THE FATE OF FOSSIL FUEL CO_2 IN THE OCEANS

MARINE SCIENCE

Coordinating Editor: Ronald J. Gibbs, *University of Delaware*

Volume 1 —*Physics of Sound in Marine Sediments*
Edited by Loyd Hampton
An Office of Naval Research symposium
Consulting Editors: Alexander Malahoff and Donald Heinrichs
Department of the Navy

Volume 2 —*Deep-Sea Sediments: Physical and Mechanical Properties*
Edited by Anton L. Inderbitzen
An Office of Naval Research symposium
Consulting Editors: Alexander Malahoff and Donald Heinrichs
Department of the Navy

Volume 3 —*Natural Gases in Marine Sediments*
Edited by Isaac R. Kaplan
An Office of Naval Research symposium
Consulting Editors: Alexander Malahoff and Donald Heinrichs
Department of the Navy

Volume 4 — *Suspended Solids in Water*
Edited by Ronald J. Gibbs
An Office of Naval Research symposium
Consulting Editors: Alexander Malahoff and Donald Heinrichs
Department of the Navy

Volume 5 — *Oceanic Sound Scattering Prediction*
Edited by Neil R. Andersen and Bernard J. Zahuranec
An Office of Naval Research symposium

Volume 6 — *The Fate of Fossil Fuel CO_2 in the Oceans*
Edited by Neil R. Andersen and Alexander Malahoff
An Office of Naval Research symposium

THE FATE OF FOSSIL FUEL CO_2 IN THE OCEANS

Edited by

Neil R. Andersen

National Science Foundation
Washington, D.C.

and

Alexander Malahoff

National Oceanographic and Atmospheric Administration
Rockville, Maryland

PLENUM PRESS • NEW YORK AND LONDON

Library of Congress Cataloging in Publication Data

Main entry under title:

The Fate of fossil fuel CO_2 in the oceans.

(Marine science; v. 6)
"Proceedings of a symposium conducted by the Ocean Science and Technology Division of the Office of Naval Research . . . held at the University of Hawaii, Honolulu, January 16-20, 1976."
Includes index.
1. Carbon dioxide—Congresses. 2. Ocean-atmosphere interaction—Congresses. 3. Chemical oceanography—Congresses. 4. Marine sediments—Congresses. I. Andersen, Neil R. II. Malahoff, Alexander, 1939- III. United States. Office of Naval Research. Ocean Science and Technology Division.
GC117.C37F37 551.4'6 77-11099
ISBN 0-306-35506-X

Proceedings of a symposium conducted by the Ocean Science and Technology Division of the Office of Naval Research on the Fate of Fossil Fuel CO_2 in the Oceans held at the University of Hawaii, Honolulu, January 16–20, 1976

A Division of Plenum Publishing Corporation
227 West 17th Street, New York, N.Y. 10011

Printed in the United States of America

Preface

The question of whether the CO_2 content of the atmosphere is increasing at a catastrophic rate is a very timely question. Will the increase in the atmospheric CO_2 content affect our climate and environment? This question cannot be answered, however, without first defining the behavior and the history of CO_2 in our total environment.

It has long been known that the carbon chemistry of the ocean must be closely related to the balance between the precipitation of $CaCO_3$ in supersaturated surface waters and its dissolution in undersaturated deep waters. This in turn is related to the biosphere, both oceanic and terrestrial, and geological processes occurring in sediments. Apparent fluctuations in this balance, as observed in the sedimentary column, have stimulated efforts recently to clarify the controlling factors which are operating in the present oceans. As this understanding has improved, it has become increasingly appreciated not only how sensitive the dynamic $CaCO_3$ precipitation/dissolution cycle may be to the injection of fossil fuel CO_2 into the oceans, but also how serious the possible consequences of such an injection may be.

$CaCO_3$ in ocean sediments is not equally distributed, either in geography or in time. Fluctuations in the productivity of the oceans and the availability of CO_2 influenced the concentration of $CaCO_3$ in the fossil record of the marine sediment. This record is further complicated by the constant drift of the oceanic plates through zones of intense productivity such as the equatorial Pacific Ocean. Given a knowledge of the geological processes involved, the past CO_2 injection rate into the oceans could be recovered from the sediment record, and historic base lines of CO_2 content developed. However, although the geological record is intriguing, the chemistry of fossil CO_2 behavior in the ocean water mass has to be understood before a complete interpretation of this record can be achieved.

A first step in quantifying the effect of fossil fuel CO_2 injections into the oceans is to estimate its effect on seawater carbonate ion concentrations, which, in turn, will strongly influence $CaCO_3$ dissolution rates. Present knowledge of marine carbonate chemistry is sufficient to do this in a general way. For example, if most of the CO_2 from burning the world's total fossil fuel reserves is eventually absorbed by the oceans, and redistributed in proportion to present inorganic carbon concentrations, the average seawater carbonate ion concentration will decrease by about 80%. It will be very surprising if this tremendous decrease is not at least partially offset by enhanced $CaCO_3$ dissolution via adjustments in the following equilibria:

$$H_2O+CO_2+CaCO_3 \rightleftarrows Ca^{++} + 2HCO_3^-$$

and

$$2HCO_3^- \rightleftarrows H_2O + CO_2 + CO_3^{=}$$

The consequence of this will be a significant increase in the inorganic carbon concentration in seawater, as well as its total alkalinity.

The impact of enhanced $CaCO_3$ dissolution can only be considered in the context of global geochemical models involving both the biosphere and the oceans as potential sinks for fossil fuel CO_2, because these are the two large reservoirs which most rapidly exchange carbon with the atmosphere. The amount of CO_2 which will be available to the oceans, therefore, depends on the future of the biosphere. Several models presently indicate that neither the biosphere nor the oceans will absorb much excess CO_2 during the next several decades. Because air-sea gas exchange is relatively rapid, the ocean surface will probably remain close to equilibrium with atmospheric CO_2 concentrations. Therefore, it is conceivable that the carbonate chemistry of the ocean surface will be determined primarily by increasing CO_2 levels in the atmosphere, and in turn, have dramatic effects on the marine biosphere and geological phenomena occurring on and in the ocean floor.

Consideration of this topic and an evaluation of the environmental problems involved go well beyond the boundaries of marine chemistry. They include physical oceanography, meteorology, terrestrial and marine biology, and geology. For example, present understanding of vertical mixing in the ocean must be improved in order to learn how fast CO_2-enriched surface waters will mix with deeper waters. It must be known what effect fossil fuel CO_2 will have on marine biological cycles, which play an important role in marine carbon geochemistry. Further, a closer look at the geological record may well provide information on how the earth has adjusted to past changes in its carbon chemistry. As a result, when initial thoughts were being given to convene a symposium to address the effects of fossil fuel CO_2 injection into the marine environment, we recognized the necessity of having individuals participate who represented a wide diversity of environmental science interest. The general structure of such a meeting was conceived to be comprised of three broad categories: the mobilization of CO_2 and resulting injections into the oceans, including the ensuing processes which would occur in the water column; the phenomena which would occur at the sea-sediment interface as a consequence of such injections; and finally, the geological implications created in the sedimentary column. As a result, the meeting was organized along these conceptual lines, with the respective session chairmen playing no small part in bringing about a meeting which saw many of the participating researchers who represented very diverse fields of science for the first time in the same room talking with one another and addressing an environmental problem of mutual concern. If the results of this meeting constitute nothing more than a continuation of the collaboration which was in some cases enhanced, and in others initiated during the deliberations, it can be considered a success.

The meeting was convened in the East-West Center at the University of Hawaii, Honolulu, Hawaii, January 16-20, 1976. It consisted of a number of Plenary Sessions and numerous individual working group sessions. Specific research recommendations arose through discussions which transpired in the latter activities. This publication represents the specific papers delivered at the symposium together with the associated discussion summaries and recommendations of the working groups, which occurred under the general guidance of the respective session chairmen.

We gratefully acknowledge the cooperation of the session chairmen and the efforts put forth by them. Without their continued dedication, our initial ideas concerning the meeting could not have been brought to fruition. We particularly want to thank Mrs. Mary-Francis Thompson, of the American Institute of Biological Sciences, whose administrative and logistical support made the entire endeavor more delightful and professional than otherwise would have been the case. Last, but by no means least, we wish to recognize our indebtedness to Mrs. Rose Marie Baldwin and Mrs. Ruth Stallings for their superb clerical assistance which made this publication possible, and their continued good humor which made the task of editing a tolerable endeavor.

Neil R. Andersen
Alexander Malahoff
Washington D. C., 1977

CONTENTS

I

CARBON IN THE ATMOSPHERE AND IN THE OCEAN

Recommendations of the Working Group on Carbon in the Atmosphere and in the Ocean

Joris M. Gieskes[1] and Ricardo M. Pytkowicz[2], Co-Chairmen

[1]*Scripps Institution of Oceanography*

[2]*Oregon State University*

This discussion will deal briefly with the recommendations made during panel discussions on "Carbon in the Atmosphere and in the Ocean." The simplified diagram in Figure 1 will be referred to and the separate topics indicated on this diagram will be addressed.

1. Man's Input of Carbon Dioxide into the Atmosphere

The future input of carbon dioxide into the atmosphere (from fossil fuels) depends on projections of future energy demand levels. The energy projections will determine

a. whether the CO_2 input will continue to rise exponentially at about 4% per year (worst case), or,

b. how much less than exponential growth can be reasonably expected at a particular future time.

These predictions, of course, will have important repercussions on future predictions, and indeed, models for different levels of fossil fuel consumption should be developed. To make these predictions, expertise must be involved, not just from geochemists, but also from specialists who concern themselves with economic and socio-economic projections.

There is strong evidence that doubling of the atmospheric level of carbon dioxide will have measurable effects on climate, although it is hardly possible to make exact projections. It should be noted that meteorologists are presently working on models and are making relevant measurements, in order to make improved climatic predictions. The importance of these predictions cannot be overstressed and further work in this area is strongly recommended.

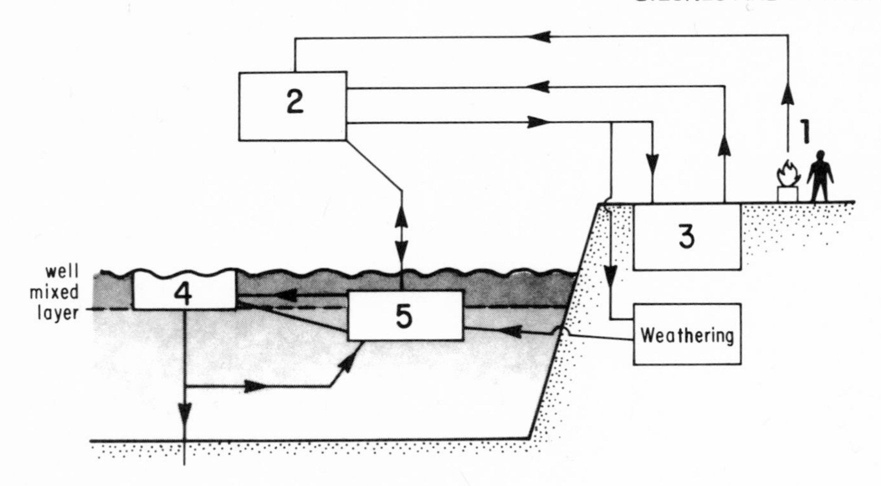

Fig. 1. Simplified carbon dioxide cycle 1) Man's input into the environment, 2) Atmospheric carbon dioxide, 3) Land biota, 4) Oceanic biota and 5) Oceanic carbon dioxide.

2. Atmospheric CO_2

It was clear during discussions of present efforts to monitor atmospheric carbon dioxide levels, that the present level of effort (i.e. at Mauna Loa, Antarctica, Point Barrow and Samoa) is an absolute minimum requirement. It was demonstrated that the presently available continuous records of the atmospheric CO_2 levels reveal seasonal, long-term, and regional variations reflecting natural processes which should be better understood.

It is desirable to have monitoring stations far removed from major sources of carbon dioxide and land biota. Seasonal variations in CO_2 levels can then be confidently interpreted as owing to exchanges with the land biota. For this, a moderate program of monitoring stable isotope C^{13}/C^{12} ratios will also be helpful, as was indicated by preliminary results reported during the conference. In addition, especially with regard to the problem of exchange of atmospheric CO_2 with the land biota, there is a need for surveying the land vegetation, possibly using satellite information, in terms of quantity and type, as well as making estimates of long-term changes of CO_2 responses of this land biota.

In order to gain more detailed information on changes in the atmospheric CO_2 levels and their causes, there is need for several more, carefully-selected monitoring stations, in addition to the temporary programs going on at Christmas Island, Fanning Island and New Zealand, which should be extended. A station in the South Atlantic Ocean is especially needed. This expansion would also provide some overlap of data, in case stations such as the one in Antarctica were to be discontinued for some reason.

Even though an expansion of the present minimum program is important, it should be carefully evaluated with regard to its priority compared with other research needs in this area.

3. Land Biota

The importance of the land biota with regard to the uptake from or release to the atmosphere of carbon dioxide has already been mentioned. For example, it was indicated that monitoring of stable carbon isotopes (C^{13}/C^{12}), for instance at Mauna Loa, would be of great use in evaluating the contribution of land biota to the seasonal changes in the atmospheric CO_2.

It is important to obtain an inventory of the areal distribution of types of land biota (forests, croplands, grasslands, etc.), possibly using satellite analyses. Changes in the CO_2 budget due to man-induced variations can then be better projected. The value of estimating possible increases or decreases in humus due to changed practices in land management should be emphasized. For this, C^{14} dating may also be informative. Continuing work on pre-"Suess effect" C^{14} levels in the atmosphere by using shells stored in musea and/or old tree rings is needed.

4. Ocean Biota

Results of preliminary experiments discussed at the meeting have demonstrated that there is some evidence that no vast increases in plant standing crops may be expected because nutrients such as nitrate appear to be factors limiting biomass increase. Some lakes and sewage lagoons might have CO_2 as a limiting nutrient, but these are quantitatively unimportant.

Further work in this area should give special attention to:

a. possible adaptations in the plant biota which may cause an increase or decrease in the carbon uptake (by a change in the C/N ratio of the protoplasm or skeletal material, for example);

b. possible effects of increased nutrient inputs by man *via* the rivers on plant productivity of the ocean. These effects may be small, but they should not be ignored; and

c. direct effects of enhanced CO_2 levels on marine animal biota, from zooplankton to higher levels in the food chain.

5. CO_2 Reservoir Exchange

a. Air-Sea Exchange

Special attention was given to this problem. Much more accurate estimates are needed to the global average exchange coefficients of CO_2 between the atmosphere and the ocean and of regional variations, so that better estimates can be made of fluxes into or out of the ocean, which is rarely in equilibrium with the atmosphere. With suitably spaced atmospheric CO_2 monitoring stations, it should be possible to establish both the global mean and regional variations in the exchanges of CO_2 between the sea and the atmosphere.

An important point considered was that an additional constraint on models for the overall exchange of CO_2 between various reservoirs would be provided by monitoring surface ocean total dissolved carbon dioxide levels. It is strongly recommended that such studies be initiated as soon as possible at certain selected ocean stations in areas with relatively stable hydrographic conditions.

b. Mixed Layer-Deep Ocean Exchange

Of great importance in the future assessment of the fate of fossil fuel carbon dioxide is the exchange of CO_2 between the mixed layer and the intermediate and deep waters of the ocean.

It is recommended that a hierarchy of models, from box models to "continuous" models of increasing complexity, be established. The most fruitful approach is to develop box models and continuous models concurrently. Assumptions about processes for box models should be put on a firmer basis than has hitherto been done. "Continuous" models are, in principle, more realistic, but they need to be refined, for example, to include topographic effects.

Steady state models of either type will make it possible to obtain information about relations among the different properties involved in the carbon cycle (alkalinity, ΣCO_2, pCO_2). Understanding of such relationships is now based mainly on chemical considerations, but it is difficult to obtain internal consistency because of the high accuracy required for each measurement. Ideally, box models and continuous models will yield a common result, thus leading to a better understanding of ongoing processes.

The transient CO_2 problem can be attacked by using continuous models including those now available. The effects of increased CO_2 concentrations introduced at high latitudes, where intermediate waters reach the surface, can be incorporated into a two-layer system with steady flow, and the transient CO_2 can be determined. One can obtain more than crude estimates from box models if the effects of important processes can be incorporated in a reliable fashion. Informational feedback from continuous models will be valuable, in order to increase the reliability of such box models.

Special attention must be given to a better determination of the boundary conditions for pCO_2, alkalinity, ΣCO_2 and other CO_2 system parameters in source areas of intermediate, deep, and bottom waters. In addition, exchange between the well-mixed surface layer and intermediate waters must be evaluated, for example from data on tritium and C^{14}. Here especially, time variations in C^{14} profiles would be of considerable value.

FOSSIL FUEL PROBLEM AND CARBON DIOXIDE: AN OVERVIEW

R. M. Pytkowicz and L. F. Small

Oregon State University

ABSTRACT

An overview of the carbon dioxide system in nature and its perturbation by man is presented in this work. Emphasis is placed upon an understanding of the factors that control this system, and on needed research, as a basis for predicting the impact of the burning fossil fuel.

INTRODUCTION

A broad look will be taken in the following discussions of the state of our knowledge of the carbon dioxide system and of the fossil fuel problem prior to this meeting. The object is to set the stage for the papers that follow in this section of the text. Our understanding and needed research will be emphasized, rather than predictions, because fossil fuel models developed thus far have not included the full range of possible pathways and feedback loops for man-injected carbon dioxide. This incompleteness has occurred not only because of mathematical difficulties encountered in treating the overall system but also because needed fundamental data were not yet available.

Numerical modelling of biogeochemical cycles is important. But, we should not be guided solely by the convention that papers should present numerical results to be worthy of publication. This is especially true in fields so ripe for speculation as the studies of broad geochemistry and of human activities. Rather, at times we should partly detach ourselves from numbers so that we may examine our premises, the gaps in our knowledge which occasionally are masked by our mathematics, and establish a perspective for needed research.

The cycle of carbon dioxide in nature before its perturbation by man shall be examined first, as the baseline for subsequent topics. Next, the natural controls for the carbon dioxide system will be considered, followed by a discussion of the impact of man through fossil fuel burning and of how the resulting increase in atmospheric carbon dioxide may be partially dissipated by uptake into various reservoirs. Finally, a brief review of the work on the fossil fuel problem will be presented before this meeting.

Only the carbon dioxide system will be discussed. This system is important, in itself, because an increase in atmospheric carbon dioxide levels may lead to an enhanced absorption of infrared radiation emitted by the earth (i.e. the greenhouse effect) and, consequently, to a warming of our environment. There are allied topics, however, namely the potential cooling casued by absorption of incoming ultraviolet radiation by clouds and industrial aerosols, which are not considered here.

THE CARBON DIOXIDE CYCLE

A view of the carbon dioxide cycle before the advent of man is shown in Figure 1. This figure is a simplified version of the model presented previously by *Pytkowicz* (1973a), with only net fluxes shown. However, the main features of the original considerations are preserved. The number of significant figures for the fluxes is used for mass balance and does not reflect the accuracy of the data. Actually, the flux values are only rough estimates. Improved values for reservoir sizes and fluxes are needed because these parameters enter into the differential equations for the perturbed carbon dioxide system.

Juvenile carbon dioxide, originating in the earth's mantle, and recycled carbon dioxide, produced by metamorphism and by oxidation, enter the atmosphere. Atmospheric carbon dioxide is then used in photosynthesis on land which, in its simplest form, can be represented by the reduction of CO_2 to carbohydrate:

$$CO_2 + H_2O \rightarrow CH_2O + O_2 \qquad (1)$$

where CH_2O symbolizes carbohydrates. Following metabolic oxidation (i.e. respiration) and decay (the reverse of equation (1)), CO_2 is released to ground waters and takes part in the chemical weathering of aluminium silicates and carbonates. This complex organic pathway for CO_2 before weathering, rather than simple uptake of atmospheric CO_2 by ground waters, is indicated because these waters are normally heavily supersaturated with CO_2 with respect to the atmosphere.

Weather processes can be represented by the generic reactions

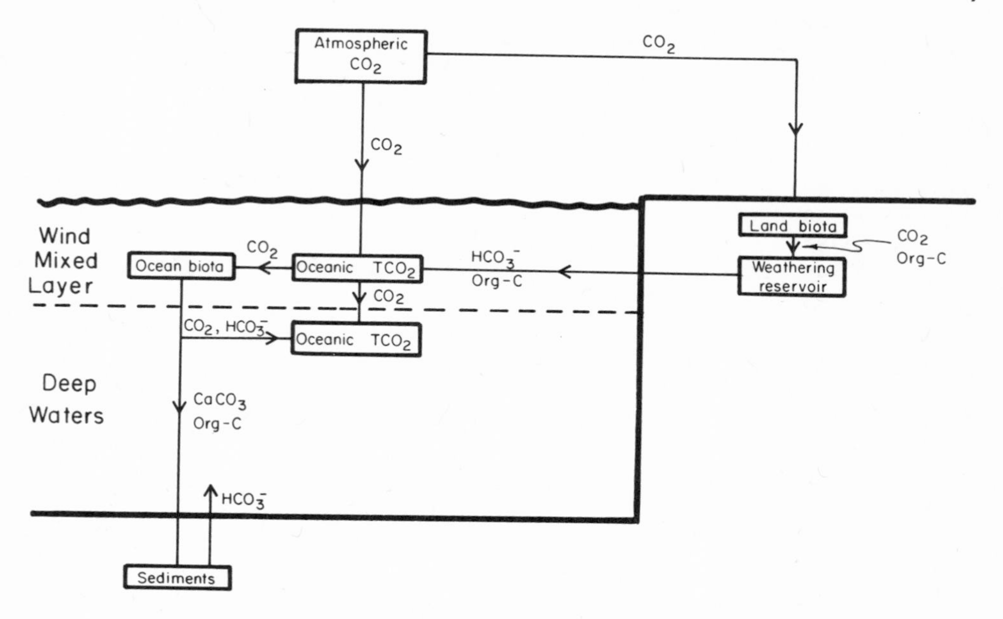

Fig. 1. Carbon dioxide cycles in nature before the advent of man; a simplified model based upon that of *Pytkowicz* (1973a). The reservoir sizes are expressed in $10^{20}gC$ and the fluxes in $10^{14}gC/yr$. The number of significant figures is necessary for mass balance purposes but does not reflect the accuracy of the data. Only net fluxes are shown.

$$Ca_xMg_{(1-x)}CO_3 + CO_2 + H_2O \rightarrow xCa^{2+} + (1-x)Mg^{2+} + 2HCO_3^- \quad (2)$$

$$\underset{(K\text{-}feldspar)}{5KAlSi_3O_8} + 4H^+ \rightarrow \underset{(illite)}{KAl_5Si_7O_{20}(OH)_4} + 8SiO_2 + 4K^+ \quad (3)$$

$$2KAl_5Si_7O_{20}(OH)_4 + 2H^+ + 5H_2O \rightarrow \underset{kaolinite}{5Al_2Si_2O_5(OH)_4} + 4SiO_2 + 2K^+ \quad (4)$$

(*Siever*, 1968; *Pytkowicz*, 1973a, 1975).

Equations (3) and (4) can be expressed in terms of CO_2 instead of H^+, with HCO_3^- appearing among the products. The use of H^+, however, renders the equations more general by including weathering by HCl. The formation of clays, such as illites from feldspars, illustrates the primary weathering of igneous rocks, while equation (4) reflects secondary weathering in soils and in the hydrosphere.

Weathering introduces cations, clays, silica, organic carbon and bicarbonate (actually carbon dioxide, carbonic acid, bicarbonate and carbonate ions, with bicarbonate predominating) to the oceans. Atmospheric carbon dioxide tends to enter the oceans at high latitudes, where low temperatures enhance its solubility and high photosynthetic rates incorporate large amounts of CO_2 as organic carbon, and leaves the oceans at low latitudes. There are, of course, regional, seasonal, and diurnal changes in this pattern (*Skirrow*, 1965). The net flux, however, is from the oceans to the atmosphere *Pytkowicz*, 1973a).

The long-term sink of CO_2, shown with a question mark in Figure 1, indicates that nature must dispose of an amount equivalent to the juvenile CO_2, or the atmosphere would have become concentrated with it throughout geologic time. Two possible long-term sinks are the mantle and the carbonate rocks. The carbonate rocks are not primordial. They were formed primarily from the weathering of plagioclase feldspars to yield calcium and bicarbonate ions, followed by the precipitation of calcium carbonate from natural solutions containing the calcium and bicarbonate ions. It is possible that such a mechanism is increasing the mass of carbonate sediments which are subsequently lithified. It should be added that marine sedimentation of carbonates at this time is essentially giogenic rather than inorganic (*Pytkowicz*, 1973b).

Within the oceans there are two carbon cycles, if carbon monoxide is neglected. The photosynthetic-oxidation cycle occurs primarily within the wind-mixed layer; the near-surface photic zone. Here, 98% of the photosynthetically-fixed carbon is released by oxidation (*Pytkowicz*, 1973a) while the remainder is oxidized in deep waters or is sedimented. The hard parts of calcareous organisms, on the other hand, primarily dissolve at depth where waters are undersaturated because of the effect of pressure upon the solubility of calcium carbonate (*Hawley and Pytkowicz*, 1969). The dissolved calcium bicarbonate is later returned to the surface at high latitudes, where deep waters upwell, and elsewhere by vertical mixing.

The two internal oceanic cycles are not closed, as part of the settling organic and carbonate carbon is incorporated into the sediments. Sediments deposited in shallow waters are returned to the weathering environment by uplift, changes in sea level, and evaporation. Deep water sediments are eventually recycled by sea-floor spreading and tectonism and undergo metamorphism while subjected to high temperatures and pressures.

The biogenically precipitated carbonates, which originated in the weathering of limestones and of dolomites, are recycled as carbonates (2.660×10^{14} *gC/yr*) (Figure 1). Those carbonates resulting from the weathering of feldspars are released as carbon dioxide (0.191×10^{14} *gc/yr*) during the reconstitution of feldspars in the metamorphic environment. Lithified organic carbon (0.300×10^{14} *g c/yr* is shown in Figure 1 to be oxidized as part of weathering for simplicity. However, part of it may actually break down and release CO_2 during metamorphism.

The overall mass in the system is balanced in such a way as to maintain a steady state before the advent of man. This steady-state view is based upon the results of many authors (e.g., *Goldberg and Arrhenius*, 1958; *Garrels and Machenzie*, 1971; *Holland*, 1972) who concluded that there have been no major trends in the composition of the oceans over millions of years. Some questions regarding this hypothesis were examined by *Pytkowicz* (1975), as new carbonates may be formed biogenically from the calcium bicarbonate originating in the weathering of plagioclase feldspars. A change in any one reservoir affects all the other ones(*Phtkowicz*, 1972a, 1973a). Still, the steady-state condition may be applicable as a first approximation to the carbon system.

NATURAL CONTROLS

It is important to understand the natural controls on the atmospheric and the marine carbon dioxide to elucidate their importance with regard to the response of the carbons system to fossil fuel carbon dioxide.

In a short time scale, (i.e. the order of years) changes in atmospheric carbon dioxide resulting from fossil fuel burning can be attenuated by uptake into the near-surface oceanic wind-mixed layer and, perhaps, into the biota of marine and land areas. *Small et al.* (1976) and *Lemon* (1976) present preliminary evidence which suggest that the biota may not be an effective sink for fossil fuel CO_2 unless the rate of addition of limiting nutrients, such as phosphate and nitrate, is accelerated by the activities of man. Also, there is a gradual transport of carbon dioxide into deep oceanic waters by advective sinking at high latitudes and by slow vertical mixing across the permanent pycnocline. These fluxes will further remove man-made CO_2 from the atmosphere. Finally, a change in atmospheric carbon dioxide may lead to changes in the rate of weathering on land and, therefore, in the river flux of weathered products. This possibility hinges, however, upon enchanced photosynthesis and subsequently on enhanced oxidation of organic matter which would increase the CO_2 content of groundwaters. These processes will be examined individually before proceeding to long-term controls.

There is a net flux of carbon dioxide from the oceans to the atmosphere. However, any increase in atmospheric content can be transmitted to the wind-mixed layer simply by decreasing this net flux; that is, by increasing the downward component of the flux. The oceanic CO_2 capacity is large because most of the CO_2 is converted to bicarbonate and carbonate ions in seawater and these ions are not in direct exchange equilibrium with the atmosphere. The wind-mixed layer, with a depth of the order of several hundred meters, contains as much CO_2 as the atmosphere. The deep oceans contain fifty-four times as much CO_2, present as TCO_2(Total carbon dioxide) = $(CO_2) + (H_2CO_3) + (HCO_3^-) + (CO_3^{2-})$.

Unfortunately, from the standpoint of assessing the significanc of fossil fuel CO_2, the change in the oceanic TCO_2 is not simply proportional to changes in pCO_2 (the atmospheric partial pressure of carbon dioxide). Rather, as can be shown from the dissociation equilibria of carbonic acid in seawater,

$$(\Delta TCO_2/TCO_2)_{sw} = B(\Delta pCO_2/pCO_2)_{atm}. \quad (5)$$

The coefficient B, the buffer factor, is slightly less than 0.1 in the surface waters of the present oceans because changes in TCO_2 are limited to values at which the pCO_2 in sea water roughly equals that of the atmosphere *Bolin and Eriksson*, 1959; *Kanwischer*, 1963; *Broecker et al.*, 1971; *Machta*, 1972; *Keeling*, 1972. Thus, if the atmospheric pCO_2 should change by 10%, then the TCO_2 of seawater would change by 1%. Because the TCO_2 of the wind-mixed layer roughly equals the atmospheric CO_2, it is concluded that, at equilibrium, 90% of the initial change in atmospheric CO_2 will remain in the atmosphere. The remaining 10% will be absorbed by the wind-mixed layer. Of course, this illustration does not include transfer of the excess CO_2 to deep waters and to the biota, as well as the possibility of enhanced weathering.

For the simplest case (i.e. no feedback control by the biota and no enhanced weathering), most authors consider that B decreases as CO_2 is absorbed and the pH decreases (e.g., *Keeling*, 1972). The values of B shown in Table 1 (*Barton, Culberson and Pytkowicz*, unpublished results), were calculated from the standard equations for the carbon dioxide-carbonate system in seawater (*Culberson and Pytkowicz*, 1968; *Mehrbach et al.*, 1973). The titration alkalinity (TA) is defined by

$$TA = (HCO_3^-) + 2(CO_3^{2-}) + (B(OH)_4^-) + (OH^-) - (H^+) \quad (6)$$

Changes in TA reflect changes in the dissociation of weak acids and bases, as well as the dissolution or removal of calcium carbonate. As the carbon dioxide-carbonate system in seawater can be expressed by four equations in six unknowns (*Pytkowicz*, 1968), any two relevant quantities, such as pH and TA, completely specify the system. The two temperatures used in Table 1 indicate the effect of latitude upon the buffer factor. A more complete treatment should also show the latitudinal effect of changes in chlorinity and in TA. The values of B for surface waters in the present oceans are about 0.090 at 17.70°C and 0.075 at 3.50°C.

In reality, the values of B which should be applied to an ocean that is changing under the influence of fossil fuel CO_2 are not certain. This uncertainty arises in part because a change in atmospheric carbon dioxide could perhaps lead to enhanced weathering if there is a change in the mass of the land biota and rate of its

TABLE 1. Buffer factor, *B*, as a function of the *pH* at a titration alkalinity of 2.331 x 10^{-3} equiv/*kg*- SW and 19.26% chor inity (*Barton et al.*,unpub. res.).

Temperature °C	*pH*	*B*	Temperature °C	*pH*	*B*
17.70	8.672	0.1323	3.50	8.590	0.1140
	498	1222		452	1005
	402	1136		349	0906
	326	1062		267	0829
	262	0999		199	0769
	207	0945		140	0722
	159	0898		089	0684
	115	0857		043	0653
	076	0822		001	0627
	040	0791		7.963	0607
	007	0764		929	0590
	7.976	0740		897	0576
	948	0718		867	0565
	921	0700		839	0557
	896	0683		813	0550
	872	0668		788	0545
	850	0655		765	0541
	828	0644		742	0539

oxidation. Accelerated weathering would in turn increase the river flux of bicarbonate into the oceans, and the excess bicarbonate could counteract the effect of CO_2 upon the *pH*. In addition, it has been suggested by *Fairhall* (1973) that a decrease in the *pH* of near-surface waters could eventually lead to their undersaturation with regard to clacium carbonate and to a dissolution of calcareous tests which, although catastrophic for some marine life, would counteract, in part, the decrease in *pH*. However, this is not certain, as calcarious organisms appear to be protected by organic coatings (*Chave and Suess*, 1967). Furthermore, the *pH* decrease possibly could be counteracted by enhanced photosynthetic activity resulting from the increase in dissolved carbon if a concomitant increase in dissolved nitrogen, or other limiting elements, occurred.

Near-surface seawater is usually at least two fold supersaturated with regard to calcium carbonate. Intermediate waters are supersaturated in most oceans, but undersaturated in the North Pacific. Deep waters are undersaturated because of the effects of high pressures and low temperatures upon the solubility of calcium carbonate (*Pytkowicz*, 1968, 1970; *Hawley and Pytkowicz*, 1969; *Ingle*, 1975).

Considerable acidification of near-surface waters would be required to ensure dissolution of calcareous organisms. However, an accelerated rate of dissolution may be expected when the extent of $CaCO_3$ undersaturation of deeper waters is increased as a result of the uptake of fossil fuel CO_2, provided there is no significant enahancement of the rate of weathering on land. In any event, an acidification of deep waters should be slow because their residence time is of the order of 1,000 years.

The effect of changes in atmospheric CO_2 upon the land and marine biotas is not fully understood, although preliminary results are available on the subject. *Machta* (1972) and *Keeling* (1972) have applied data obtained for the response of the land biota and *Pytkowicz et al.*, (in press) used preliminary results for the response of the marine biota *Small, Donaghay, and Pytkowicz* (1976) to fossil fuel models. Some lines of evidence suggest that the biota, particularly the long lived land plants, must act as a sink for a portion of the excess CO_2 due to fossil fuel burning(*Bacastow and Keeling*, 1973). Other evidence, namely the cutting of forests, suggests that land plant biomass is decreasing and cannot act as a sink for excess CO_2 (*Whittaker and Likens*, 1973; *Lemon*, 1976). Short lived land plants and phytoplankton generally are not considered to be important CO_2 sinks *Reiners*, 1973; *Small et al.*, 1977).

Understanding the role of biota as a temporary sink for increased levels of CO_2 in upper waters and on land is complicated by the fact that factors other than CO_2 (e.g. nitrogen, phosphorus, or moisture), usually limit biological production. *Riley* (1956) and others has observed that the nitrogen: phosphorus atomic ratio in marine phytoplankton found in nitrogen-depleted waters was still around the normal 16:1 ratio measured by *Redfield et al.*, (1963). This means that sufficient nitrogen becomes available through the regeneration of nutrients by oxidation, but that the nitrogen which reaches surface waters is used up as soon as it reaches the photosynthetic environment. When nitrate is added to these nitrogen-depleted waters, the biomass of phytoplankton is increased while the C:N atomic ratio in phytoplanktonic protoplasm remains 16:1. This is a classic example of nutrient limitation of population size. Similar examples can be found in freshwater and terrestrial environments.

Under population limitation by some factor other than CO_2, moderate increases in already non-limiting CO_2 levels would be expected to have little effect on plant biomass. With massive introductions of CO_2 (e.e. 5 or 10 times ambient concentrations) into nutrient-limited systems, it is not known at this time what physiological stresses may be put on the plant populations. Conceivably, increased excretion of organic carbon would be involved, or perhaps species successional changes would be induced by different levels of adaptation to the enhanced CO_2 level or to *pH* changes (*Shapiro*, 1973; *Goldman*, 1973). What direct or indirect effects increased CO_2

levels might have in various animal populations or on the whole food webs is not known. Interestingly, in certain eutrophic fresh waters, including sewage lagoons, carbon may become the limiting nutrient (*King*, 1970, 1972; *Allen*, 1972; *Wetzel*, 1965, 1972; *Kerr, et al.*, 1972). In such cases, increased atmospheric CO_2 should lead to increased plant production. However, as indicated by *Wetzel* (1972) *and Fuhs, et al.*, (1972), to emphasize carbon limitation is usually a gross oversimplification.

It is difficult to evaluate the response of weathering to atmospheric carbon dioxide. Groundwaters are already heavily supersaturated and their CO_2 content is regulated by a balance between oxidative input, loss to the atmosphere, and weathering. It is conceivable that an increased supply of carbon dioxide to groundwaters would be considerably damped by loss to the atmosphere and would have little effect upon the weathering rate.

The buffer mechanism of seawater will be considered next. This buffer mechanism must be known in order to predict future *pH* changes as the result of CO_2 uptake. *Sillen* (1961) proposed that the long term *pH* control of the oceans was caused by clay-seawater interactions. However, *Pytkowicz* (1967) showed that the alkalinity already present in the oceans can buffer the *pH* for thousands of years. Furthermore, if one considers the constant input of bicarbonate into the oceans by rivers, then the buffering capacity of the carbon dioxide system is enhanced much further. Thus, from the standpoint of short term perturbations (on a geological time scale), it is not necessary to invoke *pH* control by clays. In addition, 85% of the river bicarbonate originates in the weathering of carbonate rocks (*Pytkowicz*, 1967) so that the weathering of aluminium silicates contributes a relatively small amount of the bicarbonate buffering.

Next, for the sake of completeness, the long-term controls on the carbon dioxide system will be considered. The overall carbon dioxide system before the advent of man was probably controlled by steady states rather than equilibria, because systems in which there are gradients of chemical potential and energy (and consequently gradients of matter and energy fluxes) cannot, in principle, be at chemical equilibrium (*Pytkowicz*, 1975). Still, such systems may not be far from equilibrium, so that they can reflect the influence of 1) mineral-seawater equilibria for geochemical processes and 2) steady states for biological reactions.

The equilibrium viewpoint can be developed as follows. Wide excursions in atmospheric CO_2 are limited in the long-term sense by reactions which can be expressed in a very simplified manner (*Urey*, 1952):

$$CaSiO_3 + CO_2 = CaCO_3 + SiO_2. \qquad (7)$$

The loop expressed by Equation (7) consists of several steps (*Pytkowicz*, 1975). First, there is the primary and the secondary weathering of aluminium silicates as well as the weathering of carbonates, followed by transport of the products to the oceans. This is followed by the biogenic removal of calcium carbonate and silica from seawater, the settling of detrital clays and particulate organic matter, and the diffusion of cations from the oceans into pore waters. This diffusion may occur in the opposite direction if an excess of cations is brought to the sediments as ions absorbed on the settling solids. Finally, post-depositional reverse weathering (*Garrels*, 1965; *Siever*, 1968) occurs in submarine sediments and in metamorphic environments, and the reconstituted rocks are eventually brought back to the surface of the continents. The time scale for this cycle is of the order of 10^8 years and is not of concern for the fossil fuel problem, but is of interest for a fundamental understanding of the CO_2 system.

It may be reasoned that the atmospheric CO_2 is fixed as in Equation (7) and that the oceans are nearly at equilibrium with the atmosphere. Furthermore, the oceans may be considered to be not too far from saturation with regard to calcium carbonate. As was mentioned earlier, any two pertinent variables fix the marine carbon dioxide system. It may be concluded, therefore, that equilibria which fix the atmospheric and oceanic pCO_2 as well as the ionic product $(Ca^{2+})(CO_3^{2-})$, control the marine CO_2 system in the long run and that these equilibria are just a little perturbed by life.

It is preferred to turn the above statement around and to focus upon steady states because the existence of matter and energy fluxes suggest a steady state system in which entropy is constantly generated. Furthermore, either a system is actually at eqilibrium or it is not, and the mathematical expressions for the perturbation of steady states are quite different from those that apply to equilibria (*Pytkowicz*, 1971, 1972a).

The reasoning based upon steady states is as follows. The atmospheric pCO_2 is controlled by the biogenic settling of organic carbon and of calcium carbonate into the sediments, as shown in Figure 1. The ultimate sink is either the mantle or new carbonate rocks resulting from the weathering of plagioclase feldspars. Life is not an equilibrium process so that atmospheric CO_2, which is controlled by biological processes, reflects steady states rather than equilibria. The oceanic pCO_2 is not at equilibrium with the atmospheric one, both in the wind-mixed layer and in deep waters. Rather, it reflects the effects of thermal processes, and of the photosynthetic-oxidative cycle, as has been noted earlier. Dissolved marine carbonate is not at equilibrium with solids, as is shown for calcite in Figure 2. The values in Figure 2 were obtained from field data, the apparent dissociation constants of carbonic acid *Mehrbach et al.*, 1973), their pressure coefficients (*Culberson and Pytkowicz*, 1968), and the apparent solubility product of calcium carbonate (*Ingle et al.*, 1973). The carbonate flux out of the oceans is controlled by the biogenic settling of calcareous tests of marine organisms. Thus, in summary, the atmospheric and the marine pCO_2 are currently controlled by

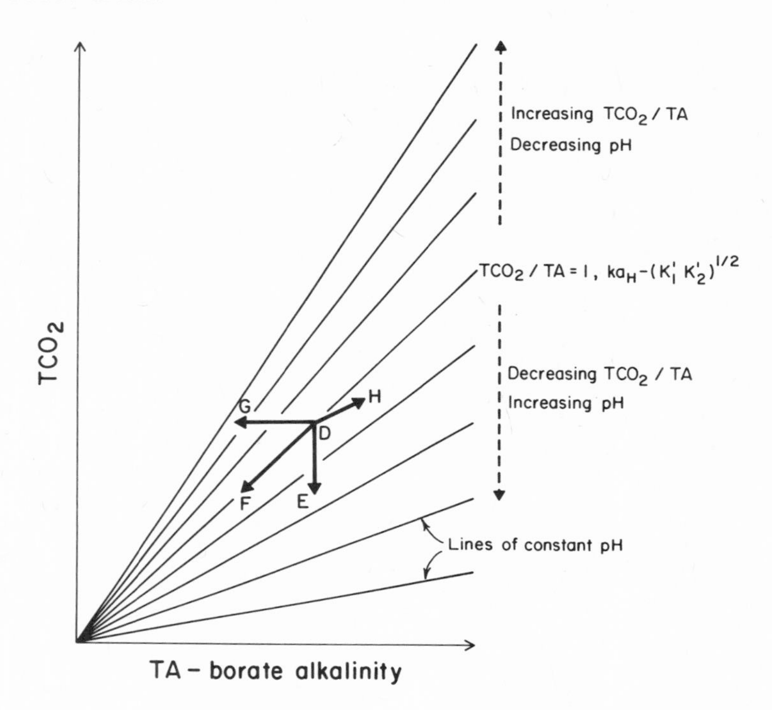

Fig. 2. Degree of saturation with regard to calcite at two oceanographic stations. The range of uncertainty in the values of the degree of undersaturation is about ±7% in terms of two standard deviations.

organisms and the same is true of the oceanic alkalinity. It is concluded, therefore, that the carbon dioxide system in the air and in the sea is controlled by biological steady states rather than by geochemical equilibria. The geochemical equilibria only impose broad limits within which the system can vary. It must be confessed, however, that at times it is felt as if it were being stated that one side of a coin is more important than the other.

It is interesting to observe that calcareous sediments persist under undersaturated waters such as those indicated in Figure 2 (*Pytkowicz*, 1970). This is confirmed by the fact that the lysocline (i.e. where the first signs of dissolution occur) is about 1-2 km shallower than the carbonate compensation depth. The presence of such carbonate sediments implies a control of dissolution rates by flow kinetics and is in contrast to the conclusion of *Li et al.* (1969).

IMPACT OF MAN

By 1966, man was already producing and releasing to the atmosphere 3.62×10^{15} gC/yr, or about 12 times as much as the natural removal rate of carbon into the sediments. Production is still increasing, but interestingly, a levelling in atmospheric CO_2 has been observed in the last few years. Should it occur, it is doubtful that a continuing exponential growth in the production of fossil fuel CO_2 could be accommodated by life and the oceans fast enough to prevent a large peak in atmospheric CO_2 (*Pytkowicz*, 1972b).

In light of the discussion of the previous section, let us examine the reservoirs that may counteract the effect of fossil fuel burning upon the atmosphere, without including the long-term geological cycle. Thoughts on this subject are summarized in Table 2 and in Figure 3. In Table 2 the time scales for the reservior

TABLE 2. Reservoirs that may attenuate the impact of fossil fuel burning upon the atmosphere, time scales for their response, and state of knowledge of their kinetics.

Reservoir	Response Time	State of knowledge of reservoir kinetics
Wind-mixed layer	Fast, years	Uptake from the atmosphere reasonably well known
Land plants and animals	Fast, years	Preliminary data on uptake, respiration, and settling rates
Weathering environment	Fast, years if it actually responds	No data on accelerated weathering kinetics
Submarine sediments	Slow, 1,000 years	Preliminary data on dissolution rates
Deep waters	Slow, 1,000 years for full response but partially effective all the time	Transfer from the wind-mixed layer fairly well known

responses and the state of our knowledge of their kinetics are presented. Both are important in order to set up realistic differential equations for the perturbed system and to calculate how effective

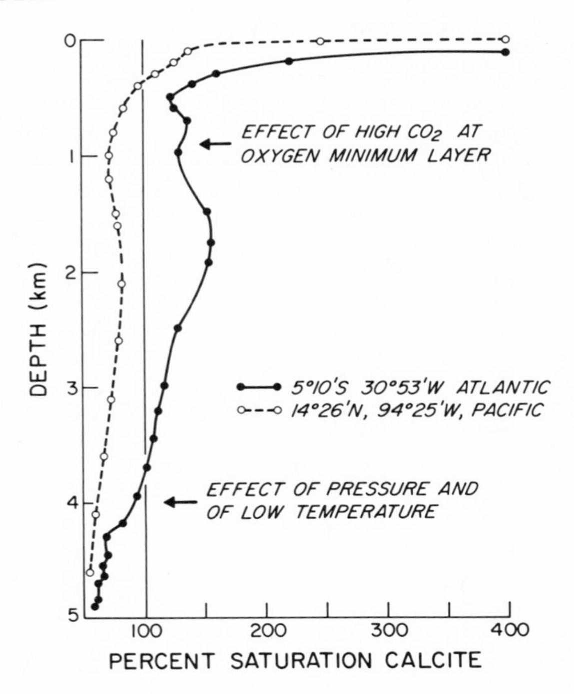

Fig. 3. Net possible pathways for fossil fuel CO_2. Chemical sub-cycles due to photosynthesis followed by metabolic oxidation, respiration, and decay as well as to calcareous test formation and subsequent dissolution are not shown.

the reservoirs are to reduce the CO_2 injected by man into the atmosphere. The long-term geological cycle is not included. The overall system has not yet been treated quantitatively by geochemists, as the state of knowledge of the relevant processes is still limited. Workers thus far have not had access to definitive data on the response of the marine biota. Similarly, geochemists have not considered the response of the weathering environment. This latter possibility, as far as is known, has been entirely neglected and perhaps, rightly so, although this point is not entirely clear.

A critical aspect of the fossil fuel problem is the prediction of the future fossil fuel consumption. Most workers thus far have concentrated on the present and the next two decades. It has been pointed out, however, that it is important to extend forecasts to about 100 years (*Pytkowicz*, 1972b). The reason for this is that, should the production of fossil fuel CO_2 continue to rise exponentially, there would come a time during the next century in which there would be only a few years in which to change energy policies to prevent a possibly catastrophic increase in atmospheric CO_2. Long-term forecasts, although crucial for evaluating the environmental impact of energy policies, leave physical and biological scientists uncomfortable and make them feel at times like gypsies looking at crystal balls. At the present time, there is not much to contribute to long-range forecasting. However, it is useful to point out that there is a fundamental decision which has to be faced in such forecasts.

In Figure 4 two types of forecasts are presented. The broken line is purely hypothetical, but represents a basic approach, such as that described by *Meadows et al.* (1972). In it one assumes that the human consumption of resources can be understood in terms of a few essential processes. If this is the case, one predicts a consumption of fossil fuels that peaks in a few decades and then decays as other needed resources such as food and metals become depleted. Of course, the size and the timing of the peak will depend upon the degree of technological optimism of the forecaster and upon an estimate of the future relative importance of alternate energy sources.

On the other hand, a simple extrapolation of present trends may be used. The solid line in Figure 4 is based upon a continuing exponential growth in production of fossil fuel carbon dioxide which accelerates at the rate of 4.5% per year. Then, by the year 2060 half of the known fossil fuel reserves will be used up and presumably production will slow down as access to the fossil fuels becomes more difficult. This type of approach involves the implicit assumption that a large number of factors, which defies definition, has controlled and will contol in the future the utilization of fossil fuels.

The exponential extrapolation does not include the possibility of catastrophes and of drastic technological and energy policy changes. It reminds one of the scientist who, to illustrate the pitfalls that occur when technological innovation is not considered, recently fed 19th century data on horse populations into a computer and concluded that by now there should be six feet of horse manure in the streets of the world.

The difficulty in making long-term forecasts is recognized but the challenge should nevertheless be accepted. Seeking clues for the processes that in the past governed the consumption of resources may possibly help. If so, loosely quoting John Masters, the truths glimmered from the past can become the half-truths that guide us in the future.

WORK ON FOSSIL FUEL BEFORE THIS MEETING

Work on the role of the oceans in the partial uptake of fossil fuels by the oceans was pioneered by *Revelle and Suess* (1957) who examined the air-seawater exchange but neglected the biota. *Bolin*

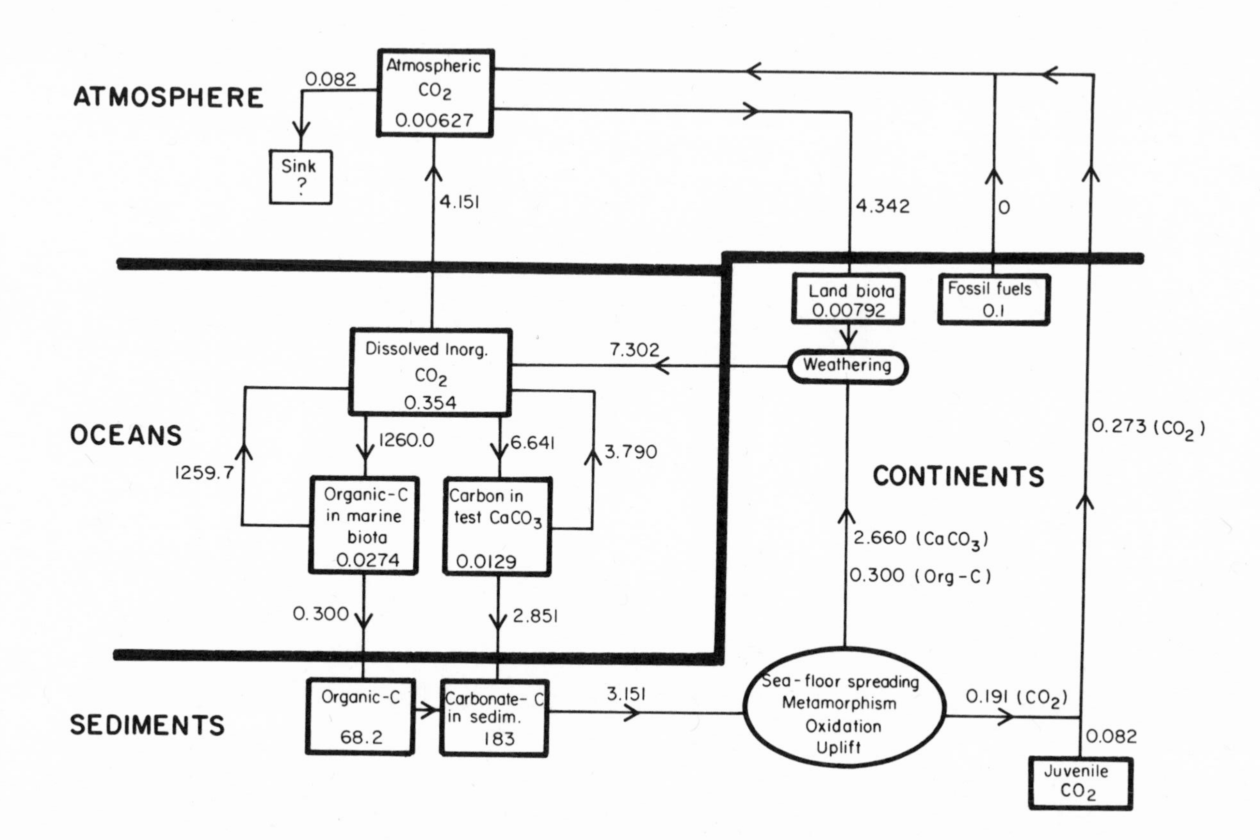

Fig. 4. Extrapolated values for the cummulative production of fossil fuels carbon dioxide, expressed in 10^{16} gC, assuming a production that continues to accelerate at 4.5% per year (solid line). Hypothetical forecast for an essentialistic (see text) model (broken line).

and Eriksson (1959) incorporated into their model the slow exchange between the wind-mixer layer and deep waters. They estimated that over half of the CO_2 generated by fossil fuel burning remained in the atmosphere.

Broecker et al. (1971) introduced an intermediate layer, the permanent pycnocline, which separates the wind-mixed layer from deep waters and slows exchange between these two reservoirs. They also considered the sinking of surface waters at high latitudes, assumed and enhanced supply of carbon to deep waters by organic matter proportional to the increase in the TCO_2 of the wind-mixed layer, and simulated the fossil fuel carbon dioxide input into the atmosphere by a step function. The land biota was neglected. *Broecker et al.* (1971) concluded that roughly 60% of the CO_2 produced up to 1960 has remained in the atmosphere.

Machta (1972) considered the atmosphere, the ocean, and the biosphere and, on the basis of data on the measured buildup of industrial CO_2 in the atmosphere, predicted atmospheric concentrations of the order of 380 ppm by the year 2000. *Keeling* (1972) carefully examined the distribution of fossil fuel CO_2 for a five-reservoir model consisting of the short and long-lived biota, the atmosphere, and the deep ocean. Like *Broecker et al.* (1971), he considered any enhancement in the downward transport of CO_2 by the marine biota proportional to the increase of TCO_2 in the mixed layer. However, none of these authors considered a possible increase in the marine biota itself, or an increase in the organic excreta of marine organisms as a sink for fossil fuel CO_2. Also, they neglected a possible enhancement in weathering rates. *Keeling* (1972) calculated the increase in the carbon content of the land biota, atmosphere, ocean surface, and deep oceans up to 1954 to be 2-4.7, 4.8-6.8, 0.9-1.4, and 0.6-0.84%, respectively, of the original sizes of these reservoirs before the advent of the Industrial Revolution.

Fairhall (1973) introduced into his model the possibility of an eventual undersaturation of near-surface oceanic waters with respect to $CaCO_3$ which could lead to the dissolution of calcareous tests, as discussed earlier in this work. However, he failed to mention that the dissolution of tests would prevent calcareous phytosynthetic organisms from taking up CO_2 during photosynthesis. *Skirrow and Whitfield* (1975) further elaborated upon the possibility of an eventual dissolution of calcareous tests due to the acidification of oceanic waters by fossil fuel CO_2 and estimated that such an event would require a tenfold increase in atmospheric CO_2. They did not consider that even then dissolution may not occur, because of protective organic coatings on the surface of the tests. *Oeschger et al.* (1975) used a model consisting of a well-mixed atmospheric box coupled to a long-term biota, an ocean surface box, and a diffusive deep ocean box, to elucidate the dynamic behavior of natural and man-made carbon dioxide. Most authors have reported on the short-term problem, up to the year 2,000. Therefore, a few years ago, very rough estimates of the long-term buildup of atmospheric and oceanic CO_2 were made to encourage long-term work on the greenhouse effect (*Pytkowicz*, 1972b). It was found that, if we are very pessimistic, we may predict a fourteenfold increase in atmospheric CO_2 in about 100 years. This was done by considering only the atmosphere and the oceans. This increase would not be permanent because uptake by deep waters and the enhanced dissolution of submarine sediments could bring the atmospheric CO_2 almost back to present values during the following millenium.

Bacastow and Keeling (1973) used a non-linear six-reservoir model, consisting of the upper and the lower atmosphere, the short-lived and the long-lived ocean biota, the wind-mixed layer, and the deep sea, to predict changes in the atmospheric CO_2 over the next hundred years. They estimated roughly a sixfold increase in atmospheric CO_2. A 4% annual growth in the production of fossil fuels was assumed and the authors estimated a sixfold to eightfold increase in atmospheric CO_2 over the pre-industrial values.

The resulting climatic changes from large CO_2 increases cannot yet be predicted because such large CO_2 changes have not yet been considered from the point of view of infrared absorption. It is possible that the rise in the average temperature of our environment would not exceed a few degrees because of the saturation in the absorption of infrared radiation by CO_2. Still, one must consider the distribution of latitudinal changes in temperature (*Manabe*, 1971) because even a relatively small warming trend could have a significant environmental impact if it led to a partial melting of the polar ice caps. The climatic problem is further complicated by possible changes in cloud cover and in the concentration of atmospheric aerosols which could change the intensity of the incoming ultraviolet and of the outgoing infrared radiation (*McCormick and Ludwig*, 1967; *Rasool and Schneider*, 1971; *Reck*, 1974; *Weare et al.*, 1974).

Further considerations regarding the impact of fossil fuel CO_2 upon the distribution of CO_2 in natural reservoirs and upon the future climate are the subject of several papers in the present volume.

ACKNOWLEDGEMENTS

This work was supported by the Oceanography Section, National Science Foundation, through NSF Grant DES 72-01631-A02 and by the Office of Naval Research Contract N00014-76-C-0067.

APPENDIX

In a conference having as one of the topics under consideration fossil fuel burning, it is appropriate to briefly review the chemistry of the carbon dioxide system in seawater, to orient those who wish to calculate the potential uptake of this gas by seawater.

The main reactions of carbon dioxide in seawater are:

$$CO_2 + H_2) \rightleftarrows H_2CO_3 \quad \text{(A-1)}$$

$$H_2CO_3 \rightleftarrows H^+ + HCO_3^- \quad \text{(A-2)}$$

$$HCO_3^- \quad H^+ + CO_3^{2-} \quad \text{(A-3)}$$

(A-1) and (A-2) are combined and one obtains equilibrium expressions

$$K_1' = \frac{ka_H(HCO_3^-)}{(H_2CO_3)} \quad \text{(A-4)}$$

$$K_2' = \frac{ka_H(CO_3^{2-})}{(HCO_3^-)} \quad \text{(A-5)}$$

which are known as the apparent dissociation constants of carbonic acid (*Buch et al.*, 1932; *Lyman*, 1956; *Mehrback et al.*, 1973). The terms in parenthesis represent concentrations and k, which is roughly 0.98 (*Hawley and Pytkowicz*, 1973), accounts for the change in liquid junction potential when electrodes are transferred from a buffer to seawater. k cancells out, within the reproducibility of pH measurements, between the determination and the application of apparent equilibrium constants (*Pytkowicz et al.*, 1974). (H_2CO_3) is the sum of the molecular carbon dioxide and the carbonic acid concentrations. The use of apparent equilibrium constants, which have been shown to be invariant for biochemical and geochemical processes of interest in seawater (*Pytkowicz et al.*, 1974), helps avoid the use of uncertain single ion activity coefficients. The concentrations are total ones rather than those of the free ions (*Pytkowicz and Kester*, 1969).

Two further equations of use in the determination of CO_2 species in seawater are:

$$TA = (HCO_3^-) + 2(CO_3^{2-}) + (B(OH)_4^-) + (OH^-) - (H^+) \quad \text{(A-6)}$$

and

$$TCO_2 = (H_2CO_3) + (HCO_3^-) + (CO_3^{2-}) \quad \text{(A-7)}$$

The equation for TA, the titration alkalinity, is obtained from the condition of electrical neutrality and TCO_2, the total carbon dioxide, represents the conversation of mass. TA changes reflect the addition of strong acids and bases as well as the addition of weak acids and bases or of their salts, such as calcium carbonate.

TA can be obtained by titration with HCl (*Culberson et al.*, 1970) and one can then calculate the carbonate alkalinity

$$CA = (HCO_3^-) + 2(CO_3^{2-}) \qquad \text{(A-8)}$$

by conventional oceanographic methods (*Skirrow*, 1965).

Equations (A-4), (A-5), (A-7), and (A-8) are a system of four equations in six unknowns, Ka_H, (H_2CO_3), (HCO_3^-), (CO_3^{2-}), CA and TCO_2, once K_1' and K_2' have been determined at the temperatures, salinities, and pressures encountered in the oceans (*Lyman*, 1956; *Culberson and Pytkowicz*, 1968; *Mehrback et al.*, 1973). One can, therefore, completely specify the carbon dioxide system in seawater by two relevant measurements and a host of inter-related pairs of measurements has been proposed in the literature.

We prefer to determine the carbon dioxide species by the use of the NBS pH and the titration alkalinity obtained by a single acid addition because this method is fast and simple, requires minimal equipment, and is accurate enough for practical purposes. Others prefer at times gasometric methods and Gran titrations also used in conjunction with apparent constants or with the modified apparent constants of *Hansson* (1973a,b) as well as with those based upon a free hydrogen ion concentration scale (*Bates*, 1975; *Bates and Macaskill*, 1975). The method of Bates is more elegant than the use of NBS buffers and may perhaps lead to more reproducible pH data.

The practical equations for the application of the pH and TA can be derived from (A-4) through (A-8) and are:

$$(H_2CO_3) = CA.ka_H/K_1' + 2K_1'K_2'/ka_H) \qquad \text{(A-9)}$$

$$(HCO_3^-) = CA.ka_H/(ka_H + 2K_2') \qquad \text{(A-10)}$$

$$(CO_3^{2-}) = CA.K_2'/(ka_H + 2K_2') \qquad \text{(A-11)}$$

where (H_2CO_3) represents $(CO_2 + H_2CO_3)$. ka_H is obtained from the measured and the relation $pH_{NBS} = -log\ ka_H$. CA is calculated from the measured TA and the equation

$$CA = TA - \frac{TB \, . \, K_B'}{ka_H + K_B} \qquad \text{(A-12)}$$

TB is the total boron known from the boron/chlorinity ratio and K_B' is the first apparent dissociation constant of boric acid. K_1', K_2', and K_B' at the in situ temperatures, salinities, and pressures, as well as ka_H under these conditions, are obtainable from the results of *Lyman* (1956), *Culberson and Pytkowicz* (1968), and *Mehrbach et al.* (1973). Note that K_1', K_2', and K_B' contain ka_H and that k, therefore, drops out of equations (A-9) through (A-12).

CA and TCO_2 can be shown to be related by

$$(ka_H)^2 - (A-1)K_1'ka_H - (2A-1)K_1'K_2' = 0 \qquad \text{(A-13)}$$

where $A = TCO_2/CA$. One can obtain from this equation sets of lines of constant pH such as those shown in Figure A-1 and vectors for the effect of oceanographic processes upon the CO_2 system. Related but more complex diagrams have been presented by a number of authors (e.g., *Revelle and Fairbridge*, 1957; *Deffeyes*, 1965).

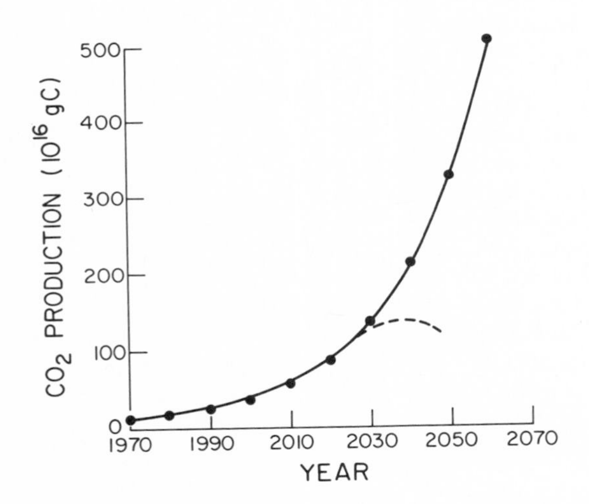

Fig. A-1. Iso-pH lines (solid ones) and process lines (heavy ones). DE corresponds to loss of CO_2 by processes such as gas exchange or photosynthesis, DF to a titration by HCl with continuous equilibration of CO_2 with a source such as the atmosphere, DG to a titration by HCl in a closed container without CO_2 loss, and DH to the dissolution of $CaCO_3$ (two equivalents of TA are added for each added mole of TCO_2). DF is only approximate as when TA becomes zero some CO_2 will still be present.

REFERENCES

Allen, H. L., 1972. Phytoplankton photosynthesis, micronutrient interactions, and inorganic carbon availability in a soft-water, Vermont Lake, In: *Nutrients and Eutrophication*, edited by G. E. Likens, pp. 63-83, Special Symposia Vol. 1, Am. Soc. Limnol. Oceanogr. Inc.

Bacastow, R. and C. D. Keeling, 1973. Atmospheric carbon dioxide and radiocarbon in the natural carbon cycle: II. Changes from A. D. 1700 to 2070 as deduced from a geochemical model, In: *Carbon and the Biosphere*, edited by G. M. Woodwell and E. V. Pecan, pp. 86-135, U.S. Atomic Energy Commission, Washington, D. C.

Barton, C., Culberson, C. and Pytkowicz, R. M. Unpublished results, Oregon State Univ. Corvallis, Ore.

Bates, R., 1975. *pH* scales for seawater, In: *The Nature of Seawater*, edited by E. D. Goldberg, pp. 313-338, Dahlem Foundation Report, West Berlin

Bates, R. G. and J. B. Macaskill, In press. Acid-base measurements in seawater, in *Analytical Methods in Oceanography*, Advances in Chemistry Series, American Chemical Society, Washington D.C.

Bolin, B. and E. Eriksson, 1959. Changes in the carbon content of the atmosphere and the sea due to fossil fuel combustion, In: *The Atmosphere and the Sea in Motion*, Rossby Memorial Volume, edited by B. Bolin, pp. 130-143, Rockfeller Institute Press, New York.

Broecker, W. S., 1971. Li, Y.-H. and T.-H. Peng, Carbon dioxide - Man's unseen artifact, In: *Impingement of Man on the Oceans*, edited by D. W. Hood, 287-324, Willey-Interscience, New York.

Buch, K., Harvey, H. W., Wattenberg, H. and S. Gripenberg, 1932. Uber das Kohlensauresystem in Meerwasser, *Rapp, Cons. Int. Explor, Mer, 79*: 1-70.

Chave, K. E. and E. Suess, 1967. Suspended minerals in seawater, *Trans. N. Y. Acad. Sci. Ser. (11), 29*: 991-1000.

Culberson, C. and R. M. Pytkowicz, 1968. Effect of pressure on carbonic acid, boric acid, and the *pH* in seawater, *Limnol. Oceanogr., 13*: 403-417.

Culberson, C., Pytkowicz, R. M. and J. E. Hawley, 1970. Seawater alkalinity determination by the *pH* method, *J. Marine Res. 28*: 15-19.

Deffeyes, K. S., 1965. Carbonate equilibria: A graphic and algerbraic approach. *Limnol. Oceanogra. 10*: 412-426.

Fairhall, A. W., 1973. Accumulation of fossil fuel CO_2 in the atmosphere and the sea, *Nature, 245*: 20-23.

Fuhs, G. W., S. D. Demmerle, E. Canelli, and M. Chem, 1972. Characterization of phosphorus-limited plankton algae (with reflections on the limiting-nutrient concept), In: *Nutrients and Eutrophication*, edited by G. E. Likens, 113-133, Special Symposia Vol 1, Am. Soc. Limnol. Oceanogr. Inc.

Garrels, R. M., Silica. 1965. Role in the buffering of natural waters, *Science, 148*: 69-70.

Goldberg, E. D. and G. O. S. Arrhenius. 1958. Chemistry of Pacific pelagic sediments, *Geochim. Cosmochim. Acta, 13*: 152-212.

Goldman, J. C. 1973. Carbon dioxide and *pH*: Effect on species succession of algae. *Science, 182*: 306-307.

Hansson, I. 1973a. A new set of acidity constants for carbonic acid and boric acid in seawater. *Deep-Sea Res. 20*:461-492.

Hansson, I. 1973b. A new scale and set of standard buffers for seawater. *Deep-Sea Res. 20*: 479-491.

Hawley, J. and R. M. Pytkowicz. 1969. Solubility of calcium carbonate in sea water at high pressure and $2^{o}C$, *Geochim, Cosmochim. Acta. 33*: 1957-1561.

Hawley, J. E. and R. M. Pytkowicz. 1973. Interpretation of *pH* measurements in concentrated electrolyte solutions, *Mar. Chem. 1*: 245-250.

Holland, H. D. 1972. The geologic history of sea water - An attempt to solve the problem, *Geochim, Cosmochim. Acta, 36*: 637-651.

Ingle, S. E., Culberson, C. H., Hawley, J. F. and R. M. Pytkowicz. 1973. The solubility of calcite in seawater at atmospheric pressure and 35% salinity, *Mar. Chem. 1*: 295-307.

Ingle, S. E. 1975. Solubility of calcite in the ocean, *Mar. Chem. 3*: 301-319.

Kanwisher, J. 1973. Effect of wind on CO_2 exchange across the sea surface. *J. Geophys. Res. 68*: 3921-3927.

Keeling, C. D. 1972. The carbon dioxide cycle: Reservoir models to depict the exchange of carbon dioxide with the oceans and with land plants, In: *Chemistry of the Lower Atmosphere*, edited by S. I. Rasool, 251-329, Plenum, New York.

Kerr, P. C., D. L. Brockway, D. F. Paris, and J. T. Barnett, Jr. 1972. The interrelation of carbon and phosporus in regulating heterotrophic and autotrophic populations in an aquatic ecosystem, Shirner's Pond, In: *Nutrients and Eutrophication*, edited by G. E. Likens, 41-62, Special Symposia, Vol. 1. Am. Soc. Limnol. Oceanogr. Inc.

King, D. L. 1970. The role of carbon in eutrophication, *J. Water Pollut. Contr. Fed. 42*: 2035-2051.

King, D. L. 1972. Carbon limitation in sewage lagoons, In: *Nutrients and Eutrophication*, edited by G. E. Likens, 98-110, Special Symposia Vol. 1, Am. Soc. Limnol. Oceanogr. Inc.

Lemon, E. R. 1977. The Panel's response to more carbon dioxide. In: *The Fate of Fossil Fuel* CO_2, N. R. Andersen and A. Malahoff (Eds.). Plenum, N. Y., (this volume).

Li, Y.-H., Takahashi, T. and W. S. Broecker. 1969. Degree of saturation of *CaCo* in the oceans, *J. Geophys. Res., 74*: 5507-5525.

Lyman, J. 1956. *Buffer Mechanism of Seawater*, Ph.D. Thesis, Univ. Calif., Los Angeles, 196 pp.

Machta, L. 1972. The role of the oceans and biosphere in the carbon dioxide cycle, In: *The Changing Chemistry of the Oceans*, edited by D. Dyrssen and D. Jagmen, 1210145, Almquist and Wiksell, Stockholm.

Manabe, S. 1971. Estimates of future change of climate due to the increase of carbon dioxide concentration in the air, In: *Man's Impact on the Climate*, edited by W. H. Matthews, W. W. Kellog, and G. D. Robinson, 249-264, MIT Press, Cambridge, Mass.

McCormick, R. A. and J. H. Ludwig, 1967. Climate modification by atmospheric aerosols, *Science, 156*: 1358-1359.

Meadows, D. H., Meadows, D. L., Randers, J. and W. M. Behrens III. 1972. *The Limits to Growth*, 205 pp., Universe Books, New York.

Mehrbach, C., Culberson, C. H., Hawley, J. E. and R. M. Pytkowicz. 1973. Measurement of the apparent dissociation constants of carbonic acid in seawater at atmospheric pressure, *Limnol. Oceanogr., 18*: 897-907.

Oeschger, H., Siegenthaler, V., Schotterer, V. and A. Gugelmann. 1975. A box diffusion model to study the carbon dioxide exchange in nature, *Tellus, 27*: 168-192.

Pytkowicz, R. M. 1967. Carbonate cycle and the buffer mechanism of recents oceans, *Geochim. Cosmochim, Acta, 31*: 63-73.

Pytkowicz, R. M. 1968. Carbon dioxide-carbonate system at high pressures in the oceans, In: *Oceanogr. Mar. Biol. Ann. Rev.*, Vol. 6, edited by H. Barnes, 83-135, Allen and Unwin, London.

Pytkowicz, R. M. 1970. On the carbonate compensation depth in the Pacific Ocean, *Geochim. Cosmochim. Acta, 34*: 836-839.

Pytkowicz, R. M. 1971. The chemical stability of the oceans, *Oregon State University Tech. Rept. 214*. Corvallis, 24 pp.

Pytkowicz, R. M. 1972a. The chemical stability of the oceans and the CO_2 system, In: *The Changing Chemistry of the Ocean*, edited by D. Dyrssen and D. Jagner, 147-152, Almquist and Wiksell, Stockholm.

Pytkowicz, R. M., 1972b. Fossil Fuel burning and carbon dioxide: A pessimistic view, *Comments Earth Sci.; Geophys., 3*: 15-22.

Pytkowicz, R. M. 1973a. The carbon dioxide system in the oceans, *Schweizer. Zeitschr. Hydrol., 35*: 8-28.

Pytkowicz, R. M. 1973b. Calcium carbonate retention in supersaturated seawater, *Am. J. Sci., 273*: 515-522.

Pytkowicz, R. M. 1175. Some trends in marine chemistry and geochemistry, *Earth Sci. Revs., 11*: 1-46.

Pytkowicz, R. M. and D. R. Kester. 1969. Harned's rule behavior of $NaCl$-Na_2SO_2 solutions explained by an ion exchange model, *Am. J. Sci., 267*: 217-229.

Pytkowicz, R. M., Ingle, S. E. and C. Mehrbach. 1974. Invariance of apparent equilibrium constants with *pH*, *Limnol. Oceanogr.*, *19*: 665-669.

Pytkowicz, R. M., Atlas, E. and C. H. Culberson. 1976 (In press). Chemical equilibrium in seawater, In: *Oceanogr. and Mar. Biology Annual Review, Vol. 14*, Barnes, H. (Ed), Hafner, New York.

Rasool, S. I. and S. H. Schneider. 1971. Atmospheric carbon dioxide and aerosols: Effect of large increase in global climate, *Science*, *173*: 138-141.

Reck, R. A. 1974. Aerosols in the atmosphere: Calculation of the critical absorption/backscatter ratio, *Science*, *186*: 1034-1035.

Redfield, A. C., Ketchum, B. H., and F. A. Richards. 1973. The Influence of Organisms on the Composition of Seawater, In: *The Sea, Vol. 2*, 26-77, Hill, M. N. (Ed.), Wiley, New York.

Reiners, W. A. 1973. Appendix: A summary of the world carbon cycle and recommendations for critical research, In: *Carbon and the Biosphere*, edited by G. M. Woodwell and E. V. Pecan, 368-392, U. S. Atomic Energy Commission, Washington D. C.

Revelle, R. and R. Fairbridge. 1957. Carbonates and carbon dioxide, In: *Treatise on Marine Ecology and Paleoecology*, Vol. 1, edited by J. W. Hedgpeth, 239-295, Geol. Soc. Am. Mem. 67, Washington D. C.

Revelle, R. and H. E. Suess, 1957. Carbon dioxide exchange between atmosphere and ocean and the question of an increase in atmospheric CO_2 during the past decades, *Tellus*, *9*: 18-27.

Riley, G. A. 1956. Oceanography of Long Island Sound 1952-1954. II. Physical oceanography, *Bull, Bingham Oceanogr. Coll.*, *15*: 15-46.

Shapiro, J. 1973. Blue-green algae: Why they become dominant, *Science*, *179*: 382-384.

Siever, R. 1968. Sedimentological consequences of a steady state ocean-atmosphere, *Sedimentology*, *11*: 5-29.

Sillen, L. G. 1961. The physical chemistry of seawater, In: *Oceanography*, edited by M. Sears, 549-581, Am. Soc. Adv. Sci., Washington D. C.

Skirrow, G. 1965. The dissolved gases - Carbon dioxide, In: *Chemical Oceanography*, Vol. 1, edited by J. P. Riley and G. Skorrow, 227-322, Academic Press, New York.

Small, L. H., P. L. Donaghay and R. M. Pytkowicz, 1977. Effects of enhanced CO_2 levels on growth characteristics of two marine phytoplankton species, In: *The Fate of Fossil Fuel CO_2*, N. R. Andersen and A. Malahoff (Eds.), Plenum, New York., (This volume).

Skirrow, G. and M. Whitfield. 1975. The effect of increases in atmospheric carbon dioxide content on the carbonation concentration of surface ocean water at 25°C. *Limnol. Oceanogr.* *20*: 103-108.

Urey, H. 1952. *The Planets: Their Origin and Development*, Yale University Press, New Haven, Connecticut. 142-149.

Weare, B. C., Temkin, R. L. and F. M. Snell, 1974. Aerosol and climate: Some further considerations, *Science, 186*: 827-828.

Wetzel, R. G., 1965. Techniques and problems of primary productivity measurements in higher aquatic plants and periphyton, *Mem. Ist. Ital. Idrobiol., Suppl., 18*: 249-267.

Wetzel, R. G. 1972. The role of carbon in hard-water marl lakes, In: *Nutrients and Eutrophication*, edited by G. E. Likens, 84-97, Special Symposia Vol. 1, Am. Soc. Limnol. Oceanogr. Inc.

Whittaker, R. H. and G. E. Likens, 1973. Carbon in the biota, In: *Carbon and the Biosphere*, edited by G. M. Woodwell, and E. V. Pecan, 281-302, U. S. Atomic Energy Commission, Washington D. C.

INFLUENCE OF THE SOUTHERN OSCILLATION ON ATMOSPHERIC CARBON DIOXIDE

Robert Bacastow

Scripps Institution of Oceanography

ABSTRACT

The atmospheric CO_2 record at the South Pole, after the seasonal effect is removed, reveals a small anomaly with a period of three to four years and an amplitude of about one part per million. The rate of change of this anomaly, and the corresponding anomaly at Mauna Loa, Hawaii, correlate well with the usual index of the Southern Oscillation. The Southern Oscillation is a large scale atmospheric and hydrospheric fluctuation involving wind strengths, ocean currents, and sea surface temperatures. Most directly, it is a fluctuation in the strength of the Pacific Trade winds, but the Southern Mid Latitude Westerlies also fluctuate in phase.

The correlations indicate that CO_2 is removed from the atmosphere at times of stronger than average Trade winds and Mid Latitude Westerlies. It is suggested that CO_2 enters the Southern Mid Latitude oceans more rapidly during times of increased Westerlies. A more complete explanation of the relationship implied by the correlation should aid our understanding of how and where fossil fuel CO_2 enters the oceans.

INTRODUCTION

The atmospheric CO_2 record at the South Pole, after the seasonal effect is removed, shows a long-term or "secular" increase with a small anomaly having a period of three to four years and a total amplitude of about one part per million (Figure 1; *Keeling et al.*, 1976a). A similar anomaly is found in the Mauna

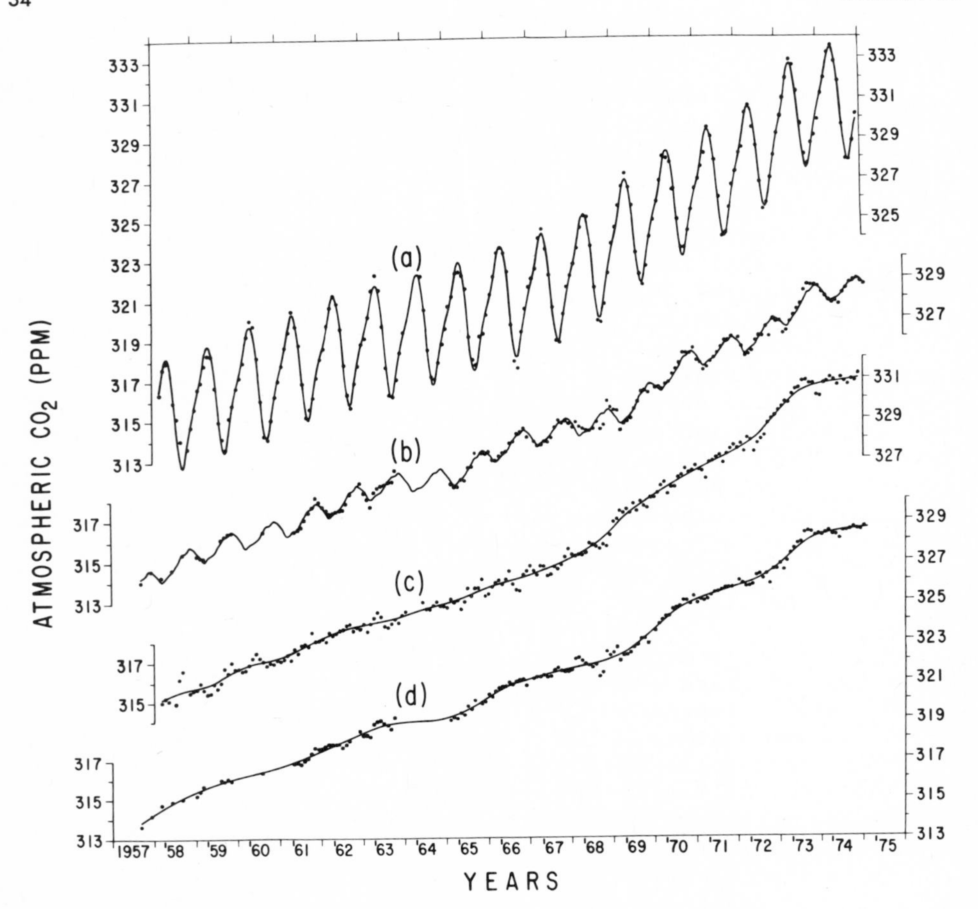

Fig. 1. Atmospheric CO_2 records at (a) Mauna Loa, Hawaii, and (b) the South Pole. The secular increases at (c) Mauna Loa, and (d) the South Pole were obtained by a procedure described in the text for removing the seasonal effects. The ordinate units are ppm CO_2 (parts per million on a dry mole fraction basis).

Loa, Hawaii, CO_2 record (*Keeling et al.*, 1976b). In this paper, the data analysis leading to the anomalies will be discussed and it will be shown that the anomalies may be connected to a meteorological phenomena called the Southern Oscillation. The cause of the connection is most likely a variation in the rate of transfer of atmospheric CO_2 to the oceans, possibly in the region of the southern mid-latitude westerlies, although other explanations are possible. A study of the anomalies may lead to a better understanding of how and where fossil fuel CO_2 leaves the atmosphere.

SEASONAL EFFECT AND SECULAR INCREASE

The seasonal effect is almost certainly due to the uptake and release of CO_2 by vegetation (*Bolin and Keeling*, 1963; *Junge and Czeplak*, 1968; and *Hall et al.*, 1975). The seasonal effect can be evaluated at a particular location by assuming that it is the same each year.

The secular increase is presumably due mainly to the input of CO_2 from fossil fuel combustion (augmented slightly by CO_2 from cement manufacture). The fossil fuel CO_2 production data do not show an anomaly similar to that in the CO_2 records. For the period of the CO_2 records, and extending back to about 1945, fossil fuel consumption per year has approximated an exponential function with an e-fold time of about 22 years (*Keeling*, 1973; and *Rotty*, 1973).

The secular increase curves in Figure 1 are spline functions (*Reinsch*, 1967) obtained by an iterative least squares fitting procedure in which the seasonal effect was evaluated by an average over identical months of the difference between the data and the spline. The stiffness of the spline was fixed by setting a parameter, similar to the statistical parameter chi-squared, to a value such that the secular increase fit appeared reasonable to the eye.

Several other methods for separating the seasonal effect and the secular increase were tried. For example; fits with the seasonal effect represented by sines and cosines with periods of six and twelve months, and the long-term increase represented by a cubic function, followed by a subtraction of the seasonal effect from the data. All methods gave essentially equivalent results for the secular increase; the spline fit was chosen because it resulted in a smooth curve which could be differentiated.

The spline secular increase was further adjusted to remove the effect of fossil fuel combustion, resulting in the anomaly curves in Figure 2. The adjustment was approximately equivalent to subtracting from the spline a linear trend, which would have been satisfactory for this study. The procedure used is based on the observation of *Ekdahl and Keeling* (1973) that a box reservoir geochemical model of the natural CO_2 system, with some reasonable approximations and an exponential CO_2 input rate, leads to a constant fraction of each year's CO_2 input remaining airborne. The CO_2 production data between 1950 and 1974 were fitted to a spline, an airborne fraction calculated between 1 Jan. 1959 and 1 Jan. 1974, and the product of this airborne fraction and the integral of the CO_2 production rate curve beginning 1 Jan. 1959 was subtracted from the spline secular increase curve. The data points in Figures 2a and 2b have been similarly adjusted for seasonal effect and fossil fuel combustion.

Irregularities in fossil fuel consumption are too small to explain the anomaly curves. The two largest differences between the CO_2 production data from 1950 and 1972 and an exponential

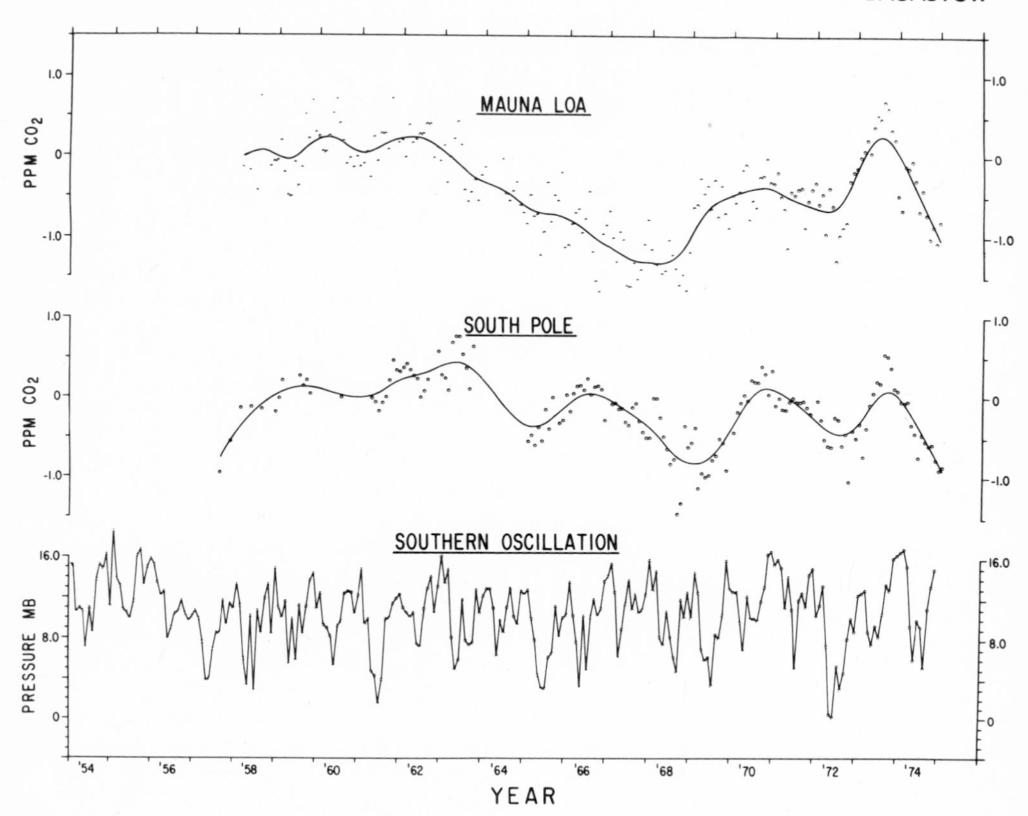

Fig. 2. Atmospheric CO_2 anomaly curves at (a) Mauna Loa, Hawaii, and (b) South Pole. The seasonal effect and fossil fuel increase were removed as described in the text. The Southern Oscillation Index, SOI, is shown at the bottom, (c). (The SOI data were assembled by W. H. Quinn, Oregon State University, Corvallis, Oregon.)

function fitted to these data are equivalent to .06 (1971, data high) and .04 (1967, data low) ppm (parts per million, mole fraction) atmospheric CO_2 per year. The total amplitude of the anomaly curves is about 1 ppm, so the change in atmospheric CO_2 associated with the anomalies is about 1 ppm in 2 years.

THE SOUTHERN OSCILLATION

The Southern Oscillation involves wind strengths, sea surface temperatures, and ocean currents (*Berlage*, 1966; and *Lamb*, 1972).

It seems to be centered in the Equatorial Pacific Ocean, but its influence is worldwide (*Berlage and de Boer*, 1960; and *Bjerknes*, 1966). The difference in average monthly barometric pressures between Easter Island and Darwin, Australia, is a suitable indicator of the Southern Oscillation, and will be called the Southern Oscillation Index or SOI (*Quinn and Burt*, 1972). Easter Island is near the center of a subtropical high pressure cell, and Darwin is representative of the Equatorial low pressure zone. The SOI is directly related to the strength of the easterly Trade Winds that blow along the Equator. When the SOI is high, the Trade Winds are stronger than average. The middle latitude westerlies along the southern flank of the subtropical anticyclone belt also tend to blow more strongly. The Humboldt Current and equatorial currents are accelerated, thereby increasing upwelling along the coast of South America and along the Equator. The Equatorial Countercurrent, which flows against the Trade Winds, is decelerated. Low SOI brings reduced Trade Winds and reduced middle latitude westerlies. The Equatorial Countercurrent flows more strongly and carries warm water to the Eastern Pacific (*Wyrtki*, 1973; and *Namias*, 1973). Low SOI is often accompanied by an El Niño, an invasion of warm water off the west coast of northern South America, which brings disaster to the fishing industry and torrential rain to a normally arid region.

DISCUSSION

An atmospheric or hydrospheric circulation phenomena should most directly effect the rate of change of the CO_2 level rather than the CO_2 level itself. For example, the transfer of CO_2 across the air-sea interface is governed approximately by an equation of the form:

$$-\frac{dp}{dt} = k(p-P) \qquad (1)$$

where p is the CO_2 partial pressure in the atmosphere, P is the CO_2 partial pressure in the ocean beneath a thin boundary layer, and k is a rate factor that depends strongly on the boundary layer thickness, and consequently on wind shear (*Skirrow*, 1965). It is, therefore, reasonable to correlate the derivatives of the anomaly curves (Figure 3) with the SOI.

First, however, the seasonal effect should be removed from the SOI and it should be smoothed so that high frequency noise does not obscure the correlations. Both the Easter Island and Darwin monthly pressure records, and their difference (Figure 2), show a strong seasonal effect, unrelated to the Southern Oscillation. This is especially clear in the Darwin pressure record because the Darwin record is less noisy than the Easter Island record. A twelve-month

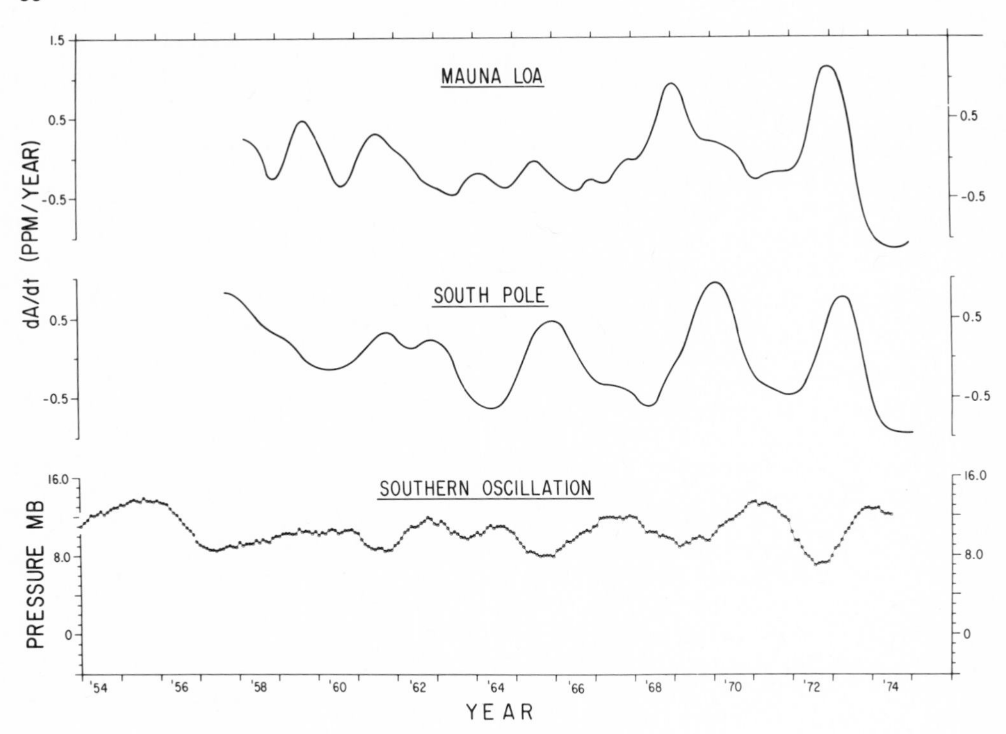

Fig. 3. Derivative anomaly curves at (a) Mauna Loa and (b) South Pole, obtained by differentiating the anomaly curves in Fig. 1. The SOI, smoothed by a twelve month moving average, is shown at the bottom for comparison.

running mean serves both to suppress noise and remove the seasonal effect.

Correlation maxima (*Box and Jenkins*, 1970; Figure 4) are then -.64, with a negative smoothed SOI lag of six months for the South Pole Derivative Anomaly, and -.55, with a negative smoothed SOI lag of 2.5 months for the Mauna Loa Derivative Anomaly. If the time series were unrelated, the cross correlation function would be expected to be zero, with a standard deviation, near zero lag, of approximately .34 and .28, respectively (*Box and Jenkins*, 1970).

The calculations were repeated with the effect of fossil fuel combustion removed by subtracting a linear function, fitted to the spline secular increase on 1 Jan. 1959 and 1 Jan. 1974, from the spline secular increase. The derivative anomaly curves and their correlations are then almost indistinguishable from corresponding curves in Figures 3 and 4.

One must use care in smoothing and differentiating data prior to correlation. Obviously, if one were to remove all frequencies but one from two time series, their correlation maxima would be approximately unity. However, in this analysis, the principal correlative features are clearly visible in the unsmoothed SOI data (Figure 2) and in the secular increase at the South Pole (Figure 3), so it does not appear that the observed correlations are an artifact of the analysis.

The nominal confidence level for the existence of a connection between the time derivatives of the anomalies and the Southern Oscillation is about 95%. However, the true confidence level is probably smaller because of the extensive data manipulation and smoothing. Longer CO_2 records should resolve the question of the existence of a connection.

The influence of the Southern Oscillation on the atmospheric CO_2 level is apparently felt 3.5 months earlier at Mauna Loa than at the South Pole. The reason may be that the South Pole is in a relatively fixed high pressure zone, isolated from the lower latitudes by circumpolar winds (*Lamb*, 1972).

Increased loss of CO_2 from the atmosphere during high SOI must be explained by increased uptake by one of the two natural reservoirs with which atmospheric CO_2 principally exchanges, the oceans and the biota. If high SOI were associated with increased rainfall over arid, but otherwise fertile land, increased vegetative uptake could explain the correlations. However, the large area of land that would have to be involved make this explanation seem unlikely.

The oceans transfer CO_2 to the atmosphere principally in regions where there is upwelling; along the Equator, along the western margins of continents, and in certain areas of the Arctic and Anarctic. The oceans absorb CO_2 from the atmosphere in other regions where the CO_2 partial pressure exerted by the oceans is less than that of the atmosphere (*Postma*, 1964; and *Keeling*, 1968).

The observed correlations may result from increased ocean uptake during high SOI in the zone of predominately westerly winds from 40° to 50°. Ocean CO_2 partial pressures there appear to be mostly less than atmospheric, although the data is very incomplete, and almost totally lacking during the southern hemispheric winter. *Kanwisher* (1963) has predicted that this zone should have a very high exchange rate for CO_2 because of its strong winds and large ocean area. The westerly winds there correlate with the easterly Trade Winds to the north, and thus with the SOI.

On the other hand, high SOI should also bring increased upwelling, at least along the northern coast of South America and along the Equatorial Pacific Ocean. The observed correlations seem to indicate that the stronger absorption by the oceans dominates the effect of the increased upwelling.

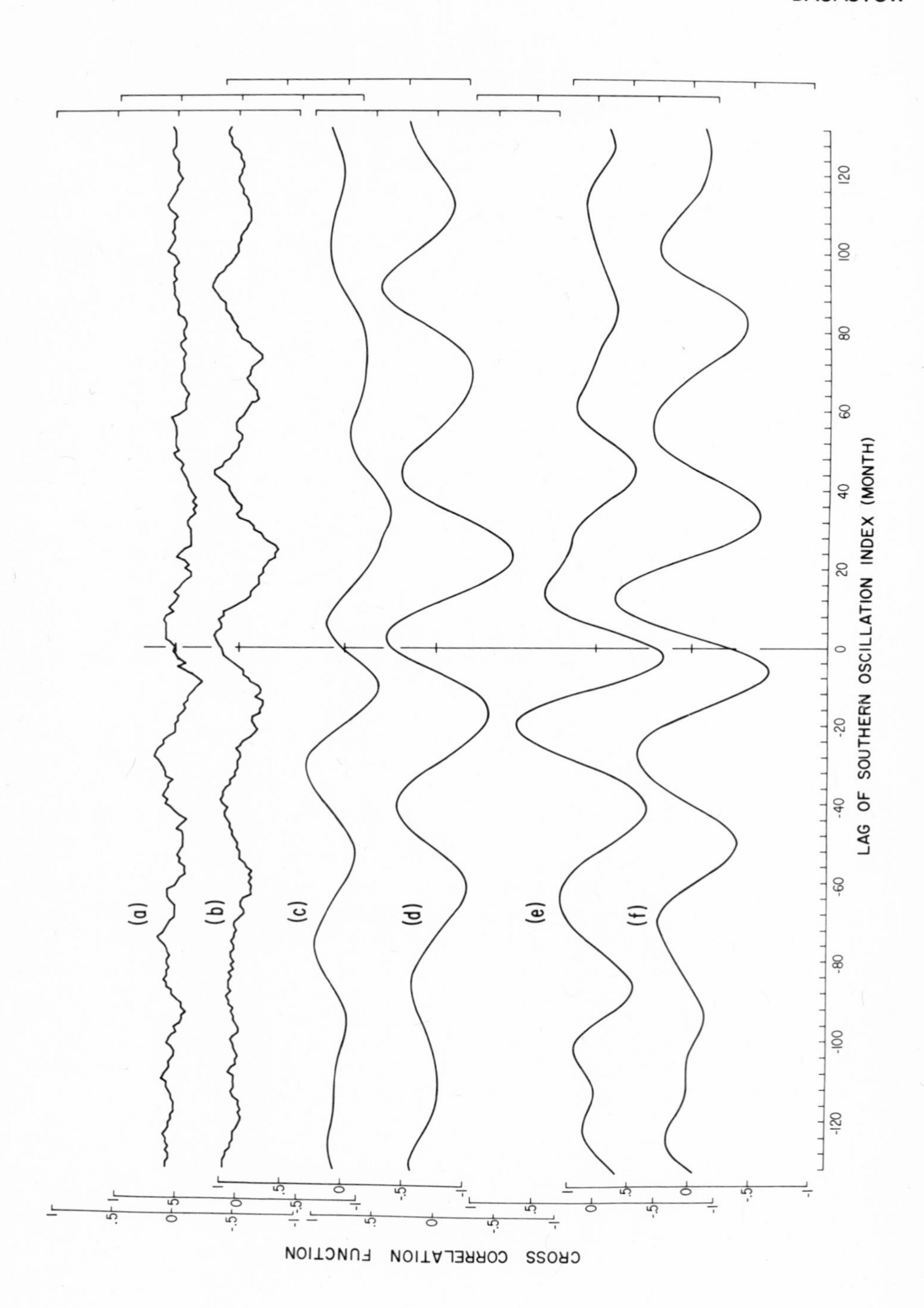
(a)
(b)
(c)
(d)
(e)
(f)
LAG OF SOUTHERN OSCILLATION INDEX (MONTH)
CROSS CORRELATION FUNCTION

Fig. 4. Cross correlation functions for Mauna Loa and South Pole CO_2 records and the Southern Oscillation Index, SOI: (a) Unsmoothed Mauna Loa Anomaly data (with missing data filled in by interpolation), and unsmoothed SOI, (b) Unsmoothed South Pole Anomaly data (with missing data filled in by interpolation), and unsmoothed SOI, (c) Mauna Loa Anomaly Curve and smoothed (twelve month moving average) SOI, (d) South Pole Anomaly Curve and smoothed SOI, (e) Derivative Mauna Loa Anomaly Curve and smoothed SOI, and (f) Derivative South Pole Anomaly Curve and smoothed SOI. All correlation functions are plotted against the lag of the SOI. Standard deviations, if each pair of curves were uncorrelated, would be, in order: .09, .12, .18, .29, .28, and .34 near zero lag.

ACKNOWLEDGEMENTS

I wish to thank C. D. Keeling for critical comments. This work was supported by the Atmospheric Sciences Division of the National Science Foundation under Grants GA-31324X and ATM74-00669.

REFERENCES

Berlage, H. P. and H. J. de Boer. 1960. On the Southern Oscillation, its way of operation and how it effects pressure patterns in the higher latitudes. *Geof. pura e appl., 46.* 329-351.

Berlage, J. 1966. The Southern Oscillation and world weather. In: *Mededel. Verhandel, Koninkl, Ned. Meteor. Inst., No. 88.* 1-152. (Staatsdrukkerij, 'S-Gravenhage.)

Bjerknes, J. 1966. A possible response of the atmospheric Hadley circulation to equatorial anomalies of ocean temperature. *Tellus, Vol. 18.* 820-829.

Bolin, B. and C. D. Keeling. 1963. Large scale atmospheric mixing as deduced from the seasonal and meridional variations of carbon dioxide. *J. of Geophys. Res., 68.* 3899-3920.

Box, G. E. P., and G. M. Jenkins. In: *Times Series Analysis,* Holden-Day, San Francisco. 370-377.

Ekdahl, C. A. and C. D. Keeling. 1973. Atmospheric carbon dioxide and radiocarbon in the natural carbon cycle: I. Quantitative deductions from records at Mauna Loa Observatory and at the South Pole. In: *Carbon and the Biosphere,* G. M. Woodwell and E. V. Pecan (Eds.), U. S. Atomic Energy Commission. 51-85.

Hall, C. A. S., C. A. Ekdahl, Jr., and D. E. Wartenberg. 1975. A fifteen-year record of biotic metabolism in the northern hemisphere. *Nature, 225.* 136-138.

Junge, C. E. and G. Czeplak. 1968. Some aspects of the seasonal variation of carbon dioxide and ozone. *Tellus, 20.* 422-434.

Kanwisher, J. 1963. On the exchange of gases between the atmosphere and the sea. *Deep Sea Res., 10.* 195-207.

Keeling, C. D. 1968. Carbon dioxide in surface ocean waters, 4. Global distribution. *J. Geophys. Res., 73.* 4543-4553.

Keeling, C. D. 1973. Industrial production of carbon dioxide from fossil fuels and limestone. *Tellus, 25.* 174-198.

Keeling, C. D., J. A. Adams, Jr., C. A. Ekdahl, Jr., and P. R. Guenther. 1976a. Atmospheric carbon dioxide variations at the South Pole. *Tellus, 28.* 552-564.

Keeling, C. D., R. B. Bacastow, A. E. Bainbridge, C. A. Ekdahl, Jr., P. R. Guenther, L. S. Waterman, and J. F. S. Chin. 1976b. Atmospheric carbon dioxide variations at Mauna Loa Observatory, Hawaii. *Tellus, 28.* 538-551.

Lamb, H. H. 1972. In: *Climate: Present, Past and Future, Vol. 1,* Methuen and Co, Ltd., London.

Namias, J. 1973. Response of the Equatorial Countercurrent to the subtropical atmosphere. *Science, 181*. 1244-1245.
Postma, H. 1964. The exchange of oxygen and carbon dioxide between the ocean and the atmosphere. *Netherlands J. of Sea Res., 2*. 258-283.
Quinn, W. H. and W. B. Burt. 1972. Use of the Southern Oscillation in weather prediction. *J. Appl. Meteor., 11*. 616-628.
Reinsch, C. H. 1967. Smoothing by spline functions. *Numerishe Mathematik, 10*. 177-183.
Rotty, R. M. 1973. Commentary on and extension of calculative procedure for CO_2 production. *Tellus, 25*. 508-517.
Skirrow, G. 1965. The dissolved gases - carbon dioxide. In: *Chemical Oceanography* (Riley, J. P. and G. Skirrow (Eds.), Academic Press, London and New York. 312-316.
Wyrtki, K. 1973. Teleconnections in the equatorial Pacific Ocean. *Science, 180*. 66-68.

Hydrogen Ions and the Thermodynamic State of Marine Systems

R. G. Bates and C. H. Culberson

University of Florida

ABSTRACT

The influence of surface pH on the exchange of CO_2 between atmosphere and ocean is considered. Different acidity functions for seawater measurements are compared, and acidic dissociation constants in the carbonate system are critically examined as a function both of temperature and salinity.

INTRODUCTION

Despite the complexity of factors influencing the transfer of carbon from its combustible source to the sea, the capacity of the ocean as a reservoir of fossil fuel carbon is profoundly influenced by the thermodynamic state of the surface water and by the partial pressure of carbon dioxide in the air mass with which the surface is in contact. Similarly, if absorption of carbon dioxide is to be a continuous process and not come gradually to a standstill, removal of dissolved carbon dioxide from the surface waters must take place. Here, ocean currents play a major role and are assisted to a significant degree by diffusion to subsurface layers. Spontaneous transport requires that a gradient in the concentration of dissolved carbon dioxide exist.

Inasmuch as dissolved carbon dioxide is weakly acidic, its concentration is also lowered by chemical interactions with bases encountered in the aqueous medium. Most important is reaction with hydroxide ion formed by the self-dissociation of water:

$$CO_2 + OH^- \leftrightarrows HCO_3^- \tag{1}$$

The bicarbonate ion formed is a weaker acid than CO_2 itself, but can nonetheless react further to some extent with OH^- to form carbonate ion:

$$HCO_3^- + OH^- \leftrightarrows CO_3^{2-} + H_2O \tag{2}$$

These two reactions add materially to the capacity of the seas to retain carbon dioxide from the atmosphere.

It is evident from equations (1) and (2) that hydroxide ion plays a primary part in the ability of seawater to accommodate dissolved carbon dioxide. Equilibrium hydroxide concentrations are quantitatively represented by the *pH* value of the medium. The approximate relationship between *pH* and the equilibrium partial pressure of CO_2 over natural seawater (with a titration alkalinity of 2.3 mmol/l) is shown in Figure 1. The same figure illustrates the increasing partial pressure of CO_2 in the atmosphere with passage of time, as taken from the measurements of Keeling (see *Bolin*, 1970). If carbon dioxide were the only acidic or basic constituent of the seawater system, the equilibrium *pH* value would lie somewhat below 6. The presence of carbonates, ammonia, borate, phosphate, and silicate from biological, atmospheric, and geological

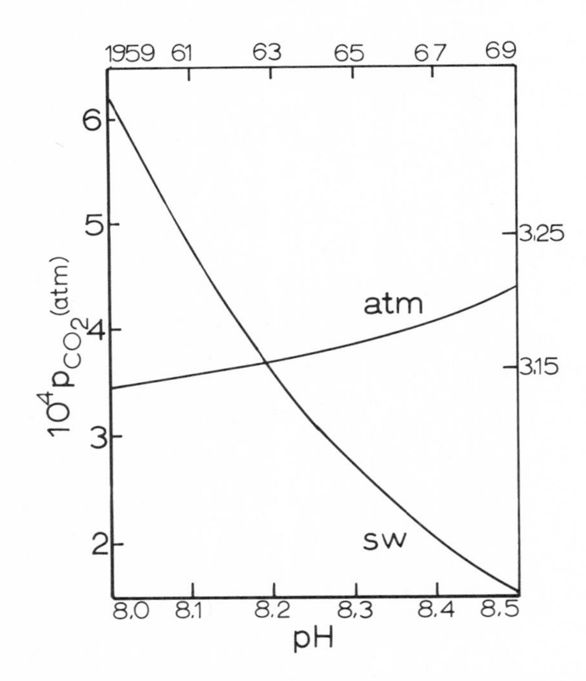

Fig. 1. Partial pressure of CO_2 in equilibrium with seawater at 25°C and various *pH* values (left and bottom scales) compared with that in the atmosphere (right and top scales).

sources, however, is sufficient to maintain the *pH* of normal surface seawater at 8, or slightly above. It may be concluded from Figure 1 that the atmosphere and the sea are nearly in equilibrium with respect to carbon dioxide. As the partial pressure of CO_2 in the atmosphere rises further, however, one may expect the absorptive process to accelerate. However, it is difficult to predict to what extent the seas will be able to temper the increase.

pH AND *pK*

The hydrogen ion concentration or *pH* value is a quantitative index of the relative concentrations of acid and its conjugate base in each of the acid-base systems present in seawater and, less directly, of speciation in certain metal-ligand systems. Formally, the proportionality constant between hydrogen ion concentration and acid/base ratio is a dissociation constant, K_c:

$$\frac{(acid)}{(base)} = \frac{(H+)}{K_c} \tag{3}$$

where parentheses indicate concentrations on either a weight or volume basis. Equation (3) acquires its form from the mass law expression for the equilibrium process $acid \leftrightharpoons base + H^+$. The simple formulation of equation (3) is however, not completely adequate for all purposes. The two main reasons for this difficulty are 1), as usually determined experimentally, *pH* values are not a unique measure of hydrogen ion concentration, and 2) K_c is not a true constant for a given acid/base system at a fixed temperature and pressure, but may vary with the composition and especially with the total ionic strength of the medium.

It is evident from equation (3) that for each definition of a hydrogen ion scale there is a corresponding definition of K. Furthermore, a particular ratio of concentrations of acid to conjugate base corresponds to a fixed difference, *pH-pK*. Thus, a *pH* scale defined in terms of hydrogen ion activity, for example, will require, for accurate results, a value of the dissociation constant different from the K_c of equation (3). For practical reasons, one is forced to define the constant in a manner consistent with practical *pH* measurements if useful assessments of acid/base ratios are to be obtained. Let us then examine various formulations of the equilibrium constant in the light of their consistency with the various scales for the measurement of hydrogen ion concentrations and activities.

Maximum constancy of the equilibrium constant as composition is varied is to be expected from the thermodynamic constant K_t, expressed in terms of activities: activity (a) equals the product of a concentration (c or m) and an activity coefficient (y or γ).

Thus, on the scale of molality (moles per kilogram of water),

$$\frac{m_{acid}}{m_{base}} = \frac{a_H \gamma_{base}}{K_t \gamma_{acid}} \tag{4}$$

The successful use of equation (4) would require an experimental measure of a_H as well as a knowledge of the activity coefficients of the acid and its conjugate base in each solution for which the acid/base ratio is desired. It is impractical to obtain this information on a routine basis.

Activity coefficients such as those on the right side of equation (4) are sensitive to changes of ionic strength and insensitive to changes of composition when a constant ionic strength is maintained, as is the case in natural seawater. Under these conditions, the "incomplete" or "apparent" dissociation constant K' is useful:

$$\frac{m_{acid}}{m_{base}} = \frac{a_H}{K'} \tag{5}$$

As will be seen in the following section, experimental scales of hydrogen ion activity have been set up on a conventional, or arbitrary basis by the National Bureau of Standards. Together with self-consistent values of K', the NBS pH scale offers a practical assessment of acid/base balance in seawater. The most common (but not only) formulations of the equilibrium constant are those designated above by the symbols K_c, K_t, and K'. The corresponding definitions of pH scales will now be considered.

PRACTICAL pH SCALES FOR SEAWATER

It is evident from the foregoing discussion that one would like to have a reliable experimental method to determine routinely pm_H, pc_H, or pa_H in seawater. The scale devised by *Sørensen* (1909) and utilized by *Buch et al.* (1932) in their classic study of the seawater CO_2 system, was intended to furnish values of the hydrogen ion concentration, pc_H. Nevertheless, it is not a precise measure of either concentration or activity. The NBS pH scale is fixed by a series of primary reference standard solutions of ionic strength (I) no greater than 0.1, whose assigned $pH(S)$ values fix a scale of pa_H (*Bates*, 1973). This is properly regarded as a scale of "conventional activity pH", inasmuch as it is based on an arbitrary choice of the activity coefficient of one species of ion. The NBS scale has been widely used to measure the pH of seawater and to determine values of K' for acid-base systems in natural and synthetic seawater (see, for example, *Skirrow*, 1965, and *Mehrbach et al.*, 1973).

The NBS *pH* scale is of general utility. It provides reproducible data and is, furthermore, the only practical scale for comparisons of the acidity of media of varying compositions and ionic strengths. If measurements are to be confined to seawater, however, recent evidence suggests that certain simplifications in *pH* measurement are possible (*Hansson*, 1973b; *Bates and Macaskill*, 1975). These rest on the effectiveness of seawater--a medium of relatively constant composition and ionic strength at a given salinity--in stabilizing the activity coefficients of solutes when small changes in the concentrations of these dissolved substances occur. In addition, the *pH* cell with glass electrode and calomel reference electrode includes a liquid junction; it appears that seawater is likewise effective in stabilizing the liquid-junction potential.

The chief disadvantage of the NBS scale for measurements of the level of acidity in seawater is the residual liquid-junction potential that is included in each experimental value of the *pH*. The *pH* cell with a reference electrode coupled to a bridge solution of concentrated potassium chloride is standardized with dilute ($I \leq 0.1$) NBS standard solutions and subsequently used to measure *pH* in seawater, the ionic strength of which may be as high as 0.7 $mol\ kg^{-1}$. The liquid-junction potential (E_j) at the interface between the seawater and the bridge solution will not be the same as between the calibrating solution and the bridge. As a consequence, the measured *pH* of the seawater sample will not fall exactly on the activity scale defined by the NBS standards but will depart from it by an amount ΔpH:

$$\Delta pH = \frac{\Delta E_j}{(RT \ln 10)/F} \tag{6}$$

where E_j is as stated above, R is the gas constant, T is temperature in degrees Kelvin, and F is the Faraday Constant.

The residual liquid-junction potential for "normal" seawater has been estimated to have a rather constant value of about $3.2mV$ (*Hawley and Pytkowicz*, 1973), and ΔpH is accordingly 0.054 *pH* unit at 25°C. Thus, measurements of *pH* in seawater vs. NBS standards do not represent pa_H exactly, and the K' values derived from them differ somewhat from those represented by equation (5). Nevertheless, $pH-pK'$ is a correct measure of the acid/base ratio (*Pytkowicz et al.*, 1974). The extensive tables of K' values are of great utility when used with *pH* based on dilute NBS standards and which includes the residual liquid-junction potential.

It is unfortunate that the NBS *pH* and the corresponding K' values in seawater do not represent well-defined thermodynamic properties of the medium, for example a_H or m_H on the one hand and K_c, K_t, or K' on the other. It has been shown (*Bates and Macaskill*, 1975), however, that the simplifications afforded by the seawater medium make it possible to define *pH* in terms of concentrations. Further, *Hansson* (1973b) has proposed that *pH* in seawater be measured by a cell with a glass electrode coupled with a silver-silver chloride reference electrode immersed in a solution matching closely the composition of seawater. In this way, one can avoid the liquid junction of the usual *pH* cell. Hansson's standard buffer solutions were assigned *pH* values based on the alteration of the glass electrode potentials when small amounts of a strong acid were added to synthetic seawater. Inasmuch as a part of the acid combines with sulfate ion in the seawater, Hansson's *pH*--designated *pH(SWS)*--corresponds to

$$pH(SWS) = -\log (m_H + m_{HSO_4}) = -\log (H^+)_t \qquad (7)$$

where $(H^+)_t$ represents the sum of the concentrations of free H^+ and that combined with SO_4^{2-}. Hence, the equilibrium constants corresponding to Hansson's *pH* values are slightly different from K_c as well as from K'. They will be designated K_{ct}.

The *pH(SWS)* can be expressed in terms of the molality of free hydrogen ion, m_H or $(H^+)_f$; β_{HSO4} (the formation constant of the HSO_4^- ion); and the molality of sulfate ion not combined with H^+:

$$pH(SWS) = -\log m_H(1+\beta_{HSO_4} m_{SO_4}) \qquad (8)$$

An essentially constant fraction of the total hydrogen ion in seawater is therefore complexed with sulfate. *Bates and Macaskill* (1975) and *Bates* (1975) have suggested that the *pH* of this constant ionic medium might be expressed more reasonably not in terms of $(m_H + m_{HSO_4})$, but rather in terms of the free hydrogen ion concentration and pm_H, thus returning to the simple definition embodied in equation (3) and to K_c. Preliminary values for the pm_H of some suitable standard buffer solutions in synthetic seawater have been presented, and a more detailed investigation is under way.

ESTABLISHMENT OF A STANDARD *pH* SCALE FOR SEAWATER

The *pH(SWS)* scale and the scale of pm_H are both fixed by standard buffer solutions in synthetic seawater matching closely the composition of natural seawater, but with borate and fluoride absent. Standard values of *pH(SWS)* or pm_H are assigned to these solutions by measurement of the *emf* (*E*) of cells without liquid-junction containing silver-silver chloride reference electrodes and an electrode responsive to hydrogen ion. Hansson's standardization procedure utilizes a glass electrode, while that in use in the authors' laboratory couples the reference electrode to the highly reproducible hydrogen gas electrode. This cell can be represented by

$$\text{Glass electrode or } Pt;H_2(\underline{g}) \;\Big|\; HCl(\underline{m}) \text{ or buffer in sw} \;\Big|\; AgCl;Ag \qquad (A)$$

Both procedures require the evaluation of a standard *emf* ($E^{\circ *}$) for the cell, referred to the standard state of "infinite dilution" of foreign solute species in the seawater medium. For this purpose, measurements of E are made with small amounts ($\underline{m}$) of HCl added to the seawater "solvent", the concentrations being adjusted so that the ionic strength and chloride molality are unchanged by the addition. Formally,

$$E^{\circ *} = E + \frac{RT \ln 10}{F} \log (H^+) m_{Cl} \gamma_{HCl}^{*\,2} \qquad (9)$$

It is evident that m_{Cl} in the seawater medium is known; furthermore, $\gamma_{HCl}^{*} = 1$ in the standard state, that is, when $m_{HCl}=0$ in the seawater, and it makes only a small contribution when the molality of added HCl is less than $0.05\ mol\ kg^{-1}$ (*Bates and Macaskill*, 1975). *Hansson* (1973b) identifies (H^+) with the stoichiometric molality of added HCl, that is, with $(H^+)_t$ in equation (7); an extrapolation to zero added HCl then yields the standard *emf* which we designate $E^{\circ *}(t)$. This extrapolation, based on measurements of K. H. Khoo in the authors' laboratory, is illustrated in the upper half of Figure 2. Although the plot is not linear in m_{HCl}, $E^{\circ *}(t)$ was found to be $0.24647V$ at 25°C.

Values of $pH(SWS)$ for reference solutions can now be calculated from measurements of E for dilute buffer solutions in the same seawater solvent by

$$pH(SWS) = \frac{E - E^{\circ *}(t)}{(RT \ln 10)/F} + \log m_{Cl} \qquad (10)$$

As indicated by equation (7), these values represent the sum of the molalities of free hydrogen ion and hydrogen in combination with the sulfate present in the seawater.

If one desires to identify (H^+) in equation (9) with the molality of free or uncombined hydrogen ion, designated m_H or $(H^+)_f$, the standard *emf* will have a different value, $E^{\circ *}(f)$. It can also be determined by a simple extrapolation if m_H in the HCl-seawater mixtures is known. Combination of equations (7) and (8) gives

$$m_H = m_{HCl}/(1+\beta_{HSO_4} m_{SO_4}) \qquad (11)$$

and shows that one must evaluate β_{HSO_4}, that is, $m_{HSO_4}/(m_H m_{SO_4})$, in

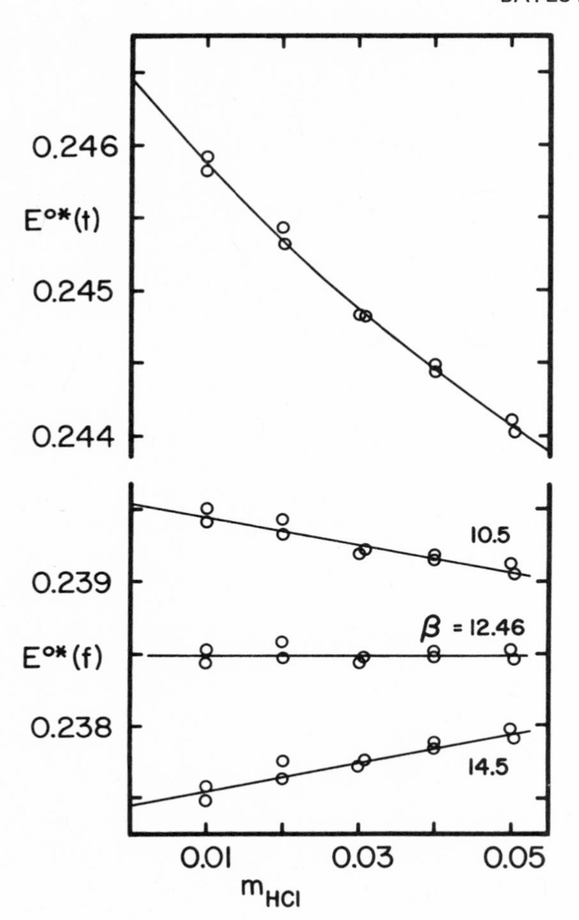

Fig. 2. The standard *emf* of cell (a) at 25°C from measurements of artificial seawater containing small amounts of added *HCl*. Salinity, 35 ‰.

seawater in order to arrive at $E^{\circ *}(f)$ by extrapolation to $m_H = 0$. Although several estimates of this constant have been reported, an independent measure can be obtained from the *emf* of cell (A) with *HCl*-seawater solutions. As shown in the lower part of Figure 2, constant values of $E^{\circ *}(f)$ at 25°C are obtained only when β_{HSO_4} is given the value 12.46 kg mol^{-1}. The intercept gives $E^{\circ *}(f)$ = 0.23849V, in excellent agreement with the value found earlier (0.23850V) from measurements in seawater without sulfate (*Bates and Macaskill*, 1975). Standard values of pm_H can now be derived from *emf* measurements for dilute buffer solutions by

$$pm_H = \frac{E - E^{o*}(\delta)}{(RT\ \ln\ 10)/F} + \log m_{Cl} \qquad (12)$$

With the use of these standard reference solutions, it appears possible to determine pm_H in seawater by means of either the usual pH cell with glass and calomel reference electrodes or with Hansson's glass-AgCl;Ag cell or cell (A), which are essentially free from a liquid-junction potential. This is because seawater, a constant ionic medium at a given salinity, is effective in stabilizing both the activity coefficients and the liquid-junction potential. The pH of a useful standard buffer solution composed of tris(hydroxymethyl)aminomethane ("Tris") and its hydrochloride in synthetic seawater is given in Table 1. Data for three scales, namely $pH(NBS)$, $pH(SWS)$, and pm_H, are listed. These values refer to the molality scale (moles per kilogram of water). Values based on moles per kilogram of seawater are all higher by 0.015 unit. From equations (10) and (12), it is evident that the difference between pm_H and $pH(SWS)$ is given by

$$pm_H - pH(SWS) = \frac{E^{o*}(t) - E^{o*}(\delta)}{(RT\ \ln\ 10)/F} \qquad (13)$$

When the respective values for $E^{o*}(t)$ and $E^{o*}(\delta)$ from Figure 2 are inserted into this equation, it is seen that the difference amounts to 0.135 unit at 25°C.

The apparent agreement between $pH(NBS)$ and pm_H is doubtless fortuitous; it is presumably due to an approximate compensation of the residual liquid-junction potential by changes in activity coefficients. A partial cancellation of this sort has been found for simulated blood serum (*Mohan and Bates*, 1975).

TABLE 1. Comparison of pH Values for Tris Buffer Solutions in Synthetic Seawater of Approximately 35 ‰ Salinity at 25°C.

Buffer Solution	$pH(NBS)$	$pH(SWS)$	pm_H
Tris(0.02m), Tris·HCl(0.02m)	8.208	8.072	8.207
Tris(0.04m), Tris·HCl(0.04m)		8.073	8.208
Tris(0.06m), Tris·HCl(0.06m)		8.074	8.209

CARBONIC ACID EQUILIBRIA IN SEAWATER

The first and second ionization constants of carbonic acid in seawater were determined by *Hansson* (1973a) and by *Mehrbach et al.* (1973). In this section we give a detailed comparison of both sets of data as a function of temperature at 35 ‰ salinity and as a function of salinity at 25°C. These are the only experimental conditions that the two sets of data have in common.

Hansson (1973a) defined K_1 and K_2 in terms of the total concentration of hydrogen ion $(H^+)_t$, while *Mehrbach et al.* (1973) expressed the equilibrium constants in terms of the conventional hydrogen ion activity on the NBS scale. The latter authors also determined the "total activity coefficient" γ_H of hydrogen ion:

$$\gamma_H = a_H(NBS)/(H^+)_t \tag{14}$$

at each of their experimental points. For comparison with the data of *Hansson* (1973a), we have converted the constants reported by *Mehrbach et al.* (1973) to K_{ct} (based on the total concentration of hydrogen ion $(H^+)_t$) by dividing their values of K_1' and K_2' by their measured values of γ_H.

The ionization constants of carbonic acid are defined by

$$K_1 = \frac{(H^+)_t(HCO_3^-)}{(CO_2+H_2CO_3)} \tag{15}$$

and

$$K_2 = \frac{(H^+)_t(CO_3^{2-})}{(HCO_3^-)} \tag{16}$$

where concentrations (indicated by parentheses) are in moles per kilogram of seawater. Both *Hansson* (1973a) and *Mehrbach et al.* (1973) determined K_1 and the product K_1K_2; K_2 was calculated from the ratio $(K_1K_2)/K_1$.

The dependence of pK_1 on salinity is shown in Figure 3. It is evident that pK_1 is a linear function of the cube root of the salinity for salinities ranging from 19 to 42 ‰. The manner in which pK_1 changes with salinity is not surprising, since it can be shown that the Debye-Hückel function $\sqrt{I}/(1+\sqrt{I})$ is also a linear function of $S^{1/3}$ for S between 19 and 42 ‰. At 25°C, the values of pK_1 found by *Mehrbach et al.* (1973) are systematically lower than those of *Hansson* (1973a), the average difference being 0.020 unit. Part of this difference is due to the fact that *Hansson* (1973a) used artificial seawater without fluoride, while *Mehrbach et al.* (1973) used natural seawater containing fluoride. The total concentration of hydrogen ion $(H^+)_t$ in the artificial seawater used by *Hansson* (1973a) is

$$(H^+)_t = (H^+)_f[1 + \beta_{HSO_4}(SO_4^{2-})_t] \quad (17)$$

while that in the natural seawater used by Mehrbach et al. is

$$(H^+)_t = (H^+)_f[1 + \beta_{HSO_4}(SO_4^{2-})_t + \beta_{HF}(F^-)_t] \quad (18)$$

The values of β_{HSO4} and β_{HF} at 25°C were found by *Culberson, Pytkowicz, and Hawley* (1970) to be 12.5 and 414 kg mol^{-1}, respectively. At 25°C and 35 ‰ salinity, the difference between the two pH scales represented by equations (17) and (18) is thus

$$(pH)_{Hansson} - (pH)_{Mehrbach\ et\ al.} = 0.010 \quad (19)$$

The values of pK_1 of Mehrbach et al. should therefore be 0.010 pK unit lower than Hansson's values. The difference in pH scales thus accounts for only half of the difference between the two sets of data for pK_1, and the remainder (0.010 pK unit) represents a systematic difference between the two sets of data.

The temperature dependence of pK_1 is shown in Figure 4. *Hansson* (1973a) expressed his values of pK_1 and pK_2 as linear functions of $1/T$, where T is the thermodynamic temperature in °K. A re-examination of his experimental data shows that a linear temperature dependence gives a much better fit of his data. The values of pK_1 of *Mehrbach et al.* (1973), when plotted against temperature, show a slight curvature. There is no thermodynamic reason to expect pK_1 or pK_2 to be linear functions of temperature over any extended range. The pK_1 values of *Harned and Bonner* (1945) in sodium chloride solutions show curvature similar to that evidenced by the data of Mehrbach et al.

The dependence of $pK_1 + pK_2$ on salinity is shown in Figure 5. A slight curvature is found when $pK_1 + pK_2$ is plotted against the cube root of the salinity. At 25°C, the values from the work of *Mehrbach et al.* (1973) are systematically lower by 0.019 unit than those of *Hansson* (1973a). Equation (19) would suggest that the presence of fluoride would cause the results to be lower by 0.020 unit. The agreement is excellent. It may therefore be said that the values of $pK_1 + pK_2$ from the two investigations are identical within experimental error at 25°C.

The temperature dependence of $pK_1 + pK_2$ at 35 ‰ salinity is also shown in Figure 5. Hansson's values are a linear function of temperature, while a plot of the data of Mehrbach et al. against temperature is slightly curved.

The change of pK_2 with salinity at 25°C is apparent in Figure 3. The experimental values are nearly a linear function of the cube root of salinity. On the average, there is no difference between the two sets of data, whereas it is expected that the effect of

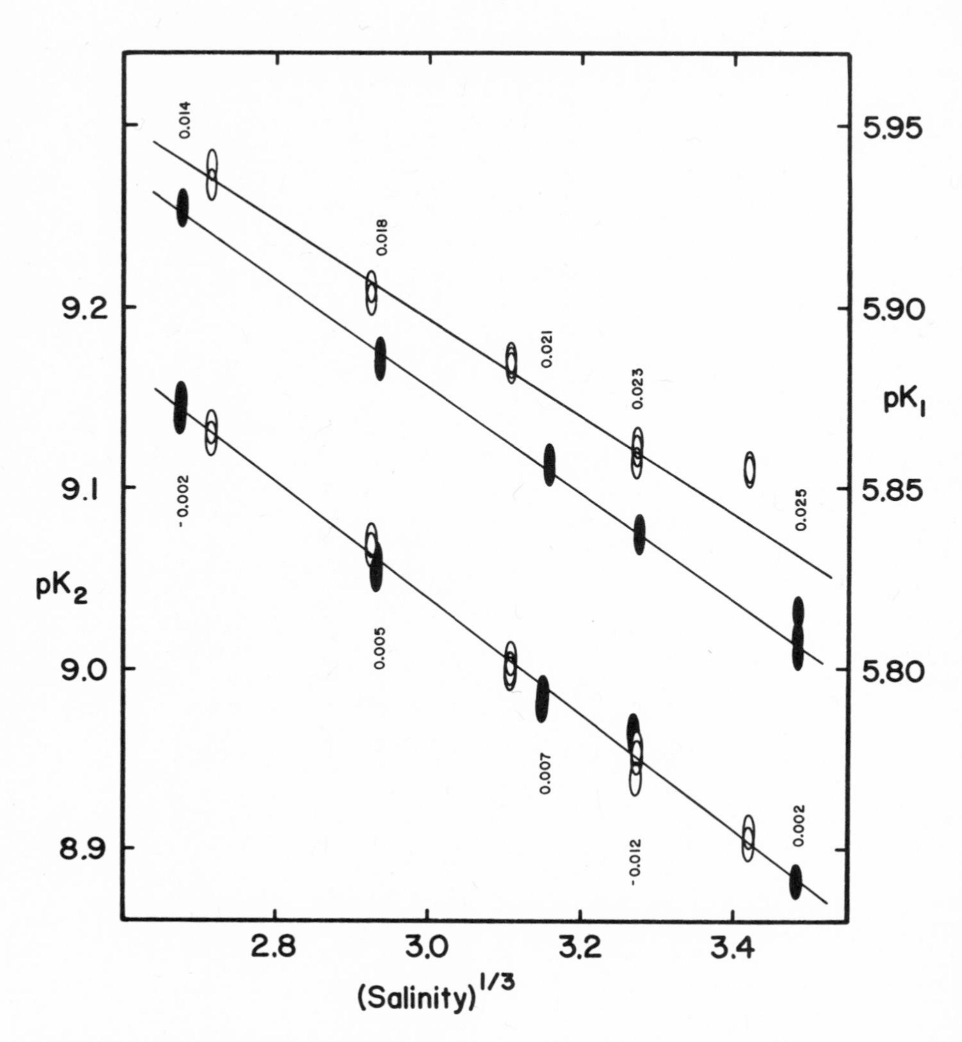

Fig. 3. Salinity dependence of pK_1 and pK_2 at 25°C. Open ovals, *Hansson* (1973a); solid ovals, *Mehrbach et al.* (1973). The lines are the least-squares lines through the experimental data, excluding Hansson's values for pK_1 at 40 °/oo salinity. For pK_2, only the line through Hansson's values is shown. The numbers above and below the ovals are the differences (*Hansson*, 1973a - *Mehrbach et al.*, 1973). The height of the ovals represents ±2 standard deviations.

fluoride on the *pH* will be such as to make the pK_2 of Hansson higher by 0.010 unit. The discrepancy is due to the fact that the two sets of values for pK_1 show a systematic difference of 0.010 unit, with Hansson's the higher, while the two values of $pK_1 + pK_2$ are in exact agreement. A high value of pK_1 results in a low value for pK_2.

The following conclusions can be reached from the above comparison. At 25°C and salinities between 19 and 42 ‰, the values of $pK_1 + pK_2$ found by *Hansson* (1973a) and by *Mehrbach et al.* (1973) are in excellent agreement, their values of pK_1 differ by 0.010 *pK* unit (Hansson higher), and their values of pK_2 differ by 0.010 *pK* unit (Hansson lower). At 35 ‰ salinity and temperatures between 2 and 35°C, the agreement of $pK_1 + pK_2$ and of pK_2 is not quite as good. The two sets of values for $pK_1 + pK_2$ differ by approximately 0.010 unit between 2 and 25°C (Hansson lower). The values of pK_1 found by Hansson are systematically higher (by 0.010 unit) than

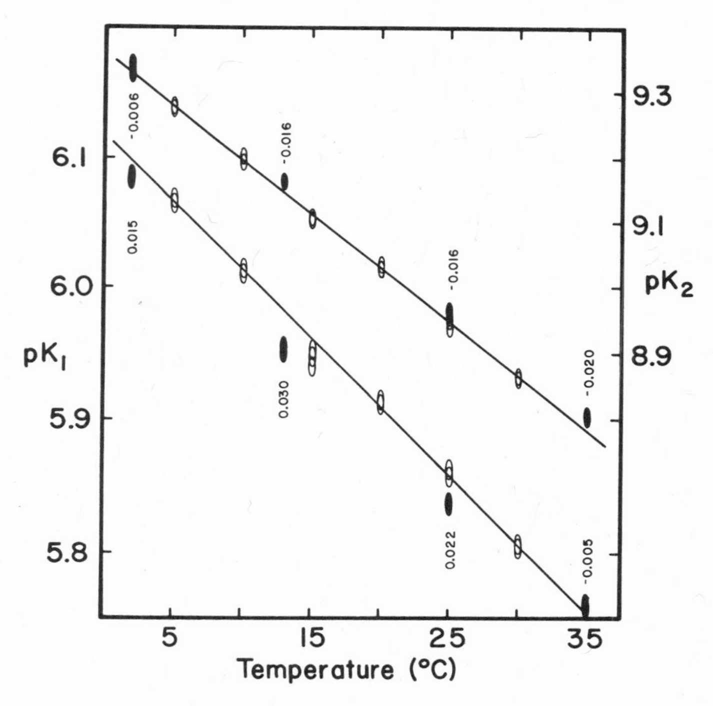

Fig. 4. Temperature dependence of pK_1 and pK_2 at 35 ‰ salinity. Open ovals, *Hansson* (1973a); solid ovals, *Mehrbach et al.* (1973). Least-squares lines through Hansson's values, excluding those at 15°C, are shown. The numbers above and below the ovals are the differences (Hansson - Mehrbach et al.) in *pK* units.

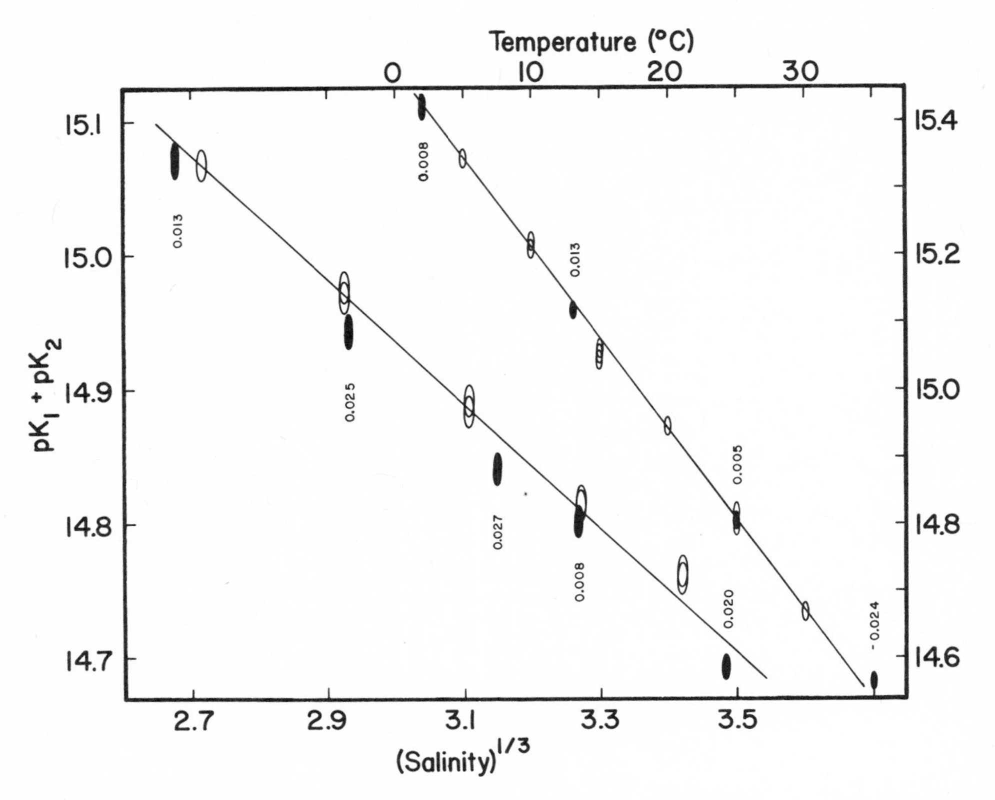

Fig 5. Salinity dependence (at 25°C) and temperature dependence (at 35 °/oo) of $pK_1 + pK_2$. Open ovals, *Hansson* (1973a); solid ovals, *Mehrbach et al.* (1973). Least-squares lines through Hansson's values, excluding that at 40 °/oo salinity, are shown. The numbers above and below the ovals are the differences (Hansson - Mehrbach et al.) in pK units.

those of Mehrbach et al., while the differences in pK_2 range from *0.02* at 2°C to *0.03* at 35°C (Hansson lower).

From the comparisons shown in Figures 3, 4, and 5, it is possible to derive smoothed values of pK_1 and pK_2 consistent within about ± 0.015 unit with the measurements of both *Hansson* (1973a) and *Mehrbach et al.* (1973). These are given, for several temperatures and a salinity of 35 °/oo, in the second columns of Tables 2 and 3. By use of the "total activity coefficients" γ_H defined by equation (14), the pK_{ct}, based on $pH(SWS)$ is converted into pK', based on $pH(NBS)$. This latter quantity is listed in the third columns of the tables, while pK_c is listed in the fourth columns. Concentrations of species i are expressed in moles per kilogram of seawater,

TABLE 2. Values of pK_1 for Carbonic Acid in Seawater (Salinity 35 °/oo) from 0 to 35°C.

t (°C)	pK_{ct}	pK'	pK_c
0	6.119	6.197	6.203
5	6.067	6.143	6.158
10	6.015	6.096	6.117
15	5.963	6.057	6.073
20	5.911	6.024	6.033
25	5.859	5.999	5.994
30	5.807	5.980	5.959
35	5.755	5.967	5.923

Concentrations in moles per kilogram of seawater. pK_{ct} calculated from the $pH(SWS)$, pK' from $pH(NBS)$, and pK_c from pM_H.

TABLE 3. Values of pK_2 for Carbonic Acid in Seawater (Salinity 35 °/oo from 0 to 35°C.

t (°C)	pK_{ct}	pK'	pK_c
0	9.362	9.465	9.446
5	9.279	9.392	9.370
10	9.196	9.322	9.298
15	9.114	9.255	9.224
20	9.031	9.190	9.153
25	8.948	9.128	9.083
30	8.865	9.068	9.017
35	8.783	9.010	8.951

Concentrations in moles per kilogram of seawater. pK_{ct} calculated from the $pH(SWS)$, pK' from $pH(NBS)$, and pK_c from pM_H.

designated M_i; hence pK_c is consistent with pM_H rather than pm_H.

Edmond and Gieskes (1970) have examined critically the results obtained by *Buch et al.* (1932) and *Lyman* (1957), deriving equations for the variation of pK_1' and pK_2' with chlorinity and temperature. Their values at 25°C and a salinity of 35 °/oo are as follows: $pK_1' = 5.967$; $pK_2' = 9.096$. The mean differences between the results given in Tables 2 and 3 and their values at the eight temperatures are 0.024 (pK_1') and 0.047 (pK_2'). For pK_1', closest agreement is obtained at the lower temperatures, whereas the pK_2' values agree best at the higher temperatures.

Tables 2 and 3 serve to illustrate the dependence of the values of pK_1 and pK_2 on the scale of *pH* selected. The differences in *pH* and *pK* evidently may amount to more than 0.1 unit. One may conclude from Figure 1 that a rather precise description of the thermodynamic state of seawater is needed to evaluate the capacity of surface water to absorb atmospheric carbon dioxide. A consistent account of the acid-base state of seawater will only be achieved when a single scale of *pH* and *pK* is universally recognized.

ACKNOWLEDGEMENT

This work was supported in part by the National Science Foundation under Grant DES75-03635.

REFERENCES

Bates, R. G. 1973. Determination of *pH*, 2nd ed., Chap. 4, John Wiley and Sons, New York. 59-104.

Bates, R. G. 1975. *pH* scales for sea water. In: *The Nature of Seawater*, edited by E. D. Goldberg, Phys. and Chem. Sci. Res. Rpt. 1, Dahlem Konferenzen, Berlin. 313-338.

Bates, R. G., and J. B. Macaskill. 1975. Acid-base measurements in sea water, In: *Analytical Methods in Oceanography*, edited by T. R. P. Gibb, Chap. 10, Adv. in Chem. Ser., No. 147, American Chemical Society, Washington. 110-123.

Bolin, B. 1970. The carbon cycle, In: *The Biosphere*, Chap. 5, W. H. Freeman, San Francisco. 47-56.

Buch, K., H. W. Harvey, H. Wattenberg, and S. Gripenberg. 1932. Über das Kohlensäuresystem im Meerwasser, *Conseil Permanent Internat. Exploration de la Mer, 79*: 1-70.

Culberson, C., R. M. Pytkowicz, and J. E. Hawley. 1970. Seawater alkalinity by the *pH* method. *J. Mar. Res., 28*: 15-21.

Edmond, J. M., and J. M. T. M. Gieskes. 1970. On the calculation of the degree of saturation of sea water with respect to calcium carbonate under *in situ* conditions. *Geochim. Cosmochim. Acta, 34*: 1261-1291.

Hansson, I. 1973a. A new set of acidity constants for carbonic acid and boric acid in sea water. *Deep-Sea Res.*, *20*: 461-478.

Hansson, I. 1973b. A new set of *pH*-scales and standard buffers for sea water. *Deep-Sea Res.*, *20*: 479-491.

Harned, H. S., and F. T. Bonner. 1945. The first ionization of carbonic acid in aqueous solutions of sodium chloride. *J. Amer. Chem. Soc.*, *67*: 1026-1031.

Hawley, J. E., and R. M. Pytkowicz. 1973. Interpretation of *pH* measurements in concentrated electrolyte solutions. *Mar. Chem.*, *1*: 245-250.

Lyman, J. 1957. Buffer mechanisms of seawater, Thesis, University of California, Los Angeles, California.

Mehrbach, C., C. H. Culberson, J. E. Hawley, and R. M. Pytkowicz. 1973. Measurement of the apparent dissociation constants of carbonic acid in seawater at atmospheric pressure. *Limnol. Oceanogr.*, *18*: 897-907.

Mohan, M. S., and R. G. Bates. 1975. Calibration of ion-selective electrodes for use in biological fluids. *Clin. Chem.*, *21*: 864-872.

Pytkowicz, R. M., S. E. Ingle, and C. Mehrbach. 1974. Invariance of apparent equilibrium constants with *pH*. *Limnol. Oceanogr.*, *19*: 665-669.

Skirrow, G. 1965. The dissolved gases - carbon dioxide, In: *Chemical Oceanography*, Vol. 1, edited by J. P. Riley and G. Skirrow, Chap. 7, Academic Press, New York. 227-322.

Sørensen, S. P. L. 1909. Enzymstudien. II. Über die Messung und die Bedeutung der Wasserstoffionenkonzentration bei enzymatischen Prozessen. *Biochem. Z.*, *21*: 131 pp.

Seasonal Patterns in Suspended Calcium Carbonate Concentrations During the Dry and Wet Seasons in the Eastern Caribbean

P. R. Betzer[1], D. W. Eggimann[1], K. L. Carder[1],
D. R. Kester[2], and S. B. Betzer[3]

[1]University of South Florida, [2]University of Rhode Island, and [3]University of Miami

ABSTRACT

The concentration of suspended calcium carbonate was measured in samples taken in shallow and deep Caribbean waters during the dry and wet seasons of 1973 and 1974. During the dry season, suspended calcium carbonate was much more abundant throughout the water column. The marked reduction in particulate carbonate levels during the wet season may be related to: 1) increases in the standing crop of zooplankton and, therefore, in the amounts of phytoplankton which are converted to rapidly-sinking fecal pellets; 2) increases in the proportion of diatoms (at the expense of coccolithophores) among the primary producers in response to the input of soluble, riverine silicon to the open ocean; and 3) a decrease in primary productivity, which is reflected in depressed numbers of coccolithophores.

INTRODUCTION

The plankton in the marine biosphere are an important link in the cycling of CO_2 in the ocean system. In fact in all oceanic areas, the activities of the marine biosphere are the dominant control over the production of calcium carbonate (*Cloud*, 1965). More specifically, examination of sediments from the deep (>1000 m) ocean show that coccolithophores and foraminifera have produced essentially all the carbonate which is found there.

Unfortunately, there are only a few open-ocean areas where any detailed seasonal studies of primary production and species composition have been made. Probably the most intensive, long-term study of temporal variations in production, species composition, and nutrients was made at Station S in the Sargasso Sea (*Ryther and Guillard*, 1959; *Menzel and Ryther*, 1960, 1961a, 1961b; *Menzel et al.*, 1963). Their results show that for all but a brief (≈2 week) period in the spring when a bloom of diatoms takes over, the coccolithophorid *Coccolithus huxleyi* is numerically dominant. These findings are consistent with those of *Hulbert* (1962a, 1962b, 1963, 1966, 1967, 1968; *Hulbert and Rodman*, 1963) who studied the species composition of plankton communities on various transects of the western Atlantic and Caribbean Sea. He found also that, in general, the euphotic zone in tropical waters was dominated by coccolithophores, the most abundant of which was normally *C. huxleyi*. While these and other published findings are sufficient to describe seasonal changes in phytoplankton productivity for the Sargasso Sea, there are not enough data to do the same for the tropical western Atlantic and Caribbean Sea.

Since essentially all calcium carbonate production in the tropical open ocean is related to the activity of coccolithophores, any model of the fluxes of carbonates and CO_2 in this part of the ocean system should take into account: 1) the distribution of coccolithophores in space and time; 2) the physical-chemical and biological factors which affect their distribution, production, and removal from the euphotic zone; and 3) whether changes in their abundance in shallow ocean waters are linked with changes in the concentrations of suspended carbonates in deep ocean waters.

We have been carrying out a study of the sources, movements, and reactions of suspended materials in the Caribbean Sea and adjacent western Atlantic during the dry and wet seasons. Because of the marked seasonal influence of the Amazon and Orinoco Rivers on the physical-chemical properties of the surface waters in this region (*Ryther et al.*, 1967; *Gade*, 1961), we would predict *a priori* that the primary producers would exhibit a different seasonal cycle than that in the Sargasso Sea.

Our initial approach to understanding the variations in suspended carbonates in these waters was to sample both shallow and deep waters during periods of relatively high and low river flow. This paper presents a comparison of suspended carbonate concentrations during these periods and also considers some of the processes which may be responsible for the changes which were observed.

MATERIALS AND METHODS

The sampling program in the eastern Caribbean and adjacent western Atlantic was carried out on two cruises of R/V TRIDENT: the first (TR 131) during the dry season (February of 1973; Figure 1); the second (TR 158-159) just after the wet season (October, November of 1974; Figure 2). Water samples were collected in 30-liter Niskin^R bottles and filtered immediately, using a closed system,

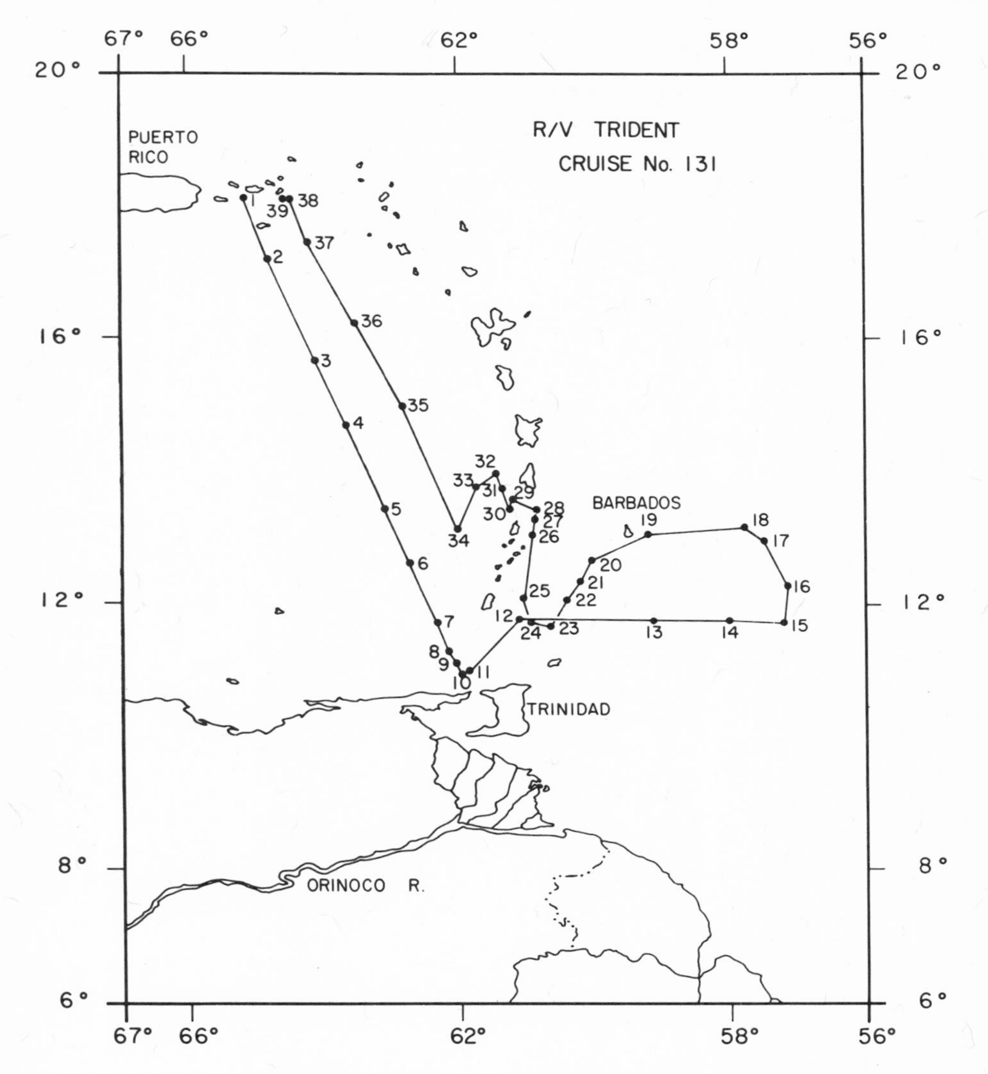

Fig. 1. Track and station locations for R/V TRIDENT cruise 131 to the western Atlantic Ocean and Caribbean Sea.

through tared, 0.4 micron pore size, 47-mm diameter, Nuclepore[R] membranes which were housed in In-Line[R], Millipore[R] filter holders. Water from the Niskin[R] bottles was gravity-fed to the filters through silicon rubber tubing. After filtration was complete, the samples were rinsed with 25 milliliters of deionized water. The filters were then removed from the filter heads with Teflon tweezers and stored in plastic vials until analyses could be initiated ashore.

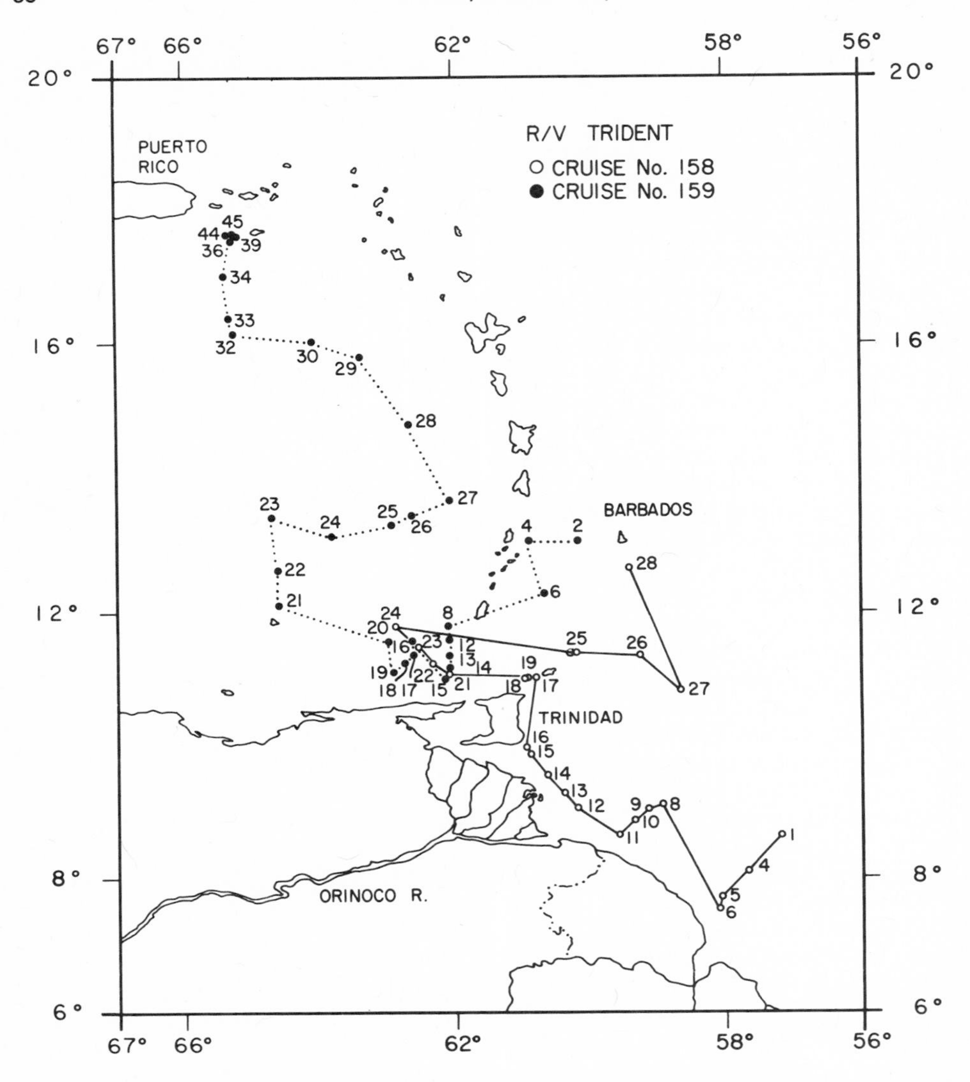

Fig. 2. Track and station locations for R/V TRIDENT cruises 158 and 159 to the western Atlantic Ocean and Caribbean Sea.

In all, 522 samples were collected for the determination of suspended particulate matter and particulate carbonates.

The dissolution and analysis of suspended carbonates was carried out in a shore-based laboratory. First, to insure the removal of any residual seawater, the samples were placed in a plastic funnel held in a clean bench and rinsed with deionized water. The samples were then removed, placed in plastic bottles, and dried under silica gel for 48 hours. After drying, they were weighed in a Mettler M-5 microbalance. To minimize static electric effects on samples weights, the microbalance had an alpha source in close proximity to the sample. When the mass of suspended material had been determined, the samples were placed in a plastic filter funnel in a clean bench and leached for about two hours with 4 milliliters of acetic acid (25% v/v) with a *pH* of 2.8. The resulting solution was then diluted to 25 milliliters with doubly deionized water and analyzed by atomic absorption spectroscopy for calcium using an air-acetylene flame. The calcium analyses were converted to the weight of calcium carbonate using a multiplication factor of 2.5. The average signal/noise ratio for the atomic absorption analyses of calcium was 9 to 1. The average analytical uncertainty (two standard deviations) for triplicate determinations of sample calcium concentrations by atomic absorption spectroscopy was 7 percent of the mean. A method blank, which included the calcium contribution from reagents, the Nuclepore^R filter, sample handling and machine noise, equaled 0.3 micrograms.

As a check on this method, several samples from TRIDENT cruise 127 to the Gulf of Mexico were examined under a microscope and found to have abundant coccoliths. These samples were leached with the acetic acid for two hours and then re-examined microscopically. No coccoliths were found in the samples after leaching. It is apparent, then, that the acetic acid does dissolve the coccoliths from natural samples of suspended material.

We also considered the potential interference of calcium contributions from clay minerals with this method. Therefore, three samples containing more than 90% clay were obtained from the plume of the Orinoco River during TRIDENT cruise 158. It was found that these samples had very low levels of weak-acid leachable calcium. Using the ratio of released calcium to aluminum in these clay-rich samples, we calculated a maximum amount of calcium which could be released from the clays in the open-ocean samples of suspended matter gathered in the eastern Caribbean based upon their aluminum concentration. The average input from the clays in the open-ocean samples would amount to approximately 0.002 micrograms/liter, which is below our detection limit (*.012 μg/mℓ*) and also well below the average calcium concentration (*0.6 μg/mℓ*) in our samples. Additional evidence that this is the case is presented in Figure 3. It is obvious from the variations in the ratio of weak-acid soluble calcium to aluminum (an indicator of clays) and from the changes in the percentage of calcium carbonate in suspended materials with depth that aluminosilicates did not release significant quantities of calcium from clays during the weak-acid leach. Therefore, it will be assumed that the contribution of calcium from clays is negligible. This finding is consistent with those of *Chester and Hughes* (1967).

Finally, we had a chance to compare results of our analyses for calcite with determinations of calcite by x-ray diffraction on nine samples collected on the West African Shelf. The average of

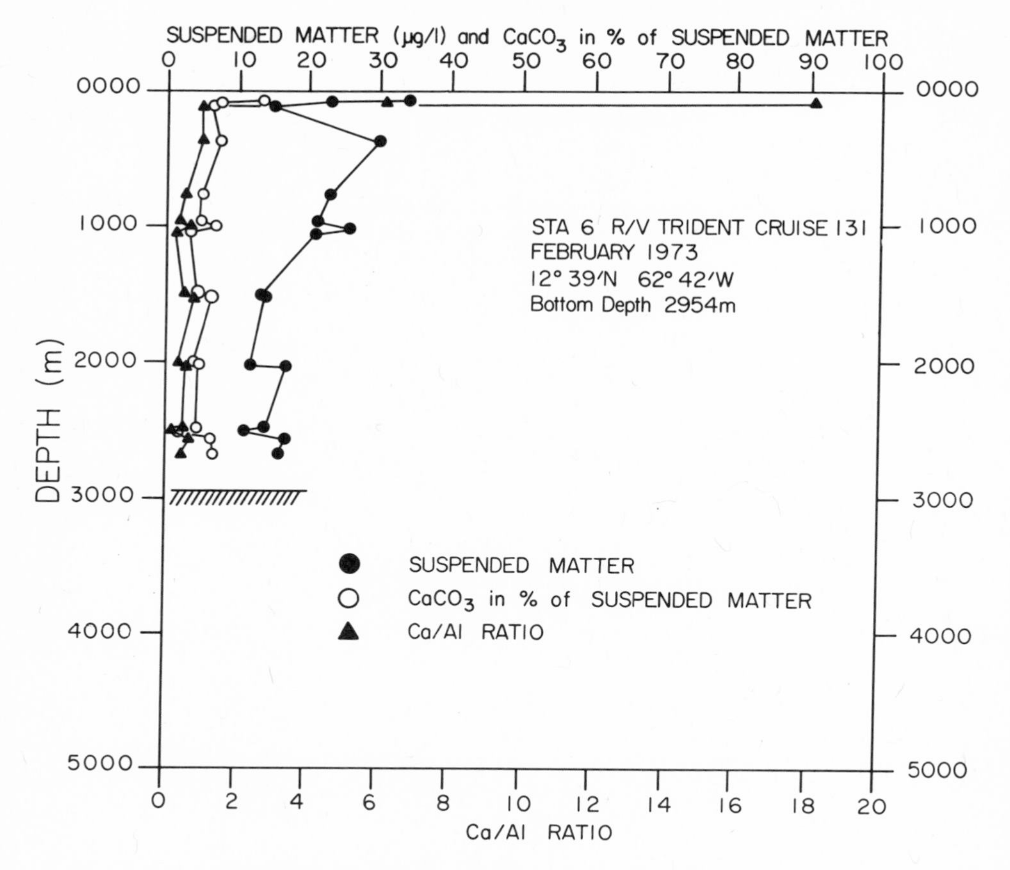

Fig. 3. Suspended particulate matter concentrations, calcium carbonate as a percentage of the suspended matter, and calcium/aluminum ratios in suspended materials from Station 6 of cruise 131.

these calcite analyses by x-ray diffraction was 3.5 ± 2.5 percent, compared to 4.2 ± 2.5 percent for the weak-acid leach (*Eggimann*, 1975). Therefore, it is apparent that the weak-acid leach gives a reasonably accurate estimate of suspended calcium carbonate in natural samples of oceanic suspended materials.

RESULTS

Representative results of calcium analyses on the weak-acid leach of samples from cruises 131, 158, and 159 are presented in Figures 4 through 8. They show that in both shallow and deep waters of the Eastern Caribbean there was significantly less suspended calcium carbonate during the wet season (TR 158 and 159)

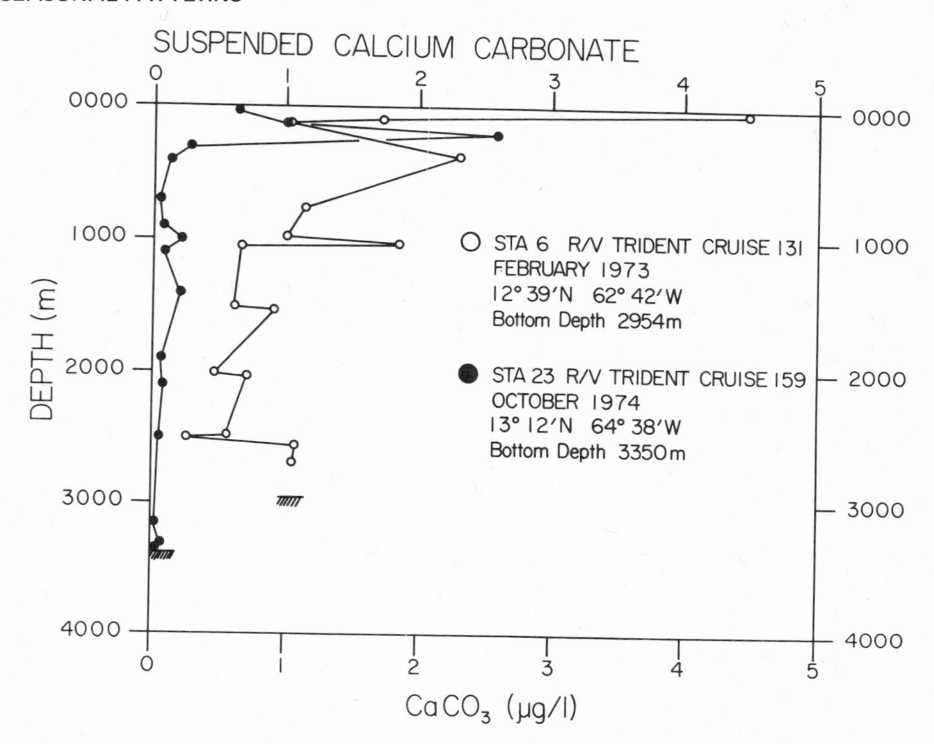

Fig. 4. Suspended calcium carbonate concentrations for Station 6 (cruise 131) during the dry season and Station 23 (cruise 159) during the wet season.

than during the dry season (TR 131). In comparing the data for the two seasons, we have considered samples gathered in surface waters (<*40 m*), the Subtropic Underwater (*75 - 200 m*), oxygen minimum zone (*300 - 500 m*), Antarctic Intermediate Water (*700 - 900 m*), and deep Caribbean water (>*1000 m*). The average suspended carbonate concentrations (in μ*g*/*ℓ*) and standard deviations for the above depths in the open eastern Caribbean during the dry season (Sta. 2 - 7 and 32 - 37 of TR 131) and the wet season (Sta. 21 - 35 of TR 159) are presented in Table 1. The differences between the means for the wet and dry seasons are significant at the 99% confidence level for all the water types. The data for the stations occupied near the Antilles chain (Sta. 25 - 31 of TR 131 and Sta. 2, 4, and 6 of TR 159) and adjacent to the Gulf of Paria (Sta. 8 - 11 of TR 131 and Sta. 22 of TR 158) showed similar changes between the two seasons. In an overall average for all areas of the eastern Caribbean, the suspended carbonate concentrations during the dry season were about two- to four-fold higher in the upper 200 meters, approximately four-fold higher between 300 and 500

TABLE 1. Average Suspended Calcium Carbonate Concentrations (μg/ℓ) in the Eastern Caribbean for Surface (<40 *m*) Water, Subtropic Underwater (75 - 200 *m*), Oxygen Minimum Water (300 - 500 *m*), Antarctic Intermediate Water (700 - 900 *m*), and Deep Caribbean Water (>1000 *m*) During the Dry (February, 1973; Sta. 2 - 7, and 32 - 37 of TR 131) and Wet (October-November, 1974; Sta. 21 - 35 of TR 159) Seasons.

	SUSPENDED CALCIUM CARBONATE*	
	Dry Season	Wet Season
Surface (<40 *m*) Water	1.24 (.66)	0.43 (.18)
Subtropic Underwater (75 - 200 *m*)	2.14 (1.55)	1.07 (.96)
Oxygen Minimum (300 - 500 *m*) Water	2.03 (.52)	0.53 (.71)
Antarctic Intermediate Water (700 - 900 *m*)	1.23 (.29)	0.14 (.14)
Deep (>1000 *m*) Caribbean Water	.93 (.48)	0.11 (.22)

* The differences between means for the dry and wet seasons at each depth were significant at the 99% confidence level.

( ) Figures in parentheses represent single standard deviations about the mean.

meters, and almost nine-fold higher below 700 meters. The difference in the concentrations of suspended calcium carbonate between the two seasons increased sharply with increasing depth.

DISCUSSION

The data we have gathered leave little question that there was a substantial reduction in the concentration of suspended carbonates throughout the eastern Caribbean between the dry season (February) of 1973 and the wet season (October) of 1974. While the change between the two seasons is not necessarily representative of those that take place every year, we can consider the physical and biological processes operating in this area which might bring it about: 1) decreases in primary productivity (from February to October) which are, in turn, reflected in lowered standing crops of primary producers; 2) changes in the relative abundance of primary producers, from a coccolithophore-dominated system during the dry season to a diatom-dominated system during the wet season;

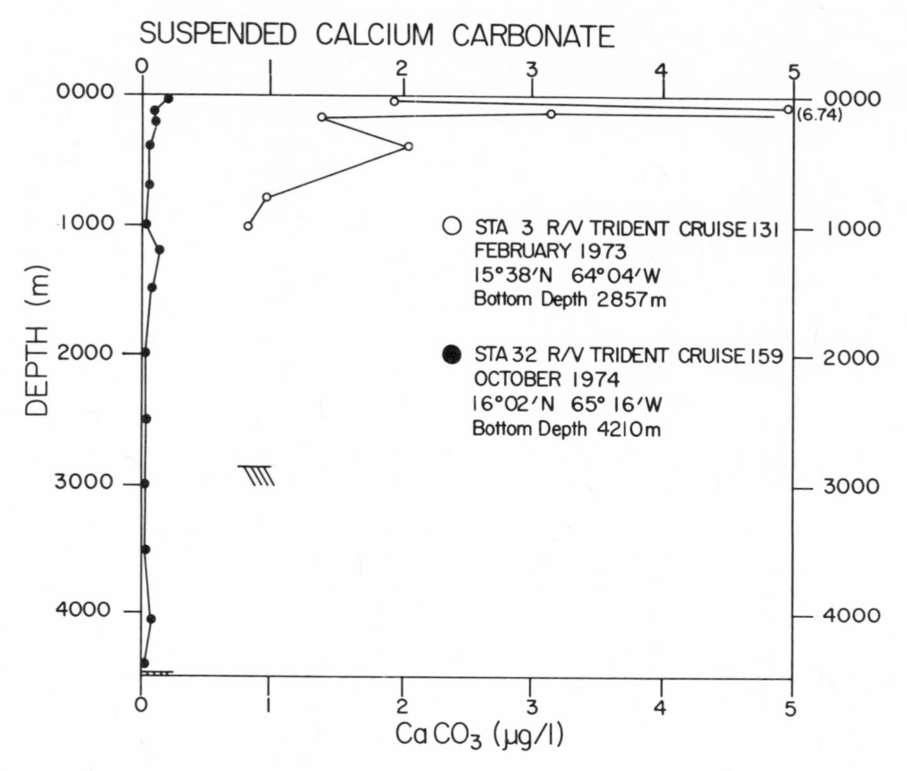

Fig. 5. Suspended calcium carbonate concentrations for Station 3 (cruise 131) during the dry season and Station 32 (cruise 159) during the wet season.

3) an increase during the wet season in the amount of coccolithophores and coccoliths which are converted into fast-sinking fecal pellets. All three processes would work to decrease the concentration of suspended carbonates during the wet season. The real question is whether any of these changes are consistent with the available biological data for this region.

Sander and Steven (1973) in a two-year study of primary productivity near Barbados found that compared to February, October and November represented months of depressed primary productivity. *Beers et al.* (1968) also studied the cycle of primary productivity in the Atlantic off Barbados over a two-year period. They too found lower primary productivity in the fall compared to the spring and summer. A decrease in primary productivity more than likely would result in decreased standing crops of primary producers in Caribbean waters. Although *Sander and Steven* (1973) present a brief summary of the more important phytoplankton groups (diatoms,

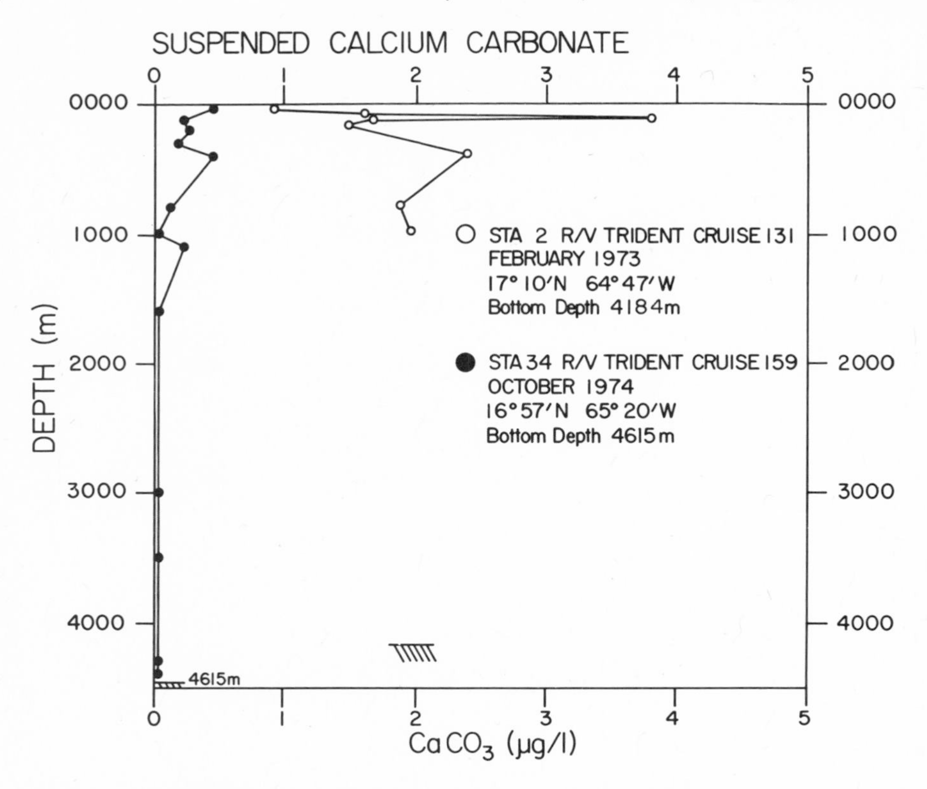

Fig. 6. Suspended calcium carbonate concentrations for Station 2 (cruise 131) during the dry season and Station 34 (cruise 159) during the wet season.

flagellates and coccolithophores) at various times near Barbados, they do not present any quantitative data on the abundance of these three groups as a function of time. We can say, however, that if standing crops of primary producers in Caribbean waters are directly related to primary productivity, our data for suspended carbonates in shallow Caribbean waters are consistent with the available biologic data.

The seasonal variation in fresh-water discharge may also influence particulate carbonate concentrations in these waters if the relative sizes of the populations of diatoms and coccolithophores in the phytoplankton are influenced by nutrient supplies. The Amazon and Orinoco Rivers together account for almost 23% of the fresh-water input to the world's oceans (*Davis*, 1964; *Oltman et al.*, 1964; *Van Andel*, 1967; *Holeman*, 1968). It is thus to be expected that at times of high discharge they would have a substantial influence upon surface salinities and nutrient concentrations in both

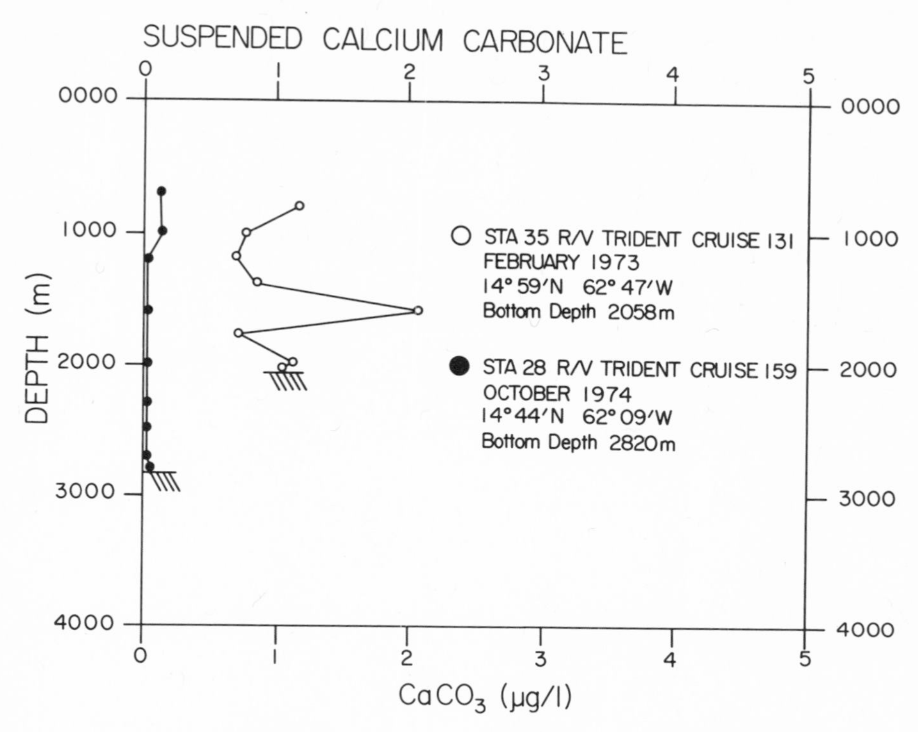

Fig. 7. Suspended calcium carbonate concentrations for Station 35 (cruise 131) during the dry season and Station 28 (cruise 159) during the wet season.

the eastern Caribbean and western Atlantic. Indeed at certain times of the year large areas (>*2.5 x* 10^6 Km^2) of the open tropical Atlantic and Caribbean have lenses of low-salinity water 30 to 40 meters thick (*Ryther et al.*, 1967; *Gade,* 1961) with high concentrations of soluble silicon. Our data, too, show the marked influence these rivers have upon the western Atlantic and eastern Caribbean. During the fall cruises (TR 158-159) in 1974, which took place several months after the point of maximum fresh-water discharge, low-salinity (<35°/oo) of Amazon and Orinoco origin was widespread. The fresh-water plumes of these two rivers were between 30 and 40 meters thick, and, in general, had elevated suspended loads and high soluble silicon concentrations. On cruise 159, the Orinoco plume was traced over 350 kilometers into the Caribbean Sea from the Dragon's Mouth. At that point (Sta. 23), the salinity at 10 meters was 33.4 °/oo, the suspended load in excess of *60* µg/ℓ, the soluble silicon concentrations 6.5 micromolar, and aluminum, 3.2% of the suspended load. For comparison, Station 5 of cruise 131 was taken near the same location during the dry season

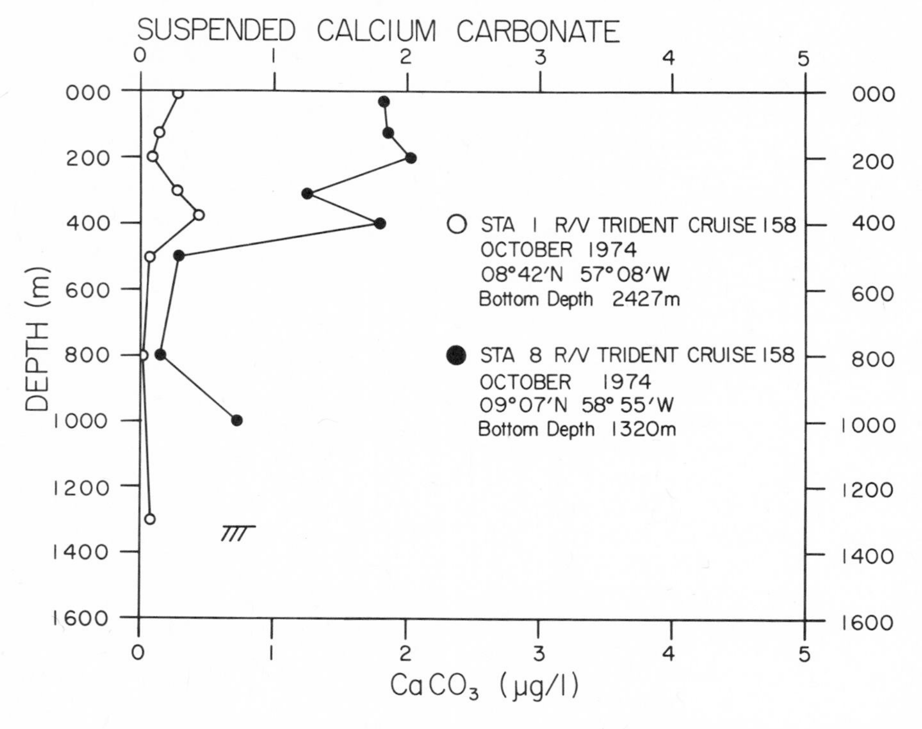

Fig. 8. Suspended calcium carbonate concentrations for Stations 1 and 8 (cruise 158) during the wet season.

and had a near-surface salinity of 36.0 ‰, a suspended load of 11 µg/ℓ, a soluble silicon concentration of 2.9 micromolar, and aluminum made up only 0.14% of the suspended matter. During this February cruise, when the discharge of both rivers was near their respective seasonal nadirs, surface salinities at all but one of 39 stations were over 35.2 ‰.

In the western Atlantic during the wet season, the Amazon and Orinoco plumes were separated by a tongue of high-salinity (>36 ‰) water which was probably part of the Guiana Current. This finding is consistent with historical data for the region (*Ryther et al.*, 1967).

The importance of the river discharge to the composition, although not necessarily the total productivity, of the plankton of these tropical waters may be largely related to the increases of soluble silicon as silicon-poor surface waters are mixed with silicon-rich river waters. A three-fold increase in soluble silicon such as we measured during the wet season might, in fact, allow large populations of diatoms (amorphous silica) to develop in water normally dominated by coccolithophores (particulate carbonates) and flagellates. If this were the case, there would be lower numbers of coccolithophores in the fresh-water lenses than in the high-salinity waters of the Guiana Current. Indeed, as shown in Figure 8, we found the concentrations of suspended calcium carbonate in surface and intermediate waters to 200 meters depth under the Amazon's fresh-water lens (Sta. 1 of TR 158) to be 11 times lower than in the high-salinity water of the Guiana Current (Sta. 8 of TR 158).

Additional evidence that there was a substantial change in the composition of the plankton populations in the low-salinity lenses comes from the measurements of amorphous silica made at Stations 25, 26, and 28 of TR 158. The technique for estimating amorphous silica has been presented by *Eggimann* (1975). Quantitative x-ray diffraction analyses, chemical analyses for particulate silicon and aluminum, and values for the silicon and aluminum content of clays are used to partition the total particulate silicon measured into clay, quartz, and amorphous (biologically-precipitated; *Siever and Scott*, 1963) fractions.

At Stations 25 and 26 (TR 158) where the 10-meter samples have high salinities (35.46 and 35.53 °/oo, respectively), intermediate dissolved silicon concentrations (*1.8* and *2.4* μM), and very low total suspended loads (*15* and *29* μg/ℓ), amorphous silica amounted to *0.74* and *1.38* μg/ℓ, respectively. Station 28, taken from the 40-meter-thick lens of low salinity (34.45 °/oo) Amazon-derived water, had slightly higher dissolved silicon (*2.8* μM), high total suspended load (*76* μg/ℓ), and an amorphous silica value of *2.03* μg/ℓ.

These data suggest that during periods of high river discharge, diatoms increase in numbers and coccolithophores decrease. We feel the major reason for such a shift in the distribution of primary producers is the increase in soluble silicon which accompanies river discharge, as this nutrient is normally in such low concentrations (<*1* μM) that it may limit the growth of diatoms in tropical waters (*Ryther and Guillard*, 1959; *Menzel and Ryther*, 1961a). In addition, coccolithophores may be adversely affected by the presence of river-derived, dissolved organic materials which are present in the low-salinity, offshore lenses (*Small*, personal communication).

Another biological factor which might have a substantial influence upon the concentration of suspended carbonates in these waters is zooplankton. There is good evidence from *Roth et al.* (1975) and *Honjo* (1975, 1977) showing that coccolithophores and coccoliths are incorporated into the fecal pellets of zooplankton and carried to deeper waters or, in cases where a resistant pellicle encases the fecal pellet, the bottom. If there were seasonal changes in zooplankton abundance, then there could be a change in the amount of carbonate material being converted into rapidly sinking (36 - 369 m/day; *Smayda*, 1970) fecal pellets, providing a means of rapid transport for coccolithophores and coccoliths from shallow waters to the bottom. In a sense, if the zooplankton were processing enough water there would be a depletion of the standing crop of coccolithophores. *Calef and Grice* (1967) studied the zooplankton associated with the Amazon's fresh-water lens in the western Atlantic. On the average, they found there were three times more zooplankton there than in adjacent higher-salinity waters. It is not unreasonable to expect, therefore, that a much greater proportion of the standing crop of coccoliths and coccolithophores are fecalized during periods of high river flow.

The increase in zooplankton numbers in the fresh-water lens of the Amazon was also accompanied by a change in fauna. *Calef and Grice* (1967) found large populations of two coastal species dominated the river-influenced (low-salinity, high-soluble silicon) offshore areas. One of these species, the decapod *Lucifer faxoni*, is much larger than the pelagic copepods normally found in these offshore areas. It is likely, therefore, that this organism produces a much larger, and therefore faster-sinking, fecal pellet than the smaller copepods. The significance of such a seasonal change to the distribution of suspended carbonates is obvious. During periods of high river discharge, much of the phytoplankton and smaller herbivorous zooplankton in the low-salinity lenses are packaged in large fecal pellets which probably sink to the bottom without significant dissolution or disintegration of the fecal pellets' protective pellicle. In contrast, the smaller, slower sinking pellets produced by open-ocean fauna during dry periods or in areas outside the low-salinity lenses probably undergo some dissolution or disintegration in moving from shallow to deeper waters (*Honjo*, 1976). In this manner, a greater proportion of coccoliths and coccolithophores would be released to intermediate and deep ocean waters during the dry than the wet season. While we were unable to locate any data on the sizes and integrity of fecal pellets produced by various oceanic and coastal copepods, there is some data (*Honjo*, 1976) which suggest that the decisive factors which determine whether a host particle reaches the bottom or is dispersed in the water column are its size and the degree to which the pellicle is degraded.

One can see from the data presented that several biological processes (changes in primary productivity, changes in species composition of photoplankton communities, changes in the rates and types of zooplankton processing) could help explain the seasonal changes in concentrations of suspended carbonates in the shallow waters of the eastern Caribbean. Inasmuch as the major biological processes are restricted to the upper several hundred meters, we are left with trying to explain how they might effect the large changes which also occurred at depth. We envision two possible changes from the dry to the wet season: a difference in transport to depth by zooplankton fecal pellets; and a seasonal formation of fast-sinking palmelloid stages.

During the dry season when zooplankton numbers have diminished, there is a reduction in the numbers of coccolithophores and coccoliths which are removed from the euphotic zone in fecal pellets.

Instead, a much greater proportion are probably physically incorporated in surface waters into organic aggregates such as the open-ocean types observed by *Riley et al.* (1964). The sizes of these materials (most are less than 50 μm) are such that most would settle much more slowly than zooplankton fecal pellets. Given the settling rates and sizes of these aggregates, it is not unreasonable to assume that we collected this type of material in deeper water with our 30-liter Niskin[R] bottles. By this explanation the dry season is envisioned as a time when organic aggregates, and to a limited extent small fecal pellets, produced in surface waters are a primary means of transporting carbonate particles to deep water while the wet season is a time when a much greater fraction of carbonate particles are pelletized, removed from the euphotic zone, and carried to the bottom in large, rapidly sinking zooplankton fecal pellets.

Another biological means of decreasing transit times for suspended carbonates through deep ocean waters, and therefore suspended carbonate concentrations, would be transport via palmelloid stages (*Smayda*, 1970). Coccolithophores are reputed to be "haters" of dissolved organic matter (*Small*, personal communication) and might react to the input of dissolved organic matter from the rivers by forming palmelloid stages. Although we are not aware of any reports which consider palmelloid formation by open-ocean coccolithophore populations, the creation of such large, rapidly-sinking masses in the euphotic zone could be another means of changing carbonate concentrations in both shallow and deep waters.

We think it is obvious from this discussion that an understanding of the delivery and cycling of carbonates in this and other ocean areas requires seasonal data on: 1) the sizes and types of aggregates which transport particles to the deep ocean; and 2) the changes which are brought to the transport processes by the marine biosphere.

ACKNOWLEDGEMENTS

The officers and crew of R/V TRIDENT, as well as the marine technicians of the University of Rhode Island, helped make this work possible. Without the incredible efforts of Chief Engineer John Symonds, who helped repair and fabricate new parts for our equipment, we would have much less to report.

This work was supported by the Office of Naval Research under contracts N00014-72-A-0363-0001 and N00014-75-C-0539 to the University of South Florida, and N00014-68-A-0215-003 to the University of Rhode Island.

REFERENCES

Beers, J. R., D. M. Stevens, and J. B. Lewis. 1968. Primary productivity in the Caribbean Sea off Jamaica and the Tropical North Atlantic off Barbados. *Bull. Mar. Sci., 18*. 66-104.

Calef, G. W. and G. D. Grice. 1967. Influence of the Amazon River outflow on the ecology of the western tropical Atlantic. II. Zooplankton abundance, copepod distribution, and a discussion of the fauna of low salinity areas. *J. Mar. Res., 25*. 84-94.

Chester, R., and M. J. Hughes. 1967. A chemical technique for the separation of ferro-manganese minerals, carbonate minerals, and absorbed trace elements from pelagic sediments. *Chem. Geol., 2*. 249-262.

Cloud, P. E., Jr. 1965. Carbonate precipitation and dissolution in the marine environment. In: *Chemical Oceanography, 2*, J. P. Riley and G. Skirrow (Eds.), Academic Press, New York. 127-158.

Davis, L. C. 1964. The Amazon's rate of flow. *Nat. His., 73*. 5-19.

Eggimann, D. W. 1975. The West African Shelf; chemical evidence of particle transport to the eastern Atlantic Basin, ms., University of South Florida.

Gade, H. G. 1961. On some oceanographic observations in the southeastern Caribbean Sea and the adjacent Atlantic Ocean with special reference to the influence of the Orinoco River. *Bol. Inst. Oceanogr. Univ. Oriente, 1*. 287-342.

Holeman, J. N. 1968. The sediment yield of major rivers of the world. *Water Resources Res., 4*. 737-746.

Honjo, S. 1975. Dissolution of suspended coccoliths in the deep-sea water column and sedimentation of coccolith ooze. In: *Dissolution of Deep-Sea Carbonates*, W. V. Sliter, A. W. H. Bé and W. H. Berger (Eds.), *Cushman Found. Foram. Research Spec. Pub. 13*. 114-128.

Honjo, S. 1976. Coccoliths: production, transportation and sedimentation. *Marine Micropaleontology., 1*. 65-79.

Honjo, S. 1977. Biogenic carbonate particles in the ocean; Do they dissolve in the water column? In: *The Fate of Fossil Fuel CO_2*, N. R. Andersen and A. Malahoff (Eds.), Plenum, N.Y., (this volume)

Hulbert, E. M. 1962a. Phytoplankton in the southwestern Sargasso Sea and North Equatorial Current, February, 1961. *Limnol. Oceanogr. 7*. 307-315.

Hulbert, E. M. 1962b. A note on the horizontal distribution of phytoplankton in the open ocean. *Deep Sea Res., 9*. 72-74.

Hulbert, E. M. 1963. The distribution of phytoplankton in coastal waters of Venezuela. *Ecology, 44*. 169-171.

Hulbert, E. M. 1966. Distribution of phytoplankton and its relationship to hydrography, between southern New England and Venezuela. *J. Mar. Res., 24*. 67-81.

Hulbert, E. M. 1967. Some notes on the phytoplankton off the southeastern coast of the United States. *Bull. Mar. Sci., 17*. 330-337.

Hulbert, E. M. 1968. Phytoplankton observations in the western Caribbean Sea. *Bull. Mar. Sci., 18.* 388-399.
Hulbert, E. M., and J. Rodman. 1963. Distribution of phytoplankton species with respect to salinity between the coast of southern New England and Bermuda. *Limnol. Oceanogr., 8.* 263-269.
Menzel, D. W., and J. H. Ryther. 1960. The annual cycle of primary production in the Sargasso Sea off Bermuda. *Deep Sea Res., 6.* 351-367.
Menzel, D. W., and J. H. Ryther. 1961a. Nutrients limiting the production of phytoplankton in the Sargasso Sea, with special reference to iron. *Deep Sea Res., 7.* 276-281.
Menzel, D. W., and J. H. Ryther. 1961b. Annual variations in primary production of the Sargasso Sea off Bermuda. *Deep Sea Res., 7.* 282-288.
Menzel, D. W., E. M. Hulbert, and J. H. Ryther. 1963. The effects of enriching Sargasso Sea water on the production and species composition of the phytoplankton. *Deep Sea Res., 10.* 209-219.
Oltman, R. E., H. O. Sternberg, F. C. Ames, and L. C. Davis. 1964. Amazon River investigations, reconnaissance measurements of July 1963. *U. S. Geol. Survey Circ.* 486.
Riley, G. A., P. J. Wangersky, and D. Van Hemert. 1964. Organic aggregates in tropical and subtropical surface waters of the North Atlantic Ocean. *Limnol. Oceanogr., 9.* 546-550.
Roth, P. H., M. M. Mullin, and W. H. Berger. 1975. Coccolith sedimentation by fecal pellets: laboratory experiments and field observations. *Geol. Soc. Amer. Bull., 86.* 1079-1084.
Ryther, J. H., and R. R. L. Guillard. 1959. Enrichment experiments as a means of studying nutrients limiting to phytoplankton production. *Deep Sea Res., 6.* 65-69.
Ryther, J. H., D. W. Menzel, and N. Corwin. 1967. Influence of the Amazon River on the ecology of the western tropical Atlantic, I. Hydrography and nutrient chemistry. *J. Mar. Res. 25.* 69-83.
Sander, F., and D. M. Steven. 1973. Organic productivity of inshore and offshore waters of Barbados: a study of the island mass effect. *Bull. Mar. Sci., 23.* 771-792.
Siever, R., and R. A. Scott. 1963. In: *Organic Geochemistry.* Pergamon Press, Oxford. p. 569.
Smayda, T. J. 1970. The suspension and sinking of phytoplankton in the sea. *Oceanogr. Mar. Biol. Ann. Rev., 8.* 353-414.
Van Andel, T. H. 1967. The orinoco delta. *J. Sed. Petrol., 37.* 297-310.

Modeling the Oceans and Ocean Sediments and their Response to Fossil Fuel Carbon Dioxide Emissions

Bert Bolin

University of Stockholm

ABSTRACT

The continuity equation for the distribution of an arbitrary tracer in a water body is integrated with due regard to the basic characteristics of the ocean circulation to deduce a multiple reservoir model for the carbon cycle. Methods for evaluating the flux of carbon between the reservoirs are presented in some detail. Awaiting results from a series of model computations, some principle characteristics are pointed out, particularly the dependence of the downward flux of carbon into the sea on the sinking of detritus and thus, of the biological production in the surface layers of the ocean. The possible role of man's input of phosphorus into coastal waters (and possibly the open sea) is assessed.

INTRODUCTION

Many attempts have been made to model the atmosphere - land biota - ocean system to describe its response to an increasing emission of fossil fuels into the atmosphere (*Bolin and Eriksson*, 1959; *Broecker et al.*, 1971; *Machta*, 1972; *Bacastow and Keeling*, 1973). All these attempts have been based on the use of simple box models, usually describing the oceans as consisting of two reservoirs, the surface layers above a thermocline placed at depths varying between 75 and 1000 m and the deep sea below. Clearly, the ocean circulation is described very poorly in this way and the approach therefore can only give order of magnitude estimates. The introduction of this kind of model dates back to *Craig* (1957) and was then used to interpret the first radio carbon data that gave an average age of the deep sea of about 1,000 years.

As an increasing amount of data for the oceans is becoming available and our understanding of the basic processes improves, it will be desirable to develop more complex models for the interpretation of these data. *Keeling and Bolin* (1967, 1968), *Broecker et al.* (1971), *Bolin* (1975) and others have proposed somewhat more elaborate models of a reservoir type. On the other hand, *Kuo and Veronis* (1973) and *Fiadeiro* (1975) have studied the tracer distribution in the deep sea with the aid of two- and three-dimensional continuous models, derived on the basis of ocean circulation theory. It is obviously necessary to make use of such knowledge about the circulation of the oceans, which has at least partly been verified by observations, in trying to understand the role of the oceans for the carbon cycle and as a sink for carbon dioxide emitted by burning fossil fuels. Our present knowledge about the chemistry of the oceans and the interactions between the ocean waters and the sediments is, however, so limited that it is still desirable to design integrated models of the kind that the box models represent, but in doing so make use of our present knowledge of ocean dynamics. It is the objective of this paper to outline a general approach in this regard and also to give some comments as to the application of such a model to a study of the carbon cycle. The model described below may well have to be modified, as work on it progresses and as knowledge of the main circulation of the oceans and the exchange between the ocean waters and the sediments increases. Even though explicit computations have not yet been completed the outline given below may be of general interest.

INTEGRATION OF THE TRANSFER EQUATION

The time rate of change of a tracer in the oceans, whose concentration is C_i, is given by the equation

$$\frac{\partial c_i}{\partial t} + \frac{1}{a\cos\phi}\frac{\partial}{\partial\lambda}(uc_i) + \frac{1}{a}\frac{\partial}{\partial\phi}(vc_i) + \frac{\partial}{\partial Z}(wc_i)$$

$$= \nabla \cdot (K_H \nabla_H c_i) + \frac{\partial}{\partial Z}\left(K_v \frac{\partial c_i}{\partial Z}\right) + J_i - \lambda^* c_i \qquad (1)$$

where u, v and w are the three velocity components along the horizontal coordinates; ϕ the latitude; λ the longitude; Z the vertical coordinate; a the radius of the earth; K_H the coefficient of horizontal eddy diffusivity; K_v coefficient of vertical eddy diffusivity; J_i denotes sources; and λ^* radioactive decay.

Kuo and Veronis (1973) have used our current knowledge of ocean dynamics as deduced theoretically to derive the distribution of oxygen below the thermocline (about 1000 m) by making simple assumptions about K_H, K_v and J_i. The deep water is formed in the Iceland-Greenland area and in the Weddel Sea, and sinks to varying depths between the bottom and the thermocline, moving then into the Atlantic, Pacific and Indian Oceans along the western boundaries. The water spreads into the major ocean basins by advection and horizontal diffusion, while at the same time slowly moving upwards and being subject to weak vertical diffusion. Good agreement between computations and observations is obtained if assuming that the injection of water into the basins as a function of the vertical is such that w increases linearly from zero at the bottom to $1.5 \cdot 10^{-5}$ cm sec^{-1} at a depth of 1000 m, $K_v = 0.6$ cm^2 sec^{-1} and $K_H = 7.5 \cdot 10^6$ cm^2 sec^{-1}. The computations further show that the vertical exchange of water is rather slow compared with the horizontal one, since otherwise the horizontal distribution of oxygen assumed as an upper boundary condition would influence the horizontal distributions at greater depths, which is not the case.

This very fact and also the finding that values for K_v, w and J_i, which are independent of λ^* and ϕ, give good results, can be taken as a justification for integrating equation (1) horizontally. Depending on the problem under consideration, integration is made over a basin or an ocean. Let

$$\overline{c}_i = \iint c_i dA \qquad (2)$$

Thus,

$$\frac{\partial \overline{c}_i}{\partial t} + \frac{\partial}{\partial Z}(w\overline{c}_i) - \frac{\partial}{\partial Z}\left(K_v \frac{\partial \overline{c}_i}{\partial Z}\right) \qquad (3)$$

(*Transfer due to water motions*)

$$= \underbrace{\oint K_H \frac{\partial c_i}{\partial n} dS + \oint v_n c_i dS}_{\text{(source)}} + \overline{J}_i - \underbrace{\lambda^* \overline{c}_i}_{\text{(decay)}}$$

The first term on the right hand side of equation (3) expresses the diffusive flux of the tracer from the lateral boundaries into the oceans, which in the case of oxygen, for example, may be taken as zero, but in the case of total carbon expresses the dissolution of carbonates from the ocean bottom. The second term on the right hand side of equation (3) is the source due to the spreading of water from the down-welling and western boundary currents into the oceans and the associated injection of the tracer (subscript n denotes the inward normal). *Kuo and Veronis* (1973) assume in their integration that the amount of water injected in this way, to be demoted by $\overline{Q}(Z)$, is the same at all depths, but that the tracer concentration may vary. The latter is obviously dependent on the manner of deep water formation, a point which will be addressed later. The source (or sink) function $J_i(Z)$ is primarily dependent on biological activity. For inorganic carbon $(i = C)$ $J_C(Z)$ depends on the decomposition of organic matter as a function of depth, which obviously is related to the availability of oxygen. In this way the oxygen and carbon cycles are interrelated. It should be emphasized at this point that $J_C(Z)$ also is dependent on the dissolution of settling carbonate shells, a process that depends on the degree of carbonate saturation of the sea water relative to the calcite or aragonite pellets that are falling through the water. We shall return to this problem, but wish at this time to merely point out that for this reason the carbon cycle cannot be understood without simultaneously considering the oxygen and calcium cycles in the sea. Also, the magnesium cycle possibly needs to be considered, but will be left out in the following. Finally, the decay term $\lambda^*\overline{C}_i$ is of concern only when dealing with radioactive tracers; in the present case the distribution of radiocarbon.

DERIVATION OF A BOX MODEL FOR STUDIES OF THE CARBON CYCLE

Equation (3) is only a function of the vertical coordinate and in this sense of the same kind as the equations used by *Wyrtki* (1962) and others. Any finite difference formulation of equation (3) is equivalent of a box model with the same number of boxes as vertical gridpoints used for the numerical formulation. It was pointed out by *Keeling and Bolin* (1967) that when merely a few gridpoints are being used, the derivatives appearing in the continuous models are very poorly described by the finite difference approximations. It should be noted that all terms in equation (3) that describe fluxes due to water motions (the two terms on the left side of the equation and the two first source terms to the right) express convergence or divergence of those fluxes into or out from the infinitesimal volume between two levels Z and $Z + dZ$. If an integration is made over a vertical interval Z_1 to Z_2, the integrated value of the tracer over the volume thus considered denoted by C_i, the following relationship is obtained:

$$\frac{\partial \overline{\overline{C}}_i}{\partial t} + \left[w\overline{C}_i \right]_{Z_1}^{Z_2} + \left[K_v \frac{\partial \overline{C}_i}{\partial Z} \right]_{Z_1}^{Z_2}$$

$$= \int_{Z_1}^{Z_2} \left[\oint K_H \frac{\partial C_i}{\partial n} dS \right] dZ + \int_{Z_1}^{Z_2} \left[\oint v_n C_i dS \right] dZ \tag{4}$$

$$+ \int_{Z_1}^{Z_2} \overline{J}_i \, dZ - \lambda^* \overline{\overline{C}}_i$$

Only the first and last terms contain the average concentration C_i, while all other terms depend on the fluxes at the boundaries and accordingly on the values of C_i at the boundaries. If it is desired to derive a box model from equation (4), those fluxes need to be expressed in terms of the concentration, $\overline{C}_i$, in the box under consideration and neighboring ones. As pointed out by *Keeling and Bolin* (1967), equation (4) is transformed to the finite difference approximation of equation (3) if a large number of boxes (or grid points) are used. However, in the case of wishing to use merely a rather limited number, the method to be used for approximating the fluxes between these boxes needs particular attention. Each term in equation (4) will be considered separately.

Convergence of Vertical Advective Flux: $\left[w\overline{C}_i \right]_{Z_1}^{Z_2}$

If w is a constant and if denoting the concentration of the box under consideration by $\overline{\overline{C}}_i^{(n)}$ and correspondingly, for the boxes below and above by $\overline{C}_i^{(n-1)}$ and $\overline{C}_i^{(n+1)}$, this term may be written as $\frac{1}{2} w \, (\overline{C}_i^{(n+1)} - \overline{C}_i^{(n-1)})$, which is the first order finite difference approximation of the corresponding term in equation (3). If, on the other hand, w varies as a function of z, as was assumed by *Kuo and Veronis* (1973) and also will be assumed here, a corresponding approximation, $\frac{1}{2} [w^{(n+1)} \cdot \overline{C}_i^{(n+1)} - w^{(n-1)} \cdot \overline{\overline{C}}_i^{(n-1)}]$ is less accurate and may introduce some systematic errors ($w^{(n+1)}$ and w^{n-1} are the values for w at the center of neighboring boxes). More accurate values for the convergence of the advective flux are obtained by using the expressing $\frac{1}{2} w_{z2} [\overline{C}_i^{(n+1)} + \overline{C}_i^{(n)}] - \frac{1}{2} w_{z1} [\overline{C}_i^{(n)} + \overline{\overline{C}}_i^{(n-1)}]$ where w_{z2} and w_{z1} are the known values of w at level z_2 and z_1 respectively.

Convergence of vertical turbulent fluxes $\left[K_v \partial \overline{C}_i/\partial Z\right]_{Z_1}^{Z_2}$

The vertical diffusivity is in reality variable, but as was done by *Kuo and Veronis* (1973), K_v will be assumed to be constant. More data and further experiments with models of the present kind needs to be carried out before a variable K_v may be included. This expression therefore is most appropriately approximated by the finite difference approximation $K_v \cdot (Z_2 - Z_1)^{-1} [\overline{C}_i(n+1) - 2\,\overline{C}_i(n) + \overline{C}_i(n-1)]$ This approximation does not hold as the bottom is approached, and the assumption that K_v remains constant is less acceptable. It is then more appropriate to consider the first term on the right hand side of equation (4) as describing the total (both vertical and horizontal) turbulent flux from the boundary into the box by interpreting K_H as the appropriate diffusion coefficient and $\frac{\partial C}{\partial Z}$ as the concentration gradient perpendicular to the bottom. For further considerations of this flux see below.

Turbulent Flux from the Boundary $\int_{Z_1}^{Z_2} \left[\oint K_H \partial C_i/\partial n dS\right] dZ$

The horizontal integration of equation (3) implies that horizontal variations of the concentration C_i will not be of concern, which implicitly implies an assumption that the horizontal advective and turbulent processes bring about a horizontally quasi-homogeneous distribution of the tracer permitting a consideration of only the vertical processes explicitly. This is only approximately true, but it has been shown by *Wyrtki* (1962) that some interesting results can be obtained in this way and will be generalized here considering a more complete model of the ocean circulation. The particular term under consideration describes the flux of the tracer from the bottom into this horizontal sheet of water or vice versa. To assess it appropriately, however, knowledge about the chemical processes of dissolution or sedimentation at the bottom and particularly which of these processes is rate limiting must be available. The following distinguishes between two extreme cases.

Chemical Reaction and Transfer Processes at the Bottom are Rate Limiting. This assumption implies that horizontal mixing (and advection) bring about an almost horizontally homogeneous distribution of the tracer and the average concentration $\overline{C}_i$ will prevail close to the bottom. This will be so for all compounds that are involved in the chemical processes at the bottom and the rate of transfer of all of these to or from the bottom may be computed from the appropriate chemical reaction. Of particular concern here is the dissolution or formation of calcium carbonate, which processes may be described in the simplest forms by the expressions.

$$CaCO_3 \rightleftarrows Ca^{++} + CO_3^{=} \quad \text{(a)}$$

$$H_2CO_3 \rightleftarrows H^+ + HCO_3^- \rightleftarrows 2H^+ + CO_3^{=} \quad \text{(b)} \qquad (5)$$

Obviously, three variables, e.g. the concentrations of calcium, bicarbonate and carbonate ions, are needed to describe these reactions. However, the concentrations of calcium and total inorganic carbon and alkalinity (cf *Keeling and Bolin*, 1967) can be used. The equilibria are qualitatively well-known and functions of pressure (or depth) which is of fundamental importance for the proper treatment of the transfers. Numerical values are, however, still uncertain. The problem is complicated by the fact that the equilibrium of (5a) is also dependent on the crystal form of the calcium-carbonate (i.e., whether it is calcite or aragonite and further, organic films into which organically formed carbonate are embedded may influence the transfer rates significantly). These microprocesses will not be considered further here, but the concept of transfer velocity (v_s) (sometimes called deposition velocity) between the water reservoir and the bottom in analogy with what is being done to describe the flux of matter between the atmosphere and an underlying land or water surface (cf e.g., *Bolin et al.*, 1974) will be introduced.

$$K_n \frac{\partial C_i}{\partial n} = v_s \, (C_i^* - C_i^e) \qquad (6)$$

where K_n is the diffusivity (molecular or turbulent) in the direction (n) parallell to the density surfaces, along which the transfer preferentially takes place; C_i^* is the concentration at some appropriately chosen distance from the bottom; and C_i^e is the concentration of the tracer being considered for which there is no net transfer in either direction between the water phase and the sediment. Most conveniently, it is desired to put $C_i^* = C_i$, which can be done if the turbulent and advective exchange processes are able to maintain a quasi-homogeneous distribution of the tracers concerned at a reasonably small distance away from the bottom. The transfer velocity then expresses primarily the effects of chemical reaction and molecular (and possibly turbulent) transfer close to the bottom, that are of importance for the exchange. It should finally be noted that there is an upper limit to v_s in order to satisfy the condition that quasi-homogeneous conditions are maintained in the interior. The mean characteristic interior transit time T_t is defined as the average time a water molecule spends in the layer under consideration (cf *Bolin and Rodhe*, 1973). Denoting further the characteristic horizontal dimension of the layer by L, the condition that

$$V_s << L/T_t \qquad (7)$$

must be imposed for intermediate depths, L is 5000 to 10,000 km and T_t of the order of several decades to 100 years. Thus V_s must be less than about $0.1 \; cm \; sec^{-1}$.

Chemical Reaction and Transfer Processes are Rapid in Comparison with the Horizontal Transfer in the Interior of the Ocean. If the chemical composition of the sediment in the depth interval Z_1, to Z_2 is similar in different parts of the oceans, the equilibria such as those expressed by equation (5) may be assumed to be valid throughout any such horizontal sheet. The integration of equation (4) would then proceed in such a way that the chemical equilibrium is determined after each time interval for which vertical transfer has been computed as given by the second and third term to the left, and the second term to the right in the equation (see further below). It should be noted, however, that the calcium carbonate deposits are found at greater depths in the Atlantic Ocean than in the Pacific Ocean, a fact that probably needs to be considered in a more accurate treatment of the interplay between the ocean and the sediment. It may then well be necessary to consider the two oceans separately, and couple them on the basis of our knowledge of the water exchange between them through the Antarctic Circumpolar Current (cf *Kuo and Veronis*, 1973).

Flux into the Interior of the Ocean from the Principal Areas of Deep Water Formation: $\int_{Z_1}^{Z_2} \left[\oint (v_n c_i)\, dS\, dZ\right]$

The line integral is extended around those regions in which deep water formation occurs, and thus, $v_n > 0$. This source term is not well-known, particularly not the vertical distribution of the deep water formation, as well as the tracer concentration C_i to be assigned to the water entering the deep ocean basins in this way. The assumption that w increases linearly with decreasing depth obviously implies that the water injections are assumed to be the same at all depths. In their study, *Kuo and Veronis* (1973) assign an oxygen concentration as observed in the Antarctic circumpolar water masses. If it is desired to deal with the tracer budget for the ocean as a whole, it is, however, necessary to relate the tracer concentration used in this term to the concentration in the surface ocean water which constitutes the source for the deep water. In the application of the present model to the carbon cycle, the concentration in the surface water may be deduced from the value which is in equilibrium with the partial pressure in the atmosphere.

Source (or sink) Term Due to Decomposition of Settling Particulate Matter: $\int_{Z_1}^{Z_2} J_i\, dZ$

A proper treatment of this term requires the simultaneous consideration of the transfer equations for all compounds that are involved in these chemical processes. In addition, an equation needs to be formulated for the concentration of particulate matter (to be denoted by C_p). With reasonable approximation it may be assumed that the transfer due to water motions are small in comparison with that due to settling. Then,

$$\frac{\partial \bar{C}_p}{\partial t} + \frac{\partial}{\partial z}(w\bar{C}_p) = J_p \tag{8}$$

or in integrated form,

$$\frac{\partial \bar{\bar{C}}_p}{\partial t} + \left[w\bar{\bar{C}}_p\right]_{Z_1}^{Z_2} = \int_{Z_1}^{Z_2} J_p \, dZ \tag{9}$$

where W denotes an appropriate settling velocity for the particulates.

To evaluate J_i requires the proper formulation of the chemical reactions involved in the decomposition of particulate matter. The simplest formulation based on *Redfield et al.* (1963) would be

$$\begin{aligned} &106\ CO_2 + 122\ H_2O + 16\ HNO_3 + H_3PO_4 \rightleftarrows \\ &\rightleftarrows (CH_2O)_{106} \cdot (NH_3)_{16} \cdot (H_3PO_4) + 138O_2 \end{aligned} \tag{10}$$

for the organic part, and the reactions of equation (5) for the dissolution of carbonate shells. It is, however, important to realize that the study of the possible changes of the carbon cycle due to the burning of fossil fuel may require considerable caution in applying these equilibrium equations. A significant shift of the carbonate and bicarbonate equilibria and an associated change of pH might well bring about a number of other perturbations of the chemical equilibria not described by the very much simplified equations (5) and (10). These would obviously also be of importance in evaluating the term ("Turbulent Flux from the Boundary") above, for which the simple equation (5) was also used. *Keeling* (1973) has included the borate ions in a more general treatment, but as has been pointed out by *Sillén* (1961), more complex reactions may also be involved. This problem will not be pursued further here, but merely its possible importance emphasized. In view of other simplifying assumptions it is hardly justified to complicate the chemical treatment further.

SOME CONSIDERATIONS OF THE CARBON CYCLE AND ITS MODIFICATION DUE TO ANTHROPOGENIC ACTIVITY

The model described above is being applied to both total carbon and radiocarbon, first to deduce the most likely set of values for the advective and turbulent fluxes based on the characteristic distributions as observed. Since the carbon cycle is interconnected with the oxygen and phosphorus cycles, the distributions of these elements add further information for the determination of ocean circulation. The computations are in a sense a generalization of the model computations presented by *Keeling and Bolin* (1968). A unique solution is, however, not obtained because of the additional degrees of freedom of the model that are introduced by the more numerous layers used for describing the tracer distribution in the ocean. Nevertheless, quite specific results can be obtained. The model then can be used for studying the time dependent case, particularly in which way emissions of fossil fuel carbon into the atmosphere will disturb the natural steady state which first has been derived. A detailed presentation of such computations will be described elsewhere.

In most models developed so far to describe the carbon cycle, the transfer of carbon into the deep sea is assumed to be accomplished by water motions, while in reality the biological formation of particulate matter, which sinks to great depths before being dissolved, represents the principal mechanism. The amount of total carbon is therefore considerably larger in the deep sea than in the surface layers. The water motions, both advertive and turbulent, on the other hand, bring about a net upward transfer of carbon, which in a steady state balances the downward transfer due to the sinking of detritus. An attempt to deal with this more complex transfer pattern was made by *Keeling and Bolin* (1967, 1968), and also by *Broecker et al.* (1971).

While *Keeling and Bolin* (1967) simultaneously considered salinity, total carbon, oxygen, phosphorus, alkalinity and radiocarbon, *Broecker et al.* (1971) essentially limited themselves to carbon, particulate matter and radiocarbon. The oceanic model in both cases basically was a crude three-box model, which obviously cannot represent well the vertical structure and circulation as outlined in the previous section. *Bolin and Keeling* (1967) computed that the downward flux of total carbon due to gravitational settling of particulate matter is 27 x 10^8 tons/year, which in the steady state is balanced by an upward flux primarily due to convective overturning at high latitudes amounting to 22 x 10^8 tons/year and by a slow meridional circulation, 5 x 10^8 tons/year (cf Figure 1). It would be of considerable interest to see whether these computed fluxes are considerably changed by using a more detailed model of the vertical structure of the ocean, and also if the transfers are modified by including the effects of sedimentation and dissolution of calciumcarbonates at the bottom. This is being done as outlined in the previous section.

It is of interest to note that *Keeling and Bolin* (1973) also derived the biologically consistent steady state flux of phosphorus, which is shown in Figure 2. It is noted first of all that the *C/P* ratio for the oceans as a whole is 400, while the *C/P* ratio in organic matter formed in the sea is about 40, *Redfield et al.* (1963). These numbers illustrate the well-known fact that phosphorus is one limiting factor for the primary production in the sea. The surface waters are rapidly depleted in phosphorus and the sinking of biogenic matter is an effective mechanism for phosphorus transfer into

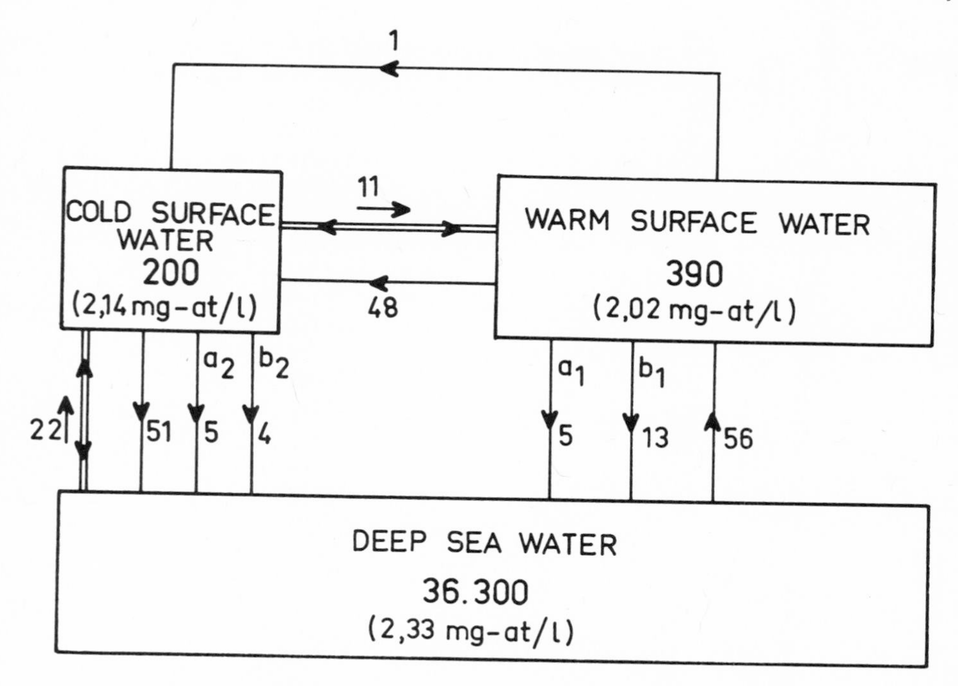

Fig. 1. Steady state transfer of total carbon as computed by *Keeling and Bolin* (1968) for a three reservoir fluxer. a) denotes flux due to settling of inorganic detritus (essentially carbonate shells), b) correspondingly denotes flux due to settling of dead organic tissue. The other arrows denote fluxes due to the water motions, where a single arrow indicates that due to a meridional circulation and the double arrow the flux caused by large scale turbulent exchange causing no net flux of water. There is finally a small flux (1 unit) due to atmospheric motions. Unit for reservoir size 10^9 ton and for fluxes 10^8 ton per year.

the deep sea. This downward transfer is balanced primarily by turbulent upward flux into the cold water reservoir in polar regions. It should be pointed out that while the downward transfer of phosphorus is only associated with the settling of organic compounds, the downward transfer of carbon is also due to settling of carbonates formed in the surface layer in association with the synthesis of biocarbon. For this reason *Keeling and Bolin* (1973) found that the ratio C/P in the total flux of particulate matter into the deep sea

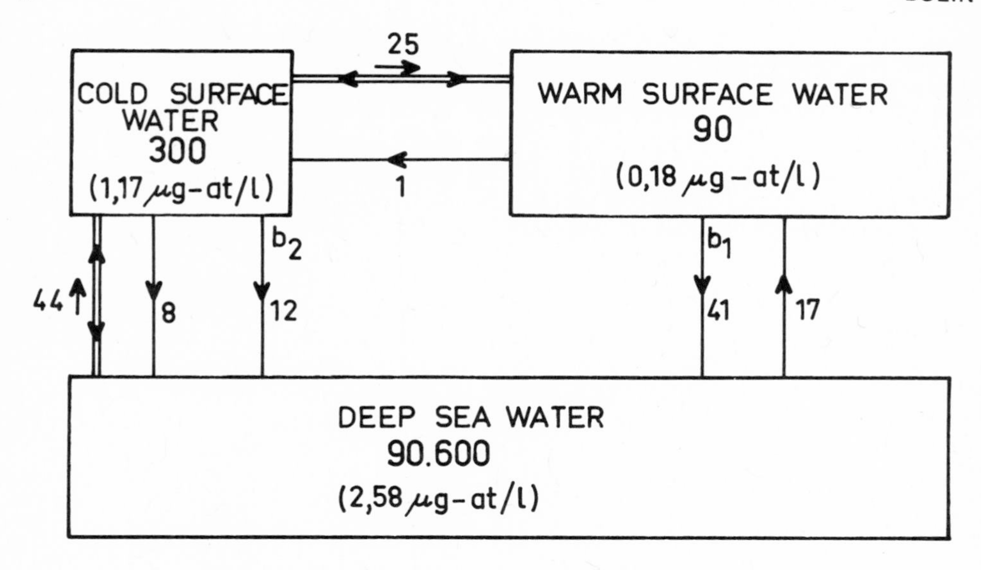

Fig. 2. Steady state transfer of phosphorus computed by *Keeling and Bolin* (1968) cf Figure 1 for notations. Unit for reservoir size 10^6 ton and for fluxes 10^6 ton per year.

is between 50 and 60. This may imply that it will be necessary to consider organic particulate matter and inorganic carbonates separately when applying the transfer equation (9). The simultaneous consideration of the phosphorus and the carbon cycles (as well as cycles for other biologically important elements) using the more detailed model outlined in the previous section is obviously of interest.

The fact that phosphorus is an important element for primary production raises the question as to whether or not man is indirectly modifying the carbon cycle by emitting rather large amounts of phosphorus into the coastal waters and the oceans. Man today extracts about 12 Tg (terragram = 10^{12} g = 10^6 tons) of phosphorus by mining, most of which (about 10 Tg/year) is used as fertilizers. About 2 Tg/year are used for other activities, particularly in detergents (cf *Pierrou*, 1976). The latter amount is added to the waste waters and transferred by rivers into lakes and the coastal waters of the oceans. Further, man handles roughly 10% of the land biota that grows per year which implies 18 Tg P/year, of which about

5 Tg represents agricultural production, mainly cereals. Most of this is returned to the soil, but 1-2 Tg/year is brought by rivers to the lakes and coastal waters of the sea. Thus, 2-4 Tg/year of phosphorus may at present be added to the fresh water systems and to the sea due to man's activities. Even though this is a rough figure, the importance of such an emission for the carbon cycle considering the fresh water systems and the sea separately should be considered. It is well-known that man in this way is fertilizing the world water systems. There are many examples of a marked increase of the biological productivity leading to eutrophication of lakes. *Stumm* (1972) gives examples from Switzerland, *Fonselius* (1972) and *Bolin* (1972) have analyzed the Baltic in this regard. In the United States the Great Lakes, particularly Lake Erie, has been seriously affected by the injection of phosphorus. Accepting a mole-ratio *C/P* in organic material of 106, corresponding to a mass ratio of 40 as given by *Redfield et al.* (1963), implies that the increased biological activity in such water systems corresponds to a withdrawal of carbon from the atmosphere in an amount 40 times the amount of phosphorus that has gone into the biota. With the present rate of phosphorus input into the hydrological system, this implies an annual carbon fixation of 80 to 160 Tg. These levels may well increase considerably in future years.

There is an additional interesting fact with regard to lakes, and even water basins of the size of the Baltic Sea, that is worth noticing in this context. As the biological activity is increased because of the additional supply of phosphorus, the oxygen demand at deeper layers for the decomposition of organic material may lead to its depletion. Further oxidation implies the breakdown of sulfates and the formation of hydrogen sulfide. The chemical balance at the water-sediment interphase in this way is radically changed. As *Fonselius* (1969) has shown for the Baltic Sea, a very significant release of phosphorus from the sediment then takes place, enriching the waters and contributing to further biological activity when transferred into the surface layers. Analyses of recently formed sediments in the Baltic Sea show *C/P* ratios much larger than the ones quoted above from *Redfield et al.* (1963). Obviously, this series of events represents a positive feed-back mechanism whereby the importance of phosphorus emissions for the carbon cycle may be enhanced.

Most of the phosphorus released into fresh-water lakes or coastal waters is utilized by organisms rather quickly and only a part reaches the oceans on time scales with which we presently are concerned. It should be noted, however, particularly at high latitudes, that in winter the primary production is markedly reduced due to low temperature and lack of sun light. During this time of the year phosphorus may be transported considerable distances.

According to *Keeling* (1973) the burning of fossil fuels and other activities presently imply a flux of carbon into the atmosphere of 4,300 Tg/year. The increase of the carbon dioxide content of the atmosphere corresponds to about 2,300 Tg/year and thus, about 2,000 Tg/year is transferred into land biota, fresh waters, the oceans and marine biota. From the discussions of the increased phosphorus flux into the hydrosphere due to man's activities it is concluded that between 100 and 200 Tg/year are transferred into the deeper water layers or sediments, possibly even more. This corresponds to 2-5% of the total industrial output, which is obviously not an insignificant amount. To the extent phosphorus is added to the water systems and recycled in the productive zones, the figure given above may even be larger. It thus may be of considerable interest to take note of our changing practices in using phosphorus and consider in more detail than has been presented here, possible implications for the carbon cycle and the likely consequences with regard to our increasing use of fossil fuels and the associated increase of atmospheric carbon dioxide. In principle this was pointed out already by *Eriksson* (1963), but some quantitative estimates are now able to be made.

REFERENCES

Bacastow, R. and C. D. Keeling. 1973. Atmospheric carbon dioxide in the natural carbon cycle: II Changes from A.D. 1700 to 2070 as deduced from a geochemical model. *Carbon in the biosphere*, U.S. Atomic Energy Commission. 86-135.

Bolin, B. and E. Eriksson. 1959. Changes in the carbon content of the atmosphere and the sea due to fossil fuel combustion. *The Atmosphere and Sea in Motion, Rossby Memorial Volume*, Rockefeller Institute Press, New York. 130-143.

Bolin, B. 1972. Model studies of the Baltic Sea. *Ambio Special Report No. 1*. 115-119.

Bolin, B. and H. Rodhe. 1973. A note on the concepts of age distribution and transit time in natural reservoirs. *Tellus*, 25. 58-62.

Bolin, B., G. Aspling, and C. Persson. 1974. Residence time of sulfur in the atmosphere as dependent on height of emission, intensity of turbulent transfer and efficiency of sink mechanisms. *Tellus*, 26. 185-195.

Bolin, B. 1975. A critical appraisal of models of the carbon cycle. Appendix 8 "Carbone dioxide processes". ICSU/WMO Global Atmospheric Research Programme - JOC GARP Publications Series No. 16. 225-235.

Broecker, W., Y. Li, and T. Peng. 1971. Carbon dioxide - Man's unseen artifact in impingement of man on the oceans. John Wiley and Son Inc., New York. 287-324.

Craig, H. 1957. The natural distribution of radiocarbon and the exchange time of carbon dioxide between the atmosphere and sea. *Tellus*, 9. 1-17.

Eriksson, E. 1963. Possible fluctuations in the atmospheric carbon dioxide due to changes in the properties of the sea. *J. of Geoph. Res., 68.* 3871-3876.

Fiadeiro, M. 1975. Numerical modeling of tracer distributions in the deep Pacific Ocean. Ph. D. Thesis. University of California, San Diego.

Fonselius, S. 1969. Hydrography of the Baltic deep basins III. Fishery Board of Sweden. Gothenburg, Report No. 23.

Fonselius, S. 1972. On biogenic elements and organic matter in the Baltic. *Ambio Special Report No. 1.* 29-36.

Keeling, C. D. and B. Bolin. 1967. The simultaneous use of chemical tracers in oceanic studies. I General theory of reservoir models. *Tellus, 19.* 566-581.

Keeling, C. D. and B. Bolin. 1968. The simultaneous use of chemical tracers in oceanic studies. II A three-reservoir model of the North and South Pacific Oceans. *Tellus, 20.* 17-54.

Keeling, C. D. 1973. Industrial production of carbon dioxide from fossil fuels and limestone. *Tellus, 25.* 174-198.

Kuo, H. H. and G. Veronis. 1973. The use of oxygen as a test for an abyssal circulation model. *Deep Sea Res. 20.* 871-878.

Machta, L. 1972. The role of the oceans and biosphere in the carbon cycle. Nobel Symposium 20. The changing chemistry of the oceans. Almqvist and Wiksell. Stockholm. 121-145.

Pierrou, U. 1976. The global phosphorus cycle, Ecological Bulletin Natural Sciences Research Council, Stockholm, No. 22. 75-88.

Redfield, A. C., B. H. Ketchum, and F. A. Richards. 1963. The influence of organisms on the composition of sea water, In: *The Sea, Vol II,* M. N. Hill (Ed.), Interscience Publishers, New York. 38 pp.

Sillén, G. 1961. The physical chemistry of sea water, In: *Oceanography,* M. Sears (Ed.), Publ. Am. Assoc. Adv. Sci. 67 pp.

Stumm, W. 1972. The acceleration of the hydrogeochemical cycling of phosphorus. Nobel Symposium 20. *The Changing Chemistry of the Oceans.* Almqvist and Wiksell, Stockholm. 329-346.

Wyrtki, K. 1962. The oxygen minima in relation to ocean circulation. *Deep Sea Res. 9.* 11-23.

THE LAND'S RESPONSE TO MORE CARBON DIOXIDE[1]

Edgar Lemon

U. S. Dept. of Agriculture

Agricultural Research Service

ABSTRACT

With a continued increase of fossil fuel carbon dioxide (CO_2) in the atmosphere there is cause for concern about future consequent climate change. Records indicate that about half the CO_2 input remains in the atmosphere and the other half is absorbed by the land and sea. No one knows the proportions or for sure the mechanisms of fixation.

On the land the initial CO_2 fixation is by photosynthesis in plants. Short term experiments indicate that photosynthesis is speeded up by more atmospheric CO_2. However, other plant processes and environmental stress may negate this initial response. The life cycle of individual plants and whole forest ecosystems could be shortened too, so that net carbon accumulation in the end is little changed.

A small change in net photosynthesis of the living biomass of the land might be quite significant however. This would depend upon the size and stability of both living and dead carbon pools. Man's influence on both is a recognized possibility yet both kind and magnitude are unknown. For example, forest cutting might be balanced by regrowth or managed agricultural production. Also soil humus losses from forest cutting or prairie plowing might be balanced by high technology agriculture where plant residues increase soil humus. Eutrophication of lakes and estuaries due to increased nutrient cycling also adds carbon to the detritus pool. So far,

---

[1]This contribution is from the U. S. Agricultural Research Service, in cooperation with the Cornell University Agricultural Experiment Station, Ithaca, N.Y. Dept. of Agronomy Series Paper No. 1156.

analysis of terrestrial plant activity through atmospheric CO_2 sampling indicates no change in the biosphere in the past decade or so. However, the method and time scale are limited.

In any event the potential for change is real. It is important. The need for more research is unquestioned.

INTRODUCTION

Two recent events have indirectly increased the public interest in the fate of fossil fuel carbon dioxide (CO_2): the escalated price of oil and the depleted grain reserves of North America. Both hit the pocketbook. The public wants to know more about the energy and agricultural economies.

Daily the newspapers elaborate on the size of fossil fuel reserves, how long they will last, etc. Also, we hear about the vagaries of the weather influencing our crops and those of the USSR.

The two economies overlap. The newspapers warn the public of the "greenhouse" effect from our adding CO_2 to the atmosphere and how this will warm the climate with dire consequence to food supply.

So far, past records do not support anyone's theory of climate change (*Landsberg*, 1976). Nonetheless serious scientists have cause for concern over the steady 1 ppm yearly increase in atmospheric CO_2. Solid evidence indicates that this represents one-half of what is man's yearly input into the air. The other half disappears into the carbon pools for the oceans and lands. How this is done and in what proportions seems unresolved at present. These are serious problems. First, we cannot rule out that future concentrations of CO_2 will not affect our climate, and second, we cannot predict what those CO_2 concentrations will be without understanding the mechanics of land and ocean uptake. Unfortunately, the problems are compounded by the fact that changes in concentrations alter land and ocean properties and processes. Still further compounding the problem is the possibility of man reducing the land pool of carbon by harvesting the tropical forests or, on the other hand, increasing the carbon pools of lakes and estuaries through speeding up inadvertent eutrophication.

The purpose here is to examine how the land will respond to increasing atmospheric CO_2. First, I will take up the problem of plant photosynthetic response to CO_2 and consequent outcome. Next I will examine the size of carbon pools on the land and speculate about their future. Finally, I will summarize our knowledge gaps and stress what needs to be done.

THE PROBLEM OF PLANT RESPONSE TO CO_2

By the land's response to CO_2 I mean land plant photosynthesis and carbon accumulation in response to added CO_2 in the atmosphere. The important problem of increased weathering of the land with more CO_2 is not considered.

All students in beginning biology learn that photosynthesis increases with more CO_2, if there is enough light and the plant system has enough nutrients, water, and a proper temperature. Logic tells us that if this is so, then more CO_2 in the earth's atmosphere should mean greater photosynthesis and in turn a greater yield or accumulation of carbon in plant life. Logic may fail us, however, because life is not that simple. Yield is a many-faceted thing and not influenced by photosynthesis alone.

Let me give an extreme example. For some time greenhouse men have been adding CO_2 in their closed system to increase yields, say, of tomatoes, lettuce, cucumbers and carnations. (By yield in this sense I mean marketable product, not necessarily carbon accumulation.) This distinction is important. The Dutch have found that more CO_2 in a glasshouse gives an earlier and bigger crop of fresh lettuce, yet on a dry matter basis (accumulated carbon) there is no difference between a CO_2 fertilized crop and a nonfertilized one. The canny Dutchmen are selling more water to the housewife packaged in green leaves. I stress this extreme case to pinpoint the complexity of forecasting what land plants will do with added CO_2.

For the annual cycle of short-lived plants, there may be little significance in their response to added CO_2 in the context of long term impact on geophysical events. What is taken up by photosynthesis in one season, much is lost by respiration before the next. However, there could be some influence on the long term carbon pool of soil humus under short lived vegetation.

Of major concern are the long lived plants, the forest and jungles where the carbon pool is large and fixed for decades, if not centuries.

In order to arrive at a "speculated" answer about the impact of more CO_2 on land plants, I must resort to indirect argument simply because there are no data on the life cycle response of trees to carbon dioxide. Much will have to be inferred from information on short lived plants or short term experiment. Some information is available from modeling.

THE HIERARCHY OF YIELD RESPONSE

We have already emphasized that while plants depend upon photosynthesis to fix carbon, the final yield of carbon accumulation is affected by many other processes. Such processes following photosynthesis are: respiration, translocation of photosynthate, the partition of photosynthate to various organs and processes, in addition to environmental limitations on the capacity of the plant to grow and use the products of photosynthesis. *Evans* (1975) gives an excellent treatise on this point of view.

I shall follow *Gifford's* (1974) outline of a hierarchical sequence of plant processes, from biochemical fixation of carbon in photosynthesis, through to final yield for two groups of plant species. His approach is most instructive because one group has a higher affinity for carbon dioxide at the initial stage of photosynthesis. Of course, this is the single most important step where more CO_2 would affect carbon accumulation.

The existence of two groups of plant species with different photosynthetic systems has been clearly recognized in the past decade. One group, the C_3 species, possess the following characteristics: 1) The first product of photosynthetic fixation of carbon is a 3 carbon carboxylic acid (phosphoglyceric acid); 2) at reduced CO_2 levels the compensation point is about 50 ppm CO_2; 3) photosynthesis is inhibited by oxygen; 4) bundle sheath cells around the vascular tissue are not well developed in the leaf anatomy. By contrast, the second group, the C_4 species have the following characteristics: 1) They initially fix carbon in a 4 carbon dicarboxylic acid (Oxaloacetic acid); 2) they have a CO_2 compensation point near zero; 3) they are not affected by oxygen; and 4) they have well-developed bundle sheath cells.

C_3 plants reputedly have lower photosynthetic potential and lower primary productivity than do C_4 plants. Maximal reported values of the process parameters of each group are compared. Thus, comparisons are for environmental conditions best suited to each group. C_4 plants thrive best in a high light environment that is warm. (Tropical grasses such as corn, sorghum, and sugar cane are C_4 plants.) C_3 plants generally thrive better at lower light and cooler temperatures. (Almost all forest species are in the C_3 group.) *Loomis and Gerakis* (1975) have examined the latitudinal distributions of maximum agricultural crop yields of the two groups. Figure 1 gives their summary hypothesis of production of C_3 and C_4 crops in relation to latitude from 0 to 65 degrees.

Now, we examine the plant processes of the two groups in the order of increasingly higher levels of biological organication.

The Primary Biochemical CO_2 Fixation Level

C_3 species initially fix carbon through a reaction between CO_2 and *RuDP* (ribulose 1,5-diphosphate) catalyzed by an enzyme, *RuDP* carboxylase. C_4 species initially fix carbon through a reaction between CO_2 and *PEP* (phosphopyruvate) catalyzed by an enzyme *PEP* carboxylase.

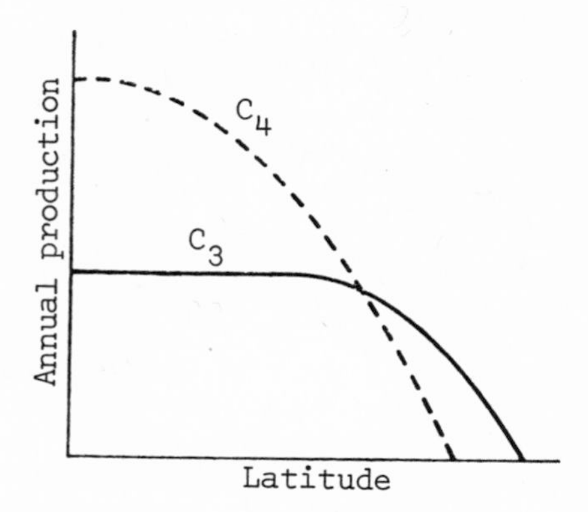

Fig. 1. Summary hypothesis of production of C_3 and C_4 crops in relation to latitude from 0° to 65°. (From *Loomis and Gerakis*, 1975).

At the biochemical level, earlier classical reaction rate constants for these two different photosynthesis pathways differed by a factor of 60 to 70. However, more recent data indicate this is only a factor of 2 to 3. The CO_2 concentration necessary to reach half the saturation rate then is 2 to 3 times greater for *RuDP* carboxylase than it is for *PEP* carboxylase. Thus, the C_4 system with its greater affinity for CO_2 enjoys a 2 to 3 fold potential advantage over the C_3 system.

The Intact Mesophyll Cell Level

At this level the differences between the two groups could be the result of differences in enzyme concentrations, or diffusional limitations on substrates and metabolites within the cell. However, the 5 fold C_4 advantage over C_3 observed at this level may be solely explained on the photorespiration basis. (If the enzyme reaction rates of the two groups are almost equal at the primary carboxylation.) C_3 plants exhibit more respiratory loss of CO_2 through a photorespirational process linked to light intensity. Significant photorespirational decarboxylation seems to occur as an inescapable part of the C_3 photosynthesis mechanism.

The Intact Leaf Level

The present concentration of CO_2 in the atmosphere is below the saturation point for normal healthy leaves. The rate of CO_2 fixation at this level is controlled in large part by gaseous diffusion of CO_2 down a concentration gradient. There are two constraints in the gas phase diffusion pathway, one external to the leaf surface, the other through the passageways (stomates) into the leaf. The external laminar boundary layer at the leaf surface damps diffusion according to the rules of aerodynamics, related in part, to the leaf geometric properties but unrelated to the C_3-C_4 grouping. The second constraint to gas diffusion, through stomates, is of profound importance.

I cannot overstate this fact for it is at the stomatal control level that environmental parameters of light, water status, temperature and CO_2 concentration have such an impact. Not only in this mechanism are there differences between the C_3 and C_4 group response to these environmental variables but overall in land plants the stomatal control of gaseous diffusion determines directly and indirectly, in no small way, what final plant yield will be.

Under optimum conditions the C_3 stomates of crops generally present a lesser barrier than those in the C_4 group, thus reducing the C_4/C_3 advantage from 5 to between 1.5 and 3 fold in the crops examined.

Daytime Net Photosynthesis of Pure Crop Stands

Photosynthesis studies have been made in pure stands of plants in which 1) clear plastic tents are placed over a representative patch of the crop, or 2) meteorological techniques of measuring gas exchange are used. Gifford cites work on four crops, one C_4 and three C_3 species. This small sampling gives a C_4/C_3 ratio of 1.5 for both techniques, down from 1.5 to 3 range at the lower intact-leaf level. It is surprising that the reduction is not greater, for the systems are now becoming really complex; with the whole plant and its many organs and processes; the crop with its various structures; and the environment with its diurnal swing. In the whole plant there are non-photosynthesizing parts that respire, as well as feedback control of leaf photosynthesis by sinks for photosynthate. At the crop level, there are differences in leaf area, leaf inclination and shape, as well as vertical and horizontal distribution. Perhaps of greatest significance is the interaction of the leaves and the diurnal swing of the sun through the sky. Since most C_4 species have a much higher CO_2 exchange rate at high light and usually do not saturate under the full sun, one would surmize that this advantage would be largely lost in a crop where most of the leaves are shaded a good share of the day. Further, only through the midday hours on clear days are the top leaves able to meet their potential in the C_4 species. There is still one more penalty to C_4 species: their stomates apparently require a higher light intensity to open (*Akita and Moss*, 1972).

Short-Term Crop Growth Rate in Pure Stands

At this level two additional factors affect the crop. During much of the daylight plants are operating at less than maximum potential because of light limitation in the morning and evening. Also, over a several day sampling period all daylight periods are unlikely to be clear. Furthermore, over 24 hour growth cycles the net exchange of CO_2 is determined in part by a drain of night respiration. Another factor may be involved in comparing maximum crop rates of C_3 and C_4 groups. That is, the crops being compared do not necessarily have an equal leaf area.

Gifford (1974) compares five C_4 species with three C_3 species having record dry matter gaines over 7 to 130 day sampling periods. Comparisons show no advantage of one group over the other. Maize and sunflower (C_4 and C_3 species) tie for top honors at 78 and 76 grams of dry matter/m^2 land area/day (roughly 35 gm C/m^2/day).

Yield

The final yield of a crop can be an economic unit such as a head of lettuce or it can be the accumulated carbon. Because economic units of yield are so different, comparisons are of limited value here.

While differences in crop growth rate influence yield at this highest and final level of organization, two factors are important to the final product; 1) partitioning of photosynthetic assimilate into the various organs and processes, and 2) the duration of crop growth. In order to make as valid a comparison between C_4 and C_3 groups, Gifford chose forage species where the foliage was removed periodically. He compared record yields of two C_3 species with one C_4 species. Annual totals gave a C_4/C_3 ratio of 2.5 to 3.5. The main reason for this high ratio is the year-around growing season for the C_4 species compared to the short season for the temperature-region C_3 forage grasses.

It was pointed out earlier that C_4 species perform best under high light and warm conditions whereas C_3 species do best under cooler conditions (*Bjorkman*, 1971). It has been assumed that a good share of this difference is due to the high temperature response of photorespiration CO_2 loss in C_3 plants. The net effect is that C_4 species have a longer growing season under conditions favoring their growth. We have already seen that in short time-sampling, yields are equal for the two groups when each is grown in its best environment. Thus, each group has equal potential rates for its optimum environment, but duration of the growing season is a dominant factor in giving C_4 plants a decided advantage in yield.

Following Gifford's approach of a hierarchy of yield processes demonstrates how metabolic advantages, prominent at the initial level of photosynthesis, disappear before final yield is achieved.

Succeeding processes influencing yield have been listed as the level of organization increased. In the wide spectrum of plants, all processes at all levels interact in a myriad of ways so that no simple universal weighing of these factors can be made.

BREEDING PLANTS FOR PHOTOSYNTHESIS

Since the photosynthesis rate is one of the components of yield, and varies widely among species and even varieties within a species, agriculturalists and foresters have attempted to improve yield in breeding programs by using short-term, leaf-gas exchange criteria (photosynthesis) for plant selection, hoping to correlate short term leaf photosynthetic rate with either economic yield or final dry matter accumulation. So far they have shown that photosynthetic rate is under genetic control and can be manipulated by breeding techniques. Unfortunately correlation of photosynthetic rate to economic yield or dry matter accumulation has eluded their efforts to date. This lack of correlation holds for maize (*Musgrave* pers. comm.); tall fescue (*Nelson, et al.*, 1975); alfalfa (*Delaney and Dobzenz*, 1974); and orchardgrass, tall fescue and timothy (*Sheehy and Cooper*, 1973). How can this be so?

Positive relationships between whole plant CO_2 uptake and yields are observed in controlled environments where conditions are near optimum, such as in glasshouses or growth chambers (*Ledig*, 1976). In the field, however, adaptations to the varying environmental factors, as well as the normal life cycle, changes photosynthetic rate in different ways among close relatives. Figure 2, from *Ledig* (1976), gives the seasonal pattern of net CO_2 uptake for second-year seedlings of European, Japanese and Siberian larch grown in Connecticut. Seasonal change in photosynthetic rank may be one reason for the frequent failure of breeding programs to relate short-time leaf photosynthetic measurement to yield or final biomass. Clearly, short-term leaf or plant samplings for photosynthesis rate is not representative of the integrated season.

Larcher (1969) emphasizes that final yield depends upon the length of "production period" as well as on CO_2 fixation rates. Failure to recognize this as a factor may defeat the plant breeder as well. Small differences in the length of the life cycles in closely related cultivars may go unrecognized such that a low photosynthesizer in the ranking may operate longer, overcoming its handicap.

A third argument that has been advanced rests on unrecognized differences in plant canopy architecture. Single leaf measurements would not account for a differing leaf-display ability to intercept light. Cultivars in a breeding program may differ sufficiently in architecture to mask photosynthetic advantage. Of course all the other factors of yield can be marshalled in the same agrument. Without a complete integrated carbon budget, difficulties arise in pinpointing cause and effect.

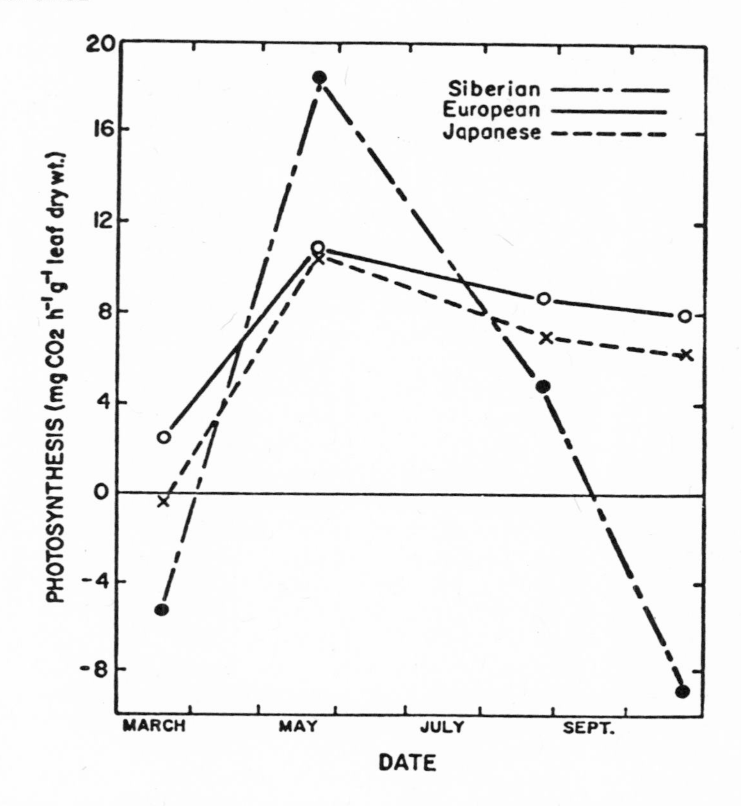

Fig. 2. Season pattern of net CO_2 uptake for second year seedlings of European, Japanese, and Siberian larch grown in Connecticut. Measured at 4000 ft-C and temperatures corresponding to the mean daily maximum. (From *Ledig*, 1976).

PHOTOSYNTHESIS RESPONSE TO CO_2

Leaf Chamber Studies

Studies abound in which intact leaves are enclosed in small chambers in order to measure short term CO_2 exchange. This is understandable because of relative ease and low cost compared to studying whole plants or even communities (especially over some extended time). Unfortunately, whole leaf studies are generally of short term, usually hours, to maintain leaf integrity. Despite relative ease of individual leaf studies, it is still tedious work and too few samples are taken to fully understand the spectrum of the variable under study. On top of this, technique is frequently faulty. Faulty technique usually resides in poor environmental control inside the enclosed leaf chamber (*Gaastra*, 1959).

If we remember that a leaf chamber measurement is on a single leaf, on a single plant, in a single environment at a single point in time, little wonder we have difficulty equating this measured photosynthetic rate to, say, dry matter gain over days, or weeks or a season. Thus these are problems with sampling in time and space as well as technique.

Despite the experimental difficulties of environmental control and, the sampling problems in space and time, the leaf chamber method of studying photosynthesis has been fruitful. Without such studies our knowledge about environmental effects on leaf photosynthesis with implication to whole plants and plant communities would be woefully lacking. To demonstrate some useful relationships fairly representative of C_3 forest species, I shall draw heavily on recent work reported by *Regehr et al.*, (1975). Figures 3, 4, 5 and 6 demonstrate the influence of light, temperature, carbon dioxide concentration, and water stress upon photosynthesis, transpiration and leaf conductance (stomate conductance) in (cottonwood) *Populus deltoides*. This tree species is a fast growing inhabitant of temperate region flood plains.

Figure 3 demonstrates that stomates open in increasing light, so that CO_2 and water vapor diffusion can increase with increasing available energy. The loss of water due to evaporation from wet cells in the leaf is an unavoidable consequence of requiring an agueous surface available to absorb CO_2 for photosynthesis. Stomates control the two-way exchange, preventing excessive water loss in time of shortage by closing. Naturally, CO_2 exchange is then throttled too.

Figure 4 shows temperature effects on these phenomena as well as on dark respiration. Dark respiration includes all respiration except photorespiration. Again, leaf conductance (or stomate control), transpiration, and photosynthesis, act in concert but with a small phase shift. Increasingly dark respiration, as well as stomate closure rob net gain in CO_2.

Figure 5 shows the influence of added CO_2 on photosynthesis and transpiration. Since transpiration is under stomatal control, the steady fall in transpiration can be interpreted as a steady closing of stomates due to CO_2 increase. Photosynthesis increases but approaches an asymptote. Figure 6 clearly demonstrates the influence of water stress on the leaf processes. More negative water potential is equated to drying of the leaf. As the leaf dries, stomates close, leaf conductance goes down, and concomitantly photosynthesis and transpiration fall.

Returning to the CO_2 effects on stomates: The response shown in Figure 5 is atypical for a C_3 species and more like a C_4. In C_4 species, CO_2 usually closes stomates so that photosynthesis response to CO_2 at high concentrations is relatively less than

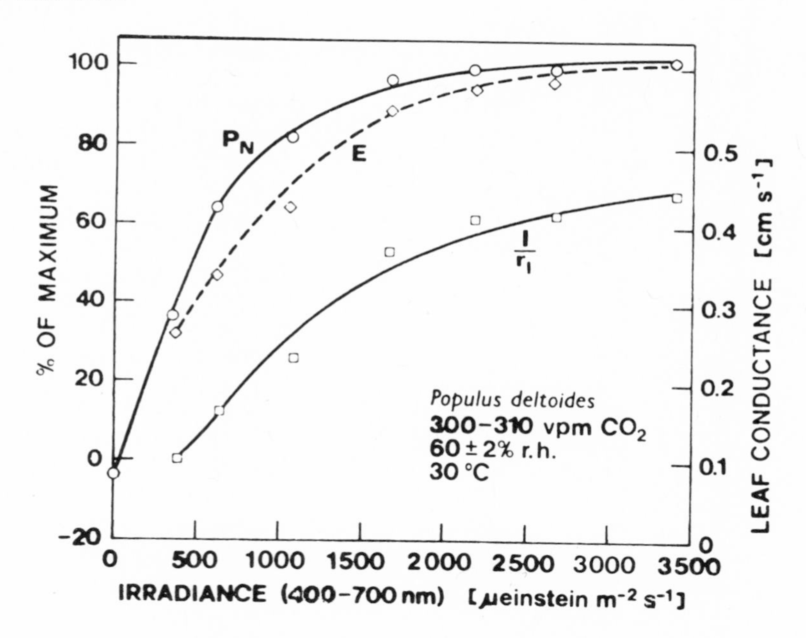

Fig. 3. Photosynthetic (p_N) and transpiration (E) rates and leaf conductance ($1/r_l$) for water vapor transfer as a function of irradiance, expressed as photosynthetically active radiation. 1500 µeinstein m^{-2} s^{-1} = 61 klx. Maximum photosynthetic rate is 26 mg dm^{-2} h^{-1}. (From *Regehr, Bazzaz and Boggess*, 1975).

plants (*Akita and Moss*, 1972, 1973). Here in the *Populus deltoides*, a doubling of CO_2 from 300 to 600 ppm increased CO_2 fixation rate by 65%.

Ludlow and Jarvis (1971) measured photosynthetic response of Sitka spruce (*Picea sitchensis*) leaves to added CO_2. Their results are more representative of C_3 species. Stomates were little affected by CO_2 and photosynthetic response was nearly linear between 0 and 600 ppm. For a doubling of CO_2 from 300 to 600 ppm, fixation rate increased 56%.

Brittain and Cameron (1973) report leaf age effects on leaf photosynthetic response to CO_2 in *Eucalyptus fastigata* in the 200 to 1400 ppm range. Juvenile leaves on week-old plants reached optimum CO_2 at 200 ppm. Juvenile and intermediate leaves of the same physical age reached optima at 400 and 500 ppm, respectively. The authors speculate that higher-than-optimum CO_2 adversely affects metabolism, ruling out stomates as a factor. The leaves gave off CO_2 at 1000 ppm CO_2.

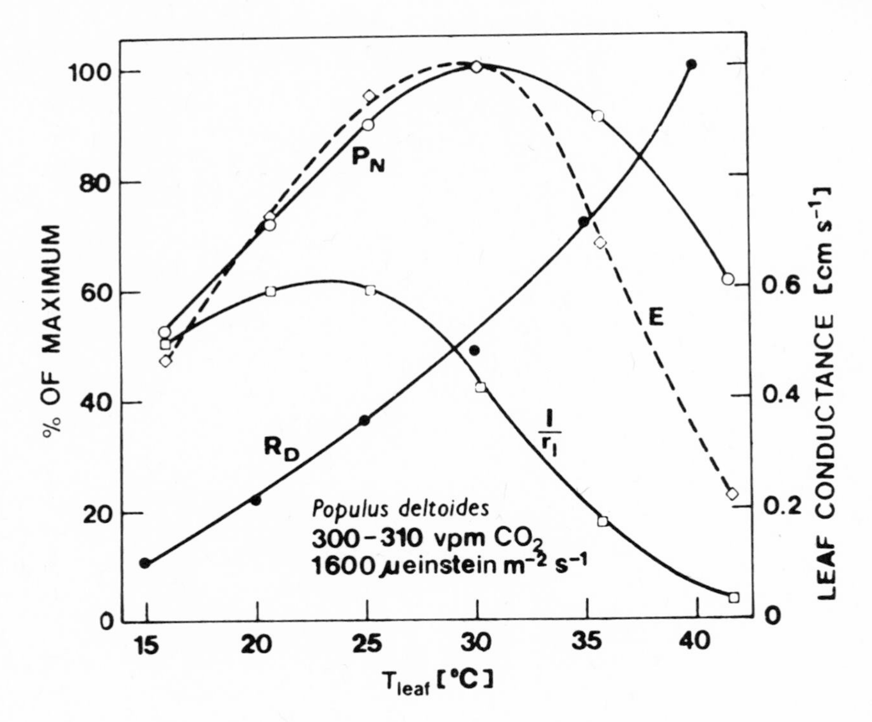

Fig. 4. Photosynthetic (P_N) and transpiration (E) rates and leaf conductance ($1/r_l$) in the light, and dark respiration rate (R_D) of single intact leaves at various temperatures. (From *Regehr et al.*, 1975).

We need to say a bit more about the environmental controls. These added complexities further damp the system to CO_2 perturbations In the tropics, the lack of available nutrient nitrogen is a significant factor for plant growth. Quite often these stresses may not necessarily damp photosynthesis per se, but hamper leaf or stem growth so in the long run the size of the photosynthesis machinery is reduced.

Figure 7, from *Boyer* (1970) clearly brings this out. For maize, soybean, and sunflower leaves, increasing leaf water potential (drying or increasing drought stress) damped both photosynthesis and leaf growth (elongation and enlargement). However, quite obviously the onset of drought stress occurred sooner and was more severe in the growth parameters compared to photosynthesis.

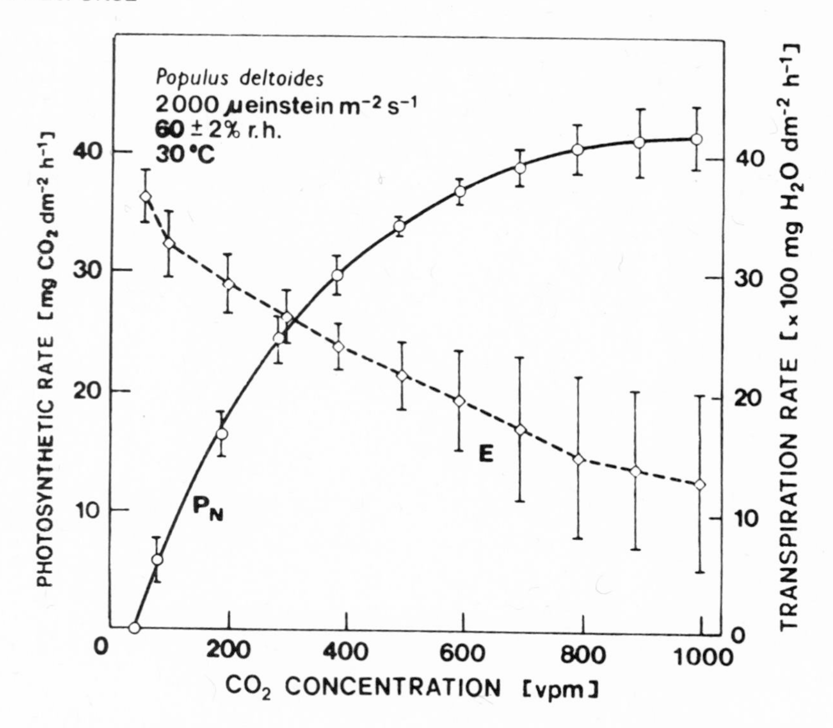

Fig. 5. Photosynthetic (P_N) and transpiration (E) rates as a function of CO_2 concentration. Vertical bars indicate one standard deviation. Leaf conductance ($1/r_1$) at compensation point, 200, 500, and 1000 vpm was 0.77, 0.54, 0.34, and 0.17 cm s^{-1}, respectively, (From *Regehr et al.*, 1975).

Usually drought stress in the temperate regions comes during the most favorable part of the growing season, (i.e., high light and high temperature) and is thus generally recognized as the single most important environmental constraint on all land productivity. It is under favorable periods of high light and high temperature that enhancement of photosynthesis by more CO_2 is most likely to occur. However, water stress at these times would damp CO_2 response the most. No definitive studies have been made on this phenomenon that I know of.

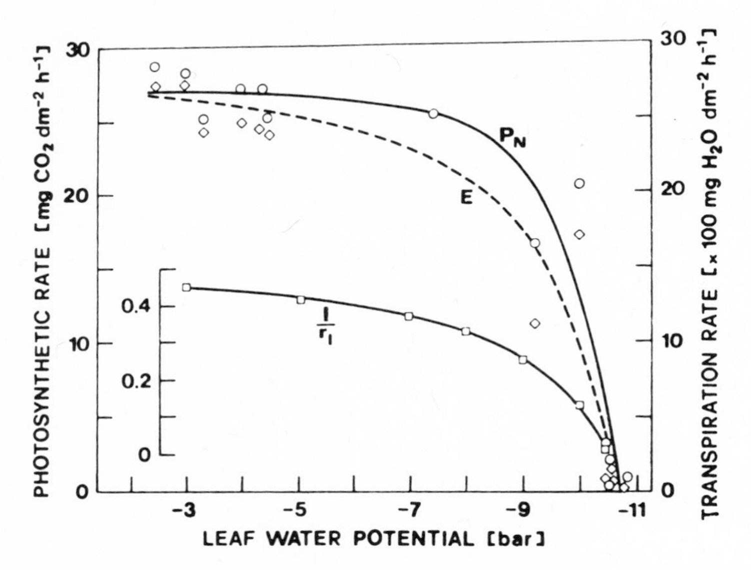

Fig. 6. Relationship between leaf water potential and photosynthetic (P_N) and transpiration (E) rates and leaf conductance ($1/r_l$ cm s^{-1}) in *Populus deltoides*. (From *Regehr et al.* 1975).

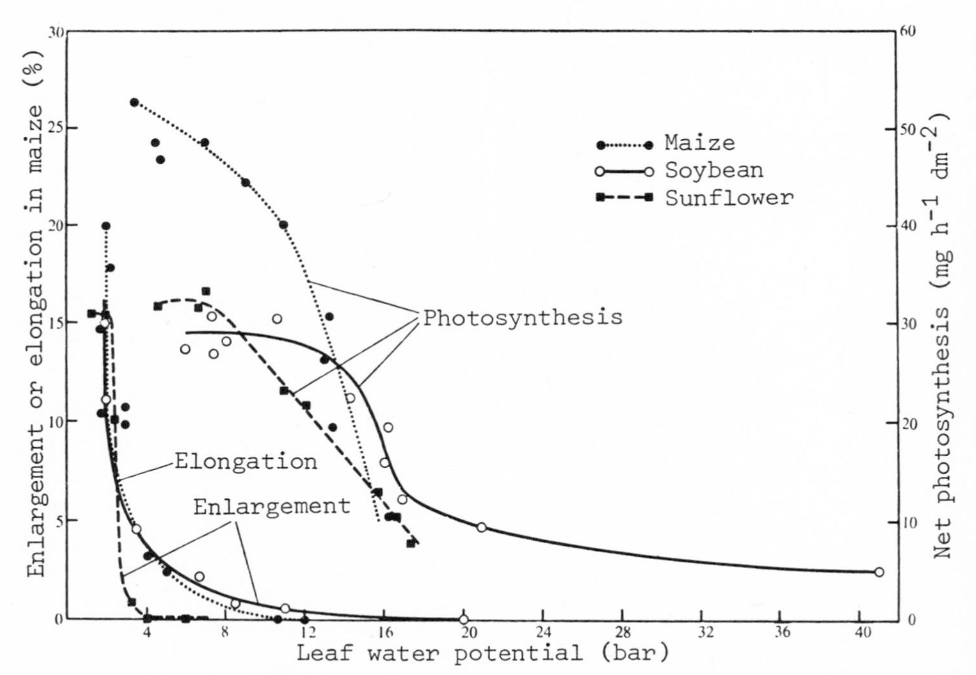

Fig. 7. Leaf enlargement and net photosynthesis rate of maize, soybean and sunflower at various leaf water potential (*Boyer*, 1970).

Field Chamber Studies

There have been a few CO_2 fertilization studies made under tents in the field, the most elegant by Thomas and Hill many years ago (see *Thomas*, 1965). Figure 8 shows sugarbeet, alfalfa, and tomato response to added CO_2 during midday under bright sunlight. All species are C_3 and responses are linear. However, nutrient defiency plays a role. Sugarbeets that were sulfur deficient had a low CO_2 fixation rate and did not respond to added CO_2. For healthy plants under high sunlight, Thomas and Hill found that doubling the CO_2 content of the air increased photosynthetic fixation 30 to 40%. Results are typical for C_3 plants.

Moss et al., (1961) report CO_2 response for maize measured under a clear plastic tent in the field. Their curvilinear response between 55 and 575 ppm gives roughly a 50% response in photosynthetic rate to a doubling of CO_2 between 300 and 600 ppm. Stomatal aperture and transpiration decreased in direct proportion to CO_2 increase. Their measured CO_2 fixation rate was asymptotic to 600 ppm. All of the results are typical of C_4 species and in accord with findings of *Akita and Moss* (1973).

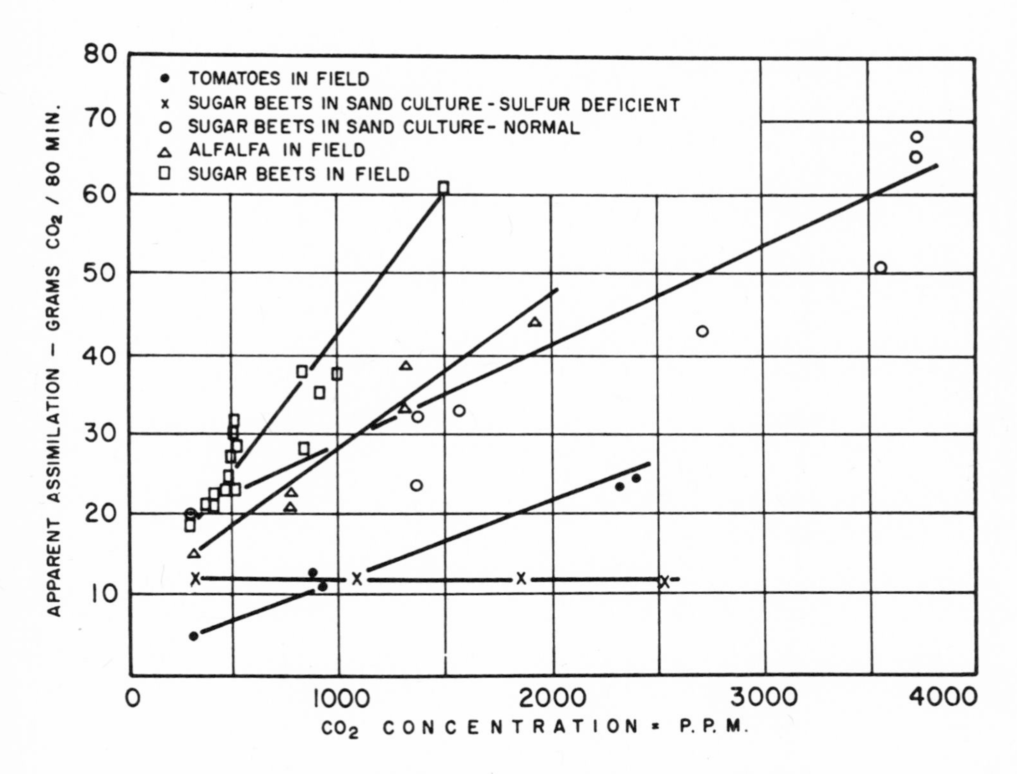

Fig. 8. Apparent assimilation by alfalfa, sugarbeets, and tomatoes in air enriched by CO_2. (From *Thomas*, 1976).

Modeling Short Term Response

A soil-plant-atmosphere model (SPAM) has been used to test the effects of CO_2 concentration and diffuse radiation changes on two crops using available data on the canopy structure and physiology of C_4 maize (*Zea mays*) and C_3 sugarbeets (*Beta vulgaris*, L.) (*Lemon et al.*, 1971; *Lister*, 1974; *Lister and Lemon*, 1976). In this study an attempt was made to demonstrate the interaction of plants and climate processes and how both may be coupled to the ambient levels of CO_2 and aerosol loading (i.e., changing the diffuse light of the atmosphere). Quantitative results bracket what may develop in the next 50 to 100 years. The simulations demonstrate the complex role that terrestrial plant life plays in climate processes.

Figure 9 is a schematic of the components of the SPAM model. The logic of the model is relatively simple: 1) to define, on the scale of the leaf surface in a plant stand how each leaf (and soil surface) responds to a given immediate climate; 2) to calculate from meteorology what that immediate climate is; 3) to calculate the specific leaf and soil response to that climate; and 4) to add up, leaf layer by leaf layer (and soil surface) the response of the whole crop to the climate input at the upper boundary.

Lister (1974) ran a factorial experiment between three levels of varying diffuse radiation (10, 30 and 60%) and two ambient CO_2 levels 315 and 400 ppm) for the C_3 and C_4 crops on a typical summer sunny day in Ithaca, N. Y. She used experimental data from leaves of each crop; a) crop structure, b) light response curves as a function of CO_2; c) temperature vs. CO_2 response curves, d) stomatal vs. CO_2 response curves, e) stomatal vs. light response curves. For each individual simulation, boundary input climate of % diffusion radiation (and direct radiation adjusted accordingly) and CO_2 concentration were held constant. The micrometeorological processes in each crop were controlled by the structure of each crop type, however.

Only a brief sample of her finds are presented here. Figures 10 and 11 illustrate the clear day response of maize and sugarbeet. Table 1 summarizes the daytime totals for net CO_2 exchange. Table 2 summarizes the daytime totals for sensible and latent heat exchange from the two crops.

The following points summarize the major findings:

(1) Net photosynthesis (primary production) increased more in sugarbeets (a C_3 carbon pathway crop) than in maize (a C_4 carbon pathway crop) with increasing levels of CO_2 in the atmosphere.

(2) When the proportion of diffuse light increased, net photosynthesis also increased up to an optimum despite a reduction in total solar radiation. The optimum for sugarbeets is somewhat higher than for maize, but decreased as CO_2 increased.

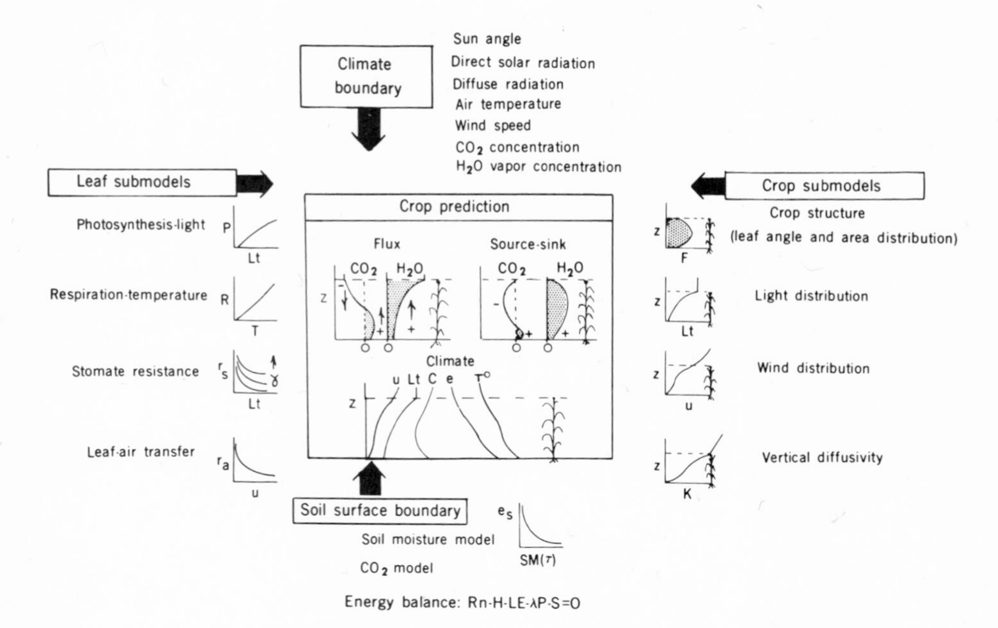

Fig. 9. Schematic summary of a mathematical soil-plant-atmosphere model (SPAM) giving required inputs, submodels, and representative daytime predictions of climate and community activity (that is, water vapor and carbon dioxide exchange). Abbreviations: height (z), wind (u), light (Lt), concentration of carbon dioxide (C), water vapor (e), air temperature (T^o), surface vapor pressure (e_s), surface soil moisture or water potential (SM[τ]), photosynthesis (P), respiration (R), leaf temperature (T), stomate resistance (r_s), minimum stomate resistance at high illuminances (γ), gas diffusion resistance (r_a), leaf surface area (F), vertical diffusivity (K), net radiation (R_n), sensible heat (H), latent heat (LE), photochemical energy equivalent (λP), and soil heat storage (S). (From *Lemon et al.*, 1971).

(3) The response of maize to more diffuse light increased when net radiation was arbitrarily increased, causing warmer leaves. A higher net radiation associated with diffuse light is common to industrial areas where the particulate matter is larger and has a higher absorption-to-backscatter ratio. Sugarbeets did not respond to increased net radiation, however. Evidently, their leaf temperatures were near optimal under the initial simulations. Also as C_3 plants, they have a lower temperature response.

(4) Increasing CO_2 had no affect on the proportions of latent and sensible heat exchange in sugarbeets. However, the proportion

of latent heat exchanged by maize decreased as a result of stomatal closure at higher CO_2.

(5) Increasing diffuse light caused a decrease in all convective heat components of both crops because of a decreased total radiation load.

(6) One speculates that terrestrial plants play a small role in damping long term atmospheric CO_2 perturbations and that increasing proportions of diffuse light due to scattering by particulate matter would slightly enhance this damping. The diffuse light effect on photosynthesis arises from a more favorable light regime in plant communities where many leaves are photosynthesizing at below light saturation levels. However, increases in diffuse sunlight, except in industrial regions, would necessarily decrease convective heat fluxes, causing a cooling.

(7) Both plant processes and climate processes evidently are coupled in ways much more complex than originally realized, such that model simulation may be the only tool available to predict outcome.

The simulations described were all for a clear cloudless day under ideal conditions. Under such conditions light levels are most favorable for photosynthesis. However, cloudless conditions are only prevalent in certain areas, usually areas whose water stress problems would severely reduce any benefit from increasing carbon

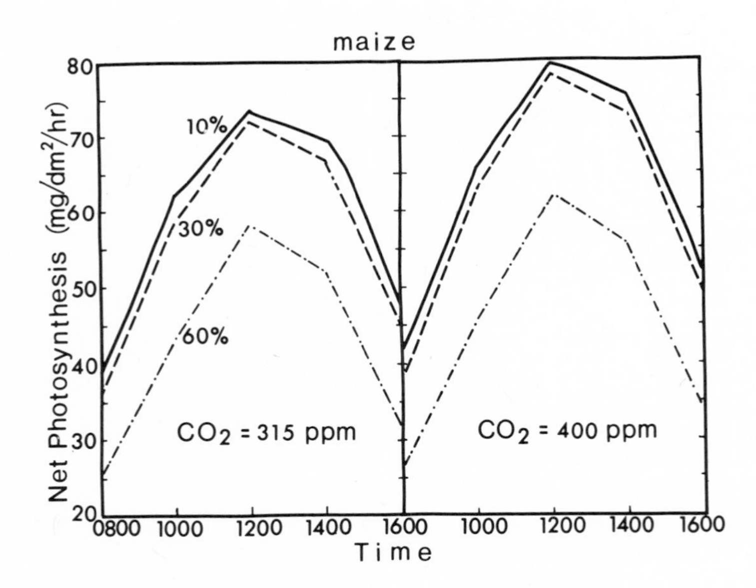

Fig. 10. Diurnal net photosynthesis for maize at 10, 30, and 60% diffuse radiation: Simulation for Ithaca, N. Y. summer radiation regime. (From *Lister and Lemon*, 1976).

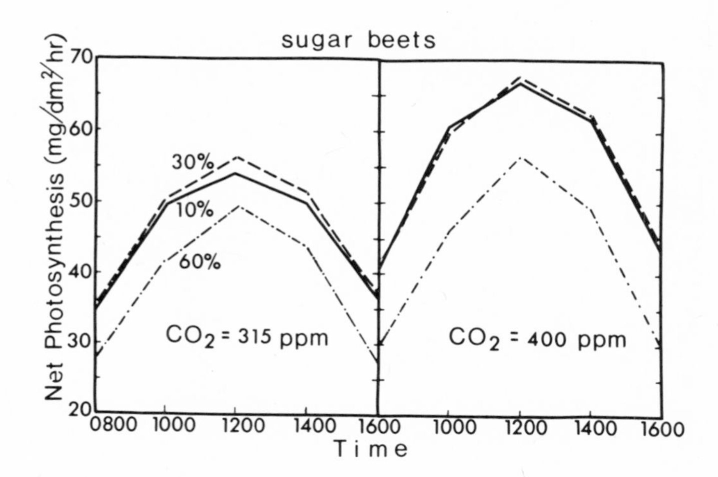

Fig. 11. Diurnal net photosynthesis for sugarbeets at 10, 30 and 60% diffuse radiation: Simulation for Ithaca, N. Y. summer radiation regime. (From *Lister and Lemon*, 1976).

TABLE 1. Simulated (SPAM) Interaction of Increased Atmospheric CO_2 and Diffuse Radiation on Net Photosynthesis.

Diffuse Radiation	CO_2 (ppm)	Net Photosynthesis ($gm/m^2/day$)		Total Radiation ($cal/cm^2/day$)
		Sugar Beets	Maize	
10%	315 (1970)	45 (1.00)	58 (1.00)	586 (1.0)
	400 (2090)	55 (1.21)	62 (1.04)	
30%	315 (1970)	46 (1.02)	55 (0.94)	385 (0.76)
	400 (2090)	54 (1.20)	59 (1.00)	
60%	315 (1970)	36 (0.82)	42 (0.72)	242 (0.41)
	400 (2090)	43 (0.94)	44 (0.75)	

TABLE 2. Simulated (SPAM) Interactions of Increased Atmospheric CO_2 and Diffuse Radiation on Latent Heat (LE) and Latent plus Sensible Heat (LE & H).

Diffuse Radiation	CO_2 ppm	LE (ly/day)		(LE & H (ly/day)	
		Sugarbeet	Maize	Sugarbeet	Maize
10%	315 (1970)	303 (1.00)	295 (1.00)	375 (1.0)	
	400 (2090)	303 (1.00)	265 (0.90)	375 (1.0)	
30%	315 (1970)	256 (0.84)	255 (0.88)	280 (0.75)	
	400 (2090)	256 (0.84)	230 (0.78)	280 (0.75)	
60%	315 (1970)	162 (0.53)	188 (0.64)	150 (0.25)	
	400 (2090)	162 (0.53)	186 (0.63)	150 (0.25)	

dioxide levels. Clouds at various altitudes and thicknesses produce different amounts of diffuse radiation and total light attenuation. Increasing aerosols under conditions other than clear or a few scattered clouds (16 to 33% diffuse radiation at Ithaca, N.Y.) would reduce both net radiation and visible radiation to a photosynthetically less advantageous level, particularly at high latitudes. Moreover, this could also shorten the growing season. This is important because lower temperature plays an important role for C_4 plants, and increasingly so for C_3 plants as CO_2 levels increase.

YIELD RESPONSE TO CO_2

Growth Cabinet Studies

I have already presented selected examples to demonstrate CO_2 effects on photosynthesis. There are also plenty of examples of CO_2 increasing economic yield of horticultural crops grown in glasshouses or growth chambers (*Witwer*, 1970). Some field crops grown in cabinets and field tents response as well (*Ford and Thorne*, 1967). It is impossible to find experimental evidence that CO_2 will increase yield or biomass of a tree species over its life cycle. There is evidence, however, that young trees do make short time dry matter gains to added CO_2 in growth chambers. Response necessarily must be measured over a short time relative to the life span of a tree and with individual young trees. *Tinus* (1972) measured a 50% greater increase in dry matter growth over a year in ponderosa pine (*Pinus ponderosa*) and blue spruce (*Picea pungens*) seedlings treated with 1200 ppm CO_2 (relative to 325 ppm). *Funsch et al.*, (1970) doubled height and diameter growth of Northern white (*Pinus strobus*) seedlings over 4 months at 1000 ppm CO_2 (compared to 300 ppm). *Siren and Alden* (1972) found that optimum CO_2 was near 300 ppm for growth of Scots pine (*Pinus sylvestres*) and Norway spruce (*Pinus abies*) in the first year. In the second year, there was no advantage to added CO_2. *Yeatman* (1971) measured an 80% increased in dry weight growth in seedlings of Jack pine (*Pinus banksiana*), Scots pine (*Pinus sylvestres*), white spruce (*Picea canadensis*) and Norway spruce (*Pinus abies*) at 1500 ppm CO_2 (compared to standard air at about 300 ppm).

How far one can extrapolate these studies to the real world is difficult to say. Intermittent or short-time response may have little bearing on final outcome and seedlings surely are not representative of other stages of maturity. An interesting experiment by *Hardman and Brun* (1971) shows the effect of intermittent response. CO_2 fertilization of soybean plants in the field had a notable effect on seed yield, if applied during the post-flowering period. Added CO_2 during the pre-flower and flowering periods had no influence on final yield. It is recognized, however, that in soybeans we are dealing with reproductive parts which may be much less significant to tree growth.

Modeling Long Time Response

Botkin, Janak and Willis (1973) report a simulation of long-time effects of CO_2 increases on a natural temperate region forest where interaction among the species plays an important role in outcome. They make the simple assumption that a percent increase in CO_2 will give an equal percent increase in the yearly increment of each tree. They simulate four levels of CO_2 increase over a 500 year time span (i.e. 10, 20, 50 and 100% CO_2 increases).

The model heavily weights competition for light among shade tolerant and shade intolerant species in a strategy for survival. Shade-intolerant species are short-lived and shade-tolerant ones are long lived. Because the annual probability of survival of an individual is related to the maximum known lifetime of its species, individuals of long-lived species have a better chance for survival than those with short lives. This strategy sacrifices productivity for persistence, and capitalizes on the older stages of a forest, thus favoring climax.

In the final outcome of the simulations,properties of birth and death mask fertilization with CO_2. Only at 50 and 100% increases in CO_2 is there significant influence on the forest. Two things happen; 1) succession is speeded so that climax is reached sooner, and 2) the standing crop is increased 40% at 100% CO_2 increase.

Climax or steady state was reached at 200 years, when CO_2 gains equal losses. The authors speculate that if enough knowledge was in hand to properly scale response over time, the sole effects of more CO_2 might be to compress the time scale and in no way change the final yield of the forest. They also go on to point out that a managed monoculture could give entirely different results. For example, in Figure 12 their spruce trees between 100 and 200 years showed a 100% increase in standing crop, at the 100% CO_2 increase level, relative to the unfertilized. Thereafter, the unfertilized trees caught up. A similar trend occurred in sugar maple. Again, it must be pointed out that linear growth response as assumed for all annual cycles leaves no chances for interactions or other constraints, either biological or environmental. Nonetheless, the authors have highlighted important factors to consider in predicting long term effects of more CO_2 on land vegetation.

THE LIVING FOREST BIOMASS

Whittaker and Likens (1973) provide recent data reproduced here as Table 3. Global estimates of primary plant production and plant biomass are given by ecosystem type. In their estimate, tropical rain forests accumulate the most carbon at a rate of 15.3 x 10^9 metric tons/year, 32% of the land total. The land plant biomass carbon in tropical rain forests is 340 x 10^9 metric tons or 41% of the land total. All land forest account for 70% of the carbon fixed on the land in a year and 93% of the total land plant biomass. Of course, there is controversy about the reality of such figures. Whittaker and Likens readily admit that data for tropical forests, in particular, are "very meager." To emphasize the diversity of estimates of this nature, I quote data given by them from various sources. For estimates of world, total, land-plant productivity published in the past six years, the range is 40 to 80 x 10^9 metric tons of carbon per year, with an average of 55 units for six esimates. Table 3 gives 48.3 units.

If there is diversity in production and biomass figures, especially for tropical rain forests, there is greater controversy about the effects of depletion of long-lived vegetation at the hands of man. There may be little doubt that old age trees are being cut. Does this necessarily mean a reduction in the carbon pool? Granted a high waste in lumbering operations, still a significant portion goes to useful lumber that might be around for a long time. Secondly, in tropical rain forests regrowth starts almost overnight. Consider the familiar sigmoid curve for growth in Figure 12. This logistic growth curve applies to a tropical rain forest, as well as the simulated results of *Botkin et al.* (1973) for the temperate forests. The old age trees of the tropics must be at climax of the curve, where CO_2 uptake barely exceeds respiration loss. Harvesting these trees and allowing regrowth or agricultural development start the system at the bottom of the curve. Either fast

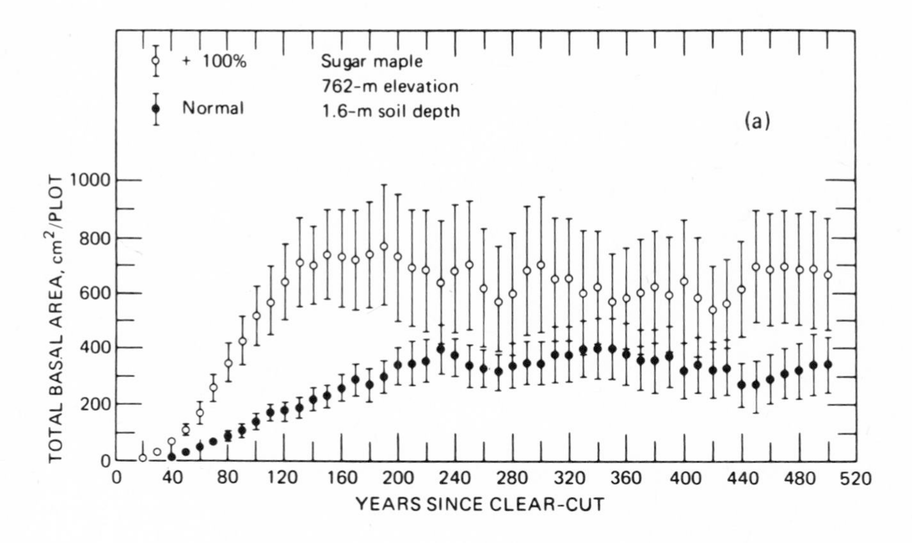

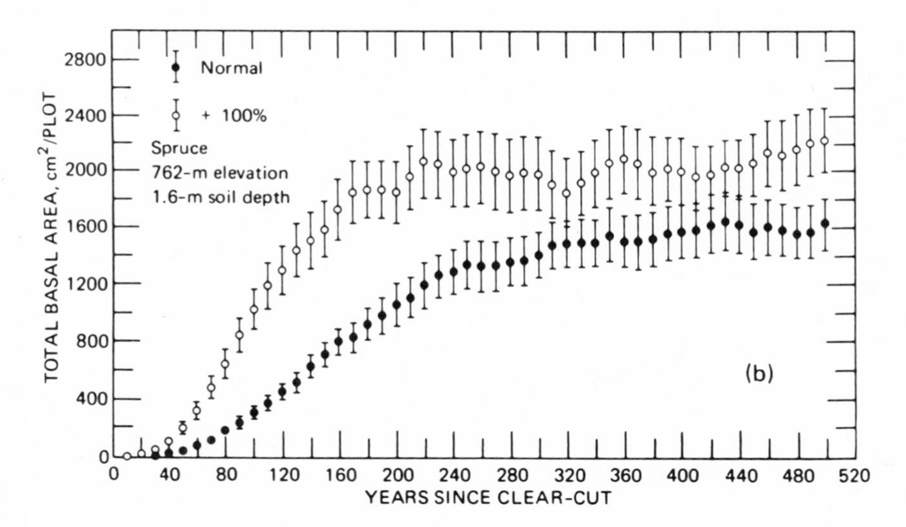

Fig. 12. Long term trends in total basal area contributed by sugar maple (a) and spruce (b) for simulated forest stands with normal and 100% greater than normal growth increment. Shown are 95% confidence intervals and means for 100 replicates. (From *Botkin, Janak and Wallis*, 1973).

TABLE 3. Primary Production and Biomass Estimates for the Biosphere*

1	2	3	4	5	6	7	8
Ecosystem type	Area 10^6 km^2 = 10^{12} m^2	Mean net primary productivity, gC/m^2/year	Total net primary production, 10^9 metric tons C/year	Combustion value, kcal/gC	Net energy fixed, 10^{15} kcal/yr	Mean plant biomass kgC/m^3	Total plant mass, 10^9 metric tons C
Tropical rain forest	17.0	900	15.3	9.1	139	20	340
Tropical seasonal forest	7.5	675	5.1	9.2	47	16	120
Temperate evergreen forest	5.0	585	2.9	10.6	31	16	80
Temperate deciduous forest	7.0	540	3.8	10.2	39	13.5	95
Boreal forest	12.0	360	4.3	10.6	46	9.0	108
Woodland and shrubland	8.0	270	2.2	10.4	23	2.7	22
Savanna	15.0	315	4.7	8.8	42	1.8	27
Temperate grassland	9.0	225	2.0	8.8	18	0.7	6.3
Tundra and alpine meadow	8.0	65	0.5	10.0	5	0.3	2.4
Desert scrub	18.0	32	0.6	10.0	6	0.3	5.4
Rock, ice and sand	24.0	1.5	0.04	10.0	0.3	0.01	0.2
Cultivated land	14.0	290	4.1	9.0	37	0.5	7.0
Swamp and marsh	2.0	1125	2.2	9.2	20	6.8	13.6
Lake and stream	2.5	225	0.6	10.0	6	0.01	0.02
Total continental	149	324	48.3	9.5	459	5.55	827
Open ocean	332.0	57	18.9	10.8	204	0.0014	0.46
Upwelling zones	0.4	225	0.1	10.8	1	0.01	0.004
Continental shelf	26.6	162	4.3	10.0	43	0.005	0.13
Algal bed and reef	0.6	900	0.5	10.0	5	0.9	0.54
Estuaries	1.4	810	1.1	9.7	11	0.45	0.63
Total marine	361	69	24.9	10.6	264	0.0049	1.76
Full total	510	144	73.2	9.9	723	1.63	829

*All values in columns 3 to 8 expressed as carbon on the assumption that carbon content approximates dry matter x 0.45.

growing pioneer species of natural regrowth or, for example, fast growing tree crops like oil palm, rubber, coffee or cocoa would rapidly bring each system to the grand period of growth on the sigmoid curve, quickly tying up carbon again.

Some experts claim that air pollution might reduce vegetative growth of the forests. There is no doubt that toxic materials produced by man harm vegetation in industrial areas. On the other hand, widespread fertilization of the landscape with nutrient elements enhances plant growth. This goes on by plan and quite inadvertently, by man and by nature. In balance, I do not feel we have clear enough answers about these various facets affecting the living forest biomass to know one way or the other the change in carbon pool size or its activity.

THE SOIL HUMUS POOL

Reiners (1973) reviews available estimates on the terrestrial detritus pool, the carbon fixed in the forest floor, mulch matting, humus, etc. Two extremes are given, 700×10^9 and 9000×10^9 metric tons of carbon, an order of magnitude difference. Comparing these figures with those in Table 3 one can see that the size of the detritus pool equals or exceeds living biomass on the land (827×10^9 tons) or total biomass of lands and seas (829×10^9 tons). Thus size and changes in the dead organic matter pool on the land could be significant to the carbon budget.

Again, many experts reason that man is depleting this pool by cutting forests or plowing grasslands, which mainly go to agricultural use. One can question this supposition.

Some evidence for raising this question is found in the levels of carbon remaining in the soil following forest clearing with continuous cropping. The belief that tropical soils are low in organic matter and that they rapidly lose what is there after forest cutting cannot be a generalization. Consider Table 4. Available experimental data from *Nye and Greenland* (1965) show surprising quantities and stability of soil organic matter under shifting agriculture in the tropics. We must not overlook the fact that nutrient fertilizers, when added, speed biomass accumulation. The lesson in Table 4 applies to our own agricultural land, as far as soil organic matter goes. Evidence is now accumulating, for example, that due to increased crop production, soil organic matter in the Corn Belt may be increasing. Large amounts of crop residue are plowed back in the soil to improve the soil tilth. Probably nitrogen fertilizer, increasing biomass, is chiefly responsible. Most striking in this regard are data from an experiment begun in 1926 on tropical laterite soils in Hawaii (*Thorne*, 1951). Under a wide spectrum of practices used in growing commercial pineapple, six treatments were by-yearly sampled for soil organic matter from 1931 to 1945. Interestingly there was no significant difference among treatments after 20 years, but soil organic matter significantly

TABLE 4. Changes in the Chemical Constitution of Tropical Soils Under Continuous Cropping

Experiment Station Data
(From Nye and Greenland - Table 10 p. 100)*

Place	Rainfall Inches/year	Cropping	Years	Depth inches	C %	N %
Ghana	82	After clearing of moist	0	0-6	2.19	0.164
		evergreen secondary forest	0	6-12	1.28	0.103
		Maize - Cassova rotation	8	0-6	1.50	0.128
			8	6-12	1.02	0.089
Trinidad		After clearing of young secondary semi-deciduous forest	0	0-6	1.02	0.13
		Rotation with maize	6	0-6	0.84	0.11
		legumes and other crops	12	0-6	0.80	0.11
Congo	70	Moist semi-deciduous forest	0	0-4	2.0	
		Bananas, groundnuts, maize, cassova	3	0-4	2.2	
Nigeria	72	Clearing and burning of	1	0-6	0.86	0.080
		50 year old moist semi-deciduous forest	1	6-18	0.44	0.037
		Oil palm with intercrops	11	0-6	0.75	0.066
			11	6-18	0.48	0.047
Ghana	50	Savana	0	0-6	0.36	0.034
			0	6-12	0.22	0.24
		Rotation of yams, cereals,	6	0-6	0.28	0.025
		groundnuts	6	6-12	0.17	0.018

*From Nye and Greenland. "The Soil Under Shifting Cultivation." Technical Comm. No. 51. Commonwealth Bureau of Soils, Harpenden. 19[illegible]

increased in all treatments over the years. The average for the six treatments was 2.3% organic matter in 1931 and 3.5% in 1945.

An important question of considerable relevance here is: At what level of agricultural technology are new lands farmed to keep pace with the population growth? Its significance rests on the influence of agriculture on the soil humus. There is reason to believe that intensive agriculture, if wisely managed, could maintain or increase soil humus instead of depleting it. Some interesting comments by *Dudal* (1975) are appropriate here. His words have bearing on the increased land needed for agricultural food production.

"I should like to place FAO's work in the general framework of developments in the last 15 years. When we compare 1958 and 1972 we see that the population of the world has increased from 2.8 billion to 3.8 billion. This is by 35 percent. During the same period the cultivated land in the world has increased approximately from 1 billion 389 million to 1 billion 480 million hectares which is an increase of about 7 percent. Fertilizer use, on the other hand, has increased from 24 million tons in 1958 to 72 million tons of plant nutrients in 1972, that is by 200 percent. If we assume that one hectare of land under average management produces an equivalent of one ton of wheat and that a 1/2 ton equivalent is needed to feed one person for one year, we find that 100 million hectares can produce food for an additional 200 million people. The remaining 800 million people have been fed, not from increased amounts of land under cultivation, but from intensification of agriculture. If we further assume that one ton of fertilizer produces an average of 8 tons of wheat equivalent, it is apparent that the 48 million tons increase in fertilizer use (1958-1972) is sufficient to produce food for up to 800 million people. This does not imply that all the additional agricultural production was due to fertilizer use but rather is indicative of other improved management practices like better seeds, more effective water control, improved land use and the application of pesticides. In other words, during the 15 year period between 1957 and 1972 only 200 million people have been fed from an expansion of agricultural land while 800 million have been fed from intensified agriculture. It is significant to note that of the additional land taken into cultivation, 70 percent was in developing countries and only 30 percent in industralized countries. On the other hand, 85 percent of the fertilizer was used in the industrialized countries compared to 15 percent in developing countries. Hence, the considerable increase of food imports by developing countries."

PERSPECTIVE

I have attempted to demonstrate the complexity of predicting response of land plants to more CO_2. Intuitively, one would guess that because CO_2 (and diffuse light) can increase photosynthetic rates (if no other factors are limiting), final yield in biomass accumulation of carbon should be greater. Certainly carefully controlled experiments in growth chambers and short-term measurements on intact leaves add weight to this logic. However, our study of the hierarchy of yield components, from the simplest level of biochemical fixation of CO_2 up through higher levels of organization to final yield opens our eyes to how increasing complexity buffers the plant system, especially at the final level of mixed forest ecosystem.

Comparing C_3 and C_4 plant species in this exercise has demonstrated how the C_4 group begins at the basic photochemical level at a 2-3 fold advantage over the C_3 group because the C_4 carboxylation enzyme has a 2-3 fold affinity for CO_2. Yet, at the short-term growth rate level this advantage has been wiped out. The tabulation below summarizes the progress of buffering the CO_2 advantage with increasing levels of organization.

Level of Organization	Ratio C_4:C_3
In vitro primary carboxylation, $K_m^{app.}$ $(CO_2 + HCO_3)$	2-3
Mesophyll resistance	5
Net photosynthesis of intact leaf	1.5-3
Net photosynthesis of crop stand at high light level (transparent tent)	1.5
Net photosynthesis of crop stand at high light level (aerodynamic)	1.5
Short-term crop growth rate	1

For lack of actual experimental data on long term CO_2 fertilization effects, I contend that the same factors that damp the C_4 advantage in photosynthesis necessarily would damp CO_2 (or diffuse light) enhancement. Two simulation studies add some support to this belief, although both have their shortcomings.

In the short-term simulation of two crops, the C_4 maize responded little to CO_2, while the C_3 sugarbeet gave roughly a 20% increase in daytime photosynthesis to a 20% increase in CO_2 under ideal growing conditions. In the long-term experiment of a mixed species forest, all C_3, there was a 40% increase in biomass carbon accumulation with a doubling of CO_2 over present ambient levels. This response is probably too high because of the 1:1 beginning assumption of increased CO_2 to annual growth increment. The speculation that CO_2 would only hasten climax jibes with experience with short lived plants. Final biomass is not necessarily increased with added CO_2, but life cycle can be decreased.

In summing up the aspects of land plant photosynthesis response to more CO_2, I take a rather conservative view. First and foremost, environmental constraints will limit plant growth. Secondly, the final yield (carbon accumulation) of a plant community is governed by so many other plant processes in addition to photosynthesis that advantages to this single process are muted. Thirdly, the life cycle of plant systems tends to be shortened with either greater photosynthesis in single plants,or faster growth of forests such that net carbon gain in the end is nil. Plants and plant communities might adapt through evolution to accommodate higher CO_2 levels to their advantage. This, however, would be a long-time process, surely not within the time frame of the next one hundred years.

Nonetheless, even a rather small increase in overall photosynthetic rate due to CO_2 might have a significant buffering effect on the anticipated, fossil-fuel, CO_2 increase in the atmosphere if the living biomass and soil humus pools are large and pool sizes are not decreasing. I have taken the devil's advocate position on the latter, arguing that man's activity may not be decreasing either the living biomass or soil humus pools because of planned or inadvertent mineral nutrient enrichment.

Is there any evidence of recent change in overall terrestrial biosphere activity? *Hall et al.*, (1975) presents interesting evidence showing no change. They have analyzed the annual highs and lows in 15 years of the Mauna Loa CO_2 samplings. Their reasoning is that the annual highs (winter net respiration) equal the annual lows (summer net photosynthesis) over the period of record (1958-1972). Since the year to year difference between each high and following low has not changed over the years there is no evidence for a reduction in net photosynthesis in the Northern Hemisphere. A similar analysis can be made for net respiration. Their case is presented in Figure 13. The authors correctly point out that a pool size decrease could be offset by a photosynthetic rate increase with more CO_2 or that the biosphere is too big to be influenced by man's activity. Of course, the period of record is short and the technique perhaps too insensitive in relation to the pool size. Yet, interestingly during this period we have had an exponential increase in fossile fuel CO_2 output, as well as an increase of tropical forest cutting.

WHAT NEEDS TO BE DONE

There can be little doubt of the need for more information on the land carbon budget. There is no question of its importance. What needs to be done?

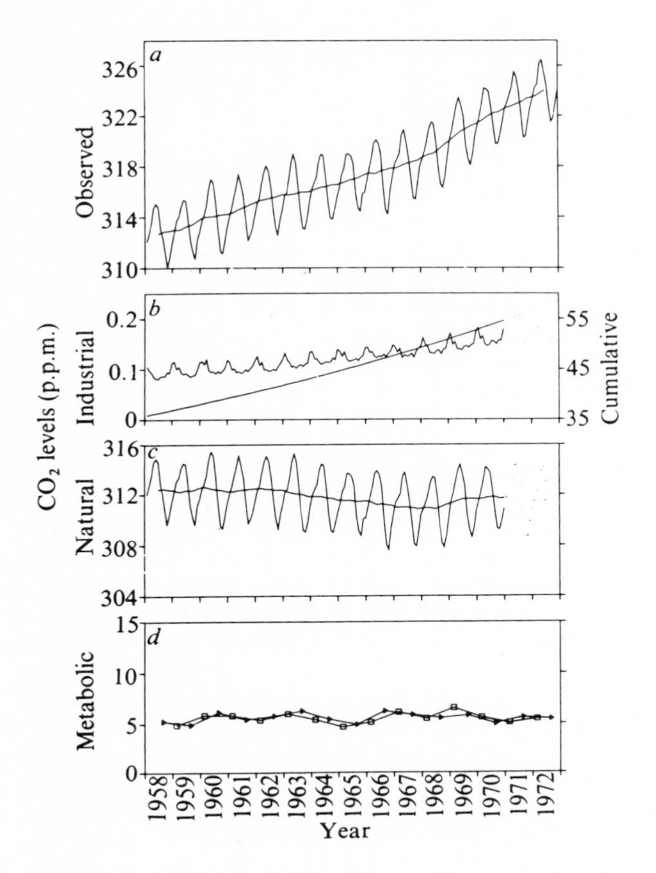

Fig. 13. a. Month-by-month variation in the CO_2 concentration of tradewind air at Mauna Loa, Hawaii. The upward sloping smooth curve is normalized for the semiannual fluctuations. b. The month-by-month addition of industrial CO_2 to the Northern Hemisphere, along with the cumulative production above the preindustrial level of 293 ppm. The past two years are extrapolated values from the least square fit of data to a third-order polynomial. c. The observed CO_2 variation corrected for net industrial additions. d. Semi-annual net ecosystem production and respiration, determined as the difference between the crests and troughs. □, Respiration by difference; Δ, photosynthesis by difference. Oceanic uptake 42%. (From *Hall, Ekdahl and Wartenberg*, 1975).

(1) Determine the life cycle response of trees to added CO_2, especially tropical forest species. While there are considerable CO_2 response data for economic yield of glass house crops, there is less for biomass yield (dry matter accumulation or net carbon fixed)

and almost none for long lived trees, especially those of the tropics. Unfortunately short-term studies may not give needed answers because final biomass is the important item of geophysical interest.

(2) Determine the interaction between environmental constraints and plant CO_2 response, especially water and mineral nutrient stress. I have reasoned that in the field when environmental conditions should be most favorable for CO_2 response (i.e. high light and high temperature) drought stress usually occurs. There is no work reporting the interaction of CO_2 and water stress.

(3) Determine by complete plant carbon budget the significance of photosynthesis (and breeding for photosynthesis) to the final biomass carbon accumulated in forest species. We all recognize photosynthesis to be a necessity for carbon accumulation, yet evidence shows how other plant processes may control final outcome in short-lived crops. Work in this area on forest species is badly needed, since photosynthesis alone is the process where CO_2 response takes place initially.

(4) Determine the significance of spacially and temporally intermittent old age forest cutting and new forest regrowth on overall carbon budget of the land (probably by modeling). Even before man, windthrow, disease, pests and fire as well as senescence imposed a natural time and space intermittency on forest tree death and regeneration. We need to know at what level of man's cutting will there be a global carbon budget influence.

(5) Determine detritus pool gains and losses, especially soil humus. This research should be easily implemented because of large collections of soil samples preserved over the years at many universities and experiment stations. From these one could determine gains or losses of carbon from the humus pool. As an aside, eutrophication of lakes and estuaries need researching too since carbon accumulation due to higher nutrient status may be significant (*Deevey*, 1973).

(6) Determine from satellite data the changes in extent and character of land vegetation over time. From this, an inventory of plant biomass on a global scale could be kept up to date.

(7) Determine the stability of net land photosynthesis and respiration in both northern and southern hemispheres. Additional stations are needed to monitor the atmospheric CO_2 content on a continuous basis. These would provide the required data for such analyses.

(8) Determine what is the wisest course of action for man. Should man do anything to promote CO_2 uptake by vegetation as a short term buffer but run the chance of catastrophic climate change in a short time period in a delayed future? Maybe it would be better for mankind to adjust slowly to gradual climate change? Since the processes are so complex that no one can predict outcome, in addition to the seriousness of the potential effects, the best course may be to eliminate the root causes of the problem through birth control and energy conservation!

REFERENCES

Akita, S. and D. N. Moss, 1972. Differential stomatal response between C_3 and C_4 species at atmospheric CO_2 concentration and light. *Crop Sci. 12.* 789-793.

Akita, S. and D. N. Moss, 1973. Photosynthetic response to CO_2 and light by maize and wheat leaves adjusted for constant stomatal aperatures. *Crop. Sci. 13.* 234-237.

Bjorkman, O., 1971. In: *Photosynthesis and Photorespiration* (Eds. M. D. Hatch, C. B. Osmond, and R. O. Slatyer). Wiley Interscience; New York. p. 18.

Botkin, D. B., J. F. Janak, and J. R. Wallis, 1973. Estimating the effects of carbon fertilization on forest composition by ecosystem simulation. In: *Carbon and the Biosphere.* Ed. G. M. Woodwell, E. V. Pecan, U. S. Atomic Energy Comm. 328-344.

Boyer, J. S. 1970. Leaf enlargement during desiccation. *Plant Physiol. 46.* 233-235.

Brittain, E. G. and R. J. Cameron, 1973. Photosynthesis of leaves of some eucalyptus species. *New Zealand J. of Botany 1.* 153-163.

Deevey, E. S., Jr., 1973. Sulfur, nitrogen and carbon in the biosphere. Ed. G. M. Woodwell, F. V. Pecan, U. S. Atomic Energy Com.

Delaney, R. H. and A. K. Dobzenz, 1974. Yield of alfalfa as related to carbon exchange. *Agron. J. 66.* 498-500.

Dudal, R., 1973. Land & Water Development Assistance. In: *Proceed: AID Soil & Water Management Workshop - Washington D. C. Feb.* 177-180.

Evans, L. T., 1975. The physiological basis of crop yield. In: *Crop Physiology.* Ed. L. T. Evans. Cambridge Univ. Press. London and New York. 327-355.

Ford, M. A. and G. N. Thomas, 1967. Effect of CO_2 correlation on growth of sugarbeet, barley, kale and maize. *Annals of Botany (N.S.) 31.* 629-643.

Funsch, R. W., R. H. Mattson, and G. R. Mowry, 1970. CO_2-supplemented atmosphere increases growth of *Pinus Strobus* seedlings. *Forest Sci. 16.* 459-460.

Gaastra, P., 1959. Photosynthesis of crop plants as influenced by light, carbon dioxide, temperature, and stomatal diffusion resistence. *Med. Landbouwhogesch., Wageningen 59.* 1-68.

Gifford, R. M., 1974. A comparison of potential photosynthesis, productivity and yield of plant species with differing photosynthetic metabolism. *Aust. J. Plant Physiol. 1.* 107-117.

Hall, C. A. S., C. A. Ekdahl, and D. E. Wartenberg, 1975. A fifteen-year record of biotic metabolism in the Northern Hemisphere. *Nature, Vol. 255, No. 5504.* 136-138.

Hardman, L. H. and W. A. Brun, 1971. Effect of atmospheric carbon dioxide enrichment at different developmental stages in growth and yield components of soybeans. *Crop Sci. 11.* 886-888.

Landsberg, H. E., 1976. How is our climate fluctuating? Presentation Amer. Assoc. Adv. Sci. Meetings, Boston, Mass.

Larcher, W., 1969. The effect of environmental and physiological variables on the carbon dioxide gas exchange of trees. *Photosynthetica 3.* 167-698.

Ledig. F. T., 1976. Physiological genetics, photosynthesis, and growth models. In: *Tree Physiology and Yield Improvem.* Ed. M.G.R. Carnell and F. T. Last. Academic Press, London and New York (In press).

Lemon, E. R., D. W. Stewart and R. W. Shawcroft, 1971. The Sun's work in a cornfield. *Science 176.* 371-378.

Lister, R. A., 1974. Effects of atmospheric carbon dioxide and particulates on plant photosynthesis. Ph.D. Thesis, Cornell Univ., Ithaca, N.Y.

Lister, R. and E. R. Lemon, 1976. Interactions of atmospheric carbon dioxide, diffuse light, plant productivity and climate processes - Model predictions. IN: *Atmosphere-Surface Exchange of Particulate and Gaseous Pollutants. ERDA Sym. Series 38:* 112-135.

Loomis, R. S. and P. A. Gerakis. 1975. Productivity of agricultural ecosystems. In: *Photosynthesis and Productivity in Different Environments. International Biological Programme Vol. 3.* Cambridge Press. London.

Ludlow, M. M. and P. G. Jarvis. 1971. Photosynthesis in Sitka spruce (*Pica sitchensis* (Bong.) Carr.). I. General characteristics. *J. Applied Ecology 8.* 925-953.

Moss, D. N., R. B. Musgrave, and E. R. Lemon. 1961. Photosynthesis under field conditions. III. Some effects of light, carbon dioxide, temperature, and soil moisture on photosynthesis, respiration and transpiration of corn. *Crop. Sci. 1.* 83-87.

Nelson, C. J., K. H. Asay and G. L. Horst. 1975. Relationships of leaf photosynthesis to forage yield of tall fescue (*Festuca arundinacea*, Schreb.). *Crop. Sci. 15.* 476-478.

Nye, P. H. and D. J. Greenland. 1965. The soil under shifting cultivation. *Tech. Comm. No. 51.* Commonwealth Bureau of Soils, Harpenden.

Regehr, D. L., F. A. Bazzaz and W. R. Boggess. 1975. Photosynthesis, transpiration and leaf conductance of *Populus deltoides* in relation to flooding and drought. *Photosynthetica 9.* 52-61.

Reiners, W. A. 1973. Terrestrial detritus and the carbon cycle. In: *Carbon and the Biosphere.* Ed. G. M. Woodwell, E. V. Pecan. U. S. Atomic Energy Comm. 303-327.

Sheehy, J. E. and J. P. Cooper. 1973. Light interception, photosynthetic activity, and crop growth rate in canopies of six temperate forage grasses. *J. of Appl. Ecology 10.* 239-250.

Siren, G. and T. Alden. 1972. CO_2 supply and its effects on the growth of conifer seedlings grown in plastic greenhouse. Institution for Skogsforyagring Research Notes NR. 37. Royal College of Forestry, Stockholm, Sweden. 15 p.

Thomas, Moyer D. 1965. Photosynthesis (carbon assimilation) environmental and metabolic relationships. In: *Plant Physiology Vol. IV A.* Ed. F. C. Steward. Academic Press, New York and London.

Thorne, M. D. 1951. The effect on pineapple yields of plowing under pineapple trash and green-manure crops. Special Report No. 25, Pineapple Research Institute of Hawaii.

Tinus, R. W. 1972. CO_2 enriched atmosphere speeds growth of ponderosa pine and blue spruce seedlings. *Tree Planter's Notes 23(1).* 12-15.

Whittaker, R. H. and G. E. Likens. 1973. Carbon in the biota. IN: *Carbon and the Biosphere.* Ed. G. M. Woodwell and E. V. Pecan. U. S. Atomic Energy Comm. 281-302.

Witwer, S. H. 1970. Aspects of CO_2 enrichment for crop production. *Trans. Amer. Soc. Agr. Eng. 13.* 249-251.

Yeatman, C. W. 1971. Genetics of jack pine seedling response to CO_2 and pollination studies. Petawawa, 1968-70. IN: *Proceedings of the Twelfth Meeting of the Committee on Forest Tree Breeding in Canada.* Universite Laval, Quebec. 101-105.

Atmospheric Carbon Dioxide and Some Interpretations

L. Machta[1], K. Hanson[1] and C. D. Keeling[2]

[1]*National Oceanic and Atmospheric Administration*
[2]*Scripps Institution of Oceanography*

ABSTRACT

The increasing amount of high quality benchmark carbon dioxide concentration data deserves critical analysis to improve our understanding of the entire carbon cycle. It is argued that the readily measured seasonal variation at benchmark stations outside of the Arctic Region can be accounted for by land photosynthesis and decay of vegetation, and that the best record at Mauna Loa shows no evidence of a changing land biosphere (although the technique is not especially sensitive). It is also argued that the time changes of the residual concentration left after long-term trend and seasonal variation have been removed could be related to sea surface temperature in the El Niño region of the eastern Pacific Ocean off South America.

INTRODUCTION

Measurements of atmospheric carbon dioxide have been performed for over 100 years. Many of the results of early observations were compared with later ones by *Callendar* (1938) and *Bray* (1959) who pointed out a significant upward trend in concentration. These early measurements were often criticized as being an inadequate base upon which to derive time trends. In 1958, *Keeling et al.* (1976a) initiated a program at the South Pole followed by a joint Scripps-Weather Bureau effort aside Mauna Loa, in the Hawaiian Islands (*Keeling et al.*, 1976b). These two stations provide the longest, mostly continuous record of atmospheric carbon dioxide using modern non-dispersive infra-red analyzer techniques based on careful gas standards. These were followed by an aircraft sampling program by *Bischof* of Sweden (*Bolin and Bischof*, 1970) and *Pearman, et al.* (1975) of Australia in 1972. The Weather Bureau (now NOAA) Program using ground based stations has expanded its network to include Point Barrow, Alaska (following earlier measurements by *Kelley* (1969)) and, more recently, American Samoa. To the "background" or "benchmark" observations one may add a very large body of data on atmospheric carbon dioxide obtained for agriculture and air pollution purposes but which may have marginal value for geophysical interpretation (e.g., *Woodwell et al.*, 1973). Finally, there are several other relatively "clean" sites at which limited records of atmospheric carbon dioxide have been obtained, such as Fanning Island in the Equatorial Pacific, Niwot Ridge on the Colorado Continental Divide, Ocean Weather Ships Papa and Charlie, and Baring Head, New Zealand.

While the role of atmospheric carbon dioxide in the global carbon cycle possesses inherent scientific interest in its own right, the possible atmospheric warming attributed to a growing concentration usually dominates any discussion of its geophysical role. The predicted increase in atmospheric carbon dioxide, as is all too well known, results from the combustion of fossil fuels and represents but one possible impact on man's environment from his plunder of past photosynthesis. While we may hope that alternative fuel resources might reduce our dependence on coal, lignite, oil and natural gas, history suggests that we ought make no such assumption in predicting the possibility of climatic effects due to a marked increase in atmospheric carbon dioxide.

This discussion will not offer yet another prediction of carbon dioxide content in the future. Many such forecasts appear in the literature (e.g., *Cramer and Myers*, 1972; *Hoffert*, 1974; *Smil and Milton*, 1974). Rather, some of the trends and periodicities in the atmospheric carbon dioxide concentrations and speculation on their significance concerning the carbon cycle will be examined.

SECULAR TRENDS

Figure 1 presents the mean monthly values (except for the Scandinavian aircraft data) of atmospheric carbon dioxide at four locations with records longer than 10 years. In the case of the Swedish aircraft measurements, Bischof has "adjusted" all data to 1 July of each year; these mid-year values are jointed by straight lines. Because of possible instrumental differences, comparisons of absolute concentrations at different locations should be viewed with caution.

The well publicized upward trend at all stations is very evident in the figure. If one calculates the expected growth in atmospheric carbon dioxide due to world-wide emissions, it turns

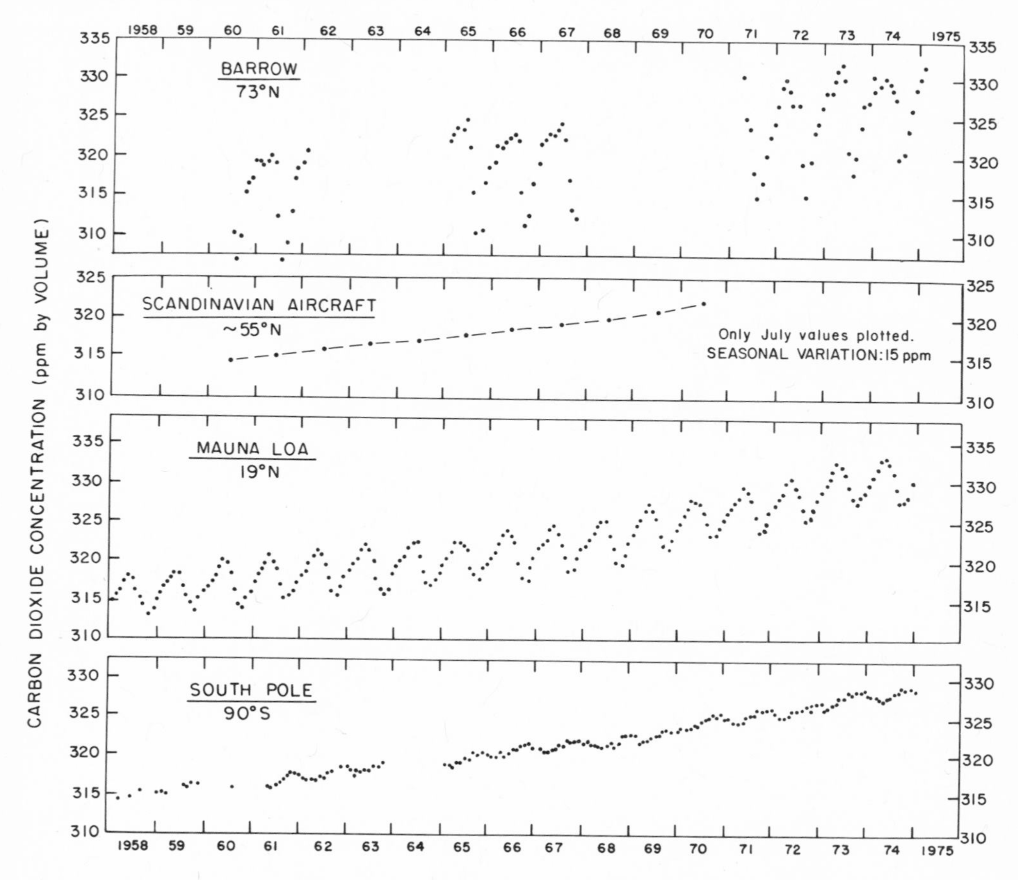

Fig. 1. The growth of atmospheric carbon dioxide concentration at four locations. Except for Scandinavian aircraft data, the points denote monthly average concentrations. In Scandinavia, the values have been referred to July using an average pattern of seasonal variability. Some data are based on intermittent flask sampling, others, including the entire record at Mauna Loa, on frequent *in situ* analyses.

out that about one half of the anthropogenic carbon dioxide remains airborne. The remainder must have entered another reservoir. This aspect will be examined in more detail later when the variations in the growth rate are discussed. The second feature, most evident at Barrow and Mauna Loa - and also appearing in the Swedish record, but not reproduced here - is the cyclical or seasonal behavior. The amplitude is largest farthest north and diminishes towards the south. Again this feature will be discussed in more detail later.

To summarize, all long-term background monitoring locations, including those not in Figure 1, display a general upward trend in atmospheric carbon dioxide. They all exhibit a seasonal variation. Finally, the year-to-year growth rate varies with time and between stations for the same period.

SEASONAL VARIABILITY

The seasonal variation is produced primarily by seasonal variations in photosynthesis and decay of organic matter, and to a lesser extent is due to seasonal variations in fossil fuel consumption and oceanic uptake and release of carbon dioxide. Seasonality, of course, increases towards the poles and is least in equatorial regions. However, the seasonal variation at the South Pole is small due to the lack of nearby land plants and remoteness from population and industry. In Figure 1, a decrease can be noted in summer and a recovery in winter when photosynthesis is minimized or absent and when fossil fuel combustion is maximized.

Figure 2 shows a time history of the amplitude of the seasonal variation at Mauna Loa. No systematic long-term changes in the amplitude over the course of 15 years stand out. It can, therefore, be speculated that there has been no detectable change in the northern hemisphere uptake and release by land vegetation - that is, in the size of the terrestrial organic carbon pool over the period of record. The seasonal amplitude trend is not an especially sensitive technique for detecting a change in the organic carbon pool. The range of variability in amplitude of Figure 2, which we feel is more likely due to either local influences or to measurement uncertainties, departs from the average amplitude by about 0.5 ppm (parts per million of dry air by volume). Using *Lieth's* (1970) estimate of net primary production of the northern hemisphere, which is about five times the withdrawl of CO_2 implied by the Mauna Loa record, a 0.5 ppm increase or decrease in amplitude in Figure 2 might be associated with a 2% change in vegetative uptake.

In Figure 3, the observed seasonal amplitude of the atmospheric carbon dioxide values for all locations for which there are data available have been plotted, together with estimates illustrating a theoretical calculation of the seasonal variation. The latter uses land net primary production values in 20° latitude bands from 50°S to 90°N and an estimate about the seasonal variation of land vegetative uptake and release of carbon dioxide within each band (*Machta*, 1972). The estimates of month by month uptake and release of carbon dioxide by vegetation, obtained independently from the observed data in Figure 2, were then introduced into a model allowing for atmospheric diffusive transport in the vertical and north-south directions. In Figure 3, the amplitude is derived from the differences between the highest and lowest monthly concentrations for

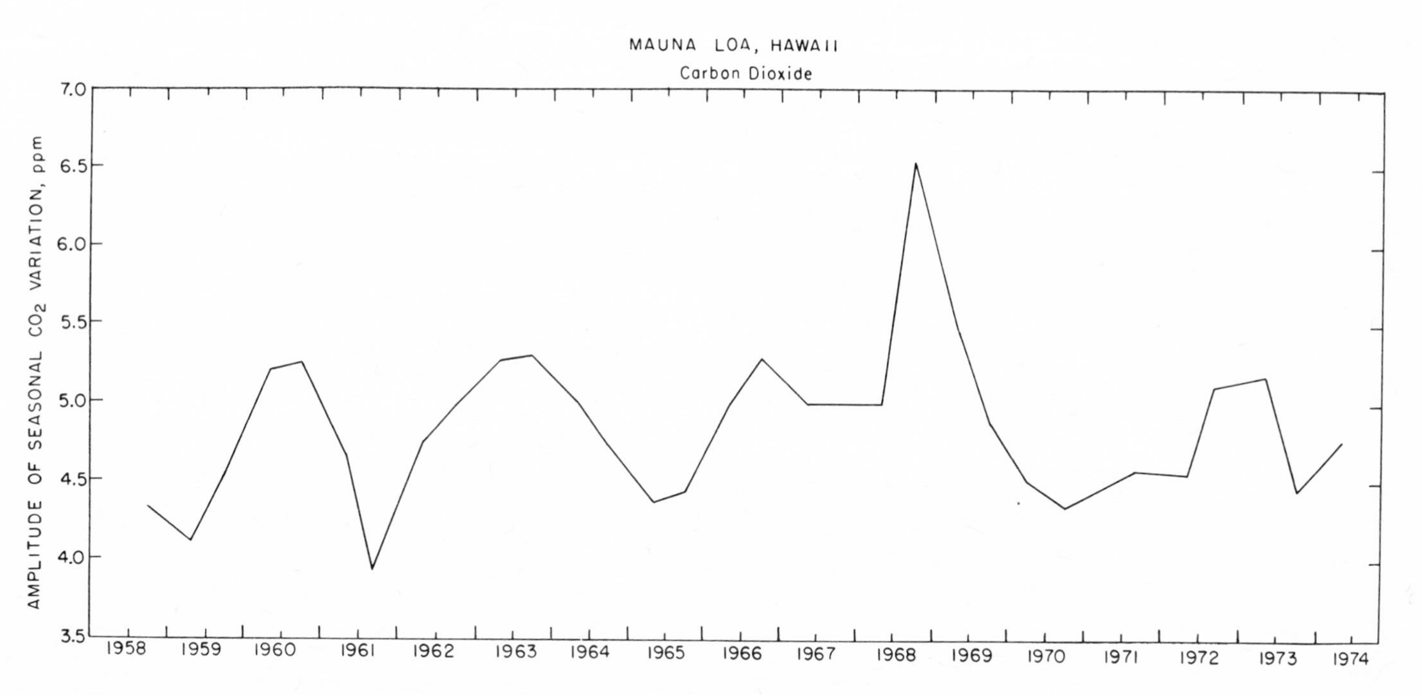

Fig. 2. The amplitude of carbon dioxide concentrations within each year at Mauna Loa. The amplitude is obtained by subtracting the minimum three-month (usually September, October and November) average from the mean of the maxima concentrations (usually March, April and May) before and after the maximum. This amplitude is plotted in October. A similar calculation is performed by subtracting the mean of the two adjacent minima from the maximum concentrations and plotted at April of each year.

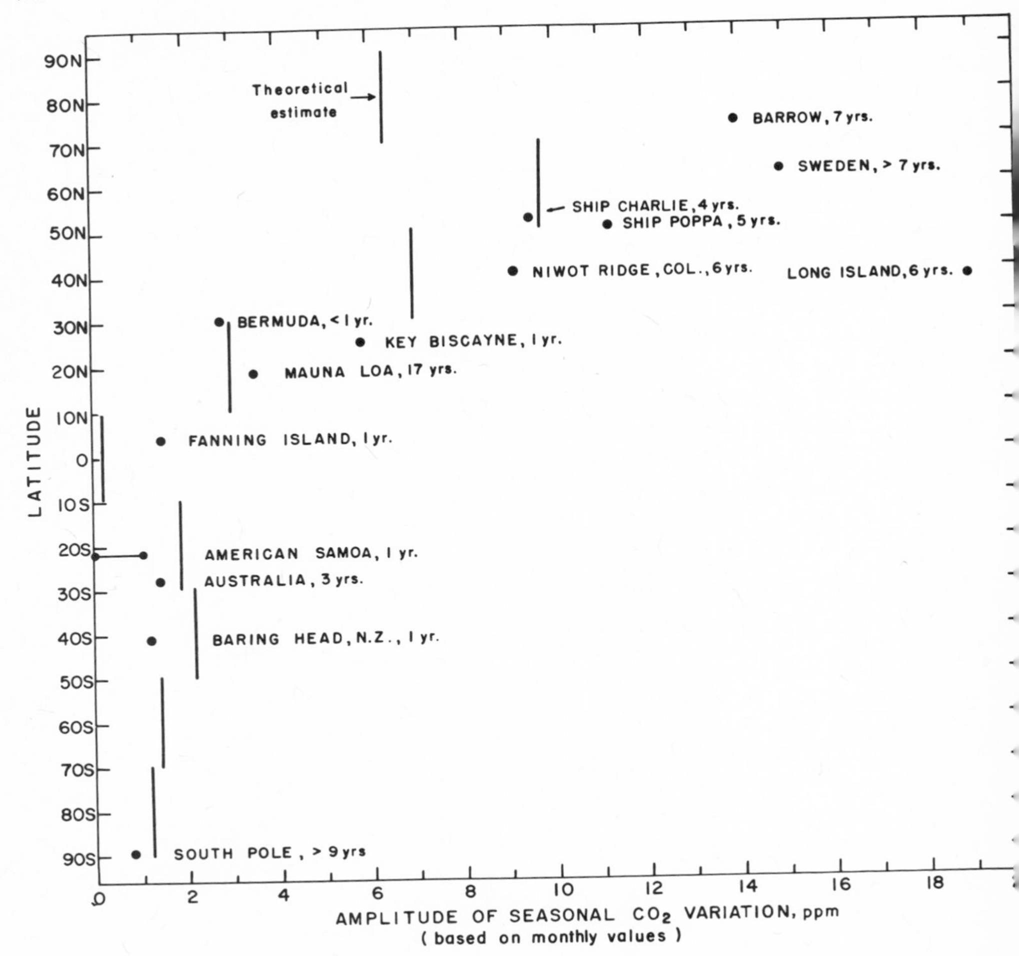

Fig. 3. Variation of amplitude of monthly carbon dioxide concentration in the course of a year, averaged over all available years with latitude (dots). The number of years of record follows the identification of the sampling point. The vertical lines provide a theoretical estimate of the same amplitude using uptake and release of carbon dioxide by the land vegetation as the sole seasonal forcing function.

both observed and model computed annual cycles, in contrast to Figure 2, where the mean three monthly minimum and maximum concentrations were employed. The month of predicted maximum and minimum concentrations were within 2 months of the average observed maximum and minimum concentration (except, perhaps, at Fanning Island and Samoa which have small amplitudes).

Figure 3 shows that the observed seasonal variation in atmospheric carbon dioxide at background stations may be predicted reasonably well from existing biological and meteorological information at all locations except at Barrow, Sweden and Long Island, NY. Here the observed amplitudes are much larger than predicted.

The ice cover of the Arctic and adjacent oceans varies seasonally. *Sanderson* (1975) has reported a decrease of ice cover in the high latitudes from over 15 to less than 7 x 10^6 km^2 between spring and late summer. If, as is suggested in the data of *Keeling* (1968), the waters of these oceans are undersaturated in carbon dioxide in the summer, their exposure to the air during the summer season will result in a sink for atmospheric carbon dioxide. This will contribute to the sharp decrease between June and September at high latitudes. On the other hand, a smaller, but uncertain, reduction in ice cover must also take place near the Antarctic continent. This may also introduce a seasonal sink in high southern latitudes. Yet, the South Pole seasonal variation in carbon dioxide is small and appears to fit the theoretical curve. According to *Coyne and Kelley* (1974, 1975), the lakes of Alaska and the tundra appear to be sources, rather than sinks, of carbon dioxide for the atmosphere during the summer. Fossil fuel usage, especially for space heating, constitutes another source of seasonality in high latitudes. Unpublished estimates by Machta, while not definitive, produce seasonal variations in ground level atmospheric carbon dioxide concentrations about an order of magnitude too small to account for the discrepancy at high latitudes in Figure 3.

The Long Island data also exhibit a larger seasonal variation (maximum in January) than expected. But, this is due to pollution from nearby human activities. It is introduced to illustrate the difficulties with a "non-background" location rather than the shortcomings of the theory.

In summary, it is argued that seasonal variation in atmospheric carbon dioxide offers no evidence of a change in the terrestrial organic carbon pool. Further, it appears that except possibly for the Arctic, the observed seasonal variation at benchmark locations can be accounted for within a few ppm, by the photosynthesis and decay of land vegetation.

THE NON-REGULAR GROWTH

Careful examination of Figure 1 reveals variations in the rate of growth of atmospheric carbon dioxide. At this point, an attempt will be made to explain the non-seasonal irregularities. To detect these features, one removes the "regular" trend and seasonal periodicity to yield residuals. These residuals will, of course, depend on the assumed mathematical form for the "regular" features. A three parameter regression equation of the form $A_0 + A_1 e^{bt}$ where t is time and A_0, A_1 and b are adjustable parameters has been assumed. The adjustable quantities, A_0, A_1 and b are computed from a least squares best fit to the observed data. The regression equation has been applied separately to each month of the year (e.g., all Januarys, all Februarys, etc.). *Bacastow* (1976) has used a spline fit and we have tried another reasonable analytical expression to remove the regular trend and periodicity. The residuals differ in magnitude between these trials but the sense of the major variations remain in all analyses.

Figure 4 compares the monthly residuals at two stations - Mauna Loa and the South Pole - with an analysis based on the same equation $(A_0 + A_1 e^{bt})$ fitted to the annual emission (monthly information not being available) values of carbon dioxide from the combustion of fossil fuels, the flaring of natural gas, and cement manufacture. Only the emission history from 1958 through 1974 has been used (*Rotty*, 1973 and 1975). The positive anomaly in 1959 and especially 1960 may be the result of questionable values of fossil fuel consumption in the Peoples Republic of China (*Keeling*, 1973).

Figure 4 suggests no significant correlation between the injection of fossil fuel carbon dioxide and the observed carbon dioxide concentrations. The zero lag correlation of annual values to Mauna Loa is 0.42 (not statistically significant). Lagging Mauna Loa carbon dioxide concentration behind inputs up to 6 years yielded smaller correlations than for zero lag. Thus, the year-to-year variations in the emission of fossil fuel carbon dioxide may contribute to the observed fluctuations in atmospheric carbon dioxide, but a substantial atmospheric variation still demands alternate explanations.

If biological or geophysical factors must indeed be invoked to account for the non-regular variations in atmospheric carbon dioxide, one looks to changes in the ocean as the cause. Thus, *Miller et al.* (1975) examined several atmospheric and oceanographic parameters as they might correlate with variations in the Mauna Loa carbon dioxide record. The month-to-month changes (with the seasonal variation eliminated) failed to correlate well with several ways of looking at sea surface temperatures of the North Pacific Ocean. However, the year-to-year changes did correlate with coefficients slightly over 0.6 between annual sea surface temperatures at about 20°N from Central America to Hawaii and the Mauna Loa yearly carbon dioxide concentrations. The correlation is statistically significant but, rather surprising, negative; that is, cooler sea surface temperatures are associated with higher carbon dioxide readings at Mauna Loa. *Miller et al.* (1975) suggested that the cooler sea surface temperatures may result from vigorous upwelling along the west coast of North America, with the water releasing its excess carbon dioxide as it warms during its southward and westward drift towards Hawaii.

It is suggested that events associated with the El Niño off the west coast of South America may exert a significant control over

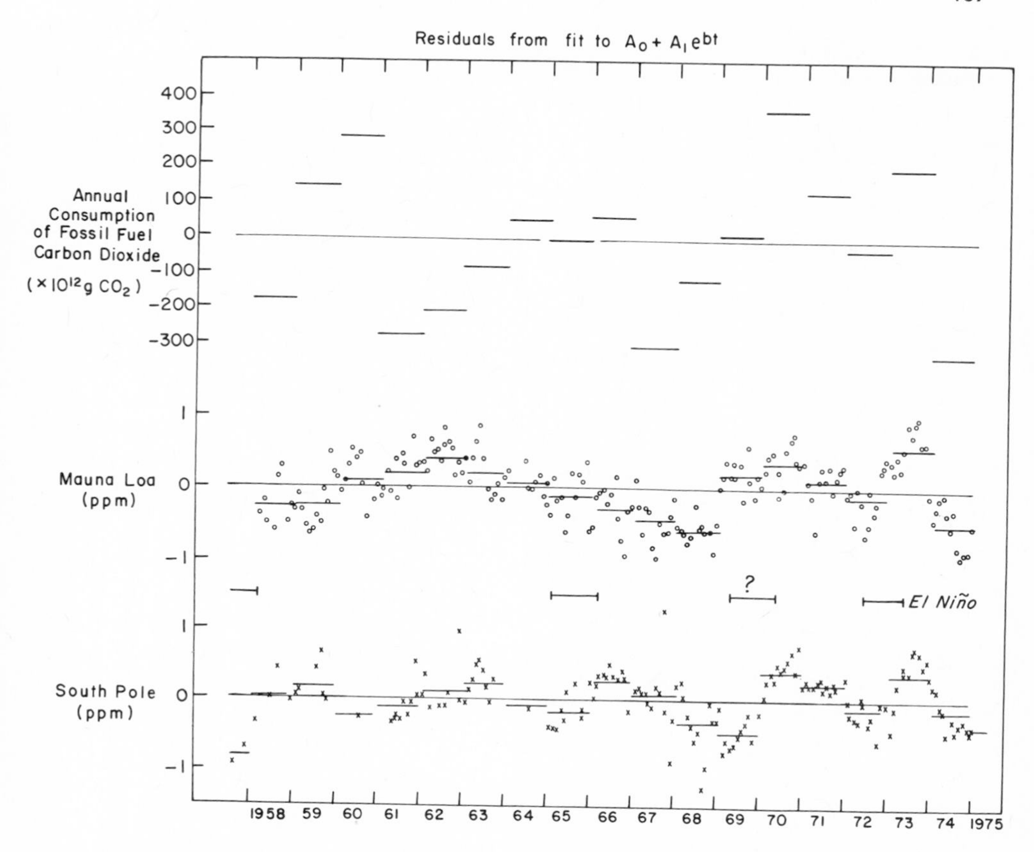

Fig. 4. Time history of residuals of monthly carbon dioxide concentrations after removing the long-term upward trend and seasonal variability at Mauna Loa and the South Pole (lower section). The horizontal lines in the upper section represent the residuals of the annual consumption (actually production) of fossil fuels after removing the long term upward trend. The periods of El Niño are also shown in the lower section. The horizontal lines in the lower section among the circles and crosses are the annual average residuals.

variations of global atmospheric carbon dioxide. Figure 4 shows four periods of much higher than normal Pacific Ocean sea surface temperatures at coastal stations off northern South America. The question mark above the 1969-1970 interval reflects the fact that it failed to qualify as a "normal" El Niño. All four sea surface temperature anomalies coincide with increasing values of the residuals at the South Pole. The last two in 1969-1970 and 1972-1973 occur at the same time as rising values of the residuals at Mauna Loa.

The lag correlations between the residuals at various stations are also suggestive of a forcing function at about the El Niño latitude. Figure 5 presents the various lag correlations between Mauna Loa, Australia (with its much more limited period of observation), and the South Pole. Computed lag corrections indicate a delay of about five or six months between Mauna Loa and the South Pole (Mauna Loa earlier), and two months between Mauna Loa and Australia (Mauna Loa again earlier). With a computer model of north-south and vertical transport, a prediction has been made as to how variable amounts of an inert tracer injected into the 0 to 10°S latitude band would be detected at the latitudes of Mauna Loa, Australia, and the South Pole, in terms of the same lag correlations The results appear as the light dashed lines of the figure and indicate that the observed lag correlations are, at least, consistent with a source at the latitude belt of the El Niño. Observed coefficients which are lower than from the model suggest that other factors for the real atmosphere also influence the uptake and release of carbon dioxide. Some of these have been discussed by *Bacastow* (1976).

The oceans must certainly exert a control over atmospheric carbon dioxide. Efforts to demonstrate this dependence from analysis of atmospheric data must be viewed as encouraging, but not yet convincing.

A DIFFERENT VIEW OF THE TREND

Figure 6 expresses the long term record at Mauna Loa in another fashion which highlights the problem of predicting future atmospheric levels of carbon dioxide. The dashed curve shows the growth of fossil fuel carbon dioxide consumption averaged over the two-year intervals centered at each dot. The growth rate of emissions to 1973 was about 5% per year. But the growth rate from 1973 to 1974 dropped to about 2% per year due to the energy crisis and a warm winter in eastern U.S. and western Europe. This rather dramatic drop in fuel consumption is not evident on the graph because of the two-year averaging procedure.

So long as the buffering factor of the oceans and the behavior and size of the biosphere remain unchanged and the carbon dioxide input growth rate remains at approximately the same exponential rate of increase, most, if not all, predictive models of carbon dioxide content of the atmosphere yield a fraction of the fossil fuel carbon dioxide remaining airborne, which remains approximately constant with time (*Ekdahl and Keeling*, 1973; *Oeschger et al.*, 1975). The observed airborne fraction appears as the solid lines on Figure 6 in which it is assumed that the changes at Mauna Loa represent the world-wide changes. A plot for the South Pole would

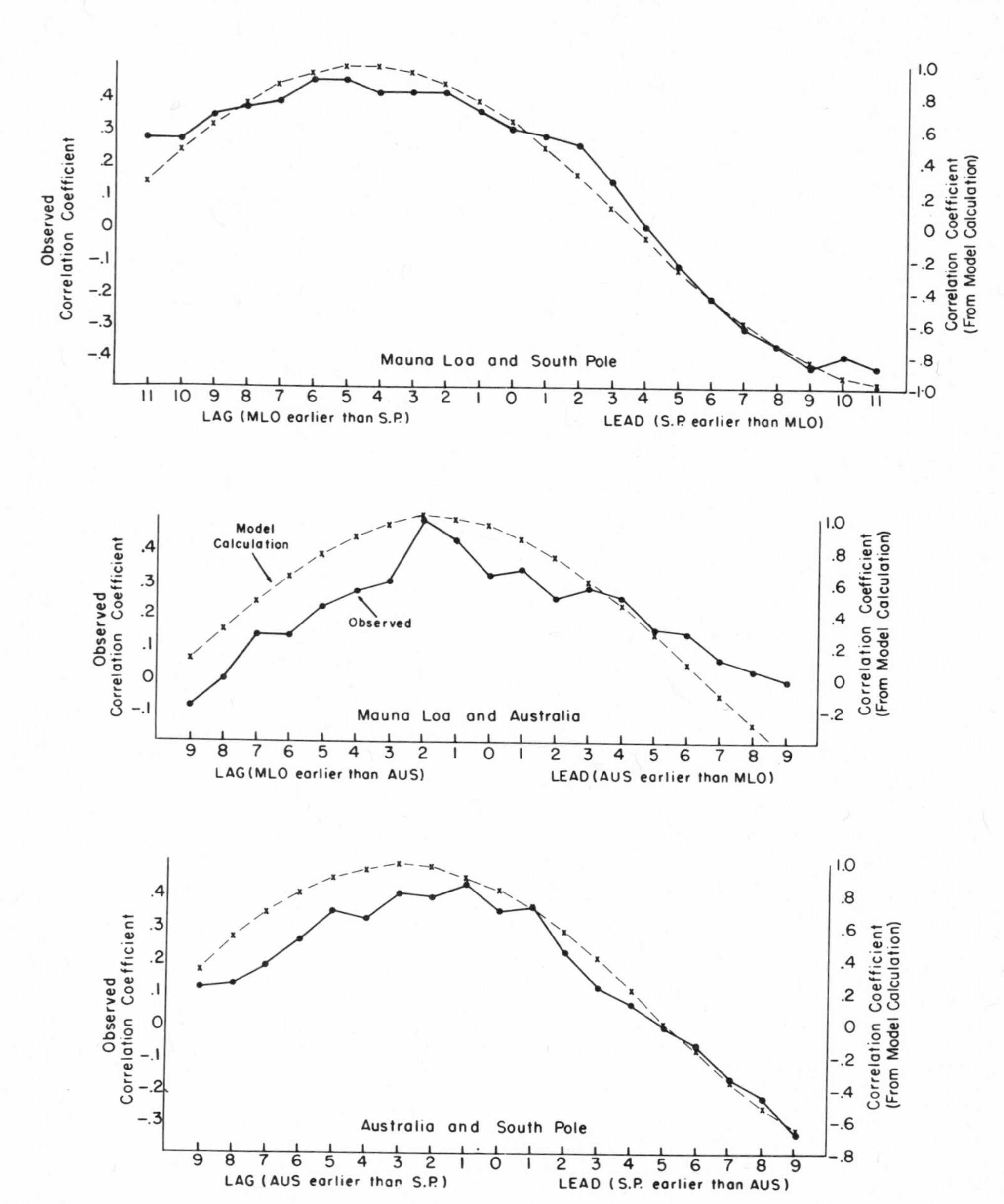

Fig. 5. Lag correlations (observed are solid lines joining dots while theoretical values are dashed lines (joining crosses) between indicated stations. The theoretical calculation assumes that the forcing function is located at 0-10°S. The observed correlation coefficients are read on the left hand scale, the theoretical correlation coefficients on the right hand scale. The lag, in months, is scaled on the horizontal axis.

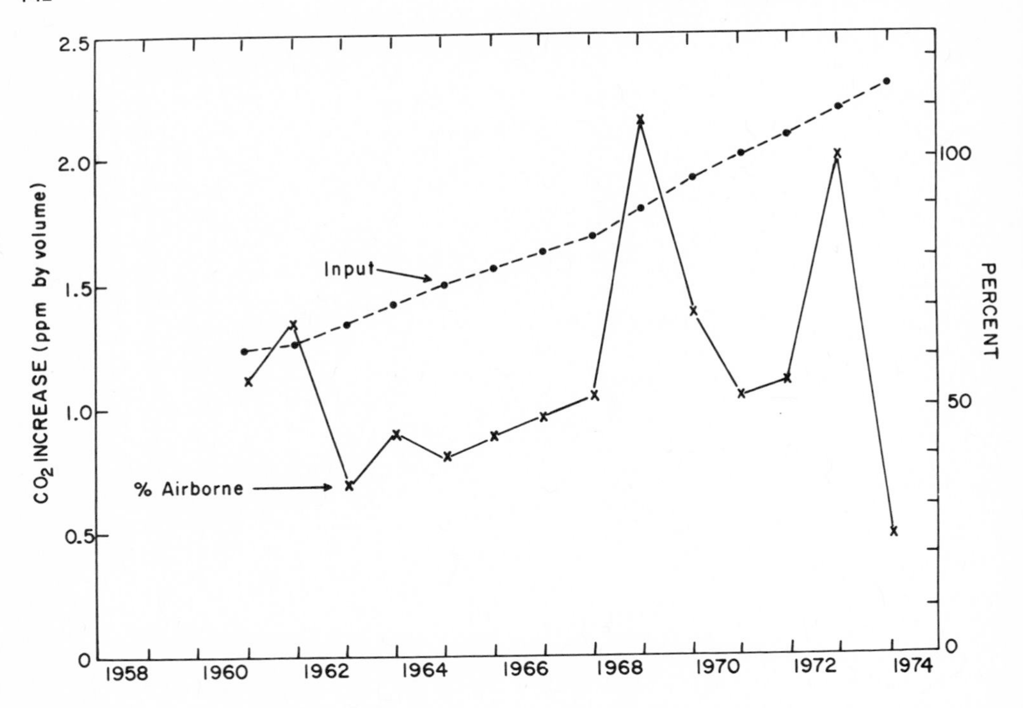

Fig. 6. The percent of the man made fossil fuel carbon dioxide remaining airborne each year (solid lines joining crosses) assuming that the observed Mauna Loa carbon dioxide growth represents the world growth. The right hand ordinate is for the percent airborne. The dashed lines joining the dots are the amounts of increase expected from the fossil fuels (and other man made sources of carbon dioxide) assuming that this carbon dioxide mixes rapidly with the entire atmosphere.

be quite similar. The fraction airborne varies from year to year with a large drop in the last year plotted on the chart.

This large drop is of particular interest. It might be attributed to a local Mauna Loa phenomenon were it not for a similar low annual growth at several other stations, namely, Barrow and the South Pole. However, not all background stations displayed a slowing down of the growth rate in 1974; for example, Ship Papa failed to show the slowing of the growth rate.

The year-to-year and longer term variation in the airborne fraction reflects the biological and geophysical control over atmospheric carbon dioxide and emphasizes the urgent need for fuller understanding of the several reservoirs for carbon dioxide and exchangeable carbon in order to acquire more confidence in long range forecasts of atmospheric carbon dioxide.

REFERENCES

Bacastow, R. B. 1976. Modulation of Atmospheric Carbon Dioxide by the Southern Oscillation. *Nature, 261.* 116-118.

Bolin, B. and W. Bischof. 1970. Variations in the carbon dioxide content of the atmosphere of the northern hemisphere. *Tellus, 22.* 431-442.

Bray, J. B. 1959. An analysis of the possible recent change in atmospheric carbon dioxide concentration. *Tellus 2.* 220-230.

Callendar, G. S. 1938. The artificial production of carbon dioxide and its influence on temperature. *Quart. J. Roy. Meteor. Soc., 64.* 223-240.

Coyne, P. I. and J. J. Kelley. 1974. Carbon dioxide partial pressures in arctic surface waters. *Limnology & Oceanography, 19.* 928-938.

Coyne, P. I. and J. J. Kelley. 1975. CO_2 Exchange over the Alaskan arctic tundra: Meteorological assessment by an aerodynamic method. *J. Appl. Ecol. 12.* 587-611.

Cramer, J. and A. L. Meyers. 1972. Rate of increase of atmospheric carbon dioxide. *Atmospheric Environment, 6.* 563-573.

Ekdahl, C. A. and C. D. Keeling. 1973. Atmospheric carbon dioxide and radiocarbon in the natural carbon cycle: 1. Quantitative deductions from records at Mauna Loa Observatory and at the South Pole. In: *Carbon and the Biosphere,* G. M. Woodwell and E. V. Pecan (Eds), U. S. Atomic Energy Commission. 51-85.

Hoffert, M. I. 1974. Global distributions of atmospheric carbon dioxide in the fossil fuel era: A projection. *Atmospheric Environment, 8.* 1225-1250.

Keeling, C. D. 1968. Carbon dioxide in surface ocean water, 4. Global distribution. *J. Geophys. Res., 73.* 4543-4553.

Keeling, C. D. 1973. Industrial production of carbon dioxide from fossil fuels and limestone. *Tellus, 25.* 174-198.

Keeling, C. D., J. A. Adams, Jr., C. A. Ekdahl, Jr., and P. G. Guenther. 1976a. Atmospheric carbon dioxide variations at the South Pole. *Tellus, 28.* 552-564.

Keeling, C. D., R. B. Bacastow, A. E. Bainbridge, C. A. Ekdahl, Jr., P. G. Guenther, L. S. Waterman, and J. F. S. Chin. 1976b. Atmospheric carbon dioxide variations at Mauna Loa Observatory, Hawaii. *Tellus, 28.* 538-551.

Kelley, J. J., Jr. 1969. An analysis of carbon dioxide in the Arctic atmosphere near Pt. Barrow, Alaska, 1961-1967. Scientific Report, Office of Naval Research, Contract N00014-67-A-01030-007, University of Washington, Seattle, Washington.

Lieth, H. 1970. Phenology in productivity studies in ecological studies, J. Jacobs, et al. (Ed.), Springer Verlag, Heidelberg, Berlin.

Machta, L. 1972. Mauna Loa and global trends in air quality. *Bull. Amer. Meteor. Soc., 53.* 402-420.

Miller, A. J., J. M. Miller, R. M. Rotty. 1975. An evaluation of the CO_2 record at Mauna Loa as baseline data for global changes. NOAA Technical Memorandum, ERL-ARL-49.

Oeschger, H., V. Siegenthaler, V. Schotterer, and A. Gugelmann. 1975. A box diffusion model to study the carbon dioxide exchange in nature. *Tellus*, *27*. 168-192.

Pearman, G. I., J. R. Garratt, P. J. Fraser, D. J. Berdsmore, and J. G. O'Toole. 1975. The CSIRO (Australia) base-line atmospheric carbon dioxide monitoring programme, Progress Report No. 3. Division of Atmospheric Physics, Aspendale, Australia.

Rotty, R. M. 1973. Commentary on and extension of calculative procedure for CO_2 production. *Tellus*, *25*. 508-517.

Rotty, R. M. 1975. A note updating carbon dioxide production from fossil fuels and cement. Institute for Energy Analysis, Oak Ridge Associated Universities, IEA (M)-75-4.

Sanderson, R. M. 1975. Changes in the area of Arctic Sea ice 1966 to 1974. *The Meteorological Magazine*, The UK Meteorological Office, 104. 313-323.

Smil, V. and D. Milton. 1974. Carbon dioxide - alternative futures. *Atmospheric Environment*, *8*. 1213-1224.

Woodwell, G. M., R. A. Houghton, and N. R. Tempel. 1973. Atmospheric CO_2 at Brookhaven, Long Island, New York. Pattern of variation up to 125 meters. *J. Geophys. Res.*, *78*. 932-940.

Climate Changes and Lags in Pacific Carbonate Preservation, Sea Surface Temperature and Global Ice Volume

T. C. Moore, Jr., N. G. Pisias and G. R. Heath

University of Rhode Island

ABSTRACT

Past increases in global ice volume (as indicated by oxygen isotopes) appear to have led increases in carbonate preservation by at least 6000 years in several cores studied from the tropical Pacific Ocean. A comparison of the carbonate-preservation and oxygen-isotope spectra in two cores shows maxima in coherence at frequencies close to the Milankovitch periods. At these frequencies the oxygen isotope record tends to lead the carbonate record by about 10,000 years. Cross correlation of the estimated sea surface temperature and the oxygen isotope records in one of the tropical Pacific cores indicates that changes in global ice volume preceded local temperature changes by about 4,000 years. In a core from the North Pacific Ocean, cross correlations of both sea surface temperature estimates and carbonate concentrations with oxygen isotopes show no lead or lag. Thus, two regions of the Pacific basin show marked differences in the fluctuations of carbonate and sea surface temperature relative to ice volume. There is no indication that changes in sea surface temperatures in the regions studied induced changes in ice volume; however, regional differences in the response time of carbonate to climatic change suggest that hydrographic as well as geochemical mechanisms are involved in glacial-interglacial changes in carbonate preservation.

INTRODUCTION

In the cycling of climate from interglacial to glacial conditions, the response of the different elements in the air-ocean-ice system is neither instantaneous nor isochronous. To understand the large changes in the earth's climate and to determine which parts of the system ultimately force changes in other parts, it is important to know the sequence of events in the ice-age cycle. Furthermore, our knowledge of this sequence of change should be globally comprehensive, for it does not necessarily follow that the order of change will be the same in one part of the globe as in another. The geologic record offers us a history of these changes, but in order to examine the sequence of change in these climatic elements through geologic time, either an extremely accurate time scale is needed for each climatic record, or indicators of each major element of the system must be found together in the same stratigraphic sections.

The deep sea is particularly well suited for such a global stratigraphic study. Not only are good sedimentary records found at nearly all latitudes in all oceans, but they also contain a comprehensive array of climatic indicators. Fossils of the planktonic flora and fauna record changes in the conditions of near-surface waters. Benthic fauna and the preservation of biogenic debris tell of changes in the character of bottom waters. Changes in the global ice volume are recorded by oxygen isotopes incorporated in calcareous tests. Pollen and other indicators of land conditions are found in hemipelagic sediments near continents; and finally, changes in wind patterns and intensity are preserved in the wind-blown detritus found in deep-sea sediments. Thus, the use of marine cores in such a study not only provides global coverage, but also a multi-channel record that bypasses the nearly insurmountable problems of inter-site correlations and obviates the usual strict requirements for a highly accurate chronostratigraphy.

The few studies that have investigated the sequence of climate change in marine cores have relied primarily on samples from the Pacific Ocean. This preference may be due in part to the assumption that here oceanographic and climatic changes are the primary controls on the concentrations of the biogenic debris and that dilution by terrigenous sediments is negligible. In particular (and for the purposes of this study), changes in carbonate concentration with time are taken to be the result of changes in the corrosiveness of bottom waters (*Berger*, 1973; *Thompson and Saito*, 1974).

In this study, three elements in the ocean-ice part of the climate system are considered: global ice volume, corrosiveness of bottom waters with respect to calcite, and sea surface temperatures. As an indicator of ice volume, the ratio of oxygen isotopes in the tests of foraminifera is used (*Shackleton and Opdyke*, 1973). To measure changes in the corrosiveness of bottom waters, both variations in carbonate content and a dissolution index based on planktonic foraminifera (*Thompson*, 1976) are used. Sea surface temperatures have been estimated using the *Imbrie and*

Kipp (1971) technique as applied to radiolaria and coccoliths. These climatic indicators were determined in three cores from the Pacific Ocean: V28-238, from the western equatorial Pacific; V19-29, from the eastern equatorial Pacific and Y6910-2, from the transitional zone of the eastern North Pacific.

Previous Work

Earlier studies of the relationship between calcite preservation and oxygen isotope records in cores have all been carried out in the tropical Pacific Ocean (*Luz and Shackleton*, 1975; *Pisias et al.*, 1975; *Ninkovich and Shackleton*, 1975; *Shackleton and Opdyke*, 1976). The first of these (*Luz and Shackleton*, 1975) compared foraminiferal dissolution indices, carbonate content and oxygen isotope curves in several cores dating to about 130,000 yr B.P. The sample spacing and sedimentation rate in these cores do not allow an accurate determination of the amount of offset in calcite preservation and oxygen isotope records; however, all cores do indicate that major decreases in global ice volume preceded the major decreases in carbonate preservation by an estimated 8,000 to 18,000 years. This same general picture was found in a longer record from the western equatorial Pacific Ocean, V28-239 (*Shackleton and Opdyke*, 1976) and in the very detailed study of a core from the eastern equatorial Pacific Ocean, V19-29 (*Ninkovich and Shackleton*, 1975; and *Shackleton*, 1977). In the latter study, it was estimated that major changes in the oxygen isotopes lead those in carbonate preservation by about 5,000 years. In a study of another core (Y69-106P) from the eastern equatorial Pacific Ocean, *Pisias et al.* (1975) avoided the assumption that there were no large variations in the accumulation rate of opal and other major non-carbonate diluents which might affect the shape of the carbonate concentration curve. Instead a minor component (quartz) with an apparently constant input was used to construct a sedimentation rate model and to calculate the variations in the accumulation of carbonate. Again, the sedimentation rate, length and quality of the record for this core did not allow a detailed analysis; however, unlike the other comparisons, in this case changes in the carbonate accumulation rates <u>preceded</u> changes in the oxygen isotope record by 5,000 to 10,000 yr. The fact that two different measures of carbonate preservation (carbonate concentration and preservation indices of the foraminiferal fauna) gave similar results (*Luz and Shackleton*, 1975) is very reassuring. However, the discrepancy between this study and that of *Pisias et al.* (1975) requires further testing of both the sedimentation model and the sensitivity of foraminiferal indices to changes in carbonate dissolution. For the cores studied in this paper, it is assumed that carbonate concentrations and solution indices do give good estimates of calcite preservation.

METHODS

In all previous studies, comparisons between the different records in individual cores have been done by "eye". Although there is little disagreement concerning the sense of the leads and lags in individual cores, the magnitude is often open to question. In order to obtain an estimate of the average offset in the two signals, a more objective approach is used in this study. First, the records of the separate indicators in each core are repeatedly correlated using successively larger lags, or displacements, of the individual records. The number of lags at which the maximum correlation occurs gives an objective estimate of the average amount that one record leads the other. These correlation coefficients are calculated over the entire time and frequency domains. In order to separate out those frequencies of oscillation at which leads and lags occur, a spectral analysis (*Pisias et al.*, 1973) is performed on each record. This is followed by a comparison of the individual spectra (cross spectral analysis; *Jenkins and Watts*, 1968), which indicates those frequencies that contain a marked amount of the variance in both records (i.e., those frequencies showing a high degree of coherence). This analysis also gives the phase relationship of the two records at any given frequency. It clearly shows the particular frequency of variation at which the leads and lags occur. This technique also offers a means of evaluating the spectra snd comparing them with those of other climatically important variables.

RESULTS

Core: V19-29, eastern equatorial Pacific
Location: 3° 35'S; 83° 18'W
Water Depth: 3158 meters
Calcite Compensation Depth (modern): 3500 meters (*Berger and Winterer*, 1974)

This core was taken on the terraced southern flank of the Carnegie Ridge (*Ninkovich and Shackleton*, 1975). The sediment is a calcareous siliceous ooze containing some ash (one distinct layer occurs at about 1330 cm). The core is 1680 cm long and, based on the ages assigned to the Emiliani oxygen isotope stages (*Shackleton and Opdyke*, 1973), has an average accumulation rate of about 6cm/10^3 years. Samples were taken at 10cm intervals for both carbonate and oxygen isotope measurements. Estimates for sea-surface temperatures based on faunal studies (*Molina-Cruz*, 1975) are available for the upper part of the core only (isotope stages 1 through 5a). The cross correlation of the carbonate and oxygen isotope records show a maximum of +.52 (Figure 1) at a lag of 3 sample intervals (6,000 yrs) with the maxima in O^{18} (global ice volume) preceding those in

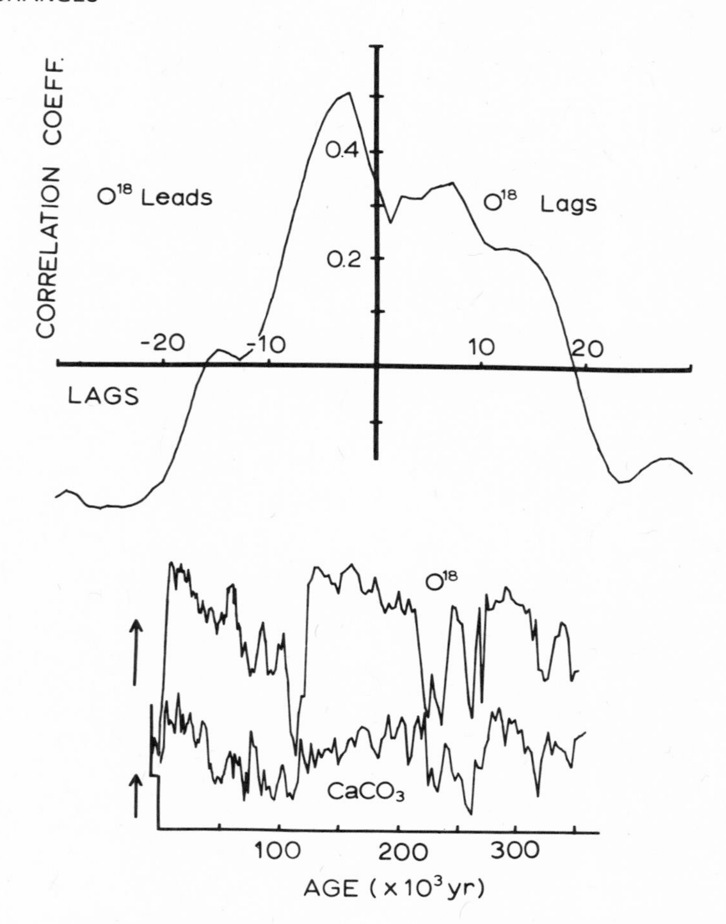

Fig. 1. Cross correlation of oxygen isotopes and carbonate concentrations in V19-29. At the bottom, plots of the two records are offset by the number of sample intervals indicated by the maximum correlation.

the carbonate record. Cross correlation of the isotope and temperature records show a maximum of -0.34, with ice maxima preceding temperature minima by about 4,000 yr (Figure 2). A second maximum of +0.50 shows increases in temperature leading increases in ice volume by 7 sample intervals (14,000 yrs); however, it is thought that this correlation is a result of the shortness of the record and may be spurious. A 7 sample offset removes stage 1 temperature changes from consideration (Figure 2) and bases the correlation primarily on fluctuations within isotope stages 2, 3 and 4. Because of the relatively sparse temperature data, spectral analysis

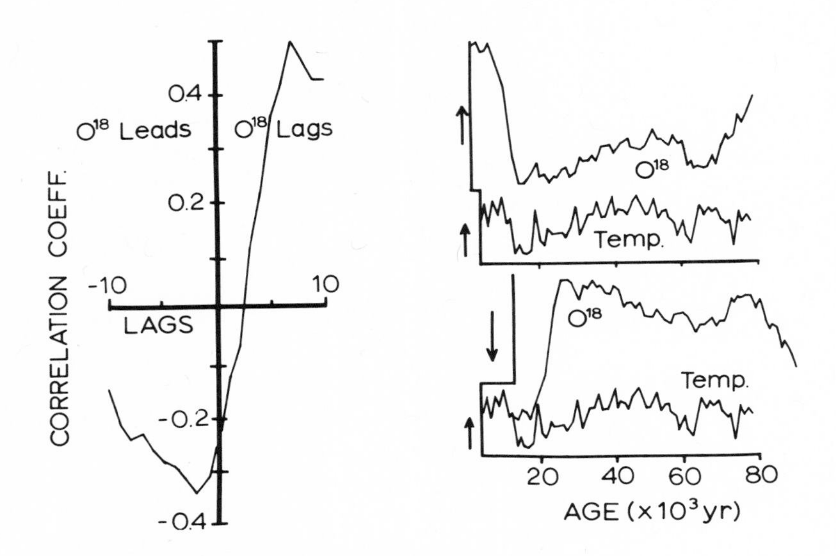

Fig. 2. Cross correlation of oxygen isotopes and estimated sea-surface temperatures in V19-29. At the right, plots of the two records are offset by the number of intervals indicated by the correlation maxima.

was conducted on the carbonate and oxygen isotope records only. Neither of the individual spectra show significant peaks. However, there is significant coherency between the two records at frequencies corresponding to periods of about 20,000, 36,000 and 100,000 yr/cycle with the 20,000 yr/cycle period showing the greatest coherency (Figure 3). The phase estimates (Figure 4) show no lag for the 100,000 yr period (0° phase angle). Oxygen isotopes lead the carbonate record at the 36,000 yr. and 20,000 yr. coherency peaks by about 9,700 yr (100° phase angle) and 9,200 yr (165° phase angle), respectively.

Core: V28-238, western equatorial Pacific
Location: 1° 01'N; 160° 29'E
Water Depth: 3120 meters
Calcite Compensation Depth (modern): 4800 meters (*Berger and Winterer*, 1974)

This core, approximately 1610 cm long, was taken on the Ontong-Java Plateau. Based on the Emiliani isotope stages, magnetic stratigraphy (*Shackleton and Opdyke*, 1973) and faunal extinctions (*Pisias*, 1976), the carbonate ooze making up this core accumulated

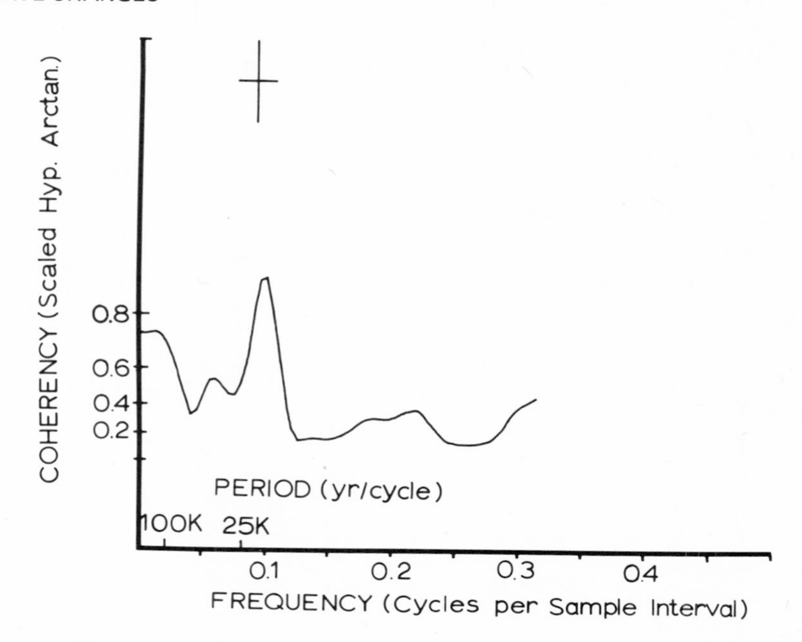

Fig. 3. Coherency between oxygen isotope and carbonate concentration spectra for V19-29. Cross indicates 0.80 confidence interval and the bandwidth. All coherency functions in this paper are plotted on a hyperbolic arctangent scale. This is done so that the calculated 0.80 confidence interval is of a constant graphic length in the figure and independent of the magnitude of the coherency estimate. The bandwidth in all figures is a measure of how much smoothing has been performed on the estimates. Estimates at two frequencies separated by more than a bandwidth are independent; at less than a bandwidth coherency estimates are not independent. The confidence interval can be used to test whether a coherency estimate at any given frequency is non-zero.

at an average rate of 1.7cm/10^3 yr. Samples were taken at approximately 10cm intervals for oxygen isotope and faunal analysis. There is little variation in the estimated sea-surface temperature and carbonate content at this site (*Geitzenauer*, 1976). Estimates of the degree of solution of samples are based on an analysis of the foraminiferal fauna (*Thompson*, 1976). The cross-correlation of the solution index and oxygen isotope curves (Figure 5) shows a maximum of -0.35 at a lag of 2 sample intervals (about 10,000 yr) with maximum ice volume (O^{18}) preceding minimum dissolution (maximum preservation). Spectral analysis of the two records show peaks near 100,000 yr/cycle and 22,000 yr/cycle. Comparison of the spectra indicate several maxima in coherence (Figure 6). Those peaks

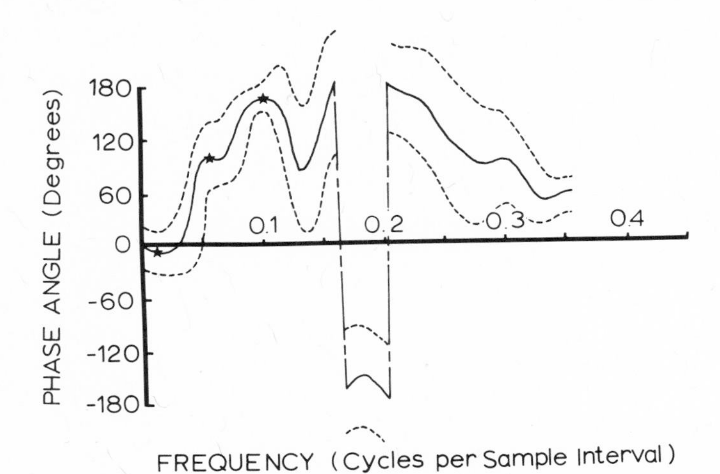

Fig. 4. Phase relationship between spectra of oxygen isotope and carbonate concentration records in V19-29 as a function of frequency. Dashed lines indicate 0.80 confidence interval. Stars indicate location of coherency maxima (Figure 3).

at frequencies greater than about 0.3 cycles per sample interval ($\leq$20,000 yr/cycle) show no phase difference in the two records. Only one well-defined peak occurs at a lower frequency, equivalent to about 26,000 yr/cycle. If a minimum in dissolution is interpreted as a maximum in preservation, the phase angle at this frequency (Figure 7) is equivalent to maximum ice volume leading maximum carbonate preservation by 10,500 years. The broad coherency peak between 100,000 and 47,000 years (Figure 6) also shows maximum ice volume leading carbonate preservation - in this case, by 15,000 to 16,000 years (Figure 7).

Core: Y6910-2, eastern North Pacific
Location: 41° 16'N; 127° 01'W
Water Depth: 2743 meters
Calcite Compensation Depth (modern): 3000 meters (*Berger and Winterer*, 1974)

This core was taken on the eastern flank of the Gorda Ridge (*Heath et al.*, 1976; *Heusser et al.*, 1975; *Moore*, 1973). The sediment is a hemipelagic silty clay with no signs of turbidite beds. The core is 1036 cm long and, based on the ages of the Emiliani oxygen isotope stage boundaries (*Shackleton and Opdyke*, 1975), has an accumulation rate of about 9cm/10^3 yr. Samples were taken at approximately 5 to 10 cm intervals for oxygen isotope, carbonate, and faunal analysis. The cross correlation of both the calcium

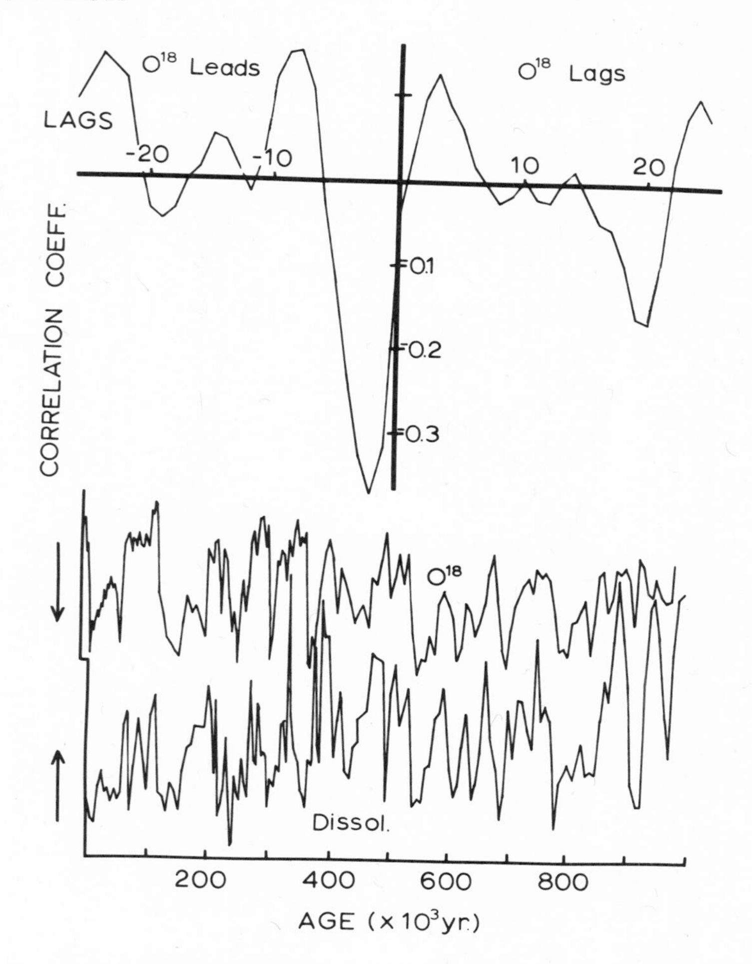

Fig. 5. Cross correlation of oxygen isotope and carbonate dissolution records in V28-238. At the bottom, plots of the two records are offset by the number of sample intervals indicated by the maximum correlation.

carbonate and sea-surface temperature records (estimated from the faunal analysis, *Moore*, 1973; *Huesser et al.*, 1975) with the oxygen isotope record show maximum correlations (+0.59 and +0.69, respectively) at zero lag (Figures 8 and 9). Thus, on the average the records are in phase. The oxygen isotope record shows no significant spectral peaks. However, the carbonate record appears to have a significant peak at a frequency corresponding to 22,500 yr/cycle. Maximum coherency between the two records occurs at a period of about 25,700 yr/cycle (Figure 10). The phase relationship

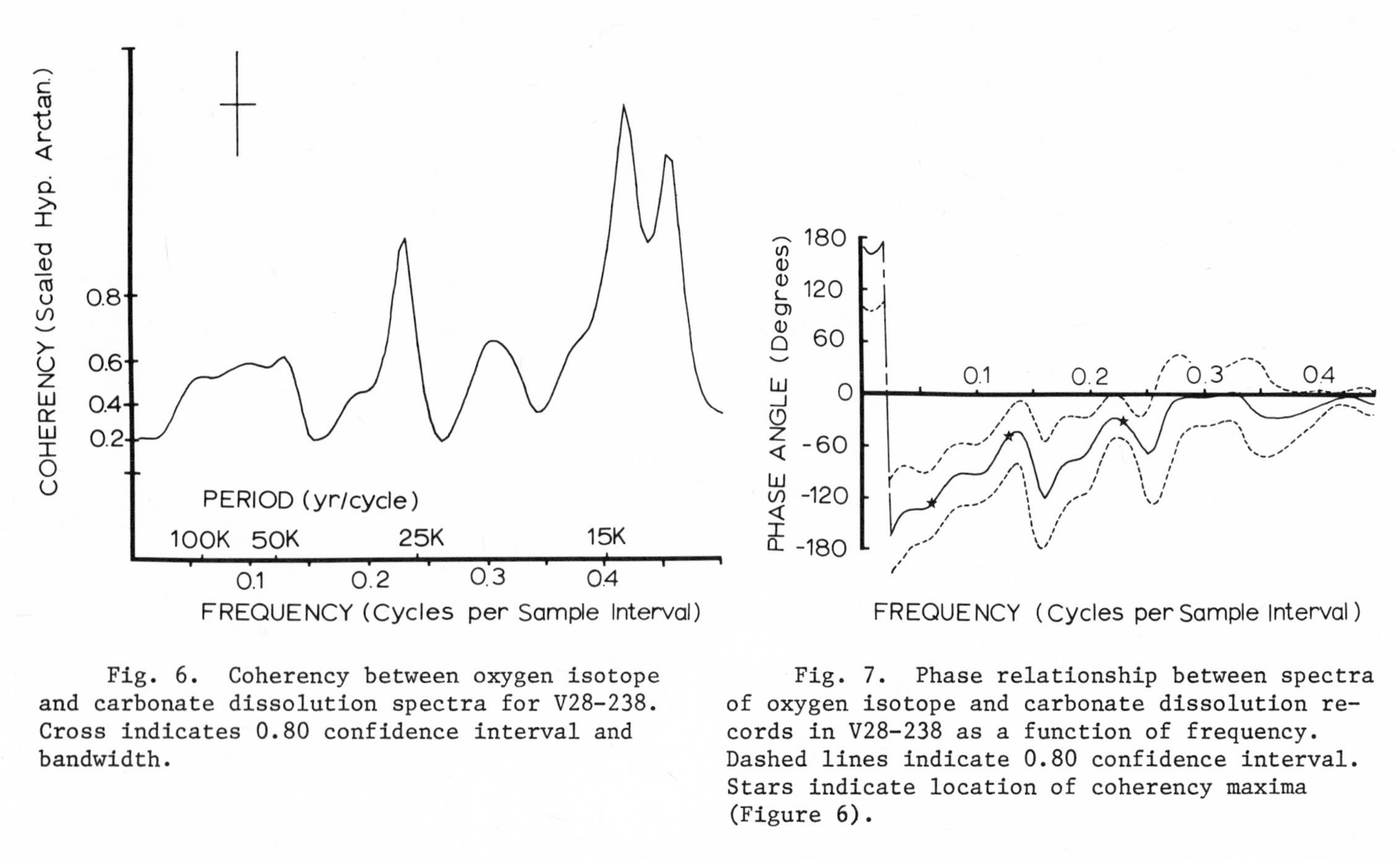

Fig. 6. Coherency between oxygen isotope and carbonate dissolution spectra for V28-238. Cross indicates 0.80 confidence interval and bandwidth.

Fig. 7. Phase relationship between spectra of oxygen isotope and carbonate dissolution records in V28-238 as a function of frequency. Dashed lines indicate 0.80 confidence interval. Stars indicate location of coherency maxima (Figure 6).

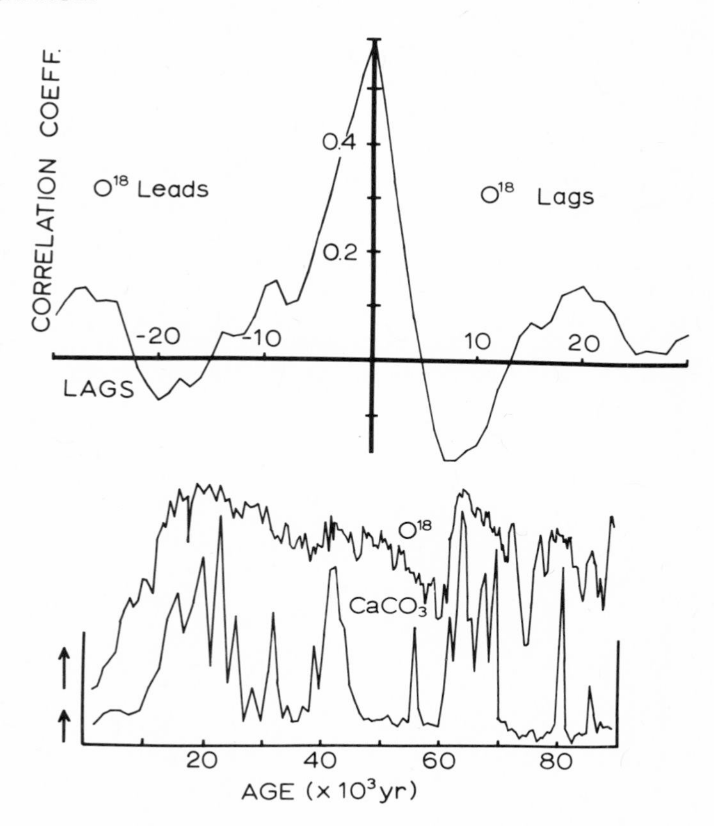

Fig. 8. Cross correlation of oxygen isotopes and carbonate concentrations in Y6910-2. At the bottom, plots of the two records are shown with no offset.

for this period (Figure 11) shows that the oxygen isotope record leads the carbonate record by about 2500 years. Spectral analysis of the temperature estimates gives an indication of a peak at 2800 years (see also *Moore*, 1973). Maxima in coherence between the temperature and isotope records (Figure 12) occurs at very low frequencies (periods >35,000 yr/cycle) and at frequencies higher than 0.3 ($\leq$2500 yr/cycle). The only coherency peak showing a phase angle significantly different from zero (Figure 13) occurs at 0.26 cycles per sample interval (about 2000 yr/cycle). At this frequency the phase relationship indicates that ice maxima precede temperature minima by about 270 yr.

A summary of the result of the cross-spectral analysis for the three cores is given in Table 1.

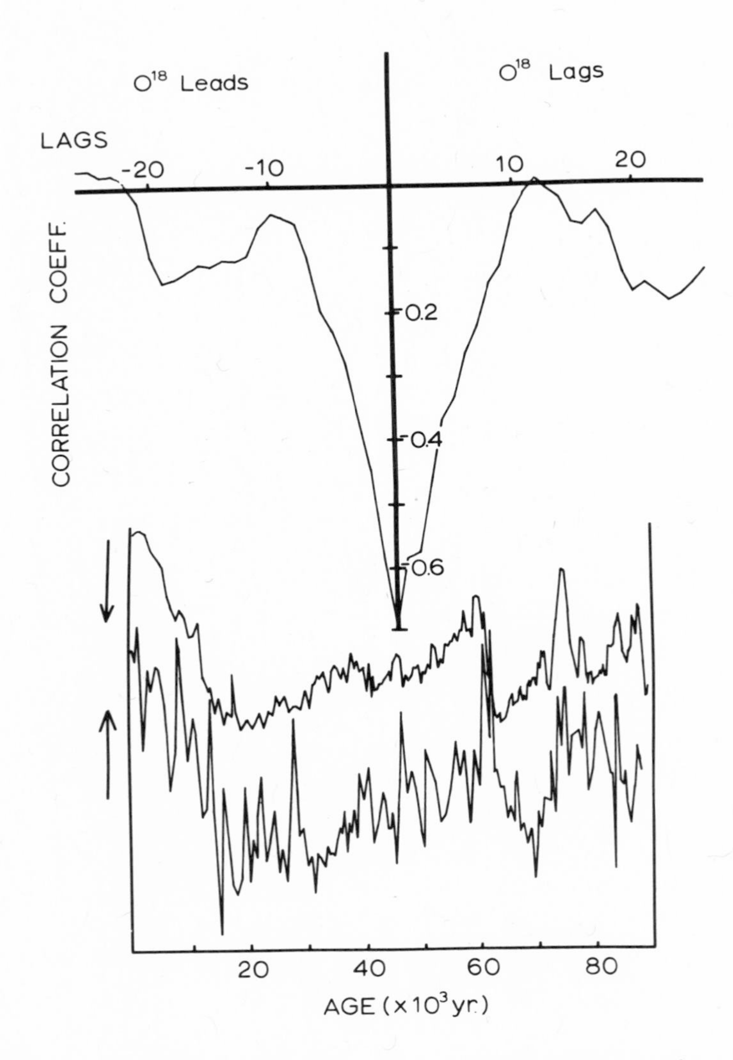

Fig. 9. Cross correlation of oxygen isotopes and estimated sea-surface temperatures in Y6910-2. At the bottom, plots of the two records are shown with no offset.

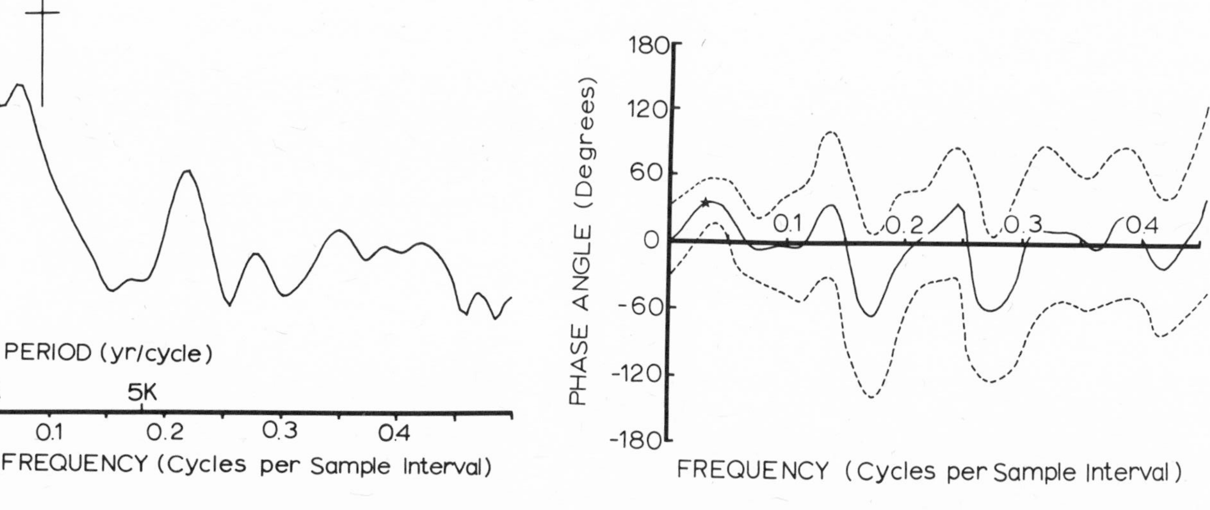

Fig. 10. Coherency between oxygen isotope and carbonate concentration spectra for Y6910-2. Cross indicates 0.80 confidence interval and bandwidth.

Fig. 11. Phase relationship between spectra of oxygen isotopes and carbonate concentration records in Y6910-2 as a function of frequency. Dashed lines indicate 0.80 confidence interval. Star indicates location of coherency maximum at approximately 26,000yr/cycle (Figure 10).

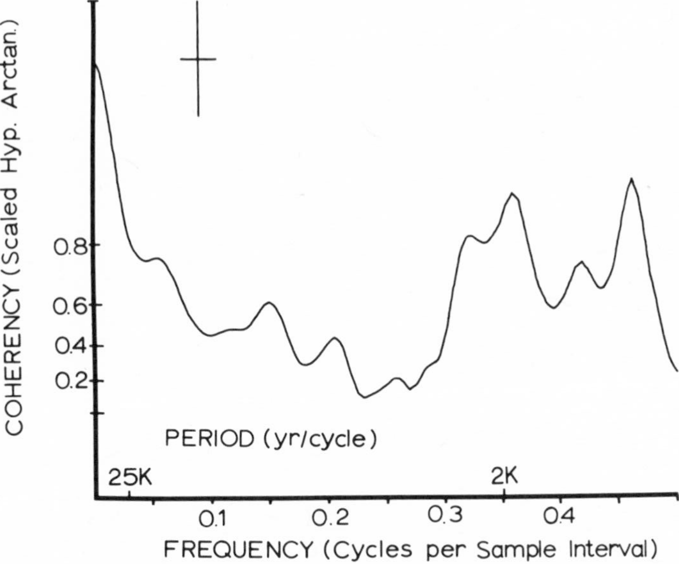

Fig. 12. Coherency between oxygen isotope and sea-surface temperature spectra for Y6910-2. Cross indicates 0.80 confidence interval and bandwidth.

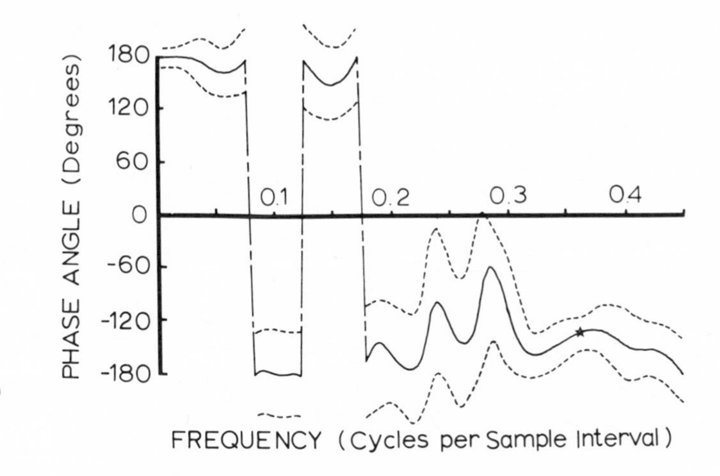

Fig. 13. Phase relationship between spectra of oxygen isotopes and sea-surface temperature records in Y6910-2 as a function of frequency. Dashed lines indicate 0.80 confidence interval. Star indicates location of coherency maximum at approximately 2,000yr/cycle (Figure 12).

TABLE 1. Summary of Lags for Maximum Correlation Coefficient in Cross-correlation Function and Frequencies of Maximum Coherencies.

Core No.	Data Pair	Lag of maximum correlation coef. 10^3 years (corr. coeff.)	Frequencies for max. Coh. function 10^3 yrs (lag in 10^3 yrs) 0^{18} leading
V19-29	0^{18} - $CaCO_3$ %	6(0.52)	100(0), 36(9.7), 20(9.2)
	0^{18} - Temperature	4(-0.38)	(temperature record too short for calculations)
V28-238	0^{18} - $CaCO_3$(sol. index)	10(-0.35)	100(15), 47(16), 26(10.5), 15(0)
Y6910-2	0^{18} - $CaCO_3$%	0(0.59)	25.7(2.5)
	0^{18} - Temperature	0(0.69)	>35(0), 2(0.27)

DISCUSSION AND CONCLUSIONS

As was pointed out in the introduction, there is no *a priori* reason to expect fluctuations in the records of different climatic indicators from different regions of the ocean to show consistent relationships. Similarly, it is likely that even at one site the exact relationship between these indicators may vary slightly from cycle to cycle (*Shackleton and Opdyke*, 1976). Depending on the quality of the stratigraphic record, however, climatic indicators in each core should show some average inter-relationship, and regional differences in such relationships may give clues to the exact process of climatic change.

The carbonate and oxygen isotope records of the tropical Pacific Ocean generally appear to be consistent. The most detailed stratigraphy in this region is that of core V19-29. Here, the increases in ice volume preceded increases in carbonate preservation by an average of 6,000 years. A slower accumulation rate and relatively wider sampling intervals may account for the longer lead time of 10,000 to 12,000 years found in V28-238 and most other cores from the tropical Pacific Ocean (*Luz and Shackleton*, 1975; *Ninkovich and Shackleton*, 1976).

Perhaps even more illuminating is the cross spectral analysis of the carbonate and oxygen isotope records. In both equatorial cores, there is considerable coherency between the spectra at frequencies very similar to the first Milankovitch period (21,000 yr/cycle), with the dominant coherency peaks in V19-29 and V28-238 at 20,000 and 26,000 yr/cycle, respectively. In both cores the phase relationship between the carbonate preservation and isotope record at this frequency shows that the oxygen isotopes lead the carbonate by 9,000-10,000 years. For the second Milankovitch period (41,000 yr/cycle), a coherency maximum is seen at 35,000 to 45,000 yr/cycle in both cores, although in V28-238 it is not a clear peak in either the power spectra or cross-spectra comparison. Just as in the 20,000 yr period, the phase angle between the two records indicates that the oxygen isotopes lead the carbonate preservation record by about 10,000 yr (9,700 to 15,600 yr). At the 90,000 yr Milankovitch period, the phase angles for the two cores are not consistent. In the east, V19-29 shows no significant difference in the carbonate concentration and oxygen isotope record at 100,000 yr/cycle. In the west, V28-238 shows a phase difference equivalent to about 15,000 yr. With the analysis of only these two cores completed, it is difficult to tell whether this disagreement is due to the differences in the two measures of carbonate preservation used in these cores, to differences in the quality of their stratigraphies, or to real regional differences in the records.

The detailed record from the North Pacific Ocean presents a quite different picture. The cross correlation of carbonate concentration and oxygen isotopes shows no lead or lag. The record is only about 80,000 years long; therefore, 100,000 and 40,000 yr/cycle periodicities are not seen in the spectral and cross spectral analyses. As in the tropical cores, there is a maximum in the coherency at about 26,000 yr. The sense of the lag is also the same. In both regions changes in the oxygen isotopes preceded changes in carbonate preservation. However, in the North Pacific Ocean the phase difference between the two records is very slight - only 2,500 yr, compared to the 10,000 yr lag seen in the tropical Pacific Ocean. The record of surface temperatures is also different in the two regions. In the eastern equatorial core (V19-29) cross correlation of the oxygen isotope and estimated temperature records suggest that the ice maxima preceded the temperature minima by about 4,000 yrs. In the North Pacific Ocean core there is no apparent lag when the two records are correlated.

Global Controls of Carbonate Preservation

Within the Pacific Ocean basin there appears to be a distinct regional difference in the way both sea-surface temperature and carbonate preservation fluctuated relative to the global ice volume. In low latitudes, the lag between the carbonate and oxygen isotope records is relatively long - on the order of the melt-back time of the northern hemisphere continental ice sheets. One explanation for this relationship lies in the geochemical balance of calcium carbonate. With the equatorward migration of the polar fronts and the lowering of sea level (*CLIMAP*, 1976) the area of the ocean available for carbonate deposition was reduced. The result, if the river influx of carbonate did not change and the ocean carbonate system was at steady state, was an increase in the amount of carbonate deposited in the deep sea at low and mid latitudes during glacial times (*Broecker*, 1971; *Berger and Winterer*, 1974; *Shackleton and Luz*, 1975). This is evidenced in Pacific Ocean sediments by enchanced preservation and increased concentrations of calcite. Under such a scenario, the response time of the ocean to change in the carbonate ion content delayed the effect on carbonate preservation in the deep sea by several thousand years, corresponding to the lag observed in the equatorial Pacific (*Broecker*, 1971).

Another mechanisms has also been proposed for this delay by *Shackleton* (1977). He suggests that increased carbon dioxide in the air-ocean system (and, therefore, greater corrosiveness of deep-ocean) resulted from the destruction of forests at the beginning of ice ages. This increase in corrosiveness suppressed the initial increase in carbonate preservation, producing the observed lag of preservation relative to increased ice volume. Similarly, the re-establishment of forests at the end of an ice age removed carbon dioxide from the air-ocean system and prolonged the interval of good carbonate preservation. Both of these proposed mechanisms should cause nearly synchronous changes in carbonate accumulation in all ocean basins. It is difficult to demonstrate these changes, however, in areas where bulk densitites, accumulation rates, and the proportion of terrigenous debris vary greatly with the rise and fall of sea level (*Broecker*, 1971).

An alternative explanation for the carbonate lag may be found in the deep ocean circulation. Recent studies strongly suggest that during the last glacial maximum the Norwegian Sea no longer contributed to the formation of North Atlantic Deep Water (NADW) (*Duplessy*, 1975) and that the deep and bottom waters in the North Atlantic Ocean were markedly different from those of today (*Streeter*, 1973; *Schnitker*, 1973). The removal of NADW from the abyssal circulation would tend to decrease the average age of waters reaching the low latitudes of the Pacific Ocean and thus cause them to be somewhat less corrosive with respect to calcite. In this scenario, the lag effect would result from the time taken by the polar front to migrate from its ice-age position to the Norwegian Sea (about 10,000 years, *Ruddiman and McIntyre*, 1973) and the time required to establish a thick, thermal blanket of ice over the Norwegian Sea (*Newell*, 1974).

In all probability, both geochemical and hydrographic mechanisms played significant roles in altering the pattern of carbonate deposition and preservation in the Pacific Ocean; for it is difficult to explain the simultaneous response of the isotope and carbonate records in the North Pacific Ocean core compared to the 6,000 to 10,000 year lag of the response in the tropical Pacific Ocean cores, without calling on a hydrographic mechanism. We hypothesize that as the global ice volume fluctuated, Antarctic deep waters, which presently overlie the site of core Y6910-2, were replaced with younger, less corrosive waters. The source of such waters most probably was the western subarctic Pacific Ocean.

Finally, based on the rather limited data available for the tropical Pacific Ocean, the lag in sea-surface temperatures relative to ice volume suggests that oceanic and perhaps atmospheric circulation in this region responded to, rather than induced, global climate changes. In the subarctic Pacific Ocean, however, all records studied appear to vary together. The only indication that global ice volume changes preceded climatically induced oscillations in the other records is found in the cross spectral analysis. The phase relationship of these records indicate that at the frequencies of maximum coherency, changes in the ice volume slightly precede changes in both carbonate content and sea-surface temperature. Thus, although there are certainly regional differences in the pattern of climate change recorded within the Pacific Ocean basin, there is no indication that changes in the character of bottom waters or decreases in tropical and subarctic sea-surface temperatures induced changes in the global ice volume.

ACKNOWLEDGEMENTS

The intellectual stimulation provided by the many discussions with CLIMAP scientists has greatly contrbuted to this study. In particular, we would like to thank N. J. Shackleton, J. D. Hays, J. Imbrie and B. Luz. We also thank A. Molina-Cruz, L. Burckle, P. Thompson and N. J. Shackleton for making their data available to us, and NSF/IDOE for making the necessary funds available (GX-28673 awarded to Oregon State University and IDO75-20358 awarded to the University of Rhode Island). Cores used in this study are in the Lamont-Doherty Geological Observatory (supported by NSF Grant GA29460 and ONR Contract N00014-67-A-1-8-0004) and Oregon State University collections.

REFERENCES

Berger, W. H. 1973. Deep-sea carbonates: Pleistocene dissolution cycles. *Jour. Foram. Res., 3:* 187-195.

Berger, W. H. and E. L. Winterer. 1974. Plate stratigraphy and the fluctuating carbonate line. *Internat. Assoc. Sedimentol. Spec. Publ., 1:* 11-48.

Broecker, W. S. 1971. Calcite accumulation rates and glacial to interglacial changes in ocean mixing, In: *The Late Cenozoic Glacial Ages,* edited by K. K. Turekian, New Haven, Yale Univ. Press. 239-265.

CLIMAP. 1976. The surface of the ice age earth. *Science, 191:* 1131-1137.

Duplessy, J. C., L. Chenouard, F. Vila. 1975. Weyl's theory of glaciation supported by isotopic study of Norwegian Core K-11. *Science, 188:* 1208-1204.

Geitzenauer, K. R., M. B. Roche, and A. McIntyre. 1976. Modern Pacific coccolith assemblages: derivation and application to late Pleistocene paleotemperature analysis, In: *Investigations of Late Quarternary Paleo-oceanography and Paleo-climatology,* edited by R. M. Cline and J. D. Hays, *Geol. Soc. Amer. Mem. 145.* 423-448.

Heath, G. R., T. C. Moore, Jr., and J. P. Dauphin. 1976 Late Pleistocene-Holocene variations in the rate of accumulation of opal, quartz, organic carbon, and calcium carbonate in the Cascadia Basin area, northeast Pacific, In: *Investigation of Late Quaternary Paleo-oceanography and Paleo-climatology,* edited by R. M. Cline and J. D. Hays, *Geol. Soc. Amer. Mem. 145.* 393-409

Heusser, L. E., N. J. Shackleton, T. C. Moore, Jr., and W. L. Balsam. 1975. Land and marine records in the Pacific northwest during the last glacial interval. *Geol. Soc. Amer. Programs and Abstracts, 7(7):* 1113-1114.

Imbrie, J. and N. G. Kipp. 1971. A new micropaleontological method for quantitative paleoclimatology: application to a late Pleistocene Caribbean core, In: *The Late Cenozoic Glacial Ages,*

edited by K. K. Turekian, New Haven, Yale Univ. Press. 71-182.
Jenkins, G. M. and D. G. Watts. 1968. Spectral analysis and its applications, Holden-Day, San Francisco. 525 pp.
Luz, Boaz and N. J. Shackleton. 1975. $CaCO_3$ solution in the tropical Pacific during the past 130,000 years, In: *Dissolution of Deep-Sea Carbonates*, edited by W. V. Sliter, A. W. H. Bé and W. H. Berger, Cushman Foundation for Foraminiferal Research, Spec. Pub. No. 13. 142-150.
Molina-Cruz, A. 1975. Paleo-oceanography of the subtropical southeastern Pacific during the late Quaternary: a study of radiolaria, opal and quartz contents of deep-sea sediments, M. S. dissertation, Oregon State University. 179 pp.
Moore, T. C., Jr. 1973. Late Pleistocene-Holocene oceanographic changes in the northeastern Pacific. *Quat. Res., 3(1)*: 99-109.
Newell, R. E. 1974. Changes in the poleward energy flux by the atmosphere and ocean as a possible cause for ice ages. *Quat. Res., 4*: 117-127.
Ninkovich, D. and N. J. Shackleton. 1975. Distribution, stratiraphic position and age of ash layer "L" in the Panama Basin region. *Earth Planet. Sci. Letters, 27*: 20-34.
Pisias, N. G. 1976. Late Quaternary variations in sedimentation rate in the Panama Basin and the identification of orbital frequencies in carbonate and opal deposition rates, In: *Investigation of Late Quaternary Paleo-oceanography and Paleo-climatology*, edited by R. M. Cline and J. D. Hays, *Geol. Soc. Amer. Mem. 145*. 375-391.
Pisias, N. G., G. R. Heath, and T. C. Moore, Jr. 1975. Lag times for oceanic responses to climatic change. *Nature, 256*: 716-717.
Pisias, N. G., J. P. Dauphin, and C. Sancetta. 1973. Spectral analysis of Late Pleistocene-Holocene sediments. *Quat. Res., 3(1)*: 3-9.
Ruddiman, W. F., and A. McIntyre. Time-transgressive deglacial retreat of polar waters from the North Atlantic. *Quat. Res., 3(1)*: 117-130.
Schnitker, D. 1973. West Atlantic abyssal circulation during the past 120,000 years. *Nature, 248(5447)*: 385-387.
Shackleton, N. J. 1977. Carbon-13 in Uvigerina: Tropical Rainforest History and the Equatorial Pacific Carbonate Dissolution Cycles. In: *Fate of Fossil Fuel CO_2*, N. R. Andersen and A. Malahoff (Eds.), Plenum, N.Y., (this volume).
Shackleton, N. J., and N. D. Opdyke. 1976. Oxygen isotope and paleomagnetic stratigraphy of core V28-239, late Pliocene to latest Pleistocene, In: *Investigation of Late Quaternary Paleo-oceanography and Paleo-climatology*, edited by R. M. Cline and J. D. Hays, *Geol. Soc. Amer. Mem. 145*. 449-464.
Shackleton, N. J. and N. D. Opdyke. 1973. Oxygen isotope and paleomagnetic stratigraphy of equatorial Pacific core V23-238: oxygen isotope temperatures and ice volumes on a 10^5 to 10^6 year scale. *Quat. Res., 3(1)*: 39-55.

Streeter, S. S. 1973. Bottom water and benthonic foraminifera in the North Atlantic - glacial-interglacial contrasts. *Quat. Res.*, *3(1)*: 131-141.

Thompson, P. R. 1976. Planktonic foraminiferal solution and progress toward a Pleistocene equatorial transfer function. *Jour. Foram. Res.*, *6 (3)*: 208-227.

Thompson, P. R. and T. Saito. 1974. Pacific Pleistocene sediments planktonic Foraminifera dissolution cycles and geochronology. *Geology 2*: 333-335.

Global Carbon Dioxide Production from Fossil Fuels and Cement, A.D. 1950 - A.D. 2000

R. M. Rotty

Institute for Energy Analysis

ABSTRACT

Industrial production of carbon dioxide has been increasing at the exponential rate of 4.3 percent almost continuously since 1860. Based on latest revised fuel use and cement manufacture data from the United Nation's Statistical Office, CO_2 production is calculated for each year, 1950 through 1974. The 1974 value is nearly 18 x 10^9 metric tons of CO_2/year (5 x 10^9 metric tons C), about 2 percent of which is a result of cement manufacture. Three scenarios for future energy demand and cement manufacture are developed and used to extrapolate the industrial carbon dioxide production to the year A. D. 2000. The selected scenarios suggest that annual CO_2 production will grow to a value between 30 x 10^9 and 53 x 10^9 metric tons by A. D. 2000, with a most probable value equivalent to 6 ppm of the atmosphere equivalent ($\sim$46.5 x 10^9 metric tons) at that time.

INTRODUCTION

Detailed calculations of industrial production of carbon dioxide from fossil fuels and cement based on United Nations (UN) data have been made by *Keeling* (1973). *Rotty* (1973) extended Keeling's calculations to include CO_2 production for two additional years (1970-1971), and included the production from the flaring of natural gas. Recently, the UN has revised the data as a consequence of new definitions of several fuel categories. This paper presents CO_2 values for the period 1950-1974 calculated on the basis of the latest UN revisions (World Energy Supplies, UN Series J; and private communication, Arthur Ramsdell, UN Statistical Office) and suggests possible quantities for future energy use and CO_2 production.

The calculation procedure developed by *Keeling* (1973) is dependent on global production data for individual fuel classifications. One of the major changes which has been made in the UN fuel data stems from the treatment of coal production from the U.S.S.R. In earlier years the bulk of the U.S.S.R. coal (which appears to be low grade) was classified as lignite. Now, the UN reports a large portion of this as "coal" after making suitable discounts for the lower heating values. The UN coal and lignite series have now been revised back to 1950. For the calculation of CO_2 production, this change makes differences of the order of one percent. However, in order to provide ease in making certain of continuity in the future, the CO_2 production was recalculated for each year (since 1950) using the revised data.

In using earlier UN data for natural gas, while adding natural gas liquids to the petroleum production, it was likely that some double counting occurred. To eliminate this possibility, the UN has revised the natural gas data to reflect the subtraction of "extraction losses" for the production of natural gas liquids. These liquids are reported separately by the UN, and in the calcu lation procedure the mass of natural gas liquids can be added to the petroleum without concern that it was also to be counted along with natural gas.

Table 1 presents CO_2 production data in the same format as earlier work (*Rotty*, 1973) with new information for 1972, 1973, and 1974. These calculations are based on latest revisions of data available from the UN. The fuel data for 1974 production is regarded as provisional, and some small changes are possible before official publication by the UN.

The total annual CO_2 production from fossil fuels and cement has almost tripled during the third quarter of the twentieth century as indicated in Table 1. This indicates an average annual growth rate of CO_2 production of 4.3%. The rate of increase has been almost steady since 1950; smaller than normal increases are usually compensated for by larger ones in adjacent years. Perhaps most significant in these data is the relatively small increase in 1974, which may or may not be indicating a trend.

Figure 1 shows the annual CO_2 production estimates from 1860 through 1974 (values for 1860-1949 are from *Keeling*, 1973). Evident in the figure is the slowing of global industrial growth during the periods of the world wars and the great depression of the early nineteen-thirties. It is remarkable that the slope of the line fitting the points has remained nearly constant before and after the interruptions. How long can the production of carbon dioxide continue to grow at 4.3%? An attempt to answer this question will be made later.

To the casual observer the two shorter straight line segments in the 1920's and 1930's appear to be forced. However, if only the

TABLE 1. Calculation of CO_2 Emissions (all entries in million metric tons, except natural gas, lines 8, 10 in 10^9 cubic meters).

	1950	1951	1952	1953	1954	1955	1956	1957	1958	1959	1960	1961	1962
1. Coal Production (low grade coal from USSR–1975 Revision)	1435.5	1511.1	1495.8	1494.2	1475.0	1598.2	1687.6	1734.7	1803.5	1876.8	1966.2	1790.0	1833.6
2. Carbon in Coal Converted to CO_2	994.8	1047.2	1036.6	1035.5	1022.2	1107.6	1169.5	1202.1	1249.8	1300.6	1362.6	1240.9	1270.7
3. Lignite Production (USSR coal/lignite–1975 Revision) incl. USSR and Ireland peat used as fuel.	418.9	457.0	475.9	507.1	541.9	589.3	612.1	651.4	669.7	679.8	692.9	713.9	724.4
4. Carbon in lignite converted to CO_2	116.0	126.6	131.8	140.5	150.1	163.2	169.6	180.4	185.5	188.3	191.9	197.8	200.7
5. Crude oil Production	520.5	590.4	621.3	657.4	688.2	771.0	839.0	884.4	906.6	978.4	1054.6	1121.7	1216.7
6. Natural Gas Liquids	19.8	22.3	24.0	26.4	26.8	29.9	31.2	31.6	32.0	35.0	37.1	40.2	42.4
7. Carbon in Petroleum & NG Liquids Converted to CO_2	415.5	471.2	496.2	525.8	549.8	615.9	669.2	704.4	721.8	779.3	839.5	893.5	968.2
8. Natural Gas (Corrected liquid extraction)	185.1	220.5	238.5	252.2	264.0	287.2	312.2	340.2	367.1	409.8	448.4	482.4	527.1
9. Carbon in Natural Gas Converted to CO_2	97.0	115.5	125.0	132.2	138.3	150.5	163.6	178.3	192.4	214.7	235.0	252.8	276.2
10. Natural gas "flared"	43.9	46.1	50.0	51.3	51.8	58.3	61.2	67.2	66.2	69.0	75.2	79.3	84.0
11. Carbon in "flared" Natural Gas Converted to CO_2	23.7	24.9	27.0	27.7	28.0	31.5	33.0	36.3	35.7	37.3	40.6	42.8	45.4
12. Cement Production	133	150	161	179	195	217	236	247	265	294	317	334	358
13. Carbon Converted to CO_2 in Cement Production	18.2	20.6	22.1	24.5	26.7	29.7	32.3	33.8	36.3	40.3	43.4	45.8	49.1
Total Carbon in CO_2	1665.2	1806.0	1838.7	1886.2	1915.1	2098.4	2237.2	2335.3	2421.5	2560.5	2713.0	2673.2	2810.3
Total CO_2 from Fossil Fuels and Cement	6106	6623	6743	6917	7023	7695	8204	8564	8880	9389	9949	9803	10305
Fraction of Atmosphere, ppm[c]	.783	.850	.865	.888	.901	.987	1.053	1.099	1.140	1.205	1.277	1.258	1.322

	1963	1964	1965	1966	1967	1968	1969	1970	1971	1972	1973	1974[a]
1. Coal Production (low grade coal from USSR–1975 Revision)	1898.9	1966.0	2012.4	2047.5	1969.1	2014.7	2058.3	2132.6	2126.5	2153.2	2163.8	2241.6
2. Carbon in Coal Converted to CO_2	1315.9	1362.4	1394.6	1418.9	1350.7	1396.2	1426.4	1477.9	1473.7	1492.2	1499.5	1553.4
3. Lignite Production (USSR coal/lignite–1975 Revision) incl. USSR and Ireland peat used as fuel.	782.2	811.3	790.3	807.4	792.5	795.4	816.9	856.5	864.7	873.4	881.2	837.4
4. Carbon in lignite converted to CO_2	216.7	224.7	218.9	223.6	219.5	220.3	226.3	237.3	239.5	241.9	244.1	232.0
5. Crude oil Production	1305.1	1409.4	1511.4	1642.3	1761.6	1924.3	2071.1	2274.5	2401.9	2532.8	2776.9	2795.1
6. Natural Gas Liquids	47.0	50.8	53.7	57.1	62.3	66.6	70.7	75.6	78.8	84.1	86.3	88.0[b]
7. Carbon in Petroleum & NG Liquids Converted to CO_2	1039.8	1122.9	1203.6	1306.8	1402.6	1531.0	1647.0	1807.2	1907.7	2012.4	2201.8	2217.1
8. Natural Gas (Corrected liquid extraction)	575.1	629.7	674.9	728.1	784.8	856.5	939.5	1037.8	1108.4	1172.1	1230.4	1309.2
9. Carbon in Natural Gas Converted to CO_2	301.4	330.0	353.6	381.5	411.2	448.8	492.3	543.8	580.8	614.2	644.7	686.0
10. Natural gas "flared"	89.5	96.9	104.2	114.9	126.0	139.4	152.2	166.6	173.1	180.4	200.6	200.9[b]
11. Carbon in "flared" Natural Gas Converted to CO_2	48.3	52.3	56.3	62.0	68.0	75.3	82.2	90.0	93.5	97.4	108.3	108.5
12. Cement Production	377	415	434	464	480	514	542	566	604	646	689	684
13. Carbon Converted to CO_2 in Cement Production	51.7	56.9	59.5	63.6	65.8	70.4	74.3	77.5	82.8	88.5	94.4	93.7
Total Carbon in CO_2	2973.8	3149.2	3286.5	3456.4	3517.8	3742.0	3948.5	4233.7	4378.0	4546.6	4792.8	4890.7
Total CO_2 from Fossil Fuels and Cement	10905	11548	12052	12675	12900	13722	14479	15525	16054	16672	17575	17934
Fraction of Atmosphere, ppm[c]	1.399	1.482	1.547	1.627	1.655	1.761	1.858	1.992	2.060	2.140	2.255	2.301

[a] 1974 Data from U.N. is provisional.
[b] Estimated by author.
[c] Moles of CO_2 per 10^6 mole of air in the earth's atmosphere.

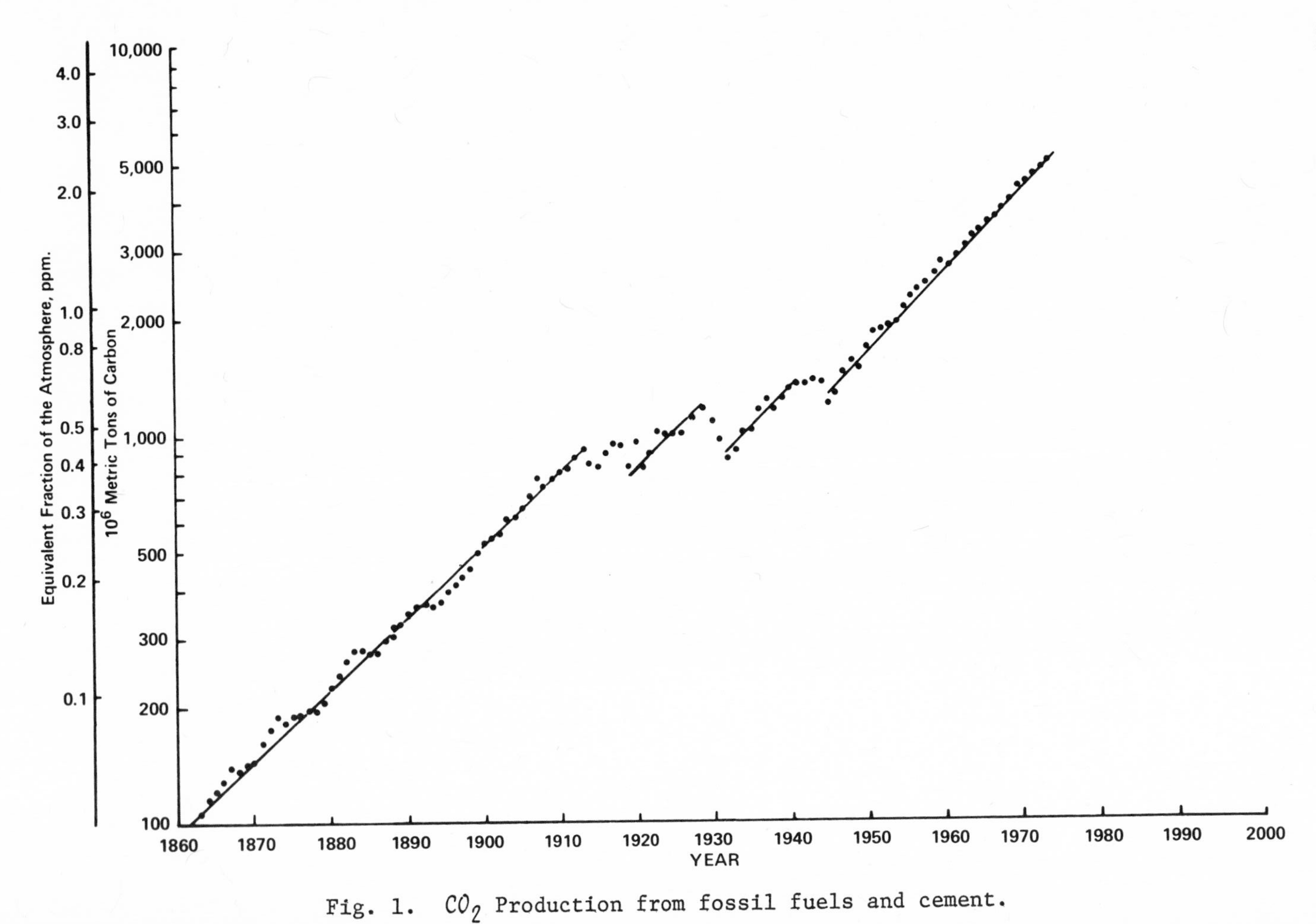

Fig. 1. CO_2 Production from fossil fuels and cement.

data for 1919 through 1929 are considered, a line similar to that shown is reasonable, and similarly for the period 1932 through 1941. Figure 1 then shows the impact on industrial activity (and CO_2 production) of major world wars and the great depression of the early 1930's.

The cumulative amount of CO_2 produced from fossil fuels and cement is presented in Figure 2. The upper set of points indicate the increase in the carbon dioxide fraction of the atmosphere which would have occurred if all the CO_2 produced since 1860 from fossil fuels and cement remained airborne. The lower set of points represents the observed increase (revised Mauna Loa data) based on an assumed value of 295 ppm in 1860. The difference between the two sets of data indicates the amount which is being taken up by the oceans (and possibly the biosphere) and placed in long-term storage.

It should be noted that although about 50% of the CO_2 produced from fossil fuels and cement seems to have found its way into reservoirs other than the atmosphere, in attempting to predict when the atmospheric concentration will reach a certain (higher) level, this fact is probably of secondary importance. The time "lag" between the "production" curve and the "observed" curve of Figure 2 is only 15 years, and is getting shorter. Predicting the year accurately in which the atmospheric fraction of CO_2 may reach, for example, 500 ppm within 15 years is most unlikely, and in any case the value <u>will</u> be reached with continued fossil energy consumption whether other natural sinks are present or not.

Any attempt to predict future global energy requirements must be predicated on a number of assumptions about factors over which we not only have no control, but also have inadequate insight. Future energy demands will depend upon at least six major factors (*Cooper and Allen*, 1975):

1) <u>People</u> - The numbers, rate of increase, and age, income, and geographic distributions.
2) <u>Economy</u> - Economic activity, rate of investment, attitude of investing public and public officials.
3) <u>Supply</u> - Heavily related to cost of energy.
4) <u>Intervention</u> - Corporate and governmental intervention, actions of cartels and even labor unions can influence energy supplies and costs and the general state of economy.
5) <u>Quality of life and aspirations</u> - Attitudes of people and willingness (or lack of it) to sacrifice in one area in order to achieve in another. For example, political pressures may be severe to improve the quality of life among those peoples of the world having a lower living standard.
6) <u>Saturation</u> - Is it possible that we are approaching the point where we are using all the energy we can for certain purposes? It is hard to conceive of energy saturation in underdeveloped parts of the world within the foreseeable

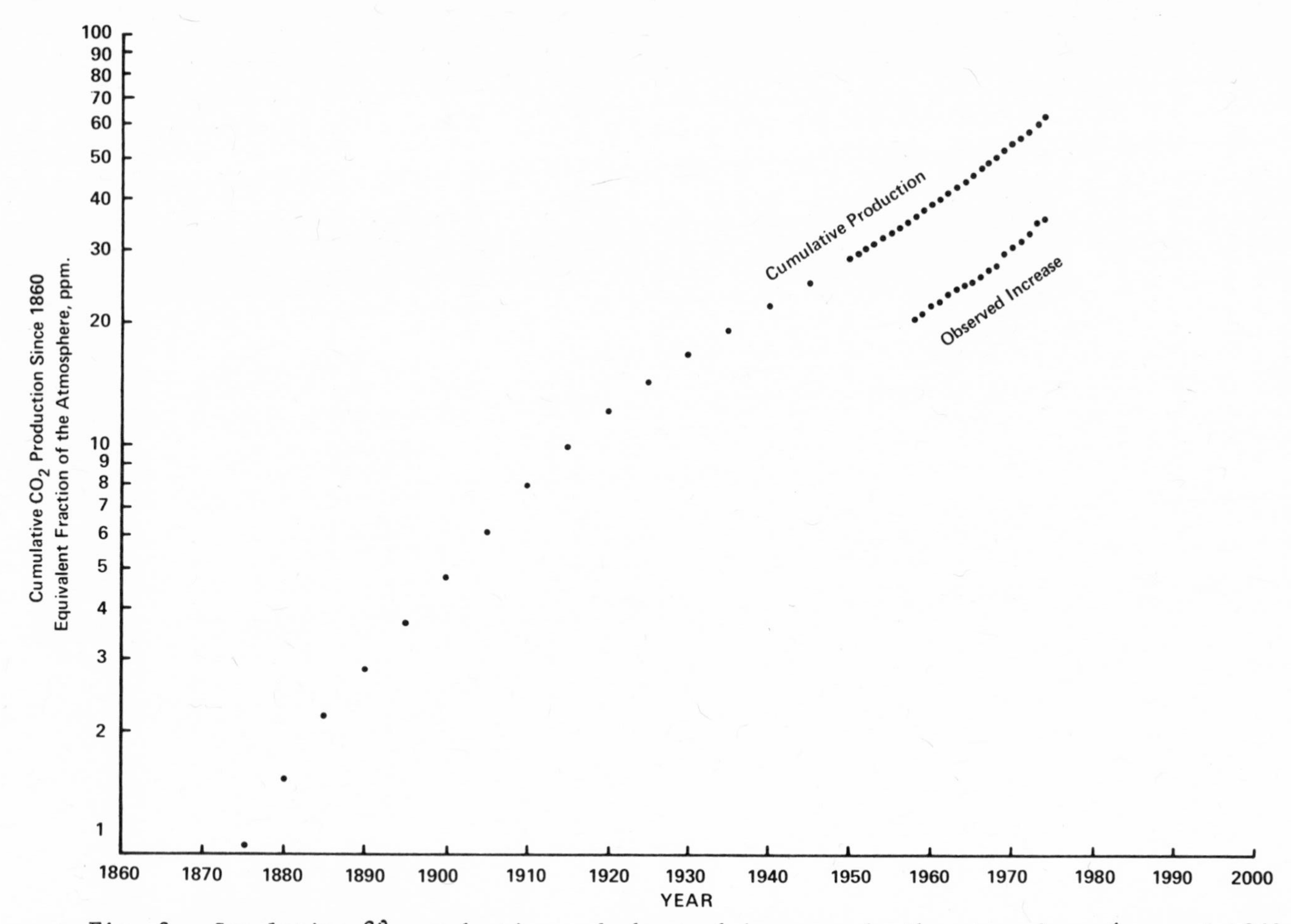

Fig. 2. Cumulative CO_2 production and observed increase in the atmosphere (assumed 1860 atmosphere fraction = 295 ppm).

future, but in the U.S. and a few other highly developed countries, it is conceivable that demands for some goods requiring energy inputs will slacken. If the period before a manufactured item becomes obsolete is lengthened, then demand is reduced.

Following are three scenarios of possible future CO_2 production, each based on a different set of assumptions. To varying degrees, factors 1, 3, and 5 are considered, but none of the three forecasts includes factors 2, 4, and 6 (except very indirectly).

Scenario No. 1 - Fuel Use Trends

In this scenario, the annual growth rate of production of each of the fossil fuels and cement was determined for the last ten years, and this growth rate was assumed to continue to A. D. 2000. This technique was not used for flared natural gas, but rather, the assumption was made that flaring will be reduced to 100×10^9 cubic meters by A. D. 2000. This procedure is outlined in Table 2.

Extrapolation of crude oil and natural gas production in this way gives estimates which exceed the probable upper limit of production (*Grove and Kristof*, 1974). Therefore, the values for oil and gas were reduced to the limits of 5500×10^6 metric tons for oil production and 2870×10^9 m^3 for gas. The energy shortfall resulting from this "arbitrary" reduction (467×10^{15}BTU) was divided equally between coal and a combination of non-fossil energy forms, (i.e., nuclear, solar, geothermal, wind, etc.). Consequently, an additional 8800×10^6 metric tons of coal must be produced and burned to make up for the oil and gas shortfall.

Scenario No. 2 - Energy Use Projections

This scenario is based on a "high energy" projection for the U.S. It is similar to the Case B projection in Economic Growth in the Future (*EEI*, 1975), and is used at the Institute for Energy Analysis in other studies. This projection provides estimates for A. D. 2000 in "quads" (10^{15}BTU) for each of the primary energy sources as indicated in the second and third columns of Table 3. For scenario No. 2, A. D. 2000 estimates of global production of coal, oil, and gas are obtained by simple scaling up in the same proportion as indicated in the projection for the U.S. This scenario places heavy reliance on electricity generated from nuclear reactors. For example, to generate 13×10^{12}kWh world-wide with oil fired boilers would require about 3500 million metric tons of oil. This suggests the serious question as to what will be the consequences if the nuclear industry is not able to expand (for either legal/political or technical reasons) to meet this need. A

TABLE 2. Annual Production, Scenario No. 1.

Fuel/Units	1974 Data	1965-1974 Growth Rate	A.D. 2000 Estimate	
			Extrapolation	Adjusted **
Coal Production/10^6mT	2241.6	1.08%	2900	11700
Lignite Production/10^6mT	837.4	0.58%	968	968
Crude Oil/10^6mT	2795.1	6.15%	13000.0*	5500
Natural Gas/10^9m^3	1309.2	6.63%	6900	2870
Natural Gas Flared/10^9m^3	200.9	NA	100	100
Cement Production/10^6mT	684.0	5.00	2385	2385

* Exceeds probable world-wide production limits.
** Adjusted to probable world-wide production limits. (See text)

TABLE 3. Construction of Scenario No. 2.

Fuel	U.S. - 10^{15}BTU		WORLD - 10^6mT or 10^9m^3		
	1975	2000	1974	2000 a	2000 b*
Coal	13.2	27.0	2241.6	4600	8900
Oil	33.8	56.0	2795.1	4600	4500
Gas	24.5	29.8	1309.2	1600	1600
Nuclear + Hydro					
Fuel Equiv.	5.5	48.1			
Elec, 10^{12}kWh(e)	0.5	4.2	(1973) 1.5	13.1	3
Lignite			837.4	1000	1000
Gas Flared			200.9	100	100
Cement			684.0	2385	2385

* Adjusted for possible limitations on nuclear power.

second version of Scenario No. 2, labeled 2b, indicates this as a possibility.

In this latter case, it is assumed that nuclear generation of electricity is limited to the amount now projected for 1980 and that this limitation is applied not only to the U.S., but also the rest of the world. Then the nuclear and hydro generation will be limited to about three trillion kWh, and the other 10 trillion kWh will likely have to be generated by burning fossil fuels if the projections of scenario 2 are to be met. Ten trillion kWh(e) requires 114 quads of energy, which, if it comes from coal, requires 4310 x 10^6 metric tons. Thus, scenario No. 2b requires a total of 8900 x 10^6mT of coal by year A.D. 2000.

Scenario No. 3 - Life Style Projections

In this scenario an attempt was made to give some consideration to the life-style and aspirations of major segments of the world's population, as well as growth in the population, and at least to some elementary issues of international politics. For this purpose the world was divided into four segments: (1) the U.S., (2) those countries having a "centrally controlled economy" (e.g., U.S.S.R., (mainland) China, North Korea, Poland, etc.), (3) other developed countries (e.g., western Europe, Japan, Australia, Canada, South Africa), and (4) the "developing" countries (e.g., Black Africa,

South America, India, Pakistan, etc.).

Individual projections as to future energy demands were made for each segment of the world as follows:

1. For the U.S. the Institute for Energy Analysis's "low energy" projection of 100 quads in A. D. 2000 was used. Based on a projected population of 250 million by that time, this will provide a per capita energy rate of 400 million BTU/capita. (Present - 1975 - comparable value is 345 million BTU/capita). Therefore, this scenario will result in a 0. % per capita growth rate in the U.S.

2. For the centrally controlled economy countries, a continued increase in energy use at 5.1% was assumed (same as in the 1960's and early 1970's). Since these countries appear to have adequate reserves to sustain such growth until well past A. D. 2000, it appears that only major upheaval of world order can substantially change this growth rate. Table 4 summarizes projections for this segment.

TABLE 4. Forecast Scenario No. 3 for "Centrally Controlled Economy Countries"

	1970	2000	Growth Rate %
Population	1.16×10^9	1.84×10^9	1.5
Energy Use, quads	53.9	249	5.1
Per Capita Energy, 10^6BTU/cap.	44.9	138	3.7

3. For other developed countries an average per capita growth rate of 2% per year for the period 1970-2000 was assumed. This value is admittedly arbitrary: it is about half the value of the growth rate of the 1960's, i.e., 4.5%. It seems more appropriate to base the projection of energy use for this segment on per capita growth rates rather than on total energy growth in order to account for life-style changes. Table 5 indicates the values used in making projections for this segment.

4. For the developing countries of the world, it was assumed that by A. D. 2000 the per capita energy use in this segment would increase to the 1970 world-wide average. This requires an 8.6% growth in energy use per year compared with a 6.7% growth rate during the 1960's. Higher growth rates are easier to attain when the total magnitude is smaller (as in the case of energy use in developing countries). Table 6 summarizes the projection for the developing countries.

TABLE 5. Forecast Scenario No. 3 for "Other Developed Countries"

	1970	2000	Growth Rate %
Population	0.52×10^9	0.70×10^9	1.0
Energy Use, Quads	55.2	135	3.0
Per capita energy, 10^6BTU/cap.	106	193	2.0

TABLE 6. Forecast Scenario No. 3 for "Developing Countries"

	1970	2000	Growth Rate %
Population	1.72×10^9	3.88×10^9	2.7
Energy Use, Quads	15.5	206	8.6
Per capita Energy, 10^6BTU/cap.	9	53	5.9

The sum of the A. D. 2000 energy demands of the four segments gives the requirement of 690×10^{15}BTU/yr. - an average global growth of 4.28%/yr., compared with a 5.4%/yr. growth during the 1960's. If the consumption of each fuel grows at 4.28%/yr., then the demand for both oil and gas will exceed the probable production capacity of 5500×10^6mT of oil and $2870 \times 10^9 m^3$ of gas. Making the same assumption here as in forecast scenario No. 1, namely that one-half of the energy shortfall from oil and gas is made up largely by nuclear power, with some geothermal, solar, wind, etc., and the other half is made up by increasing the use of coal. Thus, an additional 84.7 quads of energy must come from coal, requiring 3200 million metric tons (largely in less developed countries).

Nuclear power (if it also grows at 4.28%/yr.) will be producing 4.4 trillion kWh(e) to satisfy the growth rate and an additional 7.4 trillion kWh(e) to pick up half the shortfall in oil and gas. (Thus, the overall nuclear power requirement will be 11.8 kWh(e) - a 7.6%/yr. growth rate).

The carbon dioxide production consequences from each of the forecast scenarios is computed by the same techniques as used with the historical and current data. Table 7 shows these results, and Figure 3 shows the resulting A. D. 2000 CO_2 production estimates

TABLE 7. Calculation of Estimated CO_2 Emissions in the Year A.D. 2000 (All Entries in Million Metric Tons, Except Natural Gas, Lines 8, 10 in 10^9 Cubic Meters)

	Scenario No. 1	Scenario No. 2a	Scenario No. 2b	Scenario No. 3
1. Coal Production (low grade coal from USSR—1975 Revision)	11700	4600	8900	10000
2. Carbon in Coal Converted to CO_2	8108	3188	6168	6930
3. Lignite Production (USSR coal/lignite—1975 Revision) incl. USSR and Ireland peat used as fuel.	968	1000	1000	1000
4. Carbon in lignite Converted to CO_2	268	277	277	277
5. Crude Oil Production	5500	4600	4600	5500
6. Natural Gas Liquids	a	a	a	a
7. Carbon in Petroleum and NG Liquids Converted to CO_2	4230	3537	3537	4230
8. Natural Gas (Corrected liquid Extraction)	2870	1600	1600	2870
9. Carbon in Natural Gas Converted to CO_2	1504	838	838	1504
10. Natural gas "flared"	100	100	100	100
11. Carbon in "flared" Natural Gas Converted to CO_2	54	54	54	54
12. Cement Production	2385	2385	2385	2385
13. Carbon Converted to CO_2 in Cement Production	327	327	327	327
Total Carbon in CO_2	14491	8221	11201	13322
Total CO_2 from Fossil Fuels and Cement	53134	30144	41070	48847
Fraction of Atmosphere, ppm	6.819	3.868	5.270	6.268

[a]Included in amount estimated for crude oil production.

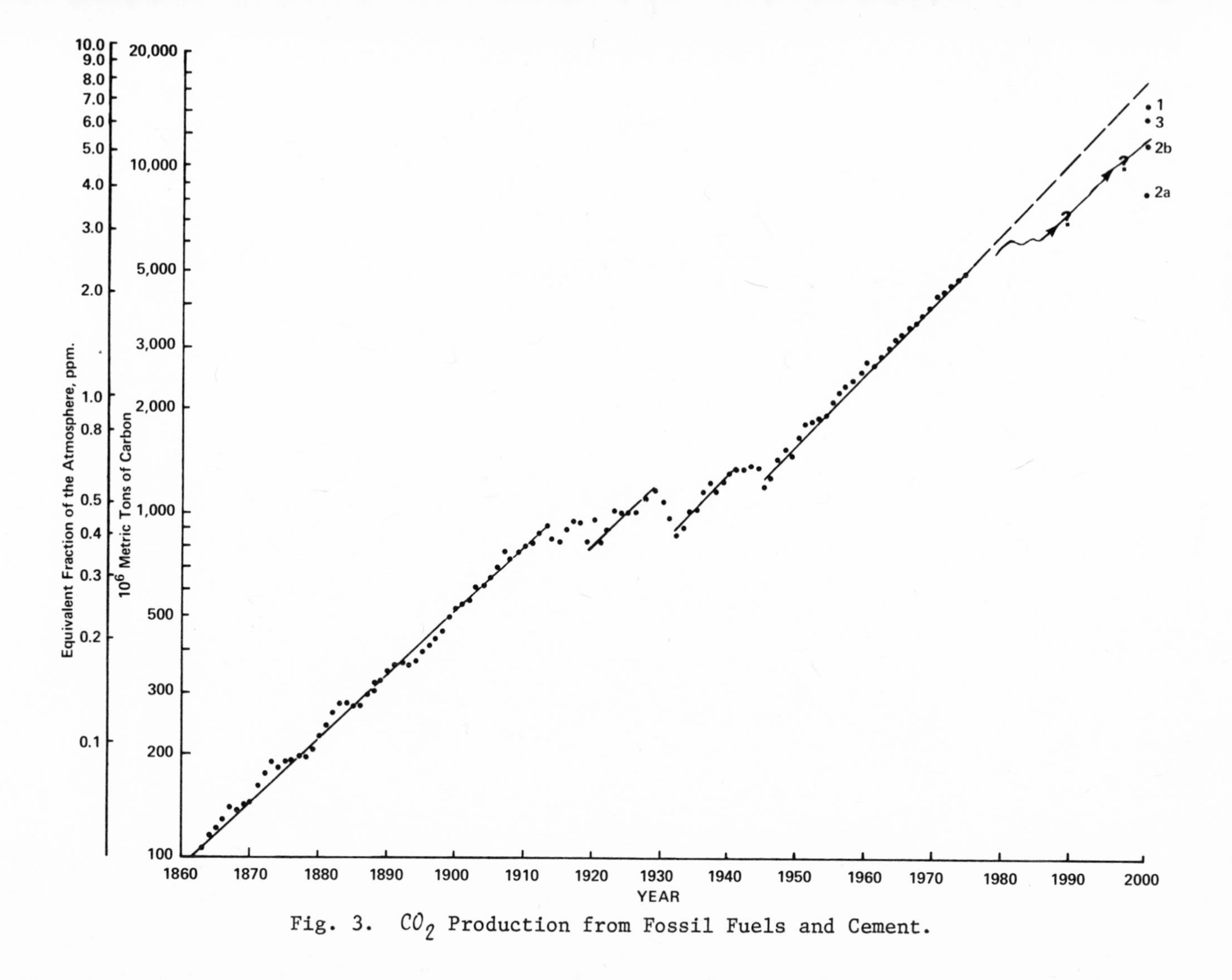

Fig. 3. CO_2 Production from Fossil Fuels and Cement.

plotted in reference to the production of the past 115 years.

Although scenario No. 2 was constructed on the basis of a "high energy" projection for the U.S., the methods used to estimate global requirements give growth at a lower rate than the other scenarios. The U.S. can achieve this "high energy" projection with a growth rate much lower than is likely in other parts of the world.

CONCLUSION

The forecast scenarios developed here are simply "first cut" approximations to give an indication of what the carbon dioxide production might be in the year A. D. 2000. From these calculations it does appear safe to assume that the CO_2 production in A. D. 2000 will be below that indicated by continued exponential extrapolation at the 4.3%/yr. rate. The range of predicted values, i.e., 4 - 7 ppm/yr., will probably include most realistic predictions. The question of the continued expansion of nuclear energy is important in this connection: the two versions of scenario No. 2 address the issue with a resulting 1.4 ppm/yr. difference. For modeling and for other studies dealing with consequences of fossil fuel CO_2 production, the value 6 ppm (12,750 x 10^6 metric tons carbon) is recommended for A. D. 2000 until better indications of future energy demands and sources are available.

Based on a 6 ppm equivalent production for the year A. D. 2000, the cumulative production from 1975 to 2000 will be just under 100 ppm, compared with the 63.5 ppm for the entire period 1860 through 1974. If half of the 100 ppm remains airborne then in the year A. D. 2000, the atmosphere will contain CO_2 in the concentration of 380 ppm.

REFERENCES

Cooper, C. L. and E. Allen. 1975. Internal memorandum, Institute for Energy Analysis, December 2, 1975.

Edison Electric Institute. 1975. Economic growth in the future, *Report of the EEI Committee on Economic Growth, Pricing and Energy Use. EEI Publication No. 75-32*, New York, New York, June 1975.

Grove, N. and E. Kristof. 1974. Oil: The dwindling treasure. *National Geographic, 145, No. 6*, June 1974.

Keeling, C. D. 1973. Industrial production of carbon dioxide from fossil fuels and limestone. *Tellus, 25, No. 2.*

Ramsdell, A. 1975. Private communications, December 1975, (Statistical Office, U.N. Department of Economic and Social Affairs, New York).

Rotty, R. M. 1973. Commentary on and extension of calculative procedure for CO_2 production. *Tellus, 25, No. 5.*

United Nations. 1975. World energy supplies, 1970-1973, Statistical Papers, Series J. No. 18, published by Statistical Office, U. N. Department of Economic and Social Affairs, New York.

EFFECTS OF ENHANCED CO_2 LEVELS ON GROWTH CHARACTERISTICS OF TWO MARINE PHYTOPLANKTON SPECIES

L. F. Small, P. L. Donaghay, and R. M. Pytkowicz

Oregon State University

ABSTRACT

The chained marine diatom Skeletonema costatum and the unicellular marine chrysophyte Isochrysis galbana were batch-grown under 0.03%, 0.15%, and 0.30% CO_2 bubbled through the growth medium. The respective pH values stabilized at 8.2, 7.6, and 7.2 and the total alkalinity remained constant. The nitrate concentration was set to ultimately limit population yields, while the other nutrients were present in overabundance. The temperature and light were within appropriate natural ranges for the algal species. Both species showed some effect of enhanced CO_2 concentrations after nitrate exhaustion from the water, if the species were not first acclimated to the new CO_2 levels (cell clumping and stickiness were noted during the stationary growth phase). Acclimation to enhanced CO_2 levels could be accomplished if CO_2 levels were increased over one cell division, however. Acclimated cells showed no effect from bubbling increased CO_2 concentrations through the water (no changes were noted in growth rates, final population yields, intracellular fractionation of nitrogen, and the particulate fractions carbon/volume (C/V), nitrogen/volume (N/V), carbon/nitrogen (C/N), and fluorescence/volume (F/V)). In our simple systems, excess CO_2 in the water was not elaborated as algal tissue.

Unbubbled control populations varied considerably from populations exposed to bubbled CO_2 treatments. Whenever the algal biomass exceeded about 10μg-at N/liter, pH values rose to 8.3 or greater, and there were often substantial changes in the particulate ratios C/V, N/V, C/N, and F/V. The pH at stationary phase (nitrate exhaustion) in unbubbled populations always exceeded 8.3, whereas pH was maintained well below 8.3 in the bubbled populations.

Although no effects of enhanced CO_2 levels were found in our nitrate-limited systems, it must be remembered that these are simple batch systems with only two phytoplankton species responding to acute CO_2 inputs. No effects of chronic CO_2 enhancement on species diversity or on physiological change were examined, and we can only speculate on the possible effects of rapid recycling of limiting nitrogenous nutrients (through zooplankton excretion) in the face of CO_2 enrichment.

One feature which emerges clearly from the results is that different algal species grown under identical conditions have entirely different F/V, C/V, N/V, and C/N ratios, and that these ratios change in a species-specific manner with changing environmental conditions. This finding has important implications in field estimates of algal biomass.

INTRODUCTION

There have been many studies on the effects of carbon dioxide on plant growth in terrestrial and freshwater systems, but relatively few in marine systems. In non-marine environments under conditions of non-limiting light, nutrients, and moisture (such as might be found in fertilized croplands, sewage lagoons, and certain freshwater lakes draining rich agricultural lands), CO_2 can limit both the rate of growth and the final plant biomass produced (see for example *Burlew*, 1961; *Eckardt*, 1968; *Likens*, 1972; and *Woodwell and Pecan*, 1973). Under very low CO_2 levels at high *pH* in some freshwater environments, algal growth rates might even become limited by the diffusional transport of CO_2 across cell membranes, rather than by intracellular photosynthetic reduction of CO_2 to organic compounds (*Gavis and Ferguson*, 1975). There has been speculation that phytoplankton species succession can be altered by changing CO_2 concentrations in fresh waters (*King*, 1970; *Shapiro*, 1973), although this idea is not universally accepted (*Goldman*, 1973). *Botkin et al.* (1973) predicted by means of a computer model that CO_2 enrichment in a forest ecosystem should favor shade-tolerant, long-lived, species and should hasten the process of succession.

In marine waters, CO_2 limitation of plant growth is rarely considered because of the enormous alkalinity of seawater. However, there is a report of CO_2 limitation of growth in *Dunaliella salina*, an algal species tolerant of high light and high salinity (*Loeblich*, 1970). Also, in some upwelling regions and other near-coast areas pCO_2 levels in the water can be half of those in the open oceans, although low pCO_2 does not necessarily imply growth limitation (*Gordon*, 1973). Instances of CO_2 limitation of marine primary production are minor, however, relative to the temporal and spatial limitation exerted by light and nutrient salts. Thus, the recent interest in CO_2 with respect to marine biota does not come as a result of the study of limiting CO_2 levels, but as a result of the possible effects upon the biota of increasing inputs of CO_2 into the atmosphere and hydrosphere. With the atmospheric CO_2 currently increasing at a rate of about 0.5% per year, and with the upper waters of the ocean acting as one of the largest short-term CO_2 sinks, questions arise as to 1) whether the oceanic primary producers can absorb some, most, or all of the CO_2 excess, given their general state of light-nutrient limitation, 2) whether enhanced photosynthetic fixation, if it occurs, automatically leads to higher plant biomass (and thus larger organic carbon pools), and 3) whether there is alteration of the species composition or of the chemical composition of the populations.

The first two questions above have been addressed by geochemists (*Broecker et al.*, 1971; *Bacastow and Keeling*, 1973) and by biologists (*Whittaker and Likens*, 1973; *Reiners*, 1973). The views are sometimes in conflict although such disagreement arises mostly from a lack of reliable data in the proper time and space scales rather than from a conflict in data interpretation. Answers to the first two questions are likely to be strongly dependent on answers to the third one, but this fact has not been generally recognized in the literature. Implicit in the third question are the potentials for alteration of whole food webs, for short-term acclimation, and for long-term irreversible genetic adaptation of biotic populations to changing CO_2 levels.

In this paper all of the questions raised on the biological effects of enhanced CO_2 levels cannot possibly be answered. But, the data base can be increased from which more realistic models can be derived. The growth rates and the elemental composition of two representative types of marine phytoplankton are particularly examined, under normal and enhanced CO_2 levels, and under realistic light, nutrient and temperature conditions. A major failing of all experimental work, including this one, is that such work cannot address the long-term chronic effects of CO_2 enhancement. Acute biotic responses and short-term acclimations must be considered. Nevertheless, such work is a start, and it is felt that some new and useful insights have been gained.

METHODS

The phytoplankton species which were used were the small, flagellated chrysophyte *Isochrysis galbana* and the ubiquitous chaining diatom *Skeletonema costatum*. It was felt that *Isochrysis* would reasonably represent small marine flagellated cells while *Skeletonema* would represent marine diatoms.

The experimental approach employed was to perform batch experiments under approximately natural conditions, but with CO_2 levels enhanced over atmospheric levels. The water temperature was maintained at $18.0 \pm 0.2^{\circ}C$ in an environmental growth chamber. This temperature is representative of surface waters of stratified temperate seas during summer. Light was made available through the sides of experimental flasks by a bank of cool-white fluorescent tubes delivering 380-400 microeinsteins/m^2/sec at the center of each empty flask. This light intensity was above light saturation of photosynthesis for the species used during the entire course of each experiment (i.e., light was never limiting). No photoperiod was employed in these initial studies. The basal growth medium for the phytoplankton was deep-ocean water with the dissolved organic matter removed by a two-week treatment with activated charcoal. The charcoal subsequently was removed by sterile filtration. Autoclaving was not done, so that alkalinity would remain as unaffected as possible. Nitrate was added to the basal medium to give the following initial concentrations of nitrogen: 46 µg-at/liter for the *Skeletonema* experiments, and 23 µg-at/liter for the *Isochrysis* experiments. Non-limiting concentrations of phosphate, silicate, iron (as Fe-EDTA), divalent sulfur (as thiourea), vitamins (thiamin, biotin, and B_{12}), and trace metals (Cu, Zn, Co, Mn, and Mo) were also added. No buffering agents were added (including bicarbonate) in order to retain the carbonate system as near normal as possible. Nutrient concentrations were calculated so that nitrate would eventually limit phytoplankton growth in all flasks, since nitrogen is generally considered to be the limiting nutrient in the ocean. The 46 µg-at NO_3-N/liter that was employed as an initial concentration for *Skeletonema* is approximately equivalent to natural deep-ocean concentrations, and is somewhat greater than surface concentrations in major coastal upwelling regions. The 23 µg-at/liter used for *Isochrysis* approximates nearshore surface concentrations in spring and during minor upwelling events.

Six flasks were used for each experimental set with each phytoplankton species. Each flask, with two liters of growth medium, was closed to the atmosphere except for a CO_2 input line and exit line with stopcock. Two of the six flasks received atmospheric concentrations of CO_2 (1X = 0.03%) by bubbling pre-filtered laboratory air through sterilized, medium-porosity frits. Two flasks received five times atmospheric CO_2 level (5X = 0.15%) from specially prepared CO_2 cylinders, with bubbling as in the 1X flasks. The final two flasks received ten times atmospheric concentration (10X = 0.30%) from cylinders, with bubbling.

Experiments were run both with and without an acclimation period for *Isochrysis* and *Skeletonema*. In the first set of experiments (in which cells were not acclimated to experimental CO_2 levels), log-growth cells were innoculated into flasks already CO_2-equilibrated at 1X, 5X and 10X levels. Carbon dioxide equilibration was achieved when *pH* values (measured at hourly intervals) stabilized (1X = *pH* 8.2, 5X = *pH* 7.6, and 10X = *pH* 7.2). This procedure provided a precise starting point for the experiments, but also resulted in the cells being subjected to an instantaneous change from atmospheric CO_2 levels (*pH* 8.2) to the experimental CO_2 levels. It was thought that immediate addition of non-acclimated cells to water with increased CO_2 levels and bubbling might place some stress on the initial cell innoculum. Thus, in a second set of experiments, the cells were acclimated to experimental CO_2 levels before actual CO_2 effects studies were begun. Cells were acclimated by slowly increasing the CO_2 bubbling into the flasks until *pH* stability was attained (1X = *pH* 8.2, 5X = *pH* 7.7, and 10X = *pH* 7.2). Total alkalinities were 2.50 $\pm$ 0.05 meq/liter for *Skeletonema* and 2.41 $\pm$ 0.05 meq/liter for *Isochrysis* throughout the time course of the studies. The length of the acclimation period was about the time required for one cell division. Then, to begin the effects studies, 20 ml of acclimated cell suspension was transferred to flasks with the same CO_2 bubbling characteristics, but with initial nutrient concentrations reestablished. Non-bubbled controls were also run. Cell fluorescence, particle numbers, particle volumes, and *pH* were measured daily during the acclimation period. With *Isochrysis*, which is always unicellular, particle numbers and volumes are exactly equivalent to cell numbers and volumes. With *Skeletonema*, the particle measurements can represent unicells and/or chains with varying numbers of cells.

A 100 ml water sample was removed at the beginning of both the *Skeletonema* and *Isochrysis* effects studies, for determination of initial nutrients, *pH* and alkalinity. Samples were drawn by closing the exit stopcock and blowing out the required sample volume. During both effects studies, the following measurements were made in each flask on a daily or twice-daily basis: *pH*; *in vivo* cell fluorescence by Aminco fluorometer; particle number, particle volume, and particle size distribution by Coulter Counter model ZBI with P-64 channelizer. All fluorometry data were normalized with a coproporphyrin standard. For the acclimated cells, daily determinations also were made of particulate carbon and nitrogen (*C* and *N*, by Carlo Erba CHN analyzer), total alkalinity, and inorganic nutrients in the water. For the non-acclimated cells, *C* and *N* measurements and alkalinities were measured only during a selected portion of the exponential growth phase, and at the end of each study.

The accounting through time of particle fluorescence, particle number, particle volume, and *C* and *N* allowed a comparison of population growth rates on the bases of these different population attributes. Changes in particle size distribution and *C/N* ratios could also be examined under the different CO_2 treatments. Particle size distributions and *C/N* ratios can be sensitive indicators of cell physiology (*Donaghay*, 1975; *Conover*, 1975). To gain further insight into physiological changes wrought by enhanced CO_2 levels under generally nitrogen-limited growth, intracellular inorganic nitrogen pools in *Isochrysis* and *Skeletonema* were measured during critical growth phases under all CO_2 treatments. *Donaghay and Small* (1975) have reported details of a total *C* and *N* budget method, of which intracellular nutrient pool measurement is a part. Intracellular inorganic nitrogen pools were determined as follows. From each flask, replicate samples were taken. A portion of each sample was gently filtered through Whatman GFF fiber filters, and the filtrates were used to determine NO_3^-, NO_2^-, and NH_3. *Donaghay and Small* (1975) determined that this filtration scheme did not break cells and gave small filter-blank values. The remaining portion from each replicate was sonicated with an Ultrasonics Sonifer to release intracellular inorganic nutrient pools. After sonication and subsequent microscopic examination to verify complete cell disruption, the samples were filtered through Whatman GFA glass fiber filters to remove the cellular debris. Debris must be removed to insure precise nutrient measurements by the Autoanalyzer. The lack of debris in the filtrate was verified with the Coulter Counter. Measurements of the inorganic nitrogen nutrients were made on the filtrates after sonication, and the differences between these values and the values determined on the non-sonicated portions of the samples were considered to be the inorganic nitrogen pools released from the cells. The inorganic nutrient techniques are specific for dissolved inorganic forms, and do not include organic fractions also released during sonication. Recent work suggests that intracellular nutrient pool sizes are related to cellular physiology under different nutritional regimes (*Donaghay*, 1975; *Lundy*, 1974; *Conover*, 1975). Complete C and N budgets have not been run in the CO_2-effects studies because of the volume of sample required and because of some uncertainties with critical measurements of dissolved organic nitrogen (DON), dissolved organic carbon (DOC), total nitrogen (TN), and total carbon (TC) in CO_2-enhanced systems. These techniques are currently being developed further. Particle fluorescence, particle number, particle volume, particle size distribution, particulate C and N, pH, and alkalinity were determined each time intracellular nitrogen pools were measured. In the *Isochrysis* and *Skeletonema* studies, two complete data sets were collected, one during exponential growth and one the day after the fluorescence growth curve had levelled off. At the latter time the cells were still healthy and suspended in the growth medium.

RESULTS

Growth Patterns

Growth curves in terms of particle volume and fluorescence for the acclimated *Isochrysis* and *Skeletonema* cultures are shown in Figure 1. For both species, particle volume and fluorescence increase log-linerly until nitrogen is depleted. After nitrogen depletion, the fluorescence begins to decrease and the volume continues to increase at a decelerating rate until a maximum volume-yield is achieved. Similar curves can be drawn on the basis of particulate carbon and nitrogen, as well as for the non-acclimated phytoplankton. The carbon-curves follow the particle volume curves

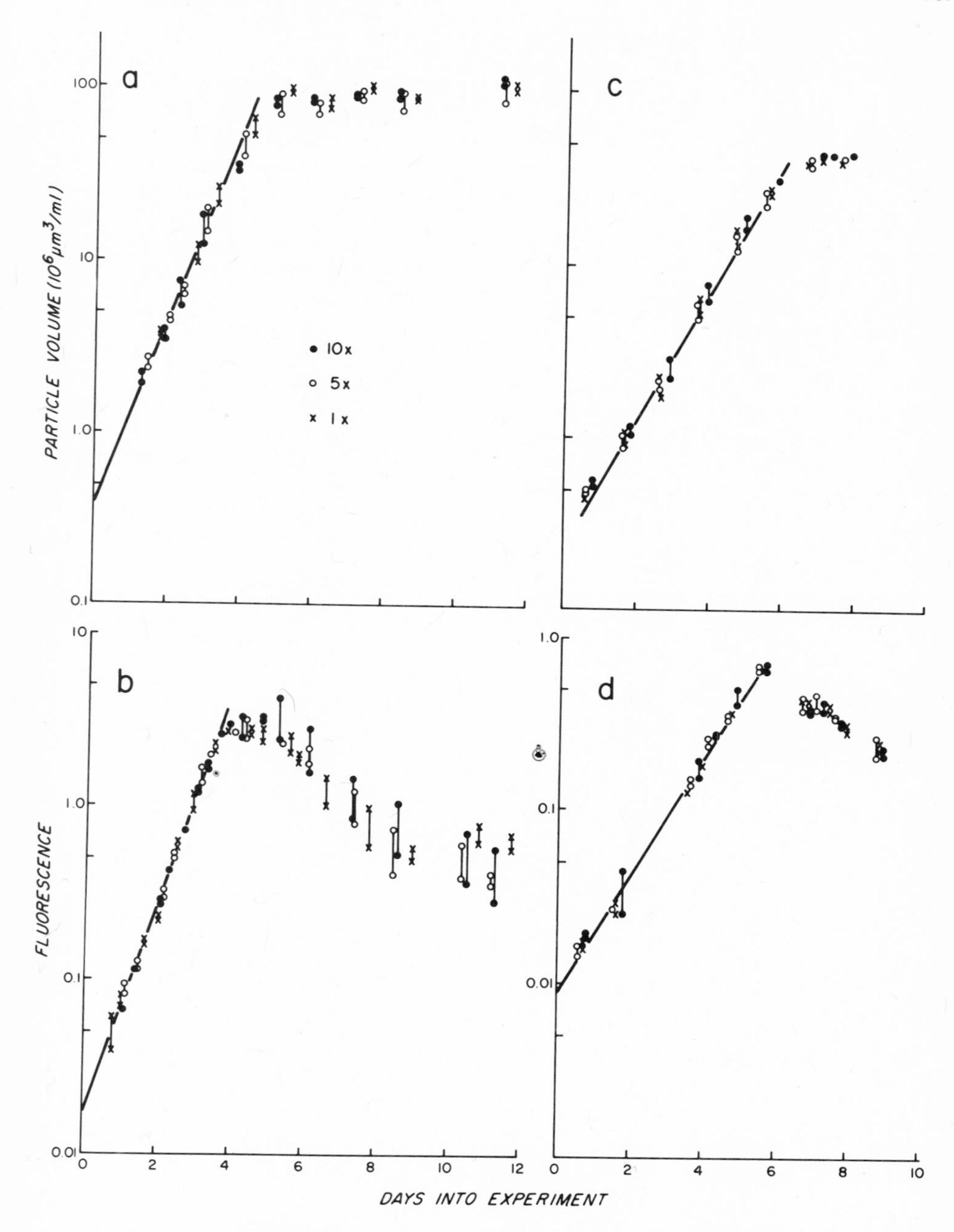

Fig. 1. Volume-based and fluorescence-based growth curves for *Skeletonema* (a,b) and *Isochrysis* (c,d), acclimated to either 0.03% (1X), 0.15% (5X), or 0.30% (10X) CO_2 concentration.

very precisely while the nitrogen-curves tend to bend over at the time of the maximum fluorescence. Growth rates were calculated from the slopes of the log-linear portions of the volume and fluorescence curves, and are summarized in Table 1 (acclimated cells) and Table 2 (non-acclimated cells). For CO_2-acclimated populations of both species, no effect of CO_2 enhancement is observed on any given set of growth rate calculations. However, volume-based growth rates are slightly higher than fluorescence-based rates at all CO_2 levels in *Isochrysis*, and slightly lower in *Skeletonema*. In the acclimated cultures, there appeared to be an effect of bubbling on *Isochrysis* fluorescence-based growth rates and on *Skeletonema* volume-based rates.

TABLE 1. Experiments with Acclimated Cells. Calculated mean exponential growth rates ($\underline{k}$ values, as day^{-1}) and final population yields (Y values) for *Isochrysis* and *Skeletonema* under 0.03% (1X), 0.15% (5X), and 0.30% (10X) CO_2 concentrations, and an unbubbled 1X control (UC). Growth rates and final yields are expressed in terms of particle volume ($\underline{k}_V$ and Y_V) and particle fluorescence ($\underline{k}_F$ and Y_F). The $\pm$ values are the ranges in replicate flasks.

			VOLUME		FLUORESCENCE	
Species	Level	*pH*	$\underline{k}_V$	Y_V	$\underline{k}_F$	Y_F
Isochrysis	1X	8.11	0.85±0.03	40.85±0.84	0.79±0.02	0.60±0.02
	5X	7.49	0.82±0.02	42.22±0.74	0.79±0.01	0.69±0.02
	10X	7.20	0.87±0.02	44.10±1.20	0.80±0.02	0.67±0.01
	UC	*	0.85±0.01	40.93±0.20	0.72±0.01	0.51±0.02
Skeletonema	1X	8.11	1.22±0.02	86.19±4.68	1.37±0.05	3.81±0.55
	5X	7.48	1.17±0.03	85.58±12.46	1.36±0.01	3.04±0.14
	10X	7.22	1.23±0.01	93.08±12.39	1.39±0.03	2.75±0.02
	UC	*	1.43±0.08	71.90±8.34	1.33±0.07	2.51±0.13

* *pH* for UC were unstable. Initially they were the same as 1X, but rose during the experiment.

TABLE 2. Experiments with Non-Acclimated Cells. Calculated mean exponential growth rates (day^{-1}) for *Isochrysis* and *Skeletonema* on the bases of particle volume (k_v) and fluorescence (k_F), under 0.03% (1X), 0.15% (5X), and 0.30% (10X) CO_2 concentrations. The ± values are the ranges in replicate flasks.

Species	CO_2 Level	pH	k_v	k_F
Isochrysis	1X	8.17	0.76±0.02	0.69±0.04
	5X	7.62	0.88±0.02	0.77±0.02
	10X	7.35	0.87±≈0	0.74±0.02
Skeletonema	1X	8.14	0.60±0.04	0.70±0.13
	5X	7.50	0.65±0.04	0.76±0.03
	10X	7.24	0.70±≈0	0.93±0.05

In contrast to the acclimated cells, non-acclimated cultures showed considerable initial growth lags (not shown) and considerable CO_2-enhancement of certain growth rates (see Table 2). For *Isochrysis*, fluorescence-based (and carbon-based) growth rates were the same under all treatments, but volume-based (and nitrogen-based) rates were greatly increased in the 5X and 10X flasks. For *Skeletonema*, fluorescence-based growth rates were larger in enhanced CO_2 levels (5X, 10X) than in air (1X). The fluorescence-based rates were greater than the volume-based rates in *Skeletonema* at all CO_2 levels, a result identical to that obtained with CO_2-acclimated *Skeletonema*. Volume-based growth rates for non-acclimated *Skeletonema* were enhanced only at the 10X CO_2 level.

For both species in the acclimated cultures there was a trend toward increasing volume yields with increasing CO_2 level (see Table 1). The trend reversed for *Skeletonema* fluorescence yields, and no trend was apparent in the *Isochrysis* fluorescence yields. It should be noted that *Isochrysis* fluorescence declined more rapidly after nitrogen depletion than *Skeletonema* fluorescence (see Figure 1). As a result, the *Isochrysis* fluorescence yields in Table 1 are based on only one maximum value and, thus, are more uncertain than the *Skeletonema* fluorescence yields.

Fluorescence yields for unbubbled controls (UC) were consistently lower than fluorescence yields for the CO_2 treatments (Table 1). *Skeletonema* volume yields in unbubbled controls also tended to be low, but *Isochrysis* volume yields in unbubbled controls could not be differentiated from those in 1X. No accurate yield estimates could be made in the non-acclimated cultures because cells began to clump and stick to the vessel walls in the latter part of the exponential growth phase.

Patterns of Particle Size

Particle volume distributions for the acclimated *Skeletonema* and *Isochrysis* cultures are shown in Figure 2. During the exponential growth phase, distributions for *Skeletonema* at 1X, 5X, and 10X were stable and were relatively unaffected by CO_2 treatment (Figure 2a). The 5X peak appears to be lower than the 1X and 10X peaks, but counting variability of *Skeletonema* chains can account for the apparent difference. In the stationary phase, the peaks of the 1X, 5X, and 10X distributions shifted to slightly smaller volumes (Figure 2b). The unbubbled control distribution was not stable, first shifting to smaller particles in early exponential phase (Figure 2a), then back to larger particles in late exponential phase. By the time stationary phase is reached (Figure 2b), particles in the unbubbled flasks average somewhat larger than those in the bubbled flasks. It must be remembered that the distributions for *Skeletonema* reflect changes in both cell size and chain length. Thus, no attempt should be made to interpret the *Skeletonema* data in terms of cell size changes, or to compare *Skeletonema* distributions with cell size distributions of *Isochrysis*.

For acclimated *Isochrysis*, a different set of distribution patterns was observed (Figures 2c,d). Distributions during the exponential phase were extremely precise and stable, so that the enhanced mean cell sizes in the 5X and 10X treatments were real (Figure 2c). Distributions of unbubbled control populations were not significantly different from 1X populations except very early in exponential phase on occasion. In the stationary phase there was a slight shift toward smaller cells in the 5X and 10X treatments, but no significant shifts in the 1X treatment and unbubbled control (Figure 2d).

Cell Physiological Patterns

Ratios of intracellular constituents often are indicative of the physiological state of phytoplankton populations under given environmental conditions. In these studies, selected ratios at natural and enhanced CO_2 levels can be examined during exponential growth (where nitrogen is not limiting) and during stationary growth

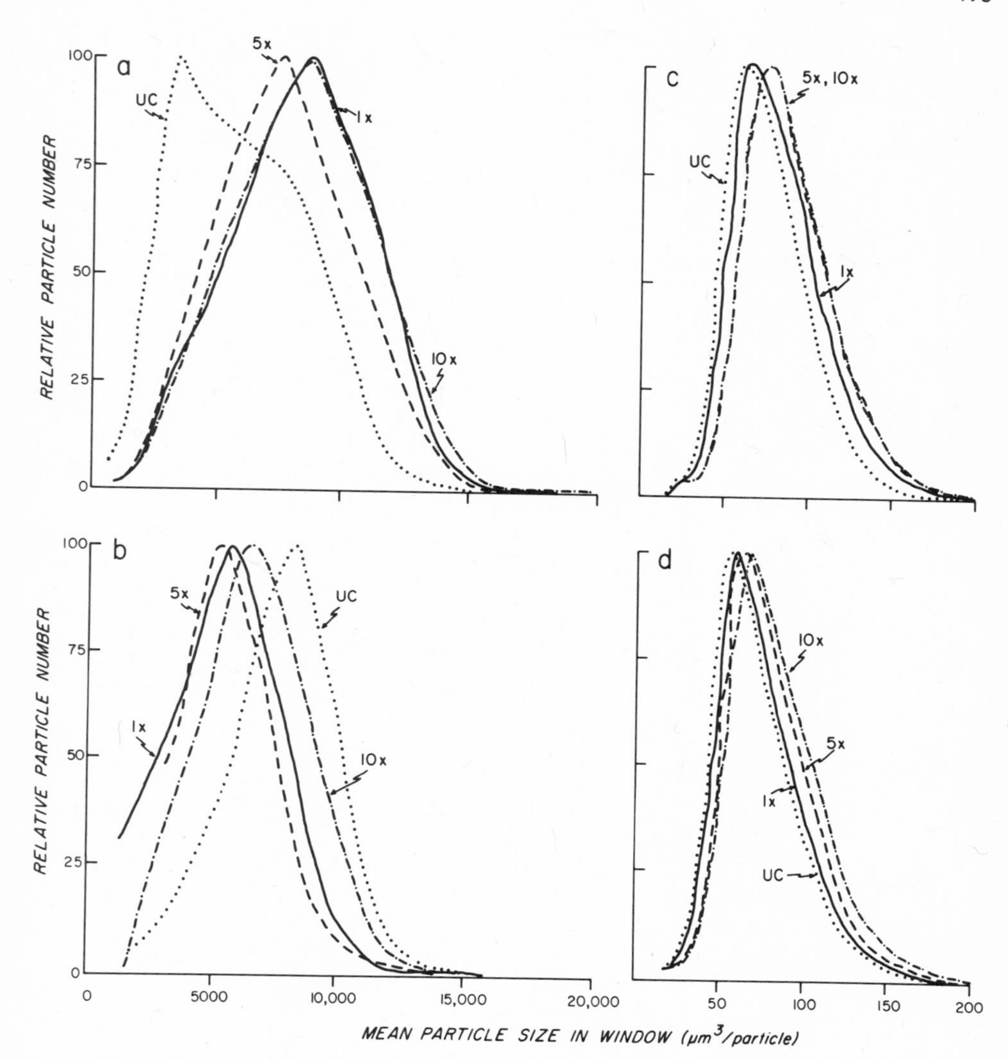

Fig. 2. Particle size distributions at 1X, 5X, and 10X CO_2 levels, and in unbubbled 1X control (UC). (a) acclimated *Skeletonema* during exponential growth; (b) acclimated *Skeletonema* during stationary phase; (c) acclimated *Isochrysis* during exponential growth; (d) acclimated *Isochrysis* during stationary phase.

phase (under nitrogen limitation). Particle fluorescence/particle volume (*F*/*V*), particulate carbon/particle volume (*C*/*V*), particulate nitrogen/particle volume (*N*/*V*), and particulate carbon/particulate nitrogen (*C*/*N*) have been selected as representative ratios.

The relationship between fluorescence and volume appears to be a particularly sensitive physiological measure. This relationship is plotted for exponential and stationary phases of the acclimated *Isochrysis* and *Skeletonema* cultures (Figure 3). Although differences in fluorescence-based and volume-based growth rates were noted earlier (Table 1), the relationship between paired fluorescence and

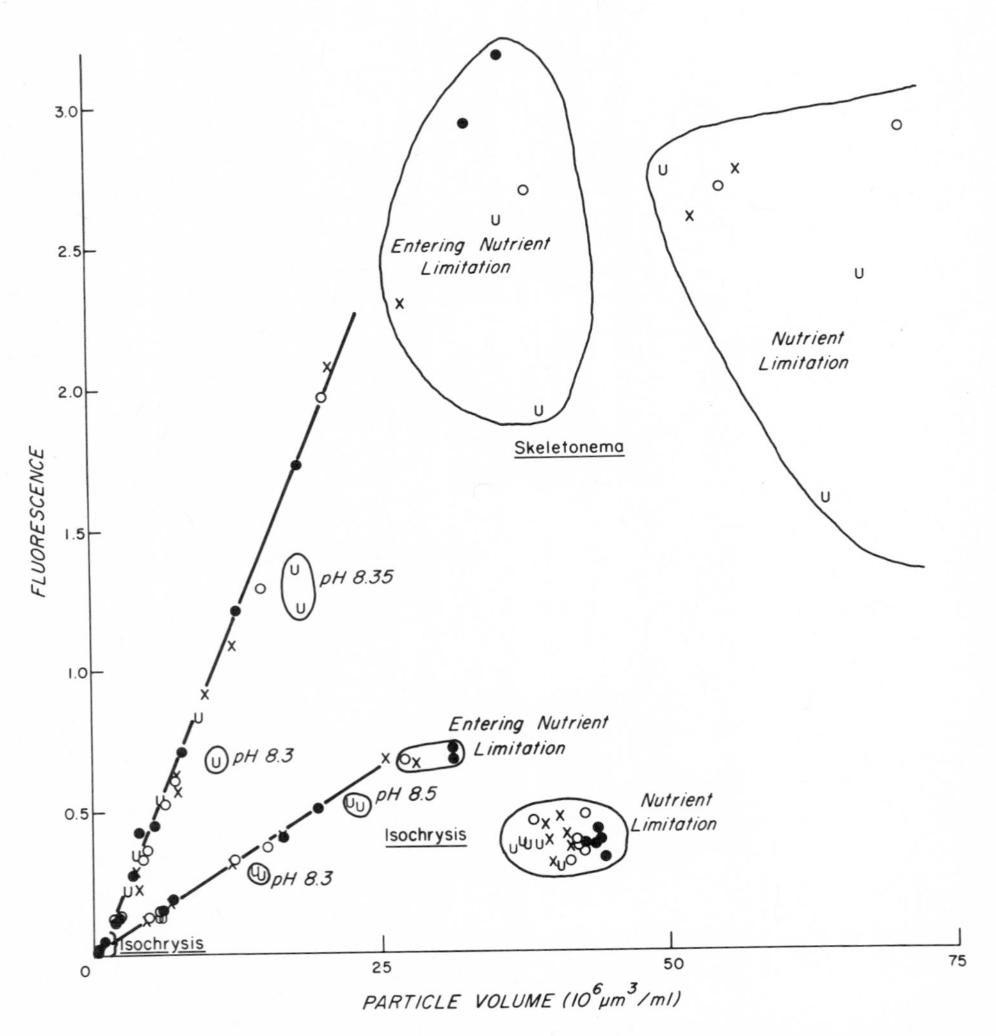

Fig. 3. Fluorescence-volume relationships during exponential and stationary phases for acclimated *Skeletonema* and *Isochrysis* at 1X, 5X, and 10X CO_2 concentrations, and for an unbubbled 1X control (UC).

volume measurements is linear during exponential growth, and appears to be unaffected by CO_2 level. The fact that the relationships are greatly different for each phytoplankton species is obvious. When the nitrogen concentration falls below 1μg-at/liter, the linear relationship begins to deteriorate. Well into the stationary phase both species show significantly less fluorescence/volume with the data being much more variable in *Skeletonema* than in *Isochrysis*.

In the unbubbled control flasks, the *pH* rose to 8.3 and higher when the phytoplankton volume exceeded about 10^7 μm^3. When this happened, the fluorescence/volume relationship during exponential growth was changed (Figure 3). The *pH* never reached 8.3 in the 1X, 5X, and 10X flasks.

Physiological Patterns

The effects of CO_2 levels on *C/V*, *N/V*, *C/N*, and *F/V* during exponential and stationary phases are shown in Figures 4 and 5. These ratios were determined on the days that complete nitrogen budgets were developed for both species, except for the *Isochrysis* exponential phase data, which were taken a day prior to the budget

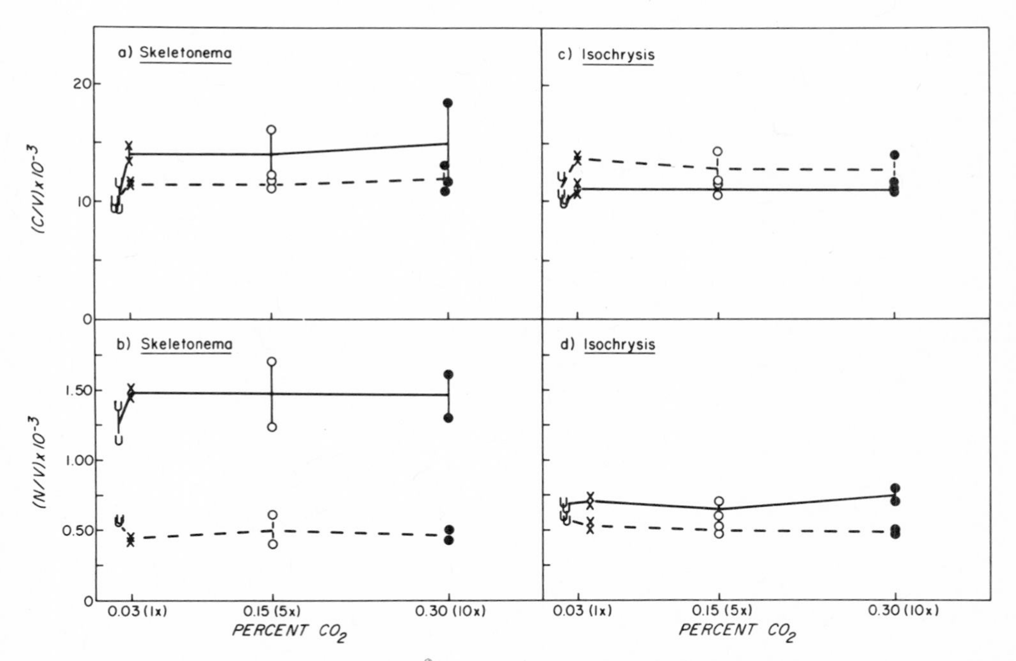

Fig. 4. Carbon/Volume (*C/V*) and Nitrogen/Volume (*N/V*) in exponential (——) and stationary (---) growth phases of *Skeletonema* and *Isochrysis* as functions of CO_2 concentration.

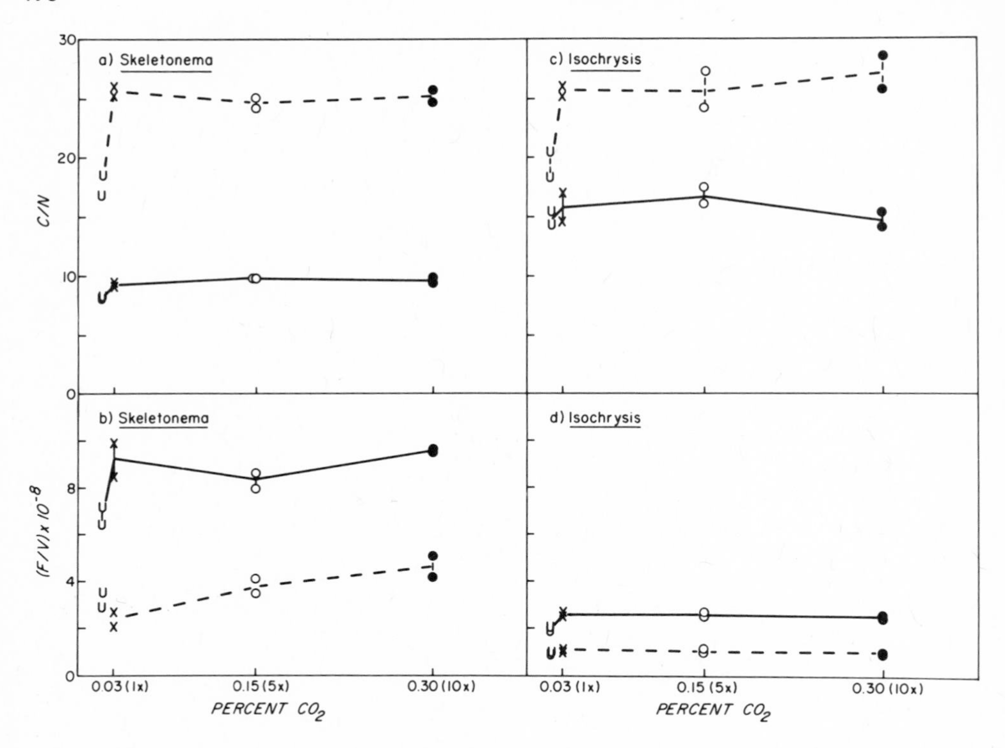

Fig. 5. Carbon/Nitrogen (*C/N*) and Fluorescence/Volume (*F/V*) in exponential (——) and stationary (---) growth phases of *Skeletonema* and *Isochrysis*, as functions of CO_2 concentration.

to insure exponential-phase cells under nitrate-sufficient conditions. Although significant differences can be observed in the exponential-phase and stationary-phase ratios for any one species, and between species in any one growth phase, there are no effects of CO_2 enhancement. However, there are effects attributable to the unbubbled controls that will be addressed later. For *Skeletonema*, both *C/V* and *N/V* are lower in the stationary phase than in the exponential phase (Figure 4a,b). The decrease in *N/V* is much more pronounced. In contrast, for *Isochrysis*, *C/V* increases slightly on entering the stationary phase (Figure 4c). The *N/V* ration for *Isochrysis* decreases in the stationary phase, a result similar to that for *Skeletonema* except that the *Isochrysis* decrease is much less pronounced (Figure 4d).

The *C/N* ratios are large at the stationary phase relative to the exponential phase (Figure 5a,c). Although the stationary-phase *C/N* values are about the same in both species, the exponential-phase values are higher for *Isochrysis* than for *Skeletonema*. As is illustrated in Figure 3, *Skeletonema* has a higher *F/V* ratio than *Isochrysis* in the exponential phase (Figure 5b,d). Care must be exercised in interpreting *F/V* ratios in the stationary phase, as the ratios must continually decrease as the stationary phase progresses (see Figure 1). The apparent increase in the *Skeletonema* stationary-phase *F/V* with increasing CO_2 level (Figure 5b) might be an artifact dependent upon how far into the stationary phase each culture had proceeded at the time of sampling.

Although no differences were observed at different CO_2 levels in the bubbled cultures, the values for the unbubbled controls were consistently different (Figures 4 and 5). One of the most striking features is that *C/V* ratios for both species in both growth phases are lower in unbubbled than in bubbled cultures (Figure 4a,c), whereas stationary-phase *N/V* ratios for species tend to be higher in unbubbled cultures relative to bubbled ones (Figure 4b,d). These relationships give rise to much lower *C/N* ratios in unbubbled stationary-phase cultures than in bubbled cultures (Figure 5a,c). It should be noted, however, that the much lower *C/N* values in unbubbled cultures somewhat reflect the fact that while particulate *N* had reached its maximum at the time of the budget, particulate *C* was still increasing towards its maximum. Late in the stationary phase, *C/N* ratios in unbubbled cultures more nearly approached those in bubbled cultures.

Nitrogen budgets for both species in the late exponential and in the early stationary phase are presented in Figure 6. The *Isochrysis* exponential-phase budget was run just after nitrate was exhausted, so that values for intracellular pools might be underestimated. Nitrogen budgets were partitioned into particulate nitrogen, inorganic intracellular NO_3^-, NO_3^- in the water, and residual ("unaccounted for") nitrogen. The "unaccounted for" fraction was determined by the difference between initial NO_3^- in the water and the sum of the other fractions. This residual fraction is composed of dissolved organic *N*, non-NO_3^- forms of inorganic *N* (such as NH_3 and urea), and any nitrogenous materials that might adhere to glassware. On the average, budgets could be balanced within 5% in the exponential phase and within 10% in the stationary phase. The increase in the residual fraction in the stationary phase involved some adhesion to the glass frits, as evidenced by coloration on those frits at the end of the experiments. Most of the nitrogen, however, clearly was in particulate form early in the stationary phases of both species.

During the exponential phase, intracellular nitrate pools comprised about 5 to 10% of the cellular nitrogen for *Skeletonema*. Nitrate pool values for *Isochrysis* in the exponential phase have little meaning because nitrate in the water had been exhausted just prior to running the *Isochrysis* exponential-phase budget. Intracellular nitrate pools were very small early in the stationary phase for both species, being close to our limits of detection. Values averaged 0.5% of the particulate nitrogen. Regardless of the

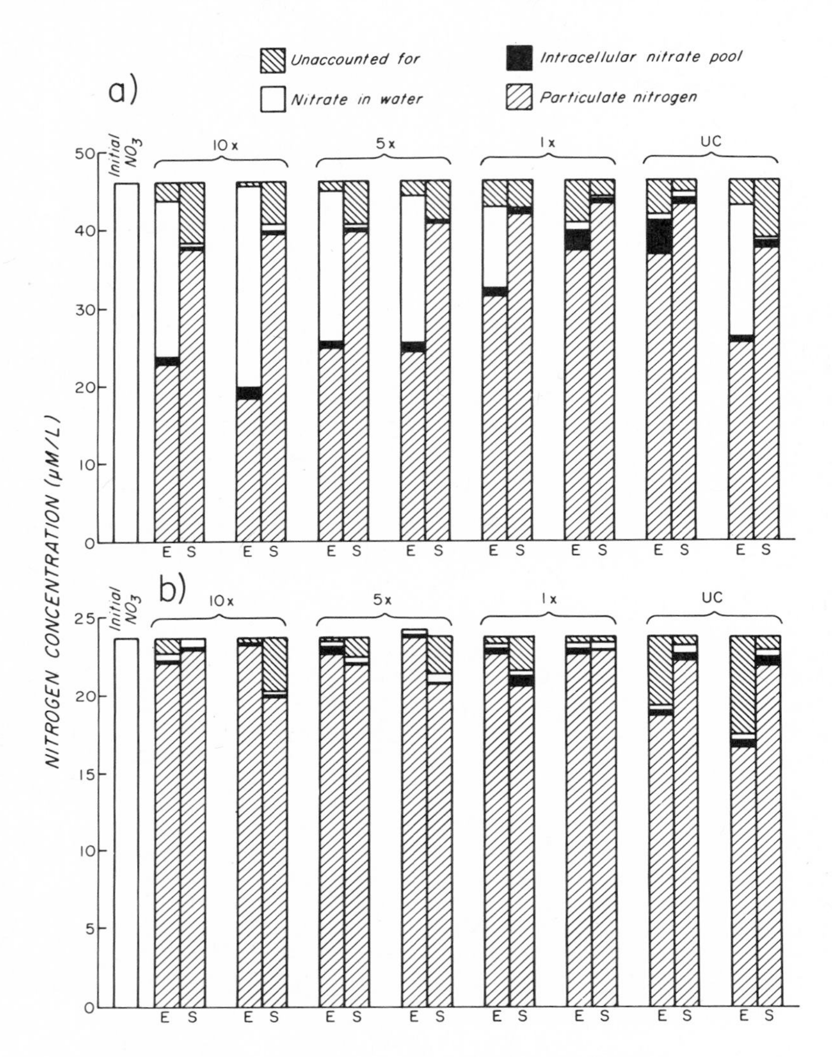

Fig. 6. Nitrogen budgets for (a) *Skeletonema*, and (b) *Isochrysis*, during exponential and stationary phases. Both replicates for each treatment are shown. Differences between exponential phase budgets (E) and stationary phase budgets (S) within any one treatment are real, but comparisons between treatments are not valid because of slightly different timing of sampling along the various growth curves. Major differences between replicates (UC during exponential phase in *Skeletonema*, for example) are also mainly sample timing effects.

growth phase or species, no CO_2 effects on pool size were observed. However, it should be noted that a significantly larger pool was measured in the *Skeletonema* unbubbled controls as NO_3^--N concentrations in the water dropped below 1μg-at/liter and *pH* was greater than 8.3.

DISCUSSION

Once the algal species were acclimated, there was no effect of increased CO_2 on exponential growth rates, yields, intracellular nitrogen fractionation, or cellular physiological ratios such as fluorescence/volume, carbon/nitrogen, nitrogen/volume or carbon/volume. There was, however, a very slight increase in *Isochrysis* cell size with enhanced CO_2 levels. These results are in general agreement with the results of *Swift and Taylor* (1966), who found no change in division rate of the coccolithophorid *Criscosphaera elongata* over CO_2 levels from 0.03% to 5.00%, and with the results of *Humphrey* (1975), who showed for a variety of marine species that maxima for growth, cell concentration, photosynthesis, and photosynthesis/respiration all occurred within the *pH* range 7.2 to 8.2 given by our CO_2 treatments.

It must be remembered that in the non-acclimated populations significant changes in growth rates and physiological features were observed. These results have been interpreted as a shock effect caused by instantaneously changing the CO_2-*pH* regime. Responses of non-acclimated cells likely are not relevant to the problem of slow CO_2 increases engendered by fossil fuel burning, but they are mentioned here as a warning in interpreting test results from short-term CO_2 experiments.

A major feature of our results, which we feel have pertinence to the ocean, is that initial nitrogen concentration in the water controls the final cellular yields in terms of both nitrogen and carbon, regardless of CO_2 concentration in the water. Since stationary-phase *C/N* ratios were independent of both CO_2 level and species in these studies, the amount of "new" (fossil-fuel-produced) CO_2 that can be absorbed by marine phytoplankton (providing our species are representative) depends on the amount of "new" nitrogen that can be brought into surface waters for use by phytoplankton. By "new" nitrogen we mean additional supplies of nitrogen above and beyond those supplied by present-day advective processes, pollution, weathering, nitrogen fixation by marine blue-green algae, etc. If, in the simplest case, all algal biomass eventually sinks through the thermocline without being eaten, then net downward transport of any algal-incorporated fossil-fuel CO_2 will be controlled by the transport of "new" nitrogen into surface waters. The addition of zooplankton herbivores and of higher trophic levels could enhance downward transport of "new" CO_2 only if animal metabolism of nitrogen (relative to carbon) is enhanced over present levels in surface waters. It is implicit in this statement that, in some fashion, "new" CO_2 present in the water would have to be linked to a change in the metabolism of the zooplankton such that the animals would alter the ratios of carbon and nitrogen removed from algal food and, as a direct result, put more carbon into fecal pellets than presently occurs. The carbon-rich fecal pellets would rapidly sink out of the surface waters. Because of the limited knowledge of the qualitative and quantitative aspects of regeneration and carbon-nitrogen transfer in trophic systems, however, it is impossible to predict if the carbon transport would be enhanced or depressed by inclusion of higher trophic levels. Because such higher levels exist in all natural systems, studies of their significance are needed.

Although the results reported herein in simple batch systems suggest no effect of enhanced CO_2 on nitrogen-based growth rates and yields, these data do not necessarily imply no effect in systems which are continually nitrogen-deficient. Nitrogen tends to exert control in marine systems in two major ways: 1) by controlling yields in static systems with initially large nitrogen concentrations (e.g., some spring blooms, which approximate batch cultures), or 2) by controlling growth rates through nitrogen fluxes in *N*-deficient systems. Control by nitrogen flux is generally important during the summer in temperate waters, and perhaps all year in some tropical and mid-ocean areas. When flux control is operating, nitrogenous nutrients are barely detectable in the water, and phytoplankton production rates are controlled by the rates of nitrogen regeneration by zooplankton and by any upward nitrogen transport across the thermocline. Understanding the effects in nitrogen-deficient, flux-controlled situations requires types of experiments other than batch culture.

Falkowski (1975) has shown that the enzyme responsible for NO_3^- uptake in *Skeletonema* is *pH*-sensitive, with a minimal activity at *pH* 7.2 (the value for 10X CO_2 level in our study), and a maximum activity at *pH* 8.0 (near 1X value of our study). However, these data likely would become significant only under conditions of nitrogen insufficiency in the water, when uptake kinetics might control phytoplankton biomass increase. If different species have different NO_3^--enzyme sensitivities to *pH*, then species competition might be affected by CO_2 enhancement in nitrogen-deficient waters. Such effects would not show up in batch culture because NO_3^- uptake does not control growth rates.

Again, despite the fact that the results of this work show no effect of CO_2 enhancement on growth rates or gross physiological attributes, it does not necessarily follow that subtle shifts in intracellular carbon metabolism have not occurred. The classic work of *Bassham and Calvin* (1957), elucidating the biochemical pathways of carbon in photosynthesis, was done with the green alga *Chlorella* under 3.0% CO_2. Under these extreme CO_2 concentrations biochemical pathways of carbon largely involve compounds of the pentose phosphate cycle. However, under natural CO_2 levels many plants convert large fractions of fixed CO_2 to glycollic acid, which can either be excreted from the cell or oxidized within the cell by the process of photorespiration (*Tolbert*, 1974). Under non-limiting nutrient and light conditions in the sea, we might expect increased CO_2 levels to damp photorespiratory effects and enhance the pentose phosphate cycle. However, such metabolic changes would have to be major in order to be discerned by the methods employed in this present investigation. In our studies, precise estimates of dissolved organics were not made, so that possible changes in metabolic carbon excretion could not be observed. Only in non-acclimated, stationary-phase cultures did it appear that considerable excretion took place to make cells clump and stick to the walls of the experimental flasks.

Throughout much of the results of this work, significant differences were observed between the unbubbled controls and the bubbled populations. The shift observed in the unbubbled *Skeletonema* particle size distribution (Figure 2) was due directly to chain breakage during frequent sampling over the time course of the experiments. Previous work with *Skeletonema* in our laboratory has shown similar particle-size shifts due to chain breakage. In systems that are constantly stirred or bubbled, a short, stable chain length develops. The other differences between unbubbled and bubbled populations,observed particularly in F/V, C/V, N/V, and C/N (Figures 3, 4, and 5), were associated with pH changes and were not due to the physical effect of the bubbling. Toward the end of exponential growth and into stationary phase in unbubbled cultures, algal biomasses were large and production rates were high enough to remove CO_2 from the water very rapidly. Such rapid removal allowed the pH to increase to 8.3 and higher. Unbubbled cultures with low biomass (in early exponential phase) experienced no pH change, and physiological ratios were the same as in the bubbled cultures with their stable pH values.

The physiological ratios not only changed when pH rose above 8.3 in unbubbled flasks, but also when nutrients became depleted in bubbled flasks with stable pH values. The F/V relationship shows a good example of these dual effects (Figure 3). It is conceivable that increasing pH in the water can be an indicator to the population of impending nutrient exhaustion. The ability of the population to adjust to low nitrogen concentration may be a function of the time period over which the adjustment is allowed to occur; i.e., the longer the time period between sensing the approach of nutrient deficiency and the actual depletion, the more effectively the population can regulate its carbon synthesis to allow a more gradual adjustment to the new environment. The problem of sufficient time for adjustment should become critical only when high nitrogen production rates are occurring, which is precisely the condition under which pH would be expected to rapidly rise above 8.3. In the pH-controlled systems of this work, cellular F/V, for example, was maintained at its non-nitrogen-deficient value until the nitrate disappeared from the water. The *F/V* ratio, as well as the other ratios, changed to values expected from highly nitrogen-deficient cultures within one day after nutrient exhaustion. In contrast, the physiological ratios of populations in *pH*-adjusting (unbubbled) systems shifted from values typical of the exponential phase to values typical of the stationary phase over a period of 5 or 6 days It appears to be carbon synthesis that is mainly suppressed in the unbubbled populations after the *pH* reaches 8.3. Almost all of the nitrogen in the system can already be accounted for as particulate *N* early in stationary phase.

A second implication of rapid, productivity-caused increases in *pH* involves the time scales over which CO_2 equilibrium can be assumed in upper waters of the ocean. In the studies reported in this communication, the *pH* began to rise in the unbubbled controls when biomass exceeded 10^7 μm^3, which is approximately equivalent to 10 μg-at particulate nitrogen/liter (a level sometimes exceeded in coastal environments). *Gordon* (1973) has recorded *pH* values up to 8.6 in natural phytoplankton blooms off the Oregon Coast. Even in the 1X and 5X bubbled *Skeletonema* cultures in this investigation, the high rates of gas dispersion were insufficient to meet the CO_2 demand of the algae in concentrations of 30 μg-at particulate nitrogen/liter (and the *pH* rose slightly for a brief period). These results indicate that, in highly productive areas, CO_2 equilibria need not be assumed during all times of year even under enhanced CO_2 levels. Such disequilibria would occur if, as in the unbubbled controls, previous populations had temporarily altered the CO_2 equilibrium. Although assumptions of CO_2 equilibration with the upper 100 meters of ocean may be satisfactory for calculations on geochemical time scales, they should be used with great caution in assessing possible impacts on, and responses of, marine phytoplankton. The whole area of short-term CO_2 equilibration would appear to require more study.

The results of these studies have further implications for field studies in general, and for attempts to detect CO_2 impacts on field populations in particular. For example, fluorescence-based growth is different than volume-based growth in both species, an indication that the use of fluorescence values as biomass indicators is inaccurate. The fact that *F/V* responds both to nitrogen depletion and to $pH \geq 8.3$, and is different for different species under identical conditions, further complicates the use of fluorescence as a biomass measure. However, in the laboratory or field in conjunction with other measurements, fluorescence-based ratios become a valuable tool for describing the physiological conditions of populations. Finally, *Skeletonema* is clearly different from *Isochrysis* by any manner of measurement: growth responses, physiological ratios, etc. Because of this, and because of the strong likelihood that other species are also different, any critical evaluation of biological effects of CO_2 enhancement in oceanic surface waters must take species composition into consideration.

ACKNOWLEDGEMENTS

We acknowledge the excellent technical support by, and discussions with, E. Jean Jensen and J. Michael DeManche.

REFERENCES

Bacastow, R. and C. D. Keeling. 1973. Atmospheric carbon dioxide and radiocarbon in the natural carbon cycle: II. Changes from A.D. 1700 to 2070 as deduced from a geochemical model, In: G. M. Woodwell and E. V. Pecan (Eds.), *Carbon and the biosphere. Proc. 24th Brookhaven Sympos. in Biology* CONF-720510, NTIS, Springfield, VA., USA. 86-135.

Bassham, J. A. and M. Calvin. 1957. The path of carbon in photosynthesis. Prentice-Hall, Englewood Cliffs, N.J., USA.

Botkin, D. B., J. F. Janak and J. R. Wallis. 1973. Estimating the effects of carbon fertilization on forest composition by ecosystem simulation, In: G. M. Woodwell and E. V. Pecan (Eds.), *Carbon and the biosphere.* CONF-720510, NTIS, Springfield, VA., USA. 328-344.

Broecker, W. S., Y. Li and T. Peng. 1971. Carbon dioxide - Man's unseen artifact, In: D. W. Hood (Ed.), *Impingement of man on the oceans.* John Wiley & Sons, Inc., New York. 287-324.

Burlew, J. S. (Ed). 1961. Algal culture from laboratory to pilot plant. Carnegie Inst. Wash. Publ. 600. Wash., D.C.

Conover, S. A. M. 1975. Partitioning of nitrogen and carbon in cultures of the marine diatom *Thalassiosira fluviatilis* supplied with nitrate, ammonium, or urea. *Mar. Biol. 32:* 231-246.

Donaghay, P. L. 1975. Population dynamics of *Skeletonema costatum* in high dilution rate chemostats. MS Thesis. Oregon State Univ., Corvallis, Ore., USA.

Donaghay, P. L. and L. F. Small. 1975. A carbon and nitrogen budget method. Am. Inst. Biol. Sci. Mtng., Corvallis, OR., USA (abstract).

Eckardt, F. E. (Ed.). 1968. Functioning of terrestrial ecosystems at the primary production level. UNESCO, Paris.

Falkowski, P. G. 1975. Nitrate uptake in marine phytoplankton: (nitrate, chloride)-activated adenosine triphosphatase from *Skeletonema costatum* (Bacillariophyceae). *J. Phycol. 11:* 323-326.

Gavis, J. and J. F. Ferguson. 1975. Kinetics of carbon dioxide uptake by phytoplankton at high *pH*. *Limnol. Oceanogr. 20:* 211-221.

Goldman, J. C. 1973. Carbon dioxide and *pH*: effect on species succession of algae. *Science 182:* 306-307.

Gordon, L. I. 1973. A study of carbon dioxide partial pressures in the surface waters of the Pacific Ocean. PhD Thesis, Oregon State Univ., Corvallis, Ore., USA.

Humphrey, G. F. 1975. The photosynthesis: respiration ratio of some unicellular marine algae. *J. exp. mar. Biol. Ecol. 18*: 111-119.

King, D. L. 1970. The role of carbon in eutrophication. *J. Water Pollut. Contr. Fed. 42*: 2035-2051.

Likens, G. E. (Ed.). 1972. Nutrients and eutrophication. *Am. Soc. Limnol. Oceanogr. Spec. Sympos. Vol. 1*. Allen Press, Inc., Lawrence, Kans., USA.

Loeblich, L. A. 1970. Growth limitation of *Dunaliella salina* by CO_2 at high salinity. *J. Phycol. 6 (suppl.)*: 9.

Lundy, D. W. 1974. Rapid equilibrium response of a marine diatom to external and internal nutrient concentrations. MS Thesis, Oregon State Univ., Corvallis, Ore., USA.

Reiners, W. 1973. Appendix: summary of world carbon cycle and recommendations for critical research, In: G. M. Woodwell and E. V. Pecan (Eds.), *Carbon and the biosphere. Proc. 24th Brookhaven Sympos. in Biology*. CONF-720510, NTIS, Springfield, VA., USA. 368-382.

Shapiro, J. 1973. Blue-green algae: why they become dominant. *Science 179*: 382-384.

Swift, E. and W. R. Taylor. 1966. The effect of *pH* on the division rate of the coccolithophorid *Cricosphaera elongata*. *J. Phycol. 2*: 121-125.

Tolbert, N. E. 1974. Photorespiration, In: W. D. P. Stewart (Ed.), *Algal physiology and biochemistry*. Univ. California Press, Berkeley, Ca., USA. 474-504.

Whittaker, R. H. and G. E. Likens. 1973. Carbon in the biota, In: G. M. Woodwell and E. V. Pecan (Eds.)., *Carbon and the biosphere. Proc. 24th Brookhaven sympos. in Biology*. CONF-720510, NTIS, Springfield, VA., USA. 281-302.

Woodwell, G. M. and E. V. Pecan. 1973. Carbon and the biosphere. *Proc. 24th Brookhaven Sympos. in Biology*. CONF-720510, NTIS, Springfield, VA., USA.

II

CARBONATE DISSOLUTION

Recommendations of the Working Group on Carbonate Dissolution

W. S. Broecker, Chairman

Lamont-Doherty Geological Observatory

If most the fossil fuel CO_2 is eventually absorbed by the oceans and reacts with marine $CaCO_3$ in a manner causing its dissolution, the oceans' total inorganic carbon content and total alkalinity will each increase by several times 10^{17} moles, or about 20%. Because the known annual alkalinity and inorganic carbon fluxes into, out of, and within the oceans are all on the order of 10^{13} to 10^{14} moles, these fluxes may take thousands to tens of thousands of years to reestablish "normal" stability within the marine carbon system as a consequence of such a perturbation. Therefore, the awesome potential for performing a truly global geochemical "experiment" with unknown consequences lasting for geological time scales presents itself, being characterized by the following questions: where will the enhanced $CaCO_3$ dissolution occur, and how fast will it occur? These questions, together with the implications of these phenomena for marine and global geochemistry, addressed briefly in the Preface of this publication, formed the basis for the research recommendations detailed in this summary of panel discussions in this session of the Symposium.

1. <u>Where Will Enhanced $CaCO_3$ Dissolution Occur?</u>

$CaCO_3$ precipitated in the open ocean first encounters undersaturated waters as it settles from the surface to the deep ocean. Thorough studies of suspended carbonates and associated organic-inorganic particulate materials are therefore necessary to estimate the primary flux of material available for dissolution. Recent studies suggest that a large fraction of particulate carbonate is transferred from shallow to deep environments by large, rapidly sinking materials which cannot be adequately sampled using standard

EDITOR'S NOTE: This discussion on the recommendations arising from this working group was put in final form to a great extent by E. Sundquist.

water-sampling techniques. Sediment traps and/or filtration pumps should thus be deployed in representative areas of the oceans. Detritus collected by these methods should be closely scrutinized for its chemical composition, mineralogy, organic content and settling behavior.

The rapid settling of carbonate particles suggests that deep-sea dissolution occurs mainly on the sea floor, rather than in the water column. One of the most basic pieces of needed information in regard to this consideration is the distribution of $CaCO_3$ in Holocene sediments. Global maps of the distribution of calcite and aragonite can be prepared from existing core collections. Using these maps and sea floor topographic information, it should be possible to estimate Holocene accumulation and dissolution rates, and how much sedimentary $CaCO_3$ is available for dissolution by fossil fuel CO_2 at various water depths.

A number of dissolution indices are now available for semi-quantitative study of the transition from undissolved to dissolved sediments. As a result, species assemblages, total carbonate content, aragonite content and other indices should be determined in closely-spaced cores which traverse dissolution transition zones. These measurements can be used to prepare detailed dissolution flux maps, which will be essential to calibrate and check dissolution models.

Several geologists at the meeting noted that sedimentary evidence suggests the possibility of $CaCO_3$ dissolution in shallow marine environments. Marine chemists, on the other hand, pointed out that surface sea water is virtually always greatly supersaturated with respect to all forms of $CaCO_3$. This discrepancy must be resolved by studying nearshore $CaCO_3$ sediments and ocean surface alkalinity -- ΣCO_2 balances. If there is any mechanism by which fossil fuel CO_2 can affect the net rate of $CaCO_3$ precipitation in the shallow ocean, the response of the marine $CaCO_3$ cycle will be much more rapid than its response via deep-sea dissolution alone.

2. How Fast Will Enhanced Dissolution Occur?

Answering this question will require research in two general areas: (1) marine $CaCO_3$ dissolution kinetics, and (2) $CaCO_3$ dissolution in sediments.

a. Marine $CaCO_3$ Dissolution Kinetics:

It is now generally accepted that dissolution of carbonates in the oceans is controlled by the degree of seawater undersaturation with respect to biogenic carbonate phases. However, the functional relationship between the degree of undersaturation and the rate of dissolution is not yet fully understood. Numerous studies have shown that carbonate dissolution in seawater is a complex process. Dissolution rates reflect the variety of carbonate particle mineralogies, compositions, shapes and sizes. The surfaces of marine carbonate particles seem to lead a life of their own, reflected in the influence of particle histories on dissolution rates. It is often difficult to compare measurements made by different laboratories using different techniques.

In order to understand the kinetics of marine $CaCO_3$ dissolution, the solubility of $CaCO_3$ in seawater must first be known. The solubilities of calcite and aragonite in seawater are poorly known at the present time. Further work is necessary to resolve disagreements concerning standard experimental conditions, temperature coefficients, and pressure coefficients. Primary consideration should be given to the effect that kinetic complications may have on equilibrium measurements. The rapid decline of dissolution and precipitation rates near equilibrium can lead to ambiguities in defining it. For example, approach via supersaturation may differ from approach via undersaturation, or dilute suspensions may differ from concentrated suspensions. *In situ* determinations are needed, in addition to laboratory measurements, and results obtained using the *in situ* carbonate saturometer are particularly encouraging. Future studies should all emphasize intercalibration among various laboratories and techniques.

$CaCO_3$ dissolution rates must similarly be studied further by a variety of intercalibrated techniques. Recent applications of *pH*-stat and saturometer methods have been promising, although there is still uncertainty as to whether the rates vary linearly or exponentially with respect to the degree of undersaturation. *In situ* measurements are much needed, and fixed-mooring experiments are anticipated to supplement the classical Peterson-Berger buoy experiment.

Both equilibrium and kinetic studies must recognize the wide variety of marine biogenic $CaCO_3$ particles. It is well known, for example, that various species of calcareous organisms are subject to different degrees of preservation in corrosive seawaters. However, it is not understood whether these differences are related to variations in dissolution rates or in solubilities. Therefore, it is necessary to measure the solubilities and dissolution rates of samples representing the full range of marine $CaCO_3$ detritus, from freshly sampled or cultured plankton shells to residual material found in partially dissolved sediments. Information regarding biogenic aragonite is particularly sparse. This information is much needed, because aragonite may be especially important in the short-term buffering of fossil fuel CO_2 injected into the ocean.

Surface phenomena are also very important in $CaCO_3$ equilibrium and kinetic measurements. Organic coatings - either absorbed or biologically inherent - have been much debated as protective barriers against dissolution in corrosive waters. It has also been proposed that dissolution rates may be inhibited by absorption and/or ion exchange involving minor constituents of seawater (e.g., phosphate). These hypotheses should be tested by careful experimentation, such as by observing the effects of various surface-active pretreatments. Surface area measurements will be required in order to evaluate rate experiments, and different surface area techniques should be evaluated critically with this purpose in mind. Above all, problematic surface effects require that special consideration be given to the history of $CaCO_3$ samples prior to their use in experiments.

For example, deep-sea organic ratings may be degraded by bacteria at warm temperatures.[1]

Many problems arise in trying to compare the results obtained by different laboratories using different techniques. The importance of intercalibration requires that careful attention be given immediately to the choice of appropriate solid and solution standards. Calcite and aragonite sources should be evaluated on the basis of their availability, purity, grain size distribution, crystallinity, and surface history. For example, commercial $CaCO_3$ (calcite) is readily available, but its preparation history is not exactly known. Its rhombs are also poorly developed, with a sloppy grain size distribution ranging only up to about 20 microns. Iceland spar calcite can be crushed and sorted to any desired grain size, but crushing introduces surface and internal strains which can strongly affect dissolution kinetics unless the grains are first painstakingly annealed. Optical grade Iceland spar is expensive and the large crystals may not be completely homogeneous. Very smooth calcite rhombs can be synthesized in the laboratory by first precipitating aragonite and then recrystallizing to calcite by stirring in hot distilled water. These rhombs are well-sorted, but no larger than about 20 microns, and the preparation is time-consuming. Laboratory precipitation seems to be the only ready source of standard aragonite, although batches are small and frequently contain unacceptable amounts of calcite. Biogenic sources of both calcite and aragonite should also be considered, because the ultimate concern is for enhanced dissolution of biogenic $CaCO_3$.

Advantages and disadvantages must be similarly weighed in choosing standards for water analyses. The establishment of solution standards may require that large amounts of solution be prepared and exchanged. Such an effort is much needed to compare laboratory solution measurements to each other and to the tremendous amount of *in situ* information obtained by the GEOSECS program.

b. $CaCO_3$ Dissolution in Sediments:

Laboratory studies of $CaCO_3$ dissolution must be directed toward ultimately formulating applicable models of dissolution in calcareou sediments. This will require a skillful summation of the full rang of dissolution behaviors displayed by the diversity of marine detritus. Moreover, sediment models must incorporate several important factors which are unique to the sedimentary environment.

[1]Editor's Note: The reader is referred to *Marine Chemistry, Vol. 6, No. 1*, 1978 for discussions on this and related topics concerning organics in seawater. It is a publication dedicated to a symposium on "Concepts in Marine Organic Chemistry" held at the University of Edinburgh, September 6-10, 1976.

Simple sediment dissolution models have been constructed on the basis of measured $CaCO_3$ dissolution rates and reasonably well known diffusion coefficients. These calculations suggest that sediment pore waters should come to equilibrium with respect to their $CaCO_3$ solids very soon after burial. However, the few relevant pore water measurements that are available suggest that changes in pore water carbonate chemistry may not conform to the simple dissolution stoichiometry assumed in the models. This discrepancy must be resolved as soon as possible because it drastically affects estimates of the time it will take sediment dissolution in pore waters to respond to changes in the chemistry of overlying bottom waters. Measurement techniques for pore waters must be improved so that their carbonate system can be determined with the accuracy and precision now attained in seawater analyses. The problems imposed by small sample volumes, extraction artifacts, and the incomplete knowledge of the ways in which pore water peculiarities affect apparent dissociation constants must be overcome. Very little is known about the extent of pore water enrichment in dissolved CO_2 and organic compounds derived from the decay of organic matter. These problems must be resolved by implementing a variety of laboratory-tested shipboard and *in situ* techniques, with enough redundancy to provide adequate checks against each other.

Sediment dissolution fluxes must pass through the sea-sediment interface, a boundary which has received the attention of physical oceanographers because of its influence on bottom water current structures. Calculations based on fluid mechanical studies suggest that the ability of overlying water to dissolve carbonate sediments may be related to the turbulence of the water. Turbulence determines the thickness of the "viscous sub-layer" just above the sea bed. Because turbulence is directly related to current velocities and sea floor roughness, more data is needed regarding the distribution and variability of these parameters. Their influence must be monitored in detailed laboratory studies and included in sediment dissolution models.

One of the least-known influences on sediment dissolution is the activity of benthonic organisms. Evidence of significant bioturbation is seen in the results of several sediment-dating techniques, primarily C^{14}. Surface sediments in regions of active deposition have apparent ages of several thousand years, and apparent accumulation rates may differ from real rates. Sedimentation rates and mechanical stirring effects must be quantified by detailed dating of box core samples using C^{14} and other tracers, such as O^{18} and fallout nuclides. Sediment mixing by bioturbation provides a means by which calcareous material can be repeatedly exposed at the sea-sediment interface, and by which saturated pore waters may be replaced by undersaturated bottom waters. These effects may strongly influence the sediments' response time to fossil fuel CO_2 oceanic injections. Benthononts may also affect sediment dissolution by releasing CO_2 during respiration and by stripping organic coatings from $CaCO_3$ particles. Intensive box coring, supplemented by time-lapse photography and benthic respiration measurements, will help to identify benthonont abundances and habits. Such information will be critical to the formulation of realistic dissolution models.

3. What Will be the Implications for Marine and Global Geochemistry?

As was pointed out in the Preface of this publication, the carbonate chemistry of the ocean surface may be determined primarily by increasing CO_2 levels in the atmosphere. As the oceans have sufficient time to mix as a whole with their CO_2-enriched surface waters, the tremendous marine carbon reservoir will in turn probably control atmospheric CO_2 concentrations. Thus, rough calculations can be performed which illustrate the possible effect of enhanced $CaCO_3$ dissolution on CO_2 in the atmosphere thousands of years from now. If all of the estimated fossil fuel reserve is burned, and if most of the CO_2 produced is redistributed in the oceans in proportion to existing inorganic carbon concentrations, the atmospheric CO_2 content will increase by about 3-fold over its preindustrial value. But, if each mole of fossil fuel CO_2 reacts with one mole of marine $CaCO_3$, the eventual atmospheric p_{CO_2} will be only around 40% higher than it was in the early 19th century.

The assumptions behind these calculations are very crude, in contrast to the complexity of estimating $CaCO_3$ dissolution effects over the full range of time scales that must be considered. As a result, research in marine chemistry, which predominates in the recommendations made here, must be accompanied by concurrent efforts in the sister environmental sciences such as physical oceanography, biology, and geology. This interdisciplinary approach cannot be underemphasized.

NEUTRALIZATION OF FOSSIL FUEL CO_2 BY MARINE CALCIUM CARBONATE

W. S. Broecker[1] and T. Takahashi[2]*

[1] *Lamont-Doherty Geological Observatory and* [2] *Queens College*

ABSTRACT

The $CaCO_3$ stored in marine sediments will ultimately neutralize the CO_2 generated by fossil fuel combustion. Details of this process are explored and a model of the early phases of this process in the western basin of the deep Atlantic Ocean is presented. The amount of $CaCO_3$ available for dissolution is derived from the calcite content of marine sediments and the extent to which these sediments are stirred by benthic organisms. The conclusion is that the available calcite is about equivalent to the CO_2 which would be released if our known resources of natural gas, oil and coal were burned. The rate at which this dissolution will proceed is estimated from the rate at which natural dissolution has proceeded during the Holocene. The conclusion is that if linear kinetics are followed, the time constant will be on the order of 1500 years. On the other hand, if the exponential kinetics found in the laboratory by Berner and Morse (1972) apply, then the time constant will be much shorter. It is possible that the dissolution process will become limited by the rate stirring of the mixed layer in deep sea sediments. The insoluble residue of clay minerals and opal built up in the upper few millimeters of the sediment as the result of dissolution must be mixed down into the underlying sediment if more than the upper few millimeters of sediment is to be attacked.

In order to model the change in the carbonate ion content of the waters in the deep Atlantic Ocean over the next few hundred years it is necessary to know:

---

* Present address: Lamont-Doherty Geological Observatory

1) the rate of chemical fuel use over this period;

2) the distribution of this CO_2 between the terrestrial biosphere, upper ocean and atmosphere;

3) the relationship between the $CO_3^=$ content of newly formed deep water and the p_{CO_2} in the atmosphere;

4) the ventilation time of the northern component of the deep water in the Atlantic Ocean and the distribution of this component in the deep mixing zone;

5) the relationship between the in situ $CO_3^=$ content of deep water and the $CO_3^=$ content necessary to trigger dissolution;

6) the relationship between the rate of dissolution of deep sea sediments and the degree of undersaturation of the bottom water.

Preliminary estimates have been made of these factors and they have been combined into a dissolution model.

INTRODUCTION

CO_2 generated by the combustion of fossil fuels will ultimately be neutralized through combination with sedimentary calcium carbonate via the reaction

$$CO_2 + CaCO_3 + H_2O \rightleftarrows 2HCO_3^- + Ca^{++} \quad (1)$$

Regardless of whether this dissolution takes place on land or in the sea, the calcium and bicarbonate ions generated will end up as part of the ocean's dissolved salt load. In this paper some of the details of this process are considered.

The primary contributor to this neutralization will be deep sea sediments. The deep sea will be gradually "acidified" by the downward mixing of surface water enriched in fossil fuel CO_2 and its sediments will become subject to enhanced calcium carbonate dissolution. Shallow marine deposits are less promising because of their relatively small area and their generally low $CaCO_3$ content, and because the water in which they are bathed is highly saturated with respect to both calcite and aragonite (see Figure 1). It is unlikely that the CO_2 content of the atmosphere will ever become great enough to bring these waters to a state of undersaturation. Although enhanced continental weathering will contribute to the neutralization, the fact that the decay of organic material maintains the CO_2 content of soil and ground waters far higher than the atmospheric equilibrium content suggests that the higher atmospheric CO_2 contents will not have a very large impact.

For the purposes of this discussion, the sediments of the sea will be divided into five categories.

1) Those lying beneath the calcite compensation depth. As these sediments contain little calcium carbonate they will not contribute significantly to the neutralization process.

2) Those lying between the calcite compensation depth and the

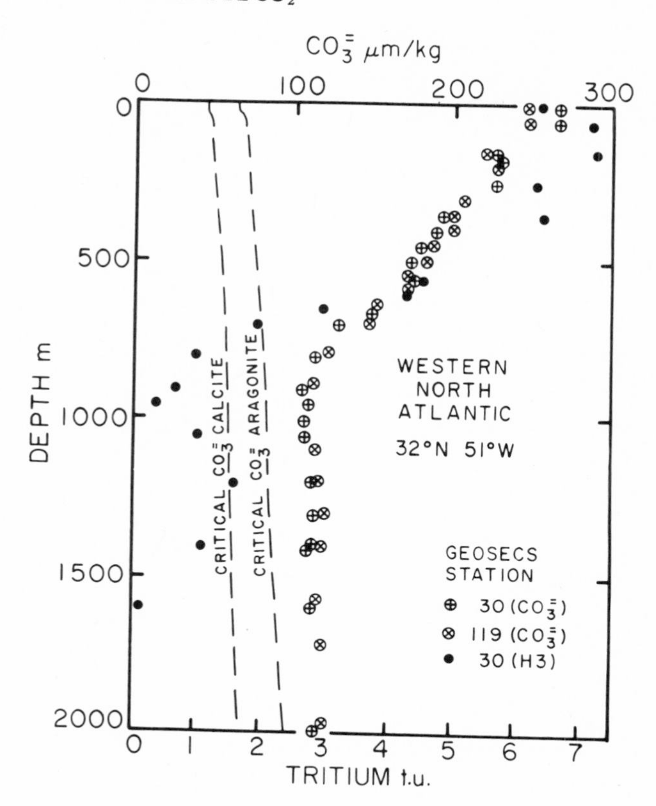

Fig. 1a. $CO_3^=$ and tritium concentration as a function of water depth in the upper water column at 32°N and 51°W in the western North Atlantic Ocean. Curves for the critical carbonate ion content for calcite and for aragonite are given for reference. They are based on those given by *Broecker and Takahashi* (in press) with a temperature correction for the decrease in solubility with increasing water temperature. The carbonate ion results were calculated from the titration data given in the GEOSECS leg reports using the constants adopted by *Takahashi et al.* (1976). At station 30 the leg report ΣCO_2 results were decreased by *15 μm/kg* to bring the leg 3 results into agreement with the pCO_2 data and with the results from the other GEOSECS Atlantic Ocean legs. The tritium results are those of *Ostlund et al.* (1974) on samples collected in Sept. 1972.

lysocline horizon (see Figure 2). These sediments lie in the transition zone between depths where natural dissolution by bottom water is small and depths where natural dissolution by bottom water removes virtually all of the calcite from the accumulating sediment (see Figure 3). These sediments still have calcite available for

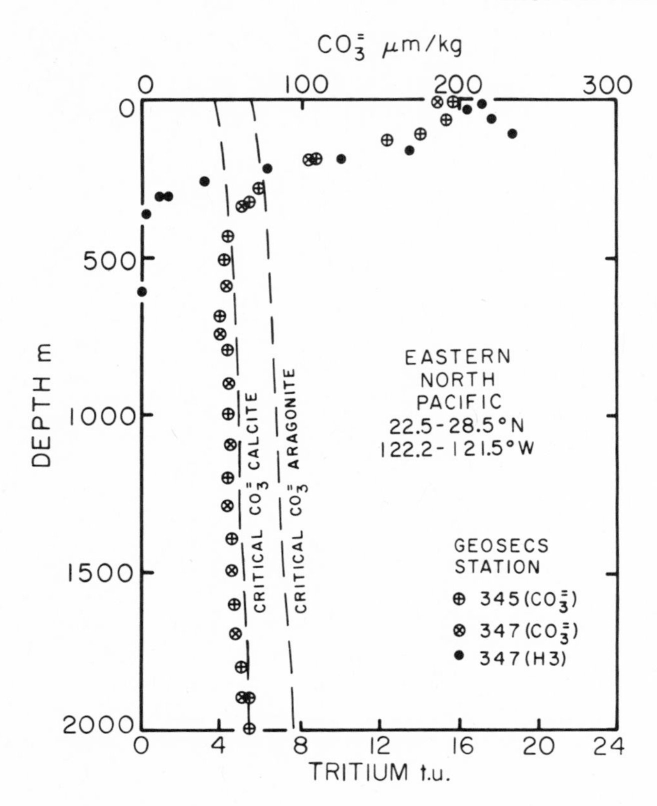

Fig. 1b. Similar results for the North Pacific Ocean (ST 345 23°N; 122°W; ST 347 29°N; 122°W). In accord with the study by *Broecker and Takahashi* (in press) the Pacific Ocean leg report ΣCO_2 results were reduced by *15 μm/kg* to bring the Pacific Ocean results into agreement with Pacific Ocean p_{CO_2} results and with the Atlanti Antarctic results. This correction increases the $CO_3^=$ concentrations by about *8 μm/kg*. The tritium results are those of *Roether* (1974) for samples collected in November 1971.

the neutralization of fossil fuel CO_2. As the bottom water with which they are in contact is already understaturated, any further decrease in $CO_3^=$ content will produce an immediate effect.

3) Those lying above the lysocline and below a water depth of 2800 meters. These sediments are located mainly on the flanks of the oceanic ridges and rises and are on the average very rich in calcium carbonate (i.e., >*80%*). Although they currently lie in non-corrosive water, as the $CO_3^=$ content of the deep sea falls they will eventually come under attack. The time elapsed before the ons of dissolution in any given area will increase with elevation above

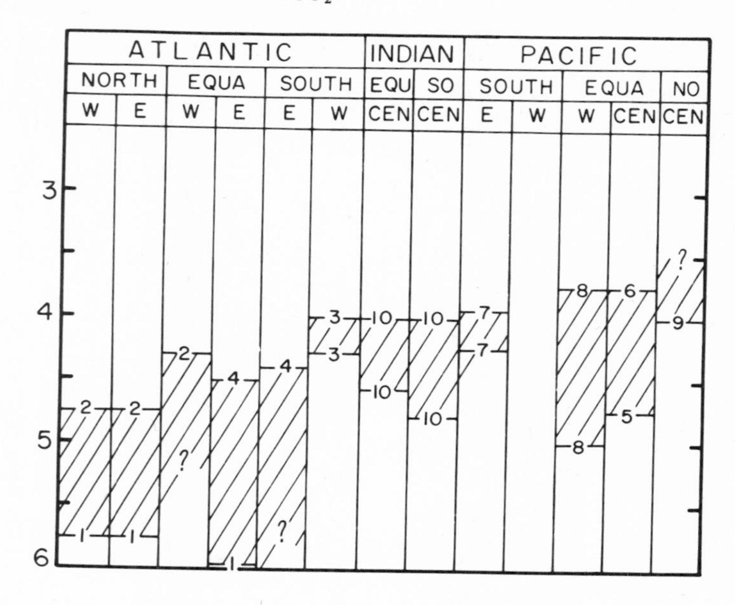

Fig. 2. Summary of lysocline and carbonate compensation depths for various oceanic regions as given by *Broecker and Takahashi* (in press). The numbers refer to the source of the information as given in the original paper.

the lysocline. (The shallower the sediment the more the $CO_3^=$ content will have to fall before dissolution is triggered.)

4) Those sediments lying above the ridge crests but beneath the 100 fathom contour. These sediments lie primarily along the continental rises and in marginal seas. They are in general low in calcite content (e.g., ∿25%) because of dilution with continental detritus. Since the acidification of the waters which they contact will occur more rapidly than that of the subridge crest waters, their contribution could be significant during the early stages of the neutralization process.

5) Shallow water sediments (e.g., <200 meter depths). The waters in this depth range will respond very quickly (i.e., a few years) to changes in atmospheric CO_2 content. Those sediments in contact with mixed layer water (i.e., upper 30-150 meters) will not, however, be subject to attack because the mixed layer is too highly supersaturated with respect to calcite and aragonite to ever become corrosive. However, as the $CO_3^=$ content drops sharply with depth below the mixed layer in some regions of the ocean, dissolution of sediments in contact with upper thermocline waters could contribute to the neutralization of fossil fuel CO_2. Assessment of this potential will prove complex. Region by region studies will be

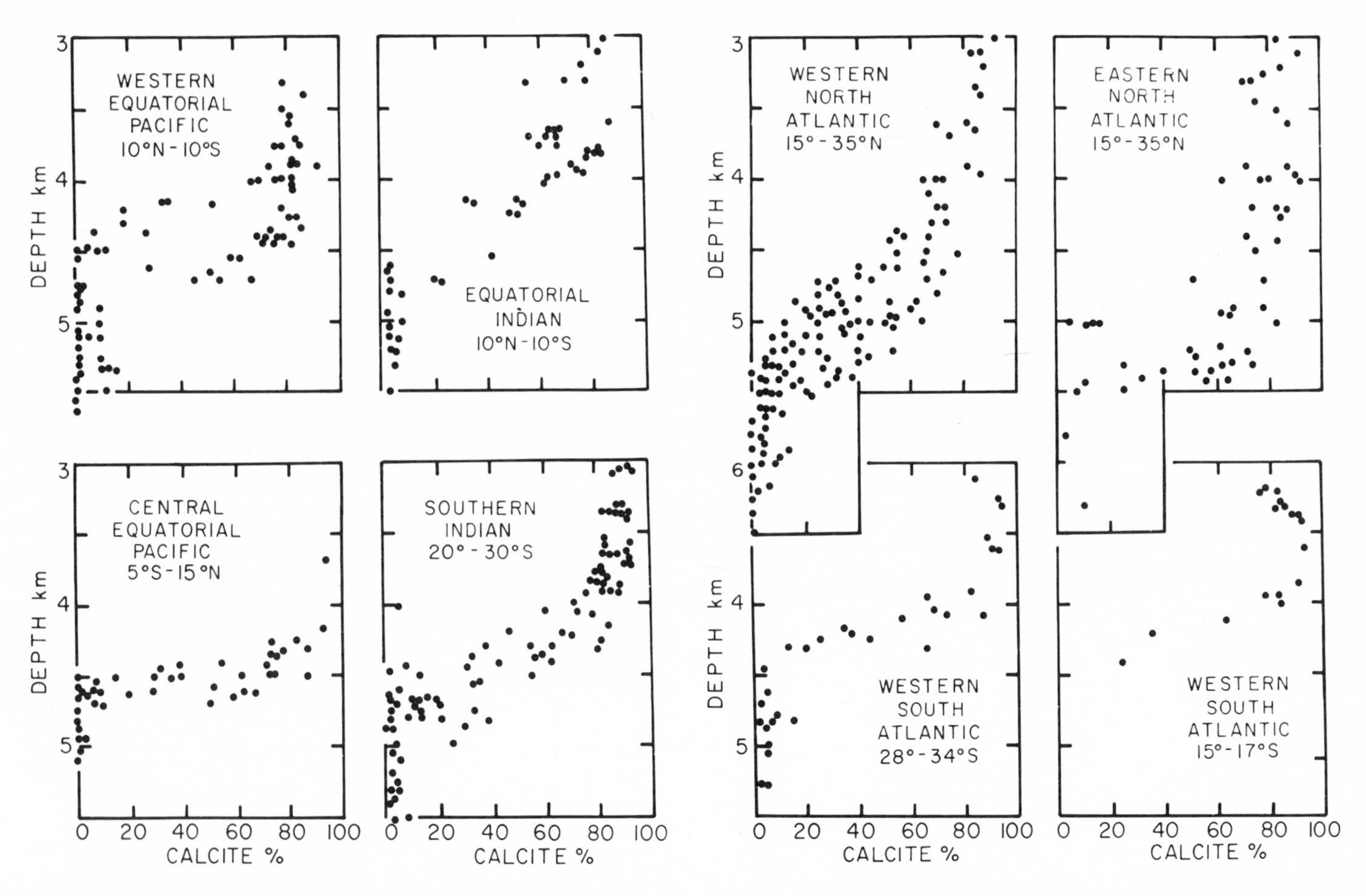

Fig. 3. $CaCO_3$ versus water depth plots as given by *Broecker and Takahashi* (in press).

necessary. Examples are given in Figure 1. Certainly on the long time scale these sediments will prove far less important than deep sea sediments in neutralizing CO_2. Their area and $CaCO_3$ content are just too low.

The depth of the lysocline varies from place to place in the world ocean ranging from 4.7 *km* in the North Atlantic Ocean to less than 3.0 *km* in the North Pacific Ocean. The thickness of the transition zone beneath the lysocline varies from 200 to 1000 meters (see Figure 2). Using this information, the area of each of these sediment types has been estimated (see Table 1) using the area versus water depth data given by *Sverdrup et al.* (1942).

As of today about 1.4×10^{16} moles of CO_2 have been generated through the burning of gas, oil and coal (*Broecker et al.*, 1971). If the production rate is increased by 2% per year, then in the year 2100, 18×10^{16} moles of CO_2 will have been produced. Our total reserves of coal, oil and gas, if burned, would produce about 50×10^{16} moles of CO_2 (*Hubbert*, 1972). How much sediment $CaCO_3$ is available for the neutralization of this CO_2? An estimate of this amount can be obtained from the following expression:

$$\text{Sediment density} \times \text{frac. } CaCO_3 \times \frac{\text{burrowing depth}}{\text{frac. non-}CaCO_3} \qquad (2)$$

TABLE 1. Sediment Area Distribution*

(Units 10^{14} m^2)

Region	Depth Range	Atlantic	Indian	Pacific	Total
Above Shelf break	<200	0.14	0.03	0.10	0.55
Shelf break to Ridge crest	200-2800	0.22	0.11	0.21	0.54
Ridge crest to Lysocline	-	0.35	0.18	0.18	0.71
Lysocline to Compensation	-	0.21	0.14	0.14	0.49
Below Compensation	-	0.14	0.30	1.15	1.59
TOTAL	-	1.06	0.76	1.78	3.60

* Based on the area versus water depth data given in Sverdrup (1942) and the lysocline and compensation depths given in Fig. 2.

The assumption is that once a carbonate-free layer one burrowing depth thick mantles the sea floor, the underlying calcium carbonate will be immune to dissolution regardless of the acidity of the over lying water. From radiocarbon data on deep sea cores it appears that the mean depth of burrowing is about 9 *cm* (see *Peng et al.*, 1977). The following average $CaCO_3$ contents are adopted for the sediment provinces mentioned above:

shelf break - ridge crest 25%
ridge crest - lysocline 85%
lysocline - compensation 40%

The corresponding amounts of available $CaCO_3$ are shown in Table 2. Two points stand out. First, the amount of "available" $CaCO_3$ is about equivalent to the amount of CO_2 locked up in recoverable fuel Second, about half of this calcium carbonate lies in the Atlantic Ocean.

BASIS FOR FUTURE DISSOLUTION RATE ESTIMATES

In order to calculate the rate at which the deep sea $CaCO_3$ will dissolve, four basic pieces of information are needed:

1) projected atmospheric CO_2 contents;
2) the carbonate ion content of newly formed deep water;
3) the rate of fossil fuel CO_2 neutralization by sediment per unit drop in the carbonate ion content of the ocean water in contact with the sediment; and
4) the ventilation times for various deep water masses.

TABLE 2. Available $CaCO_3$ in Deep Sea Sediment*

(Units 10^{16} moles)

Region	Atlantic	Indian	Pacific	Total
Shelf break to Ridge crest	0.7	0.3	0.6	1.6
Ridge crest to Lysocline	18.0	9.2	9.2	36.4
Lysocline to Compensation	1.3	0.8	0.9	3.0
TOTAL	20.0	10.3	10.7	41

* = Dry density (i.e., ~1 gm/cm^3) x Area x

$$\frac{\text{Mixing Depth (i.e., 9 cm)}}{\text{Frac. non-}CaCO_3} \times \text{Frac. } CaCO_3$$

For the decade prior to the sharp increase in oil prices, the use of chemical fuels was increasing by about 4.5% per year. Since the petroleum price hike this pace has slowed. As comprehensive energy policies have yet to be developed, any projection of chemical fuel use into the future is bound to be little more than a guess. For the calculations presented here, a 2% per year increase in the growth rate of the atmospheric CO_2 content until the year 2100 (see Figure 4) has been adopted and it has been further assumed that no further increase will occur during the following century. This choice allows a contrast of the situation during a period of exponential growth to that during a period of atmospheric stability.

As about half of the deep water in the ocean is generated at the northern end of the Atlantic Ocean, it is appropriate to take the waters in this region as an example. Table 3 summarizes the composition of a number of surface waters in the northern Atlantic Ocean. These waters are more saline and, of course, much warmer than those descending to form North Atlantic Deep Water (NADW). If these waters are cooled and freshened at equilibrium with the atmosphere, then, as shown in Table 4, they must change significantly in their ΣCO_2 content. The question is whether these waters maintain equilibrium with the atmosphere as they are cooled. The best way to answer this question is to look at newly produced deep water. As shown in Figure 5 there are three major contributors to NADW; water produced at the surface of the Labrador Sea (LSW), water produced in the Norwegian Sea which enters the Atlantic Ocean by spillage over the Denmark Straits (DSW) and water produced in the Norwegian Sea which enters the Atlantic Ocean by spillage over the Iceland-Scotland rise followed by passage through the Gibbs Fracture Zone

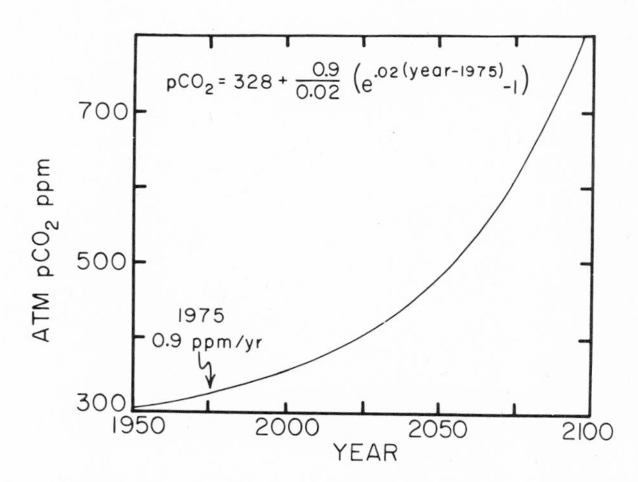

Fig. 4. Partial pressure of CO_2 versus time calculated assuming that the 0.9 ppm/yr increase rate measured by Dr. C. Keeling and his coworkers of Scripps Institution of Oceanography over the last several years will increase by 2% per year through the next century.

TABLE 3. Composition of Some North Atlantic Warm Surface Water Samples*

GEOSECS STATION NO.	24	25	26	27	28	29	30	119
T °C	11.2	20.3	21.8	22.4	25.0	25.2	26.6	21.1
S ‰	34.7	35.6	35.8	36.1	36.4	36.0	36.6	36.7
O_2 μm/kg	291	229	220	218	210	209	205	230
SiO_2 μm/kg	1.9	-	0.5	0.8	0.6	1.3	0.9	0.8
PO_4 μm/kg	0.41	-	0.09	0.03	0.02	0.05	0.04	0.04
NO_3 μm/kg	5.0	-	0.0	0.1	0.0	0.0	0.1	0.0
ALK μeq/km	2278	2308	2308	2359	2377	2357	2384	2389
ΣCO_2 μm/kg	2088	2073	2064	2085	2039	2007	2020	2040
$H_2BO_3^-$	67	79	82	90	110	112	116	113
$2CO_3^{=}+HCO_3^-$	2211	2229	2226	2269	2267	2245	2268	2276
$HCO_3^{-2}/CO_3^{=}CO_2$	1555	1345	1310	1285	1233	1233	1187	1306
HCO_3^- μm/kg	1931	1885	1872	1873	1789	1749	1752	1784
$CO_3^{=}$ μm/kg	140	172	177	198	239	248	258	246
CO_2 μm/kg	17.0	15.3	15.1	13.8	10.8	10.0	10.0	9.9
P_{CO_2} 10^{-6} atm	407	480	490	455	383	355	368	315

* Based on the titration alkalinity and total dissolved inorganic carbon data, nutrient element and dissolved oxygen and hydrographic data given in the GEOSEC leg reports. The constants used are those adopted by Takahashi et al. (in pres

(GFZW). As shown in Table 5 these three sources have similar alkalinities and ΣCO_2. All are deficient in O_2, presumably because of respiration at depth in their source regions. If the composition of these waters prior to alteration by *in situ* respiration is reconstructed by bringing the O_2 content to the atmospheric contact value (i.e., 5% supersaturated), then the CO_2 partial pressure drops to roughly the atmospheric value and the ΣCO_2 content and $CO_3^{=}$ contents become similar to that for the cooled and freshened northern Atlantic Ocean surface water (see Table 4). Hence, the contact time between the atmosphere and the cooling water appears to be long enough to bring the ΣCO_2 content to equilibrium with the atmosphere. Thus, an estimate as to how the $CO_3^{=}$ content of downwelling deep water will change with atmos-

TABLE 4. Station 119 Surface Water Freshened and Cooled at Equilibrium with the Atmosphere*

	21.1°C 36.7%	2.25°C 34.93‰
$HCO_3^{-2}/CO_3^{=} \cdot CO_2$	1306	1725
ALK μeq/km	2389	2273
$H_2BO_3^-$ μm/kg	113	65
$2CO_3^= + HCO_3^-$	2275	2209
HCO_3^-	1785	1953
$CO_3^=$	245	128
CO_2	9.9	17.5
P_{CO_2} 10^{-6} atm	315	329
ΣCO_2 μm/kg	2040	2098

* The constants used are those adopted by Takahashi et al. (in press).

pheric CO_2 content can be obtained by considering either "freshened and cooled" northern Atlantic Ocean surface water (Table 4) or "oxygenated" NADW (Table 5). An example using the latter is shown in Table 6. For a 50% increase in atmospheric CO_2 content, the $CO_3^=$ content drops by 27%. As shown in Table 7 this result is not dependent on the exact composition of the water. When the same calculation is made for warm surface water, the carbonate ion content drops by 24% for a 50% increase in atmospheric CO_2 content. In Figure 6 a plot of the fractional change in $CO_3^=$ content of "oxygenated" NADW against atmospheric CO_2 content is given.

The most reliable way to assess the rate of attack caused by a given lowering of the $CO_3^=$ content of the water in contact with the sediment is to use the Holocene record for calibration. This record is interpreted to indicate that no measurable calcite dissolution will occur until the $CO_3^=$ content drops below what is herein called the critical carbonate ion content given by the following equation (*Broecker and Takahashi*, in press).

$$[CO_3^=]_{CRIT} = 93\ e^{0.14(z - 4)}\ \mu m/kg \qquad (3)$$

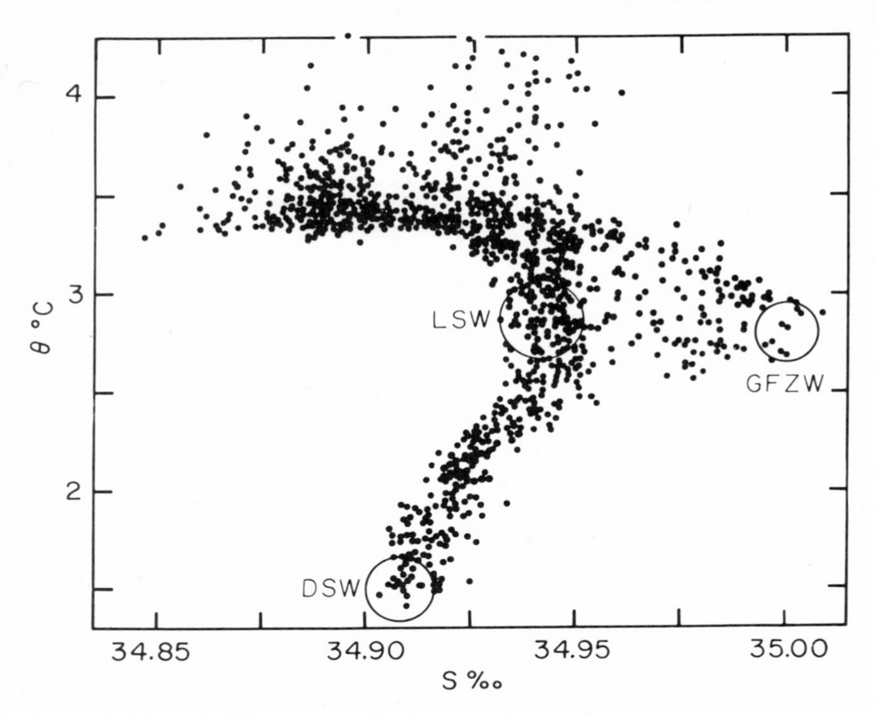

Fig. 5. Potential temperature versus salinity diagram for deep waters found at the northern end of the western basin of the North Atlantic Ocean. The end members are DSW, water spilling over the Denmark Straits; GFZW, water entering the western basin through the Gibbs Fracture Zone; and LSW, water generated at the surface of the Labrador Sea. A mixture of these three end members forms the northern component to basal North Atlantic Deep Water (NADW).

The shape of the transition zone calcite content profile can be satisfactorily explained if the dissolution rate, *R*, is assumed to follow the equation

$$R = \sqrt{f}\ 0.025\ ([CO_3^{=}]_{CRIT} - [CO_3^{=}])\ moles/m^2\ yr \qquad (4)$$

Where f is the fraction of calcite in the sediment (*Broecker and Takahashi*, in prep.)*. For small drops in $CO_3^{=}$ content (i.e., up to *15 µm/kg*) this relationship should be quite valid. For larger

*EDITOR"S NOTE: Since this paper was written, the authors have found the coefficient of dissolution, 0.025, to be a factor of two lower than that used in this contribution.

TABLE 5. North Atlantic Deep Water Components (GFZW, DSW, LSW); Their Average and Their "Oxygenated" Average+

Water Type	GFZW	DSW	LSW	NADW	NADW*
Station	24	25	26	-	-
Depth (m)	2470	4550	1990	-	-
θ °C	2.81	1.78	3.62	2.25	2.25
S ‰	35.00	34.91	34.95	34.93	34.93
O_2 μm/kg	276	288	274	280	345
SiO_2 μm/kg	13.9	-	12.0	13	-
NO_3^- μm/kg	16.1	-	16.8	17	10
$PO_4^{\equiv}$ μm/kg	1.06	-	1.16	1.14	0.66
ALK μeq/kg	2292	2290	2288	2290	2290
ΣCO_2 μm/kg	2168	2172	2168	2168	2119
$H_2BO_3^-$ μm/kg	50	48	48	49	63
$2CO_3^= + HCO_3^-$	2242	2242	2240	2241	2227
HCO_3^-	2044	2051	2046	2046	1975
$CO_3^=$	99	95	97	98	126
CO_2	25	25	25	24	18
P_{CO_2} 10^{-6} atm	436	433	463	430	310

* to 5% O_2 supersaturation assuming $\Delta O_2/\Delta\Sigma CO_2 = -1.33$

\+ The hydrographic, nutrient element and dissolved oxygen data are from the GEOSECS leg reports. The constants used are those adopted by Takahashi et al. (in press).

drops (*20-50 μm/kg*), its validity depends on whether dissolution follows linear kinetics or Morse-Berner exponential kinetics (*Morse and Berner*, 1972). As it is not yet known what constitutes the rate limiting step for dissolution, it is not possible to say for sure which type of kinetics applies. Therefore, both will be considered.

Through radiocarbon dating, a general idea of the ventilation times for most parts of the deep sea has been developed. They range up to about 1500 years for the deep water in the North Pacific Ocean (*Bien et al.*, 1965). The deep Atlantic Ocean, which is of considerable interest because of its large calcite reserves,

TABLE 6. Change in Composition of Sinking NADW with Increasing Atm. CO_2 Content*

Year	1975	2050	Δ	$\frac{2050}{1975}$
$HCO_3^{-2}/CO_3^{=} \cdot CO_2$	1725	1725	-	-
ALK μeq/kg	2290	2290	0	-
$H_2BO_3^-$ μm/kg	63	46	-17	-
$2CO_3^= + HCO_3^-$	2227	2244	+17	-
HCO_3^- μm/kg	1975	2062	+85	-
$CO_3^=$ μm/kg	126	91	-35	0.73
CO_2 μm/kg	18	27	+ 9	1.50
P_{CO_2} 10^{-6} atm	310	468	-	1.50
ΣCO_2 μm/kg	2119	2180	+61	1.029

$$\frac{\%\ \text{inc atm } CO_2}{\%\ \text{inc ocean } CO_2} = \frac{50}{2.9} = 17.2$$

* The constants used are those adopted by Takahashi et al. (in press).

is ventilated on the time scale of several hundred years (*Broecker et al.*, 1960; *Stuiver*, 1976). The presence of bomb-produced tritium and radiocarbon in upper thermocline waters demonstrates that this region of the sea is ventilated on the time scale of a few years to a few tens of years (see Figure 1).

THE TIME CONSTANT FOR SEDIMENT DISSOLUTION

A rough idea of the rate at which deep sea sediments will dissolve can be obtained. Assume that 23 x 10^{16} moles of CO_2 are generated through fossil fuel burning and allowed to come to equilibrium with the ocean. For simplicity, the ocean will be taken to be entirely NADW. In such a case the p_{CO_2} of the atmosphere would be 660 x 10^{-6} atm. The deep ocean ΣCO_2 content would then be

TABLE 7. Change in Composition of Warm Surface Water with Increasing Atm. CO_2 Content*

Year	1975	2050	Δ	$\frac{2050}{1975}$
$HCO_3^{-2}/CO_3^{=} \cdot CO_2$	1306	1306	-	-
ALK μeq/kg	2389	2389	0	-
$H_2BO_3^-$ μm/kg	114	87	-27	-
$2CO_3^= + HCO_3^-$	2275	2302	+27	-
HCO_3^- μm/kg	1785	1924	143	-
$CO_3^=$ μm/kg	245	187	-58	0.76
CO_2 μm/kg	10	15	+ 5	1.50
P_{CO_2} 10^{-6} atm	317	484	-	1.50
ΣCO_2 μm/kg	2040	2130	90	1.044

$$\frac{\%\text{ inc. atm } CO_2}{\%\text{ inc. ocean } CO_2} = \frac{50}{4.4} = 11.4$$

* The constants used in these calculations are those adopted by Takahashi et al. (in press).

120 μm/kg higher than during pre-industrial time and the $CO_3^=$ content of the deep water about half its pre-industrial value (i.e., down by about 55 μm/kg). Of the 23 x 10^{16} moles of CO_2 generated, 6.7 x 10^{16} would reside in the atmosphere and 16.3 x 10^{16} in the ocean. The amount of excess CO_2 per m^2 of ocean surface would be 640 moles.

Under such circumstances, deep sea sediments would be subjected to waters with carbonate ion deficiencies (relative to the critical carbonate ion content) ranging from about 55 μm/kg for the transition zone to between 15 and 25 μm/kg at the ridge crests. The average for all the $CaCO_3$-bearing sediments would be about 40 μm/kg. Using the dissolution coefficient of 0.025 moles/m^2 yr per μm/kg undersaturation obtained for Holocene dissolution in the ocean (*Broecker and Takahashi*, in prep.) and linear kinetics, yields an average of 1 mole/m^2 yr excess dissolution from the carbonate-bearing

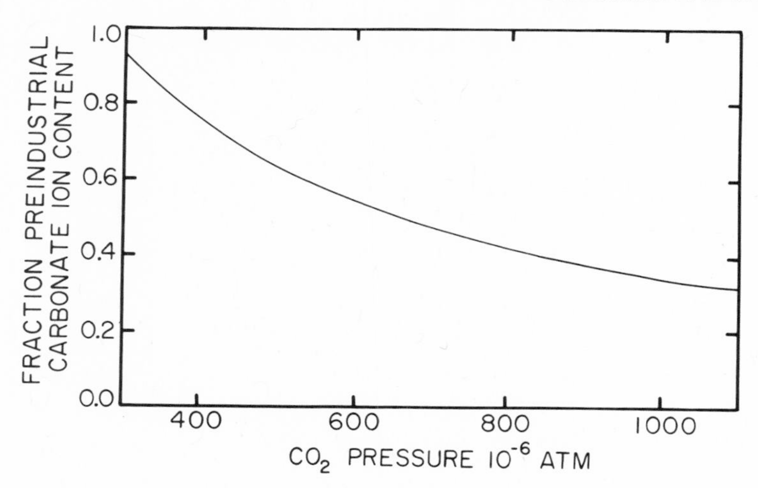

Fig. 6. $CO_3^=$ content of input northern component water (Table 6) relative to its pre-industrial value as a function of atmospheric CO_2 content. The pre-industrial $CO_3^=$ ion content is calculated to be *135* μm/kg (compared to *126* μm/kg in 1975).

sediments of the deep ocean. Since the area of deep floor bearing $CaCO_3$-rich sediment is about half the total area (see Table 1), this corresponds to 0.5 mole/m^2 yr dissolution rate as referred to the entire ocean floor. Thus, the initial rate of dissolution in this well mixed ocean would be an amount of $CaCO_3$ equivalent to one part in 1280 of the excess CO_2 in the ocean-atmosphere system. This yields a time constant (i.e., about 1280 years) for neutralization comparable to the time scale for deep ocean ventilation. Thus, any realistic model must consider dissolution and deep mixing together.

If Morse-Berner exponential kinetics (*Morse and Berner*, 1972) are used, then as shown in Figure 7, dissolution rates many times higher would be expected. They would be so high in fact, that the sediments would neutralize the excess CO_2 roughly as fast as it arrived in the deep sea.

Of course, as the dissolution proceeded, dilution of the sedimentary calcite by the buildup of a non-calcite residue would slow down the dissolution process. Eventually dissolution would become limited by the rate at which the residue was stirred into the sediment column by benthic burrowers. Thus, any realistic model of the fate of CO_2 will have to simultaneously deal with oceanic mixing interface dissolution and sediment stirring.

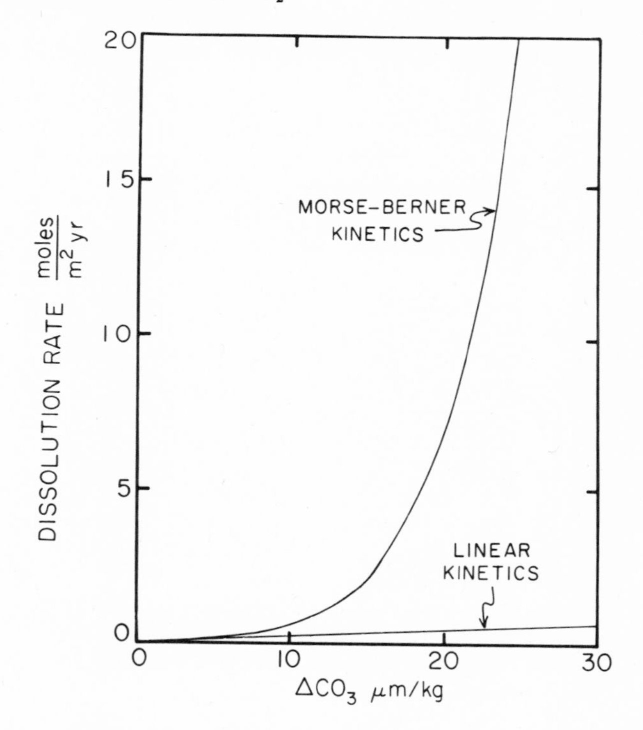

Fig. 7. Comparison of the dissolution rates for the linear kinetic model as calibrated using the Holocene $CaCO_3$ distribution with that for the exponential kinetic model as calibrated using the Holocene $CaCO_3$ distribution. The assumptions of these models are outlined by *Takahashi and Broecker* (1977).

A MODEL FOR THE DISSOLUTION OF SEDIMENT IN THE WESTERN BASIN OF THE ATLANTIC OCEAN

Since the NADW ventilation time is small compared to that for the deep Pacific Ocean, events in the Atlantic Ocean will dominate during the next hundred or so years. The deep mixing model selected here is designed with this fact in mind.

The radiocarbon data from the GEOSECS Program show that the water in the western basin of the central deep Atlantic Ocean is replaced on a time scale of about 200 years (*Stuiver*, 1976). Furthermore the "age" of the water does not show any strong dependence on depth or latitude. Thus, in modeling it is appropriate to assume that throughout the western basin the NADW component is being replaced at the rate of one part in 200 each year. The water exiting the western basin of the central Atlantic Ocean is assumed to pass in part directly to the Antarctic and in part to the eastern basin of the Atlantic Ocean.

The water below the ridge crests in the Atlantic Ocean is a mixture of NADW and Antarctic Bottom Water (AABW). The salinity section (Figure 8) through the western basin shows the geographic and depth dependence of the composition of these mixtures. The end members have salinities of 34.93% (NADW) and 34.67% (AABW). The 34.80 isohaline thus marks a 50-50 mixture of the two end members.

As AABW contains a large component of recirculated Deep Pacific Ocean Water which has not reequilibrated with the atmosphere, its $CO_3^=$ content will be slow to change. The time scale for its alteration will be many hundreds of years. Thus, for the calculations to be carried out here, the $CO_3^=$ content of AABW will be assumed to remain unchanged. Therefore, as the mixture is renewed, only the NADW component will change in $CO_3^=$ content (at least during the next hundred or so years).

Plots of $CO_3^=$ content for deep waters in the northern and southern parts of the central Atlantic Ocean are shown in Figure 9 along with the critical carbonate ion curve (i.e., the $CO_3^=$ content at which the dissolution rate becomes geologically significant). In the northern western basin where NADW dominates the deep water column, the present day crossover depth (i.e., depth where the *in situ* $CO_3^=$ content becomes less than the critical $CO_3^=$ content) is about 4700 meters. In the western basin of the South Atlantic Ocean, where AABW underrides and mixes upward into the NADW, the crossover is at a shallower depth (∿4000 meters).

Figure 10 illustrates what the situation will be when the ion content of NADW has dropped by 17 μm/kg. The crossover depth at this time will have risen to about 3700 meters throughout the Atlantic Ocean. The sediments in the transition zone between the pre-industrial lysocline and the compensation depths will be subjected at that time to much more vigorous dissolution than during the pre-industrial era. Superlysocline sediments will also become subject to attack by the acidified deep water.

In the sections that follow, an attempt will be made to model the attack of the sediments in the western basin of the Atlantic Ocean by the progressively acidified NADW. A simple box model approach is used. The NADW component of waters below 3000 meters in the western basin of the Atlantic Ocean is assumed to be well mixed. Three processes tend to change the ΣCO_2 content and $CO_3^=$ content of this water.

1) The input of new NADW progressively enriched in total dissolved inorganic carbon: The volume of water in the western basin of the Atlantic Ocean beneath 3000 meters between 30°S and 40°N is 32 x 10^{15} m^3. Of this about 75% is NADW. If this water is to be replaced once every 200 years, the flux of new NADW must be 1.2 x 10^{14} m^3/yr (i.e., about 4 Sverdrups). The $CO_3^=$ content of this input is calculated as follows.

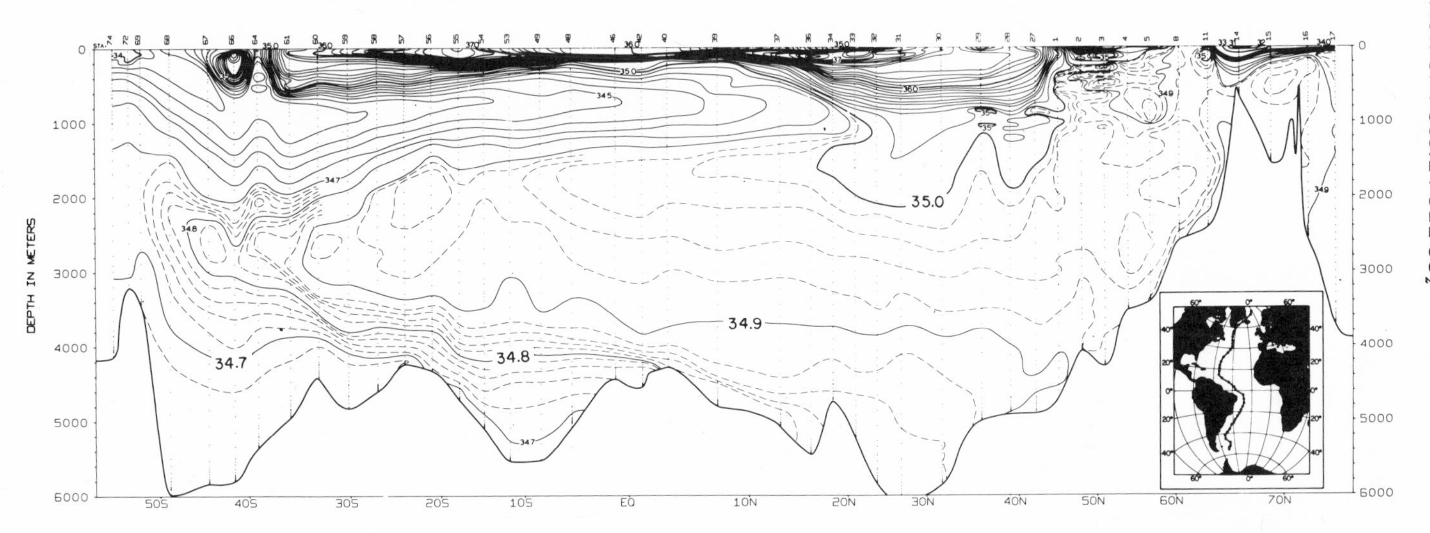

Fig. 8. Salinity section along the GEOSECS track in the western basin of the Atlantic Ocean. The mixing zone of interest to the $CaCO_3$ dissolution problem is that below 3 kilometers where AABW water ($S = 34.67^0/oo$) mixes up into NADW water ($S > 34.90^0/oo$).

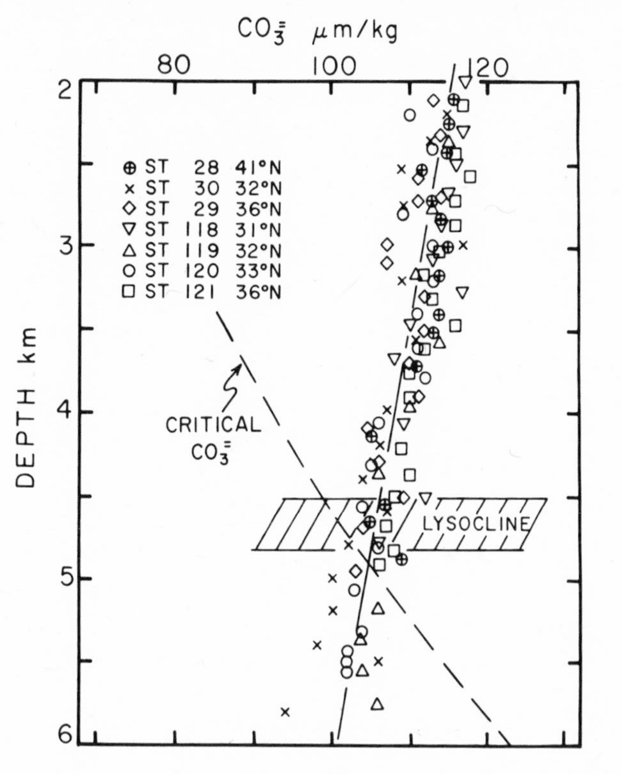

Fig. 9a. Carbonate ion content versus depth in the western North Atlantic Ocean as calculated from the GEOSECS titration data using the constants adopted by *Takahashi et al.* (in press). The lysocline depth is based on the data of *Kipp* (1976). The critical carbonate ion content curve is that of *Broecker and Takahashi* (in press).

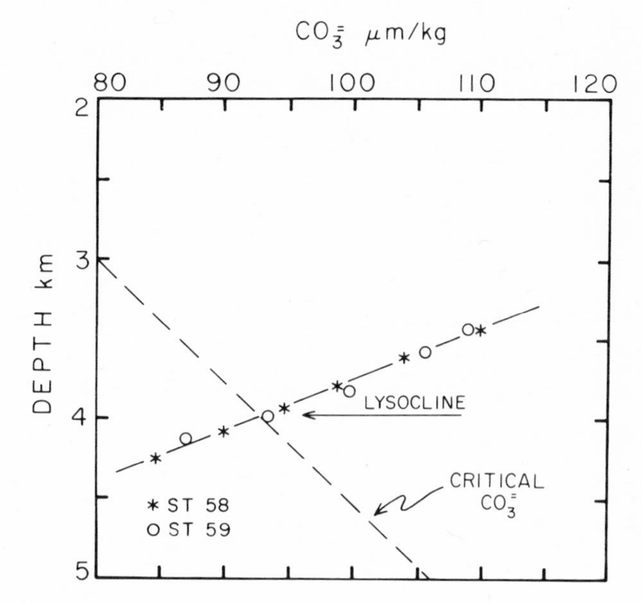

Fig. 9b. Carbonate ion content in the basal NADW-AABW mixing zone of the western South \tlantic as calculated from potential temperature using the relationship given by *Broecker and Takahashi* (in press). The critical carbonate ion content curve is given for reference.

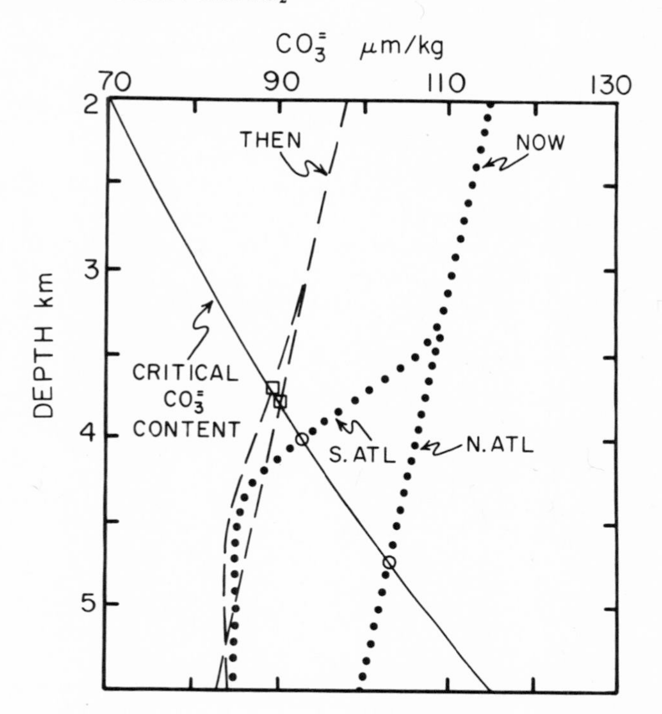

Fig. 10. $CO_3^=$ versus depth curves for the western basin of the Atlantic Ocean (dotted curves) prior to the introduction of fossil fuel CO_2 and after fossil fuel CO_2 introduction has reduced carbonate ion content of the NADW component of the mixing zone water by 17 µm/kg (dashed curves). The circles represent the present depths at which the *in situ* $CO_3^=$ content currently cross the critical value and the squares the depths at which these crossover will lie after the NADW has been "acidified" to the level where its $CO_3^=$ ion content has dropped 17 µm/kg.

$$[CO_3^=]^t_{INPUT} = \frac{[CO_3^=]^t_{INPUT\ NADW}}{[CO_3^=]^{1850}_{INPUT\ NADW}} [CO_3^=]^{1850}_{NADW} \qquad (5)$$

The ratio $[CO_3^=]^t_{INPUT\ NADW} / [CO_3^=]^{1850}_{INPUT\ NADW}$ is that given in Figure 6. In order to eliminate the subsurface processes taking place in the natural (i.e., pre-industrial) system, the actual input value is taken to be the pre-industrial mean $CO_3^=$ content of *in situ* NADW (i.e., 115 μm/kg) times this ratio. This assumes that the natural respiration processes and the natural dissolution processes continue at the same rate as in pre-industrial time. Since these processes are limited by the availability of NO_3^- and $PO_4^{\equiv}$, rather than by CO_2, this assumption is likely valid. As the alkalinity of the incoming water should remain unchanged, the carbonate system is uniquely defined by assigning the carbonate ion content of the input water.

2) Loss of excess industrial CO_2 by outflow from the western basin to the eastern basin and to the Antarctic: The excess CO_2 concentration in the exiting water is assumed to be equal to the mean excess for the water mass itself.

3) Addition of Ca^{++} and $CO_3^=$ to the deep water through excess dissolution of deep sea sediments (the natural dissolution has been taken care of through our handling of the input concentrations): For sediments below the lysocline the rate of excess dissolution at any point on the sea floor depends on the decrease in the $CO_3^=$ content of the water from its pre-industrial value. In a separate paper (*Broecker and Takahashi*, in prep.) it will be shown that for sediments moderately rich in $CaCO_3$ (i.e., >40%), dissolution proceeds at a rate of .025 moles/m^2 yr per μm/kg decrease in the $CO_3^=$ content below the critical carbonate ion content. It will be assumed that for each additional decrease of 1 μm/kg in $CO_3^=$ content the rate of dissolution of these carbonate-rich sediments will rise by 0.025 moles/m^2 year (i.e., linear kinetics). For sediments above the natural lysocline, no dissolution will take place until the $CO_3^=$ content of the water reaches the critical carbonate ion value. Once it drops below this value the same dissolution rate versus $\Delta CO_3^=$ relationship is used as for the sub-natural lysocline sediment. In the North Atlantic Ocean the lysocline will rise 100 meters per 1.7 μm/kg drop in the $CO_3^=$ content of NADW. Thus as the $CO_3^=$ content drops, both the area of sediment exposed to excess dissolution <u>and</u> the rate of the excess dissolution at any given depth will increase.

Before launching out on the model calculation, one basic assumption needs justification. Are the products of the dissolution well distributed in the deep water column? As the volume of water per unit of area of sediment contacted decreases with each depth slice down the water column, if vertical mixing does <u>not</u> occur then the impact of the dissolution process will increase strongly with water depth.

Whether or not the products of calcite dissolution are mixed throughout the water column can be best tested by considering the natural steady state. In the northern western basin nearly all the carbonate falling below about 5100 meters dissolves. In the southern western basin the corresponding depth is 4400 meters. The area of sediment nearly free of $CaCO_3$ in the western basin is about 10×10^{12} m^2. If the mean rain rate of $CaCO_3$ is taken to be 0.8

gm/cm² 10³ yrs (i.e., 8×10^{-6} moles/cm^2 yr) and if the products of this dissolution are assumed to have been confined to the volume below the top of the dissolution zone (volume of 6×10^{15} m^3) then during its residence time of 200 years the water would increase in alkalinity by *55 μeq/kg*. There would, of course, then be no alkalinity increase for waters above the dissolution zone. The observed increase is on the order of only *10 μeq/kg* (see Figure 11) and shows no strong correlation with depth or geographic location. On the other hand, if the dissolution products are mixed upward to a depth of 3000 meters (volume of 32×10^{15} m^3), then the expected alkalinity increase would be only *10 μeq/kg* -- a value roughly consistent with the GEOSECS results. The near uniformity of the alkalinity increase throughout the mixing zone beneath the Two Degree Discontinuity (TDD) suggests that the products of dissolution are not confined to the level from which they were released, but rather that they are mixed well up into the water column. Mixing along isopycnal surfaces (i.e., along the isohalines in Figure 8) could produce the observed distribution.

The model calculation is begun in the year 1900. Since the amount of CO_2 added to the NADW mass up to this time was quite small, it will be assumed that the NADW was still at its steady

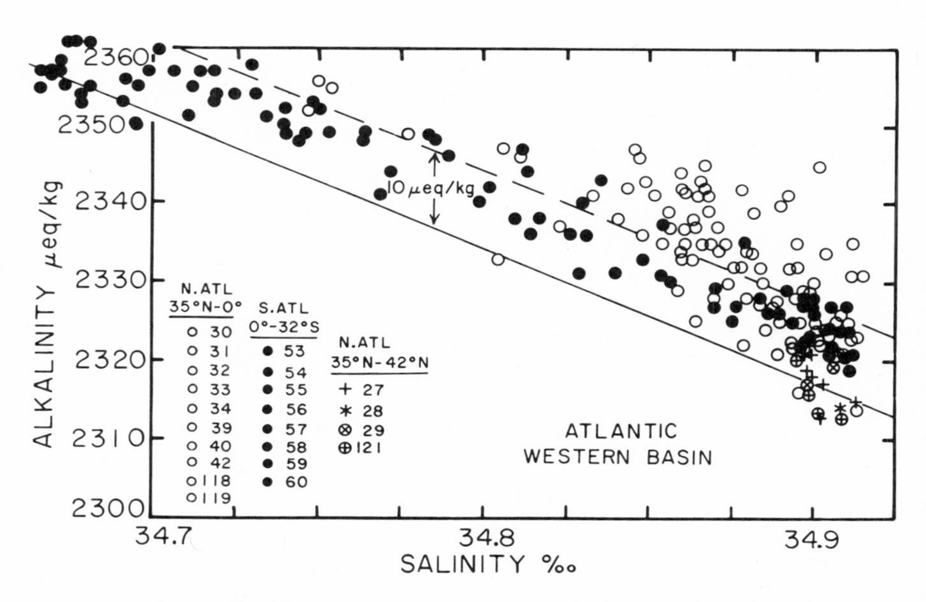

Fig. 11. Alkalinity versus salinity in the mixing zone between AABW and basal NADW. The lower solid line represents the values expected if there were no $CaCO_3$ dissolution in this mixing zone. The upper solid line represents the alkalinity expected if the products of natural $CaCO_3$ dissolution were spread uniformly through this mixing zone.

state $CO_3^=$ content (i.e., 115 m/kg) and at its mean alkalinity (2330 μeq/kg). The calculation is carried out in 10 year increments. The CO_2 pressure in the atmosphere and the composition of the input water are taken to be those for the mid-point of the decade over which the calculation is being carried out. The $CO_3^=$ content of the NADW reservoir used in the dissolution and outflow calculations is taken to be that calculated for the end of the preceding decade. At the end of each decade the net change in the ΣCO_2 content and in the alkalinity of the NADW is recomputed. The $CO_3^=$ content obtained from these values is used for the next decade.

The area of sediment in the western basin of the Atlantic Ocean lying between the lysocline and the compensation depth is about 7×10^{12} m^2. These sediments are assumed to yield 0.025 moles of excess $CaCO_3$ dissolution per square meter year for each μm/kg drop in the $CO_3^=$ ion content of NADW. The sediments beneath the compensation depth are assumed to contribute nothing. The area of sediment above the natural lysocline contributing to the excess dissolution is taken to be 0.4×10^{12} m^2 per μm/kg drop in $CO_3^=$ content of the NADW reservoir. As the difference between the critical carbonate ion content and the *in situ* carbonate ion content for this depth interval averages one half the drop in carbonate content of the NADW, a factor of 1/2 must be introduced to account for this gradient in $\Delta CO_3^=$ in the superlysocline region. Thus the dissolution rate, R, is calculated as follows:

$$R = \Delta CO_3^= (A_{L-C} + \frac{1}{2} A_{<L}) b \qquad (6)$$

where

$$\Delta CO_3^= = CO_3^= {}^{INITIAL}_{NADW} - CO_3^= {}^{t}_{NADW}$$

A_{L-C} = area of sediments between the natural lysocline and natural compensation depth

$A_{<L}$ = area of sediment above the natural lysocline which has come under attack (i.e., $0.4 \times 10^{12} \times \Delta CO_3^=$

b = coefficient of dissolution (i.e., 0.025 moles/m^2 yr per μm/kg)

The results of this calculation for the period between 1900 and 2105 are given in Table 8. This is the period where the rate of the CO_2 content increase for the atmosphere is assumed to rise

TABLE 8. Western Basin Atlantic Model Calculation Carried Out Over the Period During Which the Atmospheric CO_2 Rise Rate is Assumed to be Increasing by 2% per Year

	P_{CO_2} 10^{-6} atm	$[CO_3^=]$ Input to NADW μm/kg	$\Delta\Sigma CO_2$ Input to NADW μm/kg	$[CO_3^=]$ NADW μm/kg	$\Delta\Sigma CO_2$ NADW μm/kg	$\Delta CO_3^=$ NADW μm/kg	Increase ΣCO_2 via inflow 10^{11} moles/yr	Increase ΣCO_2 via dissolution 10^{11} moles/yr
1900-1910	294	109	5	115	0.0	0.0	6.0	0.0
1910-1920	297	108	6	115	0.2	0.1	7.2	0.2
1920-1930	300	107	8	115	0.4	0.2	9.6	0.4
1930-1940	303	106	10	115	0.7	0.3	12.0	0.5
1940-1950	308	105	13	115	1.1	0.5	15.6	0.9
1950-1960	313	104	16	114	1.6	0.7	19.2	1.3
1960-1970	320	102	19	114	2.1	1.0	22.8	1.8
1970-1980	328	101	23	114	2.8	1.3	27.6	2.3
1980-1990	338	99	27	113	3.6	1.6	32.4	2.9
1990-2000	350	97	32	113	4.6	2.0	38.4	4.0
2000-2010	365	94	38	113	5.8	2.4	45.6	4.8
2010-2020	383	91	46	112	7.1	3.0	55.2	6.0
2020-2030	405	86	55	111	8.8	3.6	66.0	7.2
2030-2040	432	82	65	111	10.8	4.4	78.0	8.8
2040-2050	465	77	75	110	13.1	5.3	90.0	10.6
2050-2060	506	72	86	109	15.8	6.3	103.2	12.6
2060-2070	555	67	97	108	18.8	7.4	116.4	16.3
2070-2080	615	61	110	106	21.9	8.5	132.0	18.7
2080-2090	689	55	125	105	25.8	9.8	150.0	21.6
2090-2100	779	49	139	104	30.2	11.3	166.8	24.9
2100-2110	888	44	157	101	35.1	13.0	188.4	28.6

by 2% per year.* The yearly increase rate goes from 0.22 ppm/yr in 1905 to 0.90 ppm per year in 1975 to 12.1 ppm in 2105. The results for the period 2105 to 2205, during which the atmospheric CO_2 content is held constant, are shown in Table 9. At the end of this period the dissolution rate of $CaCO_3$ from deep Atlantic Ocean sediment becomes about one third the input rate of excess CO_2.

---

* If, as well may be the case, the acceleration of fossil fuel use is stemmed before the end of the next century, then the carbon ion content of NADW will not drop as fast as indicated by these calculations.

TABLE 9. Western Basin Atlantic Model for Period During Which the Atmospheric CO_2 Content is Assumed to Remain Constant

	$CO_3^{=}$	$\Delta\Sigma CO_2$	$\Delta CO_3^{=}$	ΣCO_2 In	ΣCO_2 Out	ΣCO_2 Dis	ΣCO_2 Net
	← μm/kg →			← 10^{11} moles/yr →			
2100-2110	101	35	13.0	188	-42	29	175
2110-2120	100	41	14.8	188	-49	33	172
2120-2130	99	46	16.4	188	-55	36	169
2130-2140	97	51	18.0	188	-61	40	167
2140-2150	96	56	19.4	188	-67	46	167
2150-2160	94	62	20.6	188	-74	50	164
2160-2170	93	67	21.6	188	-80	52	160
2170-2180	93	72	22.4	188	-86	54	156
2180-2190	92	77	23.2	188	-92	56	152
2190-2200	91	82	23.8	188	-98	58	147
2200-2210	91	86	24.3	188	-102	60	146

$p_{CO_2}^{atm}$ = 888 ppm

ΣCO_2 input = 2282 μm/kg $\qquad$ ΣCO_2 pre-ind. NADW = 2125 μm/kg

ALK input = 2290 μeq/kg $\qquad$ ALK pre-ind. NADW = 2290 μm/kg

$CO_3^{=}$ input = 44 μm/kg $\qquad$ $CO_3^{=}$ pre-ind. NADW = 115 μm/kg

At steady state (i.e., when the NADW has come to equilibrium with this new atmospheric CO_2 content) about half of the incoming excess will be neutralized within the Atlantic Ocean. Table 10 summarizes the change in chemical composition of the input water to the NADW and of the mean NADW itself over this period. Table 11 compares the fluxes of excess CO_2 into the NADW and of $CaCO_3$ dissolution of sediments in contact with NADW with the flux of CO_2 into the atmosphere.

How long can this go on? At a $CO_3^{=}$ depletion of 20 μm/kg, the rate of dissolution is 0.5 m/m^2 yr (i.e., 0.5 mm/decade). For sediment averaging 70% $CaCO_3$, the amount of available $CaCO_3$ is about 2000 moles/m^2. Hence, the time constant for depletion is on the order of 4000 years. This is about threefold longer than the time constant for the replacement of deep Indian Ocean and Pacific Ocean waters.

TABLE 10. Summary of Composition of Input Water to the NADW and of NADW Itself Over Period Covered by the Model Calculations

		<— Input —>		<—— NADW ——>		
	P_{CO_2}	$CO_3^=$	ΣCO_2	$CO_3^=$	ΣCO_2	ALK
1850	283	115	2125	115	2125	2290
1975	328	101	2148	114	2128	2290
2105	888	44	2282	101	2160	2300
2205	888	44	2282	91	2211	2325

TABLE 11. Summary of Fossil Fuel CO_2 Fluxes Contrasting the Total Flux into the System with that Leaving the Surface Ocean for the Deep North Atlantic and with that of Fossil Fuel Induced Dissolution of Western Basin Sediments

	10^{12} moles/yr		
	Total CO_2 Influx	Flux to NADW	Flux to NADW Sediments
1850	0	0	0.0
1975	177	3	0.2
2105	2140	19	2.9
2205	-	19	6.0

To be more realistic this model would have to include:

1) The dependence of dissolution rate on the $CaCO_3$ content of the sediment.

a) If the rate limiting step is a "stagnant boundary film" at the sediment-water interface then the dependence will be quite small until very low calcite contents are achieved.

b) If the rate limiting step is the resaturation time of the sediment pore waters then the rate of dissolution will vary with the square root of the fraction calcite.

2) The dependence of the dissolution rate on $\Delta CO_3^=$. If the rate limiting step for dissolution is the release of ions from the crystal surfaces, then as shown by the experiments of *Morse and Berner* (1972) an exponential rather than linear dependence on $\Delta CO_3^=$ would have to be used. If so, dissolution will occur about an order of magnitude faster.

3) The relative rates of burrowing and dissolution. At a dissolution rate of 0.5 m/m^2 yr, in the absence of bioturbation an insoluble residue one characteristic diffusion length in thickness would be built up in a decade or so. Therefore, if burrowing occurs on the time scale of centuries, rather than years, it will quite soon become the rate limiting step.

CONCLUSIONS

The rate of deep sea sediment dissolution is certainly fast enough that this process will take place concurrently with the transfer of CO_2 from the atmosphere-shallow ocean reservoir to the deep sea reservoir. The amount of calcite "kinetically" available for dissolution is comparable to the amount of carbon locked up in recoverable fossil fuels. The major uncertainties remaining to be resolved before adequate modeling can be done are:

1) The identification of the rate limiting step for dissolution on the sea floor (i.e., a distinction between linear and exponential kinetics must be made).

2) The quantification of mixing rates of sediments on the sea floor (the mechanical eddy diffusivity as a function of depth in the sediment column must be determined).

ACKNOWLEDGEMENTS

Much of the $CO_3^=$, ΣCO_2, p_{CO_2} and alkalinity data used in this paper was generated by the GEOSECS Program. Financial support for this work was provided by a grant to Lamont-Doherty Geological Observatory from the Energy Research and Development Administration (E(11-1)2185) and by a grant to CUNY from IDOE (OCE72-06419). Lamong-Doherty Geological Observatory Contribution No. 2513.

Bien, G. S., N. W. Rakestraw and H. E. Suess. 1965. Radiocarbon in the Pacific and Indian Oceans and its relation to deep-water movements. *Limnol. Oceanogr. 10*, Supplement R25-R37.

Broecker, W. S., R. Gerard, M. Ewing and B. C. Heezen. 1960. Natural radiocarbon in the Atlantic Ocean. *J. Geophys. Res. 65*. 2903-2931.

Broecker, W. S., Y.-H. Li and T.-H. Peng. 1971. Carbon dioxide -- Man's unseen artifact, In: *Impingement of Man on the Oceans*, D. W. Hood (Ed.), John Wiley and Sons, Inc., N.Y. 297-324.

Broecker, W. S. and T. Takahashi. In press. The relationship between lysocline depth and *in situ* carbonate ion concentration. *Deep-Sea Res.*

Broecker, W. S. and T. Takahashi. In prep. A model for the sea floor dissolution of calcite.

Hubbert, M. King. 1972. Man's conquest of energy: Its ecological and human consequences, In: *The Environmental and Ecological Forum 1970-1971*, USAEC Report TID-25857.

Kipp, N. G. 1976. New transfer function for estimating past sea surface conditions from sea bed distribution of planktonic foraminiferal assemblages in the North Atlantic. *Geol. Soc. Am. Memoir No. 145*, (R. Cline and J. Hays, Eds.).

Morse, J. W. and R. A. Berner. 1972. Dissolution kinetics of calcium carbonate in sea water: II. A kinetic origin for the lysocline. *Amer. Jour. Sci. 272*. 840-851.

Ostlund, H. G., H. G. Dorsey and C. G. Rooth. 1974. Geosecs North Atlantic radiocarbon and tritium results. *Earth and Plan. Sci. Letters, 23*. 69-86.

Peng, T.-H., W. S. Broecker, G. Kipphut and N. Shackleton. 1977. Benthic mixing in deep sea cores as determined by ^{14}C dating and its implications regarding climate stratigraphy and the fate of fossil fuel CO_2. In: *The Fate of Fossil Fuel CO_2*, N. R. Andersen and A. Malahoff (Eds.), Plenum, N.Y., (this volume).

Roether, W. 1974. The tritium and carbon-14 profiles at the GEOSECS I (1969) and GOGOI (1971) North Pacific Stations. *Earth and Planet. Sci. Letters, 23*. 108-115.

Stuiver, M. 1976. The ^{14}C distribution in west Atlantic abyssal waters. *Earth Planet. Sci. Letters 32*. 322-330.

Sverdrup, H. U., M. W. Johnson and R. H. Fleming. 1942. *The Oceans: Their Physics, Chemistry and General Biology*. Prentice-Hall, New York. 1087 pp.

Takahashi, T. and W. S. Broecker. 1977. Mechanisms for calcite dissolution on the sea floor. In: *The Fate of Fossil Fuel CO_2*, N. R. Andersen and A. Malahoff (Eds.), Plenum, N.Y., (this volume).

Takahashi, T., P. Kaiteris, W. S. Broecker and A. E. Bainbridge. 1976. An evaluation of the apparent dissociation constants of carbonic acid in seawater. *Earth and Planet. Sci. Letters 32*. 458-467.

Sedimentation and Dissolution of Pteropods in the Ocean

Robert A. Berner

Yale University

ABSTRACT

Analysis of very limited data for the relative abundance of pteropods (aragonite) and planktonic foraminifera (calcite) in plankton tows suggest that approximately equal numbers of the two groups of organisms fall to the bottom in the pelagic realm. This is quantitatively verified by numerous determinations of the pteropod/planktonic foraminifera ratio in pelagic bottom sediments which have lost little or no $CaCO_3$ by dissolution. When coccoliths (calcite) and size differences are considered it is concluded that about 50% by weight of the $CaCO_3$ sedimenting to the sea floor consists of aragonite. This is a much larger number than previously believed.

Because most sea water below the thermocline is undersaturated with respect to aragonite, pteropods dissolve before burial in most pelagic sediments. However, on certain topographic highs, pteropods may accumulate to form pteropod ooze. Maximum depths for pteropod ooze are ∿500 m in the central Pacific Ocean and ∿2800 m in the Atlantic Ocean. Pteropods are also abundant in the sediments of the Mediterranean and Red Seas. Pteropods dissolve preferentially relative to foraminifera and coccoliths because of the greater solubility of aragonite, compared to calcite. This demonstrates the importance of physical chemistry in controlling the $CaCO_3$ distribution in oceanic sediments. The pteropods dissolve mainly while resting on the ocean bottom and not while settling. Because of the large flux of aragonite to the bottom, and consequently larger input of dissolved $CaCO_3$ to deep water, a $CaCO_3$ mass balance for the oceans can be obtained using the model of Broecker (1971).

INTRODUCTION

In assessing the role of fossil fuel combustion upon the carbon dioxide balance in the world ocean-atmosphere system, it is essential to understand those processes which affect the concentration of CO_2 in the surface and deep waters of the ocean. One of these processes is the secretion of calcium carbonate by planktonic organisms living in surface waters. This process converts dissolved bicarbonate ions in sea water to $CaCO_3$ plus dissolved carbon dioxide. Upon death of the organisms, their calcareous remains fall into deep water. Here, at sufficiently great depths, another major process, the dissolution of $CaCO_3$, occurs. It is the opposite of secretion and results in the conversion of dissolved carbon dioxide back to bicarbonate ions. Burial of undissolved $CaCO_3$ in sediments and mass transfer of bicarbonate and CO_2 from deep to shallow water completes the cycle. Thus, the balance between secretion in shallow water and dissolution in deep water plays an important role in the regulation of the CO_2 content and carbonate alkalinity of the oceans.

It is normally assumed that the major organisms involved in the oceanic $CaCO_3$ cycle are the planktonic foraminifera and coccolithophorids whose calcareous remains consist wholly of the mineral calcite. The reason for this assumption is that practically all of the $CaCO_3$ found in pelagic sediments consists of calcitic foraminifera and coccolith debris. However, numerous studies going back to those of the Challenger Expedition (*Murray and Renard*, 1891) have shown that, when examining plankton tows, one commonly encounters another calcareous constituent which often may exceed foraminifera tests and coccoliths in abundance. This is aragonite in the form of pteropod shells. (Other less abundant forms of aragonite include heteropod shells and various mollusc larvae.)

As has been abundantly shown (eg. *Pytkowicz*, 1965; *Berner*, 1965; *Hawley and Pytkowicz*, 1969; *Li et al.*, 1969; *Edmond and Gieskes*, 1970; *Ben-Yaakov et al.*, 1974; *Edmond*, 1974; *Takahashi*, 1975), the ocean becomes undersaturated, and increases in the degree of undersaturation with respect to calcium carbonate, with increasing depth. Since aragonite is distinctly more soluble than calcite, the shells of pteropods which have fallen to the bottom in deep water are much more readily dissolved than are accompanying foraminifera tests and coccoliths. Only in areas where the water is sufficiently shallow and, therefore, non-corrosive toward aragonite, can falling pteropod shells accumulate to any extent in bottom sediments. Average <u>maximum</u> depths for appreciable pteropod accumulation to form so-called pteropod ooze, as will be shown in this paper, range from about 2800 meters in the Atlantic Ocean to only 500 meters in the central Pacific Ocean. Thus, pteropod ooze is rare and found only on shallow parts of submarine banks, rises, and seamounts, in marginal seas (eg. the Mediterranean Sea), on the submerged slopes of oceanic islands, and occasionally on continental slopes and shelves where terrigenous and benthonic-biogenic sedimentation is limited. A map of pteropod ooze distributed in the world ocean based on occurrences cited by *Murray and Chumley* (1924), *Herman* (1971), and *Chen* (1971) is shown in Figure 1.

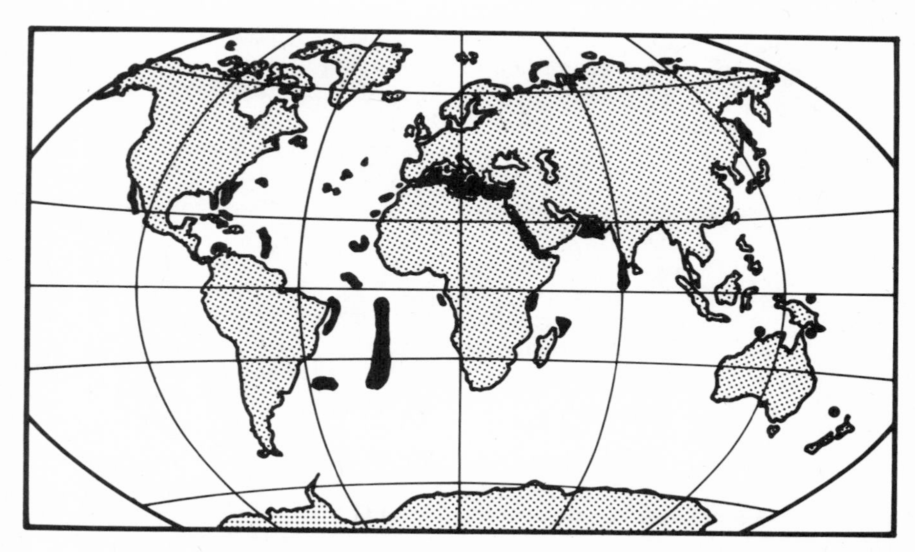

Fig. 1. Map of the distribution of pteropod ooze in the oceans. Data from *Murray and Chumley* (1924), *Chen* (1971), and *Herman* (1971).

The purpose of this paper is to demonstrate the quantitative importance of pteropod secretion and dissolution to the overall cycle of calcium carbonate in the oceans. As will be seen, neglect of the aragonite pteropod subcycle, as has been normally done in the past, has led to a serious underestimate of the amount of $CaCO_3$ secreted in shallow water and dissolved in deep water. Such neglect of aragonite has been cited by *Broecker* (1971) as a possible contributing cause to the disagreement between calculated (from dissolved carbon mass balance) and measured $CaCO_3$ accumulation rates in the present oceans. The present study will show that consideration of pteropods removes most of this disagreement.

SEDIMENTATION OF PTEROPODS

Three methods can be used to estimate the fraction of calcium carbonate delivered to the sea floor as aragonite, rather than as calcite. The first, and most direct, is the quantitative mineralogical analysis of falling particles collected on sediment traps suspended at various depths in the ocean. To the writer's knowledge such a study has not been undertaken and is sorely needed.

A second, more approximate, method is the mineralogical analysis of surface plankton tows. The mineralogical determination can be direct, via x-ray diffraction, or indirect, via plankton counts of pteropods (aragonite) and foraminifera and coccoliths (calcite). The latter technique forms the basis for the results of the present paper both for plankton tows and for bottom sediments. It should be remembered that plankton counts are only semi-quantitative because unidentifiable $CaCO_3$ grains are not counted, pteropods are generally bigger than foraminifera, and coccoliths must be enumerated on separate size fractions. Also, in the case of tows, only organisms larger than a given mesh size are collected (assuming no clogging). All these problems lead to results which are good to about a factor of two at best.

Lisitzin (1971), in an extensive study of suspended matter in oceanic surface waters, reports that x-ray diffraction techniques along with species counts were used to identify calcareous plankton. However, he gives no data concerning aragonite abundance other than the brief statement that pteropods are "distinctly subordinate" to foraminifera in surface waters. This observation is not in general agreement with those of *Adelseck and Berger* (1975) and *Berger* (1976) who have found an approximately equal abundance of pteropods and foraminifera in Pacific Ocean surface waters off Central America, nor with those of *Murray and Renard* (1891), who state that pteropods are a generally abundant constituent of plankton tows. In addition, *Bé et al.* (1971) report pteropods as "large components" or as "significant contributors" (along with foraminifera) to the ash of coarse plankton (>200 μm) collected in 342 tows from the North Atlantic Ocean. Again, no quantitative abundance data for pteropods vs foraminifera are given. The unpublished observations of *Honjo* (1976) are that pteropods are generally <u>more</u> abundant than foraminifera in plankton tows. *McGowan* (1971) presents maps of individual pteropod species abundance in terms of numbers of individual per 1000 cubic meters of sea water, but no comparable data for foraminifera are presented. Likewise *Chen and Bé* (1964) present data for pteropods in the western North Atlantic Ocean, but only in terms of relative proportions of species. In sum, it appears that pteropod aragonite may comprise a major proportion of planktonic $CaCO_3$; however, observations of co-occurring foraminifera and pteropods, as well as coccolithophorids, and bulk x-ray mineralogy should be made before any firm quantitative conclusion can be given.

Even if accurate aragonite vs calcite data for plankton were available, the results would provide only an approximation of the relative importance of aragonite vs calcite fluxes to the bottom. This is because the biological regeneration rates of pteropods are not as fast as those for foraminifera and coccoliths (*Berger*, 1976). At steady state, downward fluxes out of the surface waters must balance regeneration rates, and the flux of pteropods relative to foraminifera-plus-coccoliths, may not be equal to the ratio of pteropods to foraminifera-plus-coccolithophorids found in the standing crop of the surface water. An added complication is that pteropods undergo daily and seasonal vertical migrations (*Hardy*, 1956) which may range up to several hundred meters.

The third method of estimating the relative importance of aragonite vs calcite sedimentation fluxes, is determination of the species distribution of calcareous remains in bottom sediments which have undergone little-or-no dissolution. This is the approach used here. (Some limited data on x-ray diffraction studies of bottom sediments will also be presented.) *Herman* (1971) has counted both planktonic foraminifera and pteropods in the >74 μm fraction of sediments from the Northern Indian Ocean, the Red Sea, the Mediterranean Sea, the Caribbean Sea, and the western North Atlantic Ocean (Blake Plateau). No abundance data for pteropods vs foraminifera are given in this paper but Dr. Herman has kindly provided pteropod/foraminifera ratios to the writer from her raw data for surface sediments of the Red and Mediterranean Seas. For six Red Sea samples the pteropod/planktonic foraminifera ratio ranges from 0.4 to 1.9 with an average value of 1.2. Depths of the samples range from 768 m to 1486 m. Seven samples from the Mediterranean Sea ranging in depth from 618 to 2897 meters range from 0.25 to 1.5, with an average value of 0.8. For these two bodies of water, if no appreciable dissolution of aragonite has occurred in the sediments, the average pteropod/planktonic foraminifera ratio of sedimenting $CaCO_3$ is 1.0.

Chen (1964; 1968) provides data on the absolute abundance of pteropods in Bermuda Pedestal and Gulf of Mexico sediments, but not relative to calcitic remains. On the other hand, the results of *Berner et al.* (1976) provide abundance data for pteropods relative to planktonic foraminifera in surface sediments of the Bermuda Pedestal. In the >250 μm fraction of the two shallowest samples, which are presumed to have undergone no dissolution (see below and Figure 5), the pteropod/planktonic foraminifera ratios are 1.0 and 1.1.

Much additional data is provided by the study of *Murray and Chumley* (1924). They list the weight per cent total $CaCO_3$ and microscopically estimated per cent planktonic foraminifera in 1426 hemipelagic and pelagic sediment samples taken from the Atlantic Ocean bottom during 34 different cruises spanning the period 1857-1911. The remaining $CaCO_3$ is divided into per cent benthonic foraminifera (always a low number less than 10) and per cent "other". The other constituents (which include only microscopically identifiable grains) are described but not quantitatively evaluated. Of the 1426 samples, 40 are described as pteropod ooze meaning that "other" consists mostly of pteropods. The ratio of per cent "other" to per cent planktonic foraminifera for these samples are plotted in Figure 2 as a function of depth. If it is assumed that "other" consists entirely of pteropods, then most samples indicate a ratio of planktonic foraminifera of 1.0 $\pm$ 0.5. This number is in agreement with the studies cited above. However, this ratio may be somewhat too

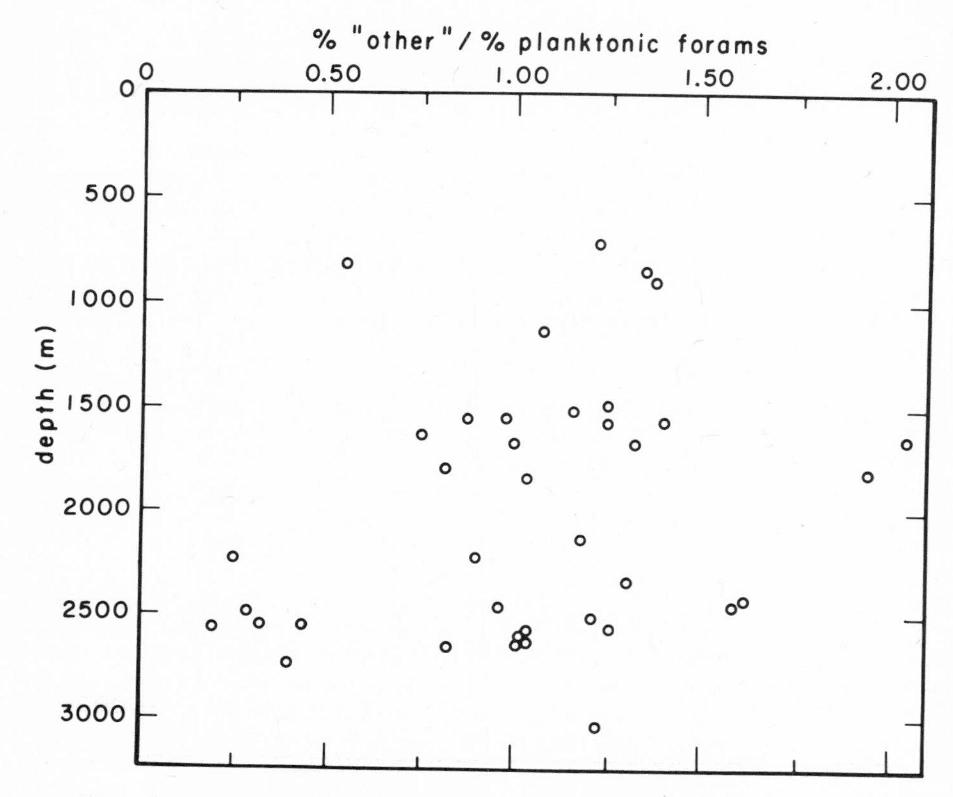

Fig. 2. Ratio of % "other"/% planktonic Foraminifera for pteropod ooze in the Atlantic Ocean. "Other" represents mainly pteropods. Data from *Murray and Chumley* (1924).

high or too low for several reasons. It may be too low because of possible dissolution, although no dissolution trend is discernable with depth. (The shallower samples which are unlikely to have undergone dissolution also show an average ratio of 1.0.) It is probably a little too high because most of the ooze samples were collected near islands or continents and contain some fragments of shallow water benthonic organisms derived from slumping. These are included under "other" along with pteropods. However, they cannot make up a major portion of the "other" category or the samples would not be designated as pteropod ooze.

In studying the Murray and Chumley data, all calcareous sediments (those containing more than 30 wt. % $CaCO_3$) taken from depths of 700-1500 meters also have been examined. At these depths, dissolution of aragonite is unlikely and dilution by slumped shallow-water debris is (hopefully) not excessive. Calculations based on

57 such samples indicate an average ratio of %"other"/% planktonic foraminifera of 0.8. This value is in fairly good agreement with that for pteropod ooze and is reasonable since almost all of the 57 samples were stated to contain pteropod remains.

An additional grouping of Murray and Chumley's data was done for all calcareous samples in the South Atlantic Ocean which were more than 1000 km from the nearest land, whether continents or islands. This was done to avoid all possibility of the inclusion of slumped coarse shallow water material. Values of %"other"/% planktonic foraminifera are shown as a function of depth in Figure 3. Note that above 2500 meters there are four samples (located on the Rio Grande Rise and the Mid-Atlantic Ridge) which exhibit a ratio of 0.9-1.4.* Here %"other" must represent entirely pelagic remains, or in other words, pteropods. The four samples are all classified as pteropod oozes.

From the above arguments it is concluded that the ratio of pteropods to planktonic foraminifera delivered to Atlantic Ocean, Mediterranean and Red Sea pelagic sediments before dissolution is approximately one. From the work of *McIntyre and McIntyre* (1971) the average concentration of coccoliths between 40°S and 40°N in (presumably undissolved) carbonate sediments of the Atlantic and Indian Ocean is about 25% of total sediment weight. If the average total $CaCO_3$ content of these sediments is 75%, the average ratio of foraminifera to coccoliths is about 2. (Pteropods are negligible in most carbonate sediments.) Combining this result with that for the pteropod-to-foraminifera ratio and assuming that pelagic carbonate consists only of planktonic foraminifera, coccoliths, and pteropods, the following average percentages for $CaCO_3$ sedimented to the bottom are: pteropods 40%, planktonic foraminifera 40%, and coccoliths 20%. This means that about 40% by weight of $CaCO_3$ supplied to deep sea sediments is supplied as aragonite. This number is probably a minimum because throughout this discussion numbers of pteropods relative to foraminifera have been used. Pteropods are normally considerably larger than planktonic foraminifera, even when fragmented, and thus, on a weight basis, they should exceed planktonic foraminifera in abundance. The conversion factor, however, is not simple because of the extreme thinness of pteropod shells and their rather variable geometry. Other complicating factors are that the proportion of coccoliths may be higher (*Honjo*, 1976) and other sources of aragonite, such as transported shallow water debris and mollusc larvae (*Thiede*, 1975), have been neglected. Obviously more data are needed but until then the writer recommends the use of 50% calcite and 50% aragonite as the mineralogical composition of $CaCO_3$ sedimenting into the deep sea.

*From the data of *Melguen and Thiede* (1974) an approximate value for the pteropod: planktonic foraminifera ratio, at 2000 m on the Rio Grande Rise, is 0.25, which is in disagreement with this data.

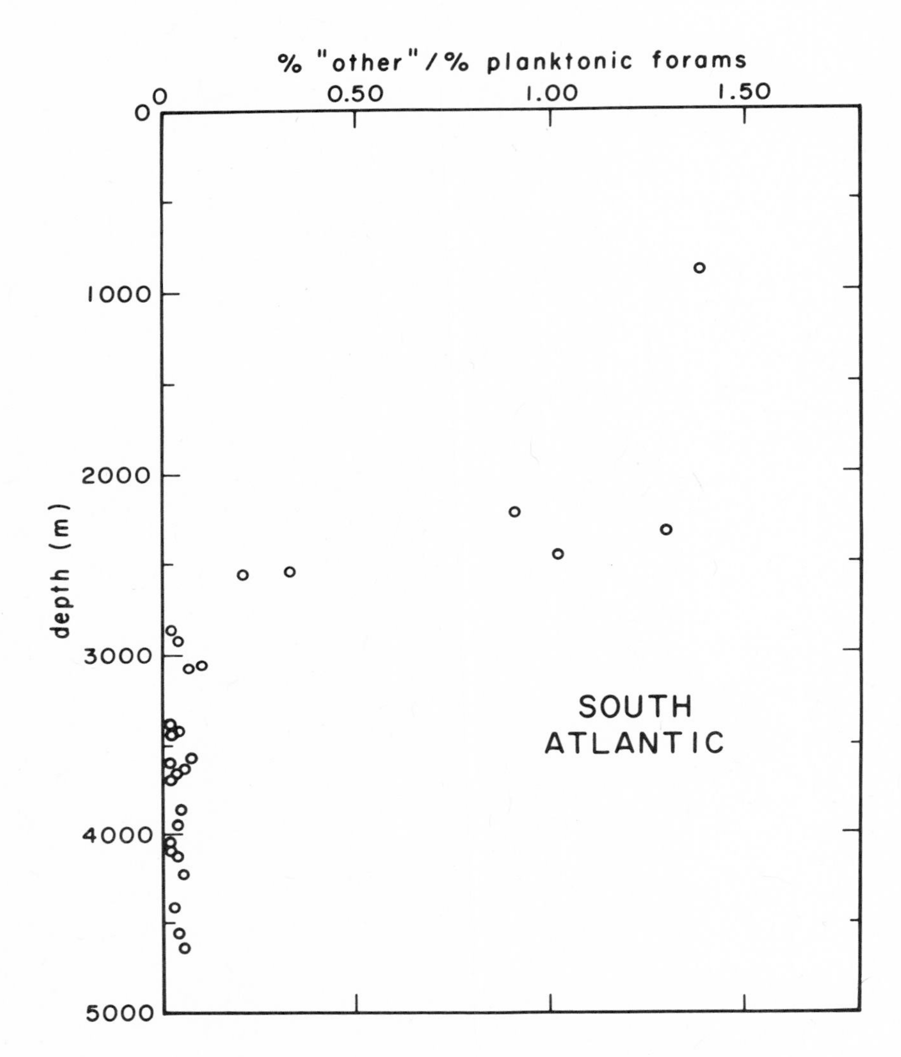

Fig. 3. Plot of % "other"/% planktonic foraminifera vs depth for all calcareous oozes (sediments containing more than 30% $CaCO_3$) in the South Atlantic Ocean which are located more than 1000 km from the nearest land. "Other" refers almost entirely to pteropods. The four samples located above 2500 meters are all classified as pteropod ooze. Data from *Murray and Chumley* (1924).

DISSOLUTION OF PTEROPODS

The reason why $CaCO_3$ in pelagic sediments does not consist of 50% aragonite is that the aragonite in most places dissolves away before burial. Most sea water below the thermocline is undersaturated with respect to aragonite. The degree of saturation can be expressed in terms of Ω_α where:

$$\Omega_\alpha = IP/K'_\alpha \tag{1}$$

The symbol IP refers to the product of calcium and carbonate ion concentrations and K'_α refers to the same product at aragonite saturation equilibrium. Representative Ω_α data for the central Pacific Ocean at the site of the *Peterson* (1966) and *Berger* (1967; 1970) experiments are shown in Figure 4. Calculations are based on *pH* measurements, temperatures, alkalinities, and salinities reported by *Berner and Wilde* (1972), and on equilibrium constants calculated from the data of *Lyman* (1957), *Culberson and Pytkowicz* (1968), *Berner* (1976), *Duedall* (1972), *Millero and Berner* (1972), *Culberson* (1972), and *Takahashi* (1975). Methods of calculation are discussed by *Takahashi* (1975). Note that the saturation, or cross-over depth where $\Omega_\alpha = 1$, falls at 350 meters and that Ω_α drops below 0.5 at only 600 meters. Dissolution in shallow water would, thus, be expected. Verification of this is provided by measurements of the rate of dissolution of pteropods suspended at various depths at the same site by *Berger* (1970). He found that pteropods suspended at 250 meters underwent no dissolution after four months, whereas those suspended at 750 meters lost over one-half their weight during the same time. Greater dissolutive weight loss was found at greater depths. If the pteropods were first treated with H_2O_2 to remove protective organic coatings, those suspended at 250 meters still showed no signs of dissolution, whereas those at 750 meters and below were completely dissolved. From the Ω_α calculations and Berger's pteropod dissolution experiment, one would not expect to find pteropods below about 500 meters in the north and central Pacific Ocean. This explains the virtual absence of pteropod ooze from this large area (see Figure 1).

A few occurrences of pteropods in Pacific Ocean pelagic sediments have been reported from topographic highs in the southwestern Pacific Ocean (*Chen*, 1971). However, compared with the Atlantic Ocean, south Pacific Ocean sediments are much rarer in pteropods because of dissolution at shallower depths. This is due, as in the case of the north and central Pacific Ocean, to a greater acidity of the water at intermediate depths. *Murray and Renard* (1891) reported that at two localities between the Fiji and New Hebrides Islands no pteropods were found in carbonate sediments at 2400 meters and 2600 meters depth although (p. 167) "they were very abundant on the surface, (ie. in plankton tows), and at a similar depth and latitude in the Atlantic they were usually present in considerable numbers."

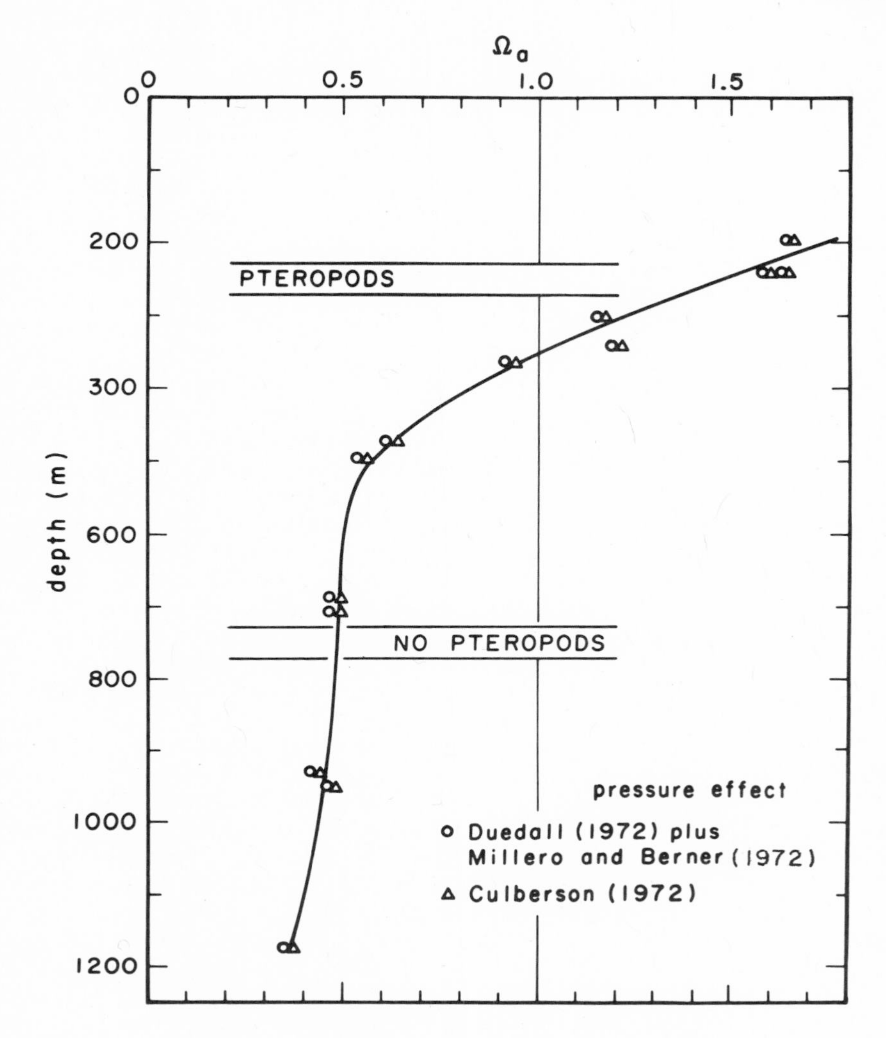

Fig. 4. Plot of the degree of saturation with respect to aragonite Ω_α, vs depth for the central Pacific Ocean at the site of the *Peterson* (1966) and *Berger* (1967; 1970) experiments (18^o49'N; 168° 31'W). The dissolution of pteropods reported by *Berger* (1970) at 750 m is designated as "no pteropods"; the lack of dissolution at 250 m is designated as "pteropods". Two sets of points are presented for Ω_α representing respectively, the pressure effect on solubility calculated from the data of *Millero and Berner* (1972) *and Duedall* (1972), and those of *Culberson* (1972). (For a discussion of pressure effects see *Takahashi*, 1975). Calculations based on the *pH*, alkalinity, salinity, and temperature data reported by *Berner and Wilde* (1972). Equilibrium constants used in the calculation are referenced in text.

In the Atlantic Ocean pteropod ooze is more common, but still a rare sediment when compared with foraminifera ooze. The principle areas of accumulation are shown in Figure 1. Depths of occurrence are shown in Figures 2 and 3. Note that the normal depth range for this sediment type is 500-2800 meters with only two occurrences cited below 2800 meters. Figure 3 shows that in the south Atlantic Ocean, on the Mid-Atlantic Ridge and Rio Grande Rise, the zone of dissolution removal appears to be between roughly 2000 and 2700 meters.

Pteropod dissolution in the North Atlantic Ocean is shown by the results of *Chen* (1964) and *Berner et al.* (1976) for the Bermuda Pedestal. In the latter study pteropod/planktonic foraminifera ratios were determined to avoid spurious changes in pteropod concentrations by dilution with coarse coral, mollusc, etc. skeletal debris supplied from shallow water by slumping. Also included was a measurement of relative aragonite/calcite ratios, in the fine grained fraction of the sediments, using x-ray diffraction. The fine fraction is believed to emanate by suspension transport, from the shallow Bermuda platform, of ground-up reef material (*Chave et al.*, 1962; *Friedman*, 1965). Results of pteropod/planktonic foraminifera ratios and fine grained aragonite/fine grained calcite ratios are shown in Figure 5. Included are x-ray data for shallower sediments from the study of *Chave et al.* (1962). Note that a decrease in pteropods with depth more-or-less parallels a decrease in fine-grained aragonite. This is interpreted as selective dissolution of pteropods and fine-grained aragonite relative to planktonic foraminifera and fine-grained calcite. The zone of removal is roughly 1600-3000 meters.

The state of saturation of sea water with respect to aragonite near Bermuda is shown in Figure 6. Calculations are based on the alkalinity and *pH* measurements of *Berner et al.* (1976) and on GEOSECS data. Methods of calculation and equilibrium constants are the same as those used in the calculation of Figure 4. Note that the aragonite disappearance depth zone shown in Figure 5 falls at a range of Ω_α values from about 0.60 to 0.80. These values are distincly less than one, and illustrate that the simple attainment of undersaturation is insufficient, in itself, to bring about dissolution. Kinetic problems such as surface reactivity, protective organic coatings (*Chave and Suess*,1970) or adsorbed inorganic inhibitors (*Berner and Morse*, 1974; *Morse*, 1974) combine to enable the persistence of aragonite in weakly undersaturated water above 1600 meters and to produce a gradual disappearance below this depth. Some of the breadth of the zone of disappearance may be due to slumping of pteropods from above, but this is considered unimportant because of the relative lack of shallow water coral-mollusc debris in most of the samples from this zone that were selected for study (*Berner et al.*, 1976).

Looked at in a broader perspective, however, thermodynamics is a principal guide to aragonite dissolution in the oceans. Aragonite disappears at much shallower depths than calcite because of its greater solubility. Thus, theories for calcium carbonate dissolution in the ocean that place chemical controls (whether thermodynamic or

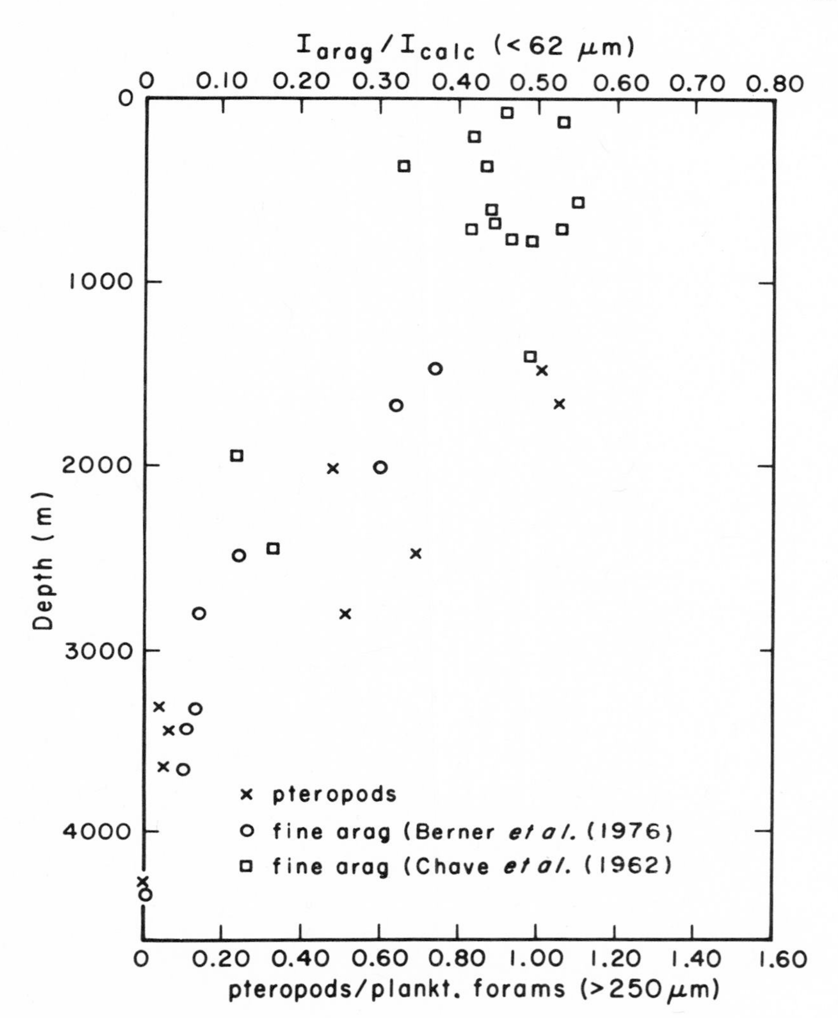

Fig. 5. Pteropods/planktonic foraminifera and fine-grained aragonite/fine-grained calcite vs depth in Bermuda Pedestal sediments. $I_{arag.}$ refers to the integrated x-ray diffraction peak area; $I_{calc.}$ refers to low-Mg calcite only. Data denoted by squares from *Chave et al.* (1962). Other data from *Berner et al.* (1976). After *Berner et al.* (1976).

kinetic) in a minor role (eg. *Edmond*, 1974) run into major problems when both aragonite and calcite are considered. Since aragonite constitutes a major component of $CaCO_3$ falling to the bottom, as shown in the previous section, its dissolution can no longer be ignored. Many previously perplexing appearances of excess alkalinity in waters of intermediate depth (where calcite does not dissolve) may be due to the dissolution of pteropods.

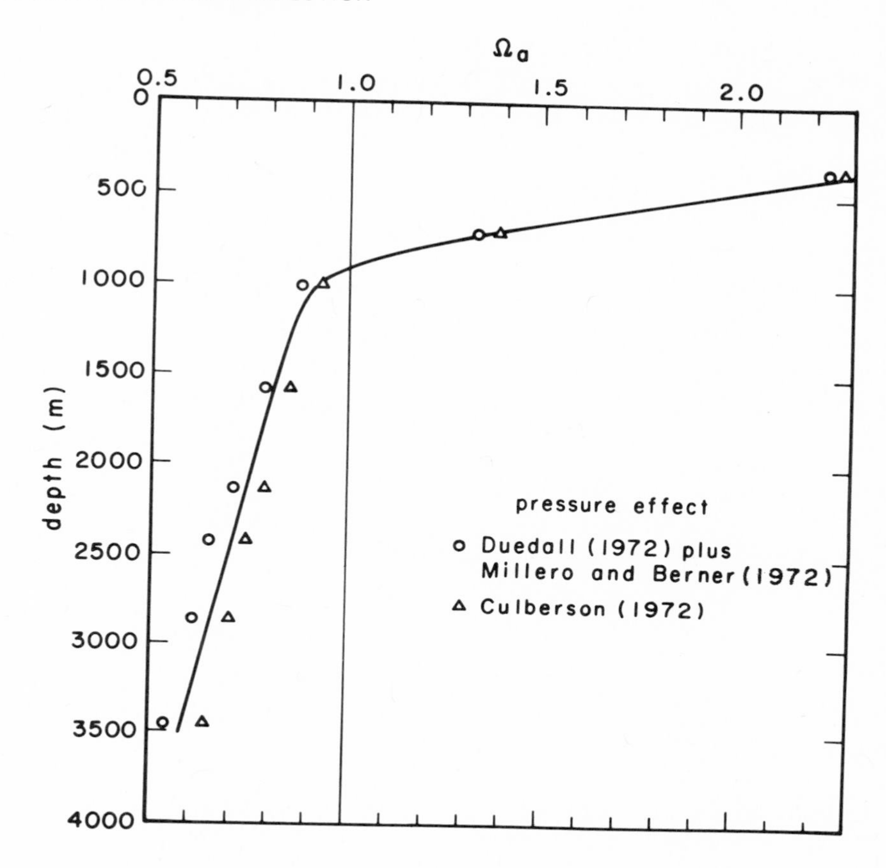

Fig. 6. Plot of the degree of saturation with respect to aragonite, Ω_α, vs depth for the North Atlantic Ocean near Bermuda. Two sets of points are presented for Ω_α representing, respectively, the pressure effect on solubility calculated from the data of *Millero and Berner* (1972) and *Duedall*, (1972), and those of *Culberson* (1972). (For a discussion of pressure effects see *Takahaski*, 1975). Calculations are based on GEOSECS data (Station 121) and equilibrium constants referenced in text. Station 121 is located at 35°59.5'N; 68°0.0'W.

Petropods probably undergo most dissolution while sitting on the bottom and not while settling out. They are large enough that settling times are only a few days to a few weeks (*Lisitzin*, 1971). *Adelseck and Berger* (1975), through the use of deep plankton tows, have demonstrated the existence of relatively undissolved, sedimenting pteropods in deep (3000 meters) highly undersaturated, Pacific

Ocean seawater. Furthermore, *Morse* (1976) has found pteropods on the uppermost surface of a sediment box core, taken from about 3000 meters depth on the mid-Atlantic Ridge, below which in the sediment they disappear due to dissolution. A similar dissappearance below the top few cm has also been found by *Melguen and Thiede* (1974) for cores taken on the Rio Grande Rise at depths greater than 3000 meters.

The importance of aragonite dissolution can be shown by means of *Broecker's* (1971) model for the $CaCO_3$ budget of the oceans. In this model the rates of calcite accumulation, aragonite accumulation, and total $CaCO_3$ dissolution are related by the $CaCO_3$ mass balance expression:

$$A_o(R_a + R_c) = A_oR_s + (A_aR_a + A_cR_c) \quad (2)$$

where: A_a = area of sea floor covered by aragonitic (pteropod) sediment

A_c = area of sea floor covered by calcitic(foraminifera-coccolith) sediment

A_o = area of total sea floor (for a given ocean)

R_c = average rate of sedimentation of calcite

R_a = average rate of sedimentation of aragonite

R_s = average rate of dissolution of aragonite plus calcite (for a given ocean)

Values of R are in terms of mass per unit area per unit time. Equation (2) assumes implicity that accumulation rates and sedimentation rates are equal for areas of calcite-rich and aragonite-rich sediments, or in other words, that no dissolution of the minerals occurs in their respective principal areas.

Since the area of aragonite (pteropod) accumulation is negligibl as compared to that of calcite or the whole ocean floor, $A_o - A_a \approx A_o$. Using this assumption and solving for R_c, we obtain:

$$R_c = \frac{(R_s - R_a)}{(1 - A_c/A_o)} \quad (3)$$

Broecker (1971) states that for the Atlantic Ocean $A_c/A_o = 0.65$ and for the Pacific-Indian Ocean system $A_c/A_o = 0.40$. He also gives average values for R_s, based on independent mass balance calculations for the Atlantic and Pacific-Indian Oceans. The values are given in Table 1. Using this data, and the assumption that $R_a = 0$, in other words that aragonite sedimentation is negligible, *Broecker*

TABLE 1. $CaCO_3$ mass balance model of *Broecker* (1971) with and without inclusion of aragonite sedimentation. All R values in $mg\ cm^{-2}yr^{-1}$.

Parameter	Source	Atlantic	Pacific
R_s	*Broecker* (1971)	1.8	1.5
R_c	*Broecker* $CaCO_3$ model with $R_a = 0$	5.0	2.5
R_c	Same model with $R_a = R_c$	1.3	0.9
R_c	Observed rates from *Broecker* (1971)	1.0	0.5

(1971) obtained values for R_c which are much greater than measured accumulation rates for calcite-rich sediments. This is shown in Table 1. He then stated that one of the principal reasons for disagreement may have been due to the neglect of aragonite sedimentation. Indeed, distinctly different results are obtained if the values from the present study for pteropod aragonite sedimentation are used. From the arguments presented above one can assume that $R_a \approx R_c$. In this case equation (3) reduces to:

$$R_c = \frac{R_s}{(2 - A_c/A_o)} \tag{4}$$

Values of R_c calculated from equation (4) are also shown in Table 1. Much better agreement with observed R_c values is now obtained. This indicates that, when corrected for aragonite sedimentation using the data of the present study, *Broecker's* $CaCO_3$ mass balance can actually be demonstrated. Small remaining errors can be attributed to the many other assumptions of the model.

ACKNOWLEDGMENTS

Special thanks go to Dr. Yvonne Herman-Rosenberg for permission to use unpublished data on total pteropod abundances in the Mediterranean and Red Seas. The writer has benefited greatly from discussions with K. K. Turekian, W. H. Berger, E. K. Berner, S. Honjo, and W. S. Broecker. Research supported by NSF Grant DES 73-00365 A01. Special acknowledgment goes to the executive committee of GEOSECS for permission to use unpublished data for station 121.

REFERENCES

Adelseck, C. G. and Berger, W. H., 1975. On the dissolution of planktonic foraminifera and associated microfossils during settling and on the Sea Floor, *Cushman Found. for Foraminifera Research*, Spec. Publ. No. 13, 70-81.

Bé, A. W. H., Forns, J. M. and Roels, O. A., 1971. Plankton abundance in the North Atlantic Ocean, *in Fertility of the Sea*, ed. J. D. Costlow, Gordon & Breach, N.Y., 17-50.

Ben-Yaakov, S., Ruth, E. and Kaplan, I. R., 1974. Carbonate compensation depth: relation to carbonate solubility in ocean waters, *Science, 184*: 982-984.

Berger, W. H., 1970. Planktonic foraminifera: selective solution and the lysocline, *Marine Geology, 8*: 111-138.

Berger, W. H., 1967. Foraminiferal ooze: solution at depth, *Science, 156*: 383-385.

Berger, W. H., 1976. Scripps Institution of Oceanography, pers. comm.

Berner, R. A., 1965. Activity coefficients of bicarbonate, carbonate, and calcium ions in sea water, *Geochim, Cosmochim, Acta, 29*: 947-965.

Berner, R. A., 1976. The solubility of calcite and aragonite in sea water at atmospheric pressure and 34.5^{o}/oo salinity, *Am. Jour. Sci., 276*. 713-730.

Berner, R. A., Berner, E. K., and Keir, R. S., 1976. Aragonite dissolution on the Bermuda Pedestal: its depth and geochemical significance, *Earth Planet. Sci. Letters, 30*. 169-178.

Berner, R. A. and Morse, J. W., 1974. Dissolution kinetics of calcium carbonate in sea water IV. Theory of calcite dissolution, *Am. Jour. Sci., 274*: 108-134.

Berner, R. A. and Wilde, P., 1972. Dissolution kinetics of calcium carbonate in sea water I. Saturation state parameters for kinetic calculations, *Am. Jour. Sci. 272*: 826-839.

Broecker, W. S., 1971. Calcite accumulation rates and glacial to interglacial changes in oceanic mixings, in *Late Cenozoic Glacial Ages*, edited by K. K. Turekian, Yale, Univ. Press, New Haven, 239-265.

Chave, K. E., Sanders, H. L., Hessler, R. R., and Neumann, A. C., 1962. Animal-sediment interrelationships on the Bermuda Slope and in the adjacent deep sea, *Office of Naval Research Contract Nonr-1135(02) Final Report*, 17pp.

Chave, K. E. and Suess, E., 1970. Calcium carbonate saturation in seawater; effects of dissolved organic matters, *Limnol. and Oceanog., 15*: 633-637.

Chen, C., 1964. Pteropod ooze from Bermuda Pedestal, *Science, 144*: 60-62.

Chen, C., 1968. Pleistocene pteropods in pelagic sediments, *Nature, 219*: 1145-1149.

Chen, C. and Bé, A. W. H., 1964. Seasonal distribution of euthecosomatous pteropods in the surface waters of five stations in the western North Atlantic, *Bull. Mar. Sci. Gulf and Caribbean, 14*: 185-220.

Culberson, C. H., 1972. Processes affecting the oceanic distribution of carbon dioxide, Ph.D. Dissertation, Oregon State Univ., 178pp.

Culberson, C. H., and Pytkowicz, R. M., 1968. Effect of pressure on carbonic acid, boric acid, and the *pH* in sea water, *Limnol. and Oceanog., 13*: 403-417.

Duedall, I. W., 1972. The partial molal volume of calcium carbonate in sea-water, *Geochim. Cosmochim. Acta, 36*: 729-734.

Edmond, J. M., 1974. On the dissolution of silicate and carbonate in the deep sea, *Deep Sea Research, 2l*: 455-480.

Edmond, J. M. and Gieskes, J. M. T. M., 1970. On the calculation of the degree of saturation of sea water with respect to calcium carbonate under in situ conditions, *Geochim, Cosmochim, Acta. 34*: 1261-1291.

Friedman, G. M., 1965. Occurrence and stability relationships of aragonite, high-magnesian calcite, and low-magnesian calcite under deep sea conditions, *Bull. Geol. Soc. America, 76*: 1191-1196.

Hardy, A. C., 1956. The Open Sea: The World of Plankton, Collins, London, 335 pp.

Hawley, J. and Pytkowicz, R. M., 1964. Solubility of calcium carbonate in sea water at high pressures and 2°C., *Geochim, Cosmochim. Acta, 33*: 1557-1561.

Herman, Y., 1971. Vertical and horizontal distribution of pteropods in quarternary sequences, *in the Micropaleontology of Oceans*, edited by B. M. Funnell and W. R. Riedel, 3-74.

Honjo, S., 1976. Woods Hole Oceanographic Institution, pers. Comm.

Li, Y. H., Takahashi, T. and Broecker, W. S., 1969. Degree of saturation of $CaCO_3$ in the oceans, *Journ. Geophys. Res., 74*: 5507-5525.

Lisitzin, A. P., 1971. Distribution of carbonate microfossils in suspension and in bottom sediments, in *The Micropaleontology of Oceans*, edited by B. M. Funnell and W. R. Riedel, 197-218.

Lyman, J., 1957. Buffer mechanism of seawater, Ph.D. Dissertation, Univ of California, Los Angeles, 196 pp.

McGowan ,J. A., 1971. Oceanic biogeography of the Pacific, in *The Micropaleontology of Oceans*, edited by B. M. Funnell and W. R. Riedel, 3-74.

McIntyre, A. and McIntyre, R., 1971. Coccolith concentrations and differential solution in oceanic sediments, in *The Micropaleontology of Oceans*, edited by B. M. Funnell and W. R. Riedel, 253-262.

Melguen, M. and Thiede, J., 1974. Facies distribution and dissolution depths of surface sediment components from the Vema Channel and the Rio Grande Rise (Southwest Atlantic Ocean), *Marine Geology, 17*: 341-353.

Millero, F. J. and Berner, R. A., 1972. Effect of pressure on carbonate equilibria in sea water, *Geochim. Cosmochim, Acta, 36:* 92-98.

Morse, J. W., 1974. Dissolution kinetics of calcium carbonate in sea water V. Effects of natural inhibitors and the position of the chemical lysocline, *Am. Jour. Sci. 274:* 638-647.

Morse, J. W., 1976. Florida State Univ. pers. comm.

Murray, J. and Chumley, J., 1924. The deep sea deposits of the Atlantic Ocean, *Trans. Roy. Soc. Edinburgh, 54:* 1-252.

Murray, J. and Renard, A. F., 1891. Report on deep-sea deposits based on the specimens collected during the voyage of H.M.S. Challenger in the years 1872 to 1876, Longmans, London, 525 pp.

Peterson, M. N. A., 1966. Calcite: rates of dissolution in a vertical profile in the Central Pacific, *Science, 154:* 1542-1544.

Pytkowicz, R. M., 1965. Calcium carbonate saturation in the ocean, *Limnol. and Oceanog., 10:*220-225.

Takahashi, T., 1975. Carbonate chemistry of seawater and the calcite compensation depth in the oceans, *Cushman Foundation for Foraminifera Research Spec. Publ. No. 13,* 11-26.

Thiede, J., 1975. Shell and skeleton-producing plankton and nekton in the eastern North Atlantic Ocean, *Meteor. Forsch. Ergeb. C. 20.* 33-79.

On the Carbonate Saturation - Dissolution Rate Relationship

Edward A. Boyle* and John M. Edmond

Massachusetts Institute of Technology

ABSTRACT

Although the re-cycling rate of calcium carbonate is usually presumed to be dependent on the degree of saturation of the deep ocean, the functionality of this relationship is not known. The saturation-dissolution relationship could be determined by measurements of recent carbonate sedimentation rates and separation of the effect of productivity variations from the effect of variations in the degree of saturation. Preliminary analysis suggests that the sediment dissolution rate is highly sensitive to small (i.e. 10%) shifts in the near-bottom saturation value.

INTRODUCTION

A fundamental aspect of the carbonate system is the evidence of fluctuations in carbonate sedimentation due to time variations in seawater chemistry (e.g. *Luz and Shackleton*, 1975). An understanding of the feedback mechanism balancing the carbonate cycle would make it possible to decode the sediment record and make quantitative estimates of past variations in ocean composition.

Carbonate sedimentation is controlled by surface productivity variations and the dissolution of sediments in the deep sea. Recent data (*Edmond*, 1974; *Takahashi*, 1975) demonstrate that the onset of significant dissolution occurs well below the horizon where water becomes undersaturated with respect to calcium carbonate. The preservation of carbonate sediments is thus related to the degree of saturation

---

* Present Address: University of Edinburgh

$$\Omega = 100 \left[\frac{[Ca^{++}]\ [CO_3^{=}]}{Ksp} \right] \quad (1)$$

by a dynamic, rather than a static equilibrium process. Laboratory dissolution experiments (*Morse and Berner*, 1972) and mooring experiments (*Peterson*, 1966; *Milliman*, 1975) can provide important constraints on the functionality of the saturation-dissolution relationship. However, the rate information derived from these experiments cannot be applied in any simple fashion to the dissolution of actual carbonate sediments; the physical regimes are certainly different, as may be the nature of the carbonate material. Field evidence on the functional relationship between dissolution and saturation is necessary for modeling the response of the ocean to environmental changes.

TABLE 1. Evidence for Near-Bottom Dissolution

Type of Evidence	Reference
Coccolith assemblages in sediments resemble surface diversity whereas deep water column diversity is markedly lower.	*Honjo*, 1975
Foraminifera found at sediment surface below regional compensation depth show a wide range of preservation states.	*Adelsack and Berger*, 1975
Similarity of the distribution of dissolved silicate and alkalinity throughout the world ocean despite marked differences in their saturation chemistry.	*Edmond*, 1974

Several lines of evidence (Table 1) suggest that the dissolution of carbonate particles may occur mainly at or near the sediment-water interface. By this interpretation, it is the near-bottom saturation value that is important to the dissolution of carbonate particles. The sedimentation rate of calcium carbonate is then

$$r_s = r_p - r_d \ (\Omega_B, \text{ other factors}) \tag{2}$$

where r_s = rate of sedimentation of $CaCO_3$.

r_p = rate of productivity of $CaCO_3$.

r_d = rate of dissolution of $CaCO_3$.

and Ω_B = near-bottom saturation factor.

The problem is to determine r_d as a function of the saturation factor. Formulating the problem in this fashion leads to approaching its solution by a study of variations in the sedimentation rate, where the degree of saturation varies and the productivity and the "other factors" remain constant. It is then necessary to define explicitly how this experiment should be carried out.

DISCUSSION

We can directly determine the saturation factor only for the present-day ocean. Therefore, the dissolution function must be determined by studying present-day sedimentation rates. In practice this involves considering the sedimentation rate since the beginning of the last interglacial. However, derivation of the sedimentation rate for this short interval is hampered somewhat by the mixing of the surface sediments by organisms. As *Peng et al.* (1977) show elsewhere in this volume, detailed oxygen isotope and C^{14} studies can provide a rate which, despite mixing, is subject to only a small error relative to the total rate variations across the compensation horizon.

The simplest way to insure constant surface productivity would be to locate the study in a central gyre, far from upwelling areas. By this criterion, a latitudinal section at about 25° S across the East Pacific Rise would perhaps be the best location. Such a plan resembles the approach of *Broecker and Broecker* (1974) in their study of sedimentation rate variations in a section at 16° S.

Figure 1 illustrates three depth profiles of the calcite saturation factor in the Tropical Pacific Ocean. The saturation factor was calculated for preliminary GEOSECS data by the method of *Edmond and Gieskes* (1970) using the solubility pressure-dependence of *Duedall* (1972). Uncertainties in the constants used in this calculation only shift the saturation scale a short distance, or compress or expand it slightly; such corrections will not affect the following discussion. The profiles in Figure 1 were used to estimate the bottom saturation factor for the study area considered

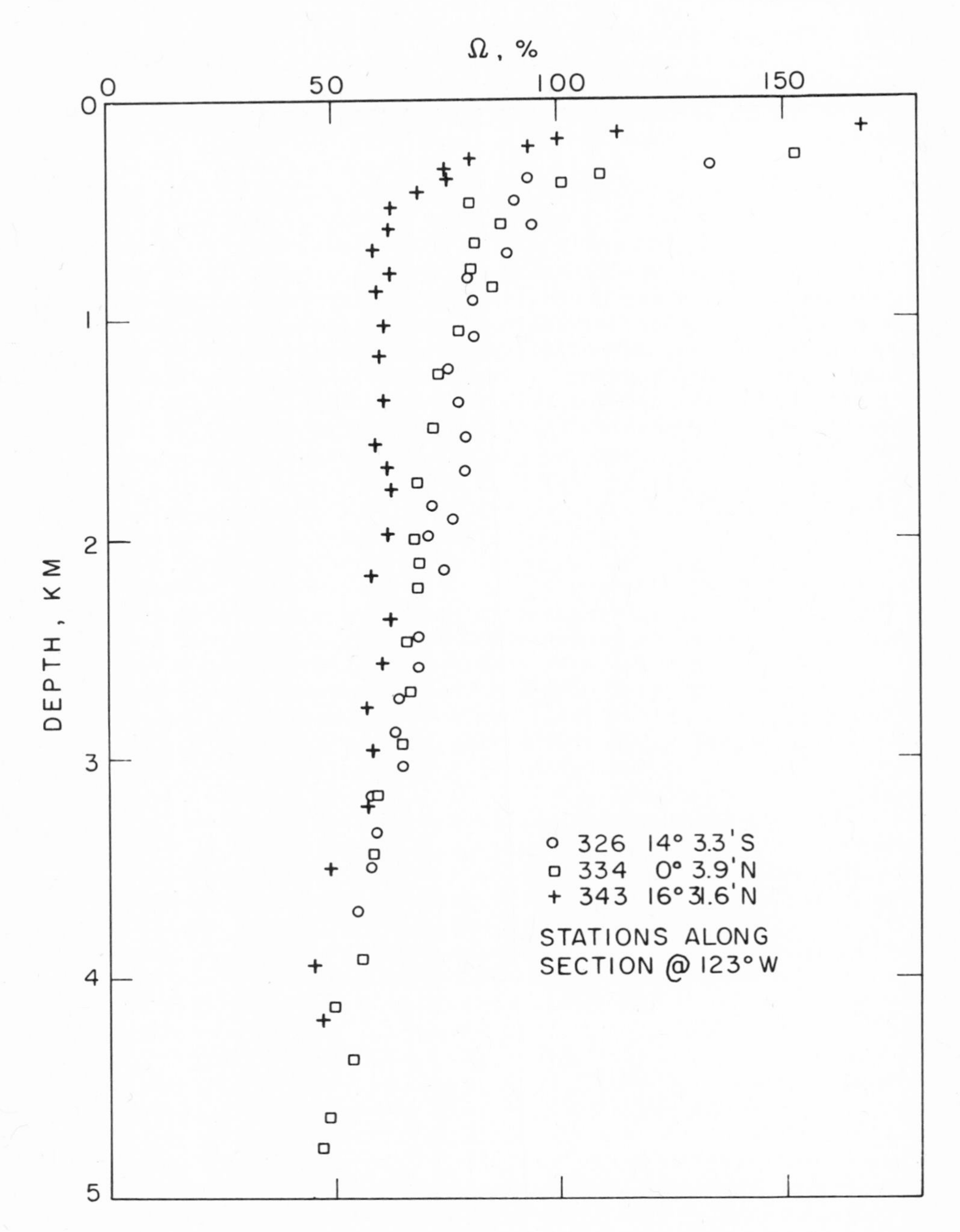

Fig. 1. Depth profiles of calcite saturation for three GEOSECS stations in the eastern Tropical Pacific Ocean.

by *Broecker and Broecker* (1974). Their sedimentation rate data is plotted vs. the estimated saturation factor in Figure 2. The functionality of the curve is not well constrained by this data, but it is evident that there is a large change in sedimentation rate for a small change in the bottom saturation factor.

Locating the study in a region of low productivity limits the dissolution relationship to higher values of the saturation factor. Since some of the most interesting aspects of carbonate dissolution occur at lower saturation values, such as under the equatorial upwelling region, it is desirable to locate the experiment in a region of high productivity. Presuming that by some independent means contours of constant carbonate productivity can be established (e.g. by silica sedimentation rates, surface nutrient levels, productivity levels, productivity measurements, or measurements of particle flux), then both the dissolution rate function and productivity variations can be determined. This is illustrated in Figure 3. Measuring the sedimentation rates at the contour intersections provides redundant information for 4 independent equations in 5 unknowns:

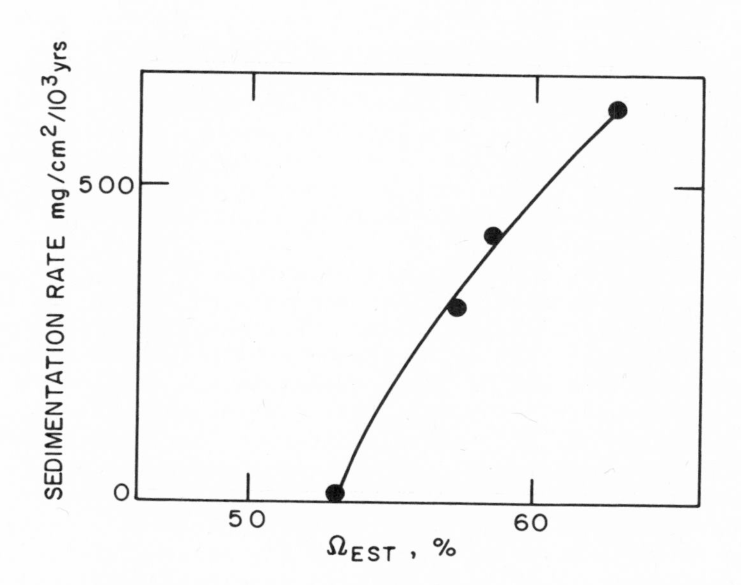

Fig. 2. Carbonate sedimentation rate vs. estimated calcite saturation at 16°S in the eastern Pacific Ocean. Rate data from *Broecker and Broecker* (1974).

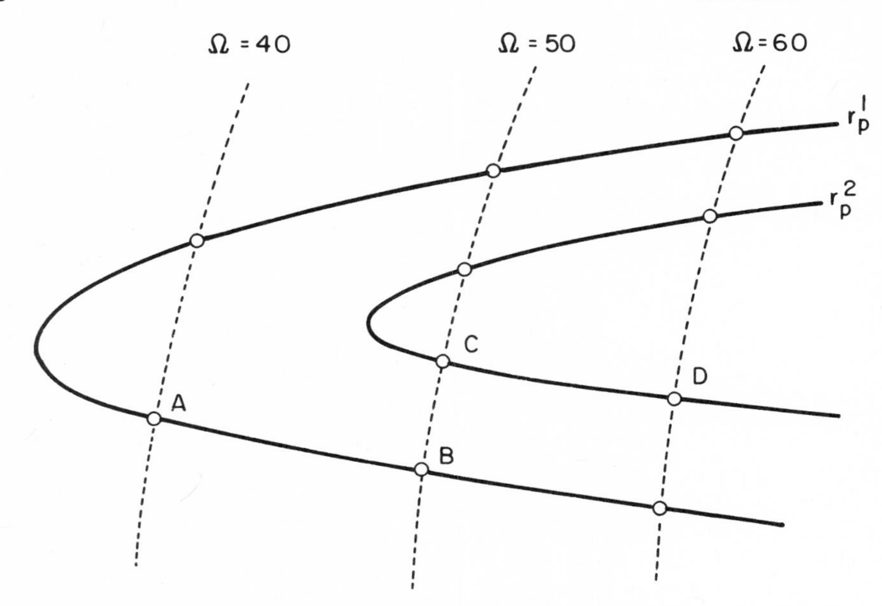

Fig. 3. Map showing hypothetical contours of constant productivity and constant calcite saturation factor.

$$r_s\ (r_{p1},\ \Omega = 40) = r_{p1} - r_d\ (\Omega = 40) \tag{3a}$$

$$r_s\ (r_{p1},\ \Omega = 50) = r_{p1} - r_d\ (\Omega = 50) \tag{3b}$$

$$r_s\ (r_{p2},\ \Omega = 50) + r_{p2} - r_d\ (\Omega = 50) \tag{3c}$$

$$r_s\ (r_{p2},\ \Omega = 60) = r_{p2} - r_d\ (\Omega = 50) \tag{3d}$$

The unknowns are the productivity and dissolution rates. The redundant information is provided by the measured sedimentation rates. For each additional saturation or productivity contour, one more independent equation and one more unknown is added. By assuming or independently measuring the magnitude of one productivity rate, the system is completely determined and a "best-fit" procedure will provide estimates for the remaining productivity and dissolution rates. This formulation assumes that the dependence of the dissolution rate on other factors, including the productivity rate itself, is secondary to its dependence on the bottom saturation factor. Once the functionality of r_d (Ω) has been determined by the fitting procedure, the effect of other factors can be investigated by mapping the difference between the measured sedimentation rates from the best fit and comparing the distribution of any anomalies with the distribution of those other factors.

This approach resembles that of *Berger* (in press) who mapped estimated sedimentation rate contours vs. latitude and depth for the Eastern Tropical Pacific Ocean. His sedimentation rate contours are replotted against estimated bottom saturation value in Figure 4. Again, there are large variations in sedimentation rate with small saturation value intervals. With more detailed information on productivity contours and recent sedimentation rates, similar maps could be used to determine the dissolution/saturation relation as outlined above.

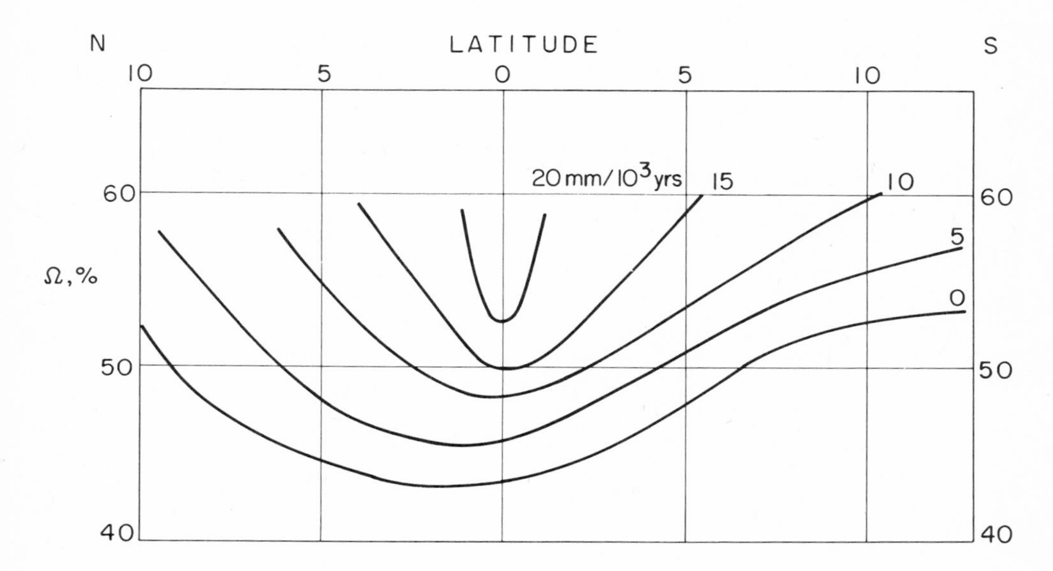

Fig. 4. Contours of carbonate accumulation rate vs. calcite saturation factor and latitude in the eastern Tropical Pacific Ocean. Data derived from *Berger* (in press).

SUMMARY

The variation in the dissolution rate as a function of the near-bottom carbonate saturation factor can be determined by normalization to the present-day sedimentation rate-saturation distribution. The preliminary evidence currently available demonstrates that the dissolution rate is extremely sensitive to small shifts in the saturation values occurring near the bottom of the Pacific Ocean. Once this function has been determined for the present it can be used to interpret the sediment record, quantitatively, in terms of changes in the bottom saturation factor and to produce constraints for models predicting the response of the ocean to the input of fossil fuel carbon dioxide.

REFERENCES

Adelsack, C. G. and W. H. Berger, 1975. On the dissolution of planktonic foraminifera and associated micro-fossils during settling and on the sea floor, In: *Dissolution of Deep-Sea Carbonates* (eds. Sliter, Bé, and Berger), Cushman Foundation for Foraminifera Research, Special Publication No. 13, p. 70.

Berger, W. H. Sedimentation of Deep-Sea Carbonate: Maps and Models of Variations and Fluctuations, In: *Marine Plankton and Sediments* (eds. Riedel and Saito) Micropaleontology Press (in press).

Broecker, W. S. and S. Broecker, 1974. Carbonate Dissolution in the Western Flank of the East Pacific Rise, In: *Studies in Paleo-oceanography* (ed. Hay),Society of Economic Paleontologists and Mineralogists, Special Publication No. 20, p. 44.

Duedall, I. W., 1972. The partial molal volume of calcium carbonate in seawater, *Geochim. et. Cosmochim. Acta 36*: 729-734.

Edmond, J. M., and J. Gieskes, 1970. On the calculation of the degree of Saturation of Seawater with respect to calcium carbonate under *in situ* conditions, *Geochim. et. Cosmochim. Acta 34*: 1261-1291.

Edmond, J. M., 1974. On the dissolution of carbonate and silicate in the deep ocean, *Deep-Sea Res., 21*: 455.

Honjo, S. 1975. Dissolution of suspended coccoliths in the deep-sea water column and sedimentation of coccolith ooze, In: *Dissolution of Deep-Sea Carbonates* (eds. Silter, Bé, and Berger), Cushman Foundation for Foraminiferal Research, Special Publication No. 13, p. 114.

Luz, B. and N. J. Shackleton, 1975. $CaCO_3$ solution in the tropical east Pacific during the past 130,000 years In: *Dissolution of Deep-Sea Carbonates* (eds. Silter, Bé, and Berger) Cushman Foundation for Foraminiferal Research, Special Publication No. 13, p. 142.

Milliman, J. D., 1975. Dissolution of aragonite, Mg-calcite, and calcite in the North Atlantic Ocean, *Geology, 3*: 461-462.

Morse, J. W. and R. A. Berner, 1972. Dissolution kinetics of calcium carbonate in sea water: II A kinetic origin for the lysocline, *Amer. J. Sci., 272*: 840.

Peterson, M. N. A. 1966. Calcite: rates of dissolution in a vertical profile in the central Pacific, *Science 154*: 1542.

Peng, T. H., W. S. Broecker, G. Kipphut and N. Shackleton, 1977. Benthic mixing in deep sea cores as determined by 14C dating and its implications regarding climate stratigraphy and the fate of fossil fuel CO_2. In: *The Fate of Fossil Fuel CO_2*. N. R. Andersen and A. Malahoff (Eds.), Plenum, N.Y., (this volume)

Takahashi, T. 1975. Carbonate Chemistry of Sea Water and the Calcite Compensation Depth in the Oceans In: *Dissolution of Deep Sea Carbonates* (eds. Sliter, Bé, and Berger), Cushman Foundation for Foraminiferal Research, Special Publication No. 13, p. 11.

BIOGENIC CARBONATE PARTICLES IN THE OCEAN; DO THEY DISSOLVE IN THE WATER COLUMN?

Susumu Honjo

Woods Hole Oceanographic Institution

ABSTRACT

Excess CO_2 on the earth should eventually be neutralized in the ocean by the skeletal remains of calcite and aragonite produced by planktonic organisms particularly foraminifera, coccolithophores and pteropods. The remineralization rates of biogenic carbonate, which controls the alkalinity adjustment of the ocean, is a function of their absolute dissolution rates. It is important to clarify the process and locality, how and where major remineralization occurs in the changing oceanic environment. The residence time of planktonic foraminiferan tests and pteropod shells is too short to undergo the significant dissolution while they are settling through the water column. On the other hand, the residence time and specific surface area of coccoliths is very large and their skeletons should be completely dissolved in the water column before arriving at the ocean floor. However, well preserved suspended coccoliths in undersaturated deep water have been collected. This paradox has been explained by invoking a two stage transportation hypothesis; first coccoliths descend with high speed within the fecal pellets of grazers; secondly, discrete coccoliths sink very slowly after having been shed from the host fecal pellet in the deep sea. Carbonate in fecal pellets are protected from undersaturated sea-water and their residence time in the water column is short through the process of fast sinking. Dissolution of carbonate particles does not take place within the grazer's guts, biodegraded fecal pellets disintegrate while sinking. Thus a number of coccoliths are suspended in undersaturated water and they are *probably the major source of in situ calcite remineralization in the open ocean. The rates were as high as 3 to 12 mg $m^{-3}yr^{-1}$ (if Peterson's (1966) dissolution profile was applied) in a 5,000 m water column in the North Equatorial Pacific. These rates are probably some of the highest in the oceans. The annual rates of remineralization were estimated to be less in the waters of colder climatic zones. Under lower temperature, microbial activity is subdued and pellets were less intensely biodegraded. Consequently, the shedding rate of coccoliths from pellets was low and resulting in less in situ remineralization.*

INTRODUCTION

Biogenic calcium carbonate, secreted by planktonic organisms in the open ocean, is a major part of the gigantic CO_2 biogeochemical cycle. Myriads of calcareous skeletal remains are preserved in a temporary sedimentary sink whose dissolution acts as a control on the alkalinity of ocean water. Growing excess CO_2 on this planet (*Pales and Keeling*, 1965), which is partly generated by the combustion of fossil chemical fuels (*Broecker et al.*, 1971), should eventually be neutralized in the ocean by these biogenic carbonate particles.

The dissolution of calcium carbonates can occur in many ocean environments. Deep water, particularly in the Pacific Ocean, is undersaturated with respect to calcite and aragonite. If the major dissolution of carbonates occurs in shallow layers of the water column, the alkalinity response of sea water is relatively rapid. On the other hand, if the dissolution of biogenic carbonate occurs only at the sea-bottom or in the sediment at great depths, the neutralized water mass takes a long time to be exchanged with surface water by diffusion and advection. This geochemical "impedance", which is equivalent to the oceanic turnover period, could be an important time lag affecting the rate at which CO_2 concentration increases and is neutralized.

The purpose of this review is to examine whether biogenic carbonate particles undergo dissolution while sinking through the undersaturated water column, and if not, to examine other regimes where dissolution could occur. As elaborated in the text, it does not appear that the major dissolution of biogenic carbonates takes place in the water column while they sink from the productive layer surface waters toward the bottom (excluding resuspended carbonate particles in the near bottom layer). However, some coccoliths should dissolve in relatively shallow water in the Pacific Ocean because the saturation depth is so shallow (*Takahashi*, 1975). A budget for biogenic carbonate dissolution which has been estimated for highly undersaturated water is submitted in this paper.

There are serious quantitative uncertainties in the basic information used to estimate the carbonate biogeocycle in the ocean. In particular, our knowledge of carbonate production rates by planktonic organisms, particularly by foraminifera and pteropods, is at best sparse. Secondly, the available information about the absolute dissolution rates of biogenic (and inorganic) carbonates appears unreliable. As a matter of fact, *Peterson* (1966) and *Berger's* (1967) experiment have provided the only coherent data on the dissolution rates of inorganic calcite and planktonic foraminiferan tests in undersaturated water (also *Milliman*, in press). The development of surface properties of carbonate particles involves complex processes originating from epitaxic biomineralization. Thus the application of dissolution rates to physico-chemically determined surface areas is not straightforward and may not be correct. It was found that the study of accelerated sinking by aggregated coccoliths, particularly by zooplankton fecal pellets, is important in understanding the residence time of these small but abundant carbonate particles in the undersaturated water column (*Honjo*, 1975; *Roth et al.*, 1975). This phenomena requires a detailed study of coccolithophore production and grazing, for which the quantitative relation is not well known. Under the present state of the art, therefore, an accurate estimation of mass balance for production-dissolution-delivery is difficult. The budgets presented in this paper are therefore preliminary.

BIOGENIC CALCIUM CARBONATE PRODUCTION IN THE OPEN OCEAN

In the open oceans, coccolithophorids and foraminifera constitute the principal agents of calcium carbonate deposition in the form of low magnesium calcite (*Milliman*, 1974). Relatively large populations of pteropods that precipitate low strontium aragonites are found in the ocean but the mass of carbonate fixation is said to be small compared to the former. Heteropods and species of gastropod genus *Janthina* and some species of cephalopods also precipitate aragonite but they are minor contributions (*Lowenstam*, 1974).

Coccoliths

Coccolithophorids live almost exclusively in the upper 350 m (*Okada and Honjo*, 1973). Their depth range is limited to the uppermost few tens of meters in the Subarctic zone. Maximum densities of coccolithophorids are observed in the 40 to 80 m layer in the Transitional zone, Central Gyre, and Equatorial zones, coinciding with the local chlorophyll maximum (*Honjo and Okada*,1974; *Honjo*, 1976a). Numbers of coccospheres are found throughout the deep water column except in the Subarctic zone, but there are usually less than 100 cells per liter (*Okada and Honjo*, 1973).

The standing crop of calcite as coccoliths in living coccolithophorids is approximately 350 mg m^{-2} (estimated value integrating the density through 200 m photic water column with section of 1 m^2) in the Equatorial, 100 mg m^{-2} in Central Gyre and as high as 1000 mg m^{-2} at a station in the North Pacific 50°N, 155°W measured in August. The calcite mineral productivity (as coccoliths) by coccolithophorids can be roughly estimated in tropical and subtropical waters where the seasonal fluctuation of biomass is relatively small and the turn-over period is presumed to be less irregular than the area of higher latitudes. In the Equatorial Zone of the Pacific Ocean the annual production is approximately 10 grams m^{-2} y^{-1} if the year-round turn-over period is 10 days (*Honjo*, 1975). In Gyre water the calcite production will be only a few grams, but it can be higher than 10 grams m^{-2} y^{-1} in the Subarctic and Transitional Zones.

Planktonic Foraminiferan Tests

It is difficult to estimate calcite production by planktonic foraminiferan tests because their concentration and turn-over rates are not known with reasonable precision. Planktonic foraminifera are found mainly in the upper 350 m and their geographic abundance strongly reflects oceanic fertility (*Berger*, 1969; *Bé and Tolderlund*, 1971). Published records indicate that abundance ranges from fewer than 1 specimen per m^3 of water in the Central Gyre, to greater than 10 specimens in m^3 in high productivity areas such as western sides of continents (*Bradshaw*, 1959; *Bé and Tolderlund*, 1971; *Tolderlund and Bé*, 1971). Several year's mean annual standing crop of planktonic foraminifera in surface waters of the Subarctic, Transitional and Central Gyre (Bermuda) of the Atlantic was 5.6, 4.0, 3.9 (3.6) per m^3. A 200 μm mesh plankton net was mostly used for sampling in those studies). *Adelseck and Berger* (1975) reported rather high densities in the Equatorial Pacific; 9 to 19 individuals per m^3 (using a 153 μm mesh net). It is evident that counts of living planktonic foraminifera drastically increase if collection is made by smaller meshed nets or large volume pumping (*Bishop et al.*, 1976). The approximate world average concentration (*Berger*, 1969, statistically normalized to the catch by a 158 μm net) is 10 individuals m^{-3} (in surface layers). Roughly, the difference in concentration between the Pacific and Atlantic is significant and the concentration in the Indo-Pacific appears to be at least twice as much as Atlantic (*Berger*, 1969, fig. 3). The delivery rate of planktonic foraminiferan tests at the Equatorial deep sea floor of the Pacific is about 1 x 10^5 m^{-2} y^{-1} (*Adelseck and Berger's* calcuation in 1975 based upon *Arrhenius*, 1962; *Berger*, 1971) and the turn-over rate must be of the order of 10 times a year.

Berger (1969) presented oceanwide statistics which indicate that below 500 m the concentration of living and empty tests is roughly equal. He also estimated that the turn-over period was about equal to the time which a test required to sink 500 m; a few days. According to *Berger and Soutar's* (1967) estimate for the Santa Barbara Basin in August, the average turnover period for several planktonic foraminiferan species was approximately 50 days, giving an annual turnover rate of 7.

Berger (1971) estimated the turnover time was 4 to 10 days applying various methods. The standing crop of planktonic foraminifera in the Santa Barbara Basin in August was approximately 2 x 10^3 m^{-2} (estimated from *Berger and Soutar*, 1967, table 1, p. 1496) and an annual production of tests approximately 10^4 m^{-2} (1.4 x 10^4 tests m^{-2} yr^{-1}). If we use the average weight of foraminiferan tests from the 250 μm to 175 μm fraction, which is approximately 6 μg, the annual production of calcite by those foraminifera is of the order of 100 mg m^{-2}. In a hypothetical situation, if all the foraminiferan tests grew larger than 250 μm the estimated annual calcite production rate would still be less than a gram per m^2, an order of magnitude less than the production rate of coccolith's calcite. Research on the *in situ* standing crop and turn-over rates warrant further study.

Berger and Soutar's (1968) estimates for reproduction rates are valid only for the larger (> 150 μm) specimens. However, smaller tests are also abundant oceanic foraminiferan ooze sediments and are thus an important part of $CaCO_3$ sedimentation. *Berger and Piper* (1972) reported that approximately 20% of "coarse" laminae and 90% of "fine" laminae consist of between 62 and 125 μm diameter (the average of each fraction is illustrated in D through A in Fig. 3, p. 280, *Berger and Piper*, 1972). *Berger* (1969) reported that relatively large standing crops (more than 25 individuals m^{-3}) of *Globorotalia bulloides* in surface (0-50 m) Gulf of California waters (March) were mostly smaller specimens. Moderate numbers of these individuals were found from 100 to 200 m (a few individuals per m^{-3}) but they were mostly large specimens (Fig. 5, p. 16). This indicates that small tests outnumber the large tests in a living community. If smaller tests divide faster than the larger ones, the contribution by the former would be significant.

Pteropod Shells

Euthecosomation pteropods (shelled pteropods) are ubiquitous in the open oceans but are essentially restricted to the euphotic layer. Pteropods and heteropods undergo diurnal depth migrations (*McGowan*, 1960; *Chen and Bé*, 1964). Species distribution is latitudinal like many other oceanic zooplankton species (*McGowan*, 1960). Some species such as *Clio pyramidata* and its varieties are common, with wide temperature tolerance ranges from 7.0 to 27.8°C (*van der Spoel*, 1967). Euthecosomates have not been found in the high Arctic, Central Baffin Bay, the Ross Sea and the Weddell Sea. Inland or marginal seas such as the Baltic and Black Sea that lack pteropods probably because of their low salinity (*van der Spel*, 1967).

Pteropod concentrations were reported to be 1 to 10 individuals per m^3 (*Lowenstam*, 1974). *Adelseck and Berger* (1975) found 5 to 11 individuals of pteropods per m^3 in the surface layer of the Equatorial Pacific Ocean. The amount of aragonite in pteropod shells varies over 2 orders of magnitude (*Honjo*, in press) and there is a complete lack of information on turn-over rates, making estimates of aragonite production by pteropods almost impossible.

ABSOLUTE DISSOLUTION RATES OF CARBONATES

Dissolution of particles in the water column depends on: (1) the absolute dissolution rate of each mineral phase; (2) the surface area in contact with water; and (3) the residence time of the particles in undersaturated water.

A serious discrepancy exists between the dissolution rates of calcite measured in the laboratory (*Morse and Berner*, 1972) versus those measured in the field (*Berger*, 1967). Even different laboratory measurements of the apparent solubility products of inorganic calcite crystals do not agree well (*Morse and Berner*, 1972; *Ingle et al.*, 1973; *McIntyre*, 1965). Generally speaking the laboratory calcite dissolution rate was over 100 times faster than *Berger's* (1967) *in situ* dissolution rate for planktonic foraminifera tests and 30 times faster than *Peterson's* (1966) *in situ* dissolution rate for inorganic calcite (*Morse*, 1976), when they are compared under the conditions equivalent to the Ω value of 0.50, on a $mg/cm^2/yr$ basis.

Peterson machined spar-calcite into spheres which were subsequently exposed to sea water. Weight loss was normalized to surface area and thus an absolute rate versus depth profile was obtained. Ground surfaces have a rather rough topography and appear cracked when observed under SEM. Therefore, the exposed surface area is larger than that calculated from the diameter of a sphere. The dissolution of polished or lapped calcite surfaces by acid solutions was significantly different from untreated crystalline surfaces or ground surfaces. If the initial site of solution takes place at so called "kinks and corners" (*Berner and Morse*, 1974), the physical stress on the surface which is imparted during the milling process will change the dissolution properties of the sample.

Only *Berger's* (1966) experiment is available for the *in situ* dissolution rates of biogenic calcite (planktonic foraminiferan tests). Measurements of *in situ* dissolution rates of coccoliths have not been tried because of experimental difficulties. Very little is known about biogenic aragonite dissolution rates. *Berger* (1966, 1970) retained samples in a permeable container covered by 64 μm mesh synthetic gauze. Whether this retainer prevented water circulated and increased the Ω value resulting in the decrease of sample dissolution rates has been controversial (*Morse*, 1976). *Milliman's* (1975) experiment involves a similar situation. *Berger's* (1967, 1970) *in situ* dissolution rates were given as percentage of weight loss per day. Applying specific surface area (N_2/He BET method, reviewed by *Grab et al.*, 1975) of planktonic foraminifera tests from moderate to warm water assemblage in two fractions, > 250 μm and 250 to 125 μm (Table 1), the absolute dissolution rates of planktonic foraminiferan tests in the Pacific Ocean from 1,000 m to 2,500 m is 6 to 4 μg cm^{-2} yr^{-1} (0.02% per day, *Berger and Piper*, 1972, p. 283, table 8). The average dissolution rate of calcite spheres for the same depth range measured by *Peterson* (1966) was 30 μg cm^{-2} yr^{-1}. The discrepancy can best be explained by the difference in effective area for dissolution between solid sphere and tests. The sea water circulation through the interior of foram chambers is highly inefficient, but N_2/He BET measurement does not discriminate such areas; i.e. areas where light gases might freely enter. The non-critical surface of a test (inside chambers and pore surfaces) may be several (5 to 7) times larger than the effective surface for dissolution. A similar conclusion was reached by *Morse* (1976) from a laboratory dissolution study of planktonic foraminiferan tests.

SURFACE PROPERTIES OF BIOGENIC PARTICLES

Surface Area of Particles

As shown in Table 1, biogenic calcium carbonate particles have large specific surface areas because of their complex architecture and porous structure. The specific surface area of coccoliths is an order of magnitude larger than planktonic foraminiferan tests and pteropod shells. A BET method does not discriminate against the surface area covered by foreign material. One can question whether the surface represented by the adsorption sites of N_2, He or Kr molecules are identical to the kinetic reaction sites for dissolution. A BET surface of *Berner and Morse* (1974) standard calcite was often larger by a factor of 2 or 3 (*Morse*, 1976) than the estimate based upon topographic information by SEM observation for the same sample. Despite these uncertainties, BET methods still appear to be the best first approximation of the area of biogenic calcite particles to be exposed to sea water.

Zooplankton tests or shells are not exposed directly to the water but usually formed inside of soft tissue while they are living. Typically, foraminiferan tests or radiolarian skeletons are covered by protoplasmic material. The interior of molluscan shells is

TABLE 1: Specific Surface Area of Carbonate Samples

Sample	Specific surface (m^2 g^{-1})	Method
Calcite sphere (with ideal surface)	0.00247	calculated
Berner-Morse calcite	0.553	Kr BET (1)
Morse's aragonite	1.40	Kr BET (2)
Pelagic carbonate sediment	3.60	Kr BET (3)
Planktonic foram tests		
> 64 μm	0.877	Kr BET (4)
125 μm-250 μm	1.70	1p N_2 BET (5)
< 250 μm	1.50	1p N_2 BET (6)
Pteropod	1.04	1p N_2 BET (7)
Coccoliths		
*Emiliania huxleyi**	10.442	3p N_2 BET (8)
*Cruciplacolithus neohelis**	9.629	3p N_2 BET (9)
Cyclococcolithus leptopora	4.3	SEM (10)

* Coccoliths were extracted from laboratory culture (1) through (4), after *Morse and de Kanal*, (1976). (5) to (9), *Honjo and Erez* (in prep). 1p: One point BET method. 3p: 3 points BET. Helium was used for the carrier. (10) Average surface area in this species was estimated by SEM photographs. Species assemblage of (5) and (6) is *G. tumida*, 25%; *G. inflata*, 30%; and *G. ruber*, 45%. Measured total of 864 and 473 mg, respectively.

generally filled by soft organs and usually a periostracum covers the outside surface. After death the hard tissue is not immediately available for dissolution until the covering soft tissue is biodegraded. The term organic coating is usually applied to a hypothetical refractory film which is believed to cover the surface of recent, semifossil and fossil biogenic particles. Foreign organic material suspended in the oceanic water column may be adsorbed on particulate surfaces, slowing or inhibiting dissolution. The inhibitory effect of dissolved organic matter (*Chave and Suess*, 1970) on carbonate dissolution in the water column has not been confirmed (*Morse*, 1973).

Planktonic Foraminiferan Tests

In many benthonic foraminifera, individuals empty their tests of soft tissue during schizogomy and gamogony cycle (for example, *Grell*, 1958). Knowledge of the life history of planktonic foraminifera was extremely limited until *Bé and Anderson* (1976) observed the gametogenesis of two planktonic species; the test was emptied after the gametes swam away, leaving the empty shell to descend through the water column. Small residues of soft tissue remain on sinking tests (*Berger and Soutar*, 1970) and may prevent immediate contact with sea water until biodegradation removes such layers. Biodegradation is more active in warmer (shallower) water but it may be significantly slower in deep and cold water because microbial activity is subdued (*Jannasch and Wirsen*, 1973).

It is not clear that thin and refractory organic coatings universally exist and, if they do, how they effect the dissolution of sinking tests in undersaturated water on the surface of planktonic foraminiferan tests. During a mooring experiment, *Berger* (1967) found that H_2O_2 treated specimens dissolved significantly faster below the lysocline than untreated tests. However, electron micrographs of planktonic foraminiferan wall structure suggest the presence of an organic membrane (for examples, *Towe and Cifelli*, 1967; *Bé and Hemleben*, 1970; *Towe*, 1971). The increased dissolution after H_2O_2 treatment may not have simply resulted from the removal of an organic material but rather could have resulted from a broadening of available surface area by a loosening of wall structures. Amino acids were reported to represent from 0.02 to 0.04% (by weight) of foram tests from deep sea core tops (*King and Hare*, 1972). It is possible that the amino acids they found were delivered from proteinacious membranes analogous to three organic matrices metazoan's use for hard tissue mineralization. Interaction of these organic materials with biocrystals and their effect on dissolution warrant further study.

Coccoliths

Coccoliths appear more resistant to dissolution than foraminiferan tests in the deep sea sediments. A hypothetical long lived refractory organic coating has been postulated by paleontologists (for examples, *Cita*, 1970; *McIntyre and McIntyre*, 1971).

One of the primary functions of coccolith plates is probably as ballast to adjust the best depth of survival (*Honjo*, 1975). Thus the physiologic significance of carbonate hard tissue is entirely different in coccolithophores and planktonic foraminifera. No refractory film-like material has been observed under SEM or FESEM on the surface of collected or cultured coccoliths. It appears that the distal side of a coccolith is directly exposed to sea water while the coccolithophore is living. (As an exception, the "palmelloid stage" of coccolithophores (*Bernard*, 1963) has an organic film covering cell aggregates. We have frequently observed this in laboratory cultures of *Cruciplacolithus neohelis* whose life cycle is unusual. The significance of palmelloid formation of coccolithophores in the open sea is unknown.)

Inhibition of dissolution by the organic matrix which serves as a template for epitaxic growth of coccolith calcite (*Wilbur and Watabe*, 1963) remains a possibility. Coccoliths are formed in the cell's Golgi apparatus (*Klaveness*, 1972). The membrane matrix is formed on the proximal side of the coccolith-to-be inside the vesicle, with dorsoventral symmetry. The template appears to accompany the complete coccolith when extruded to the cell surface (*Wilbur and Watabe*, 1962). Recalcification of coccoliths reveals a stainable organic matrix of coccoliths appears to be entirely composed of carbohydrates (*Westbroek et al.*, 1973), made by uronic acid (up to 25%), simple sugars, and polysaccharide-sulfates (*de Jong*, 1975) associated with ribose (*Mopper and Degens*, 1972). It is uncertain whether the polysaccharide has an intra- or extra-crystalline location and if it actually impedes dissolution in undersaturated water.

Pteropod Shells

The trophic level of pteropods appears relatively high in the oceanic food webb. The causes of depth, predators, and production cycles of euthecosomation pteropods are not known. Therefore, there is no reliable information on the manner in which pteropod shells are delivered from shallow to deep water. In case of "natural death" remnant tissue may prevent immediate dissolution of shell interiors while they descend through the water column. *Vinogradov* (1961) reported that the experimental sinking speeds of dead pteropods were between 910 to 2,270 m, measured on the specimens ranging 0.5 to 4.17 mm. His data indicate that pteropods shells remain several days before reaching the bottom and it is assumed that complete biodegradation of tissue does not occur before they reach the deep sea floor (Votnintsev, 1948; Skopinstev, 1959 in *Vinogradov*, 1961). Some species are known to attack other pteropods and clean the shell by eating soft tissue while they are still suspended but such modes of predation may not be universal. However, *Vinogradov* (1961) reported that below 2,000 m practically all (suspended) pteropods shells are empty, while below 3,000 m not a single shell with the remains of mollusk was found at a station in the Equatorial Indian Ocean. He explained this by biodegradation of the tissue during a prolonged residence time to local picnocline and turbulence. When pteropods are eaten by fish with strongly acid stomachs, the shell may be destroyed. However, predation by fish may be minor in the open ocean. The large number of juvenile shells, many less than 100 μm, which were collected by sediment trap (*Wiebe et al.*, 1976) suggest high juvenile mortality.

Other Dissolution Inhibitor; Suspended Organic Matter

In scanning electron microscope studies, we have occasionally observed a thin amorphous fouling of film which prevents the dissolution of calcite microcrystals (such as Berner and Morse's standard calcite) when directly exposed to water on a deep mooring line (*Honjo*, in press). In addition, particulate residues on Nuclepore filters from samples of subsurface sea water which were collected often retain amorphous matter similar to the film mentioned above. The "marine snow" phenomena (*Costin*, 1971; *Kajihara*, 1971) may be a cause of such deposits but no further information is available at present.

RESIDENCE TIME AND PENETRATION DISTANCE

Planktonic Foraminiferan Tests

We assume that the sinking velocity of a planktonic foraminiferan test in the ocean is comparable to laboratory sinking rates measured on tests picked (up) from the sediment. The estimated residence time of tests in a water column of 5,500 m is on the order of weeks to a month (5.2 days for the fraction larger than 250 μ, 12.3 days for 250 to 177 μm, 40.9 days for 125 to 62 μm, after *Berger and Piper*, 1972). If we apply the BET specific surface area (Table 1) and absolute dissolution profile of spar calcite measured by *Peterson* (1966), the estimated dissolution during a 5000 m descent is 2.5 to 3.5% for a test of larger than 250 μm and 7.5 to 8.5% for 250-150 μm fraction. On the other hand, if Berger's dissolution rate for planktonic foraminiferan tests is applied, the 250 μm fraction dissolves 0.7% compared to 1.7% for the 250-177 μm fraction. As explained before, this difference might be due to the relatively large proportion of enclosed surface area in the larger test. *Adelseck and Berger* (1975) noticed no signs of dissolution on large, recently arrived foraminiferan tests collected at the sediment surface below the lysocline in the Equatorial Pacific.

The sedimentation processes of small foraminifera (smaller than 64 μm) in the ocean is not clear. Individuals in early stages of division are very common in many calcareous core samples from a variety of depths and oceans. *Früterer* (1974) reported that a large number of small planktonic foraminifera tests are present in a so-called "coccolith" fraction of sediment from off the West African shore. These small tests were usually ignored by researchers and omitted from faunal statistics, or simply assigned to similar forms such as *Thoracosphaera* (*Roth*, pers. comm.).

Bishop et al. (1976) collected 800 to 2,000 planktonic foraminifera m^{-3} (mixture of living and empty tests), of less than 53 μm, between the surface and 388 m at an Equatorial Atlantic station using a large volume submersible pump (*Bishop and Edmond*, 1976). Also, the density of tests larger than 53 μm which they reported was far larger than any number previously published: 5,800 (32 m) to 230 (388 m) per m^3. A reasonable explanation of those data has not been obtained yet.

The sinking speed of small spherical planktonic foraminiferan tests with diameters of 30 μm and 10 μm is estimated to be 6 x 10^2 μm sec^{-1} (from *Lerman et al.*, 1975, fig. 1, p. 24) and their residence time in a 5,500 m water column is 105 and 1600 days, respectively. If we apply Berger's dissolution profile (*Berger*, 1967) at Peterson's mooring site, the weight loss of a sinking 300 μm spheric specimen is 14% and compared to 200% (it is totally dissolved at 4,500 m) for a 10 μm diameter specimen. The biodegradation of remaining tissue would be more complete in the smaller specimen because of its greater residence time, therefore they may dissolve more but they also have a greater chance of being predated and protected in fecal pellets (*Honjo and Harbison*, in press).

Pteropod Shells

Dissolution of pteropods appears much dependent upon the persistence of the organic tissues which cover the aragonitic shell. If "naked shells" under 2,000 m observed by *Vinogradov* (1961) is universal, minor dissolution may take place while they sink. Suppose a) the dissolution rates of pteropod aragonite is twice as fast as foraminifera, b) the BET surface is 75% of the over 250 μ fraction of foraminifera (Table 2), and c) the residence time of a shell after being exposed to sea water is three days, approximately a half percent of shell can be dissolved before they reach the bottom.

Coccoliths

From a purely Stokesian point of view, a coccolith sinks very slowly (Table 2). A typical residence time is on the order of 100 years in the deep sea. The weight of a single coccolith amounts to only several picograms, and it has a large surface area (Table 1). Since the morphology of coccoliths generally involve no enclosed volumes which can retain stagnant pore fluids (although some holococcoliths such as *Scyphosphaera* and *Thoracosphaera* have enclosed space), the whole surface of a coccolith is in contact with sea water.

If *Peterson's* (1966) dissolution rates are applicable to coccoliths, it takes only a few months to disappear at a rate of 30 μg cm^{-2} yr^{-1} (the average of Peterson's dissolution rates from 1,000 m to 3,500 m (Fig. 1, *Peterson*, 1966)), while a coccolith sinks only several tens of meters (Table 2). Consequently, undersaturated deep-sea water should contain no coccoliths. This should be particularly true in the North Pacific where undersaturated waters approach the surface (*Takahashi*, 1975). This means that no coccoliths can reach the bottom *via* normal sinking, therefore, no coccoliths should be found in the sediment deeper than the calcite undersaturation depth. However, abundant, excellently preserved coccoliths were found suspended throughout Pacific deep water, even where the water was strongly undersaturated (*Honjo*, 1975). Coccoliths are a major constituent of deep sea sediment and intact specimens are abundant in the sediment at great depth.

PROTECTION AND TRANSPORTATION OF CARBONATE PARTICLES BY FECAL PELLETS

A large number of well preserved coccoliths were found suspended throughout undersaturated deep Pacific water (*Honjo*, 1975). The distribution of suspended coccoliths species in deep water reflected the species assemblages in the overlying euphotic zone. No selective depletion of more delicate coccolith species was evident and the original species composition and nannofloral boundaries in the photic layer were well preserved throughout the underlying sea water column (*Honjo*, 1975). Considering the dissolution rates of carbonates and advection of small particles by deep-water movement, the sinking speed of coccoliths should be at least 3 orders of magnitude faster than the speed measured in the laboratory. *Honjo* (1976b) explained this using a two stage sinking model; first the coccoliths sink rapidly enclosed in fecal pellets, then after "lateral injection" of sloughed off coccoliths from their host pellets in deeper water, their sinking speed is reduced to normal (0.15 m day^{-1}) and they quickly dissolve *in situ*.

TABLE 2. Surface Area and Sinking Rates of Individual Particles

	Weight (μg)	Individual surface area (μm^2)	Sinking velocity[1] @ 20°C ($\mu m\ sec^{-1}$)
Planktonic foraminifera[(1)]			
150-250 μm	6	10×10^6	$2\text{-}10 \times 10^3$
> 250 μm	22	30×10^6	$10\text{-}15 \times 10^3$
Coccoliths			
E. huxleyi	8×10^{-6}	3.2×10	1.6[(2)]
C. neohelis	6×10^{-6}	3.6×10	1.8
C. leptopora	330×10^{-6}	41×10	15

(1) Approximate sinking rates obtained from Fig. 1 of *Berger and Piper* (1972) and *Honjo* (in press).

(2) *Honjo*, 1976b.

Grazing pressure on phytoplankton is very high in the open ocean oligotrophic communities (for example, *Menzel and Ryther*, 1962). A large variety of filter feeders efficiently graze upon phytoplankton. *Menzel* (1974) indicated that 90% or more of living phytoplankton are eaten while living. The majority of undigestable tissue of phytoplankton such as coccoliths or diatoms thus end up in a fecal pellets.

Fecal Transportation Observed in Field and Laboratory

A sediment trap set at 2,200 m on the floor of the Tongue of the Ocean, Bahamas (*Wiebe et al.*, 1976) collected large numbers of fecal pellets, some of which were packed full with excellently preserved coccoliths including delicate species such as holococcoliths and *Umbelicosphaera irregularis* as well as other hard-tissue of phytoplankton (*Honjo*, 1975). Many small (juvenile) planktonic foraminiferan tests and "*Thoracosphaera*" (less than 40 μm) were also in the pellets. A typical pellet contained about 10^5 coccoliths, or approximately 1 μg of $CaCO_3$ (*Honjo*, 1975). They sank at the rate of 100 m day^{-1} and reached the deep ocean floor within a month (*Wiebe et al.*, 1976). The majority of the pellets found in the trap were assumed to have been produced by pelagic copepods. Copepods are important because of their abundance and their low trophic levels in pelagic ecosystems.

A variety of copepods (*Calanus finmarchicus*, *Centropages typicus* and *Acartia tonsa*) were fed a phytoplankton mixture containing laboratory cultured coccolithophores, free coccoliths, and inorganic carbonate crystals (less than 10 μm) of calcite and aragonite. The copepods produced fecal pellets collected by sedimentation trap (*Honjo and Roman*, in press).

Does Dissolution Occur in the Zooplankton Gut?

The calcite and aragonite crystals ingested by laboratory copepods and incorporated into their fecal pellets were completely intact and showed no etching when observed under SEM (*Honjo and Roman*, in press). Thus the inside of fecal pellets appears not acidic and the *pH* is probably similar to that of ambient seawater. Field and laboratory evidence suggests that dissolution of biogenic calcite does not therefore occur in zooplankton guts.

Uptake Mechanism and Destruction of Food

The spacing of the secondary maxiella of mature and copepodite stage marine copepods appears too coarse to collect coccolithophores efficiently. However, laboratory cultures of marine copepods including *C. finmarchicus*, a large (2.5-6 mm) cosmopolitan species, eat not only coccospheres but even discrete coccoliths. Scanning electron microscope observations of the mouth organs of *A. tonsa* suggested that small particles such as coccoliths or aragonite microcrystals are glued by mucus material to accrete them into a manageable size. No *E. huxleyi* coccoliths were found mechanically destroyed or damaged. Delicate coccolith specimens of *Syracosphaera* were usually completely intact in natural fecal pellets (*Honjo*, 1976b). When food material is plentiful in the culture, coccospheres were not destroyed by ingestion. On the other hand, many diatoms such as *Chaetoceros lorenzianus* or *Coscinodiscus* were found mechanically crushed into quasi-uniform sized fragments of 20 to 30 μm in fecal pellets (*Smayda*, 1970; *Marshall and Orr*, 1956).

Protective Role of Pellets

It was found that pellets produced by copepods in the laboratory (*Honjo*, 1976) are covered by a thin "peritrophic" membrane of "pellicle" (*Forster*, 1953; *Gauld*, 1957). The pellicle is important in protecting the contents from being sloughed off, and provides a chemical barrier while smoothing the surface of a pellicle, reducing drag and increasing sinking velocity (*Honjo*, 1976b). The porosity of laboratory cultured copepod fecal pellets ranged between 40 (when fine aragonite crystals were fed) to 90%. After being processed by a critical dehydration technique and when viewed under SEM, the space between particles was filled in by densely formed thread-like material. This material is probably mucus that is viscid when wet. It would insulate coccoliths and other particulates in a pellet from interacting with sea water. However, it appears water soluble so that the protection only lasts (as long as) the insoluble pellicle coverage is intact.

Biodegradation and Dispersion of Coccoliths

Newly produced fecal pellets are immediately colonized by bacteria, most actively from the surface of pellets. (According to TEM study in this section, the bacterial contents inside of fecal pellets were less significant.) In warm water, the pellicle soon is covered by microbial colonies and rapidly degraded to expose the pellet contents. This results in increased surface roughness and slower sinking speeds. The rate of pellicle degradation is roughly proportional to the water temperature. In an exposure experiment at 25°C, the pellicle disappeared within several minutes. The pellet content falls off the exposed pellet surface and pellets were seen to crack into pieces. When the temperature is low, the speed of biodegradation is slowed so that at 5°C in natural surface sea water, pellets remained intact for over 20 days (*Honjo and Roman*, in press).

These studies suggest that pellets produced in warm water have a small chance of penetrating the water column, being disintegrated into small pieces which are possibly re-grazed by zooplankton in shallow water. When pellets produced in cold water such as below the thermocline, or in high latitude water, they have a better chance of sinking through the water column with their contents intact. When they reach deep cold water, bacterial activity decreases drastically (*Jannasch and Wirsen*, 1973). Some of the raining pellets and their fragments are possibly regrazed by deep sea zooplankton.

Pellets with biodegraded pellicles shed their contents such as coccoliths and diatom frustules while they sink. The existence of abundant, excellently preserved suspended coccoliths in under-saturated deep water can be explained by such "lateral injection" of coccoliths from sinking pellets whose pellicles have been damaged by biodegradation (*Honjo*, 1976b).

Protective Fecal Pellets; Other Than of Copepods

There are many other efficient filter feeders that live in the open ocean. The filtering efficiency of salps (tunicates) is high and they are common in pelagic water. In fact, many workers of the Woods Hole Oceanographic Institution have encountered large, densely populated patches of salps particularly in Slope Waters and the Tongue of the Ocean (*Wiebe and Harbison*, pers. comm). Again, no dissolution, destruction and etching of carbonate particles was found after ingestion and excretion by salps. Fecal pellets of salps usually contain large number of clay particles, small planktonic foraminiferan tests, radiolarians and silico-flagellates. Coccoliths and coccospheres were excellently preserved. Salp fecal pellets are covered by pellicles which are thicker than those produced by copepods (*Honjo and Harbison*, in prep.). Because of their large size, the sinking rates of salp fecal pellets are one or two orders of magnitude greater than copepod fecal pellets. An average sized salp (5 cm) filters over 50 cc of water per minute (*Harbison*, pers. comm.). Besides such normal filter feeders, "net catchers" such as the corollarian tunicates, or appendicularian pteropods are often observed in warm water. They produce a web screen up to several feet wide to collect small particulates such as coccoliths and take them up along with their own screen. It is assumed that the filtering efficiency of pyrosome colonies would be high.

DISSOLUTION OF BIOGENIC CARBONATE PARTICLES IN WATER COLUMN

The above discussion indicates that most biogenic carbonate particles, including coccoliths, are not likely to be appreciably dissolved in the water column while settling. Dissolution of adult planktonic foraminiferan tests is minor because of their high sinking rates and protection by remaining organic tissue. Field observations (*Adelseck and Berger*, 1975) established that adult tests arrive at the ocean bottom with no detectable dissolution. It would be safe to conclude that the dissolution of planktonic foraminiferan tests in the water column is insignificant. Dissolution of pteropods shells during settling remains unclear because of a lack of ecological information. However, considering their sinking velocity and possible organic tissue protection, dissolution while settling is highly unlikely (Table 3). The majority of coccolithophorids are eaten by grazing filter feeders. Invertebrate filter feeders do not dissolve carbonate particles (even partially) during ingestion and excretion. Coccoliths are included in fecal pellets which sink rapidly to the bottom. Small planktonic foraminiferan tests, particularly the fraction less than 50 μm, remain a problem. It is presumed that some of them are grazed by zooplankton and transported by fecal pellets in a fashion analogous to coccolithophores.

Dissolution of biogenic carbonate can occur in undersaturated deep water when coccoliths are released at depth into undersaturated water by the breakage of fecal pellets. Perhaps the dissolution of free coccoliths in undersaturated water is a unique cause of *in situ* carbonate mineralization in the open ocean environment. The suspended coccoliths which originate by fecal pellet disintegration are dissolved if they are "injected" into undersaturated sea-water. Estimates of remineralization of such suspended coccoliths depends on the absolute rate of dissolution of a coccolith calcite and the standing crop at the given depth, assuming that the supply is in steady state (*Goreau and Honjo*, in prep.). Annual turnover rate is the inverse number of the residence time which can be estimated from the specific surface area of coccoliths (9.0 m^2 g^{-1} or 0.11 g m^{-2} in average) and the annual *in situ* dissolution rate of coccolith calcite. Since the latter is not available at present, Peterson's dissolution profile was used as a first order approximation.

For example, the annual turnover rate of suspended coccoliths in the Equatorial Pacific was estimated between 1 and 5 km every 250 to 1,000 m. *In situ* dissolution rates were calculated at each depth-range applying the estimation of standing crop of calcite as suspended coccoliths given in fig. 6, p. 124, (*Honjo*, 1975). Obtained Ji values were integrated through the water column. Thus the annual remineralization rate was estimated as 3 g m^{-2} at 5°N, 155°W (where the length of the undersaturated water column is unusually long). The estimated annual production of coccoliths at this station was 9 g m^{-2} yr^{-1} (*Honjo*, 1976b).

The density of suspended coccoliths in the deep water column of the Central Gyre was generally an order of magnitude less than that of the Equatorial Zone. This should be due to the smaller

TABLE 3. Dissolution of Individual Specimen; Estimate

	Residence time days to sink 5000 m	Dissolution in this time interval (%)	Penetration distance
Planktonic foraminifera			
125-250 μm	29	1.8	any water column
> 250 μm	6	0.7	any water column
			any water column
Coccoliths			
E. huxleyi	36×10^3	3.9×10^4	82 m*
C. neohelis	32×10^3	3.2×10^4	90 m*
C. leptopora	6×10^3	850	1,250 m*

* Distance survived when the dissolution rate of 0.03 mg cm^{-2} y^{-1} (Peterson's rate between 500 m to 3,000 m) was applied.

production of coccolithophores in the productive layer (*Okada and Honjo*, 1973). Significant remineralization of coccoliths would occur in the Transitional zone where the density of suspended coccoliths was seasonably higher than the Equatorial Zone. The estimated remineralization rate at 1,000 m in the Transitional Zone is approximately 1.5 mg m^{-3} yr^{-1} at the end of the summer. The year round rate is assumed to be smaller.

Although high standing crops of coccoliths were found in the euphotic layer of the Subarctic zone (monopolized by *E. huxleyi*), the suspended coccoliths were two orders less than the density in the Equatorial water column. This may be due to the cold surface water (12°C to 14.5°C). Also, biodegradation of fecal pellets will be severely subdued in the water below the well developed thermocline at 40 m where the water temperature drops to 5°C within 50 m. Because of their less-biodegraded, intact pellicles, the majority of these fecal pellets probably arrive at the bottom without shedding a large fraction of their coccoliths while they penetrate the deep-sea water column. Considering the year round water temperature, suspended coccoliths should be rare and annual remineralization of biogenic calcite in the water column limited in the cold climate zone, despite the undersaturation. This would not violate the observation of a depressed CCD at lower latitude (*Berger and Winterer*, 1975) because relatively small remineralization of calcite in warm water regions can be offset by other factors such as overall carbonate production rates.

The year-round production of coccolithophores in the photic layer of the Equatorial zone of the Pacific is high. The fecal pellets of grazers in such warm water in 5°W, 155°W station (up to 28°C to 50 m, *Marumo et al.*, 1970) are more likely to be biodegraded. Consequently, there is a good chance that fecal pellets will be disintegrated and coccoliths shed from sinking pellets. These factors result in high densities of suspended coccoliths in the underlying water column (particularly in the water column from 400 m to 3000 m deep, *Honjo*, 1975). The sea water in this area was undersaturated from very shallow depths throughout the deep water (*Takahashi*, 1975). The estimated remineralization rates were 12 to 3 mg m^{-3} yr^{-1} at the 1,000 m layer in the stations in the North Equatorial zone. This layer appears to have the largest *in situ* remineralization rate along the 155°W longitudinal transect. The remineralization rate below the lysoclinal depth in this area is estimated to be less than a few mg m^{-3} yr^{-1}; the delivery rates are small at this depth. Thus the Equatorial Pacific probably has one of the highest (data insufficient for the Indian Ocean) *in situ* remineralization rates for calcium carbonates in the world oceans. Because of very deep saturation depths, the remineralization of calcite is probably insignificant in the Atlantic even in the Equatorial area.

COMPARISON TO OTHER GEOCHEMICAL ESTIMATES

The dissolution of carbonates in the water column has been investigated by a variety of geochemical methods. Using titration alkalinity data is one way to approach the problem. Some titration alkalinity profiles have indicated that alkalinity in the deep ocean (*Edmond and Gieskes*, 1970; *Culberson and Pytkowitz*, 1970) is essentially conservative and is only affected significantly by dissolution of calcium carbonate at the interface (*Gieskes*, 1974). *Tsunogai* (pers. comm.) however, indicated that more precise titration alkalinity data from some of the GEOSECS stations in the Pacific show small increases with depth. He suggested that if this positive value was due to the *in situ* dissolution of carbonates, at least 2 to 4 g of calcium per m^2 carbonate should remineralize while in the water column.

The vertical distribution of calcium in the Equatorial South Pacific can be better explained when the J value is 4×10^{-5} mM yr^{-1}, which is equivalent to the dissolution of 15 g of $CaCO_3$ m^{-2} yr^{-1} for 2 n undersaturated water column 4.5 km in length (*Tsunogai*, 1973).

The *in situ* dissolution rate of $CaCO_3$ can also be estimated by ΣCO_2 profiles. *Craig's* (1969) J value (rate of *in situ* remineralization) for ΣCO_2 was 4×10^{-3} cc kg^{-1} yr^{-1}, when W (upwelling velocity) is 5 m yr^{-1} (J/W = 0.80 cc kg^{-1} km^{-1}). According to *Craig and Weiss* (1971) J was 0.24×10^{-3} mM/kg yr^{-1} (H/W = 0.047 mM kg^{-1} km^{-1}), equivalent to 9 mg CO_2 m^{-3} yr^{-1} or 5 mg $CaCO_3$ m^{-3} yr^{-1}. The fraction of organic carbon was 75% (*Craig and Weiss*, 1971) to 70% (*Brewer et al.*, 1975), therefore the J value due to the dissolution of $CaCO_3$ was a few mg m^{-3} yr^{-1}. These geochemical estimates coincide (within the same order of magnitude) with the estimate which was derived from the dissolution of suspended coccoliths in undersaturated sea water.

ACKNOWLEDGEMENTS

I would like to thank T. Takahashi, S. Tsunogai, R. Berner, A. Bé, H. Lowenstam, M. Roman, R. Scheltema, L. Madin and M. Goreau for their constructive discussions. I am grateful to T. Goreau, J. Erez, G. P. Lohmann, W. A. Berggren and K. O. Emery for their critical reading of the manuscript and valuable suggestions. My thanks are due to S. Bernardo for editorial assistance.

This investigation was made possible through financial support by the Biological Oceanography Program, and partially by the Marine Geochemistry Program, Oceanography Section of the National Science Foundation under Grants DES71-00509 and OEC76-2122, respectively. Contribution No. 3751 from the Woods Hole Oceanographic Institution.

REFERENCES

Adelseck, C. G. and W. H. Berger. 1975. On the dissolution of planktonic foraminifera and associated microfossils during settling and on the sea floor. Sliter, W., Be, A. W. H. and Berger, W. H., eds. *Cushman Found. Foram. Res. Sp. Pub. No. 13.* 70-81.

Arrhenius, G.. 1962. Sediment cores from the East Pacific. *Rep. Swed. Deep Sea Exped. (1947-1948), Parts 1-4, 5.* 1-288, 152.

Bé, A. W. H. and C. Hamleben, 1970. Calcification in a living planktonic foramifer, *Globigerinoides sacculifer* (Brady). *Neues. Jahrb. Geol. Pal. Abh. 134.* 221-234.

Bé, A. W. H. and D. S. Tolderlund. 1971. Distribution and ecology of living planktonic foraminifera in surface waters of the Atlantic and Indian Oceans. In: *The Micropalenotology of Oceans.* Cambridge University Press. 105-149.

Bé, A. W. H. and O. R. Anderson. 1976. Gametogensis in planktonic foraminifera. *Science.* In press.

Berger, W. H. 1967. Foraminiferal ooze: Solution at depths. *Science, 156.* 383-385.

Berger, W. H. 1969. Ecologic patterns of living planktonic foraminifera. *Deep Sea Research. 16.* 1-24.

Berger, W. H. 1970. Planktonic foraminifera: selective solution and the lysocline. *Marine Geology, 8.* 111-138.

Berger, W. H. 1971. Sedimentation of planktonic foraminifera. *Marine Geology, 11.* 325-358.

Berger, W. H. and A. Soutar. 1967. Planktonic foraminifera: Field experiment on production rate. *Science, 156.* 1495-1497.

Berger, W. H. and A. Soutar. 1970. Preservation of planktonic shells in an anaerobic basin off California. *Bull. Geol. Soc. Amer. 81.* 275-282.

Berger, W. H. and D. J. Piper. 1972. Planktonic foraminifera: differential settling, dissolution and redeposition. *Limno. Oceanogra. 17.* 275-287.

Berger, W. H. and E. L. Winterer. 1974. Plate stratigraphy and the fluctuating carbonate line. IN: *Pelagic Sediments on Land and in the Oceans.* Hus, K. J. and H. Jenkins (Eds.). *Int. Asso. Sedimentologists. Spec. Pub. 1.*

Bernard, F. 1963. Vifesse de chute en mer des amas palmeloides de *Cyclococcolithus.* Ses Consequences pour le cycle vital des mers chaudes. *Pelagos 1.* 1-34.

Berner, R. A. and J. W. Morse. 1974. Dissolution kinetics of calcium carbonate in sea water: IV. Theory of calcite dissolution. *Am. Jour. Sci. 274.* 108-134.

Berner, R. A., E. K. Berner and R. S. Keir. 1976. Aragonite dissolution on the Bermuda Pedestal: Its depth and geochemical significance. *Earth Planetary Sci. Letters.* In Press.

Bishop, J. K. B. and J. M. Edmond. 1976. A new large volume filtration system for the sampling of oceanic particular matter. *Jour. Mar. Res.* In Press.

Bradshaw, J. S. 1959. Ecology of living planktonic foraminifera in the North and Equatorial Pacific Ocean. *Contrib. Cushman Foundation Foramin. Res. 10.* 25-64.

Brewer, P. G., G. T. F. Wong, M. P. Bacon, and D. W. Spencer. 1975. An oceanic calcium problem? *Earth Planet. Sci. Letters. 26.* 81-87.

Broecker, W. S., Y. H. Li and T.-H. Peng. 197. Carbon dioxide - man's unseen artifact. IN: *Impingement of man on the oceans.* D. W. Hood (Ed.) *Wiley-International Science.* 287-324.

Chave, K. E. and E. Suess. 1970. Calcium carbonate saturation in sea water: Effects of dissolved organic matter. *Limnol. Oceanogr. 15.* 633-637.

Chen, C. and A. W. H. Bé. 1964. Seasonal distribution of euthuosomatous pteropods in surface waters of five stations in the Western North Atlantic. *Bull. of Mar. Sci. of the Gulf and Caribbean, 14.* 186-220.

Cita, M. B., 1970. Paleoenvironmental aspects of DSDP Legs I-IV(1). A. Farinucci, (Ed.) *Proceedings of Second Planktonic Conference Rome, 1.* 251-285.

Craig, H. 1971. Abyssal carbon 13 in the South Pacific. *Jour. Geophy. Res. 75.* 691-695.

Craig, H. and R. F. Weiss. 1971. The GEOSECS 1969 intercalibration station: Introduction, hydrographic features and total CO_2-O_2 relationships. *Jour. Geophys. Res. 75.* 7641-7647.

Costin, J. M. 1971. Visual observations of suspended particle distribution at three sites in the Caribbean Sea. *Jour. Geophys. Res., Vol. 75.* 4144-4150.

Culberson, C. and R. M. Pytkowicz. 1970. Oxygen-total carbon dioxide correlation in the eastern Pacific Ocean. *Jour. Oceanogr. Soc. Japan, 26.* 95-100.

de Jong, E. W. 1975. Isolation and characterization of polysaccharides associated with coccoliths, a paleobiochemical study. Ph.D. Dissertation, University of Leiden.

Edmond, J. M. and J. M. Gieskes. 1970. On the calculation of the degree of saturation of seawater with respect to calcium carbonate under *in situ* conditions. *Geochim. Cosmochim. Acta, 34.* 1221-1291.

Forster, G. R. 1953. Peritrophic membranes in the caridea (Crustacea decapoda). *J. Mar. Biol. Assoc. U. K., 32.* 315-318.

Fruterer, D. 1974. Plankton organisms as sediment contributors in the fine-grained fraction of surface sediments off West Africa. Symposium *Marine Plankton and Sediments and 3rd Planktonic Conf., Kiel. Abst.* p. 25.

Gauld, D. T. 1957. A peritrophic membrane in calanoid copepods. *Nature, 179.* 325-326.

Gieskes, J. M., and T. M. Gieskes. 1974. The alkalinity-total carbon dioxide system in seawater. IN: *The Sea, Vol. 5.* 123-151. Goldberg, E.D. (Ed.) p. 895. John Wiley and Sons, New York.

Grab, R. L., M. A. Kaiser and M. J. O'Brien. 1975. Specific surface area analyses; application and limitations, Part 1. *Ame. Lab.*, June. 13-41.

Grell, K. G. 1958. Studien zum differenzierungsproblem an foraminiferen. *Naturwissenschaften, 45.* 3-32.

Hamano, M. and S. Honjo, 1969. Qualitative analysis of the organic matter measured in Oligocene coccoliths. *Jour. Geol. Soc. Japan, 75.* 607-614.

Honjo, S. 1975. Dissolution of suspended coccoliths in the deep-sea water column and sedimentation of coccolith ooze. W. Sliter, A. W. H. Bé and W. H. Berger (Eds.) *Cushman Found. for Foram. Res. Pub. No. 13.* 115-128.

Honjo, S. 1976a. Biogeography and provincialism of living coccolithophorids in the Pacific Ocean. Oceanic Micropaleontology, A. T. Sramsay (Ed.) Academic Press, London. (In press).

Honjo, S. 1976b. Coccoliths; production, transportation and sedimentation. *Marine Micropaleontology, 20.* Academic Press

Honjo, S. and H. Okada. 1974. Community structure of coccolithophorids in the photic layer of the Mid-Pacific Ocean. *Micropaleontology, 20.* 209-230.

Ingle, S. E., C. H. Culberson, J. E. Hawley, and R. M. Pytkowicz. 1973. The solubility of calcite in sea water at atmospheric pressure and 35% salinity. *Marine Chem. 1.* 295-307.

Jannasch, H. W. and C. O. Wirsen. 1973. Deep-sea microorganisms: *in situ* response to nutrient enrichment. *Science, 180.* 641-643.

Kajihara, M. 1971. Settling velocity and polarity of large suspended particles. *Journ. Oceanogr. Soc. Japan, 27.* 158-162.

King, K., P. E. Hare. 1972. Amino and composition of the test as a taxinomic character for living and fossil planktonic foraminifera. *Paleontology 18.* 285-293.

Klaveness, D. 1972. *Coccolithus huxleyi* (LOHMANN) KAMPTNER: 1 - Morphological investigation on the negative cell and the process of coccolith formation. *Protistologica, 8.* 335-346.

Klaveness, D. 1973. The microanotomy of *Calyptrosphaera sphaeroidae*, with some supplementary observation on the motile stage of *Coccolithus pelagicus*. *Norw. J. Bot. 20.* 151-162.

Lerman, A., D. Lal, and M. F. Dacey. 1975. Stokes' settling and chemical reactivity of suspended particles in natural waters. IN: *Suspended Solids in Water*, R. J. Gibbs, (Ed.). 17-47.

Lowenstam, H. A. 1974. Impact of life on chemical and physical processes. IN: *The Sea, Vol. 5.* 715-796. E. D. Goldberg (Ed.) John Wiley and Sons, New York.

Marshall, S. M. and A. P. Orr. 1956. On the biology of *Calanus finmarchicus*, IX, Feeding and digestion in the young stages. *Marine Biol. Assoc. United Kingdom Jour., 35.* 587-603.

Marshall, S. M. and A. P. Orr, 1962. Food and feeding in copepods. *Rapp. Cons. Explor. Mer.*, *152*. 92-98.

McGowan, H. A. 1960. The systematic distribution and abundance of Euthecosomata of the north Pacific. Ph.D. dissertation, Univ. of California San Diego, La Jolla, Calif.

McIntyre, A., and R. McIntyre. 1971. Coccolith concentrations and differential solution in oceanic sediments. IN: *The Micropaleontology of Oceans*. B. M. Funnell and W. R. Riedel, (Eds.) Cambridge University Press, 253-262.

McIntyre, W. G. 1965. The temperature variation of the solubility product of calcium carbonate in sea water. *Fisheries Res. Board Canada Manuscript Rept. Series*, *200*. 153 p.

Menzel, D. W. 1974. Primary productivity dissolved and particulate organic matter, and the sites of oxidation of organic matter. IN: *The Sea, Vol. 5*. 659-678. E. D. Goldberg, (Ed.) 895 pp. John Wiley and Sons, New York.

Menzel, D. W. and J. H. Ryther. 1962. Zooplankton in the Sargasso Sea off Bermuda and its relation to organic production. *Jour. Cons. Perm. Inter. Expl. Mer.* *26*. 250-258.

Milliman, J. D. 1974. Marine Carbonate. Springer-Verlag, Berlin. p. 375.

Milliman, J. D. 1975. Dissolution of aragonite, Mg-calcite and calcite in the North Atlantic Ocean. *Geology*, *3(8)*. 461-462.

Mopper, K. E. T. Degens. 1972. Aspect of the biochemistry of carbohydrates and proteins in organic environments. *Tech. Rep. WHOI 72-68*. 118 pp.

Morse, J. W. 1973. The dissolution kinetics of calcite: A Kinetic origin for the lysocline. Ph.D. dissertation, Yale Univ., 118 pp. (Univ. Microfilms, Ann Arbor, Mich.).

Morse, J. W. 1976. Dissolution kinetics of calcium carbonate in sea water VI: The near-equilibrium dissolution kinetics of calcium carbonate-rich deep sea sediments. *Amer. Jour. Sci.* In Press.

Morse, J. W. and R. A. Berner. 1972. Dissolution kinetics of calcium carbonate in sea water. II. A kinetic origin for the lysocline. *Am. Jour. Sci.* *272*. 840-851.

Morse, J. W. and J. de Kanel. 1976. A simple and inexpensive method for surface area determination by the Kr BET method. *Anal. Chem.*, In press.

Okada, H., and S. Honjo. 1973. The distribution of oceanic coccolithophorids in the Pacific. *Deep Sea Res.* *20*. 355-374.

Pales, J. C. and C. D. Keeling. 1965. The concentration of atmospheric carbon dioxide in Hawaii. *J. Geophys. Res.* *70*. 6053-6076.

Peterson, M. N. A. 1966. Calcite: rates of dissolution in a vertical profile in the central Pacific. *Science*, *154*. 1542-1544.

Roth, P. H., M. M. Mullin and W. H. Berger. 1975. Coccolith sedimentation by fecal pellets; Laboratory experiments and field observations. *Geol. Soc. Amer. Bull.*, *86*. 1079-1084.

Smayda, T. J. 1970. The suspension and sinking of phytoplankton in the sea. *Oceanogr. Mar. Biol. Annu. Rev. 8.* 353-414.

Takahashi, T. 1975. Carbonate chemistry of sea water and the calcite compensation depth in the oceans. IN: *Dissolution of deep-sea carbonates: Cushman Found. Foram. Research. Spec. Pub. No. 13.* W. V. Sliter, A. W. H. Bé, and W. H. Berger (Eds.) 11-26.

Tolderlund, D. S. and A. H. W. Bé. 1971. Seasonal distribution of planktonic foraminifera in the western North Atlantic, *Micro-paleontology, 17.* 297-329.

Towe, K. M. 1971. Lamellar wall construction in planktonic fora-minifera. IN: *Proceedings of the Second Planktonic Conference, Rome, 1970, 1.* A. Farinacci (Ed.) 1213-1224.

Towe, K. M., R. Cifelli. 1967. Wall ultrastructure in the cal-areous foraminifera: crystallographic aspects and a model of calcification. *Jour. Paleon. 41.* 742-762.

Tsunogai, S., H. Yamahata, S. Kudo and O. Saito. 1973. Calcium in the Pacific Ocean. *Deep Sea Res., 20.* 717-726.

van der Spel, S. 1967. Euthecosomata, a group with remarkable developmental stages (gastropoda, pteropoda). *J. Noorcluign en Zoon N.V. gurinchem.* pp. 375.

Vinogradov, M. Ye. 1961. Food sources of the deep-water fauna. Speed of decomposition of dead Pteropoda. (Trans. from Russian) *Am. Geophys. Union. 136/141.* 39-42. Dokl. Akad. Nauk SSSR Okeanol.

Westbroek, P., E. W. de Jong, W. Dam and L. Bosch. 1973. Soluble intercrystalline polysaccharides from coccoliths of *Coccolithus huxleyi* (LOHMANN) KAMPTNER (T). *Cal. Tiss. Res. 12.* 227-238.

Wiebe, P. H., S. H. Boyd, and C. Winget. 1976. Sedimentological trap for use of above sea-floor with preliminary results of its use in the Tongue of Ocean, Bahamas. *Jour. Mar. Res.* In press.

Wilber, K. M. and N. Watabe. 1963. Experimental studies on calcification in molluscs and the alga *Coccolithus huxleyi*. *Ann. N. Y. Acad. Sci. 109.* 82-111.

$\delta^{13}C$ Variations in Marine Carbonate Sediments as Indicators of the CO_2 Balance Between the Atmosphere and Oceans

P. M. Kroopnick[1], S. V. Margolis[1], and C. S. Wong[2]

[1]University of Hawaii, [2]Institute of Ocean Sciences,

Victoria, British Columbia, Canada

ABSTRACT

A model is presented to explain the departures from equilibrium of the ocean-atmosphere CO_2-^{13}C system with respect to temperature, latitude and water mass characteristics. Carbon isotope data suggests that invasion and evasion of CO_2 to and from the sea is at least partially controlled by the organic productivity of a water mass. Surface water $\delta^{13}C$ values are determined by exchange with the atmosphere and by the amount of carbon removed versus that replenished. Deeper water $\delta^{13}C$ values are controlled by the relative rates of upwelling of deep water, by in situ respiration, and by the dissolution of settling particles.

The record of $\delta^{13}C$ fluctuations for the last 75 million years in oceanic surface and bottom waters is recorded in the calcareous remains of benthic and planktonic foraminifera and in nannofossils. The assumption of equilibrium deposition of carbon isotopes by these organisms is shown to not always be the case. Nevertheless, the trend of $\delta^{13}C$ values derived from these three organisms are parallel for the majority of the Late Cretaceous and Cenozoic, perhaps indicating that they are responding to ocean-wide productivity variations. The exact nature of the processes responsible for $\delta^{13}C$ variations in fossil carbonate sediments, however, is still unknown, but increased understanding of $\delta^{13}C$ variations in the present-day oceans and atmosphere will eventually aid in understanding and interpreting the significance of past fluctuations in the CO_2-^{13}C-$CaCO_3$ system.

INTRODUCTION

The isotopic composition of dissolved inorganic carbon in surface and near-surface water is controlled by several processes: (1) a steady-state quasi-equilibrium with the atmosphere; (2) *in situ* production of CO_2 in the water column; (3) the oxidation of organic matter and dissolution of carbonate; and (4) the overall rate at which CO_2 from deep water is introduced into surface water (*Kroopnick*, 1974a, 1974b). Most of these processes respond to climatic change, although the respective time constants and interrelationships may be complex and variable from ocean to ocean. Variation in the mean surface water temperatures may also change the isotopic composition of each reservoir. Latitudinal variations of temperature, ΣCO_2, its $\delta^{13}C$ and the $\delta^{13}C$ of atmospheric CO_2 are discussed here. The data will be used to map the departures of the ocean-atmosphere CO_2-^{13}C system from equilibrum. In addition, the deep-sea sedimentary record of $\delta^{13}C$ variations of benthic and planktonic foraminifera and calcareous nannofossils will be examined in order to assess whether such data can be used in light of our knowledge of the present-day system to interpret past fluctuations in the CO_2 system.

In the present-day oceans, the $\delta^{13}C$ of dissolved inorganic carbon varies from about 2°/oo at the surface to below -0.5°/oo in the oxygen minimum (∿500m) of the North Pacific Ocean. Bottom water values are generally between -0.2 and +0.2 (*Kroopnick*, 1974a and references cited therein). In accordance with these observations, surface dwelling plankton should have higher $\delta^{13}C$ values, while benthic foraminifera should have lower $\delta^{13}C$ values due to a deeper and colder habitat. Carbon isotope variations in benthic and planktonic foraminifera have received significant attention in recent years with the increased thrust in paleoclimatological studies of deep-sea sediments. $\delta^{13}C$ values are obtained during $\delta^{18}O$ paleotemperature determinations, but their significance has largely been ignored because of a lack of understanding. *Saito and Van Donk* (1974), *Savin et al.* (1975), *Douglas and Savin* (1975), *Savin and Douglas* (1977), *Sommer and Matthews* (1975) and *Shackleton and Kennett* (1975a) have all published $\delta^{13}C$ data on foraminifera and discussed their interpretations. Calcareous nannoplankton carbon isotope data has been presented by *Margolis et al.* (1975).

Several of the above workers have proposed that $\delta^{13}C$ values of planktonic foraminifera generally reflect their depth habitat. Theoretically, if one could find a suite of calcareous organisms that calcified in equilibrium with respect to carbon and oxygen isotopes at various depths throughout the water column, then a paleo-$\delta^{13}C$ profile and a paleothermocline could be reconstructed. However, the assignment of depth habitats to extinct species is not an easy task (see *Berger*, 1969; *Savin and Douglas*, 1973; *Shackleton et al.*, 1973; *Goodney et al.*, 1975). Temperature changes and food availability cause vertical migration in zooplankton, which along with horizontal advection, could alter such depth rankings.

The oxygen isotopes of many species of foraminifera appear to respond directly to temperature (and $\delta^{18}O$ of water). By analyzing several species from the same core, reliable depth habitats have been determined (*Emiliani*, 1954; *Savin and Douglas*, 1973). However, the $\delta^{13}C$ data for planktonic foraminifera can show large, species-dependent, non-equilibrium fractionation effects (*Shackleton et al.*, 1973; *Savin and Douglas*, 1973; *Goodney et al.*, 1975). Large fractionations have also been observed within a single species as a function of size or age *Savin and Douglas*, 1977). On the other hand, calcareous nannoplankton appear to assimilate inorganic carbon with a $\delta^{13}C$ equal to that of their environment (*Goodney et al.*, 1975; *Margolis et al.*, 1975).

Analyses of fossil calcareous nannoplankton and benthic foraminifera indicate that the effects of past climatic fluctuations on the carbon isotope distributions in the oceans are about equal to the present-day vertical gradient between surface and deep waters (about 2°/oo) (*Margolis et al.*, 1975; *Savin et al.*, 1975; *Douglas and Savin*, 1975). If thermodynamics were to prevail, the temperature dependence of the $\delta^{13}C$ of $CaCO_3$, precipitated under equilibrium conditions, would be 25°C/°/oo. It is shown in this paper that the record of $\delta^{13}C$ variations during the last 75 million years indicates that a maximum shift in the $\delta^{13}C$ of calcareous plankton is about 2°/oo. Since sea surface water temperature change cannot have exceeded 25°C, the mechanism for the formation of biogenic carbonates cannot involve equilibration between the calcite phase and dissolved inorganic carbon. Thus the $\delta^{13}C$ values for fossil carbonates more closely resemble the $\delta^{13}C$ values for the dissolved inorganic carbon reservoir, and hence, should reflect either the growth habitats of marine plankton, the productivity of the surrounding surface waters, or perhaps the rates of vertical mixing, rather than temperature. Species-dependent carbon isotope fractionation effects have also been shown to occur during the incorporation of dissolved bicarbonate. Species-dependent biogenic fractionation of oxygen isotopes between bicarbonate and water is not as great a problem, and thus, $\delta^{18}O$ can be used as an indicator of paleotemperatures.

In the following section a model for the CO_2 and $\delta^{13}C$ distributions in the ocean-atmosphere system will first be presented, and then $\delta^{13}C$ variations in benthic and planktonic organisms contained in deep-sea sediments from Late Cretaceous to Late Cenozoic will be discussed.

EXPERIMENTAL METHODS AND SAMPLES SELECTED

Sea-Water and Atmospheric Samples

The samples discussed here were collected during April and May 1970 on the Pacific leg of the HUDSON-70 Expedition. The Pacific Ocean was traversed from 63°S to 57°N along longitude 150°W. Sea water samples were collected from Nansen bottles, immediately poisoned with mercuric chloride and returned to the laboratory in greased ground-glass stoppered reagent bottles (*Wong*, 1970). The ΣCO_2 (total dissolved inorganic carbon) was measured by gas chromatography (*Weiss*, 1970). Separate aliquots from the same bottle were extracted for isotopic analysis by stripping with nitrogen gas and collecting the dissolved gases in a vacuum system (*Kroopnick*, 1974a).

The atmospheric CO_2 samples were collected in pre-evacuated 5 ℓ glass flasks with greased ground-glass stoppers. Water vapor was removed at the time of collection by passing the air through silica gel. The isotopic results presented here have been corrected for their N_2O concentration (*Craig and Keeling*, 1963; *Craig*, pers. comm.).

Deep-Sea Sediments

Carbon and oxygen isotope studies of sediments from D.S.D.P. site 284 (Figure 1), drilled during Leg 29 at a water depth of 1068 meters, were initiated by *Shackleton and Kennett* (1975a). This site contained well-preserved calcareous biogenic components and was well suited for isotopic investigation to evaluate the nature of paleoclimates during the last 6 million years. Benthic foraminifera of the genus *Uvigerina* were the first to be isotopically analyzed from this site by *Shackleton and Kennett* (1975a) and the changes in their oxygen and carbon isotopes were found to coincide with changes in planktonic foraminiferal faunas considered by *Kennett and Vella* (1975) to reflect changes in Late Cenozoic surface water temperatures. The paleomagnetic stratigraphy for this site is given in Figure 2.

Monospecific samples of three species of planktonic foraminifera representing what was believed to be different depth habitats (*Globigerina quinqueloba*, *Neogloboquadrina pachyderma* and *Globigerina bulloides*) were selected for isotopic analysis from the same horizons as the benthic foraminifera and nannofossils (*Kennett et al.*, in prep.). Only well-preserved nannofossil samples showing little or no evidence of secondary overgrowths during scanning electron microscopic examination were isotopically analyzed. Slight differences in the selection of sample depths (Figures 9 and 10) are due to changes in the relative abundances of the planktonic foraminifera and in the preservation of coccoliths. Where planktonic foraminifera and coccoliths could not be selected for analysis from the same sediment sample, the closest possible sample containing either sufficient numbers of individuals of the respective species or coccoliths lacking overgrowth was selected.

Analyses were also performed on calcareous nannofossils from D.S.D.P. sites 277, 279 and 281 and from Eltanin core 55-26. The procedures used for separation and isotope analysis of fossil carbonates have been described elsewhere (*Shackleton and Opdyke*, 1973; *Margolis et al.*, 1975). Briefly, relatively pure but polyspecific

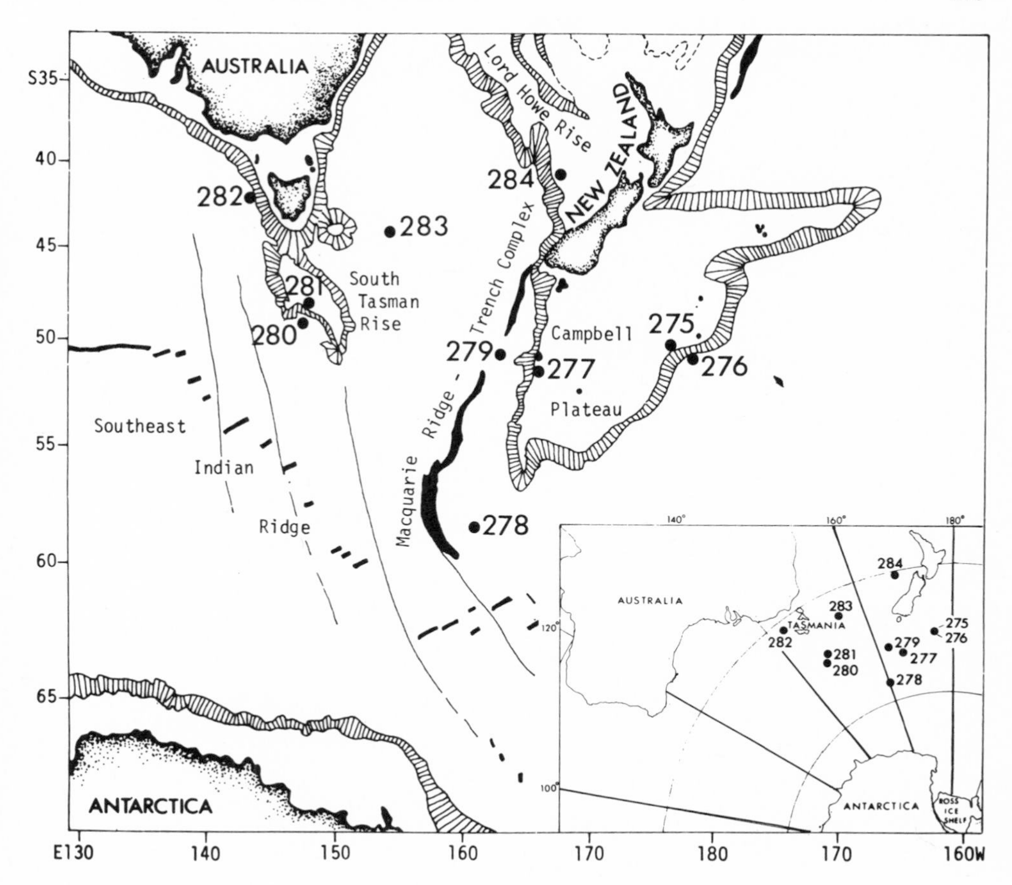

Fig. 1. Locations of sites drilled during D.S.D.P. Leg 29.

calcareous nannofossil fractions consisting of isolated coccoliths, coccospheres and discoasters were separated from the <44 μm fraction of the sediment samples by using short centrifuge techniques. Reproducible isotopic results were achieved by roasting the sample prior to reaction with 100% H_3PO_4. The CO_2 gas was then analyzed on a Nuclide 3-60 RMS mass spectrometer with a precision of ± 0.05‰ (*Goodney and Kroopnick*, 1977). The data are given as the per mil deviations (δ value) of the $^{13}C/^{12}C$ ratio relative to the PDB-carbon standard, which corresponds approximately to average limestones (*Craig*, 1957) as in equation (1).

$$\delta^{13}C = \left[\frac{^{13}C/^{12}C_{(sample)}}{^{13}C/^{12}C_{(standard)}} - 1\right] x\ 1000 \qquad (1)$$

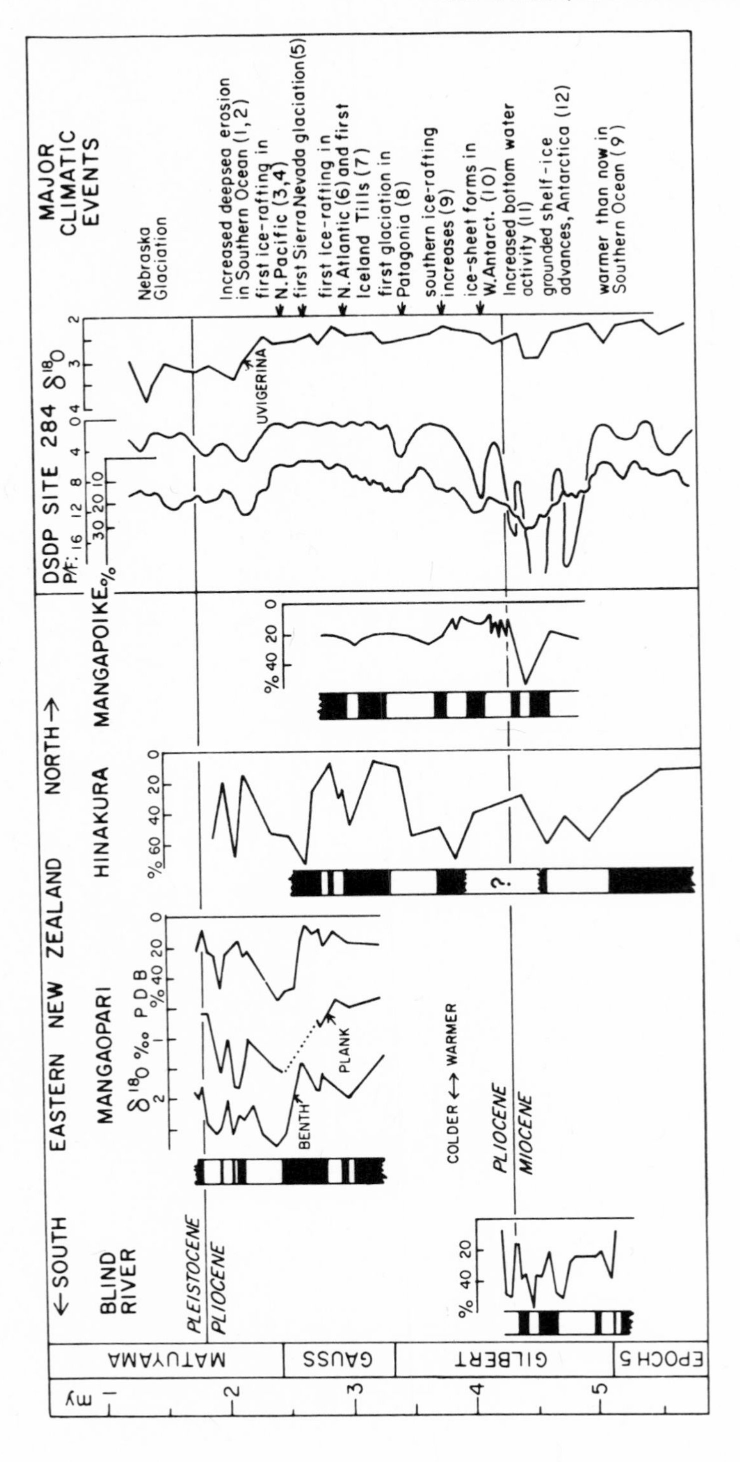
←SOUTH
EASTERN NEW ZEALAND
NORTH→
BLIND RIVER
MANGAOPARI
HINAKURA
MANGAPOIKE
DSDP SITE 284 δ18O
P/F %
MAJOR CLIMATIC EVENTS
Nebraska Glaciation
Increased deepsea erosion in Southern Ocean (1,2)
first ice-rafting in N.Pacific (3,4)
first Sierra Nevada glaciation (5)
first ice-rafting in N.Atlantic (6) and first Iceland Tills (7)
first glaciation in Patagonia (8)
southern ice-rafting increases (9)
ice-sheet forms in W.Antarct. (10)
Increased bottom water activity (11)
grounded shelf-ice advances, Antarctica (12)
warmer than now in Southern Ocean (9)
UVIGERINA
δ18O ‰ PDB
BENTH
PLANK
COLDER ↔ WARMER
PLEISTOCENE
PLIOCENE
PLIOCENE
MIOCENE
MATUYAMA
GAUSS
GILBERT
EPOCH 5
my

Fig. 2. Paleomagnetic correlations and chronology of DSDP Site 284 and eastern New Zealand marine paleoclimatic sequences and chronology of selected late Cenozoic major climatic events in other regions. Paleomagnetic stratigraphy for the Mangaopari section after Kennett et al. (1971); for the Hinakura section after Linert et al. (1972); and for the Mangapoike and Blind River sections after Kennett and Watkins (in press). Oxygen isotopic records for the Mangaopari section determined on benthonic and planktonic foraminifera are after Devereaux et al. (1970) and for DSDP Site 284 based on the benthonic foraminifera *Uvigerina* are after Shackleton and Kennett (this volume). Percentage curves for each section represent frequency oscillations in the cool foraminifer *Neogloboquadrina pachyderma*. Increased percentages (toward the left) reflect colder intervals. Oscillations in the *N. pachyderma/Globigerina falconensis* ratio (P/F) are also shown for Site 284. Paleomagnetic epochs shown at the left and chronology are after Cox (1966). Major climatic events recorded in other regions shown at right: 1=Watkins and Kennett, 1972; 2=Fillon, 1972; 3=Kent et al., 1971; 4=Echols, 1973; 5=Curray, 1966; 6=Berggren, 1972; 7=McDougall and Wensink, 1966; 8=Mercer, 1973; 9=Blank and Margolis, in press; 10=Mercer, 1972; 11=Kennett and Brunner, 1973; 12=Hayes, Frakes et al., 1973. (From *Kennett and Vella*, 1975).

RESULTS AND DISCUSSION

Surface Water

Pacific Ocean surface water sections for temperature, total dissolved inorganic carbon dioxide (ΣCO_2), $\delta^{13}C$ of the ΣCO_2 and $\delta^{13}C$ of atmospheric CO_2 obtained along 150°W for April-May 1970 are presented in Figures 3-6. The temperature data shows the expected warming as one approaches the equator (Figure 3). The subtropical convergence is seen as a break in slope at about 40°S and 40°N latitude. The equatorial divergence brings cooler water to the surface at the equator.

The ΣCO_2 data (Figure 4) similarly reflect the major hydrographic features of the Pacific Ocean. Near the equator, local effects cause significant scatter in the data. The $\delta^{13}C$ of the ΣCO_2 data (Figure 5) indicate that ^{13}C-depleted carbon is being added at high latitudes and along the equator, causing the $\delta^{13}C$ to decrease. Kroopnick (1974b) showed that the water upwelling along the polar front reflects the contributions of organic carbon and calcium carbonate added to deep water by oxidation and dissolution. Organic carbon has a $\delta^{13}C$ of between -20 and -25°/oo while $CaCO_3$ precipitated in equilibrium with surface water will have a $\delta^{13}C$ between 0 and +2.5°/oo. A plot of $\delta^{13}C$ versus ΣCO_2 was used by Kroopnick (1974b) to determine that 83% of the increase of the ΣCO_2 in deep water was caused by the oxidation of entrained organic matter.

The high latitude data presented here is consistent with that model, but, between 40°S and 40°N, $\delta^{13}C$ and ΣCO_2 are not covariant. Within the two central gyres the surface $\delta^{13}C$ value is controlled by a combination of the local processes of photosynthesis, respiration, calcification and atmospheric exchange, rather than the larger scale mixing of Atlantic Ocean deep water and Pacific Ocean surface water, which occurs near the polar front.

The $\delta^{13}C$ of atmospheric CO_2 at various latitudes is shown in Figure 6. The most striking feature is the decrease in $\delta^{13}C$ in the Northern Hemisphere. The $\delta^{13}C$ of the ΣCO_2 (Figure 5) is also skewed toward lower values in the north. In general, organic carbon and fossil fuels have $\delta^{13}C$ values of between -10 and -30°/oo. If the average $\delta^{13}C$ (Figure 6) of the Southern Hemisphere atmospheric CO_2 is taken as -7.1°/oo (fall season) and that of the Northern Hemisphere as -8.0°/oo (spring season) and one assumes that CO_2 with a $\delta^{13}C$ of -25°/oo is being added to the atmosphere in the Northern Hemisphere, then to account for the isotopic shift with latitude, the atmosphere in the Northern Hemisphere must have 5% more organically produced CO_2 than the Southern Hemisphere. The source of this CO_2 could be either the combustion of fossil fuel during the winter, or the decay of organic material produced during the previous summer. Further measurement of the $\delta^{13}C$ of atmospheric CO_2 at various seasons are required before the full impact of the seasonal release of biogenic carbon on the atmosphere can be evaluated.

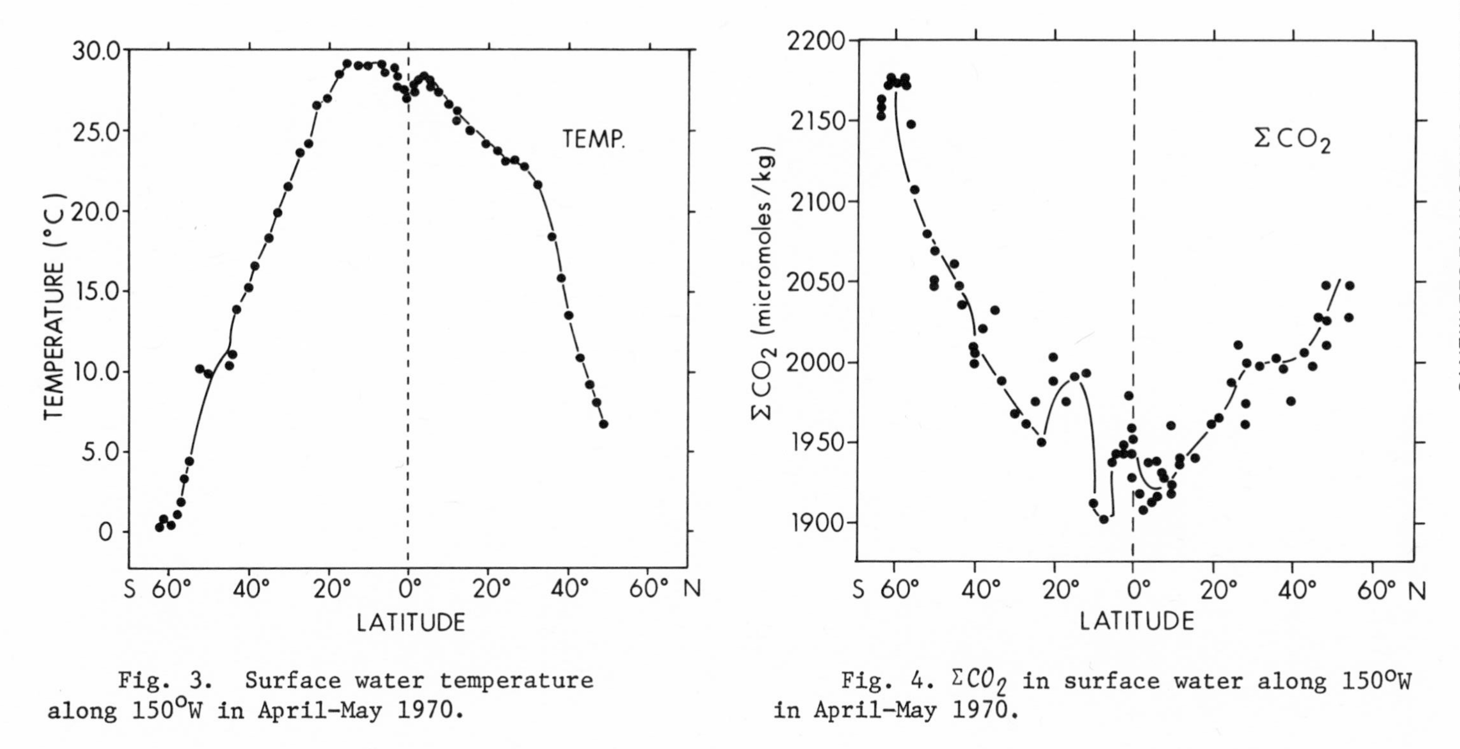

Fig. 3. Surface water temperature along 150°W in April-May 1970.

Fig. 4. ΣCO_2 in surface water along 150°W in April-May 1970.

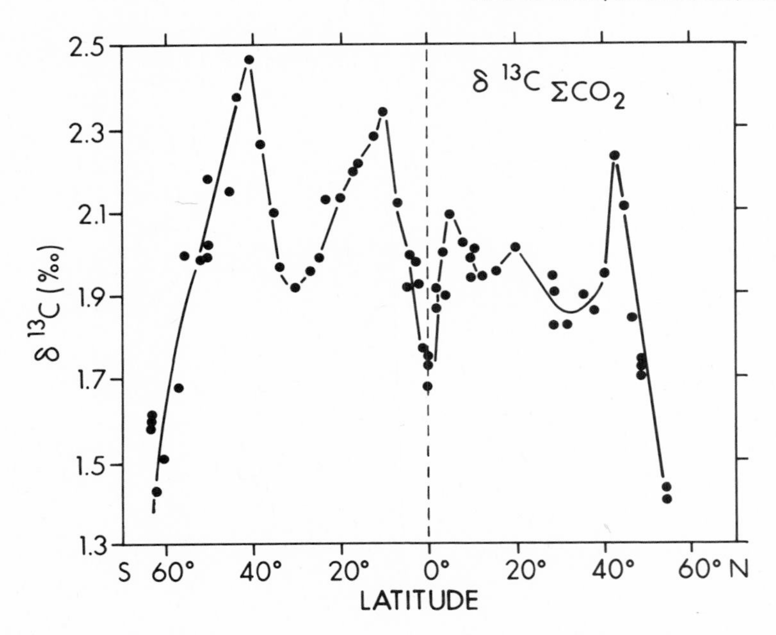

Fig. 5. $\delta^{13}C$ of ΣCO_2 along 150°W in April and May 1970.

The departure of ocean surface water from equilibrium with respect to the CO_2 system can now be calculated. *Kroopnick* (1974b) showed that the $\delta^{13}C$ and ΣCO_2 values of mid to high latitude samples (39° to 60°S) corresponded to a quasi-steady-state equilibrium with atmospheric CO_2 at a mean temperature of 10°C. To perform that calculation, he was required to assume a constant atmospheric P_{CO_2}, and a constant *pH* for oceanic surface water. In light of the data presented here, the restrictions required for the previous model can now be removed.

Given the latitudinal variations of temperature, ΣCO_2 and $\delta^{13}C$, the only assumption required now is a constant atmospheric CO_2 level (i.e., 322 ppm). Given these four variables, the carbonate system is completely determined and the chemical speciation at each sampling site can be calculated. The isotopic fractionation for equilibrium between gaseous CO_2, dissolved CO_2, bicarbonate ion and carbonate ion have been measured by several workers (*Mook et al.*, 1974) and is also completely determined given the four quantities listed above (*Kroopnick*, 1974b).

For atmospheric CO_2 in equilibrium with the carbonate system in surface sea water, the expected $\delta^{13}C$ of atmospheric CO_2 will be calculated now as a function of latitude. Considering the mass balance relationship for ^{12}C and ^{13}C, the following equation can be derived:

$$\delta_{\Sigma CO_2} = \delta_b \cdot F_b + \delta_c \cdot F_c + \delta_{aq} \cdot F_{aq} \quad (2)$$

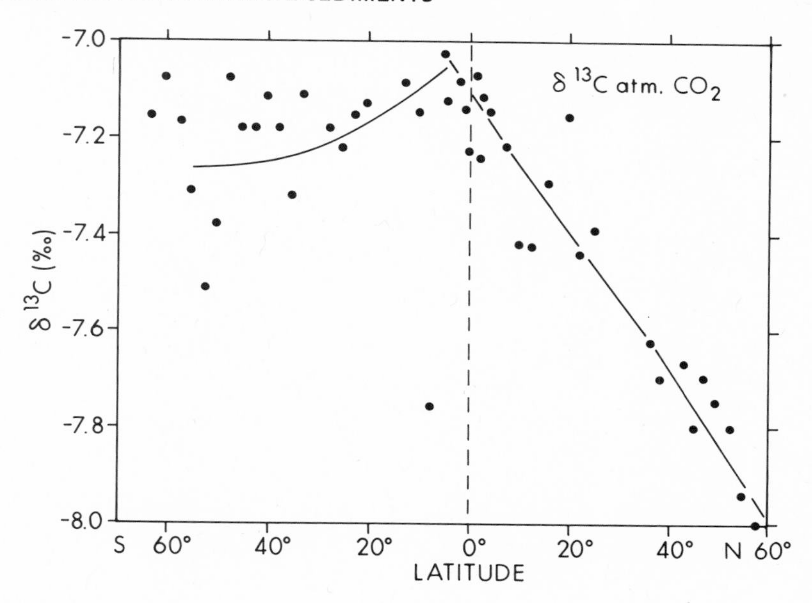

Fig. 6. $\delta^{13}C$ of atmospheric CO_2 along 150°W in April and May 1970.

where $\delta_{\Sigma CO_2}$ is the ^{13}C of the total dissolved inorganic carbon, F_i is the percentage abundance for each of the "i" species $\{HCO_3^-, CO_3^=, CO_2(aq)\}$, and δ_i is its $\delta^{13}C$ value. For each species,

$$\delta_i = \delta_{CO_2(g)} + \varepsilon_i \tag{3}$$

where ε_i is the isotopic enrichment relative to atmospheric CO_2. Substituting for δ_i in equation (2), and rearranging terms, it is found that,

$$\delta_{atm \cdot CO_2} = \delta_{\Sigma CO_2} - F_b \cdot \varepsilon_b - F_c \cdot \varepsilon_c - F_{aq} \cdot \varepsilon_{aq} \tag{4}$$

The difference between the measured values of the $\delta^{13}C$ of atmospheric CO_2 and the calculated values is shown in Figure 7. While considerable smoothing of the data occurred naturally during this calculation, obviously bad data points were rejected during interpolation in order to present a coherent picture of the carbonate systems approach to equilibrium in surface waters.

At high latitudes, the measured $\delta^{13}C$ of atmospheric CO_2 is enriched by up to 2°/oo in $\delta^{13}C$ relative to the calculated equilibrium value. The upwelling of deep water at the polar front brings waters enriched in ΣCO_2 and depleted in $\delta^{13}C$ to the surface. A similar enrichment occurs at the equator where upwelling also occurs. It is generally accepted that at these latitudes a net

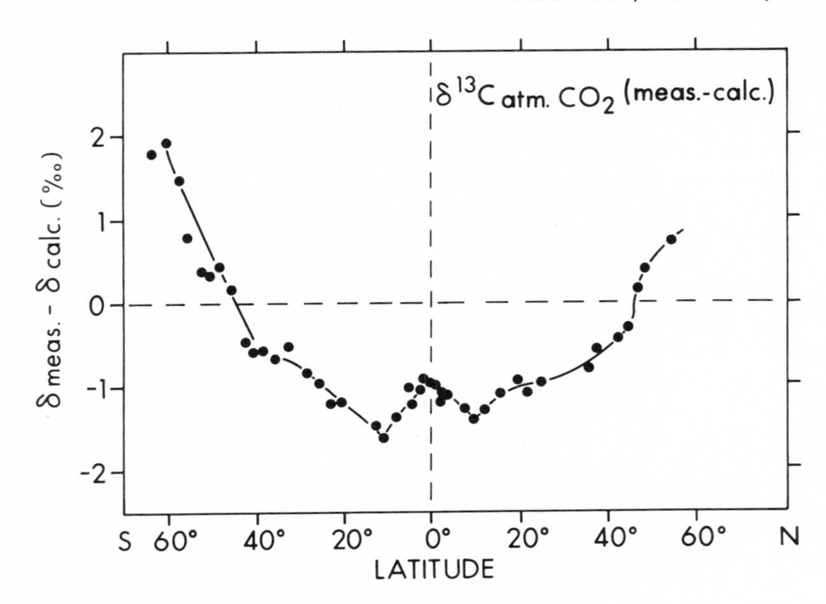

Fig. 7. ($\delta^{13}C$ meas. - $\delta^{13}C$ calc.) for atmospheric CO_2 along 150°W in April-May 1970.

transfer of CO_2 into the ocean is occurring. Carbon-12 must then be entering the sea in preference to ^{13}C. This preferential uptake of ^{12}C is known to occur during organic fixation with the residual CO_2 being enriched by ∿25°/oo and during the equilibrium solvation of $CO_2(g)$ to $CO_2(aq)$ (enrichment - 1.1°/oo). The latter effect has already been included in the calculated value.

Mid-latitude samples are depleted in ^{13}C relative to equilibrium. In these areas photosynthesis is low and the net biological effect is for ^{13}C-depleted organic carbon to be added to the atmosphere by respiration. Thus, the carbon isotope data indicates that the invasion and evasion of CO_2 to and from the sea is at least partially controlled by the productivity of the appropriate water mass. The relative importance of biological and chemical mechanisms of CO_2 exchange cannot be fully evaluated until seasonal records of productivity, ΣCO_2, and $\delta^{13}C$ are available.

Deep Water

As discussed previously, the $\delta^{13}C$ values for fossil carbonate are expected to reflect the $\delta^{13}C$ value of the bicarbonate reservoir in which the calcification occurred. Thus, a knowledge of the depth habitats of ancient planktonic organisms might allow reconstruction of the $\delta^{13}C$-depth profiles during past eras. However, before considering the fossil record of $\delta^{13}C$ variations, the causes of the vertical distribution of $\delta^{13}C$ as observed today must first be discussed. ANTIPODE 15 STA. 6 at 17°S; 172°W represents a "typical" mid-latitude Pacific Ocean site (Figure 8). Note the broad thermocline in which the temperature drops from 27°C at the surface to 9°C at 500 meters. The deep dissolved oxygen minimum is poorly developed here but is sufficient for a calculation to show that between 1 and 3.4 km, *in situ* respiration is consuming 0.04 μmole oxygen per kg per year (*Craig et al.*, 1972). In accordance with previous observations (*Kroopnick*, 1974a), about 0.03 μmole kg^{-1}/yr of CO_2 is produced during respiration and an additional 0.01 μmole kg^{-1}/yr of $CO_3^{=}$ is produced by the dissolution of $CaCO_3$ tests. Thus, in the Pacific Ocean deep water, ΣCO_2 increases at about the same rate that oxygen decreases. As discussed earlier, the addition of organic carbon lowers the observed $\delta^{13}C$ value. Thus, the $\delta^{13}C$ profile shown in Figure 8 has a broad minimum at 2.5 km. The bottom water below the "benthic front" is also depleted in ^{13}C since it is derived from "aged" North Atlantic Ocean water.

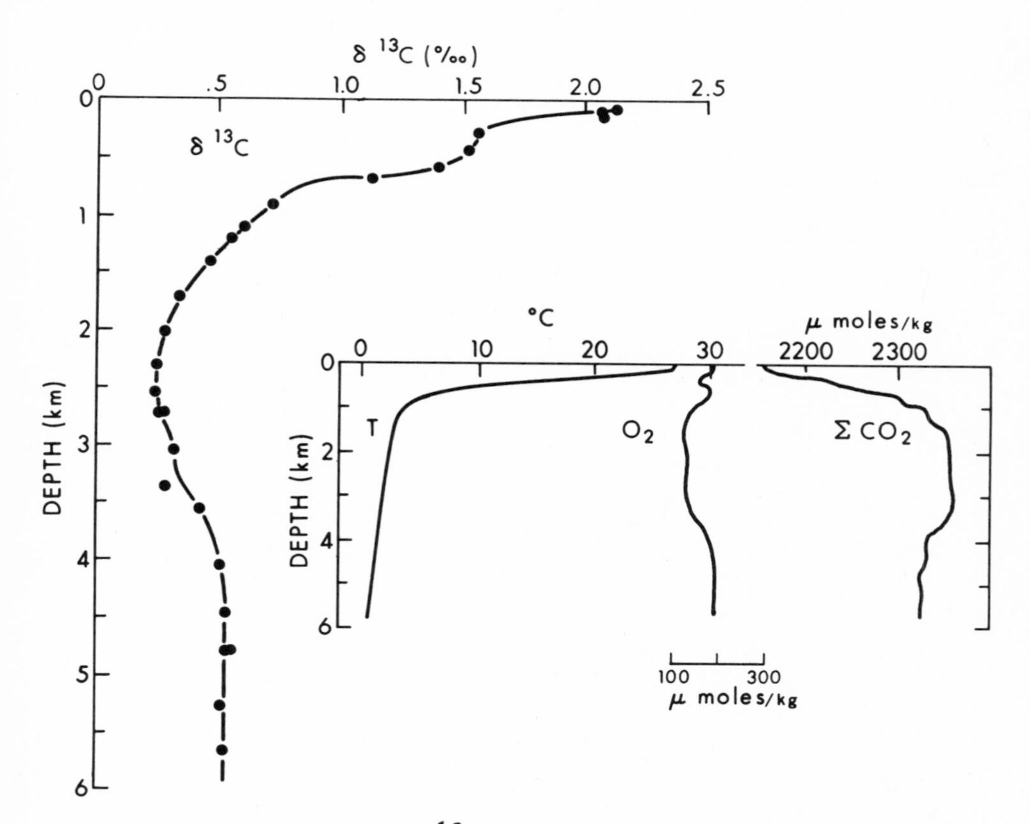

Fig. 8. Composite of $\delta^{13}C$ of ΣCO_2 and temperature, dissolved oxygen and ΣCO_2 vs. depth for antipode 15 station 6. (17°S - 172°W).

Most planktonic organisms live within the upper 400 meters where the temperature and $\delta^{13}C$ gradients are the steepest. The $\delta^{13}C$ value at the surface is determined by exchange with the atmosphere and by the amount of carbon removed versus that replenished (i.e., net productivity). The gradient in the deep water is controlled by the relative rates of upwelling of deep water, *in situ* respiration and solution of falling particles. The first variable (upwelling) is controlled by the thermohaline circulation, while the other two are proportional to productivity. The response times for these processes vary between approximately 1000 years and 1 year. Plankton with the highest $\delta^{13}C$ values would occur in surface waters and lower $\delta^{13}C$ values indicate a deeper habitat. Post-depositional exchange of surface species with bottom water could lead to an erroneous habitat assignment, but this problem can be minimized by careful examination of the samples.

ISOTOPIC INVESTIGATIONS OF CALCAREOUS NANNOFOSSILS FROM LATE MESOZOIC AND CENOZOIC SEDIMENTS

The results of laboratory culturing of three species of coccolithophorids (*Dudley*, 1976; *Goodney*, 1977) show evidence of species-dependent, non-equilibrium deposition of oxygen isotopes in coccolith calcite. All culture experiments to date, nevertheless, show a temperature dependence of oxygen isotopes in coccoliths. This fact has also been substantiated by analysis of calcareous nannofossils and associated benthic and planktonic foraminifera from deep-sea cores.

Fig. 9. Summary plot of oxygen isotope data for benthic foraminifera (squares), planktonic foraminifera (triangles), and nannofossils (circles and dots). Data are from: *Savin et al.* (1975), *Douglas and Savin* (1975), *Saito and Van Donk* (1974), *Margolis et al.* (1975). Solid dots with dark line in Upper Cretaceous are nannofossils from E55-26.

- ⧳ Site 305 Nannofossils, Shatsky Rise
- o Site 47 Nannofossils, Shatsky Rise
- ϕ Site 167 Nannofossils, Magellan Rise from *Douglas and Savin* (1975)
- Δ Site 167 Planktonic foraminifera
- ⍦ Site 167 Benthic Assemblages

Benthic and planktonic foraminifera data older than 60 million years are from *Savin et al.* (1975) and *Saito and Van Donk* (1974). The carbon and oxygen isotope record of calcareous biogenic components contained in sediments from D.S.D.P. sites 279, 281 and 277 are from *Shackleton and Kennett* (1975b) using monospecific foraminiferal samples, and *Margolis et al.* (1975a) using mixed assemblages of nannofossils. Those data for sites 47, 167, and 305 are from *Douglas and Savin* (1975), using mixed assemblages of benthic and planktonic foraminifera and calcareous nannofossils.

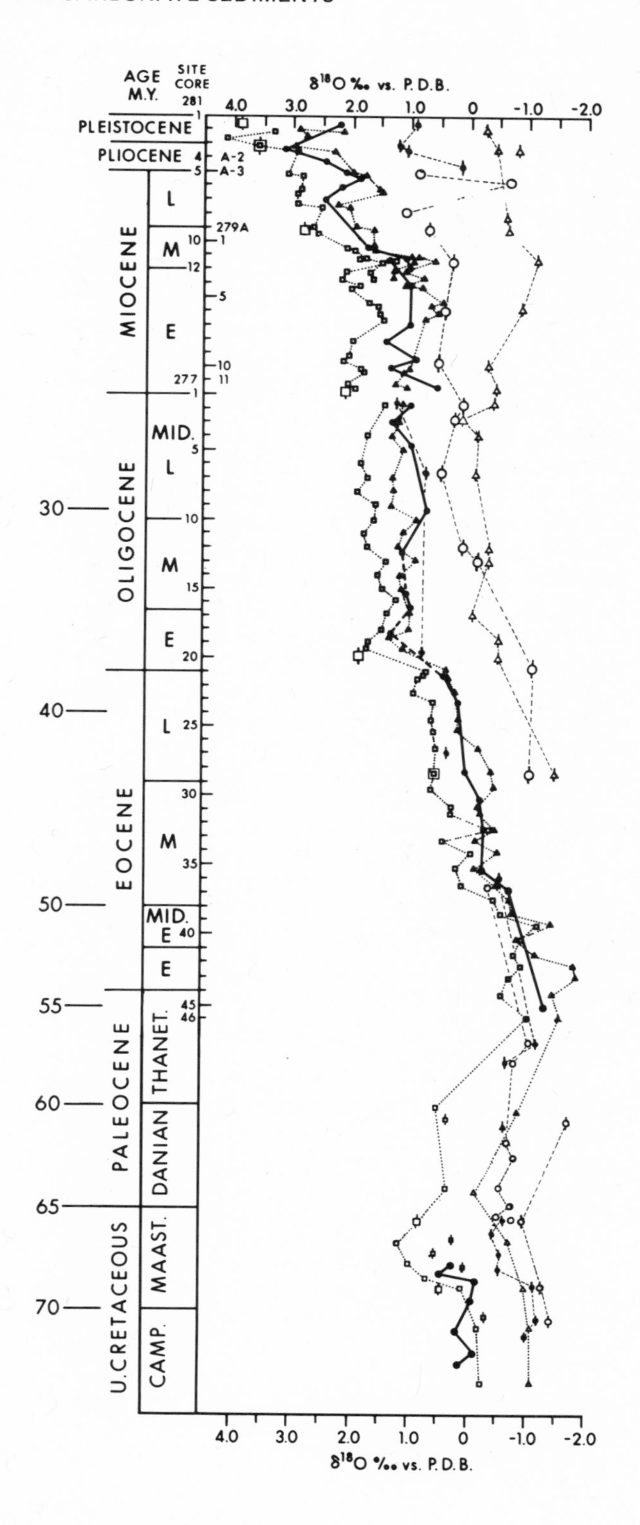

AGE M.Y.
SITE CORE
281
$\delta^{18}O$ ‰ vs. P.D.B.
4.0
3.0
2.0
1.0
0
-1.0
-2.0
PLEISTOCENE
PLIOCENE
A-2
A-3
279A
277
MIOCENE
L
M
E
OLIGOCENE
MID.
L
M
E
EOCENE
L
M
MID. E
E
PALEOCENE
THANET.
DANIAN
U.CRETACEOUS
MAAST.
CAMP.
30
40
50
55
60
65
70

Margolis et al. (1975) isotopically analyzed moderately well-preserved nannofossils contained in three D.S.D.P. cores of Cenozoic age from the Southern Ocean (Figure 9). These nannofossils have a $\delta^{18}O$ curve that closely parallels the planktonic foraminifera curve, although nannofossil $\delta^{18}O$ values are occasionally slightly higher than the associated planktonic foraminifera. The majority of nannofossil samples from site 277 show $\delta^{18}O$ values that are equal to or slightly lower than those of planktonic foraminifera, indicating preservation of surface water $\delta^{18}O$ values after burial in sediments of up to middle Eocene in age. Diagenetic alteration rendered separation of relatively pure nannofossils difficult in older sediments, and the isotopic data show evidence of re-equilibration with bottom waters, in the form of calcite overgrowths on the coccoliths.

The $\delta^{13}C$ profile for calcareous nannofossils from these cores also shows a tendency to parallel the foraminiferal curves (Figure 10). The progressive increase in $\delta^{13}C$ values from the benthic and planktonic foraminifera to the nannofossils at sites 279A and 277 suggests that carbon isotopes reflect the $\delta^{13}C$ of the surrounding media and may be indicative of the water depth during growth (*Saito and Van Donk*, 1974). The $\delta^{13}C$ of the ΣCO_2 in present-day South Pacific Ocean waters varies from about +2.0°/oo at the surface to around +5.0°/oo in bottom waters (*Kroopnick*, 1974a).

The effects of dissolution and secondary encrustation with resultant isotopic re-equilibration of calcareous nannofossils are related to either water depth or subsequent burial history. In present-day oceans, planktonic foraminifera show slight dissolution effects below 1000m, with appreciable dissolution occurring below 3000m in the Central Pacific Ocean. Etching, fragmentation, and dissolution features are evident on coccoliths found below 3000m, and further deterioration increases rapidly between 3500 and 4800m and apparently can be produced at or near the sediment-water interface (*Roth and Berger*, 1975). Among the cores discussed here, only sediment samples from site 279A now lie below a water depth of 3000m and should contain the most evidence for dissolution. The isotope data from site 279A (3341m), when compared with sediments from sites 277 and 281 (1214 and 1591m), do not show any evidence of extensive encrustation.

Encrustation of the nannofossils during or after burial should produce higher $\delta^{18}O$ values because of colder bottom water temperatures and lower $\delta^{13}C$ values because of incorporation of ^{13}C-depleted organic carbon (*Kroopnick*, 1974a). Several samples analyzed do show possible indications of re-equilibration with deeper water when compared to planktonic foraminifera. Detailed electron micro-

Fig. 10. Summary plot of $\delta^{13}C$ for benthic foraminifera, planktonic foraminifera, and nannofossils. Symbols used, sites, and references used are the same as in caption for Figure 9.

Benthic foraminifera
Planktonic foraminifera
Calcareous nannofossils

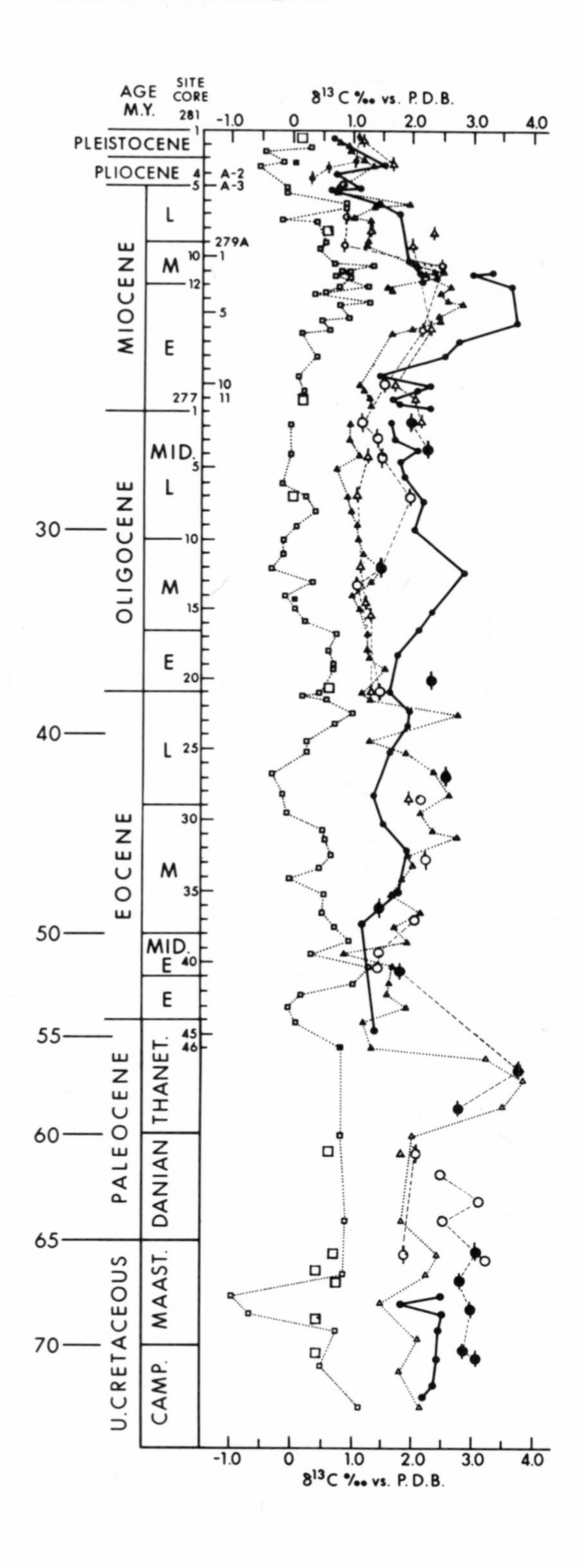
AGE
M.Y.
SITE
CORE
281
δ13C ‰ vs. P.D.B.
-1.0
0
1.0
2.0
3.0
4.0
PLEISTOCENE
PLIOCENE
MIOCENE
OLIGOCENE
EOCENE
PALEOCENE
U.CRETACEOUS
L
M
E
MID.
THANET.
DANIAN
MAAST.
CAMP.
A-2
A-3
279A
277
30
40
50
55
60
65
70

scopic examinations of nannofossils from site 277, core 28, and site 281, core 6, which have lower $\delta^{13}C$ and higher $\delta^{18}O$ values than planktonic foraminifera (Figures 9 and 10) reveal that most of the nannofossil remains in these cores exhibit abundant evidence of dissolution and secondary calcite overgrowths and contain no whole coccospheres. However, samples where $\delta^{13}C$ values for nannofossils are higher than those for planktonic foraminifera, and $\delta^{18}O$ values are lower, such as from site 277, cores 4 and 9, contain numerous whole coccospheres and coccoliths which are well-preserved. Thus, samples of nannofossils must be subjected to electron microscopic studies of the degree of preservation if meaningful paleoceanographic data are desired.

Examination of the Eocene portion of site 277 (Figure 10) reveals a divergence in $\delta^{13}C$ values for benthic and planktonic foraminifera between cores 22 and 34. Individuals of the species *Globigerapsis* index were used to obtain the planktonic foraminiferal values from those samples, whereas in other samples mixed planktonic species have been used. It has been found by several researchers (*Shackleton et al.*, 1973; *Savin and Douglas*, 1973, 1977) that there are definite species-dependent departures from isotopic equilibrium in planktonic foraminifera, so that surface water $\delta^{13}C$ values cannot be derived from planktonic foraminiferal measurements even if the depth habitat of the species is known (and it is not for extinct species). However, polyspecific samples of calcareous nannofossils may provide a more reliable indication of surface $\delta^{13}C$ changes than planktonic foraminifera.

Site 284

Figures 11 and 12 are the oxygen and carbon isotopic curves for benthic foraminifera, nannofossils and the three species of planktonic foraminifera analyzed from this site. A more thorough discussion of these data is presented in *Kennett et al.* (in prep.). The most noticeable features of the four $\delta^{18}O$ curves, however, are: (1) The nannofossil $\delta^{18}O$ values with few exceptions, are isotopically heavier than the planktonic foraminifera. (2) Occasionally values for *G. bulloides* are heavier than the nannofossil values; for instance, in the two shallowest and the two deepest samples. *G. bulloides* also shows the widest range in $\delta^{18}O$ values of the three planktonic foraminifera, with fluctuating depth or isotopic rankings from lightest of the three in one sample to the above mentioned cases where they were isotopically heavier than nannofossils. On the other hand, *G. quinqueloba* consistently shows the lightest oxygen isotopic values, which indicates that it is the shallowest calcifying species of these three, occupying the warm surface waters. *N. pachyderma's* isotopic trend closely parallels that of *G. quinqueloba*, and in a number of samples is isotopically lighter than the surface water dwelling thin-shelled foraminifera. Disregarding *G. bulloides*, the nannofossil isotope curve is consistently 1.0 $\pm$ 0.5^{o}/oo heavier than the lightest planktonic

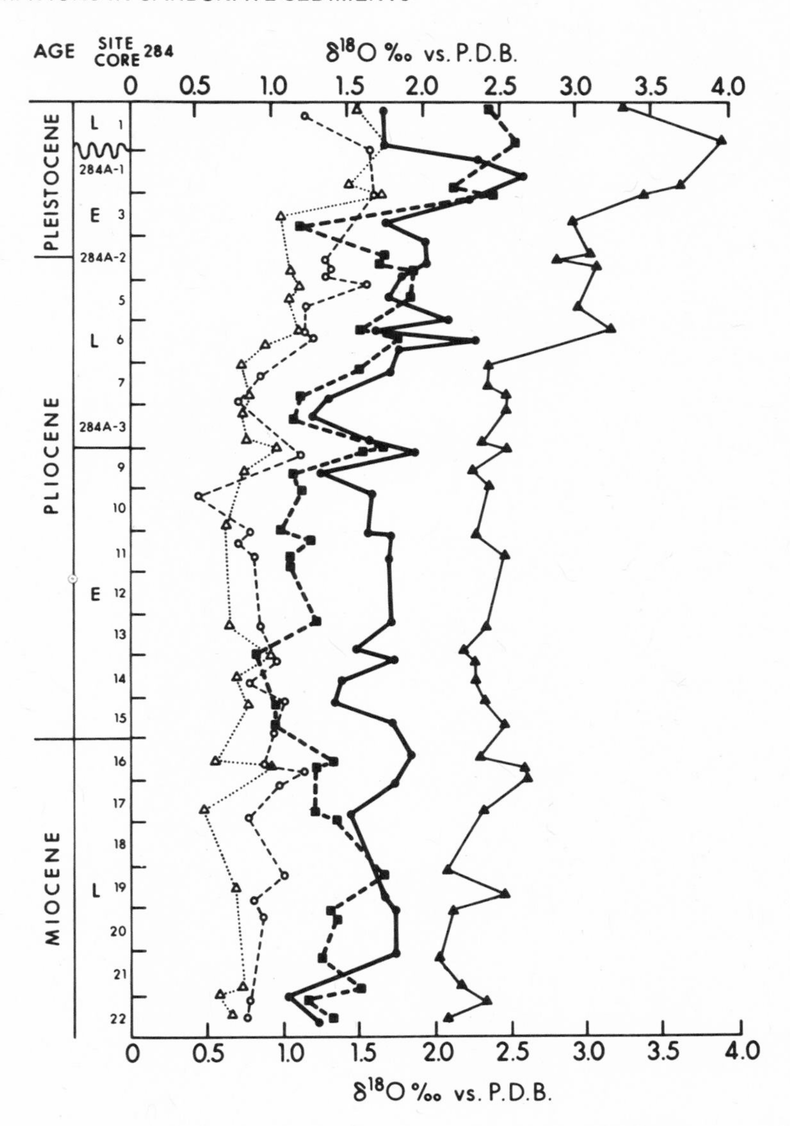

Fig. 11. Plot of $\delta^{18}O$ for site 284, solid triangles are benthic foraminifera (*Uvigerina*) from *Shackleton and Kennett* (1975b). Solid line with dark dots are calcareous nannofossils. Solid squares with dashed line are *G. bulloides*, open triangles are *G. quinqueloba* and open circles are *N. pachyderma*. (From *Kennett et al.*, in prep.).

foraminifera, and occupies a position about equidistant between the planktonic and benthic foraminifera curves. The isotopic trends from the Middle Pliocene to the Late Pleistocene seem to be synchronous

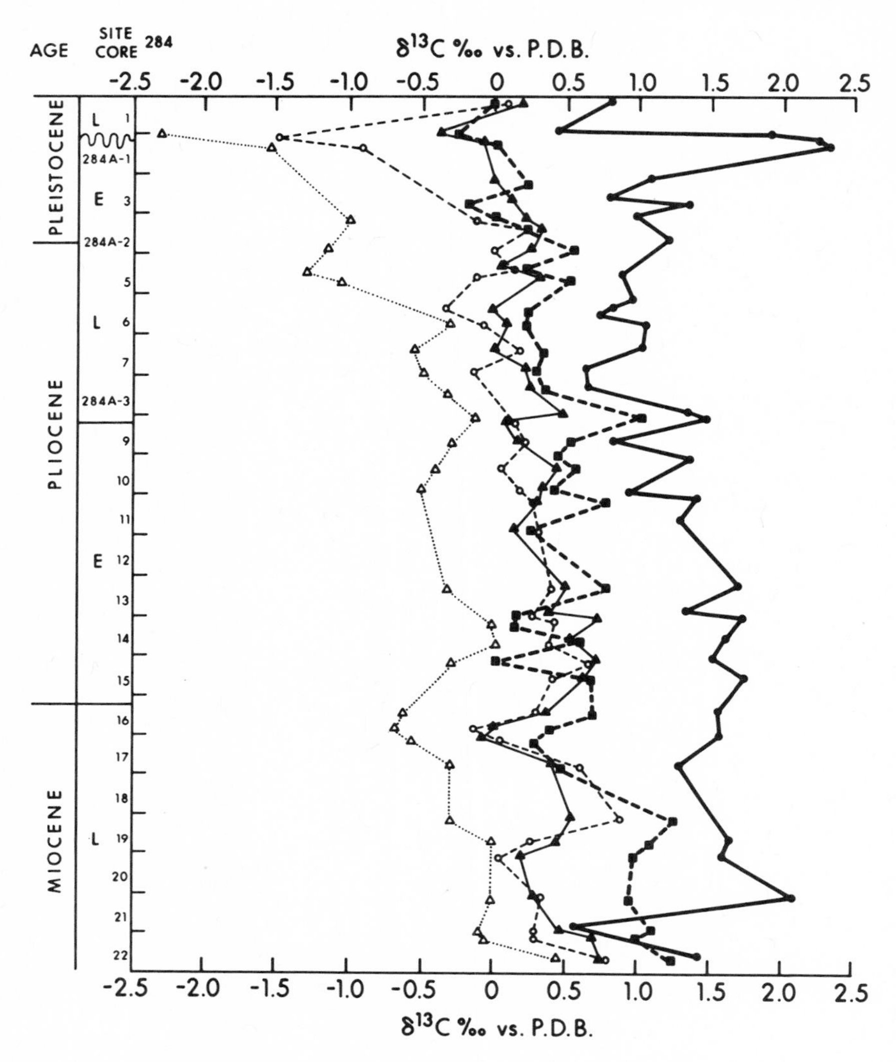

Fig. 12. $\delta^{13}C$ vs. PDB for site 284. Symbols are the same as Figure 10. (From *Kennett et al.*, in prep.).

with respect to major excursions, taking into consideration the 1.0 $\pm$ 0.5‰ positive offset of the coccolith data. *Kennett et al.* (in prep.) discuss the paleoclimatic significance of these excursions. It is difficult to explain this offset in terms of diagenetic overgrowths on the coccoliths since the section shown represents carbonate deposition above the lysocline. *Douglas and Savin* (1975) and *Savin et al.* (1975) and *Goodney* (1977) have also observed this 1°/oo shift between coccolith and foraminifera samples.

At this time, planktonic foraminifera grown in controlled laboratory cultures have not been isotopically analyzed, so that either one has to assume isotopic equilibrium or somehow relate isotopic temperatures of Recent planktonic foraminifera to present-day surface temperatures in order to determine if isotopic equilibrium is attained. Several workers (*Van Donk*, 1970; *Shackleton et al.*, 1973) have reported that some species of planktonic foraminifera show non-equilibrium fractionation during growth of their tests. Correction factors have been calculated by *Savin et al.* (1975) to permit estimation of surface-water temperatures from planktonic foraminiferal multi-specific assemblages, which have taken into consideration (1) differences between the isotopic temperature of the assemblage and the shallowest dwelling species; and (2) the difference between the isotopic temperature of the shallowest species and the actual surface water temperature estimated by proportion from (1). Similar correction factors may have to be used in order to estimate surface temperatures from oxygen isotope data of nannofossil assemblages. Our preliminary coccolithophorid culturing data indicate that non-equilibrium deposition of oxygen isotopes may explain the 1°/oo difference between fossil coccoliths and associated planktonic foraminifera (*Dudley*, 1976; *Goodney*, 1977).

Carbon isotopes for the calcareous plankton from site 284 (Figure 12) show a remarkably wide range of values. Coccolith $\delta^{13}C$ values are consistently heavier than all planktonic foraminifera, with the exception of one point near the bottom of the section. The coccolith carbonate $\delta^{13}C$ values are also consistent with a model proposing isotopic equilibration with surface water ΣCO_2, mainly HCO_3^- (*Kroopnick*, 1975). However, the carbon isotopes of the three species of planktonic foraminifera show a wide departure from one another; *G. bulloides* perhaps more closely approaching the nannofossil curve than the other foraminifera, and *N. pachyderma*, except for a large negative excursion in the Late Pleistocene, exhibiting $\delta^{13}C$ values similar to *Uvigerina*. *G. quinqueloba*, presumably the shallowest dwelling of the planktonic foraminifera, with the lightest $\delta^{18}O$ values, is the most negative in $\delta^{13}C$. This observation is not consistent with the depth stratification model of calcareous biota, where organically deposited carbonate should reflect the $\delta^{13}C$ of the ΣCO_2. Evidently, *G. quinqueloba* is either fractionating carbon isotopes, or perhaps utilizing a more negative reservoir of carbon in secreting its test.

The $\delta^{13}C$ of the nannofossils shows a large positive excursion during the Early Pleistocene immediately below the disconformity with Late Pleistocene sediments (Figure 12). This positive excursion is synchronous with a large negative excursion in $\delta^{13}C$ values for *G. quinqueloba* and *N. pachyderma* (Figure 12) and a positive excursion in both surface and bottom water oxygen isotopic values.

Late Cretaceous and Cenozoic Oxygen and Carbon Isotope Data

Figures 9 and 10 represent a compilation of published and unpublished oxygen and carbon isotope data on benthic and planktonic foraminifera, and associated calcareous nannofossils contained in deep-sea sediments in the world's oceans. The spread in planktonic oxygen isotope values is due to the latitudinal distribution of the cores, representing equatorial, temperate, and high latitude water masses. A noticeable feature of the $\delta^{18}O$ curves are the values for benthic foraminifera. During the majority of the time span represented here, deep-sea bottom water isotope values show significantly less latitudinal effects than surface water values, except perhaps for data from the Upper Cretaceous. Both the nannofossil and planktonic foraminifera curves parallel the benthic curves for the majority of the time. As one goes back in time, the surface to bottom temperature gradient becomes less pronounced, and the difference between nannofossil and planktonic foraminifera isotope values is smaller. Figures 9 and 10 should be used strictly in a qualitative manner because they represent analyses performed by several laboratories all using different techniques of specimen selection, preparation and isotopic analyses. Nannofossil oxygen isotope values obtained from the sub-Antarctic core Eltanin 55-26 of Upper Cretaceous age show values similar to benthic foraminifera from lower latitudes (Figure 9), indicating that the thermohaline circulation system must have been active at that time as it is at present (*Margolis et al.*, 1977 in press).

The plot of $\delta^{13}C$ values for the last 75 million years from the same samples (Figure 10) shows a continuation of the trends discussed above, and those presented by *Margolis et al.* (1975). Nannofossils are usually heavier in $\delta^{13}C$ than associated foraminifera, except where there is evidence of secondary diagenetic overgrowths which contain lighter carbon. In this case, corresponding oxygen isotopes are usually heavier, representing bottom water influence. The data exhibit good parallelism of trends, even as far back as the Upper Cretaceous. Where coccolith $\delta^{13}C$ values are heavier than those for associated planktonic foraminifera, it lends credence to using the corresponding nannofossil oxygen isotope value as a paleoceanographic indicator. Other features of the $\delta^{13}C$ curves are parallel excursions in all curves occurring over several million year periods. There is a negative excursion during the middle Maastrichtian, where benthic foraminifera attain their lightest values in the time series shown; there is a corresponding decrease in $\delta^{13}C$ values for planktonics but not as significant as compared to changes elsewhere in the planktonic curve. There is also a well-defined positive (heavy) excursion in the planktonic $\delta^{13}C$ during the Middle to Late Paleocene. Unfortunately, there are no corresponding data for benthic foraminifera from this interval. Another positive excursion occurs in the late Early to early Middle Miocene where $\delta^{13}C$ values for benthic and planktonic organisms become heavier than the rather consistent Late Oligocene and Early Miocene values. After this positive excursion there is a general trend towards lighter $\delta^{13}C$ values, corresponding to the Late Cenozoic cooling and ice build-up shown by the oxygen isotopes, which are associated with intensification of high latitude and eventually middle latitude Pleistocene glaciation. *Shackleton* (1977) has associated decreases in $\delta^{13}C$ in benthic foraminifera with glacial periods as discussed below.

At the present time we can only speculate as to the significance of these long-term fluctuations in $\delta^{13}C$ of carbonate secreting benthos and plankton. One possible explanation for the Paleocene and Miocene positive (heavy) excursions could be that there were transgressions of the oceans over the continents at those times, increasing the area of continental slopes where isotopically light organic carbon would be buried in the sediment. The oceans at that time would thus be enriched in ^{13}C, which would then be reflected as more positive $\delta^{13}C$ in calcareous tests. Other possible causes for excursions in carbon isotope values would be: increased weathering rates of fossil carbon containing sediments on land; changes in the $\delta^{13}C$ profile in the oceans caused by increased upwelling, which in turn would increase organic productivity in surface waters and increase the rate of burial of light carbon; and volcanism which would introduce isotopically heavier CO_2 gas into the atmosphere (heavier than organic carbon).

Shackleton (1977) has shown that carbon isotopes of monospecific samples of benthic foraminifera are more depleted in ^{13}C in Late Pleistocene glacial specimens than Recent (interglacial) specimens. He suggests that dissolved CO_2 in the oceans was lighter during the last glacial period than now by about 0.5°/oo. A possible source of light carbon would be the terrestrial biosphere which may contain as much as 2×10^{18} g of carbon with a composition of about -25°/oo. If half of this amount were added to the oceans, it would cause a change of about -0.3°/oo in dissolved CO_2. According to *Shackleton* (1977), changes in the development of tropical rainforests and temperate forests in Northern Eurasia and North America caused by the advance of the continental ice-sheets could result in the transfer of this isotopically light carbon between the oceans and the terrestrial biosphere. This is a plausible explanation for Late Pleistocene fluctuations in $\delta^{13}C$; however, it would be difficult to evaluate the effect of terrestrial biosphere fluctuations with respect to long-term Cenozoic fluctuations in biogenic carbonate $\delta^{13}C$. There is, however, a general trend of heavier carbon isotopes with lighter (warmer) oxygen isotopes, and lighter carbon with heavier (cooler) oxygen isotopes, shown in the major excursions in Figures 9 and 10.

Other than stating that there are a number of possible processes that would influence the carbon isotopic composition of deep-sea biogenous carbonates, we will not speculate any further until we understand the processes which control the mechanism of incorporation and the reservoirs of carbon which are utilized by calcifying organisms. As mentioned earlier, however, it appears that nannofossils may provide an important key to our understanding of carbon isotopes in the ocean since well-preserved samples do not show the species-dependent fractionations discussed earlier. The carbon isotope data presented in this paper consistently show the nannofossil samples to be more positive than foraminifera data, although the observed trends are often parallel. The $\delta^{13}C$ of the most Recent nannofossil samples from sites 281 and 284 are +.6 and +.8, respectively. These samples are definitely not contemporary but still can be interpreted as resembling today's surface $\delta^{13}C$ of ΣCO_2 value of about +1.9°/oo (*Kroopnick*, 1974b). $\delta^{13}C$ analysis of coccoliths from present-day surface waters and from surface sediments, such as those presented by *Goodney* (1977), will yield a better comparison.

Calcite inorganically precipitated from sea water is enriched in ^{13}C by 0.9°/oo at 25°C (*Rubinson and Clayton*, 1969) and by 1.7°/oo at 2°C. The mechanism(s) for organic carbonate production are not known but two obvious sources of carbon are the HCO_3^- pool (97% of the ΣCO_2) and dissolved $CO_3^=$. If these are the sources of carbon, clearly "vital" effects are playing a large role in determining the $\delta^{13}C$ of foraminifera shells. Another possible carbon source is dissolved organic matter which, if applicable, further complicates the problem. On the other hand, the calcareous nannofossil data presented here suggest that coccoliths incorporate HCO_3^- directly from the surrounding water mass. This is surprising since as was discussed earlier, the oxygen isotopes are not deposited in equilibrium with the surrounding water either.

In conclusion, it has been shown that the response of the ocean and the atmosphere to climatic and paleoceanographic changes and their significance relative to changes in the CO_2 budget can be studied via the carbon and oxygen isotope record in fossil carbonate sediments. Further studies, however, are needed to help us predict the ocean's response to climatic and man-induced changes in the present-day CO_2-^{13}C-$CaCO_3$ system.

ACKNOWLEDGEMENTS

This work is supported by NSF Grants DES75-19386 and OCE76-05057 to S. V. Margolis and P. M. Kroopnick. We thank N. Shackleton and J. P. Kennett for providing the oxygen and carbon data on planktonic foraminifera from site 284, which is the subject of a more detailed paper with them (*Kennett et al.*, in prep.). H. Craig and R. Weiss of Scripps Institution of Oceanography kindly allowed us to use some of their unpublished ΣCO_2 and $\delta^{13}C$ data. We thank R. D. Bellegay for his assistance in the collection and analysis of the HUDSON-70 CO_2 samples.

REFERENCES

Berger, W. H. 1969. Ecologic patterns of living planktonic foraminifera. *Deep Sea Res., 16.* 1-24.

Craig, H. 1957. Isotopic standards for carbon and oxygen and correction factors for mass spectrometric analyses of carbon dioxide. *Geochim. Cosmochim. Acta, 12.* 133.

Craig, H. and C. D. Keeling. 1963. The effects of atmospheric nitrous oxide on the measured isotopic ratio of atmospheric carbon. *Geochim. Cosmochim. Acta, 27.* 549-551.

Craig, H., Y. Chung, and M. Fiadeiro. 1972. A benthic front in the South Pacific. *Earth Planet. Sci. Lett., 16.* 50-65.

Douglas, R. and S. Savin. 1975. Oxygen and carbon isotope analyses of Tertiary and Cretaceous microfossils from Shatsky Rise and other sites in the North Pacific Ocean. In: *Initial Reports of the Deep Sea Drilling Project, 32.* 509-520.

Dudley, W. C. 1976. Paleoceanographic application of oxygen isotope analyses of calcareous nannoplankton grown in culture. Ph.D. dissertation, University of Hawaii. 168 pp.

Emiliani, C. 1954. Depth habitats of some species of pelagic foraminifera as indicated by oxygen isotope ratios. *Amer. J. Sci., 252.* 149-158.

Goodney, D. E., W. C. Dudley, M. A. Mahoney, and S. V. Margolis. 1975. Paleoclimatic implications of the relationship between Late Cenozoic oxygen isotopes and explosive volcanic ash. *Amer. Geophys. Union,* Fall Annual Meeting.

Goodney, D. E. and P. Kroopnick. In Press. Current techniques for the collection and processing of stable isotope data. In: *Proceedings of International Conference on Stable Isotopes;* Lower Hutt, New Zealand; August 1976; D.S.I.R. Special Bulletin.

Goodney, D. E. 1977. Non-equilibrium fractionation of the stable isotopes of carbon and oxygen during precipitation of calcium carbonate by marine phytoplankton. Ph.D. dissertation, University of Hawaii. 146 pp.

Kennett, J. P., and P. Vella. 1975. Late Cenozoic planktonic foraminifera and paleoceanography at D.S.D.P., site 284 in the cool subtropical South Pacific, D.S.D.P. Leg 29, In: *Initial Reports of the Deep Sea Drilling Project, 29.* 760-800.

Kennett, J. P., N. J. Shackleton, S. V. Margolis, D. E. Goodney, W. C. Dudley, and P. M. Kroopnick. In prep. Late Cenozoic oxygen and carbon isotope history and volcanic ash stratigraphy: D.S.D.P. site 284, South Pacific.

Kroopnick, P. 1974a. The dissolved O_2-CO_2-C^{13} system in the eastern equatorial Pacific. *Deep Sea Res., 21.* 211-227.

Kroopnick, P. 1974b. Correlations between C^{13} and ΣCO_2 in surface waters and atmospheric CO_2. *Earth Planet. Sci. Lett.*, *22*, #*4*. 397-403.

Kroopnick, P. 1975. Respiration, photosynthesis and oxygen isotope fractionation in oceanic surface water. *Limnol. Ocean.*, *20*. 988-992.

Margolis, S. V., P. M. Kroopnick, D. E. Goodney, W. C. Dudley and M. A. Mahoney. 1975. Oxygen and carbon isotopes from calcareous nannofossils as paleoceanographic indicators. *Science*, *189*. 555-557.

Margolis, S. V., P. M. Kroopnick, and D. E. Goodney. In press. Cenozoic and late Mesozoic paleoceanographic and paleoglacial history contained in circum-Antarctic deep-sea sediments. *Marine Geology*.

Mook, W. G., J. C. Bommerson and W. H. Staverman. 1974. Carbon isotope fractionation between dissolved bicarbonate and gaseous carbon dioxide. *Earth Planet. Sci. Lett.*, *22*. 169-176.

Perch-Neilsen, C. and A. R. Edwards. 1975. Calcareous nannofossils from the southern Southwest Pacific, D.S.D.P. Leg 29. In: *Initial Reports of D.S.D.P.*, *29*. 469-540.

Roth, P. H. and W. H. Berger. 1975. Distribution and dissolution of coccoliths in the south and central Pacific. In: *Dissolution of Deep-Sea Carbonates*, W. V. Sliter, A. W. H. Bé and W. H. Berger (Eds.), Cushman Foundation Special Publication No. 13, Allen Press, Lawrence, Kansas. 87-113.

Rubinson, H. and R. N. Clayton. 1969. Carbon 13 fractionation between aragonite and calcite. *Geochim. Cosmochim. Acta*, *33*. 997-1004.

Saito, T. and J. Van Donk. 1974. Oxygen and carbon isotope measurements of Late Cretaceous and Early Tertiary foraminifera. *Micropaleo.*, *20*. 152-177.

Savin, S. and R. G. Douglas. 1973. Stable isotope and magnesium Geochemistry of recent planktonic foraminifera from the South Pacific. *GSA Bull.*, *84*. 2327-2342.

Savin, S. M., R. G. Douglas and F. G. Stehli. 1975. Tertiary marine paleotemperatures. *GSA Bull.*, *86*. 1499-1510.

Savin, S. M. and R. G. Douglas. 1977. Depth stratification of planktonic foraminifera. *J. Foram. Res.*, (in press).

Shackleton, N. J. and N. D. Opdyke. 1973. Oxygen isotope temperatures and ice volumes on a 10^5 year and 10^6 year scale. *J. Quat. Res.*, *3*. 39-55.

Shackleton, N. J., H. Wiseman and A. Buckley. 1973. Non-equilibrium deposition between seawater and planktonic foraminifera tests. *Nature*, *242*. 177-179.

Shackleton, N. J. and J. P. Kennett. 1975a. Paleotemperature history of the Cenozoic and initiation of Antarctic glaciation: oxygen and carbon isotope analyses in D.S.D.P. sites 277, 279, 281. In: *Initial Reports of the Deep Sea Drilling Project*, *29*.

Shackleton, N. J. and J. P. Kennett. 1975b. Late Cenozoic oxygen and carbon isotopic changes at D.S.D.P. site 284: Implications for glacial history of the Northern Hemisphere and Antarctic, D.S.D.P. Leg 29. In: *Initial Reports of the Deep Sea Drilling Project, 29*. 801-808.

Shackleton, N. J. 1977. Carbon-13 in *Uvigerina*: tropical rain-forest history and equatorial Pacific dissolution cycles. In: *The Fate of Fossil Fuel CO_2*, N. R. Andersen and A. Malahoff (Eds.), Plenum, N.Y., (this volume).

Sommer, M. A. II and R. K. Matthews. 1975. Carbon-13 stratigraphy in deep-sea cores. *G.S.A. Annual Meeting*, Salt Lake City.

Van Donk, J. 1970. The oxygen isotope record in deep-sea sediments, Ph.D. dissertation, Columbia University, New York. 228 pp.

Weiss, R. F. 1970. Dissolved gases and total inorganic carbon in seawater: distribution, solubilities, and shipboard gas chromatography. Ph.D. thesis, Scripps Institution of Oceanography, U. C., San Diego.

Wong, C. S. 1970. Quantitative analysis of total carbon dioxide in seawater: a new extraction method. *Deep Sea Res., 17*. 9 pp.

The Carbonate Chemistry of North Atlantic Ocean Deep-Sea Sediment Pore Water

John W. Morse

University of Miami

ABSTRACT

The carbonate chemistry of twenty cores taken in the central North Atlantic Ocean has been studied in order to elucidate the processes controlling the diagenesis of calcium carbonate in deep-sea sediments. Two major problems make the interpretation of the results uncertain: lack of knowledge of the extent to which reactions take place between the calcium carbonate in the sediment and the pore water during recovery and short-term storage; and the large differences presently reported in values for the equilibrium ion product of calcite.

If there is little reaction between the calcium carbonate in the sediment and the pore water during the core recovery and short-term storage, three important conclusions can be drawn from the data presented in this study. Sediments below 3000 meters water depth are more under-saturated with respect to calcite than is the overlying water. Very little reaction takes place between the calcium carbonate in these sediments, and the pore waters in the upper meter of sediment, which could be due to rapid coating of the calcium carbonate grains after deposition at the sediment-water interface. In sediments above 3000 meters water depth, small amounts of aragonite in the upper few centimeters of sediment exert a strong influence on the carbonate chemistry of the pore waters.

INTRODUCTION

In recent years factors controlling the deposition of calcium carbonate in the deep sea have been intensely investigated. The investigations have shown that deposition continues to depths where water overlying the sediment in which the calcium carbonate is accumulating is undersaturated with respect to both calcite and aragonite (e. g., see *Takahashi*, 1975). Studies based on the data collected in the GEOSECS Program indicate that over large areas of the Atlantic and Pacific Oceans the undersaturation at the foraminiferal lysocline (*Broecker and Takahashi*, 1977) and the calcite compensation depth (*Takahashi*, 1975) are close to constant. Assuming similar productivity in the overlying water, this suggests that there is a rather constant relationship between dissolution and undersaturation in much of the ocean.

Although it appears at this time that the relationship of the chemistry of the water column to the carbonate content of adjacent sediments is well known, little is known about the conditions controlling the dissolution of calcium carbonate at or immediately below the sediment-water interface. It is reasonable to presume that equilibrium is reached at some depth in the sediments between calcium carbonate (low magnesium calcite) and the pore water. The work of *Hammond* (1974) on JOIDES cores suggests that such equilibrium is eventually reached, but does not indicate the rate or path of the process of equilibration. A knowledge of the equilibration process is crucial for a complete understanding of calcium carbonate deposition in the deep sea.

EXPERIMENTAL METHODS

In this investigation twenty cores were studied. They were collected on three cruises in a band running from the center of the Mid-Atlantic Ridge to the center of the Canary Basin (25° to 30^oN). One core was collected in the Nares Abyssal Plain. The locations are given in Table 1, and shown in Figures 1 and 2.

Sediment samples from within 1.5 meters of the sediment-water interface were collected with either hydroplastic or "boomerang" corers. Samples from greater sediment depths were collected on the first cruise with a piston corer. Immediately upon recovery, all cores were refrigerated at close to *in situ* temperature (2^oC) for approximately 24 hours.

Pore water was extracted from the sediment by squeezing at 2^oC. On the first two cruises, a hydraulic squeezer with Millipore pre-filters and 0.4 μ filters was used at a maximum squeezing pressure of 3500 p.s.i. On the third cruise, a bank of nylon, gas (N_2) powered squeezers was used. These squeezers contained precombusted glass fiber filters, and 0.4 μ Nuclepore after-filters.

pH

The *pH* was measured using precooled "punch in" electrodes

TABLE 1. Location of Coring Sites.

Core	Latitude(N)	Longitude(W)	Water Depth(m)
I-1	25° 37.3'	24° 14.2'	5150
I-2	26° 14'	25° 15'	5246
I-3	26° 23'	25° 13'	5244
I-4	27° 15'	27° 43'	5317
I-5	28° 15'	30° 42'	5129
I-6	28° 36'	34° 38'	5939
I-7	29° 12'	38° 29'	4754
I-8	29° 45'	38° 38'	3689
I-9	29° 53'	38° 28'	4357
I-10	29° 46'	40° 53'	4226
I-11	29° 58'	40° 54'	3767
I-12	29° 57'	41° 42'	4208
II-1	26° 19'	44° 43'	4078
II-2	26° 19.7'	44° 43.6'	4163
II-3	26° 19.1'	44° 44.5'	4105
II-4	26° 19.1'	44° 44.5'	4105
II-5	26° 50.8'	60° 13.6'	6254
III-1	25° 56.3'	44° 42.1'	2900
III-2	25° 46.8'	44° 42.0'	2980
III-3	25° 51.6'	40° 46.5'	4930

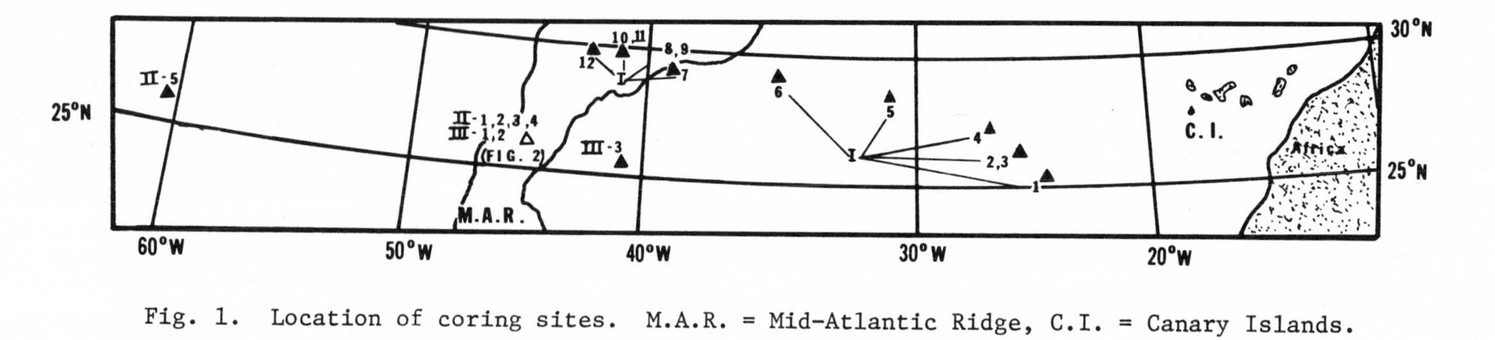

Fig. 1. Location of coring sites. M.A.R. = Mid-Atlantic Ridge, C.I. = Canary Islands.

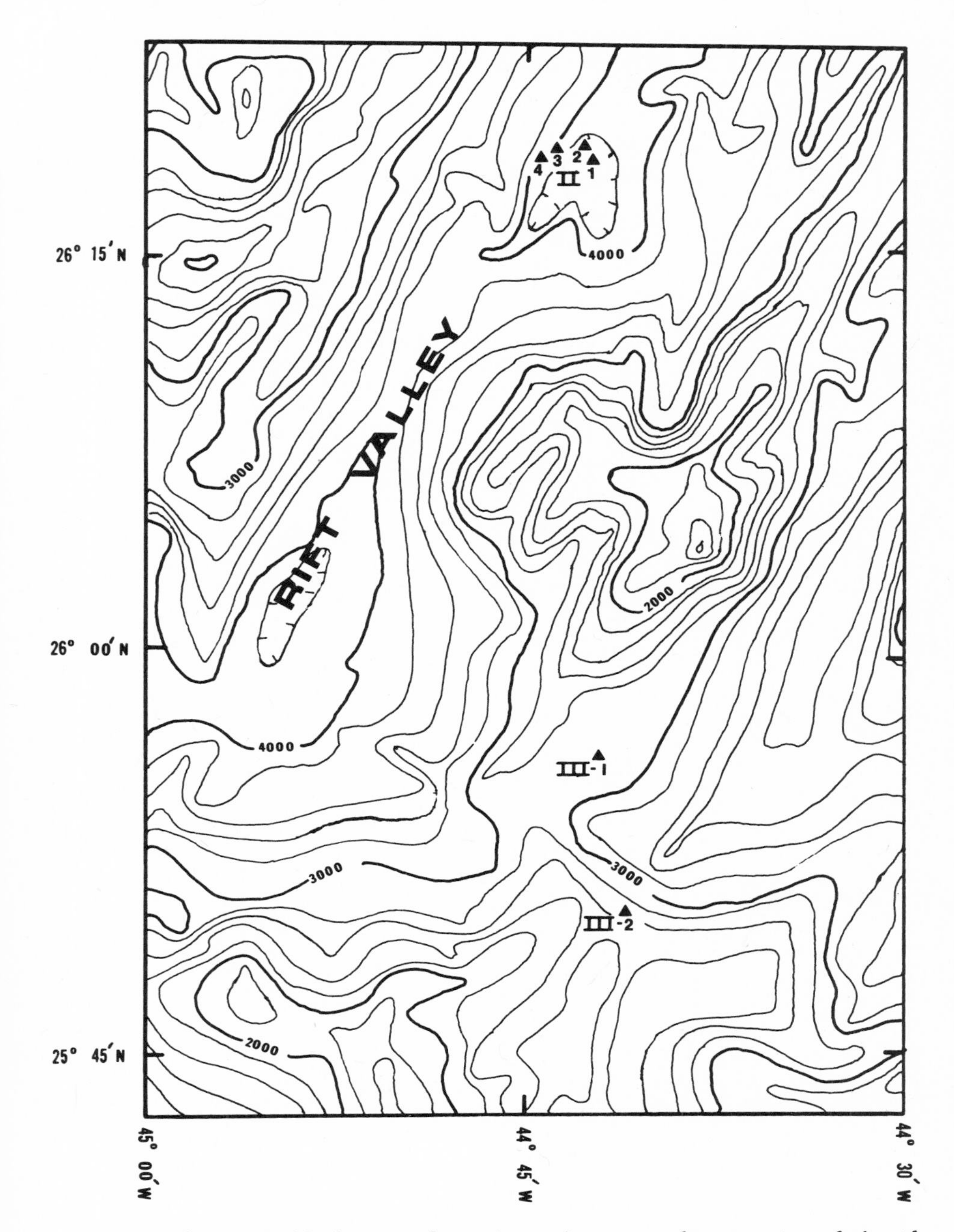

Fig. 2. Detailed map of coring sites on the crest and in the central rift valley of the Mid-Atlantic Ridge (after *McGregor and Rona*, 1975). Depths on contour lines are in meters.

standardized with a buffer solution chilled to the sediment temperature (2°C). *Gieskes* (1974) reports the precision of this method as ± 0.08 *pH* units.

Alkalinity

The titration alkalinity was determined by potentiometric titration immediately following pore water extraction. On the first cruise, values of the titration alkalinity were taken as the inflection point of a plot of volume of titrant versus millivolts. Precision was approximately 2%. On the other two cruises, the Gran plot method was used and a precision of better than 1% was obtained.

Calcium

Compleximetric titration with EGTA (*Tsunogai et al.*, 1968) as modified by *Gieskes* (1974) was used for calcium determination. Copenhagen IAPSO standard seawater was used as the standard. Strontium was not determined, and only a standard correction for average seawater strontium concentration was made. Values of calcium concentration were normalized to 34.5 o/oo salinity based on calcium to chloride ratios. Calcium values reported from the second cruise are based on analysis of pore water which had been stored untreated for approximately ten months. Calcium values from the third cruise are based on the analysis of pore water which had been titrated for alkalinity; measurements were carried out within three months of collection. In carrying out the calcium analyses a precision of better than 2% was obtained on the standard seawater, better than 1% on deep seawater, and only about 4% on a combined pore water sample. The reason for the variability in precision is not known, but it may be due to some unknown organic interference. Since the total variation in calcium concentration was only about ± 7%, it is the author's opinion that not much significance should be attached to variations in calcium at this time.

Chloride

Chlorinity was determined for normalization of calcium values by the standard Mohr titration with $AgNO_3$ and K_2CrO_4 indicator. Copenhagen IAPSO standard seawater was used as the standard. A precision of better than 1% was obtained.

Weight % $CaCO_3$

Weight percent calcium carbonate was determined by acid extraction of calcium. Calcium was analyzed on the samples from the first cruise by atomic adsorption spectroscopy, and on the samples from the other cruises by the previously described EGTA titration method. In this set of analyses a precision of approximately 3% was obtained for the total procedure.

Analytical results are given in Table 2. Alkalinity, *pH* and weight percent calcium carbonate were determined on the samples from the first cruise. Alkalinity, *pH*, weight percent calcium carbonate, and calcium and chloride concentrations were determined on most samples from the second and third cruises.

CALCULATION OF SATURATION STATE

The saturation state of pore waters with respect to calcite has been calculated for both extraction and *in situ* conditions. The *pH* values measured at 2°C (near *in situ* temperature) were corrected to *in situ* conditions, using the pressure corrections of *Culberson and Pytkowicz* (1968). Titration alkalinity (A_T) was converted to carbonate alkalinity (A_C) by subtraction of borate alkalinity for both extraction and *in situ* conditions, as described by *Culberson and Pytkowicz* (1968). Equation (1) was used to calculate the saturation state.

$$\Omega = \left[\frac{K_2'}{K_C'} \quad \frac{Ac\ m_{Ca^{++}}}{2K_2' + 10^{-pH}} \right] \qquad (1)$$

Where:

Ω = the saturation state defined as the ratio of the concentration product of calcium and carbonate ions to the equilibrium ion product of calcite in seawater (K_C')

K_2' = the second dissociation constant of carbonic acid in seawater

$m_{Ca^{++}}$ = total molality of calcium ions (free ions plus ion pairs)

The value of K_2' at 2°C of *Lyman* (1957) was used and corrected for pressure in the *in situ* calculations according to the results of *Culberson and Pytkowicz* (1968).

At present, the major uncertainty in the calculation of saturation state is in the equilibrium ion product for calcite. Major differences exist in the experimentally determined and calculated values for the absolute value of K_C', and for the effects of temperature and pressure. The values of *Ingle et al.* (1973) and *Berner* (1976) for K_C' at 2°C were used for the calculation of saturation state under extraction conditions. Only the K_C' value of *Berner* (1976) was used for *in situ* saturation state calculations.

TABLE 2. Measured Sediment $CaCO_3$ Related Data

Core	Sampling Interval(cm)	A_T (meq/l)	pH (2°C)	Ca x 10^2 (moles/l)	Weight % $CaCO_3$
I-1	0-4	1.69	8.01	-	52
	4-9	2.01	7.74	-	64
	86-90	2.21	7.78	-	54
	170-173	2.38	7.78	-	43
I-2	0-4	2.10	7.81	-	58
	20-24	1.97	7.74	-	36
I-3	0-4	2.16	7.79	-	54
	4-8	2.00	7.78	-	33
	50-54	2.37	7.85	-	39
	86-90	2.74	7.75	-	38
I-4	0-3.5	2.07	7.70	-	47
	7-12	2.06	7.72	-	29
	24-28	2.27	7.70	-	37
I-5	0-4	2.14	7.71	-	78
	4-8	2.16	7.73	-	74
I-6	0-76	1.97	7.75	-	56
	1000-1004	2.27	7.64	-	64
I-7	0-4	2.22	7.95	-	84
	100-104	2.45	7.91	-	85
I-8	0-4	2.32	7.88	-	89
I-9	0-5	2.30	7.84	-	86
	50-54	2.47	7.76	-	79
I-10	0-4	2.25	7.76	-	89
	42-46	2.41	7.75	-	80
I-11	0-4	2.38	7.76	-	88
	43-47	2.55	7.87	-	86
I-12	0-4	2.30	7.80	-	85
II-1	0-4	2.27	7.83	1.08	67
	4-8	2.22	7.87	1.02	70
	8-14	2.23	7.90	1.05	76
	14-18	2.20	7.88	1.00	76
	18-22	2.22	7.88	1.02	76
	22-26	2.19	7.85	1.04	74

TABLE 2 (Continued)

Core	Sampling Interval(cm)	A_T (meq/l)	pH (2°C)	Ca x10^2 (moles/l)	Weight % $CaCO_3$
II-1	26-30	2.18	7.88	0.96	49
	33-37	2.13	7.84	1.04	58
	41-50	2.13	7.84	1.04	-
	70-75	2.18	7.87	1.02	54
	95-100	2.16	7.90	1.01	55
	100-120	2.15	7.86	1.02	52
	120-127	2.16	7.83	1.03	53
II-2	0-2	2.54	7.89	-	-
	2-4	2.69	7.90	-	-
	4-6	2.25	7.92	-	-
	6-8	2.26	7.91	-	-
	8-10	2.19	7.91	-	-
	10-14	2.18	7.91	-	-
	14-16	2.20	7.91	-	-
	16-20	2.17	7.89	-	-
	23-27	2.11	7.87	-	-
II-3	0-8	2.21	7.76	-	-
II-4	0-4	-	7.91	1.06	-
	4-8	2.23	7.98	-	-
	8-12	-	7.94	1.04	-
	12-16	2.29	7.98	-	-
	16-20	-	7.92	1.04	-
	20-25	2.26	7.94	-	-
	25-30	-	7.93	1.05	-
	30-35	2.21	7.95	-	-
	35-42	-	7.92	-	-
II-5	0-4	2.38	7.78	-	2
	4-8	2.36	7.74	-	2
	8-12	2.43	7.74	-	2
	12-16	2.38	7.72	-	2
	20-25	2.46	7.76	-	2
	30-35	2.53	7.72	-	3
	70-80	2.77	7.67	-	6
	90-100	3.04	7.61	-	8
	120-130	3.36	7.68	-	-
III-1	0-2	2.59	8.48	1.03	82
	2-4	2.44	8.48	1.06	85
	4-6	2.35	8.21	1.08	-
	6-8	2.26	8.31	1.08	89
	8-10	2.29	7.89	-	-
	10-12	2.39	8.00	1.00	86
	12-14	2.43	7.86	1.00	-
	14-16	2.52	7.96	-	84
	16-18	2.52	7,89	0.98	-
	18-20	2.54	8.04	1.04	85
	20-22	2.54	8.06	1.00	-
	22-24	2.48	8.07	-	-
	24-28	2.49	8.04	1.00	87

TABLE 2 (Continued)

Core	Sampling Interval(cm)	A_T (meq/l)	pH (2°C)	Ca x 10^2 (moles/l)	Weight % $CaCO_3$
III-1	34-38	2.47	7.98	0.98	87
	48-52	2.62	7.99	1.04	83
	70-74	2.62	8.09	-	88
III-2	0-2	2.36	8.05	1.00	84
	2-4	2.40	8.16	1.04	89
	4-6	2.39	8.26	1.04	-
	6-8	2.40	8.29	1.04	89
	8-10	2.45	7.94	1.00	-
	10-12	2.47	7.96	0.98	85
	12-14	2.52	7.95	1.09	-
	14-16	2.53	7.96	1.00	83
	16-18	2.58	7.90	1.04	-
	18-20	2.57	7.92	1.00	73
	20-22	2.57	7.93	1.07	-
	22-24	2.47	7.96	1.08	-
	24-28	2.60	7.94	1.00	88
	38-42	2.69	7.98	1.10	98
	52-56	2.63	7.94	-	83
	76-86	2.62	7.92	1.04	90
III-3	0-2	2.85	7.93	1.05	78
	2-4	2.40	7.91	1.04	76
	4-6	2.42	7.87	1.04	-
	6-8	2.35	7.84	1.05	70
	8-10	2.23	7.86	1.02	-
	10-12	2.22	7.88	0.97	64
	12-14	2.27	7.89	1.02	-
	14-16	2.32	7.91	0.96	62
	16-20	2.27	7.89	0.96	-
	20-24	2.23	7.93	-	-
	24-28	2.28	7.94	0.98	68
	28-32	2.27	7.98	0.95	-
	45-47	2.26	7.79	0.99	55
	60-62	2.30	7.86	0.97	41
	90-92	2.27	7.85	1.00	37
	140-142	2.34	7.98	1.09	81

The pressure correction was carried out via equation 2.

$$\log \frac{(K_C')_P}{(K_C')_1} = \frac{-\Delta V^*}{2.3RT} (P-1) \quad (2)$$

Where: ΔV^* = the partial molal volume change for the dissociation of calcite to calcium and carbonate ions in seawater
T = absolute temperature
R = gas constant
P = pressure at *in situ* depth

Values of ΔV^* are presently in dispute (e.g., see, *Takahashi*, 1975). In this paper the ΔV^* values based on the measurements and estimates of *Duedall* (1972) and *Millero and Berner* (1972) are used with the temperature coefficient of *Culberson and Pytkowicz* (1968). This value of ΔV^* is in good agreement with the most recent best estimate of ΔV^* by *Millero* (1976).

Due to uncertainty in the validity of the measured calcium concentrations, and the lack of data on calcium concentration on many of the samples, the average seawater concentration of calcium at the observed salinity has been used in all saturation calculations. From this consideration, along with the other analytical uncertainties, a total analytical uncertainty in the saturation state of approximately $\pm$ 15% (non-statistical) is probable. The results of these calculations are given in Table 3.

THE SATUROMETER PROBLEM

Perhaps the biggest problem associated with assessing the saturation state of calcium carbonate in the pore waters of deep-sea sediments is that the calcium carbonate may respond to the changes in pressure and temperature that occur during core recovery and short-term storage. In the worst possible case, the response time would be fast enough so that if the pore water were in equilibrium with the calcium carbonate on the sea floor, it would be able to maintain equilibrium with the pore water as the pressure and temperature conditions changed. If this were the case, the calculated saturation state under the conditions of pore water extraction would be 1. The percent of samples in various saturation state ranges is shown in Figure 3. Samples from cores collected above 3000 meters water depth were excluded for reasons which will be discussed later. Saturation states calculated using the value for K_C' of *Berner* (1976) or of *Ingle et al.* (1973) show a wide range of values, indicating that equilibrium is not being maintained in all samples. The observation that some values are close to equilibrium does not necessarily mean that they have reacted during recovery and storage since they could be coincidental with this model.

If the reaction rate of the calcium carbonate with the pore

TABLE 3. Calculated Sediment $CaCO_3$ Related Data

		2°C, 1 atm.				In Situ		
Core	Sampling Interval (cm)	pH	A_c (meq/1)	Ω_B	Ω_I	pH	A_c (meq/1)	Ω_B
I-1	0-4	8.01	1.64	0.82	1.16	7.79	1.63	0.28
	4-9	7.74	1.98	0.55	0.78	7.52	1.98	0.19
	86-90	7.78	2.18	0.66	0.93	7.56	2.18	0.23
	170-173	7.78	2.35	0.72	1.02	7.56	2.35	0.25
I-2	0-4	7.81	2.07	0.67	0.95	7.59	2.06	0.23
	20-24	7.74	1.94	0.54	0.76	7.52	1.94	0.19
I-3	0-4	7.79	2.13	0.67	0.95	7.57	2.12	0.23
	4-8	7.78	1.97	0.60	0.85	7.56	1.97	0.21
	50-54	7.85	2.33	0.83	1.17	7.63	2.33	0.28
	86-90	7.75	2.71	0.78	1.10	7.53	2.71	0.27
I-4	0-3,5	7.70	2.04	0.52	0.74	7.48	2.04	0.17
	7-12	7.72	2.03	0.54	0.76	7.50	2.03	0.18
	24-28	7.70	2.24	0.58	0.82	7.48	2.24	0.19
I-5	0-4	7.71	2.11	0.55	0.78	7.49	2.11	0.19
	4-8	7.73	2.13	0.58	0.82	7.51	2.13	0.20
I-6	0-76	7.75	1.94	0.56	0.79	7.51	1.94	0.16
	1000-1004	7.64	2.25	0.50	0.71	7.40	2.24	1.15
I-7	0-4	7.95	2.17	0.96	1.36	7.75	2.17	0.34
	100-104	7.91	2.41	0.97	1.37	7.71	2.40	0.35
I-8	0-4	7.88	2.28	0.87	1.23	7.72	2.28	0.40
I-9	0-5	7.84	2.26	0.79	1.12	7.66	2.26	0.31
	50-54	7.76	2.44	0.71	1.00	7.58	2.44	0.28
I-10	0-4	7.76	2.22	0.65	0.92	7.58	2.22	0.27
	42-46	7.75	2.38	0.68	0.96	7.57	2.38	0.29
I-11	0-4	7.76	2.35	0.69	0.97	7.60	2.35	0.30
	43-47	7.87	2.51	0.93	1.31	7.71	2.51	0.41
I-12	0-4	7.80	2.27	0.72	1.02	7.62	2.26	0.31
II-1	0-4	7.83	2.24	0.76	1.07	7.66	2.23	0.33
	4-8	7.87	2.18	0.81	1.14	7.70	2.18	0.35
	8-14	7.90	2.19	0.87	1.23	7.73	2.18	0.37
	14-18	7.88	2.16	0.82	1.16	7.71	2.16	0.35
	18-22	7.88	2.18	0.83	1.17	7.71	2.18	0.36

TABLE 3 (continued)

		2°C, 1 atm.				In Situ		
Core	Sampling Interval (cm)	pH	A_c (meq/l)	Ω_B	Ω_I	pH	A_c (meq/l)	Ω_B
II-1	22-26	7.85	2.15	0.77	1.09	7.68	2.15	0.33
	26-30	7.88	2.14	0.81	1.14	7.71	2.14	0.35
	33-37	7.84	2.09	0.73	1.03	7.67	2.09	0.31
	41-50	7.84	2.09	0.73	1.03	7.67	2.09	0.31
	70-75	7.87	2.14	0.80	1.13	7.70	2.14	0.34
	95-100	7.90	2.12	0.84	1.19	7.73	2.11	0.36
	100-120	7.86	2.11	0.77	1.09	7.69	2.11	0.33
	120-127	7.83	2.13	0.72	1.02	7.66	2.12	0.31
II-2	0-2	7.89	2.50	0.97	1.37	7.73	2.42	0.43
	2-4	7.90	2.65	1.05	1.48	7.74	2.64	0.46
	4-6	7.92	2.21	0.91	1.29	7.76	2.20	0.40
	6-8	7.91	2.22	0.90	1.28	7.75	2.21	0.40
	8-10	7.91	2.15	0.87	1.23	7.75	2.14	0.38
	10-14	7.91	2.14	0.87	1.23	7.75	2.13	0.38
	14-16	7.91	2.16	0.88	1.24	7.75	2.15	0.39
	16-20	7.89	2.13	0.83	1.17	7.73	2.12	0.36
	23-27	7.87	2.07	0.77	1.09	7.71	2.07	0.34
II-3	0-8	7.76	2.18	0.64	0.90	7.69	2.17	0.34
II-4	4-8	7.98	2.18	1.03	1.46	7.81	2.18	0.44
	12-16	7.98	2.24	1.06	1.50	7.81	2.24	0.46
	20-25	7.94	2.22	0.96	1.36	7.77	2.21	0.41
	30-35	7.95	2.16	0.96	1.36	7.78	2.16	0.41
II-5	0-4	7.78	2.35	0.71	1.00	7.52	2.34	0.20
	4-8	7.74	2.33	0.65	0.92	7.48	2.33	0.18
	8-12	7.74	2.40	0.67	0.95	7.48	2.40	0.19
	12-16	7.72	2.35	0.63	0.89	7.46	2.35	0.18
	20-25	7.76	2.43	0.71	1.00	7.50	2.43	0.20
	30-35	7.72	2.50	0.67	0.95	7.46	2.50	0.19
	70-80	7.67	2.75	0.66	0.94	7.41	2.74	0.19
	90-100	7.61	3.02	0.63	0.89	7.35	3.02	0.18
	120-130	7.68	3.34	0.82	1.15	7.42	3.33	0.23
III-1	0-2	8.48	2.44	3.12	4.41	8.35	2.42	1.64
	2-4	8.48	2.28	2.92	4.13	8.35	2.27	1.55
	4-6	8.21	2.27	1.72	2.43	8.07	2.26	0.90
	6-8	8.31	2.16	2.00	2.83	8.18	2.15	1.05
	8-10	7.89	2.25	0.87	1.23	7.76	2.24	0.46
	10-12	8.00	2.34	1.15	1.63	7.87	2.33	0.60
	12-14	7.86	2.39	0.87	1.23	7.73	2.39	0.46
	14-16	7.96	2.47	1.12	1.58	7.83	2.47	0.59
	16-18	7.89	2.48	0.96	1.36	7.76	2.48	0.51
	18-20	8.04	2.48	1.33	1.88	7.91	2.48	0.70
	20-22	8.06	2.48	1.38	1.95	7.93	2.48	0.73

TABLE 3 (continued)

		2°C, 1 atm.				In Situ		
Core	Sampling Interval (cm)	pH	A_c (meq/1)	Ω_B	Ω_I	pH	A_c (meq/1)	Ω_B
III-1	22-24	8.07	2.42	1.38	1.95	7.94	2.44	0.73
	24-28	8.04	2.43	1.30	1.83	7.91	2.43	0.68
	34-38	7.98	2.42	1.14	1.61	7.85	2.42	0.60
	58-52	7.99	2.57	1.24	1.75	7.86	2.57	0.65
	70-74	8.09	2.56	1.52	2.15	7.96	2.55	0.80
III-2	0-2	8.05	2.30	1.26	1.78	7.92	2.30	0.66
	2-4	8.16	2.33	1.60	2.26	8.03	2.32	0.84
	4-6	8.26	2.30	1.93	2.73	8.13	2.29	1.02
	6-8	8.29	2.30	2.05	2.90	8.16	2.29	1.08
	8-10	7.94	2.41	1.04	1.47	7.81	2.40	0.55
	10-12	7.96	2.42	1.09	1.54	7.83	2.42	0.57
	12-14	7.95	2.47	1.09	1.54	7.82	2.47	0.57
	14-16	7.96	2.48	1.12	1.58	7.83	2.48	0.59
	16-18	7.90	2.54	1.01	1.43	7.77	2.54	0.53
	18-20	7.92	2.53	1.05	1.48	7.79	2.53	0.55
	20-22	7.93	2.53	1.07	1.51	7.80	2.52	0.56
	22-24	7.96	2.42	1.09	1.54	7.83	2.42	0.57
	24-28	7.94	2.56	1.10	1.55	7.81	2.55	0.58
	38-42	7.98	2.64	1.24	1.75	7.85	2.64	0.65
	52-56	7.94	2.59	1.12	1.58	7.81	2.58	0.59
	76-86	7.92	2.58	1.07	1.51	7.79	2.57	0.56
III-3	0-2	7.93	2.81	1.19	1.68	7.71	2.80	0.40
	2-4	7.91	2.36	0.96	1.36	7.69	2.35	0.33
	4-6	7.87	2.38	0.86	1.22	7.65	2.38	0.29
	6-8	7.84	2.31	0.81	1.14	7.62	2.31	0.28
	8-10	7.86	2.19	0.80	1.13	7.64	2.19	0.27
	10-12	7.88	2.18	0.83	1.17	7.66	2.18	0.28
	12-14	7.89	2.23	0.87	1.23	7.67	2.23	0.30
	14-16	7.91	2.28	0.92	1.30	7.69	2.27	0.32
	16-20	7.89	2.23	0.87	1.23	7.67	2.23	0.30
	20-24	7.93	2.19	0.93	1.32	7.71	2.18	0.32
	24-28	7.94	2.24	0.97	1.37	7.72	2.23	0.33
	28-32	7.98	2.22	1.05	1.48	7.76	2.22	0.36
	45-47	7.79	2.23	0.70	0.99	7.57	2.23	0.24
	60-62	7.86	2.26	0.82	1.16	7.64	2.26	0.28
	90-92	7.85	2.23	0.80	1.13	7.63	2.23	0.27
	140-142	7.98	2.29	1.08	1.53	7.76	2.29	0.37

Ω_B calculated using K_C' of Berner (1976).

Ω_I calculated using K_C' of Ingle and others (1973).

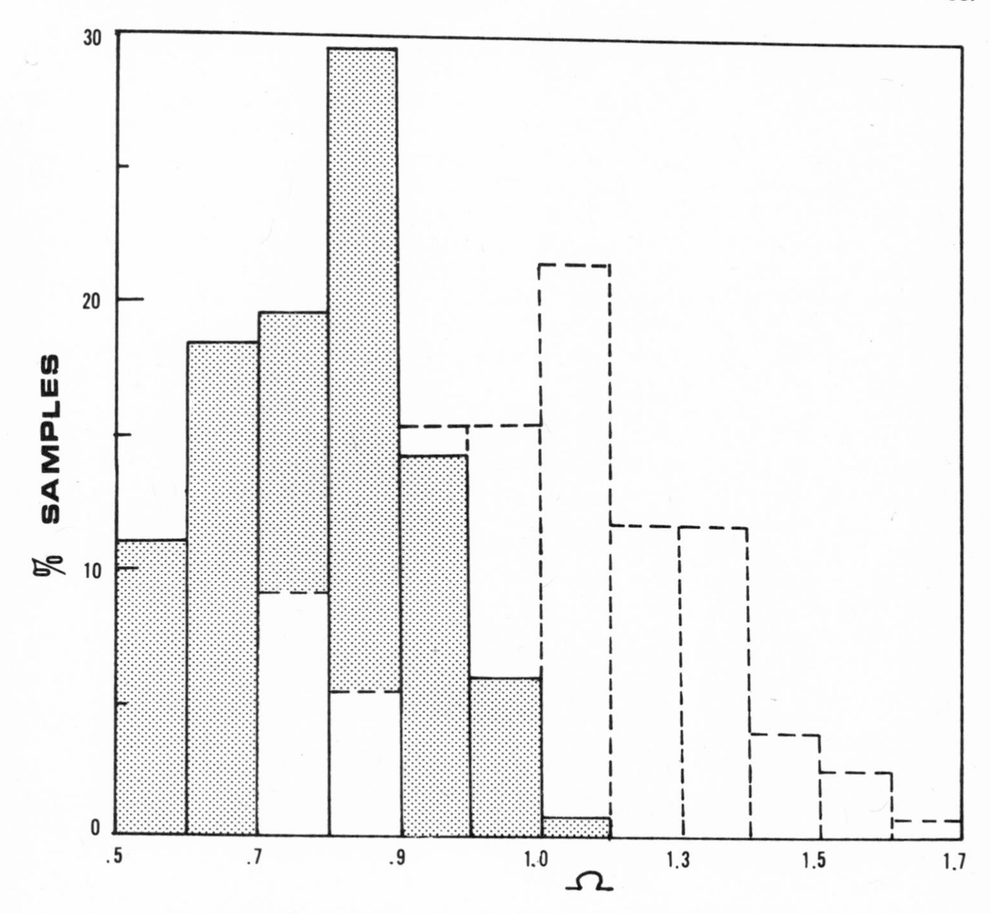

Fig. 3. The percentage of samples from cores collected below 3000 meters water depth in various saturation ranges (relative to calcite). The solid line-shaded histograms are for saturation states calculated using the K_C' of *Berner* (1976). The dashed line-open histograms are for saturation states calculated using the K_C' of *Ingle et al.* (1973).

water was not fast enough to keep up with the changing conditions, and the calcium carbonate was in equilibrium with the pore water before coring, an extration saturation state always greater than 1 would be expected. If the value for K_C' of *Berner* (1976) is used to calculate the saturation state, approximately 80% of the samples are undersaturated beyond analytic uncertainty. This means that under *in situ* conditions the pore waters would be highly undersaturated (see Table 3), and that if there were any saturometer effect it would minimize the calculated undersaturation. However, if the value for K_C' of *Ingle et al.* (1973) is used to calculate the saturation state, most of the samples are supersaturated under extraction conditions. Again, this does not necessarily mean that reaction has taken place during recovery and short-term storage since it could be coincidental with the model.

One line of evidence in general support of the validity of these measurements comes from measurements of *pH* made on pore water samples from the North Atlantic Ocean by *Sayles* (pers. comm., 1976). His pore water samples were collected using an *in situ* sampler. The use of an *in situ* sampler, while not allowing close interval sampling, eliminates the "saturometer problem". The results that were obtained indicate that the *pH* of the pore water is always less than that of the overlying water. A range of *pH* from 7.7 to 7.9 under *in situ* conditions was found. In this study, 56% of the measured *pH* values, when corrected to *in situ* conditions, fall within this range, and none are above 7.9. If a range of *pH* from 7.6 to 7.9 is used, 74% of the measured *pH* values fall within the range. Since *pH* is a very sensitive indicator, these results, while far from conclusive, are supportive of the argument that a major "saturometer effect" is not occurring.

Throughout this discussion of the "saturometer effect" it has been necessary to separate the cores taken off the ridge crest from the two cores taken on the ridge crest between 2900 and 3000 meters depth. The cores from the ridge crest are significantly less undersaturated, when corrected to *in situ* conditions, than are the other cores and they show much greater variation in saturation state. The top 2 centimeters of the shallowest core gave the only *in situ* supersaturation value (1.64) encountered in core tops in this study. A possible explanation for this, which is supported by the observation of weak aragonite X-ray diffraction peaks in these cores, is that aragonite dissolution is occurring on and within the very uppermost part of the cores and significantly modifying the general calcium carbonate chemistry. This could account for the "anomalous" behavior of these cores.

CHEMICAL TRENDS WITHIN CORES

A wide variety of trends, with increasing depth below the sediment-water interface, has been found for both *pH* and alkalinity. In most of the cores no strong trends for *pH* with depth are apparent. In the two shallowest cores, there is a definite decrease in *pH* with increasing depth. Several alkalinity trends exist. In some cores (e.g., I-3), a definite minimum occurs near the sediment-water interface. In others, the alkalinity is relatively constant (e.g., II-1) or while varying shows no definite trend (e.g., III-1). The only core which showed a strong increase of alkalinity with depth was core II-5 (see Figure 4), which is also the core which had the least calcium carbonate.

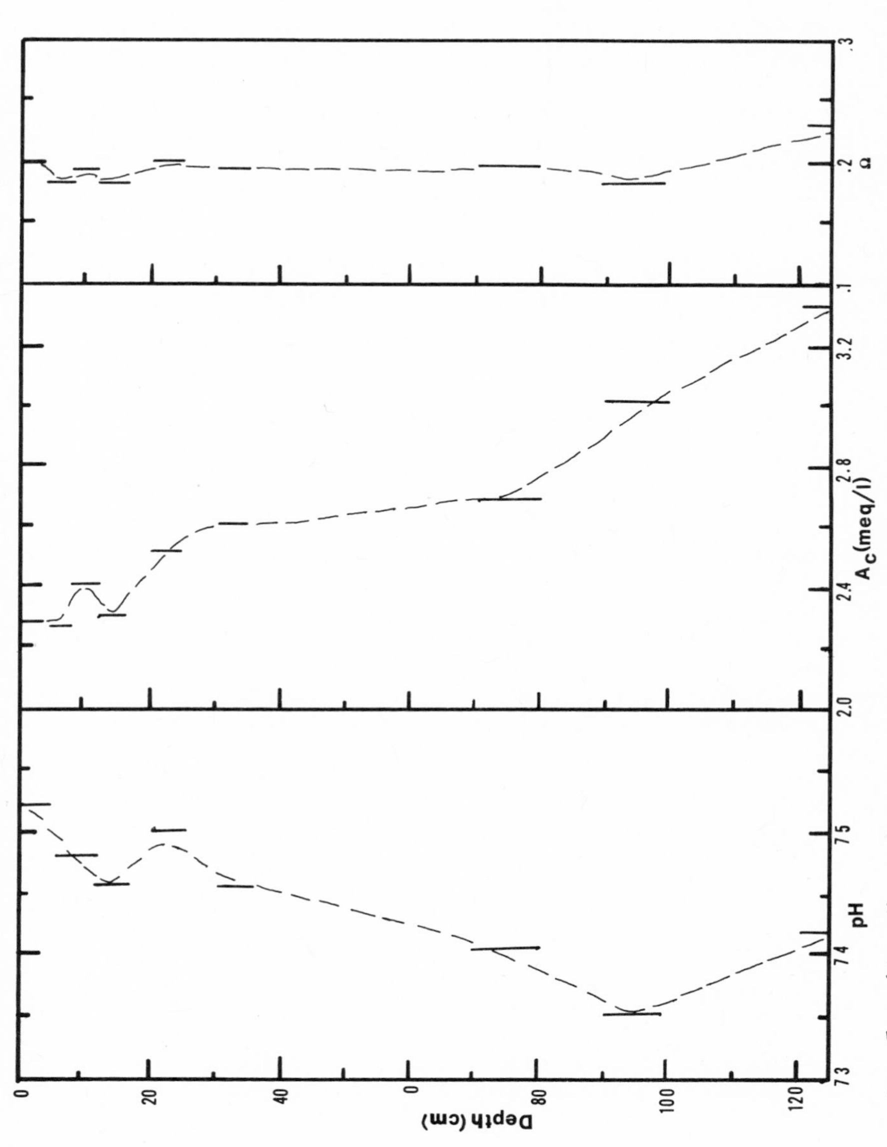

Fig. 4. *In situ pH*, carbonate alkalinity, and saturation state (Ω) found in core II-5. Note that while both *pH* and alkalinity show relatively large variations, the saturation state is close to constant.

For the cores collected below 3000 meters water depth, the saturation state, while varying between cores, has a range of only 10% within a given core, except for core III-3, where the range in saturation state is 19%. Considering that these cores exhibit quite different *pH* and alkalinity trends, widely varying calcium carbonate contents both within and between cores, and are all highly undersaturated (if the K_C' of *Berner* (1976) is used), the constancy in saturation state within a given core is quite perplexing. It may indicate that soon after deposition a saturation state is arrived at which remains fixed for long periods of time. The two cores from above 3000 meters water depth have more complex trends in their saturation states. In the shallower core, there is a large decrease in saturation state in the upper 8 centimeters; below this depth the saturation state is variable with a range of 36%. The core from 2980 meters exhibits a complex saturation state profile in which the saturation state increases in the top 8 centimeters to near equilibrium and then abruptly decreases and remains constant to within 12%. As discussed previously, the behavior of the shallow cores may be due to the presence of aragonite.

CHEMICAL TRENDS WITH WATER DEPTH

The chemical trends with water depth are presented in Figure 5. The saturation state in the water column is based on the data of *Takahashi* (1975) at 35°N in the Atlantic Ocean corrected to be consistent with the saturation state calculations for the sediment. The saturation state used for the sediments is the saturation state in the top 3.5 to 5 centimeters. In cases where data were available for both 0 to 2 and 2 to 4 centimeters, the average value for the two intervals was used. The "hollow" data points are for cores taken in a sediment pond in the central rift valley of the Mid-Atlantic Ridge (see Figure 2). These sediments contained many glass and rock fragments.

Perhaps the most important observation that can be made from Figure 5 is that with the exception of the two shallowest cores, the saturation state of the pore water is always significantly less than that of the overlying water. There is also no apparent relationship between the saturation state in the water column and the saturation state of pore water. Another observation which can be made from Figure 5 is that there is no apparent relationship between the calcium carbonate content of the sediment and its saturation state. Both of these observations lend further support to the argument that a "saturometer effect" is not strongly modifying the results.

It does not seem possible at this time to reconcile the differences between the saturation state of the water column and the sediments. If the primary problem were in the measurement of *pH* in the sediments, a consistent offset or an irregular offset related

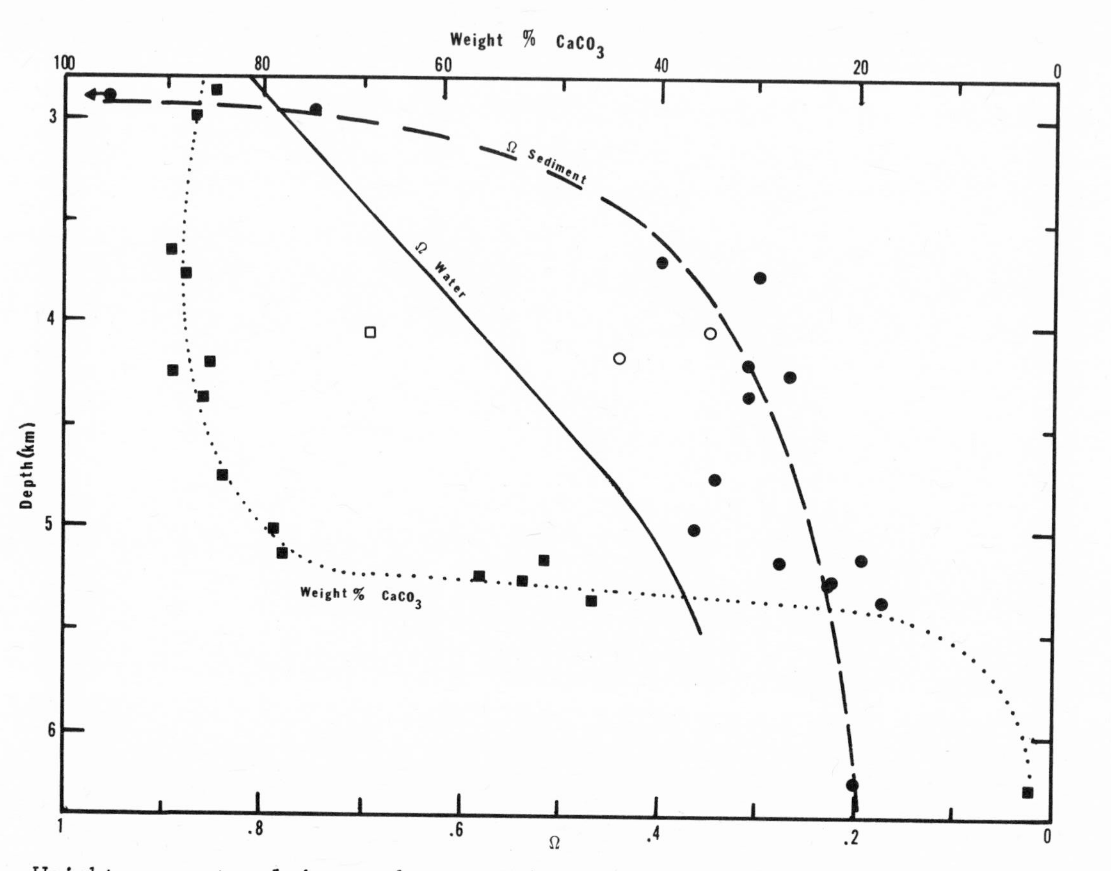

Fig. 5. Weight percent calcium carbonate and saturation state (Ω) in top 3.5 to 5 centimeters of cores and saturation state in the water column as a function of water depth. The saturation state of the water is based on the findings of *Takahashi* (1975) modified to be consistent with the constants used in calculation of the sediment saturation state.

to sediment composition arising from liquid junction potential effects would be expected. Since neither type of offset is observed, it is unlikely that *pH* measurement is the source of the differences. Simply changing the values of constants used in the saturation state calculation also fails to explain the differences, since the same constants were used to calculate the saturation state in the water column and in the sediment. Any attempt to change the constants so that the pore water is in equilibrium with calcite results in a supersaturated state persisting in the water column to depths at least in excess of the calcium carbonate compensation depth.

CONCLUSIONS

At this time it is not possible to draw definite conclusions about the carbonate chemistry of deep-sea sediment pore waters. There are two primary reasons for this. First, it is not possible to determine to what extent, if any, reaction between calcium carbonate and pore water takes place during core recovery and short-term storage. If significant reaction does take place, then the shipboard measurements do not accurately reflect the *in situ* saturation states. The second problem is that at this time it does not seem possible to determine which, if either, of the recently reported equilibrium ion products for calcite is correct. Since at 2°C the K_C' of *Berner* (1976) is approximately 40% greater than that reported by *Ingle et al.* (1973), the conclusions drawn from use of the different values for K_C' are quite different.

If future research shows that there is no significant reaction between calcium carbonate in the sediment and pore water during core reovery and short-term storage, then the generally accepted model for calcium carbonate accumulation will have to be largely revised. Instead of the rate of equilibration of pore water with calcium carbonate being the dominant controlling mechanism for accumulation, the rate of surface coating of calcium carbonate particles will be the controlling mechanism.

ACKNOWLEDGEMENTS

The author wishes to thank the crews and scientific parties of the NOAA ships R/V DISCOVERER and RESEARCHER and the R/V AKADEMIK KURCHATOV for their help in the collection and processing of cores. Sampling was carried out as part of the Trans-Atlantic Geotraverse Program, P. Rona, Director. Helpful comments and discussion were provided by Drs. R. A. Berner, W. S. Broecker, R. Scott, T. Takahashi and K. K. Turekian. Financial support was provided by NSF Grants DES74-02352-A03 and DES74-12369.

REFERENCES

Berner, R. A. 1976. The solubility of calcite and aragonite in seawater at atmospheric pressure and 34.5% salinity. *Amer. Jour. Sci.* (In press)

Broecker, W. S. and T. Takahashi. 1977. The relationship between lysocline depth and *in situ* carbonate ion concentration. *Deep-Sea Research.* In press.

Culberson, C. H. and R. M. Pytkowicz. 1968. Effect of pressure on carbonic acid, boric acid, and the *pH* in seawater. *Limnol. Oceanogr., 13.* 403-417.

Duedall, I. W. 1972. The partial molal volume of calcium carbonate in seawater. *Geochim. Cosmochim. Acta, 36.* 729-734.

Gieskes, J. M. 1974. Interstitial water studies Leg 15; Alkalinity, *pH, Mg, Ca, Si,* PO_4 and NH_4, In: Heezen, B. C., I. B. MacGregor, et al., *Initial reports of the deep sea drilling project, 20,* Washington, D. C.

Hammond, D. E. 1974. Interstitial water studies Leg 15; A comparison of major element and carbonate chemistry data from sites 147, 148 and 149, In: Heezen, B. C., I. B. MacGregor, et al., *Initial reports of the deep sea drilling project, 20,* Washington, D. C.

Ingle, S. E., C. H. Culberson, J. E. Hawley and R. M. Pytkowicz. 1973. The solubility of calcite in seawater at atmospheric pressure and 35 o/oo salinity. *Mar. Chem. 1.* 295-307.

Lyman, J. 1957. Buffer mechanisms of seawater, Ph.D. dissertation, University of California, Los Angeles, Calif. 1960.

Millero, F. J. and R. A. Berner. 1972. Effect of pressure on carbonate ion equilibrium in seawater. *Geochim. Cosmochim. Acta, 36.* 92-98.

Millero, F. J. 1976. The effect of pressure on the solubility of calcite in seawater at 25°C. *Geochim. Cosmochim. Acta.* (In press)

McGregor, B. A., and P. A. Rona. 1975. Crest of the Mid-Atlantic Ridge at 26°N. *J. Geophys. Res., 80.* 3307-3314.

Takahashi, T. 1975. Carbonate chemistry of seawater and the calcite compensation depth in the ocean. In: *Dissolution of Deep-Sea Carbonates,* W. V. Sliter, A. W. H. Bé and W. H. Berger, Eds., Cushman Foundation for Foraminiferal Research Special Pub. No. 13. 11-26.

Tsunogai, S. M. Nishimura and S. Nakaya. 1968. Complexometric titration of calcium in the presence of larger amounts of magnesium. *Talanta, 15.* 385 pp.

THE EFFECT OF BENTHIC BIOLOGICAL PROCESSES ON THE CO_2 CARBONATE SYSTEM

Allen Z. Paul

Lamong-Doherty Geological Observatory of

Columbia University

ABSTRACT

Benthic organisms effect the dissolution of calcium carbonate tests and particles through the following activities: the production of CO_2 by respiration increases the solubility of calcium carbonate; digestive processes strip organic coatings from calcium carbonate tests and shells allowing the undersaturated water to reach their surfaces; and bioturbation keeps the upper layer of sediments mixed and thus prevents calcium carbonate from being buried rapidly and lost to the system.

Bioturbation is probably the most important factor and includes the following processes: organisms moving over or through the sediment; benthos eating the sediment and egesting feces or pseudofeces, or rejecting certain size grains; tube or burrow building activities of infauna; and sediment filling open burrows. Although benthic populations are not high there may be adaptations that permit processing of large quantities of sediment allowing the organisms to obtain their required nutrition from low organic carbon deposits. The meiobenthos (organisms smaller than 0.5mm) may play an important role in the upper 1-2cm in deep-sea sediments.

INTRODUCTION

In the upper waters $CaCO_3$ is utilized by organisms, such as foraminiferans and coccolithophorids, to form their shells or tests, which upon the death of the organism sink slowly to the ocean floor. $CaCO_3$ is injected into the ocean by terrestrial weathering of Ca-containing rocks, possibly by dissolution of calcareous tests as they fall to the bottom (*Pytkowicz*, 1973), and from deep-sea sediments (*Edmond*, 1974). Sediments containing more than 30% $CaCO_3$ cover approximately one-third to one-half of the deep ocean floor (*Revelle and Fairbridge*, 1957; *Rezak*, 1974). These deposits are an important factor in the recycling of $CaCO_3$ and the activities of benthic organisms are necessary in the CO_2 cycle.

RESPIRATION

Respiration by benthic animals produces CO_2, which in turn increases the rate of solution of carbonate tests. One study consisting of 10 measurements of *in situ* community respiration at 1850 m (*Smith and Teal*, 1973) showed that biological respiration is two orders of magnitude lower at this depth than in shallow water (e.g. 0.50 ml/m^2/hr compared to over 40 ml/m^2/hr). But, because porous carbonate sediments contain an abundance of microorganisms (*Meadows and Anderson*, 1968) and are rich in organic aggregates (*Suess*, 1968), a higher population of benthos, and therefore a higher respiration rate, may be attained at depth. Consequently *in situ* measurements of respiration need to be made in deep-sea deposits before an accurate evaluation of the contribution of benthic metabolic activity to CO_2 production and eventual test dissolution can be achieved. *Craig* (1971) modeled the total CO_2-^{14}C-^{13}C-O_2-alkalinity system and concluded that an active, although slow, metabolism does take place in the deep-sea.

REMOVAL OF ORGANIC COATINGS

It is known that organic coatings on $CaCO_3$ shells and tests inhibits their dissolution during descent through the water column and after they settle on the bottom. Deposit feeding benthic organisms strip this encrusting matter exposing the tests to the action of the undersaturated water. *Johnson* (1974) recently studied particulate matter at the sediment-water interface in coastal environments in a novel way. He pointed out that most studies of animal-sediment relations have tried to associate bulk properties of the sediment with the distribution of particular species of organisms, and that the conventional methods of analyses, the drying and sieving of a sample, will destroy fecal pellets and change the original state of coagulation of the sediment. To avoid this, *Johnson* (1974) conducted a microscopic study of fresh marine deposits attempting to identify materials present from a biological point of view. Using a variety of histological stains he concluded that 61% of the particles at the surface of the sediment were potential food sources to the benthos, although bulk analyses revealed only a few percent organic carbon by weight.

BIOTURBATION

Sediment deposited on the ocean floor is mixed by animals living in it and on it, this process being termed bioturbation (*Richter*, 1952). The abundance of biogenic structures on deep-sea sediments has been shown in innumerable bottom photographs. *Ewing and Davis* (1967) set forth a classification scheme based on the geometry of the tracks, trails, holes, and mounds observed. *Heezen and Hollister* (1971) show hundreds of bottom pictures with a variety of organisms and their traces (Figures 1 and 2).

Rowe et al., (1974) placed a time-lapse camera at a depth of 360m at the head of the Hudson Canyon for 42 hours. The camera was set to take a picture about every 4 seconds. 526 bottom photographs were obtained. Recognizable benthic organisms were seen in the $1m^2$ field during about 3 hours of the total time. A crab, an ophiuroid, an octopus, a burrowing decapod crustacean and a small worm were recognized. The crustaceans made the deepest tracks and it appeared that the ophiuroid and worm probably had a smoothing effect. They concluded that the organisms were a major contributing factor to the high rates of erosion and sediment reworking in this area of the Hudson Canyon.

Fig. 1. Bottom relief resulting from biologic activity. (Taken by B. C. Heezen from the D. S. V. Turtle). By permission of author.

Fig. 2. An asteroid beginning to burrow in sediment. Note sediment covering edges of arms. (Taken by B. C. Heezen from the D. S. V. Turtle). By permission of author.

Heezen and Hollister (1971) used bottom photographs to deduce the rate of movement of holothurians. The fecal casts of large elasipod sea cucumbers have a characteristic shape and are easily recognized. They calculated that the holothurians egest feces about every 10m and probably twice a day (*Crozier*, 1918) so that their rate of movement is about 50cm per hour. *Rowe* (1974) estimated that the feces have a density of about $2gm/cm^3$ and a volume of about $25cm^3$, indicating that 100g of sediment must pass through their gut per day. In the deep, sea animals of this size occur about once per $100m^2$, and they turn over 1g of sediment per square meter per day.

Charles Darwin (1881) established that burrowing earthworms play an important role in mixing the top layers of soil. Similar effects can be seen in any shallow water environment inhabited by marine burrowers such as worms, clams, and crabs. However, it was not until recently (*Bramlette and Bradley*,1942) that it has been demonstrated that mud-feeding animals play a significant part in reworking sediment, even on the abyssal sea floor. Burrow structures and unpattered mixed layers in deep-sea cores show that intense biologic reworking of abyssal sediments occurs on a worldwide basis (*Ekdale*, 1974).

There have been a number of investigations of the methods by which benthic infauna rework, or mix, the sediment. One of the earliest studies of these animal-sediment relationships (*Rhoads*, 1963) determined the rate of reworking by a clam (*Yoldia limatula*, Figure 3) found in Buzzards Bay, Massachusetts, and Long Island Sound.

Y. limatula feeds with more than half of its shell buried in the substrate. Sediment is brought into the mantle cavity by a pair of feeding palps which extend beneath the surface. Ciliary action transports particles to a sorting area near the mouth where large grains are rejected, moves them to the mantle, and then intermittently ejects them from the excurrent siphon. Smaller particles are diverted to the mouth, passed through the gut, and the undigested portion is egested in compact fecal pellets. This clam changes the character of the substrate by transferring material from below the surface to mounds above the surface, by binding some of the sediment into fecal pellets, and by its motion through the substrate as it moves from one area to another.

In laboratory experiments conducted at controlled temperatures *Rhoads* (1963) collected the ejected sediment and measured its volume. Temperature had a strong effect on the reworking rate. Using mean monthly temperatures and the known density of *Y. limatula*, it was claculated that the sediment deposited yearly in Buzzards Bay (2.3mm/yr) is turned over twice each year and that deposited in Long Island Sound is turned over several times per year, by this clam alone.

Rhoads (1967) also examined some polychaete worms to determine the rate and depth of their sediment reworking. *Pectinaria gouldi* constructs a conical tube of agglutinated sand grains. This 5-6cm long tube is buried in the sediment with its posterior end extending

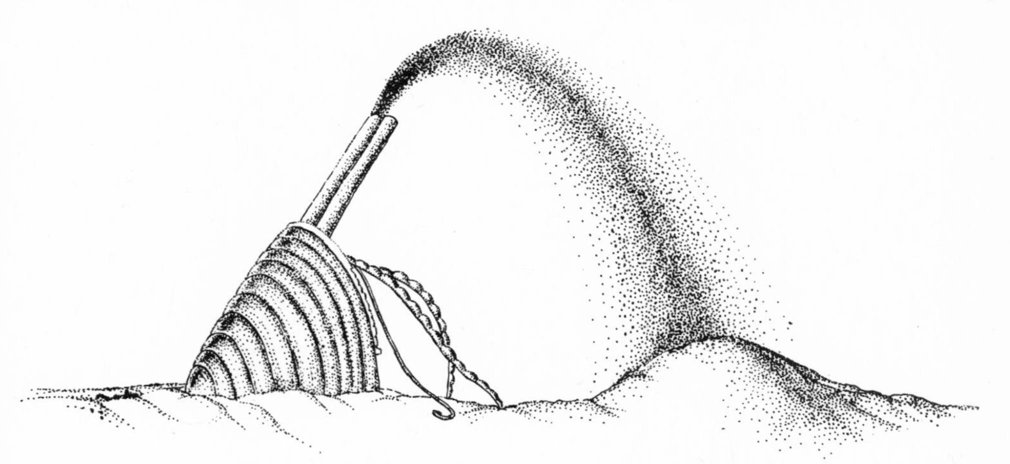

Fig. 3. Feeding and sediment reworking behavior of the clam, *Yoldia limatula*.

just above the surface. The worm ingests particles smaller than 1mm at its anterior end and rejects larger particles, concentrating them at about 6cm below the sediment-water interface. It was experimentally determined that *P. gouldi* would completely rework 670cm^2 to this depth in a year.

Another worm, *Amphitrite ornata*, builds a U-shaped burrow over 0.5cm in diameter, about 30cm long and approximately 20cm deep. Its numerous ciliated tentacles spread in a circle about 10-15cm in diameter at the oral end of the tube. These tentacles bring particles to the center, where larger grains are rejected, building a mound of coarse granules. Smaller grains are ingested, passed through the digestive system, and feces are egested at the anal end. Thus, each worm produces two mounds, one of larger particles and one of fine grained feces. Because these mounds were easily eroded by current action the rate of turnover could not be determined accurately. However, it was considered to be high.

Rhoades and Young (1970) studied a burrowing holothurian, *Molpadia oolitica*, to determine its sediment reworking capabilities. These organisms burrow vertically into the sediment with the anal end extending above the sediment surface and the mouth about 20cm below. Sediment is ingested and passed up through the gut with the uningested portion forcefully ejected as a fecal sludge from the cloaca. A cone-like mound is built around the anus with slumping taking place at the edge of these cones which are 10-30cm in diameter and 2-3cm high. These cones and depressions form a micro-relief that influences the distribution of other organisms. Tubes of a suspension feeding polychaete, *Euchone incolor* are found in heavy concentrations only on the mounds and are absent, or are in low density, in the depressions between mounds. *Rhoads and Young* (1970) were not able to calculate reworking rates in their study, but report that *Yamanouchi* (1927) noted that a similar species, *Caudina chilensis*, passes about 160g of sediment/day through its digestive system.

The "man-in-the-sea" Projects Tektite I and Tektite II offered a unique opportunity to study bioturbation *in situ* for a total time of about 80 days (*Clifton and Hunter*, 1973). One of their experiments consisted of making ripple-like structures, each 100cm long, 5cm high, and 20cm apart in an area where there was no bottom wave action. At the end of 37 days no ripple marks were discernible. They also implanted colored sand layers 2mm thick at 3-5cm depth and 10-12cm depth in four locations. Each site was sampled with a box corer every 4 days, which showed that mixing in the four sites varied. However, during the 36 day experiment, the uppermost centimeter was well mixed, with some burrows extending as deep as 4cm. Their previous study (*Clifton and Hunter*, 1973) showed mixing of the uppermost 5cm and some burrows as deep as 20cm. In nearly every core there was a thoroughly mixed upper layer, overlying a partly mixed zone, which in turn was overlying a zone in which mixing occurs only within individual burrows. The thickness of these layers varied from site to site and it was noticed that in some locations burrowers moving primarily in a horizontal direction did most of the reworking, while in other sites organisms moving vertically dominated the mixing. Taking all their data into account, they determined that 225,000 kg of sand was reworked yearly in a strip 10m wide bordering the 500m long reef front near the habitat.

In an investigation of sediment-fauna interaction in deeper water (2800-2900m) off the Washington-Oregon coast, *Griggs et al.*, (1969) collected piston cores for the study of burrows, and organisms for identification of probable burrowers. Three, 6m long, cores from the axis of the Cascadia Channel showed alternating layers of olive green silts and gray clays. Three of the layers in each core showed numerous, distinct burrows to depths of 25cm and the other 6 layers showed burrows only in the upper 5-10cm. The type and size of burrows were correlated with the organisms collected and it was postulated that: various sizes of polychaete worms could be responsible for all size burrows seen (0.1mm-20mm); burrowing pelecypods were responsible for burrows about 10mm in size; and holothurians and echinoids created burrows about 20mm in diameter.

Clarke (1968) examined burrow frequency in abyssal cores by using color variations seen within the sediment, which he called "mottles", and used as an indication of post-depositional animal activity. He noticed a range of 20-50 mottles per 10cm in carbonate rich layers, while elsewhere in the cores there was a range of 0-15 mottles per 10cm. This is additional evidence that there may be faunistic differences between carbonate and non-carbonate deposits.

Recently, *Gerard* (pers. comm.) placed two bottom ocean monitors (BOM) on the sea floor at 5000 m depth in the North Equatorial Pacific Ocean. Time-lapse cameras were one of the pieces of equipment attached to the monitors, approximately 1200 photographs have been examined. One camera took pictures every 4 hours, the other every 2 hours.

The area shown in each frame was about 1.9 m^2, and a number of organisms moved past the camera. Elasipod holothurians were the most common organisms seen. Other recognizable animals were crabs and sea urchins. The holothurians moved in a typical meandering pattern at about 10 cm/hr leaving a slight trail that could only be seen by extremely close examination. Fecal piles left by a passing holothurian began to disintegrate in about a month, but were still easily recognizable. The crabs, most of whom appeared in only one frame and whose speed therefore could not be determined, left the most visible tracks of perhaps 1 cm deep indentation where their walking appendages touched the sediment. The tracks that were made then tended, over a period of weeks, to disappear, probably due to infaunal activity.

It appears that at least 35-50% of the visible area was changed by animal activity over a period of six weeks. However, it seems to be a steady state condition with certain areas showing more perturbation as others become smoother. The depth in the sediment to which these changes take place remains unknown, but future experiments with the BOM should elucidate this point.

It is well known that there are fewer organisms per m^2 in the deep ocean than in shallow water and, therefore, is assumed that their effect can not be important. However, these bottom dwellers, that in general live on sediments low in organic carbon, may well be adapted to processing extremely large quantities of material in order to extract sufficient food. An abyssal clam, *Abra profundorum*, shows morphological changes in its feeding palps, gills, and hind-gut, all modifications thought to enable it to handle, in the words of the investigators, (*Allen and Sanders*, 1966) "vast quantities" of bottom sediment.

To this point, only the effect of the macrobenthos, organisms larger than 0.5mm have been discussed. The meiobenthos were recently also found to have a profound effect on lebensspuren (*Cullen*, 1973). It was noticed that tracks and trails left in aquaria after individual macrobenthonents had been removed disappeared after the termination of some experiments. New, controlled experiments were initiated and it was found that all tracks and trails were obliterated after 10-14 days. The sediment was examined and found to have a vigorous population of meiobenthic fauna consisting mainly of ostracods and nematodes. Even though these animals are small and may not be present in large numbers in the deep-sea, their bioturbational effects in this low energy environment with its concomitant low sedimentation rates may well be significant.

CONCLUSIONS

1. The respiration of benthic organisms produces CO_2 that will increase the rate of dissolution of $CaCO^3$.
2. Benthonents can strip organic coatings from $CaCO_3$ particles making their surfaces available to the action of the corrosive water.
3. Bioturbation results from: animals moving over or through the sediment; deposit feeders eating sediment and egesting feces; burrow or tube building activities of organisms; and sediment filling in open burrows.
4. Macrofaunal deposit feeders that may have adaptations to process large quantities of sediment low in nutritive value, and the meiobenthos in general probably keep the reworking rate high, even in the deep sea.
5. There are indications that carbonate depositions support a more active burrowing fauna than non-carbonate sediments.

ACKNOWLEDGMENTS

Contribution No. 2495 from Lamont-Doherty Geological Observatory of Columbia University, Palisades, New York 10964. I would like to thank W. S. Broecker and C. Garside for helpful conversations and criticisms of this paper. This work was partially funded by the National Oceanic and Atmospheric Administration contract number 03-6-022-35141.

REFERENCES

Allen, J. A., and H. L. Sanders, 1966. Adaptations to abyssal life as shown by the bivalve *Abra profundorum* (Smith), *Deep-Sea Res.*, *13*, 1175-1184.

Bramlette, M. N., and W. G. Bradley, 1942. Geology and biology of North Atlantic deep-sea cores between Newfoundland and Ireland. I. Lithology and geologic interpretations, *U. S. Geol. Survey, Profess. Papers No. 196A*, 1-34.

Clarke, R. H., 1968. Burrow frequency in abyssal sediments, *Deep-Sea Res.*, *15*, 397-400.

Clifton, H. E., and R. E. Hunter, 1973. Bioturbation rates and effects in carbonate sand, St. John, U.S. Virgin Islands, *J. Geol.*, *81*, 253-268.

Craig, H., 1971. The deep metabolism: oxygen consumption in abyssal ocean water, *J. Geophys. Res.*, *76*, 5078-5086.

Crozier, W. J., 1918. The amount of bottom material ingested by holothurians (Stichopus), *J. Exp. Zool*, *26*, 379-389.

Cullen, D., 1973. Bioturbation of superficial marine ingested by interstitial meiobenthos, *Nature*, *242*, 323-324.

Darwin, C., 1881. The formation of vegetable mould through the action of worms, John Murray, London.

Edmond, J. M., 1974. On the dissolution of carbonate and silicate in the deep ocean, *Deep-Sea Res.*, *21*, 455-480.

Ekdale, A. A., 1974. Geologic history of the abyssal benthos: evidence from the trace fossils in the Deep-Sea Drilling Project Cores, Ph.D. thesis, Rice University, Houston Texas, 156pp.

Ewing, M., and R. A. Davis, 1967. Lebenssupren photographed on the ocean floor, In: *Deep-Sea Photography*, 259-294.

Griggs, G.B., A.G. Carey, and L.D. Kulm, 1969. Deep-sea sedimentation and sediment-fauna interaction in Cascadia Channel and on Cascadia Abyssal Plain, *Deep-Sea Res.*, *16*, 157-170.

Heezen, B. C., and C. D. Hollister, 1971. The face of the deep, Oxford University, Press, N.Y.

Johnson, R. G., 1974. Particulate matter at the sediment-water interface in coastal environments, *J. Mar. Res.*, *32*, 313-330.

Meadows, P. S., and J. G. Anderson, 1968. Microorganisms attached to marine sand grains, *J. Mar. Biol. Ass. U.K.*, *48*. 161-176.

Pytkowicz, R. M., 1973. The carbon dioxide system in the ocean, *Schweizer, Zeitschr. Hydrol.*, *35*, 8-28.

Revelle, R., and R. Fairbridge, 1957. Carbonates and carbon dioxide, *Geological Society of America Memoir*, *67*, 239-296.

Rezak, R., 1974. Deep-sea carbonates, In: *Deep-Sea Sediments*, 453-461.

Rhoads, D. C., 1963. Rates of sediment reworking by *Yoldia limatula* in Buzzards Bay, Massachusetts, and Long Island Sound, *J. Sediment. Petrol.*, *33*, 723-727.

Rhoads, D. C., 1967. Biogenic reworking of intertidal sediments in Barnstable Harbor and Buzzards Bay, Massachusetts, *J. Geol.*, *75*, 461-476.

Rhoads, D. C. and K. K. Young, 1970. The influence of deposit-feeding organisms on sediment stability and community tropic structures, *J. Marine Res.*, *28*, 150-176.

Richter, R., 1952. Fluidal-Textur in Sediment-Gesteinen and über Sedifluktion überhaupt, Notizbe. Hess. L.-Amt Bodenforsch., 3, 67-81

Rowe, G. T., 1974. The effects of the benthic fauna on the physical properties of deep-sea sediments, In: *Deep-Sea Sediments*, 381-400.

Rowe, G. T., G. Keller, H. Edgerton, N. Staresinic, and J. MacIlvaine, 1974. Time lapse photography of the biological reworking of sediments in Hudson Submarine Canyon, *J. Sediment. Petrol.*, *44*, 549-552.

Smith, K. L., and J. M. Teal, 1973. Deep-sea benthic community respiration: an *in situ* study at 1850 meters, *Science*, *179*, 282-283.

Suess, E., 1968. Calcium carbonate interaction with organic compounds, Ph.D. thesis, Lehigh Univ., Bethlehem, Pa. 153pp.

Yamanouchi, T., 1927. Some preliminary notes on the behavior of the holothurian, *Caudina chilensis* (J. Muller), *Sci. Rep. Tohoku Univ.*, *(Ser.4) 2*, 85-91.

BENTHIC MIXING IN DEEP SEA CORES AS DETERMINED BY ^{14}C DATING AND ITS IMPLICATIONS REGARDING CLIMATE STRATIGRAPHY AND THE FATE OF FOSSIL FUEL CO_2[1]

T.-H. Peng[1], W. S. Broecker[1], G. Kipphut[1] and N. Shackleton[2]

[1]*Lamont-Doherty Geological Observatory, Columbia Univ., and* [2]*University of Cambridge*

ABSTRACT

A sereis of radiocarbon dates on Indian Ocean core V19-188 reveals that this core was vertically mixed to a depth of 9 cm. Core top ^{14}C dates from a number of other areas also indicate that the sediment has been mixed to a depth ranging from 6 to 18 cm. This mixing has important implications with regard to the fate of fossil fuel CO_2. A numerical model has been developed to study the effect of sediment mixing on radiocarbon dates, carbonate concentrations, oxygen isotopes and fauna abundances. This model indicates that the deviation of apparent ^{14}C ages from the true age ranges from 530 years to 1800 years for core V19-188, with the maximum deviation at the climatic transition depth.

INTRODUCTION

It is the consensus of many who have studied deep sea sediments that they are mixed to a depth on the order of 5-30 cm by biological activity and the bottom currents (*Arrhenius*, 1963; *Berger and Heath*, 1968; *Hanor and Marshall*, 1971; *Ruddiman and Glover*, 1972; *Guinasso and Schink*, 1975). The most convincing evidence for sediment mixing is based on the dispersion of volcanic ash and microtektite layers (*Ruddiman and Glover*, 1972; *Glass*, 1969, 1972; *Glass et al.*, 1973). A review of some of the biological mechanisms producing this turbation is presented by *Paul* (1977). Models of the turbation process have been proposed by *Berger and Heath* (1968) and by *Guinasso and Schink* (1975). In this paper yet another line of evidence is reported for the extent of benthic sediment mixing based on radiocarbon dating.

To evaluate the effect of sediment mixing on radiocarbon dates and on some of the climatic indices in the deep sea sediments, a numerical version of the Berger-Heath mixing model has been developed. Core V19-188 from the Indian Ocean was chosen for this mixing study. This core was subjected to a variety of measurements including radiocarbon dating, carbonate concentrations, fauna abundances and oxygen isotopes.

THE EXTENT OF BENTHIC MIXING AS DEDUCED FROM ^{14}C DATING

Core V19-188 was taken from the Mid-Indian Ridge (06°52'N; 60°40'E) at a depth of 3356 meters during the 1963 cruise of the R/V VEMA. Radiocarbon measurements were made on bulk carbonate at various depths in the core. Using 0.95 times the $^{14}C/^{12}C$ ratio of carbon dioxide obtained from NBS oxalic acid as a control (*Broecker and Olson*, 1959), ages were calculated for the sampled segments. The results of radiocarbon dates are listed in Table 1 and are plotted in Figure 1 with least square best fit lines. A change in the slope of the straight line is noticed at the depth of 23 cm (10,200 years). This indicates that the sedimentation rate decreases from 3.5 cm per 1,000 years during the last glacial to 2.3 cm per 1,000 years during the present interglacial. A similar break in sedimentation rate across the transition from the last glacial to the current interglacial (Termination I) has been reported by *Broecker et al.* (1958).

TABLE 1. Radiocarbon Ages for Core V19-188

LDGO Sample No.	Sample Depth (cm)	Mid-point (cm)	Age (yr)
1383F	0-3	1.5	2,960±150
1391T	5-7	6.0	3,450±150
1391A	7.5-9	8.25	3,630±160
1391P	11-13	12.0	4,920±180
1391Q	19-22	20.5	9,000±280
1391B	24-26	25.0	10,840±360
1991R	30-33	31.5	12,730±340
1391S	37-39	38.0	14,000±490
1391C	41-43	42.0	15,690±750
1391D	47-49.5	48.25	17,780±600
1383G	54-56	55.0	18,600±880
1391E	61-63	62.0	21,700±1550

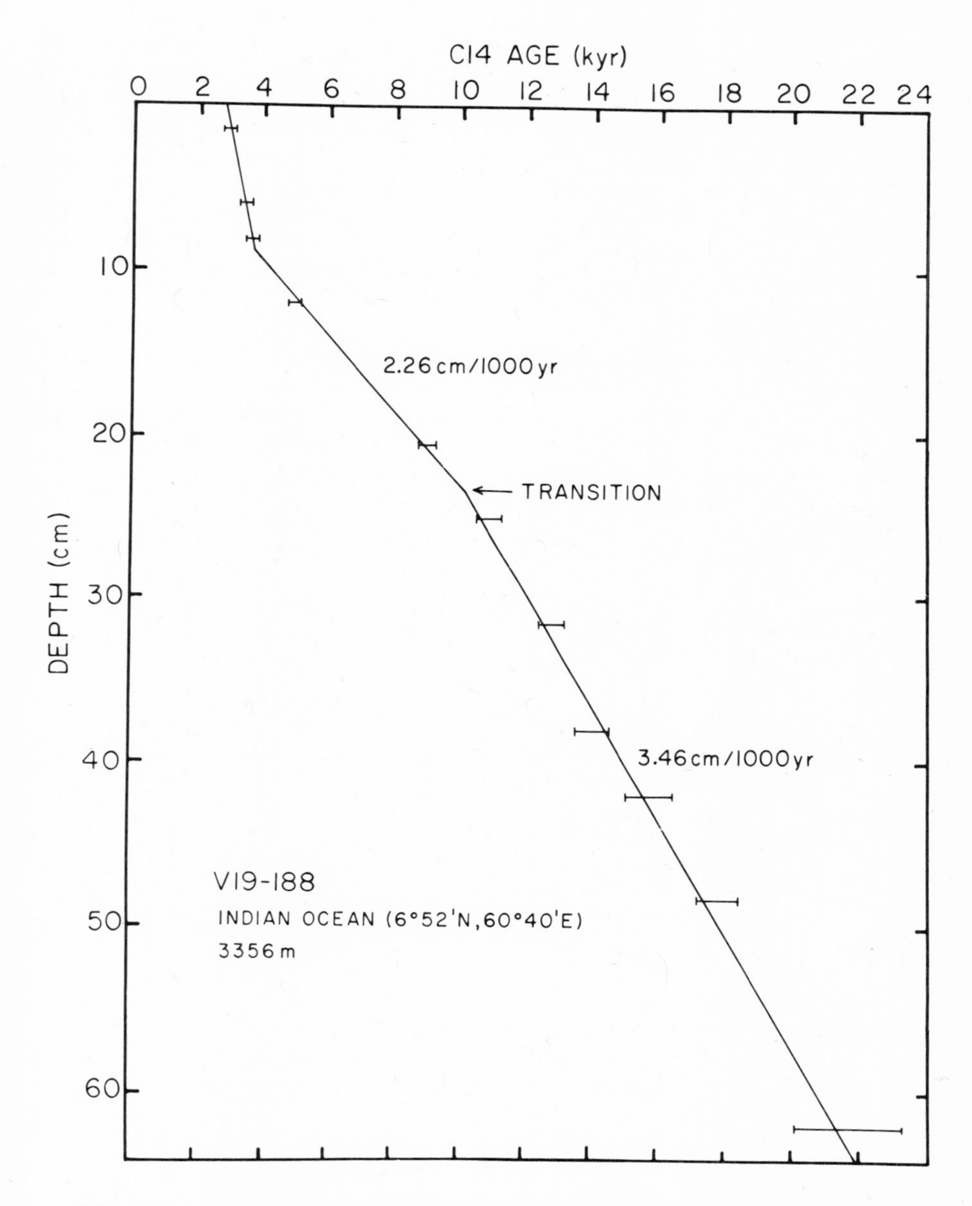

Fig. 1. Radiocarbon dates for core V19-188 (Mid-Indian Range, 06°52'N, 60°40'E). The apparent transition from glacial to interglacial sedimentation rate took place at 23 cm (10,200 yrs).

The core top ^{14}C age is 3,000 years instead of zero age. In fact, three samples between 0 and 9 cm appear to have similar ages. This is a clear indication of sediment mixing. The mean age in this top section is 3,350 years. This is not unusual. Almost all core top ages, found in the literature, of the cores with sedimentation rates slower than 5 cm/10^3 yr are older than 3,000 years (see Table 2). The classical interpretation has been that this high age reflects missing core top material. It is easy to conceive of mechanisms that might remove the top few centimeters of the piston core (impact splash, drag along sides of core pipe, extrusion loss, etc.). However, as long as part of the sediment mixed layer is preserved, the dated ^{14}C age should be close to the mean age of the mixed section. Thus, even if a few centimeters of material have been lost, the magnitude of the core top age carries information regarding the depth of mixing.

In order to understand the relationship between the core top age and the depth of mixing, a relationship has been derived between these two parameters and the Holocene sedimentation rate. It is based on the mass balance of radiocarbon in the mixed layer as shown in Figure 2. The input flux of radiocarbon to the layer can be expressed as

$$\left.\frac{^{14}C}{C}\right)_{SO} \times R_{CaCO_3} \qquad (1)$$

where $^{14}C/C)_{SO}$ is the ratio of ^{14}C atoms to ^{12}C atoms in the surface ocean and R_{CaCO_3} is the carbonate deposition rate ($gm/cm^2\ 10^3$ yrs). the loss of radiocarbon from the layer is by carbonate burial beneath the base of the mixed layer and by the radioactive decay of ^{14}C within the layer. If carbonate dissolution is assumed to be negligible, the total output flux can be given as

$$\left.\frac{^{14}C}{C}\right)_{ML} \times R_{CaCO_3} + \left.\frac{^{14}C}{C}\right)_{ML} M \rho f_{CaCO_3} \lambda \qquad (2)$$

where $^{14}C/C)_{ML}$ is the ^{14}C ratio in the mixed layer, M is the thickness of the mixed layer, ρ is the density of the sediment, f_{CaCO_3} is the fraction of carbonate and λ is the ^{14}C decay constant. At steady state we have

$$\left.\frac{^{14}C}{C}\right)_{SO} R_{CaCO_3} = \left.\frac{^{14}C}{C}\right)_{ML} R_{CaCO_3} + \left.\frac{^{14}C}{C}\right)_{ML} M \rho f_{CaCO_3} \lambda \qquad (3)$$

Solving for M we obtain

TABLE 2. Summary of Mixed Layer Thickness Derived from Core Top ^{14}C Ages and Holocene Sedimentation Rates

Core No.	Sed. Rate[a] (cm/10^3 yr)	Core Top Age (yr)	Mixed Layer[b] Thickness M (cm)	Reference
ERDC92-BX-2	1.6	4440	9	Unpublished LDGO ^{14}C dates
V16-36	1.8	5940	16	Stuiver, 1969
A240-ML	2.1	3850	10	Rusnak et al., 1963
A180-74	2.2	3630	10	Broecker et al., 1958
RC14-35	2.2	4200	12	Unpublished LDGO ^{14}C dates
V19-188	2.3	3000	8	This work
V29-183K	2.5	3860	12	Ruddiman and McIntyre, 1973
A180-73	2.5	2960	9	Ericson et al., 1961
A179-4	2.7	3950	14	Ericson et al., 1961
V12-66	2.8	4320	16	Stuiver, 1969
RC9-150	3.3	3200	13	Unpublished LDGO ^{14}C dates and Conolly, 1967
MG60-17	3.3	1515	6	Ostlund et al., 1962
A172-6	3.5	3700	16	Ericson et al., 1961
CP-28	3.7	4870	25	Rusnak et al., 1964
C-16	4.1	3400	17	Thommeret and Thommeret, 1966
C-18	4.6	3050	17	Thommeret and Thommeret, 1966
MG60-4	7	1770	14	Ostlund et al., 1962
MG58-6	8	1660	15	Rusnak et al., 1963
R10-10	9	4260	48	Ericson et al., 1961
MG57-11	10	845	9	Rusnak et al., 1963
MG57-18	10	300	< 4	Rusnak et al., 1963
MG60-20	10	2425	28	Ostlund et al., 1962
MG60-18	12	1360	18	Rusnak et al., 1963
A179-15	13	850	12	Ericson et al., 1961

[a]Sedimentation rates were calculated from the submixed layer Holocene ^{14}C dates by dividing the thickness of overlying sediment by the ^{14}C age. Except for the box cores (ERDC92-BX-2 and V29-183K) and the well dated piston core V19-188, 5 cm has been added to take into account the probable loss of material during coring. The uncertainty in these rates (and hence also in the computed mixed layer depths) generally lies in the range of 10 to 25%.

[b]To be included in this list the calculated thickness of the mixed layer must be internally consistent with the core top ^{14}C date, namely, the core top point must lie within the mixed layer. For this reason we have not included the following cores: A254-BR-C (Rusnak et al., 1964), MG60-14, MG61-1, MG61-2 (Ostlund et al., 1962), LJF-71 (Hubbs and Bien, 1967), and A180-48 (Ericson et al., 1961).

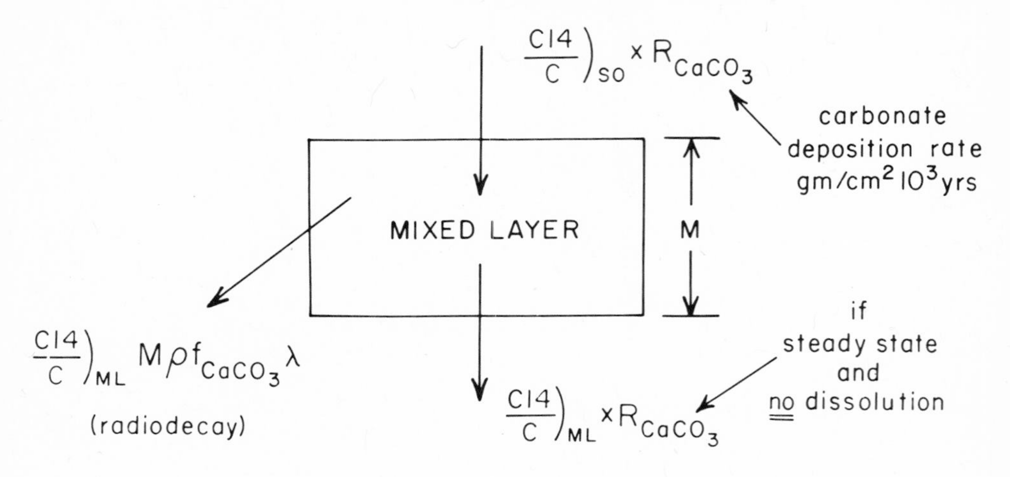

$$R_{CaCO_3}\left[\frac{C14}{C}\Big)_{SO} - \frac{C14}{C}\Big)_{ML}\right] = \frac{C14}{C}\Big)_{ML} M\rho f_{CaCO_3}\lambda$$

total deposition rate cm/10^3yrs

$$M = \frac{R_{CaCO_3}}{\rho f_{CaCO_3}\lambda}\left[\frac{C14/C)_{SO}}{C14/C)_{ML}} - 1\right] = \frac{S}{\lambda}(e^{\lambda t} - 1)$$

Fig. 2. Radiocarbon mass balance at the steady state in the sediment mixed layer.

$$M = \frac{R_{CaCO_3}}{\rho f_{CaCO_3}\lambda}\left(\frac{{}^{14}C/C)_{SO}}{{}^{14}C/C)_{ML}} - 1\right) \tag{4}$$

Since the total sediment deposition rate S (cm/10^3 yrs) is given by

$$S = \frac{R_{CaCO_3}}{\rho f_{CaCO_3}} \tag{5}$$

and the radiocarbon decay equation is

$$\left.\frac{^{14}C}{C}\right)_{ML} = \left.\frac{^{14}C}{C}\right)_{SO} e^{-\lambda t} \qquad (6)$$

where t is the core top age, the above equation (4) can be rewritten as

$$M = \frac{S}{\lambda} (e^{\lambda t} - 1) \qquad (7)$$

M values have been calculated from the core top age and the Holocene deposition rate for the 24 cores listed in Table 2. With the exception of three cores (CP-28, R10-10, and MG60-20) the values of M range from 6 to 18 cm. Although unlikely, it is possible that the entire mixed layer was lost during coring. If so, the calculated thickness of sediment mixed layer is only an upper limit.

For core V19-188, the calculated M is 8.4 cm whereas the radiocarbon profile indicates that a mixed layer of 9 cm is present in this core. This internal consistency implies that no more than a centimeter or two of material was lost during the coring process.

Carbonate dissolution above the lysocline (as in core V19-188) is considered to be insignificant. The effect of dissolution would be to increase the thickness of the mixed layer calculated from the core top age and Holocene sedimentation rate. If 10% carbonate dissolution is assumed, the calculated M would be 10.2 cm rather than 8.4 cm for core V19-188.

For comparison of the mixed layer thicknesses derived from ^{14}C core top ages with those derived from some other methods, a brief review of mixing depth is proper. *Laughton* (1963) studied bottom photographs from stations in the North Atlantic Ocean and the Gulf of Mexico. He concluded that the sediments were mixed to a mean depth of about 5 cm. In his studies of mixing across unconformities, *Arrhenius* (1963) stated that worms may burrow as deep as 20-30 cm, but that the mean mixing depth was about 5 cm. *Donahue* (1971) examined photographs of freshly split cores and noted that burrows were evident to distances of 10-30 cm away from contacts between sediments of different colors. He stated that the mixing zone in oceanic sediments ranged from 10-30 cm with a mean value of 20 cm. *Guinasso and Schink* (1975) calculated mixing parameters from microtektite data by using their mixing model and obtained mixed layer thicknesses ranging from 17 cm to 48 cm. Under the assumption of instantaneous mixing in the mixed layer of Berger and Heath's model, the same set of microtektite data gave the thickness of the mixed layer as ranging from 8 to 20 cm. From this review, it is evident that the thickness of the mixed layer is on the order of 5 to 30 cm. The ^{14}C-derived values of 6 to 18 cm are certainly consistent with previous work. We conclude that a mixed layer of 10 cm is a good approximation for sediment mixing.

Support for mixing of this order also comes from an examination of over fifty oxygen isotope records obtained as a stratigraphic contribution to the CLIMAP project (*CLIMAP*, 1976). The distribution of cores is shown in Figure 3. In many parts of the ocean the record of oxygen isotopic changes in planktonic foraminifera is primarily a reflection of changing ocean isotopic composition. To the extent that this is true, the record, if unaffected by post-depositional mixing, would be similar in all sediment cores. This is even more likely if benthonic foraminifera are analyzed. In reality, mixing ensures that the isotopic extremes are lost in slowly accumulating sediment. In Figure 4 we plot the maximum range in oxygen isotopic composition seen in each core studied against average accumulation rate for the core. For high accumulation rates and benthonic foraminifera, the range is, within analytical limits, a constant 1.8 per mil. Among planktonics there is more scatter, reflecting the fact that there is some temperature contribution present in a few of the areas studied (for example, core V28-14 shows a range of over 2.5‰, because surface temperatur at that site in the southern Norwegian Sea was substantially colder 18,000 years ago than it is today). In slowly accumulating sediments having a sedimentation rate of less than 1 cm per thousand years, we see half or even less than half the true glacial-interglacial isotopic range.

Using the model described below the glacial-interglacial isotopic range expected for various sedimentation rates has been calculated. Curves for mixing depths of 6, 10, and 14 cm are shown in Figure 4. The mixing depth of 12±4 cm necessary to explain the isotopic data is consistent with that based on ^{14}C dating.

IMPLICATIONS REGARDING THE FATE OF FOSSIL FUEL CO_2

The CO_2 generated by the combustion of fossil fuel will be in part neutralized by the dissolution of $CaCO_3$ in deep sea sediments. This process will be limited by the build up of a non-carbonate residue at the sea water-sediment interface. Once this residue obtains a thickness of a few millimeters the dissolution process will be slowed to a negligible pace, in the absence of mechanical stirring of the sediment. If this is so, the sediments will be capable of neutralizing only a small fraction of the CO_2 that is likely to be produced. On the other hand, if benthic stirring to a depth of 10 cm operates on the time scale of decades, then the residue will be homogenized throughout the mixed layer and the dissolution will proceed until non-carbonate residue accumulates one mixed layer in thickness. In this case, the amount of per unit area of sea floor available for dissolution, H, will be given by the expression

$$H = \rho M \frac{\delta}{1 - \delta} , \qquad (8)$$

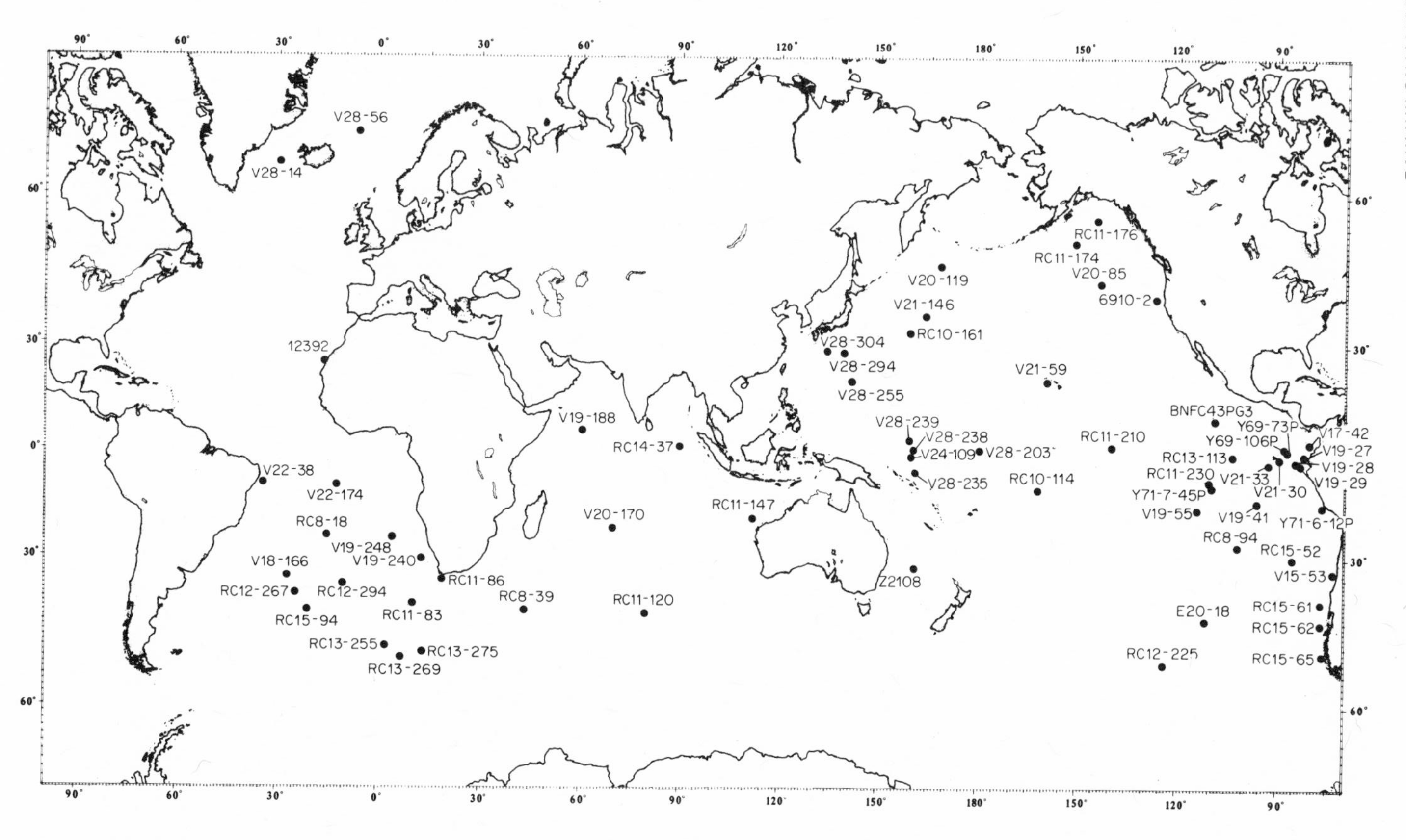

Fig. 3. Locations of CLIMAP cores. The maximum ranges in oxygen isotopic composition of these cores since the last glacial period are plotted in Figure 4.

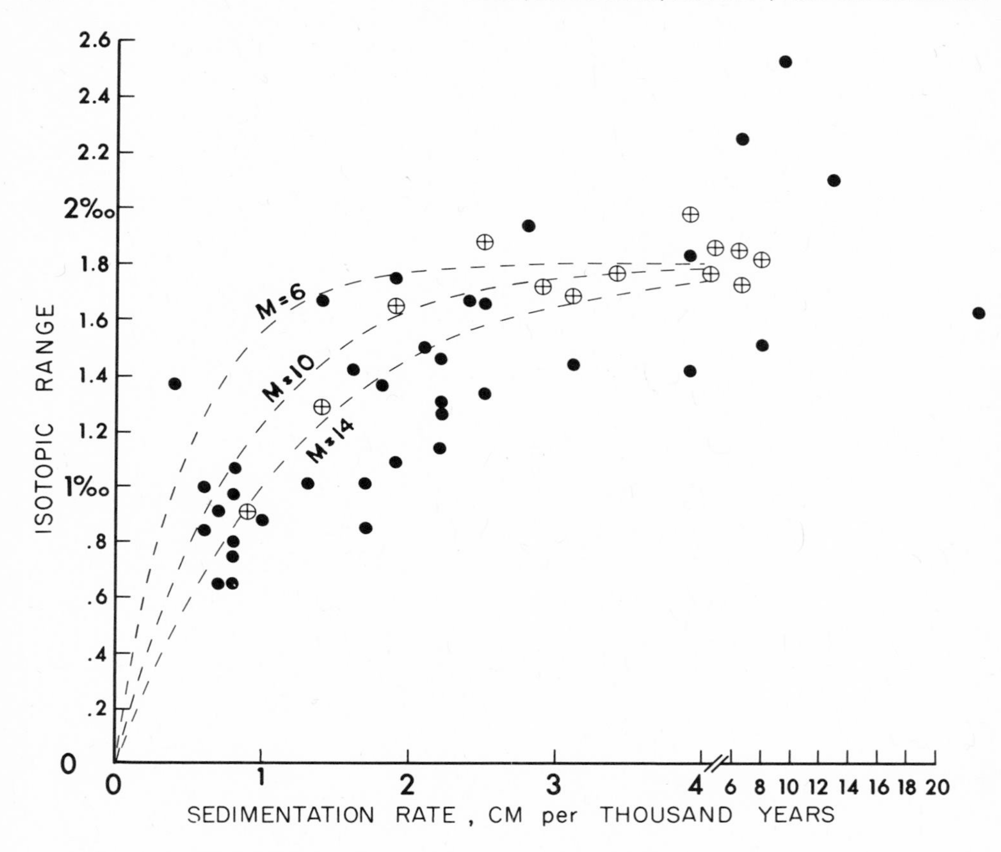

Fig. 4. The maximum oxygen isotopic range as a function of sedimentation rates. The solid circles are planktonic foraminifera, and the open circles are benthonic foraminifera. The dashed lines are calculated from the mixing model with various mixing depth M.

where f is the fraction of $CaCO_3$ in the sediment, M is the mixed layer thickness and ρ the dry density of the sediment. Taking M to be 10 cm, ρ to be 1 gm/cm^3, the amount of $CaCO_3$ available in a sediment for which f is 0.9 would be 90 gm/cm^2. For a sediment with 50% $CaCO_3$ the value would be 10 gm/cm^2 and in one with 10% the amount would be only 1 gm/cm^2. The total amount of $CaCO_3$ available for dissolution has been estimated to be about equivalent to the CO_2 locked up in our fossil fuel reserves (*Broecker and Takahashi*, 1977).

IMPLICATIONS TO CLIMATE STRATIGRAPHY

The correlation of deep sea sediment stratigraphy in the late Pleistocene (up to 40,000 years B.P.) relies heavily on radiocarbon chronology. What is the magnitude of deviation of the observed radiocarbon dates from the real radiocarbon dates caused by the sediment mixing? How will the interaction between sediment mixing and variations in sedimentation rates affect the true distribution of climatic indicators in the sediment such as the depth distribution of fauna, of carbonate content and oxygen isotope ratios? To answer these questions, a mixing model is needed that can evaluate the mixing effect on a continuously accumulating sediment pile whose input parameters change with time.

A number of quantitative models of vertical sediment mixing have been proposed (*Goldberg and Koide*, 1962; *Berger and Heath*, 1968; *Guinasso and Schink*, 1975). *Berger and Heath* (1968) described the effect of mixing on the vertical distribution of a species after its sudden appearance and disappearance. In their eddy diffusion model, *Guinasso and Schink* (1975) estimated the rate of stirring within the mixed layer to be between 1 and $100\ cm^2\ Kyr^{-1}$ by using the distribution of impulse tracers such as microtektites. Although the Guinasso and Schink model is capable of treating the redistribution of continuous inputs, to date these authors have considered only instantaneous inputs. To study the results of continuous inputs of material a numerical model is needed which can be more readily applied to various complex situations.

The mixed layer of the model is divided into M sublayers. For convenience, each sublayer is assigned a thickness of 1 cm. Sediment accumulation is approximated by the instantaneous addition of sublayers separated by time intervals of $1/S$, where S is the sedimentation rate (in $cm/10^3$ yrs). During the time interval between the addition of two successive sublayers the mixed layer is well stirred and ^{14}C decay takes place. Each time a new sublayer is added the lowest existing sublayer is removed and becomes an isolated entity, thereafter immune from the effects of mixing. The mixing rate is assumed to be fast enough so that the material in the mixed layer is homogeneous at the time the bottom sublayer is removed from the model.

As is now being used, the model is simply an extension of the *Berger and Heath* (1968) approach. There would be no difficulty, however, introducing finite mixing rates between sublayers. If such rates were uniform to the depth of the mixed layer, the model would become an extension of the *Guinasso and Schink* (1975) approach. In future work we will be inclined, however, to the approach which will abandon the concept of a fixed mixed layer and introduce a mixing coefficient which decreases with depth in the core. This could readily be done by introducing depth-dependent transfer rate coefficients between successive sublayers.

In modeling the glacial to interglacial transition it is

assumed that steady state glacial conditions have been achieved in the core at the time of Termination I. To do this the modeling is commenced at a depth well below Termination I. Layers are added one by one using the mean glacial $^{18}O/^{16}O$ ratio, faunal abundance and $CaCO_3$ content. The $^{14}C/^{12}C$ ratio of the $CaCO_3$ is that of present day surface ocean water. In this way a steady state ^{14}C content in the "GLACIAL" mixed layer is achieved by the time Termination I is reached. Termination I is assumed to be rectangular. In other words, the last glacial layer is added with mean glacial characteristics and the first post-glacial layer with mean Holocene characteristics. This assumption is made not because it is believed that this is necessarily the case, but rather as the logical starting point for such a study. If the transition between sediment bearing full glacial and that containing full interglacial characteristics can be explained entirely by mixing, then it can be said that the actual change must have been sufficiently close to rectangular so as not to have produced any gradations beyond those explainable by mixing alone.

In order to model the radiocarbon distribution in a core, proper thicknesses of the mixed layer, the depth of the transition between glacial and interglacial conditions, the glacial rate of sedimentation, and the interglacial rate of deposition must be chosen. As shown before, the core top material is well preserved in the core V19-188. The best estimate of the mixing depth in this core is 9 cm. The glacial rate of 3.5 cm/10^3 yrs is assumed to be unaffected by the mixing process. The other two parameters are derived by iteration. The best combination of these variables for core V19-188 is as follows:

transition depth 20 cm

interglacial sedimentation rate 2.0 cm/10^3 yrs

The resulting model derived apparent ^{14}C age versus depth plot is shown in Figure 5, along with the actual ^{14}C ages determined for this core. The agreement is satisfactory.

From the sedimentation rates the ^{14}C age versus depth curve which would have been obtained if no mixing had occurred can be reconstructed. The deviation of apparent ^{14}C ages after mixing from the true ages before mixing is shown in Figure 6. The maximum deviation appears at the transition depth. The apparent ^{14}C age of the sediment is 1,800 years younger than the true age. At the bottom of the mixed layer, the apparent age is 1,200 years younger than the true age. Beyond one mixing depth below the transition level, the deviation remains at a constant value of 530 years.

The Holocene sedimentation rate obtained by modeling (i.e., 2.0 cm/10^3 yrs) is about 15% lower than that obtained from the best line fit to the submixed layer Holocene ages (i.e., 2.3 cm/10^3 yrs).

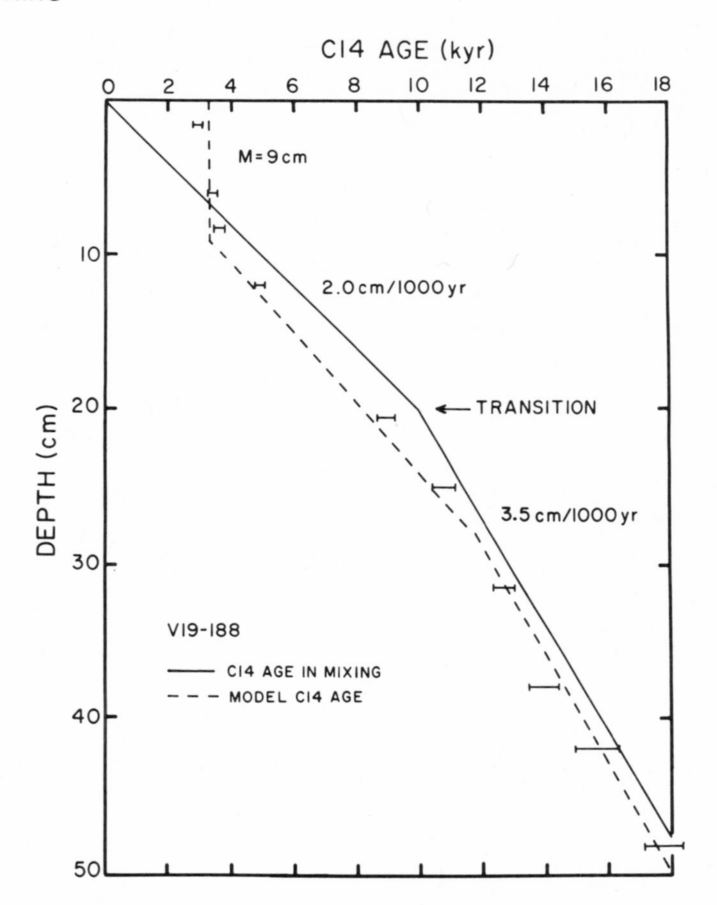

Fig. 5. Effect of mixing on ^{14}C dates. The following parameters are used: (1) mixed layer thickness = 9 cm, (2) transition depth = 20 cm, (3) glacial sedimentation rate = 3.5 cm/1000 yrs, (4) interglacial sedimentation rate = 2.0 cm/1000 yrs. The solid line is ^{14}C age without mixing and the dashed line with mixing.

The age of the transition obtained by modeling is 10,000 years at a depth of 20 cm, as opposed to 10,200 years at a depth of 23 cm obtained by the method used by *Broecker et al.* (1958) (i.e., the intersection of straight line segments joining the Holocene points and the glacial points on a ^{14}C age versus depth plot).

Figure 7 shows the results of model calculation for the expected $CaCO_3$ distribution (using a mean glacial value of 69% and a mean interglacial value of 84%). Although the fit is not bad, a better one would have been achieved if the transition

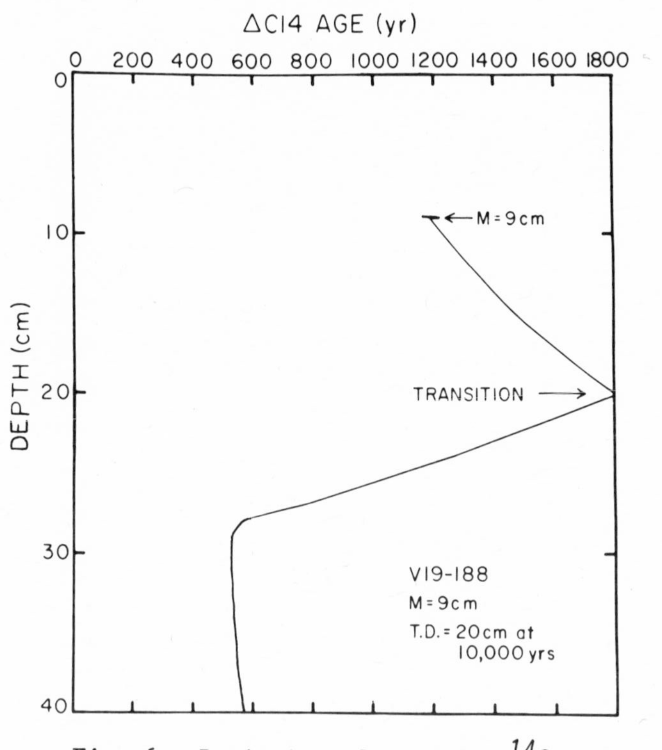

Fig. 6. Deviation of apparent, ^{14}C ages from the true ages caused by the sediments mixing process. $\Delta^{14}C$ age = true age - apparent age.

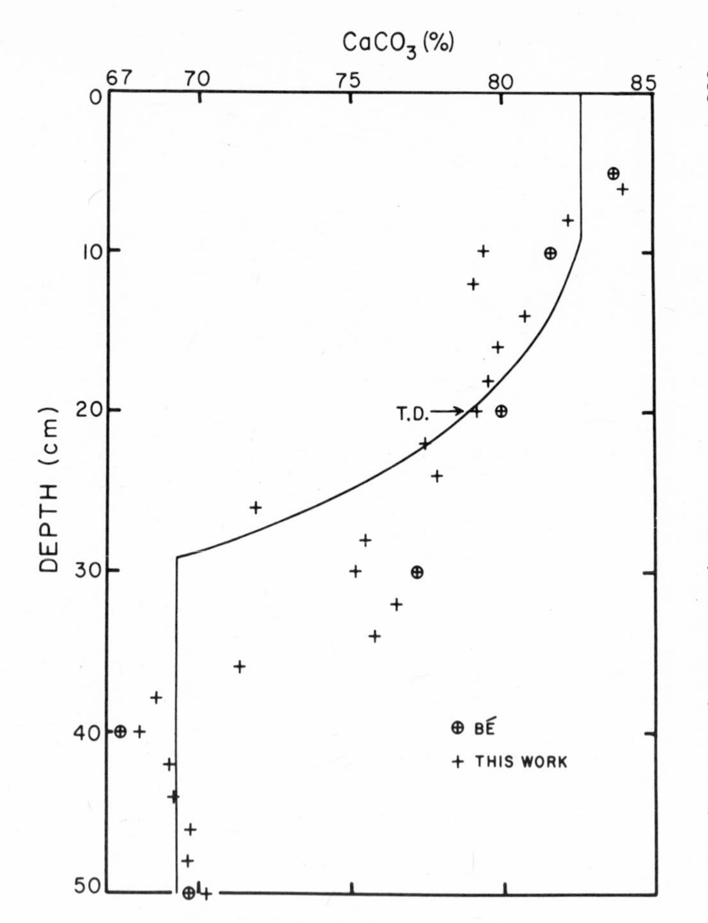

Fig. 7. $CaCO_3$ distribution in core V19-188. The curve indicates the expected $CaCO_3$ distribution from the model calculation by assuming M = 9 *cm* and the transition depth (T.D.) = 20 *cm*.

depth were moved down a few centimeters and the mixing depth increased by a few centimeters.

In Figure 8 a similar plot is given for the $^{18}O/^{16}O$ ratio in *G. sacculifer*. From coarse fraction data and faunal abundance data (see Table 3) it is estimated that the glacial abundance of this species (speciments per unit volume of sediment) is about 6.2 times lower than its post-glacial abundance. In this case the model fits the limited amount of data available quite well.

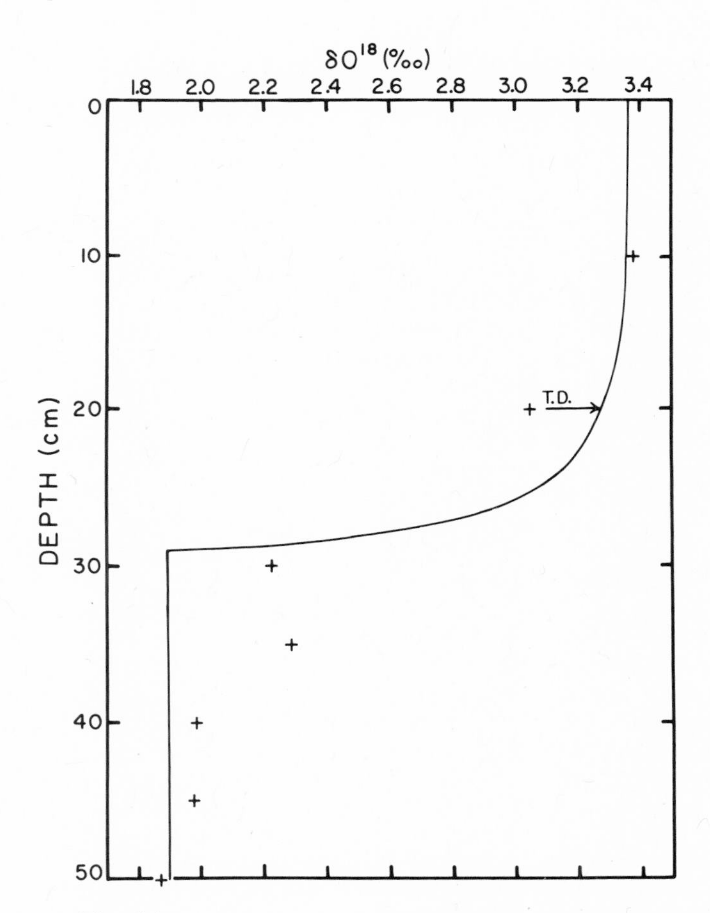

Fig. 8. $^{18}O/^{16}O$ ratio in *G. sacculifer* for core V19-188. The curve indicates the expected $\delta^{18}O$ variations from the model calculation (M = 9 cm, transition from $\delta^{18}O$ of 1.9°/oo for glacial to 3.4°/oo for post-glacial took place at 20 cm and the ratio of interglacial to glacial abundance, per gram of sediment, of *G. sacculifer* is 6.2).

TABLE 3. *G. sacculifer* Abundance in Core V19-188*

	Holocene	Glacial
Coarse Frac. (>149μ)%	27	13
G. sac./Forams >149μ	.14	.04
G. sac. (>149μ)%	3.7	0.6

$$\text{Ratio } \frac{(G.\ sac.)_H}{(G.\ sac.)_G} = 6.2$$

* Data provided by Allan Bē, Lamont-Doherty Geological Observatory

Had the mid-point of the climatic change been estimated from the mid-point of the $CaCO_3$ shift and the apparent ^{14}C ages, a result of about 11,500 years would have been obtained. Had this been done for the mid-point of the $^{18}O/^{16}O$ change, an age of about 11,000 years would have been obtained. The best estimate from our model is about 10,000 years. Thus, whereas the effects are somewhat self compensating, they are not completely so. Biases of 1,000 years can easily creep in.

Using the same information used for the analysis of core V19-188, a theoretical oxygen isotope stratigraphy has been calculated (Figure 9) to show the apparent leads and lags resulting if mixing is superimposed on a 40,000 year rectangular climatic cycle. The sedimentation rate is taken to be $2\ cm/10^3$ years so that the warm period, as shown in the figure, proceeds for 20,000 years before the next cold period of 20,000 years starts. The model curve (with $M = 10\ cm$) shows that the oxygen isotope changes very rapidly to the warm period value. A much slower change occurs as the climate becomes colder. If the mid-point of the observed ^{18}O transition had been taken to be the time of the climatic event, it would have been sooner than the true one for the transition from cold to warm, and later than the true one for that from warm to cold. Apparent leads or lags between various climatic indices could easily be generated in this way. Also the apparent shape of climatic changes derived from a single index could be quite misleading.

The model adopted here is undoubtedly too simple in that it assumes:

1) A rectangular change from glacial to interglacial conditions.

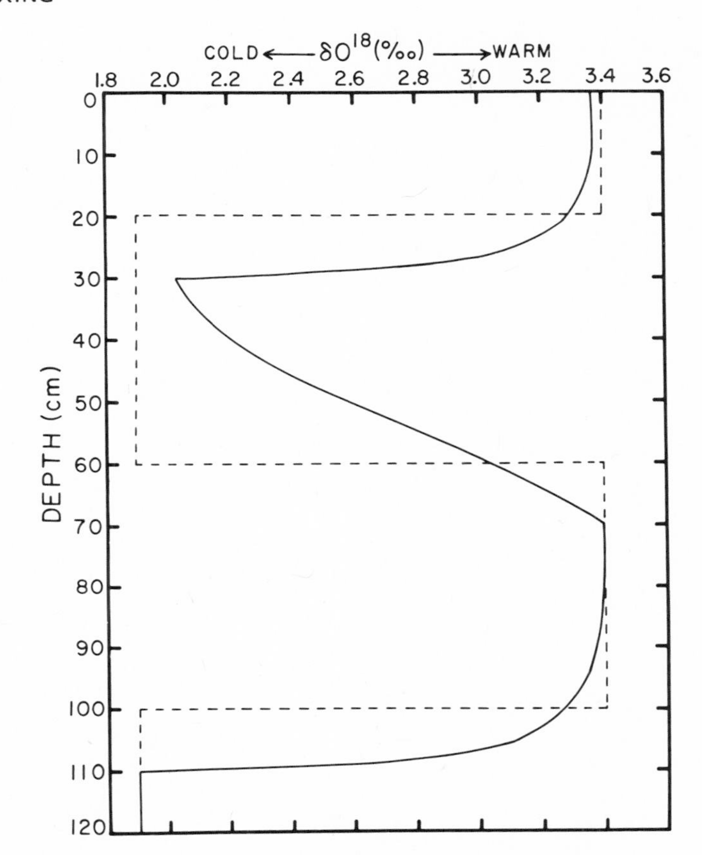

Fig. 9. Model of $^{18}O/^{16}O$ stratigraphy (*G. sacculifer*) for a 40,000 year climatic cycle. The dashed line is the distribution of $\delta^{18}O$ without mixing, and the solid line is the $\delta^{18}O$ distribution with mixing. The sedimentation rate is assumed to be 2 cm/10^3 yrs. The following mixing parameters are used: M = 10 cm, $\delta^{18}O$ cold = 1.9%, $\delta^{18}O$ warm - 3.4%, *G. sacculifer* warm/*G. sacculifer* cold = 6.2, and transitions at 20, 60, and 100 cm.

2) No phase lag between the change in sedimentation rate, in $^{18}O/^{16}O$ ratio, in fauna abundance, and calcium carbonate content.
3) Rapid stirring within a mixed layer of finite thickness.
4) No change in mixed layer thickness, sedimentation rate, calcium carbonate content, $^{18}O/^{16}O$ ratio, and faunal abundances during Holocene time.

However, as the model does take into account the first order effects, there is no doubt that it does portray the kind of errors which might occur if the conventional treatment of ^{14}C ages and property changes is used. The mixing effects can produce apparent leads and lags. They can also produce systematic age biases.

This study suggests that layer by layer ^{14}C dating of deep sea cores from core top to about 20,000 years must be done on a sizable number of cores before realistic second generation models can be constructed.

ACKNOWLEDGEMENTS

We thank A. Bē and L. Lott for providing ^{14}C dating material and some unpublished data on coarse fraction, faunal abundance and $CaCO_3$ distribution in core V19-188. N. Evensen aided in computer programming and M. Friedmann in laboratory measurements. M. Zickl typed the manuscript and P. Catanzaro drafted the figures. Financial support came from the National Science Foundation Grant DES 74 21412 and by Grant E(11-1) 2185 from the Energy Research and Development Administration to Lamont-Doherty Geological Observatory.

REFERENCES

Arrhenius, G. O. S. 1963. Pelagic sedimentation, In: *The Sea, vol. 3,* M. N. Hill, Ed., John Wiley, New York. 655-727.

Berger, W. H. and G. R. Heath. 1968. Vertical mixing in pelagic sediments. *J. Marine Res. 26(2):* 134-143.

Broecker, W. S. and E. A. Olson. 1959. Lamont radiocarbon measurements VI. *Am. J. Sci. Radiocarbon Suppl. 1:* 111-132.

Broecker, W. S., K. K. Turekian and B. C. Heezen. 1958. The relation of deep sea sedimentation rates to variations in climate. *Am J. Sci. 256:* 503-517.

Broecker, W. S. and T. Takahashi. 1977. Neutralization of fossil fuel CO_2 by marine calcium carbonate, In: *The Fate of Fossil Fuel* CO_2, N. R. Andersen and A. Malahoff (Eds.), Plenum, N. Y., (this volume).

CLIMAP. 1976. The surface of the ice-age earth. *Science 191:* 1131-1144.

Conolly, J. R. 1967. Postglacial-glacial change in climate in the Indian Ocean. *Nature 214:* 873-875.

Donahue, J. 1971. Burrow morphologies in north-central Pacific sediments. *Mar. Geol. 11:* M1-M7.

Ericson, D. B., M. Ewing, G. Wollin, and B. C. Heezen. 1961. Atlantic deep-sea sediment cores. *Bull. Geol. Soc. Am. 72:* 193-286.

Glass, B. P. 1969. Reworking of deep-sea sediments as indicated by vertical dispersion of the Australasian and Ivory Coast microtektite horizons. *Earth Planet. Sci. Letters 6:* 409-415.

Glass, B. P. 1972. Australasian microtektites in deep-sea sediments, Antarctic Oceanology II, The Australian-New Zealand sector. *Antarctic Res. Ser. Vol. 19*, B. E. Hayes, Ed., AGU, Washington, D. C. 335-348.

Glass, B. P., R. N. Baker, D. Storzer and G. A. Wagner. 1973. North American microtektites from the Caribbean Sea and their fission track age. *Earth Planet. Sci. Letters 19*: 184-192.

Goldberg, E. D. and M. Koide. 1962. Geochronological studies of deep-sea sediments by the thorium-ionium method. *Geochim. Cosmochim. Acta 26*: 417-450.

Guinasso, N. L. and D. R. Schink. 1975. Quantitative estimates of biological mixing rates in abyssal sediments. *J. Geophys. Res. 80*: 3032-3043.

Hanor, J. S. and N. F. Marshall. 1971. Mixing of sediment by organisms, In: *Trace Fossils*, B. F. Perkins, Ed., School of Geoscience, Louisiana State University, Baton Rouge. 127-135.

Hubbs, C. L. and G. S. Bien. 1967. La Jolla natural radiocarbon measurements V. *Radiocarbon 9*: 261-294.

Laughton, A. S. 1963. Microtopography, In: *The Sea, vol. 3*, M. N. Hill, Ed., John Wiley, New York. 437-472.

Ostlund, H. G., A. L. Bowman, and G. A. Rusnak. 1962. Miami natural radiocarbon measurements I. *Radiocarbon 4*: 51-56.

Paul, A. Z. 1977. The effect of benthic biological processes on the CO_2 carbonate system. In: *The Fate of Fossil Fuel* CO_2, N. R. Andersen and A. Malahoff (Eds.), Plenum, N.Y., this volume.

Ruddiman, W. F. and L. K. Glover. 1972. Vertical mixing of ice-rafted volcanic ash in North Atlantic sediments. *Geol. Soc. Am. Bull. 83*: 2817-2836.

Ruddiman, W. F. and A. McIntyre. 1973. Time-transgressive deglacial retreat of polar waters from the North Atlantic. *Quaternary Research 3*: 117-130.

Rusnak, G. A., A. L. Bowman, and H. G. Ostlund. 1963. Miami natural radiocarbon measurements II. *Radiocarbon 5*: 23-33.

Rusnak, G. A., A. L. Bowman, and H. G. Ostlund. 1964. Miami natural radiocarbon measurements III. *Radiocarbon 6*: 208-214.

Stuiver, M. 1969. Yale natural radiocarbon measurements IX. *Radiocarbon 11*: 545-658.

Thommeret, J. and Y. Thommeret. 1966. Monaco radiocarbon measurements II. *Radiocarbon 8*: 286-291.

MODELLING THE INFLUENCE OF BIOTURBATION AND OTHER PROCESSES ON CALCIUM CARBONATE DISSOLUTION AT THE SEA FLOOR

D. R. Schink and N. L. Guinasso, Jr.

Texas A&M University

ABSTRACT

Profiles of dissolved and particulate carbonate in abyssal sediments have been calculated and the associated fluxes determined. The fractions of $CaCO_3$ in accumulating sediments were then evaluated. The calculations considered: effect of depth on solubility; rate of dissolution linearly decreasing as saturation approaches; rate of bioturbation; rate of supply of carbonate sediment; input rate of non-calcareous sediment; rate of bioturbation; and impedance of a non-turbulent boundary layer.

These calculations indicate (1) that the "carbonate compensation depth" would have a narrow depth range even if the dissolution rate increased only linearly with departure from saturation; (2) that when carbonate accumulation is nil, pore waters may still saturate with carbonate ion; (3) that increases in the supply flux of carbonate by a factor of two will deepen the compensation level by more than a kilometer; (4) that bioturbation has much less influence on the carbonate system that it has on interstitial silica; and (5) that bottom currents or turbulence can have a very significant influence on the calcium carbonate accumulation patterns.

INTRODUCTION

Silica and calcium carbonate constitute nearly all of the deep sea's biogenous sediment. These sediments are similar in many respects. Both result from shell formation by planktonic organisms near the sea surface, and subsequent shell deposition on the sea floor. Where the rate of supply exceeds the rate of dissolution biogenous sediments accumulate. Where solution exceeds supply they disappear.

The ocean is everywhere undersaturated in silica, but surface waters are supersaturated with respect to calcite and aragonite, the dominant biogenic calcium carbonate minerals. Because the ocean is more nearly saturated with respect to carbonate minerals, they accumulate over much broader regions. Since calcium carbonate solubility is strongly pressure dependent, the calcium carbonate distribution is closely related to depth. The transition from calcium carbonate accumulation can be quite sharp (*Bramlette*, 1961 *Broecker and Broecker*, 1974), but the depth of transition varies. *Berger and Winterer* (1975) have mapped the level of the "carbonate compensation surface" between 3 and 5.5 km depth--deeper in the Atlantic, shallower in the Pacific, especially at the edges or high latitudes.

The degree of calcium carbonate saturation in sea waters at depth can be calculated from measurements made on these waters after they are brought to the surface (*Berner*,1965; *Pytkowicz*, 1965; *Li, Broecker and Takahashi*, 1969; *Hawley and Pytkowicz*, 1969; and *Broecker and Takahashi*, 1977). Although the calculations are fairly involved, there is now general agreement that the carbonate compensation surface lies below the depth at which sea water becomes undersaturated. Thus, calcareous sediments like siliceous sediments accumulate in undersaturated waters. This accumulation process then represents a dynamic imbalance where input exceeds solution. Solution is relatively rapid compared to accumulation and should continue after burial until all solids were dissolved unless the solution reaction ceased. Dissolution might cease either due to saturation of the surrounding fluids as suggested by *Broecker and Broecker* (1974) or due to formation of protective coatings (*Chave*, 1965; *Chave and Suess*, 1970). Similar arguments apply to siliceous accumulations.

We hypothesize that saturation of pore water is a necessary element in carbonate accumulation, and that the concentration profile of dissolved calcium carbonate (in particular, carbonate ion) in pore water will determine the fraction of deposited $CaCO_3$* that accumulates. The effect of a dissolving particle on pore water concentration will be much greater if that particle is buried rather than simply resting on the surface. Dissolution products from a surface particle escape by diffusion through a thin, non-turbulent boundary layer of water to be removed by rapid eddy diffusion in the overlying waters. Dissolution products of a particle buried deep in the sediment must diffuse through the complex pore structure

*Throughout this paper we have considered only calcite. Similar arguments apply to aragonite, but depths of interest will be less.

of the solids and fluids matrix, then through the boundary layer to final removal into the body of the ocean.

In the dissolution of silica the depth of a dissolving particle has profound effect upon the resultant dissolved silica gradient. $CaCO_3$ however is always near saturation. As a result, only small concentration differences are possible, and the zone of variation must be narrow if strong gradients are to be formed. In the carbonate system dissolution processes must be confined to just a few centimeters. As a consequence, bioturbation rates will have less effect on carbonate accumulation than on silica profiles. However they do affect the system to some degree.

A MODEL FOR VERTICAL DISTRIBUTION OF CARBONATE IN SEDIMENTS

Model Equations

Calculation of carbonate profiles--for both solid and dissolved phases--can provide considerable insight to the way this system operates. In order to perform such calculations we have been forced to adopt a number of simplifications in the system. We will try to identify each of these, so their influence can be considered when examining the final result.

Our model closely resembles that employed for silica (*Schink, Guinasso, and Fanning*, 1975). In that study the distribution of mass within a unit volume of sediment (solid + fluid) was examined, and the rate of change in dissolved concentration was balanced against the diffusive flux and the production by dissolution. The rate of change in reactive solid concentration was balanced against the redistribution by bioturbation and the loss by dissolution. Effects of advection due to accumulating sediments were omitted in that study and the effects of adsorption were eliminated by assuming steady-state conditions. No significant changes in porosity were found within the zone of interest and, thus porosity was treated as a constant.

To evaluate vertical distributions under conditions where solids accumulate at a significant rate, the advance of the sediment water interface must be considered. The model of *Schink, Guinasso, and Fanning* (1975) has been modified here to include this advection. The effects of adsorption have again been neglected. *Berner* (1976) has pointed out that this may lead to errors, even in the steady-state condition, whenever advection is important. Nevertheless, we are not familiar with carbonate adsorption experiments and have therefore adopted this simplification. The equations then become:

$$\phi \frac{\partial c}{\partial t} = \phi D_c \frac{\partial^2 c}{\partial x^2} - \phi v \frac{\partial c}{\partial x} + K_B B \left(1 - \frac{c}{c_s}\right) \quad (1)$$

$$\frac{\partial B}{\partial t} = \frac{\partial}{\partial x}\left(D_B \frac{\partial B}{\partial x}\right) - v \frac{\partial B}{\partial x} - K_B B \left(1 - \frac{c}{c_s}\right) \tag{2}$$

where

ϕ = porosity (dimensionless)

c = concentration in the aqueous phase of dissolved species of interest (μ mole cm^{-3})

c_s = saturation concentration of species of interest (see discussion following)

t = time (kyr)

D_B = biological mixing coefficient (cm^2 ky^{-1})

D_c = aqueous diffusion coefficient for porous media. $D_c = D_m/\Theta^2$ where D_m is the diffusion coefficient in aqueous phase ($cm^2 kyr^{-1}$) and Θ represents tortuosity (dimensionless)

x = depth in sediments increasing downward and defined with respect to the upward-moving sediment-water interface (cm)

v = rate of apparent movement relative to the sediment-water interface of a particle or associated fluid; sediment accumulation rate (cm kyr^{-1})

K_B = rate of dissolution of solid phase (kyr^{-1}) (see following discussion)

B = concentration of dissolvable solid per volume of bulk sediment (μ mol cm^{-3})

In general, we here treat these terms as in the previous work, but a brief review is included for convenience. As in the previous work we assume steady-state so that the left hand side of both equations become zero. Note that c is the concentration of dissolved $CO_3^=$ in the aqueous phase, while B represents the concentration of solid $CaCO_3$ in a volume of bulk sediment (solids + fluids).

Boundary Conditions and Method of Solution

The boundary conditions for the solid phase are

$$F_B = - D_B \frac{\partial B}{\partial x} + vB; \qquad x=0 \tag{3}$$

and

$$vB = -D_B \frac{\partial B}{\partial x} + vB; \quad x=x_b \tag{4}$$

The first equation specifies the flux of solids across $x = 0$ as equal to the input supply rate (F_B) of $CaCO_3$. The second equation specifies that the flux at the bottom of the region of study is given by vB, i.e. there are no gradients in B at the bottom of the model. This implies that the flux through the bottom of the solution region is only by the advection associated with the accumulating sediment.

To model the effect of the benthic boundary layer on diffusion of the dissolved phase, the calculation is extended for c into the water column above the sediments. In the water column the concentration is described by

$$\frac{\partial}{\partial x} D \frac{\partial c}{\partial x} = 0 \tag{5}$$

where D is an eddy diffusivity coefficient specified as a function of height above the bottom. We describe this function later.

At some level in the water column, the boundary condition for the aqueous phase is given by requiring that c equal c_w, the concentration in the overlying water. The exact height of this point is not important as long as it is away from the boundary region.

Joining the water column region to the sediment region is accomplished by equating fluxes across $x = 0$, or

$$D_m \left(\frac{\partial c}{\partial x}\right)_w = \phi \; D_c \left(\frac{\partial c}{\partial x}\right)_s \tag{6}$$

The left hand side is the flux in the water region; the right hand side the flux in the sediment region. This equation implies a discontinuity in the gradient of c across the interface. At the bottom of the solution region the boundary condition is

$$vc = - D_c \frac{\partial c}{\partial x} + vc \tag{7}$$

Again this implies that there is no gradient at $x = x_b$ or, equivalently, that the only flux across x_b is that associated with advection.

The equations are solved using a procedure similar to that used by *Schink, Guinasso and Fanning* (1975). Equations (1) and (2) are placed in finite difference form. An initial sedimentation rate, υ, made up of all the non-carbonate components plus 50% of the flux of solid $CaCO_3$ to the sea floor is assumed. An iterative procedure is then used to calculate c and B as functions of x. Next, a new sedimentation rate is calculated from the value of B at the bottom of the solution region and this value is compared to the one used to make the calculation. The value of υ is adjusted and the equations are solved again for B and c. This procedure is repeated until the value for the sedimentation rate converges. For all of the calculations reported here a value of .3 cm kyr^{-1} was used for the non-carbonate sedimentation rate.

It has been pointed out that these equations have the form of a damped harmonic oscillator and that our iterative procedure for solving them might lead to oscillating solutions which are not the sought after steady-state solutions. The harmonic oscillator equation takes the form

$$\alpha\ddot{y} + \beta\dot{y} + \gamma y = \delta \tag{8}$$

where the dots imply differentiation with respect to the independent variable, x. The general solution for (8) is

$$y = R\, exp\left\{\left[\frac{-\beta}{2\alpha} + \frac{(\beta^2-4\alpha\gamma)^{1/2}}{2\alpha}\right]\right\}x + S\, exp\left\{\left[\frac{-\beta}{2\alpha} - \frac{(\beta^2-4\alpha\gamma)^{1/2}}{2\alpha}\right]\right\}x \tag{9}$$

where R and S are constants.

Oscillatory solutions can be expected when the argument of the exponential terms are complex numbers, that is when $\beta^2 - 4\alpha\gamma < 0$. For the equation describing the concentration of the aqueous phase: $\alpha = D_c$, $\beta = -\upsilon$, and $\gamma = -\frac{BK}{c_f}$. Since γ is always negative and β^2 and α are always positive, we can expect $\beta^2 - 4\alpha\gamma$ always to be positive so that oscillatory solutions are not possible. A similar analysis shows that the equation for the solid phase concentration will not have oscillatory solutions; and in fact, such oscillations do not appear in our calculated solutions.

The Benthic Boundary Layer

A chemical substance diffusing through the sea floor leaves a region where fluxes are controlled by molecular diffusion and passes into a region where fluxes are controlled by turbulent motions. Specifying the nature of the transition zone becomes one of specifying $\mathcal{D}$ as a function of height above the bottom in (5).

Wimbush and Munk (1970) distinguish two regions in the vicinity of the sea floor: 1) just above the interface is the viscous sublayer where momentum is transferred predominantly by viscous forces; 2) a logarithmic transition zone connects this layer to the sea above. They considered the height of the viscous sublayer to be about 12 ν/u^* where u^* is the friction velocity and ν is the kinematic viscosity of water. The friction velocity is the square root of the shear stress divided by density and is a measure of the magnitude of the current speed at the sea floor.

Morse (1974a) estimated chemical fluxes by assuming that diffusion takes place only by molecular diffusion across the entire viscous sublayer. However, *Wimbush* (1976) has pointed out that this is not correct. The viscous sublayer is distinguished from the zone above because the turbulent momentum flux has become less than molecular momentum flux. This does not mean that turbulent fluxes do not exist--only that they are less effective than viscosity in transporting momentum to the sea floor. Although momentum transport is dominated by viscosity in the viscous sublayer, chemical transport can still be dominated by turbulent motions over most of this region because (in sea water at 2°C) chemical diffusivity (~ 1 x 10^{-5} cm sec^{-1}) is much less than the kinematic viscosity (~ 1.7 x 10^{-2} cm^2 sec^{-1}). The "diffusion sublayer" is therefore much thinner than the viscous sublayer.

Deissler (1954) compiled a variety of data on heat, mass and momentum transfer in turbulent flow, and, using dimensional analysis, developed a model to fit the data. Using this model the thickness of the diffusion sublayer has been estimated to be

$$\ell_{\mathcal{D}} = \frac{24\nu}{u^* \, Sc^{1/3}} \qquad (10)$$

where $\ell_{\mathcal{D}}$ is the diffusion sublayer thickness

ν is the kinematic viscosity

u^* is the friction velocity

and Sc is the Schmidt Number ($Sc = \nu/\mathcal{D}_m$)

For sea water at 2°C, $Sc \simeq 3000$ and the height of the diffusion sublayer is only 14% of the height of the viscous sublayer.

In our calculations we assume that transport is by molecular diffusion in the diffusion sublayer and that eddy diffusion obeys a mixing rate law above this layer. Accordingly, $\mathcal{D}$ in Equation (5) becomes

$$D = \begin{cases} D_m & 0 < x < \ell_D \\ ku^* x & x > \ell_D \end{cases} \quad (11)$$

where k is the von Kármán constant (0.4) and x is the height above the bottom.

Wimbush (1969) measured u^* at a location in the deep Pacific Ocean and found it to vary with the tidal cycle. Numerical values for u^* varied from about 0.4 to 0.05 cm sec^{-1}. This corresponds to a diffusion sublayer thickness 0.63 to 5 mm. Any diffusion gradient established when u^* was small would be obliterated during a tidal cycle; the viscous boundary layer would be reduced to its least extent every six hours. Transient turbulent effects occasionally erode even this minimum boundary layer. Accordingly, the effective boundary layer for chemical diffusion is a few millimeters at most. In our calculations we have used values of u^* of 0.2 cm sec^{-1} and 0.3 cm sec^{-1} which give diffusion sublayer thicknesses of 1.25 mm and 0.83 mm.

Concentration

In the model describing silica dissolution (*Schink, Guinasso and Fanning*, 1975) there is a fortunate simplification that the dissolved and solid species may both be described by the same molecular formula (albeit not exactly correctly) and one mole of solid dissolves to one mole of solute. Calcium carbonate is more complex, forming two dissolved species upon dissolution which have different concentrations, $CaCO_3 \rightleftarrows Ca^{++} + CO_3^{=}$. Moreover the carbonate ion may further react in a variety of ways, e.g. forming bicarbonate, or ion associations.

In the model for silica, c_s simply represented the saturation concentration of dissolved SiO_2 and the $(c_s - c)/c_s$ term described the slowing of dissolution as saturation was approached. In order to pursue the analogy in a carbonate system, c_s should be replaced here with the apparent solubility product $K'_{sp} = [Ca^{++}]\,[CO_3^{=}]$. But fortunately the $[Ca^{++}]$ is effectively constant in sea water and pore water, so the ratio $I.P./K'_{sp}$ (where $I.P.$ = ion product) can be simplified from $[Ca^{++}]\,[CO_3^{=}]$ in situ $/[Ca^{++}]\,[CO_3^{=}]_{sat.}$ to $[CO_3^{=}]\;/\;[CO_3^{=}]_{sat.} = c/c_s$.

Moreover the dissolution of one mole of solid $CaCO_3$ can be regarded as producing one mole of $CO_3^{=}$ since so little dissolved CO_2 is present either in sea water or pore water that the reaction

$$CO_2 + CO_3^{=} + H_2O \rightleftarrows 2\ HCO_3^{-} \quad (12)$$

cannot proceed to a significant extent, and will have negligible effect on the $[CO_3^{=}]$. Decomposition of organic matter will produce

CO_2 and thereby consume $CO_3^=$. Carbonate sediments accumulate only to the extent that the arriving flux of particulate carbonates exceed the production of CO_2 from organics. This effect will be discussed in a later section. Complex ion formation reactions can be neglected because their effects are incorporated into the K'_{sp} (apparent solubility product).

Broecker and Takahashi (1977) discuss the uncertainties in K'_{sp} and $\Delta\bar{V}^*$ (partial molal volume). Possible K'_{sp} values for calcite range from 4.8×10^{-7} to 7.6×10^{-7} (*Hawley and Pytkowicz*, 1969; *McIntyre*, 1965; *Ingle et al.*, 1973; *Berner*, 1976). For most of our calculations we have selected an intermediate value of 7×10^{-7} for 1 atmosphere pressure at 2°C without any attempt to judge the merits of the various experimental determinations.

Measurements of $\Delta\bar{V}^*$ show similar uncertainty. We have adopted the *Millero and Berner* (1972) value for calcite $\Delta\bar{V}^* = -39$ cm^3/ mol. Actually the Millero and Berner value is for 25°C. *Edmond and Gieskes* (1970) have estimated that $\Delta\bar{V}^*$ changes about 0.23 cm^3/mol °C. Based on their evaluation the correct value at 0°C would be -45 cm^3/mol. Since $\Delta\bar{V}^*$ was measured at sea level and extrapolated to 6000 m without accounting for compressibility effects, and since *Broecker and Takahashi* (1976) have estimated the *in situ* value to be -31 cm^3/mol, we consider the -39 cm^3/mol volume adequate for qualitatively evaluating the effects of pressure on solubility. Then taking $[Ca^{++}] = 0.010$ M we have estimated $[CO_3^=]_{sat.}$ as a function of depth by integrating

$$\frac{\partial\ (\ln K'sp)}{\partial P} = -\frac{\Delta\bar{V}^*}{RT} \tag{13}$$

Assuming the density and temperature of sea water to be constant with depth ($\rho = 1.025$ gm/cm^3, $T = 275$°K) and $\Delta\bar{V}^*$ to be constant with pressure, provides us with the expression

$$c_{\delta} = \frac{K'sp\ 10^6}{[Ca^{++}]} \exp\ (Z\ \Delta\bar{V}^*/828T) \tag{14}$$

where c_{δ} is in micromoles/cm^3 and z is depth in meters.

Various calculations of *in situ* $[CO_3^=]$ (e.g., *Li et al.*, 1969; *Broecker and Takahashi*, 1977) suggest that this concentration is not strongly depth dependent. Therefore, the sea water value of $[CO_3^=]$ has been taken as a constant with depth at a value of 100 µ M. Figure 1 compares the assumed sea water concentrations with calculated $[CO_3^=]_{sat}$.

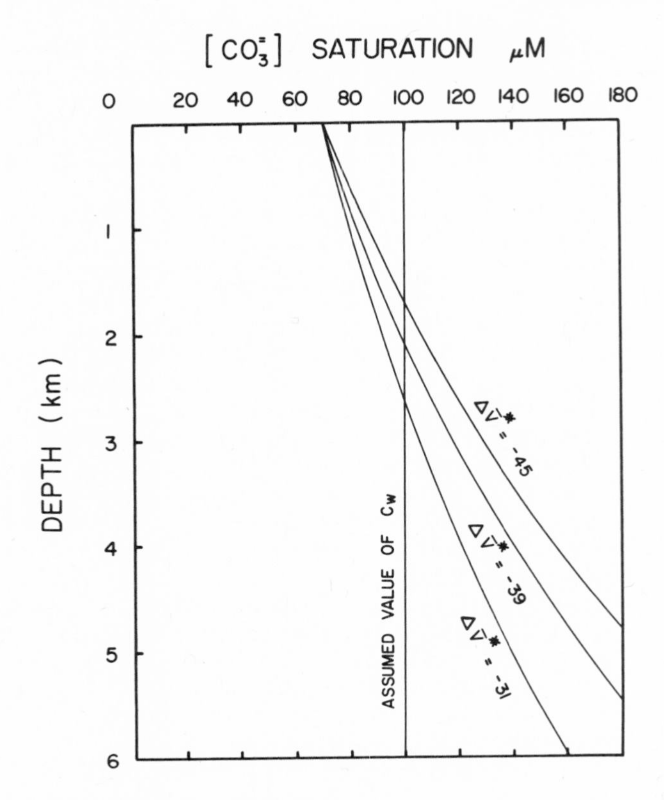

Fig. 1. Variation with depth of the saturation concentration for $CO_3^=$ based on $K'_{sp} = 7 \times 10^{-7}$ at the surface $T = 2^oC$ (constant) $[Ca^{++}] = .01$ M and $\Delta\bar{V}^* = -39$ cc/mole for the calcite dissolution reaction. Similar curves for $\Delta\bar{V}^* = -31$ cc/mole and -45 cc/mole are shown for comparison. Our assumed deep water carbonate ion concentration is also indicated.

Dissolution

The amount of $CaCO_3$ available for dissolution at the sea floor obviously depends upon the amount present, but the dependence is not trivially simple. Dissolution is a surface process and the rate will be related to the surface available; this is not necessarily the surface measured by nitrogen adsorption. The carbonate surface may have "pores" which can admit adsorbing N_2 but which are too small to participate effectively in the dissolution process. Surface coatings may remove some material from contact with the solvent.

In the face of these complexities we will treat dissolution as a rate reaction related to mass of carbonate rather than surface available. Since the calcareous material is seldom spherical, often complex, this assumption is not so bad as it might seem. It merely suggests that specific surface (i.e. reactive surface area per mass of carbonate) remains constant.

The dissolution rate of $CaCO_3$ has been studied by *Morse* (1974b,c); *Morse and Berner* (1972); and *Berner and Morse* (1974). They found rapid acceleration of dissolution as c/c_s increasingly departed from unity. The point at which acceleration occurred was shifted by phosphate so that more phosphate delayed onset of rapid dissolution. In the range of 1-6 μM phosphate the dissolution rate increases slowly as c/c_s decrease from 1.0 to 0.6 or 0.5.

The experiments of *Morse and Berner* (1972) do not focus on these slower dissolution rates, but the data suggest that dissolution rates may be approximated by a linear decrease as the solution approaches saturation. Within the concentration zone where $c/c_s >$ 0.5 the dissolution may be approximately represented by our dissolution term $K_B\ (c/c_s - 1)B$. K_B was very roughly evaluated by extrapolating Morse and Berner's data to $c/c_s = 0$ and converting from a rate dependent on surface area to a first order rate constant by assuming a specific surface of 2 $cm^2\ mg^{-1}$ for carbonaceous sediments.

Figure 2 is redrawn from *Morse and Berner* (1972). It shows the extrapolation to $c/c_s = 0$. $K_B(mg/cm^2yr)$ seems to be between 0.09 and 0.45. Using the higher value (0.45 mg $cm^{-2}yr^{-1}$) times the assumed surface area we estimate K_B at 900 kyr^{-1}. Admittedly this is a crude estimate. *Honjo* (1977) has measured specific surface areas on coccoliths as much as ten times our assumed values, so our choice of K may be too low. On the other hand the extrapolated value of dissolution rate may be equally overestimated, and the question remains as to whether N_2 adsorption areas are relevant.

Because the appropriate rates are uncertain, we have also made some parallel calculations with substantially higher and lower dissolution rates. We later present evidence that the value selected may be a fairly good choice. Whether or not this is the case, calculations of the effect of solution rates and other variables on the pattern of carbonate sedimentation can prove instructive.

Takahashi and Broecker (1977) employ a solution rate constant ($K = 1/\tau$) that is dimensionally similar but has a different meaning. The product of their K with the fraction of carbonate in the sediment and with the Δc value gives their dissolution term. The two rate constants differ primarily by the factor B/c_s, which depends on depth, porosity and percent carbonate. For a depth of 4 km their rate constant is about 1.4 yr^{-1} when transformed to our system, or about 50% greater than our central value of 0.9 yr^{-1}.

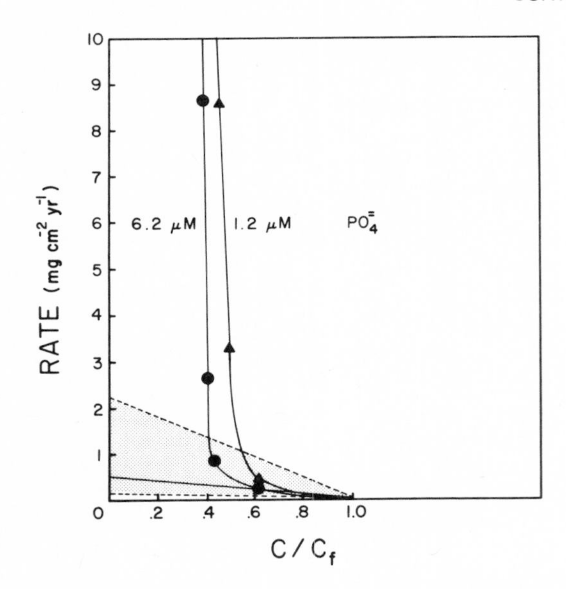

Fig. 2. Selection of a dissolution rate constant based on Morse and Berner's data but using only the near-saturation rates and extrapolating to zero carbonate.

Bioturbation

The effect of bioturbation has been represented by a vertical eddy diffusion coefficient over a mixing length. The quantitative evaluation of bioturbation rates was discussed by *Guinasso and Schink* (1975). Two different "tracers" gave differing answers. Microtektites suggest bioturbation rates of 0.5 to 3 cm^2/kyr and plutonium redistribution suggested rates of 100 to 400 cm/kyr. Since the tektite distributions represent average mixing to a depth of roughly one-half meter in the sediments over millenia of mixing and burial, whereas the plutonium has been stirred over just about a decimeter during just a decade or so, we would interpret the faster rate as appropriate to the upper few centimeters of sediment. Most of our calculations have focused on bioturbation rates of 300 cm/kyr acting to a depth of 30 cm.

Obviously an eddy diffusion coefficient is far from the ideal description of sediment mixing, and striking inhomogenieties should be expected. Moreover, the rate of bioturbation will probably vary with the food supply. Seemingly, coastal waters undergo mixing of solids at rates up to a thousand times faster than those in the deep sea. The influence of surface fertility on bottom sediment mixing deserves further study.

Rate of Accumulation

The vertical axis is fixed to the sediment-water interface with x increasing downward. As sediment accumulates, solid and aqueous phases both appear to be moving downward as viewed from the advancing reference system. Measurements made in this work suggest that porosity seldom varies strongly in this zone. As a result, it has been treated as a constant. With constant porosity the velocities of solids and fluids are identical. Although some evidence suggests that "irrigation" effects associated with bioturbation may significantly influence pore water distributions of some shelf sediment, little evidence is seen of irrigation influencing pore water velocities in the deep ocean floor.

To estimate the percentage of $CaCO_3$ in sediment, a constant influx of inert material has been assumed (0.3 cm/kyr in these calculations) and this has been added to the undissolved calcareous material at each level. The density of calcareous and inert materials were both taken as 2.6 g/cm^3 and both sediment types were assigned the same porosities (ϕ = 0.80) and tortuosities (θ = 1.15).

$CaCO_3$ Supply Rate

The rate at which calcareous materials reach the sea floor will vary with depth and with surface productivity. The deeper the sea floor, the greater will be the chance for complete dissolution as the particle falls. However, it would appear that much of the calcareous material arrives at the floor in fecal pellets (*Honjo*, 1976). In that case the relative success of such delivery must be somewhat depth related, but the relation may not be a strong one.

Obviously the rate of supply of calcareous material will often be greater than the rate of accumulation. Accumulation rates well above the compensation level might be used to estimate the incoming flux. *Broecker and Broecker* (1974) found 650 mg cm^{-2} kyr^{-1} accumulating in their shallowest core. Based on the change in alkalinity of deep sea water as it ages, *Li et al.* (1969) suggested a mean production rate of 2 g/cm^2 kyr. They estimated that on the (worldwide) average 85% of this total redissolves.

In this work, several flux rates around 2 g cm^{-2} kyr^{-1} have been adopted. Assuming some $CaCO_3$ dissolves while settling, a normal input rate of 1.5 g cm^{-2} kyr^{-1} was adopted as a central value for calculations.

RESULTS OF THE CALCULATIONS

Figure 3 shows several profiles of carbonate ion in pore waters and of calcite in sediments under varying conditions. It had been expected that calcareous sediments would accumulate when pore water

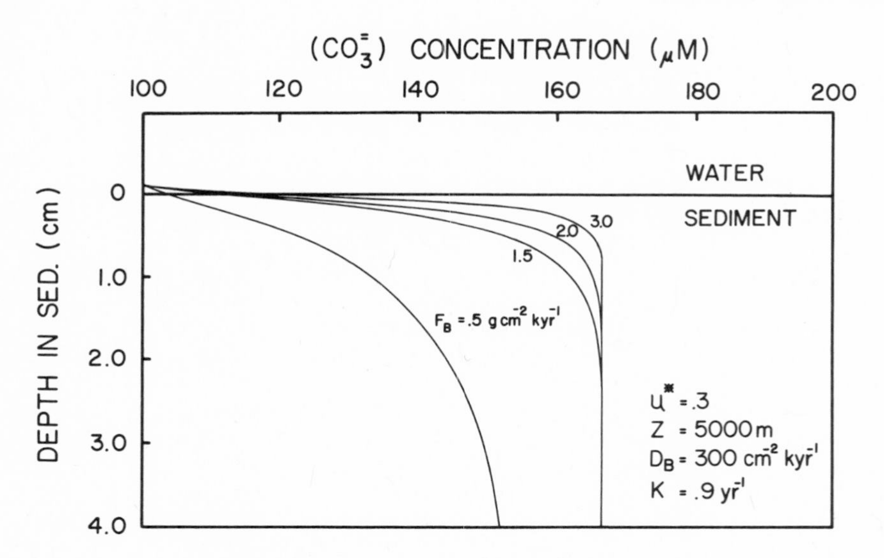

Fig. 3.a. Carbonate ion concentration as a function of depth for four different input fluxes, F_B. Water depth is 5000 m.

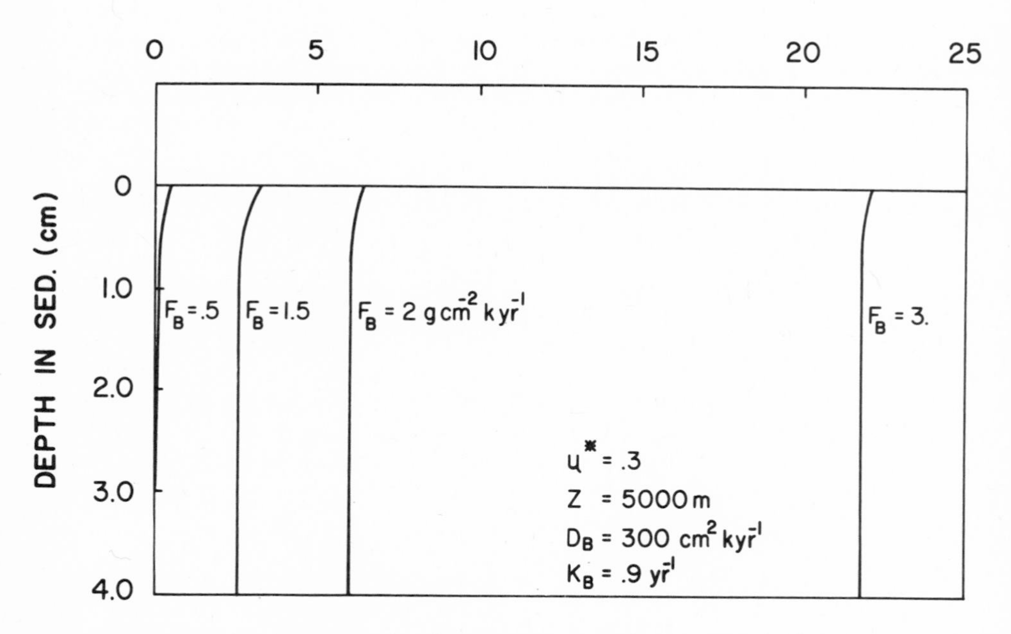

Fig. 3.b. Percent, $CaCO_3$ in sediment for same conditions as (a). No $CaCO_3$ accumulates for F_B = .5 g cm^{-2}kyr^{-1}.

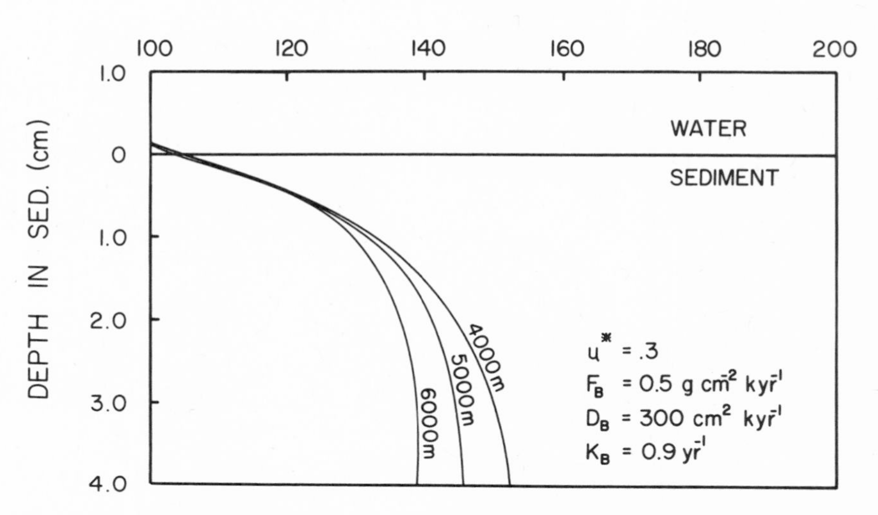

Fig. 3.c. Carbonate ion concentration in sediment for a low flux rate of .5 g $cm^{-2}kyr^{-1}$. At 4000 m pore water saturates and some $CaCO_3$ accumulates. At 5000 m and 6000 m the pore water does not saturate and all $CaCO_3$ dissolves.

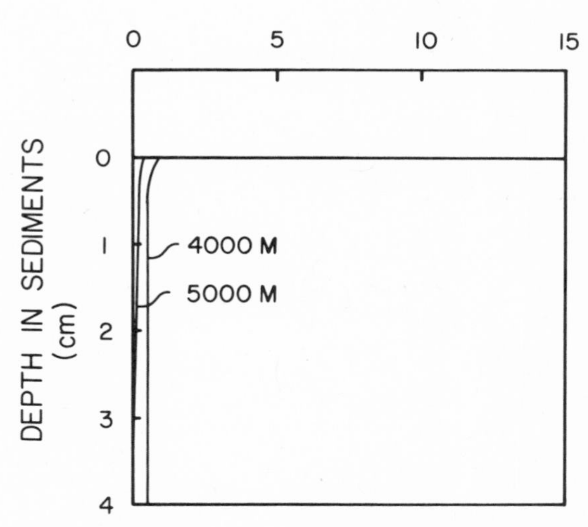

Fig. 3.d. Percent $CaCO_3$ in sediments for same conditions as (c).

Fig. 3. Some profiles of carbonate ion concentration or particulate carbonate in pore water and bulk sediment, respectively.

was saturated, and would not accumulate when pore water remained undersaturated. This expectation was fulfilled. But, when the pore waters went to saturation, a broad range of $CaCO_3$ accumulation rates resulted. When the saturation condition was quickly and easily achieved, a minority of deposited material redissolved. When saturation was achieved at several centimeters depth, accumulating carbonates formed a much smaller fraction of the sediments.

Calculations made in this work suggest that carbonate accumulation, at concentrations sufficient to be considered calcareous ooze, depends not just on the saturation of pore waters, but rather on the vertical extent of the undersaturated zone in pore water. This finding has been anticipated by a number of previous investigators.

The zone of undersaturation is most strongly influenced by the depth and the rate of supply of calcareous material. Figure 4 shows the $CaCO_3$ content of accumulating sediments at various depths where all conditions are identical, except the input flux. A doubling of the $CaCO_3$ rain rate deepens the carbonate compensation depth (taken as 50% $CaCO_3$ in sediments) by 1150 meters. Figure 4 shows that a compensation depth will exist in the ocean even when carbonate dissolution rates vary linearly with the degree of saturation.

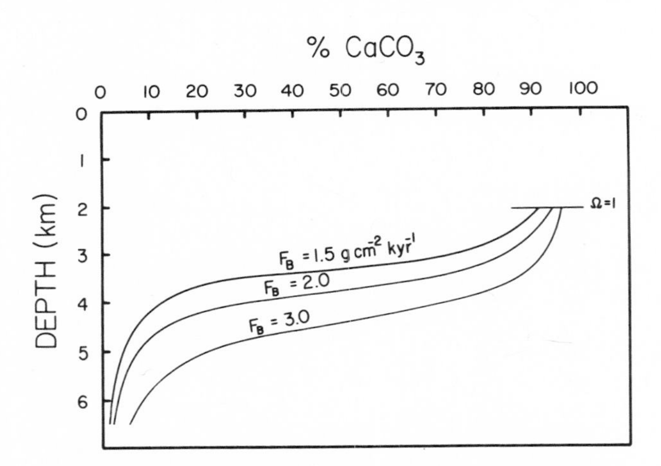

Fig. 4. Fraction of $CaCO_3$ in sediments versus depth for three different input fluxes. Dissolution rate, K_B = 0.9 yr^{-1}, bioturbation rate D_B = 300 $cm^2 kyr^{-1}$, bottom water carbonate concentration is 100 μM, porosity = .80 and non-carbonate sedimentation rate = 0.3 cm kyr^{-1}.

Figure 5 shows the effect of varying the rate "constant" for dissolution. The rate of solution is clearly an important parameter in the carbonate system, but major variations in solution rate (e.g.

five-fold) have less effect than a doubling of the input flux. Increasing K_B by a factor of 10 from 0.09 yr^{-1} to 0.9 yr^{-1} raises the compensation depth by 1000 m. Increasing K_B from 0.45 to 4.5 yr^{-1} raises the compensation depth only 400 m. As K_B increases, the dissolution rate becomes less important in controlling the compensation depth.

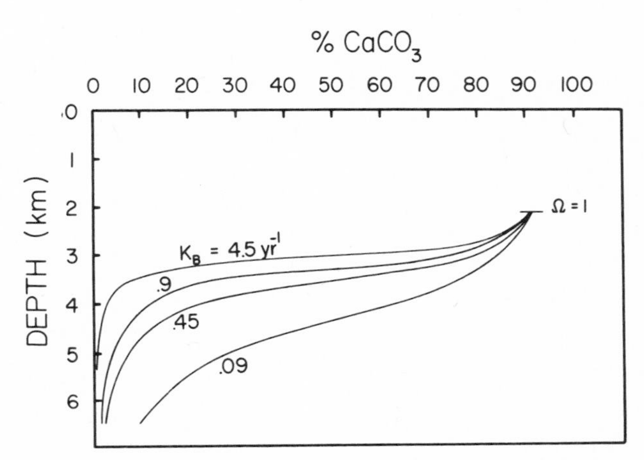

Fig. 5. Fraction of $CaCO_3$ in sediments versus depth using four different dissolution rate constants. Other conditions as in Figure 4 with input flux F_B = 1.5 g $cm^{-2}kyr^{-1}$.

Figure 6 shows the effect of two different bioturbation rates on calcareous accumulation. In *Schink, Guinasso and Fanning* (1975) the importance of bioturbation as an influence on interstitial silica profiles was emphasized. Bioturbation is most effective on profiles where the reacting substance is undersaturated. It has been found that even where carbonate accumulation is slight, the pore waters will saturate. Interstitial carbonate profiles are therefore only slightly influenced by bioturbation rates. At great depths, where pore waters are undersaturated, increasing the bioturbation rate has the effect of raising the interstitial carbonate concentration. Therefore, the dissolution rate is decreased, allowing for more carbonate to be buried in the sediment.

Bottom currents enter our model through the parameter u^* which controls the impedance of the boundary layer to diffusion. Figure 3 indicates that a substantial part of the gradient across the sediment interface extends into the water column, even at relatively high values of u^*. Increasing the bottom current increases the amount of the gradient within the sediments, causing more solution, and raising the compensation depth. Figure 7 shows that a 50% increase in the friction velocity will raise the compensation depth 650 m. The magnitude of bottom currents has a major influence on the carbonate compensation depth.

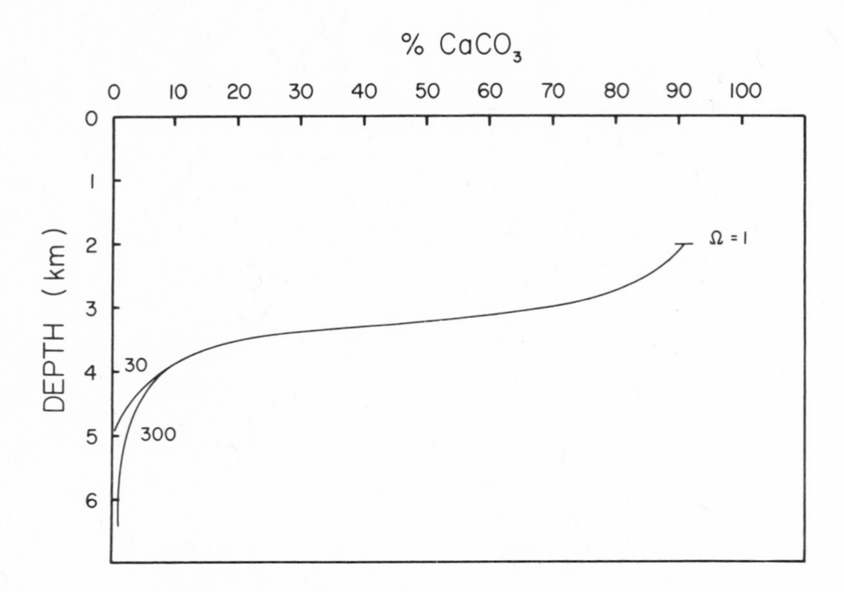

Fig. 6. Fraction of $CaCO_3$ in sediments versus depth for two different bioturbation rates, D_B = 30 or 300 cm^2kyr^{-1}. Other conditions as in Figures 4 and 5.

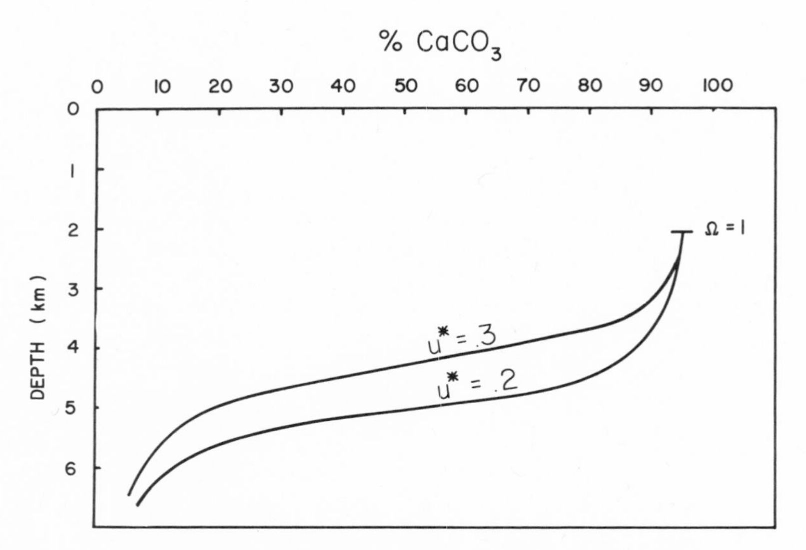

Fig. 7. Fraction of $CaCO_3$ in sediments versus depth for two different boundary layer conditions. Larger $u*$ values imply greater bottom current or turbulence, F_B = 3.0 g $cm^{-2}kyr^{-1}$. Other conditions as in Figure 4.

The slopes of the lines between 20% and 70% $CaCO_3$ have been determined from Figures 4, 5, and 7. Table 1 compares the slopes for various changing conditions. From Table 1 it is seen that the calculated carbonate distribution patterns, as expressed by these slopes, are remarkably similar to the observed slopes based on actual samples as reported by *Biscaye, Kolla and Turekian* (1976). They found values generally around 15 meters/percent carbonate, with a range from 3-22 m/%.

TABLE 1. Slopes of the lines defining the variation in carbonate content of sediments as a function of depth, for various sets of conditions.

a. Dissolution Rate

F_B	K_B	u^*	slope
g cm^{-2}kyr^{-1}	yr^{-1}	cm sec^{-1}	m/%
1.5	0.09	0.3	29
	0.45		14
	0.90		12
	4.50		7

b. Friction Velocity

F_B	K_B	u^*	slope
3.0	0.90	0.2	14
		0.3	21

c. Input Flux

F_B	K_B	u^*	slope
3.0	0.90	0.3	21
2.0			13
1.5			12

Table 1.a. shows that dissolution rate has the greatest effect on the carbonate vs. depth slope. Very slow solution generates greater slopes than those normally observed. The dissolution rate selected in this work generates a slope very near the values commonly observed by *Biscaye, Kolla and Turekian* (1976). Table 1.a. and Figure 5 also suggest that increasingly rapid dissolution has a diminishing effect on the carbonate accumulation pattern as expressed by the

slope in Figure 5. That is because the diffusion sublayer becomes relatively more important as dissolution becomes more rapid.

Table 1.b. shows the effect of this sublayer on the carbonate vs. depth patterns. The lower $u*$ values (i.e. slower bottom currents generate a smaller slope in the curve. Table 1.c. shows that greater input fluxes of carbonate produce larger slopes in the carbonate vs. depth curves, but that the slopes of these curves change less dramatically than their depths. Obviously these slopes can also be affected by changing the carbonate ion concentration with depth, or by changing the fraction of non-carbonate materials.

The calculations that have been performed utilized coefficients which were selected from relevant studies of the various effects without ad hoc adjustments to make the model fit observations. The calculations, for the most part, match observed distributions, except that our depths are usually too shallow, at least with respect to the Atlantic. This may be the result of selecting too large a $\Delta\overline{V}*$. Figure 1 compared the carbonate saturation values for our selected $\Delta\overline{V}* = -39$ cm^3/mol. with the value of -31 cm^3/mol., suggested by *Broecker and Takahashi* (1977). Figure 8 compares the carbonate distribution vs. depth when the $\Delta\overline{V}*$ values differ in this manner. If $CO_3^=$ = 110 µM had been used for the calculations, the curves in Figures 4 - 8 would have been shifted downward by 1000 m. Our calculated values might also be too shallow because a K'_{sp} of 7×10^{-6} was selected; a lower value would increase the depth at which $\Omega = 1$, and shift the curves in Figures 4 - 8 toward deeper depths.

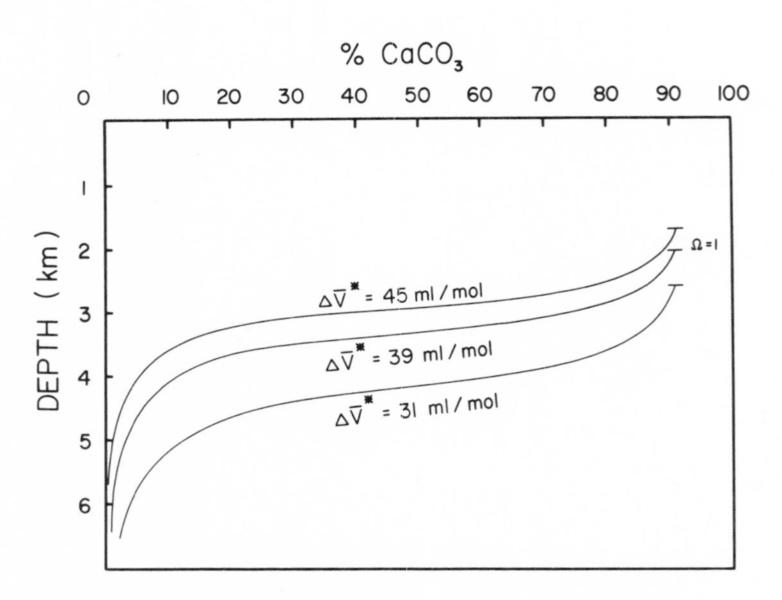

Fig. 8. Fraction of $CaCO_3$ in sediments versus depth calculated using three different values for the change of partial molal volume during dissolution of carbonate in sea water.

Figure 9 shows the effect on these calculations of selecting different values of K'_{sp} (1 Atm.). This parameter is presumably constant. However, adoption of different measurements of the actual value can produce strikingly different calculated carbonate distributions. The large variations introduced by this range of uncertainty emphasize the importance of determing K'_{sp} accurately and precisely.

Although a wide range of input parameters have been employed in a great variety of combinations, none of the calculated profiles showed a substantial change in the fraction of $CaCO_3$ with depth in the sediment at any given location. We must now agree with *Broecker and Broecker* (1974) that the variations they observed (e.g. Core V19-66) are the result of variations in conditions of dissolution over the period of accumulation. This is not what we expected to show.

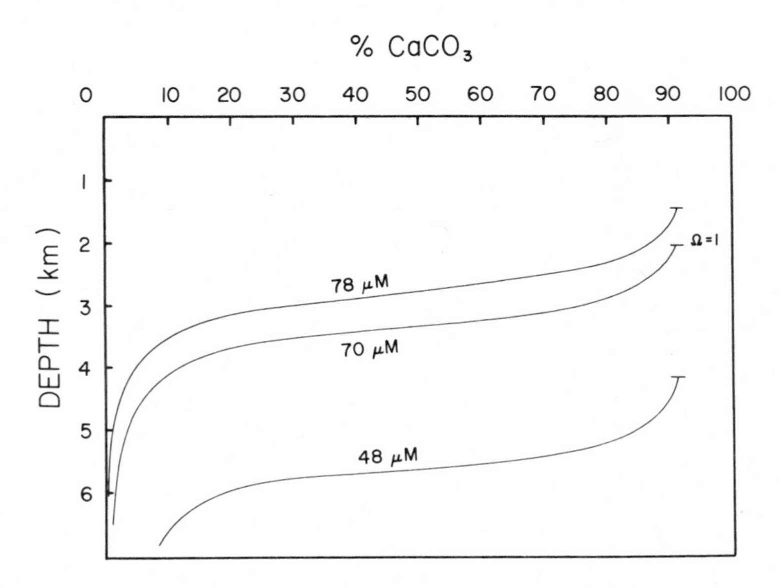

Fig. 9. Fraction of $CaCO_3$ in sediments versus depth for various values of K'_{sp} at one atmosphere. The numbers on the figure refer to the saturation concentration of carbonate ion at one atmosphere. Other conditions are the same as in Figure 4.

Effects of Organic Matter

The discussion so far has ignored the effect of organic carbon in the sedimentary rain, except to note that organic films might markedly reduce the dissolution rate. Organic material will decompose--with the aid of bacteria--in the well-oxygenated abyssal sediments. The result will be delivery of CO_2 to the overlying waters, and to the extent that the organic material has been buried, its decomposition will influence the carbonate gradient.

There will be a source of CO_2 for the reaction $CO_3^= + CO_2 + H_2O \rightleftarrows 2HCO_3^-$. The net effect will be almost a mole for mole removal of $CO_3^=$ by CO_2. *Broecker* (1974) suggests a 2:1 ratio for organic: inorganic carbon in pelagic detritus. If such a ratio persisted into the sea floor, it would cause almost all of the calcareous sediment to dissolve. Presumably much of the organic carbon in the ocean is more readily attacked than is the inorganic carbon, so that organic decomposition occurs mainly in the water column or at the interface where it will have little effect on the dissolved carbonate profile. If bioturbation processes succeed in burying organic matter before it decomposes, this profile of decomposing carbon could effectively subtract an equivalent amount of $CaCO_3$.

Surface productivity then does not depress the compensation level unless this productivity enhances the $CaCO_3$ delivery more effectively than it enriches the organic content of the sediment. (Obviously unreactive organic matter can be excluded from this argument.) Possibly the depression of the compensation depth in equatorial regions and the rise in the compensation depth in high latitudes can be attributed to the relative organic to inorganic content of the resultant detritus. Warm waters produce $CaCO_3$ rich populations whereas the colder waters emphasize the production by non-calcareous organisms.

Morse and Berner (1972) have shown that phosphate inhibits dissolution rates. Since decomposing organic matter contains phosphate it might be tempting to suggest that organic decomposition inhibits solution. This would not be correct. The ratio of CO_2: $PO_4^{\equiv}$ is so great and the retardation effects of phosphate so limited at the concentrations encountered in pore water (2-5 μM) that this second-order effect can be neglected.

CONCLUSIONS

All of the effects noted in our calculations have been suggested before--usually by several authors. Our calculations serve only to reinforce previous conclusions and perhaps to de-emphasize some less important factors. We consider the compensation depth to occur where rate of $CaCO_3$ dissolution almost equals rate of supply. At this level carbonate still saturates the pore water, however, the majority of the carbonate dissolves. Above this level accumulation of calcareous ooze proceeds, but with substantial dissolution. Changes in the dissolution rate constant are not necessary to achieve this pattern.

Only slight variations of $CaCO_3$ content versus depth in the core can be derived under steady-state conditions. A small enrichment at the surface is to be expected. The degree of this enrichment, as well as the depth of it, increase with increasing bioturbation rates. Variation by more than a few percent in $CaCO_3$ content with depth in the core must be the result of variations in environmental conditions over the period of accumulation. Such changes could have been caused by variations in the saturation concentration due to temperature or pressure, by variations in the $CO_3^=$ content of the overlying water, by changes in the supply rate of $CaCO_3$ or the ratio of $CaCO_3$ to organic carbon, by changes in near-bottom turbulence, or by changes in the non-calcareous sedimentary accumulation rate. We do not believe that variations in bioturbation rates, phosphate content, or deep water temperature changes of 1°C could significantly alter the $CaCO_3$ content of accumulating sediments. However, detailed calculations have not been made to support these last arguments.

The importance of boundary layer effects on the solution flux of carbonate suggest that processes in this zone must receive much more careful attention.

ACKNOWLEDGEMENTS

This work was supported by NSF Grant # OCE 75-21275 and ONR Contract # N00014-75-C-0537. We are grateful to Mark Wimbush for setting us on the right track with respect to the boundary layer zone. The authors benefited from a talk on this same subject by W. S. Broecker given at the IUGG meeting, Grenoble, Sept. 1975. It should also be noted that the Broecker and Takahashi paper (Deep Sea Research) was presented at this (Honolulu) meeting, and we had the benefit of a preprint of that work.

REFERENCES

Berger, W. H. and E. L. Winterer, 1974. Plate stratigraphy and the fluctuating carbonate line, *Spec. Publs., Int. Ass. Sediment* *1*, 11-48.

Berner, R. A., 1965. Activity coefficients of bicarbonate, carbonate, and calcium ions in seawater. *Geochim. Cosmochim. Acta 29*, 947-965.

Berner, R. A., 1976. Solubility of Calcite and Aragonite in seawater at Atmospheric Pressure and 34.5% Salinity. *Amer. Jour. Sci.* *276*, 713-730.

Berner, R. A. and J. W. Morse, 1974. Dissolution kinetics of calcium carbonate in seawater. IV. Theory of calcite dissolution, *Am. Jour. Sci. 274*, 108-134.

Biscaye, P. E., V. Kolla and K. K. Turekian, 1976. Distribution of calcium carbonate in surface sediments of the Atlantic Ocean, *J. Geophys. Res. 81*, 2595-2604.

Bramlette, M. N., 1961. Pelagic sediments, In: *Oceanography* Sears, M. (Ed.) *Publs. Am. Ass. Advmt. Sci. 67*, 345-366.

Broecker, W. S., 1974. *Chemical Oceanography*, Harcourt-Brace-Jovanovich, Inc., New York, p. 42.

Broecker, W. S. and S. Broecker, 1974. Carbonate Dissolution on the Western Flank of the East Pacific Rise, In: *Studies in Paleo-Oceanography*, (Ed. by William W. Hay) *Soc. of Eco. Paleontologists and Mineralogists. Spec. Publ. No. 20*, 44-57.

Broecker, W. S. and T. Takahashi, (In press) 1977. The relationship between lysocline depth and *in situ* carbonate ion concentration, *Deep-Sea Res.*

Chave, K. E., 1965. Carbonates: Association with Organic Matter in Surface Seawater, *Science 148*, 1723-1724.

Chave, K. E. and E. Suess, 1970. Calcium carbonate saturation in seawater: Effects of dissolved organic matter, *Limn. and Ocn. 15*, 633-637.

Deissler, R. G., 1954. Analysis of turbulent heat transfer, mass transfer, and friction in smooth tubes at high Prandtl and Schmidt numbers, *National Advisory Committee for Aeronautics, Technical Note 3145*, 53 pp.

Edmond, J. M. and J. M. T. M. Gieskes, 1970. On the calculation of the degree of saturation of seawater with respect to calcium carbonate under *in situ* conditions, *Geochim et Cosmochim., Acta, 34*, 1261-1291.

Guinasso, N. L., Jr. and D. R. Schink, 1975. Quantitative estimates of biological mixing rates in abyssal sediments, *J. Geophys. Res. 80*, 3032-3043.

Hawley, J. and R. M. Pytkowicz, 1969. Solubility of calcium carbonate in seawater at high pressures and 2°C., *Geochim. Cosmochim. Acta 33*, 1557-1561.

Honjo, S., 1976. Coccoliths; Production, Transportation, and Sedimentation, *Marine Micropaleontology, 1*, 65-79.

Honjo, S. 1977. Biogenic carbonate particles in the ocean; and do they dissolve in the water column? IN: *The Fate of Fossil Fuel CO_2*, N. R. Andersen and A. Malahoff (Eds.), Plenum, N. Y., (this volume).

Ingle, S. E., C. H. Culberson, J. Hawley, and R. M. Pytkowicz, 1973. The solubility of calcite in sea water. *Marine Chemistry, 1*, 295-307.

Li, Y.-H., T. Takahashi and W. S. Broecker, 1969. The degree of saturation of $CaCO_3$ in the oceans, *J. Geophys. Res. 74*, 5507-5525.

McIntyre, W. G., 1965. The temperature variation of the solubility product of calcium carbonate in sea water. *Fisheries Research Board of Canada, Manuscript Report Series, 200*, 153.

Millero, F. J. and R. A. Berner, 1972. Effect of pressure on carbonate equilibria in seawater, *Geochim. Cosmochim. Acta 36*, 92-98.

Morse, J. W., 1974a. Calculation of diffusive fluxes across the sediment-water interface, *Jour. of Geophys. Res. 79*, 5045-5048.

Morse, J. W., 1974b. Dissolution kinetics of calcium carbonate in seawater. III. A new method for the study of carbonate reaction kinetics, *Am. Jour. Sci. 274*, 97-107.

Morse, J. W., 1974c. Dissolution kinetics of calcium carbonate in seawater. V. Effects of natural inhibitors and position of the chemical lysocline, *Am. Jour. Sci. 274*, 638-647.

Morse, J. W. and R. A. Berner, 1972. Dissolution kinetics of calcium carbonate in seawater. II. A kinetic origin for the lysocline. *Am. Jour. Sci. 272*, 840-851.

Pytkowicz, R. M., 1965. Rates of inorganic calcium carbonate precipitation, *J. Geol. 73*, 196-199.

Schink, D. R., N. L. Guinasso, Jr., and K. A. Fanning, 1975. Processes affecting the concentration of silica at the sediment-water interface of the Atlantic Ocean, *Jour. of Geophs. Res. 80*. 3012-3031.

Takahashi, T., and W. S. Broecker, 1977. Mechanism of calcite dissolution on the deep sea floor. In: *The Fate of Fossil Fuel* CO_2, N. R. Andersen and A. Malahoff (Eds.). Plenum, N.Y.

Wimbush, A. H. M. H., 1969. Dissertation, Univ. of Calif., San Diego.

Wimbush, A. H. M. H., and W. Munk, 1970. The benthic boundary, In: *The Sea, Vol. 4.* (Ed. by A. E. Maxwell) John Wiley, New York, 731-758.

CARBON-13 IN UVIGERINA: TROPICAL RAINFOREST HISTORY AND THE EQUATORIAL PACIFIC CARBONATE DISSOLUTION CYCLES

N. J. Shackleton

University of Cambridge and Lamont-Doherty Geological Observatory

ABSTRACT

Benthonic foraminifera in late Pleistocene deep-sea cores show significant variation in $\delta^{13}C$ with depth in sediment. This, and the report by Sommer et al. (in prep.) of $\delta^{13}C$ variations in planktonic foraminifera, indicate that $\delta^{13}C$ in dissolved oceanic CO_2 undergoes a significant change in a few thousand years. This is in apparent contradiction to the estimated 300 ka residence time for carbon in the ocean. It is suggested that this is a consequence of changes in the terrestrial plant biomass, which has a $\delta^{13}C$ of about -25°/oo. Postulated changes in world vegetation, particularly in tropical rainforests during the Late Pleistocene, were sufficient to produce a change of the magnitude observed. Rapid expansion of forests between 13 ka and 8 ka ago may have resulted in the striking accumulation of aragonite pteropods in Atlantic Ocean sediments of that age. Rapid deforestation during an interglacial-glacial transition probably caused the intense carbonate dissolution which is observed in Equatorial Pacific Ocean sediments deposited over this interval. The current rate of injection of fossil fuel CO_2 into the atmosphere is substantially greater than the rate at which it was added during post-interglacial aridification in the tropics.

INTRODUCTION

Since 1973 oxygen isotope determinations in foraminifera have been conducted in association with the CLIMAP project (*CLIMAP*, 1976) on over fifty deep-sea cores from all the oceans covering a wide latitudinal and depth range. In a selection of these cores, carbon isotopic composition was also measured. A coherent pattern of variation has gradually emerged, primarily from the cores in which benthonic foraminifera were analyzed. It appears that deep-sea CO_2 available for test formation was depleted in ^{13}C in glacial, relative to interglacial times. Although this may have been caused in part by increased oxidation of isotopically light organic matter at the sea floor in some areas, it seems unlikely that this was a general occurrence. The similarity between carbon isotope records from planktonic foraminifera in the subantarctic, and benthonic foraminifera at low latitudes over nearly half a million years, indicates that the $^{13}C/^{12}C$ ratio variations affected the entire ocean.

It would appear at first glance, that changes in the ^{13}C content of the ocean CO_2 reservoir are only possible on a 100 *ka* scale because of the long residence time of carbon in the ocean (Table 1). However, current opinions of the history of world vegetation indicate that the mass of this ^{13}C-depleted reservoir has undergone substantial fluctuations, with the ocean acting both as a sink and a source for carbon. This is a convenient geological analogue of the present transfer of fossil fuel CO_2 to the oceans. It suggests an explanation for the pulses of carbonate dissolution in the equatorial oceans which occur at interglacial-glacial transitions (*Luz and Shackleton*, 1975), and for the excellent carbonate preservation, including pteropod aragonite in the Atlantic Ocean, in sediments deposited during the last glacial-interglacial transition (*Diester-Haass, et al.*).

TABLE 1. Carbon Reservoir Time Constants.

Dissolved carbon in ocean	300 *ka*	(steady-state residence time: *Broecker*, 1974, p.195)
$\delta\ ^{13}C$ in ocean dissolved CO_2	300 *ka*	(derived from same assumption)
$CO_3^{=}$ in seawater	15 *ka*	(response time: *Broecker*, 1974, p.195)
Ocean deep water	1 *ka*	(mixing time: *Gordon*, 1975)

PLEISTOCENE CLIMATIC RECORD: TEMPORAL FRAMEWORK

Because there have been major glaciations in the Pleistocene, during which enormous ice sheets spread over northern North America and much of northern Europe, which were separated by warmer intervals similar to the present, it has become usual to divide Pleistocene time into glacial and interglacial ages. Since by definition these together fill the whole of Pleistocene time (i.e., ∿ *1.5 Ma*) and since fewer than a dozen local stages are defined in most areas, one might reach the misleading conclusion that on the average each age lasted over 100 *ka*. In the marine Pleistocene, using the oxygen isotope record, 19 stages have been defined in sediments deposited during the past 700 *ka* (*Emiliani*, 1955; *Shackleton and Opdyke*, 1973). These ages, averaging 35 *ka* in duration, were characterized by alternately larger and smaller ice volumes. However, even this degree of refinement does not give an adequate portrayal of the actual climatic record. *Shackleton* (1969) showed that the last interglacial, in the sense that the word is used by palynologists (i.e., the time during which temperate forests flourished in Northwest Europe), lasted only about 10 *ka*, and occupied only the fraction of age 5 from about 125 to 115 *ka* ago. Recent work (*Mesolella et al.*, 1969; *Hays, Imbrie and Shackleton*, 1976) has substantiated this model and has drawn attention to the very strong 22 *ka* cycle in Pleistocene climate records. Unfortunately, this climatic signal is below the stratigraphic resolution allowed by sediment mixing in many parts of the ocean, as has been revealed by the reconstruction of a detailed glacio-eustatic sea-level record (*Mesolella et al.*, 1969; *Chappell*, 1974a, b) and by analyzing oxygen isotope records in high sedimentation rate cores (*Ninkovich and Shackleton*, 1975; *Hays, Imbrie and Shackleton*, 1976). It should be noted that in a few of the 22 *ka* cycles, only the'interglacial' portion would correspond to an interglacial in the European pollen record. However, probably no palynologically defined interglacial would span more than the 'interglacial' portion of one 22 *ka* cycle.

At present there is little direct evidence to enable the Pleistocene history of the tropical continents to be related to this model. On the other hand, there is no evidence to support *Suggate* (1975) in his assertion that the climatic change marking the end of the last interglacial in norther Europe did not have a marked effect in the tropics. Here, the study of *Hays and Perruzza* (1972) is significant, because they showed a dominant period of about 20 *ka* in the inferred fluctuations in the flux of Saharan dust into the ocean. Their peak 8, during which only small quantities of Saharan dust blew into the ocean, was equivalent to substage 5e, and probably had about a 10 *ka* duration. *Damuth* (1975) shows a similar result; in his terminology carbonate peak P9 corresponds with the brief interglacial substage 5e. In short, it is highly likely that even in the tropics the glacial interval from about 115 *ka* to 105 *ka* ago (substage 5d of *Shackleton*, 1969) approached a 'glacial' (i.e, arid) extreme. Although stage 5 (125 *ka* to 75 *ka* ago) when seen homogenized in slowly accumulating sediment seems interglacial by comparison with stage 6, it in fact encompasses most of the range of Pleistocene climates.

THE PLANT BIOSPHERE AND ITS HISTORY

The world biomass has been estimated to contain 8.4×10^{17} g carbon. The asociated humus in soil is thought to contain up to 50% more carbon than the living biomass (*Whittaker and Likens*, 1975). Thus, a reasonable estimate for the total carbon would be 2×10^{18} g carbon (Table 2). It is my contention that this quantity has varied immensely during the Pleistocene.

Today, more than half of the total carbon is contained in the tropical rainforests and the tropical seasonal forests, which are distributed in Africa, South America and Southeast Asia (*Dansereau*, 1957). The history of the African tropical rainforests was reviewed recently by *Hamilton* (1976). In Hamilton's reconstructions the total forest area in tropical Africa at the arid extreme about 20 *ka* ago (*Street and Grove*, 1976) was very roughly one-third of its present area. About 8 *ka* ago when Lake Chad and many other lakes were near their maximum size, *Hamilton* (1976) shows northward extensions in forest area which would have added about half the present area.

A map of the tropical forest refuges of South America during an arid phase (*Vuilleumier*, 1971) shows an even greater proportional reduction in rainforest than that suggested for Africa. *Van der Hammen* (1972) quoted evidence from several areas showing that the tropical forests were extensively replaced by savanna during glacial arid phases, stressing the evidence for the age of this phase, which was lacking in *Vuilleumier's* (1971) review (see also *Damuth and Fairbridge*, 1970). Recently, *Prance* (1973) has gathered a considerable amount of phytogeographic support for *Vuilleumier's* (1971) reconstructions of forest refuges. The refuges shown by *Prance* (1973, Figure 24) are not as small as those proposed earlier. Nevertheless, they represent a very considerable reduction in forest area as compared with the present. *Tricart* (1975), who has examined some of the areas where it is known that tropical forests are rooted in dunes from the last glacial, also supports this notion.

The third major tropical rainforest area encompasses New Guinea, Java, Sumatra, Borneo and Malaysia. *Bowler et al.* (1975) have shown that there was a northward movement of the Australian arid area at the last glacial maximum resulting in a shift of vegetational belts to the north, and that at the same time the tree line in New Guinea was substantially lower than today. Although the Southeast Asian tropical land area was significantly increased

TABLE 2. Carbon Reservoirs, Their $\delta\ ^{13}C$ Values, and Changes in Them.

Reservoir	$\delta\ ^{13}C$	Mass (as carbon)
living plant biomass	-25°/oo (*Craig*, 1953)	8.4×10^{17} *g* (*Whittaker & Likens*, 1975)
"humus"	-25°/00 (from above)	12×10^{17} *g* (*Whittaker & Likens*, 1975)
marine biomass	-23°/oo (*Sackett et al.*, 1965)	1.8×10^{15} *g* (*Whittaker & Likens*, 1975)
ocean dissolved CO_2	0°/oo (*Kroopnick et al.*, 1970)	3.5×10^{19} *g* (*Broecker*, 1974, p. 182)
atmospheric CO_2	-7°/oo (*Craig*, 1954, p. 116)	6×10^{17} *g* (*Broecker*, 1974, p. 182)
fossil fuel CO_2 year 2000 (cumulative)	-24°/oo (*Craig*, 1953, p. 76, coal)	2×10^{17} *g* (*Rotty*, 1977, projection)
	rate of change (as carbon)	
ocean dissolved CO_2	10^{14} *g*/year (throughput: *Broecker*, 1974)	
terrestrial biomass (net primary productivity)	1.15×10^{17} *g*/year (*Whittaker & Likens*, 1975)	
terrestrial biomass (geological rate of change)	10^{14} *g*/year (this paper)	
fossil fuel	5×10^{15} *g*/year (1974 burning: *Rotty*, 1977)	

by the glacio-eustatic lowering of sea level, the reconstructions of *Nix and Kalma* (1972) suggest that this additional area was covered by woodland, rather than rainforest, and that the area of closed forest on New Guinea was appreciably less than today. They postulated an appreciably greater area of closed tropical forest 8 *ka* ago than exists today. *Verstappen* (1975) has reviewed the evidence from Malaysia, stressing the effects on vegetation and landform development of lower precipitation and a longer dry season during glacials. Thus, in the third great tropical rainforest area, the rainfall variations documented (e.g., by *Kershaw*, 1974), may have caused variations in the area of tropical rainforests by as much as a factor of two, although several regions have scarcely been investigated.

In conclusion, not only do the three major areas of tropical rainforests show evidence for a synchronous diminution associated with aridity at the last glacial maximum (*Williams*, 1975), but in addition, there is evidence from all three regions that the tropical forests significantly exceeded their present area during the early part of the Holocene in association with a 'pluvial' episode. In all three areas, published reconstructions have indicated a variation by at least a factor of two.

The changes in forest extent in northern Eurasia and northern North America are almost as striking. For Eurasia they have been summarized by *Frenzel* (1968). His reconstructions of Eemian (interglacial) and Weichselian (glacial) vegetation again clearly indicate that the area of forest was reduced by at least a factor of two during glacials. The areas of ice and tundra are quite well established for this region so that reconstructions of former forest cover are more accurate than in the tropics.

In North America the extensive ice cover substantially reduced the area of forests during the glacial period, although well south of the ice front the change in forest cover may not have been so dramatic.

Summarizing, it seems not unlikely that the total terrestrial plant biomass and associated humus increased by a factor of three between about 14 *ka* and 8 *ka* ago, the change being largely due to the enormous changes in tropical rainforest extent caused by a change from arid to pluvial conditions, and to the reforestation of Eurasia and North America. Since 8 *ka* ago it may have fallen to roughly two-thirds of its maximum. In some areas, man has contributed to the recent reduction.

It is perhaps worth remarking that as the vegetation changes or is replaced by desert, essentially all the carbon, both living and in the soil, is oxidized. Only a minute fraction will be preserved to form a geological deposit. This is, of course, well known to Pleistocene geologists who need to locate these deposits to make their reconstructions.

FACTORS CONTROLLING CARBON ISOTOPE VARIATIONS IN OCEANIC CARBONATES

As with the oxygen isotopes, the carbon isotopes show a temperature-dependent fractionation when carbonate is deposited in isotopic equilibrium with water containing dissolved CO_2, $CO_3^=$ and HCO_3^-. Several attempts have been made to determine the various fractionation factors and their temperature dependences, of which, that by *Emrich, Ehhalt and Vogel* (1970) is the most comprehensive. Use of their data would in principal enable an estimation of $\delta\,^{13}C$ for carbonate deposited in isotopic equilibrium in any part of the ocean from a knowledge of the carbon isotopic composition of the available inorganic carbon, and the temperature. The temperature effect is small; about 0.03°/oo per degree C (*Emrich et al.*, 1970).

The carbon isotopic composition of the available inorganic carbon varies from place to place in the open ocean, largely as a result of the fact that the production of organic carbon usually results in a large isotopic fractionation. Typical $\delta\,^{13}C$'s for marine organic carbon at low latitudes averages about -23°/oo (*Sackett, Eckelmann, Bender and Be,* 1965; *Kroopnick*, 1974). Organic productivity close to the ocean surface leaves the remaining inorganic carbon isotopically enriched (ie., $\delta\,^{13}C \sim +2^o/oo$). The gradual oxidation of falling organic debris results in a profile of carbon isotopic composition that varies with depth and whose relation to dissolved oxygen and total inorganic carbon, in the same profile, forms the basis of several fascinating studies (*Deuser and Hunt*, 1969; *Craig*, 1970; *Kroopnick, Deuser and Craig*, 1970; *Kroopnick*, 1974). Typical values in deep water are +1.1°/oo in the North Atlantic Ocean, +0.6°/oo in the South Pacific Ocean, about 0.0°/oo (P.D.B.) in the Equatorial Pacific Ocean, and -0.1°/oo in the North Pacific Ocean (*Duplessy*, 1972; *Kroopnick*, 1971). This trend reflects the slow passage of deep water to the south in the Atlantic Ocean and thence north in the Pacific Ocean with the gradual depletion in ^{13}C of the dissolved CO_2 by the addition of the oxidation products of organic debris sinking from the ocean surface.

CARBON ISOTOPE VARIATIONS IN FORAMINIFERA

The ^{13}C content of foraminifera is measured by mass spectrometry in the same manner as the ^{18}O content (*Shackleton and Opdyke*, 1973). Measurements are expressed in the δ notation; where δ ($^o/oo$) is defined by:

$$\delta = 1000 \; \frac{^{13}C/^{12}C \text{ sample}}{^{13}C/^{12}C \text{ standard}} - 1 \qquad (1)$$

and the standard is P.D.B. belemnite, calibrated through the NBS-20 standard using the calibration of *Craig* (1957).

Analytical precision, as determined by replicate analyses of small aliquots of standard carbonates, is about ± 0.07°/oo (1-δ). Taking all aspects of sampling and analysis into account, the precision with which the ^{13}C content of the foraminifera in a particular core section may be estimated on the basis of a single analysis is probably about ± 0.12°/oo.

Since the capability for analyzing carbon isotopic composition in foraminifera has existed for over twenty years, it is reasonable to ask why it is that so little research has been done in this area. *Broecker* (1971a) investigated changes in the ocean carbon/phosphorus ratio with time by considering the carbon isotopic composition of planktonic and benthonic foraminifera. However, his inferences are somewhat tenuous because (1) different benthonic foraminiferal species show different departures from carbon isotopic equilibrium with sea water bicarbonate (*Shackleton*, unpublished data); (2) this is also true for different planktonic species (*Shackleton and Vincent*, in prep.); and (3) in the near-surface water where the carbon isotope gradient is steep, uncertainty in the depth of carbonate secretion precludes the estimation of the carbon isotopic composition at the surface.

Recently *Sommer, Matthews and Shackleton* (in prep.) have obtained correlatable carbon isotopic records for planktonic foraminifera from several cores. However, there are areas where a quite different pattern, or no perceptible pattern, emerges, perhaps as a result of the complications mentioned above. ^{13}C data obtained from benthonic foraminifera are presented here; hopefully their interpretation is subject to fewer complications.

Until recently very few monospecific samples of benthonic foraminifera from cores had been analyzed isotopically. *Duplessy, Chenouard and Reyss* (1974) and *Shackleton and Opdyke* (1973) obtained the first continuous records through the last glacial cycle. *Shackleton* (1974) showed that *Uvigerina* are close to isotopic equilbrium for oxygen and yield reproducible values for $\delta^{18}O$. *Ninkovich and Shackleton* (1975) published detailed oxygen isotope records based on this genus (Figure 1). Oxygen and carbon isotope measurements for several cores (Table 3) are given in Tables 4 to 7, and several points can be made. There is a trend towards isotopically lighter carbon in the Pacific Ocean relative to the Atlantic Ocean. The inter-ocean difference in *Uvigerina* $\delta^{13}C$ is more or less equivalent to the difference in dissolved CO_2 $\delta^{13}C$ discussed above. In another genus, *Planulina wuellerstorfi*, $\delta^{13}C$ shows a trend which is not proportional to the $\delta^{13}C$ change in dissolved CO_2, from which it is deduced that another unknown factor contributes to variation in the ^{13}C content of the test of *P. wuellerstorfi*.

The *Uvigerina* $\delta^{13}C$ is also lighter in the eastern Equatorial Pacific Ocean (core V19-29) than in the western Equatorial Pacific Ocean

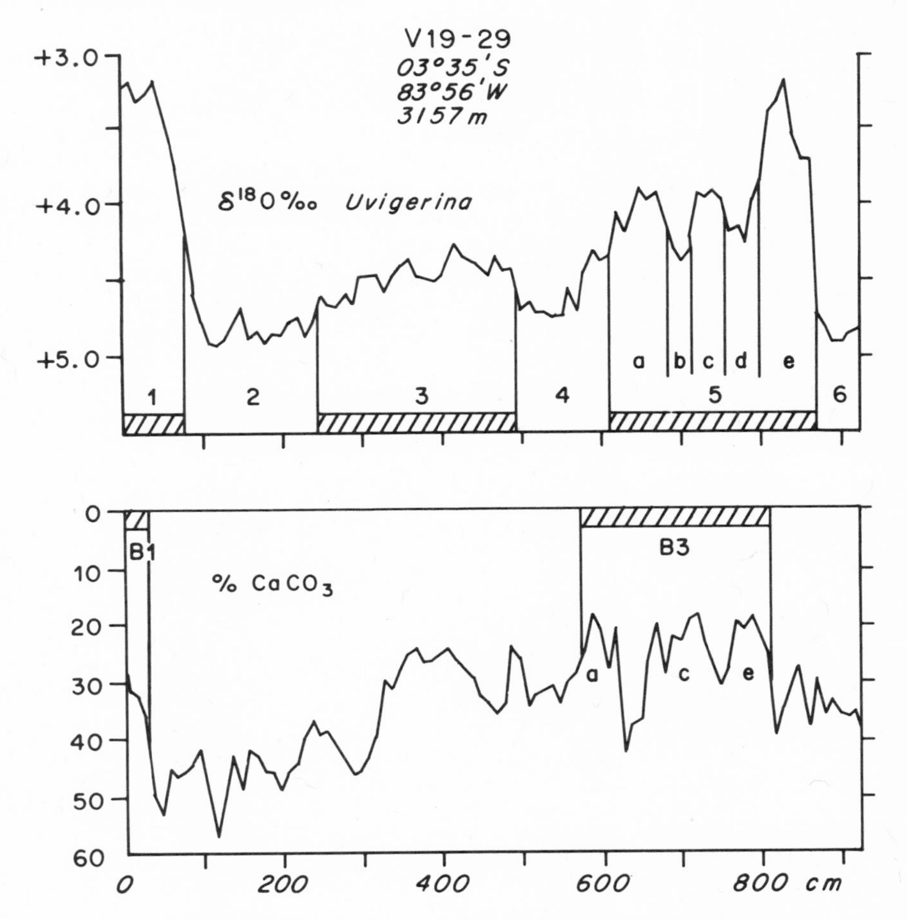

Fig. 1. Oxygen isotope and carbonate per cent record for core V19-29 (from *Ninkovich and Shackleton*, 1975). The section spans about 140 *ka*.

(core V28-238) as would be expected from the aging history of the deep water mass and from the additional oxidation of organic matter below the fertile surface waters of the Panama region. It is concluded that a study of ^{13}C distribution in the fossil benthonic foraminifera will provide a valuable additional tool for the study of water mass movements and for the location of regions of former high productivity.

Finally and most significantly, it is noticed that glacial samples have a $\delta\ ^{13}C$ several tenths of a part per mil lighter than Holocene samples. In Figures 2, 3 and 4 three continuous ^{13}C records are shown. In Table 7 are data from a few other cores in which there

TABLE 3. Location of Cores.

Core	Latitude	Longitude	Depth (m)
12392	25°10'N	16°50'W	2573
V22-174	10°04'S	12°49'W	2630
V19-29	03°35'S	83°56'W	3157
V28-238	01°01'N	160°29'E	3120
RC11-120	43°31'S	79°52'E	3193
RC12-267	38°41'S	25°47'E	4144

are only isolated measurements for the Holocene and glacial periods. As may be seen from Figure 3, in core V22-174 the $\delta^{13}C$ record of planktonic species is parallel to that of benthonic species. This implies that a change in the $\delta^{13}C$ of ocean-dissolved CO_2 is reflected in these data. The whole ocean contains about 3.5×10^{19} g carbon ($\delta^{13}C$ 0°/oo). The admixture of 10^{18} g carbon by destruction of plant biomass ($\delta^{13}C$ -25°/oo) would change the ^{13}C content of the ocean CO_2 by about 0.7 per mil. If neutralized by the dissolution of marine carbonate, this figure would be halved. Thereafter, since the response time of CO_2 in the ocean is about 300 *ka* (Table 1), isotopic steady state would not be approached before a new continental climatic regime reversed the change.

CARBON ISOTOPES IN PLANKTONIC FORAMINIFERA

As indicated by *Sommer et al.* (in prep.) the $\delta^{13}C$ record of shallow-living planktonic species varies in detail from area to area. The reason for this is exemplified by the following. In core V19-29 on the Carnegie Rise (*Ninkovich and Shackleton,* 1975) the ^{18}O content of *Uvigerina* is about 1.6°/oo greater at the glacial maximum than today. This figure is an estimate of the interglacial-to-glacial change in ocean water $\delta^{18}O$. In this area, *Globigerinoides sacculifer,* which lives relatively near the surface, registers a similar ^{18}O difference. Additionally, glacial samples of both *G. sacculifer* and *Uvigerina* are depleted in ^{13}C, in accordance with the hypothesis that in the glacial period $\delta^{13}C$ in ocean CO_2 was more negative. However, the deeper-living species *Neogloboquadrina dutertrei* and *Globorotalia menardii* register smaller ^{18}O variations. Therefore, they must have lived in warmer water during the glacial than they do today (*Shackleton,* 1968), presumably because they are morphologically adapted to live in water of a particular density; the glacial ocean was saltier and hence

TABLE 4. Oxygen and Carbon Isotope Data, Core V22-174.

	$\delta^{18}O$			$\delta^{13}C$		
depth (cm)	*G. ruber*	*G. sacculifer*	*U. auberiana*	*G. ruber*	*G. sacculifer*	*U. auberiana*
0	-1.23	-0.92	+3.07	+1.62	+2.18	+0.48
10	-1.21	-0.74	+3.23	+1.25	+2.03	+0.56
20	-1.01	-0.59	+3.09	+1.30	+1.90	+0.28
30	-0.73	-0.14	+3.44	+1.16	+1.73	+0.03
40	-0.22	+0.22	+3.99	+0.98	+1.65	+0.09
43	-0.18	+0.41	-	+1.09	+1.74	-
47	+0.01	+0.53	+4.74	+1.00	+1.90	+0.11
50	+0.12	+0.64	+4.73	+0.89	+1.69	+0.14
53	+0.28	+0.49	+4.74	+1.09	+1.90	+0.04
57	+0.31	+0.81	+4.71	+1.18	+2.04	+0.24
60	+0.29	+0.73	+4.50	+0.96	+1.89	+0.24
63	+0.32	+0.66	+4.58	+1.00	+1.82	+0.15
67	+0.08	+0.66	+4.61	+0.88	+1.87	+0.06
70	+0.08	+0.53	+4.60	+0.93	+1.82	+0.28
80	+0.02	+0.51	+4.52	+1.10	+1.39	+0.17
90	+0.02	+0.31	+4.31	+0.92	+1.81	+0.26

TABLE 5. Oxygen and Carbon Isotope Data, Meteor Core 12392.

depth (cm)	$\delta^{18}O$			$\delta^{13}C$		
	Uvigerina peregrina	Planulina wueller-storfi	Melonis pompilloides	Uvigerina peregrina	Planulina wueller-storfi	Melonis pompilloides
2-4.5		+2.69			+1.01	
7.5	+3.34	+2.80		+0.19	+0.80	
12.5-15		+2.58			+1.02	
22.5-25	+3.16	+2.56		+0.26	+0.99	
30-32.5	+3.05			+0.10		
50-52.5	+3.82			-0.91		
60-62.5	+4.28	+3.32		-0.77	+0.22	
70-72.5	+4.43			-0.95		
75-77.5			+4.38			-1.00
80-82.5	+4.58	+3.62	+4.20	-0.91	+0.23	-1.07
90-92.5	+4.67			-1.08		
100-102.5	+5.28	+4.57	+4.93	-0.94	+0.37	-0.82
110-112.5			+4.85			-0.93
120-122.5	+5.30	+4.56	+5.06	-1.04	+0.23	-0.94
130-132.5			+4.76			-0.99
140-142.5		+4.51			+0.25	
150-152.5			+4.52			-0.96
160-162.5			+4.57			-0.95
170-172.5	+4.64		+4.68	-1.13		-0.98
180-182.5	+4.95		+4.69	-1.21		-1.05
190-192.5	+4.93			-1.30		
200-202.5	+5.18			-1.20		
210-212.5	+4.78			-1.08		
220-222.5	+5.05			-0.99		

TABLE 5. Oxygen and Carbon Isotope Data, Meteor Core 12392 (Continued)

depth (cm)	$\delta^{18}O$			$\delta^{13}C$		
	Uvigerina peregrina	Planulina wueller-storfi	Melonis pompilloides	Uvigerina peregrina	Planulina wueller-storfi	Melonis pompilloides
230-232.5			+4.46			-1.02
240-242.5	+4.59		+4.56	-0.93		-0.96
250-252.5	+4.78			-1.06		
260-262.5	+4.76			-1.00		
270-272.5	+4.81		+4.44	-1.29		-1.07
280-282.5	+4.51		+4.32	-1.23		-1.01
290-292.5			+4.41			-0.96
300-302.5	+4.60			-0.87		
310-312.5	+4.62			-1.14		
320-322.5	+4.66			-1.14		
330-332.5	+4.66			-1.19		
340-342.5	+4.51			-1.07		
350-352.5			+4.15			-1.21
360-362.5			+4.23			-1.10
373-375.5			+4.23			-0.93
380-382.5						
383-385.5			+4.36			-0.88
393-395.5	+4.51		+4.35	-0.88		-1.02
400-402.5			+4.33			-0.97
413-415.5	+4.54			-1.05		
420-422.5	+4.63		+4.32	-0.72		-1.00
450-452.5			+4.27			-1.02
460-462.5			+4.15			-1.15
473-475.5			+4.17			-1.27

TABLE 5. Oxygen and Carbon Isotope Data, Meteor Core 12392 (Continued).

depth (cm)	$\delta^{18}O$			$\delta^{13}C$		
	Uvigerina peregrina	Planulina wueller-storfi	Melonis pompilloides	Uvigerina peregrina	Planulina wueller-storfi	Melonis pompilloides
480-482.5			+4.22			-1.19
493-495.5			+4.27			-0.80
503-505.5			+4.19			-1.22
513-515.5	+4.76			-1.34		
523-525.5			+4.44			-1.31
533-535.5	+4.75			-0.25		
543-545.5	+4.67			-0.82		
553-555.5	+4.82			-0.91		
563-565.5	+4.55			-0.71		
573-575.5	+4.47			-0.94		
585-585.5	+4.60			-0.59		
590-595.5	+4.30			-0.79		
603-605.5	+4.37			-0.41		
613-615.5		+3.46			+0.74	
623-625.5		+3.33			+0.88	
633-635.5		+3.16			+0.82	
653-655.5	+4.23			-0.31		
663-665.5		+3.70			+0.67	
673-675.5		+3.60				
683-685.5		+3.27			+0.92	
693-695.5		+3.29			+0.73	
702.5-705		+3.11			+0.91	
713-715.5		+2.97			+0.66	
723.5-725.5		+3.23			+0.32	

TABLE 5. Oxygen and Carbon Isotope Data, Meteor Core 12392 (Continued).

	$\delta^{18}O$			$\delta^{13}C$		
depth (cm)	Uvigerina peregrina	Planulina wuellerstorfi	Melonis pompilloides	Uvigerina peregrina	Planulina wuellerstorfi	Melonis pompilloides
733.5-735.5		+3.25			+0.59	
743-745.5		+3.11			+0.36	
753-755.5	+4.35			-0.06		
763-765.5		+3.48			+0.59	
773-775.5	+4.04			+0.01		
782.5-785		+3.13			+0.73	
793-795.5	+3.75			+0.03		
803-805.5	+3.92			0.00		
813-815.5		+2.42			+0.70	
823-825.5	+3.17	+2.51		-0.38	+0.67	
833-835.5		+2.34			+0.28	
842-846	+4.12			-0.51		
855-857.5	+4.54			-1.03		
863-865.5	+4.52			-1.27		
875-877.5	+4.69			-1.47		
883-885.5	+4.94			-1.59		
893-895.5	+4.88			-1.42		
903-905.5	+4.95			-1.45		
913-915.5	+4.89			-1.42		
923-925.5			+4.58			-1.41
933-935.5	+5.11			-1.69		
943-945.5			+4.48			-0.16
952-954.5			+4.71			-1.41
960-963	+4.96		+4.80	-1.53		-1.40

TABLE 6. Oxygen and Carbon Isotope Data for *Uvigerina peregrina*, Core Y71-6-12MG.

depth (cm)	$\delta^{18}O$	$\delta^{13}C$
0-1	+3.51	-0.74
6-7	+3.58	-0.92
12-13	+4.55	-1.04
18-19	+4.68	-1.09
24-25	+4.73	-1.30
30-31	+4.57	-1.30
36-37	+4.80	-1.39
42-43	+4.73	-1.34
48-49	+4.71	-1.33
54-55	+4.49	-1.27
60-61	+4.35	-1.30
66-67	+4.42	-1.27
72-73	+4.72	-1.20
78-79	+4.30	-1.37
84-85	+4.35	-1.33
90-91	+4.10	-1.20
96-97		
	+4.17	-1.25
102-103		
108-109	+4.64	-1.39
114-115	+4.38	-1.04
120-121	+4.27	-0.98
126-127	+3.83	-1.11
132-133	+3.88	-1.15
138-139	+4.07	-1.30
144-145	+4.01	-1.04
150-151	+4.03	-1.14
156-157	+3.62	-1.31
162-163	+4.06	-1.21
168-169	+3.81	-1.50
174-175	+4.02	-1.01
182-183	+4.53	-1.09

TABLE 7. Oxygen and carbon isotope data, other cores.

core	depth (cm)	species	$\delta^{18}O$	$\delta^{13}C$
RC11-120	5	Uvigerina peregrina	+3.52	-0.09
RC11-120	75	Uvigerina peregrina	+5.22	-0.92
RC12-267	2-3	Uvigerina peregrina	+3.31	-0.13
RC12-267	70-71	Uvigerina peregrina	+4.84	-0.88
V19-29	10	Uvigerina proboscidea	+3.25	-0.85
V19-29	10	Uvigerina proboscidea	+3.42	-0.65
V19-29	190	Uvigerina proboscidea	+4.92	-1.45
V19-29	200	Uvigerina proboscidea	+4.87	-1.49
V28-238	35	Uvigerina peregrina	+4.76	-0.76

In each core a sample from the recent and from the last glacial maximum are compared (see Hays, Lozano, Shackleton & Irving (1976) for subantarctic cores RC11-120 and RC12-267, Ninkovich & Shackleton (1975) for Panama Basin core V19-29, and Shackleton & Opdyke (1973) for Western Equatorial Pacific core V28-238 in which the recent section contained insufficient Uvigerina for reliable analysis).

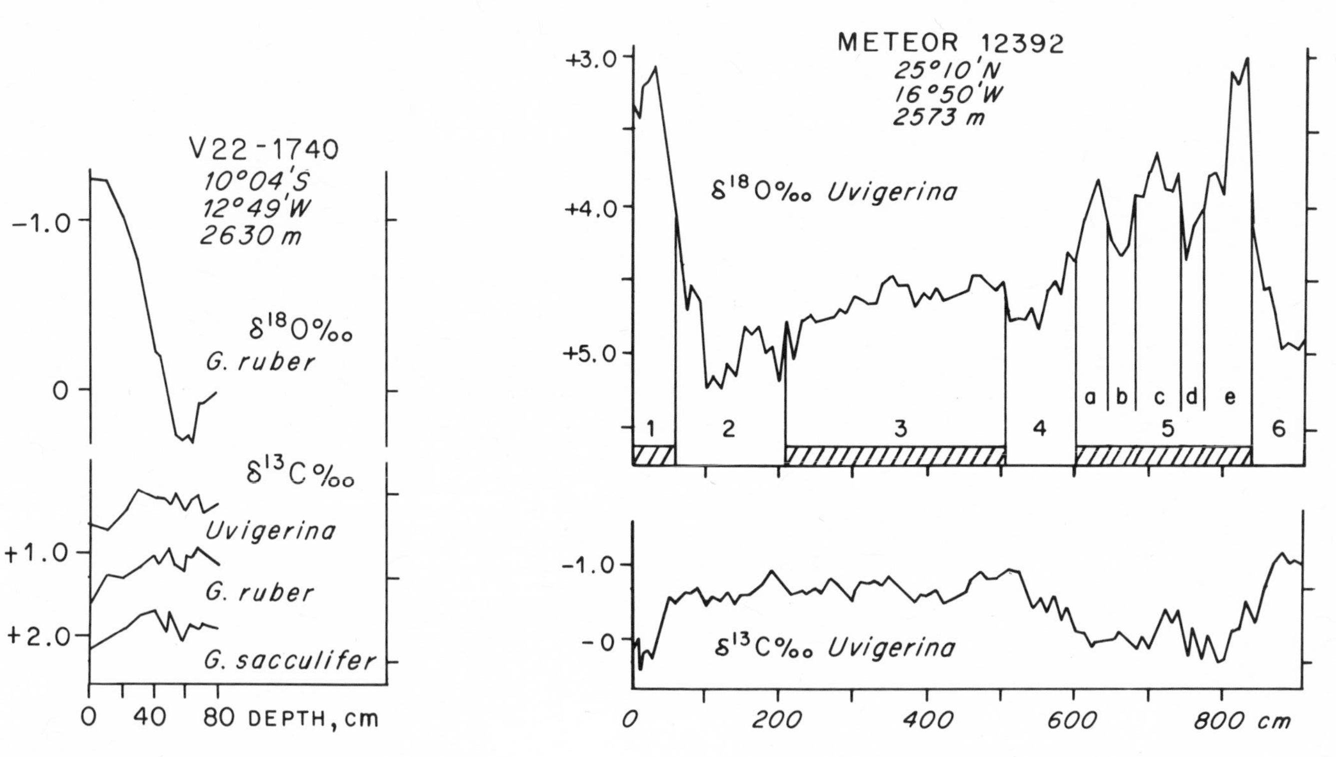

Fig. 2. Oxygen isotope and carbon isotope record of core V22-174. The section illustrated covers about 25 *ka*.

Fig. 3. Oxygen isotope and carbon isotope record of core 12392. The section spans about 140 *ka*.

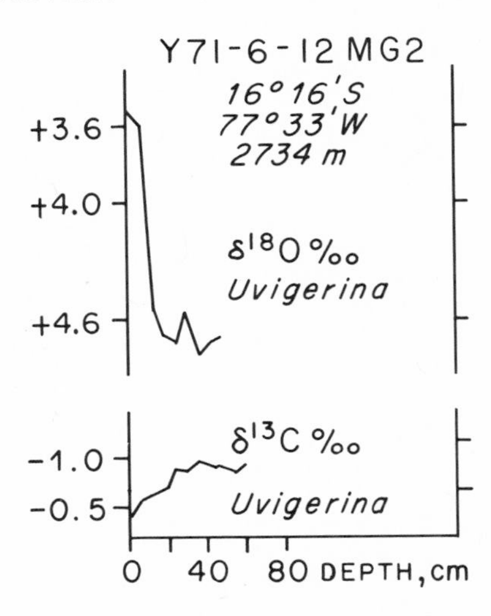

Fig. 4. Oxygen isotope and carbon isotope record of core Y71-6-12MG2. The section illustrated spans about 20 *ka*.

denser due to the storage of fresh water as ice sheets. Compensation was achieved by living nearer the ocean surface. The carbon isotope values for these two species support this observation; they show higher ^{13}C content reflecting the ^{13}C enrichment towards the surface. This vertical migration prevents using $\delta\ ^{13}C$ in these species to monitor the ^{13}C content of ocean deep CO_2. However, it has been shown by *Shackleton and Vincent* (in prep.) that *Globorotalia scitula gigantea* secretes the bulk of its carbonate in very cold, deep water (in the region studied, in the Antarctic Intermediate Water). At this depth, there is a low vertical ^{13}C gradient. It is, therefore, significant that in the core selected for glacial-interglacial comparison, *G. scitula gigantea* was found to be lighter in ^{13}C in the glacial sample. Perhaps the true variation in ocean ^{13}C content would be best obtained by analyzing this species, or another living below the oxygen minimum, where the ^{13}C profile is essentially constant over a wide depth range.

CARBONATE DISSOLUTION

It is well known that in the eastern Equatorial Pacific Ocean, the carbonate content of sediment cores displays quasi-cyclic fluctuations (*Arrhenius*, 1952). Broadly speaking, these fluctuations may be related to climatic changes (*Arrhenius*, 1952; and *Hays, Saito, Opdyke and Burckle*, 1969), higher percentages of carbonate being found in glacial age sediments. *Arrhenius* (1952) interpreted the variations primarily in terms of productivity changes. However, workers more recently have tended to the view that changes in dissolution rate are the primary controlling mechanism (e.g., *Broecker*, 1971b; *Berger*, 1973) and that changes in productivity remain to be demonstrated to be of significant effect.

Until recently, a proper investigation of these important features was hampered by a misleading concept of the Pleistocene climatic record, and by the absence of any independent stratigraphic scale against which the carbonate fluctuations could be studied in detail. The important study by *Broecker* (1971b) assumed a roughly square-wave climatic driving function with a 100 *ka* period, and assumed synchroneity between the climatic changes and the carbonate responses. As it turns out, this climatic model is too simple to reveal the mechanism involved. Both the climatic record derived from $\delta\,^{18}O$ measurements, and the $CaCO_3$ record, contain important energy with a 22 *ka* period (*Luz*, 1973; *Ninkovich and Shackleton*, 1975).

The establishment of a detailed oxygen isotope stratigraphy in the Pacific Ocean (*Shackleton and Opdyke*, 1973) has permitted the temporal relationship between climatic changes and dissolution events to be investigated. It is now known that episodes of intense dissolution occurred at interglacial-glacial transitions, rather than being centered on interglacials (*Luz*, 1973; *Luz and Shackleton*, 1975; *Ninkovich and Shackleton*, 1975; *Shackleton and Opdyke*, 1976). This has been viewed as a lead-lag phenomenon, and *Moore et al.* (1977), taking this viewpoint, have calculated that changes in the oxygen isotope ratio (reflecting terrestrial ice volume and hence extent of glaciation) lead changes in $CaCO_3$ content in core V19-29 by an average of 6 *ka*. *Luz and Shackleton* (1975) suggested that the carbonate dissolution lag reflected the response of the ocean carbonate ion system to a change elsewhere; for example, increased $CaCO_3$ deposition at high latitudes or in the North Atlantic Ocean (cf. *Olausson*, 1967). Here an alternative model is proposed, that it is in reality controlled directly by the injection of CO_2 into the ocean as a result of the destruction of terrestrial vegetation, and that the measured lag approximates one-fourth of the dominant 22 *ka* period in the climatic record.

The possibility of as much as 10^{18} *g* carbon being added to the ocean (as CO_2) in 10 *ka* has been discussed above. At most, this could dissolve about 8×10^{18} *g* $CaCO_3$. This figure must be compared with the observed calcite dissolution event.

Berger (in press) discussed oceanic dissolution in terms of the variation with depth in the actual carbonate dissolution rate, and cited a figure equivalent to an increase in dissolution rate of $2\ g/cm^2/ka/km$ of depth. Approximately 20% of the ocean is in the depth range 2 to 3 *km*, so that a Carbonate Compensation Depth (CCD) movement of 200 *m* (*Berger*, in press) might transfer 4% of the ocean floor, about $1.4 \times 10^{17}\ cm^2$, from above to below the lysocline. This would result in the dissolution of 3×10^{18} *g* carbonate during 10 *ka*, assuming that the $CaCO_3$ input to the top of the 1 *km* dissolution transition is 2 $g/cm^2/ka$. This calculation, though somewhat crude, indicates that each dissolution pulse involved the dissolution of less carbonate than could have been dissolved by means of the mechanism postulated.

It is concluded that climatically induced changes in the terrestrial biosphere have been of sufficient magnitude and rate to give rise to the observed carbonate dissolution cycles. Apart from the inherent interest in explaining a well known phenomenon, this conclusion clearly points to the necessity for modelling the response of the ocean to a changing CO_2 flux. The response of the oceans to Pleistocene changes will never be understood if only steady-state conditions are modeled.

DEGLACIATION, PLUVIALS AND PTEROPODS

Just as the concept of tropical aridity at the time of glacial maximum is becoming widely known, so the timing of maximum tropical wetness is becoming better appreciated. It is becoming increasingly clear that the so-called pluvial, with accompanying high lake levels in tropical Africa, began around 13 *ka* ago and was maximal around 9 *ka* ago (*Street and Grove*, 1976). In Africa, tropical forests were considerably more extensive 8 *ka* ago than today (*Hamilton*, 1976), so that the accumulation of the terrestrial biosphere occurred well within 10 *ka*, and the rate of removal of CO_2 from the ocean, via the atmosphere to the biomass, was as great as the rate of its injection at the beginning of a glacial.

To some extent, this would have had the opposite effect to injecting CO_2; there would have been an interval of diminished dissolution on the sea floor. However, there is one respect in which the effect is worthy of special remark. The extraction of CO_2 would be from warm surface waters, whereas its uptake would be in cold waters. Thus, during the period of forest growth, Atlantic Ocean waters could have become supersaturated to a substantially greater depth, and it is possible that this gave rise to the pulse in pteropod abundance seen in sediments of the deglacial interval (*Diester-Haass, Schrader and Thiede*, 1973). However, this conclusion is somewhat speculative because (1) *Chen* (1968) reported a peak in the abundance of pteropods at the glacial maximum in sediments in the Caribbean Sea; (2) in some cores (e.g., M8058B) there are two pteropod peaks (*Kudrass*, 1973); and (3) a clear correlation between the $\delta^{13}C$ change with the pteropod pulse is not yet proved. It is worth noting that in many of the cores examined by *Luz and Shackleton* (1975) the best preservation (lowest Solution Index) was found in the sections representing deglaciation. Thus, although carbonate preservation, especially in the Atlantic Ocean, is affected by several factors in addition to the one discussed here (e.g., AABW production), there may have been times when injection or removal of CO_2 was the dominant factor, so that the effect occurred essentially synchronously in both oceans.

SUMMARY AND INTERPRETATION

Figure 3 shows the oxygen and carbon isotope record of core 12392 (25°10.3'N; 16°50.7'W, 2573 *m*). The two records, though expressed as if for *Uvigerina*, are in fact based on three different genera: *Uvigerina peregrina*, *Melonis pompilloides*, and *Planulina wuellerstorfi*, as no single species is sufficiently abundant throughout the core. On the basis of areas of overlap between species isotopic differences have been estimated between *Uvigerina peregrina* and the other two species, and these differences have been used to obtain a 'correction factor', to estimate the record that would have been obtained had *Uvigerina* been abundant throughout. The stratigraphy of the core is indicated by the stages and substages of *Emiliani* (1955) and *Shackleton* (1969).

Substage 5e is believed to correlate with the Eemian interglacial in Northwest Europe (*Shackleton*, 1969). By analogy with the Holocene, it may be assumed that tropical forests were extensively developed. Sea level may have been a little higher than today (*Veeh*, 1966). By substage 5d, only 10 *ka* later, the sea had dropped 60 *m* (*Steinen, Harrison and Matthews*, 1973; *Chappell*, 1974b), ice had accumulated on North America, dust was blowing off the Sahara onto the Atlantic Ocean, loess was blowing over the remains of soil formed under a temperate forest in Czechoslovakia (*Kukla and Koci*, 1972) and isotopically light carbon dioxide had been released into the atmosphere at a rate of at least 10^{14} *g C* per year. This rendered $\delta^{13}C$ in oceanic dissolved CO_2 more negative and caused a dissolution event in the Pacific Ocean. This is the carbonate minimum denoted B3e in Figure 1. The ^{13}C event is clearly seen in the records shown by *Sommer et al.* (in prep.).

Judging from the carbon isotope record, tropical vegetation recovered in substage 5c and did not suffer catastrophic destruction again until the end of stage 5. At this time an even more striking carbon isotope change is seen, and the dissolution even (B3a in Figure 1) is even manifested in the Equatorial Atlantic Ocean (*Crowley*, 1975). Again the carbon isotope event is evident in the records shown by *Sommer et al.* (in prep.).

Figure 3 shows little change between 70 *ka* and the most recent revegetation at the end of the last glacial. Although the records shown by *Sommer et al.* (in prep.) for planktonic foraminifera show fluctuations in this interval, they seem not to be synchronous and may not depict massive biomass changes but may represent local events.

Between about 15 *ka* and 5 *ka* ago the increasing plant biomass pumped CO_2 out of the ocean (via the atmosphere) again at a rate exceeding 10^{14} *g C* per year, giving rise to exceptionally good preservation of carbonate and particularly of aragonitic pteropods.

It is difficult to study the past few thousand years because of the difficulty in recovering the topmost sediment intact, but there is some indication of a late Holocene isotopic and dissolution even (denoted B1 in Figure 1), perhaps reflecting man's influence as well as that of partial tropical aridification.

ACKNOWLEDGEMENTS

This research was supported by N.E.R.C. grant GR3/1762. I am indebted to M. A. Hall for operating the VG Micromass 602C mass spectrometer on which the analyses were made; to Lamont-Doherty Geological Observatory for a Senior Visiting Research Fellowship for 1974-75; to L.D.G.O. for sediment samples taken and curated with the support of the Office of Naval Research (N00014-75C-0210) and the National Science Foundation (DES72-01568-A04); to Oregon State University for use of their core collection; and to E. Siebold and J. Thiede for samples from Meteor core 12392.

I am grateful to many of my CLIMAP colleagues for stimulating discussions of Pleistocene sediments and climates; to A. T. Grove and F. A. Street and to A. C. Hamilton for permission to quote their unpublished manuscripts, and to J. M. Bowler, P. L. Carter, J. Chappell, R. W. Fairbridge, A. Hamblin, A. P. Kershaw, D. Walker, and R. G. West for discussions on the continental record. I am grateful to Anne Boersma for assistance and enthusiastic discussion in the world of benthonic foraminifera. Discussions with W. S. Broecker and a very careful review by S. M. Savin contributed substantially to the final manuscript, as did the assistance of Heather Jarman. Contribution No. 2494 of Lamont-Doherty Geological Observatory.

REFERENCES

Arrhenius, G. 1952. Sediment cores from the East Pacific. *Swedish Deep-Sea Expedition Reports* 5. 6-227.

Berger, W. H. 1973. Deep-sea carbonates: Pleistocene dissolution cycles. *Journal of Foraminiferal Research* 3. 187-195.

Berger, W. H. In press. Sedimentation of deep-sea carbonate: maps and models of variations and fluctuations. In: *Marine Plankton and Sediments*, W. R. Riedel and T. Saito (Eds.). New York: Micropalaeontology Press.

Bowler, J. M., G. S. Hope, J. N. Jennings, G. Singh, and D. Walker. 1975. Late Quaternary climates of Australia and New Guinea. *Quaternary REsearch, 6*. 359-394.

Broecker, W. S. 1971a. A kinetic model for the chemical composition of sea water. *Quaternary Research 1*. 188-207.

Broecker, W. S. 1971b. Calcite accumulation rates and glacial to interglacial changes in ocean mixing. In: *The Late Cenozoic Glacial Ages*, K. K. Turekian (Ed.), Yale University Press. 239-265.

Broecker, W. S. 1974. *Chemical Oceanography*. New York: Harcourt Brace Jovanovich.

Chappell, J. 1974a. Geology of coral terraces, Huon Peninsula, New Guinea: a study of Quaternary tectonic movements and sea-level changes. *Geological Society of America Bulletin 85*. 553-570.

Chappell, J. 1974b. Relationships between sea levels, ^{18}O variations and orbital perturbations, during the past 250,000 years. *Nature 252*. 199-202.

CLIMAP. 1976. The surface of the ice-age earth. *Science 191*. 1131-1137.

Chen, C. 1968. Pleistocene pteropods in pealgic sediments. *Nature 219*. 1145-1148.

Craig, H. 1953. The geochemistry of the stable isotopes of carbon. *Geochimica et Cosmochimica Acta 3*. 53-72.

Craig, H. 1954. Carbon 13 in plants and the relationships between carbon 13 and carbon 14 variations in nature. *The Journal of Geology 62*. 115-149.

Craig, H. 1957. Isotopic standards of carbon and oxygen and correction factors for mass-spectrometric analysis of carbon dioxide. *Geochimica et Cosmochimica Acta 12*. 133-149.

Craig, H. 1970. Abyssal carbon 13 in the South Pacific. *Journal of Geophysical Research 75*. 691-695.

Crowley, T. J. 1975. *Fluctuations of the Eastern North Atlantic Gyre during the Last 150,000 Years*. Ph.D. Thesis, Brown University.

Damuth, J. E. 1975. Quaternary climate change as revealed by calcium carbonate fluctuations in western Equatorial Atlantic sediments. *Deep-Sea Research 22*. 725-743.

Damuth, J. E. and R. W. Fairbridge. 1970. Equatorial Atlantic deep-sea arkosic sands and ice-age aridity in tropical South America. *Geological Society of America Bulletin 81*. 189-206.

Dansereau, P. 1957. *Biogeography, an Ecological Perspective*. New York: Ronald Press Co.

Deuser, W. G. and J. M. Hunt. 1969. Stable isotope ratios of dissolved inorganic carbon in the Atlantic. *Deep-Sea Research 16*. 221-225.

Diester-Haass, L., H.-J. Schrader, and J. Thiede. 1973. Sedimentological and paleoclimatological investigations of two pelagic ooze cores off Cape Barbas, North-West Africa. *"Meteor" Forschungsergebnisse C, No. 7*. 19-66.

Duplessy, J. C. 1972. *La géochimie des isotopes stables du carbone dans la mer, CEA-N-1565*. Paris: Commissariat à l'Energie Atomique.

Duplessy, J. C., L. Chenouard, and J. L. Reyss. 1974. Paléo-températures isotopiques de l'Atlantique Equatorial. *Colloques Internationaux du Centre National de la Recherche Scientifique 219*. 251-258.

Emrich, K., D. H. Ehhalt, and J. C. Vogel. 1970. Carbon isotope fractionation during the precipitation of calcium carbonate. *Earth and Planetary Science Letters 8*. 363-371.

Emiliani, C. 1955. Pleistocene temperatures. *Journal of Geology 63*. 538-578.

Frenzel, B. 1968. The Pleistocene vegetation of northern Eurasia. *Science 161*. 637-649.

Gordon, A. L. 1975. General ocean circulation. In: *Numerical Models of Ocean Circulation*, Washington, D. C., National Academy of Science. 39-53.

Hamilton, A. C. 1976. The significance of pattern of distribution shown by forest plants and animals in tropical Africa for the reconstruction of paleoenvironments: a review. *Palaeoecology of Africa 9*. 63-97

Hays, J. D., J. Imbrie, N. J. Shackleton. 1976. Variations in the earth's orbit: pacemaker of the ice ages. *Science, 196*, NY. 1121-1132.

Hays, J. D., J. Lozano, N. Shackleton, and G. Irving. 1976. Reconstruction of the Atlantic Ocean and western Indian Ocean sectors of the 18,000 B.P. Antarctic Ocean. In: *Investigation of Late Quaternary Paleoceanography and Paleoclimatology*, R. M. Cline and J. C. Hays (Ed.). *Geological Society of America Memoir 145*. 337-372.

Hays, J. D. and A. Perruzza. 1972. The significance of calcium carbonate oscillations in eastern Equatorial Atlantic deep-sea sediments for the end of the Holocene warm interval. *Quaternary Research 2*. 355-362.

Hays, J. D., T. Saito, N. D. Opdyke, and L. H. Burckle. 1969. Pliocene-Pleistocene sediments of the equatorial Pacific: their palaeomagnetic, biostratigraphic, and climatic record. *Geological Society of America Bulletin 80*. 1481-1514.

Kershaw, A. P. 1974. A long continuous pollen sequence from north-east Queensland, Australia. *Nature 251*. 222-223.

Kroopnick, P. 1971. *Oxygen and Carbon in the Oceans and Atmosphere: Stable Isotopes as Tracers for Consumption, Production, and Circulation Models*. Ph.D. Thesis, University of California at San Diego.

Kroopnick, P. 1974. The dissolved O_2-CO_2-^{13}C system in the eastern equatorial Pacific. *Deep-Sea Research 21*. 211-227.

Kroopnick, P., W. G. Deuser, and H. Craig. 1970. Carbon 13 measurements on dissolved inorganic carbon at the North Pacific (1969) GEOSECS station. *Journal of Geophysical Research 75*. 7668-7671.

Kudrass, H.-R. 1973. Sedimentation am Kontinentalhang vor Portugal und Marokko im Spätpleistozän und Holozän. *"Meteor" Forschungsergebnisse C, No. 13*. 1-63.

Kukla, G. J. and A. Kocí. 1972. End of the last interglacial in the loess record. *Quaternary Research 2.* 374-383.

Luz, B. 1973. Stratigraphic and paleoclimatic analysis of Late Pleistocene tropical southeast Pacific cores (with an Appendix by N. J. Shackleton). *Quaternary Research 3.* 56-72.

Luz, B. and N. J. Shackleton. 1975. $CaCO_3$ solution in the tropical east Pacific during the past 130,000 years. *Cushman Foundation for Foraminiferal Research Special Publication 13.* 142-150.

Mesolella, K. J., R. K. Matthews, W. S. Broecker, and D. L. Thurber. 1969. The astronomical theory of climatic change: Barbados data. *Journal of Geology 77.* 250-274.

Moore, T. C., Jr., N. G. Pisias and G. R. Heath. 1977. Climate Changes and Lags in Pacific Carbonate Preservation, Sea Surface Temperature and Global Ice Volume. In: *The Fate of Fossil Fuel* CO_2, N. R. Andersen and A. Malahoff (Eds.), Plenum, N.Y. (this volume).

Ninkovich, D. and N. J. Shackleton. 1975. Distribution, stratigraphic position and age of ash layer "L", in the Panama Basin region. *Earth and Planetary Science Letters 27.* 20-34.

Nix, H. A. and J. D. Kalma. 1972. Climate as a dominant control in the biogeography of northern Australia and New Guinea. In: *Bridge and Barrier: The Natural and Cultural History of the Torres Strait,* D. Walker (Ed.). Department of Biogeography and Geomorphology Publication BG/3, A.N.U. Canberra, Australia. 61-92.

Olausson, E. 1967. Climatological, geoeconomical and paleo-oceanographical aspects on carbonate deposition. *Progress in Oceanography 4.* 245-265.

Prance, G. T. 1973. Phytogeographic support for the theory of Pleistocene forest refuges in the Amazon Basin, based on evidence from distribution patterns in Caryocaraceae, Chrysobalanaceae, Dichapetalaceae and Lecythidaceae. *Acta Amazonica 3.* 5-28.

Rotty, R. M. 1977. Global carbon dioxide production from fossil fuels and cement, A.D. 1970 - A.D. 2000. In: *The Fate of Fossil Fuel* CO_2, N. R. Andersen and A. Malahoff (Eds.), Plenum, N.Y. (this volume).

Sackett, W. M., W. R. Eckelmann, M. C. Bender, and A. W. H. Bê. 1965. Temperature dependence of carbon isotope composition in marine plankton and sediments. *Science 148.* 235-237.

Shackleton, N. J. 1969. The last interglacial in the marine and terrestrial records. *Proceedings of the Royal Society (London) B 174.* 135-154.

Shackleton, N. J. 1974. Attainment of isotopic equilibrium between ocean water and the benthonic foraminifera genus *Uvigerina*: isotopic changes in the ocean during the last glacial. *Colloques Internationaux du Centre National de la Recherche Scientifique 219.* 203-210.

Shackleton, N. J. 1968. Depth of pelagic foraminifera and isotopic changes in Pleistocene oceans. *Nature 218.* 79-80.

Shackleton, N. J. and N. D. Opdyke. 1973. Oxygen isotope and palaeomagnetic stratigraphy of Equatorial Pacific core V28-238: oxygen isotope temperatures and ice volumes on a 10^5 year and 10^6 year scale. *Quaternary Research 3*. 39-55.

Shackleton, N. J. and N. D. Opdyke. 1976. Oxygen isotope and palaeomagnetic stratigraphy of Equatorial Pacific core V28-239, Late Pliocene to Latest Pleistocene. In: *Investigation of Late Quaternary Paleoceanography and Paleoclimatology*, R. M. Cline and J. D. Hays (Ed.). *Geological Society of America Memoir 145*. 449-464.

Shackleton, N. J. and E. Vincent. (In prep) Oxygen and carbon isotope studies in recent foraminifera from the Mozambique Channel region. (Also G.S.A. *Abstracts with Programs*)

Sommer, M. A., R. K. Matthews, and N. J. Shackleton. (In prep) Carbon isotope stratigraphy in the marine deep sea record.

Steinen, R. P., R. S. Harrison, and R. K. Matthews. 1973. Eustatic low stand of sea level between 125,000 and 105,000 B.P.: evidence from the sub-surface of Barbados, West Indies. *Geological Society of America Bulletin 84*. 63-70.

Street, F. A. and A. T. Grove. 1976. Late Quaternary lake level fluctuations in Africa: environmental and climatic implications. *Nature 261*. 385-390.

Suggate, R. P. 1975. When did the Last Interglacial end? *Quaternary Research 5*. 246-252.

Tricart, J. 1975. Existence de périodes sèches au Quaternaire en Amazonie et dans les régions voisines. *Revue Géographique Dynamique 23*. 145-158.

van der Hammen, T. 1972. Changes in vegetation and climate in the Amazon Basin and surrounding area during the Pleistocene. *Geologie en Mijnbouw 51*. 641-643.

Veeh, H. H. 1966. Th^{230}/U^{238} and U^{234}/U^{238} ages of Pleistocene high sea level stand. *Journal of Geophysical Research 71*. 3379-3386.

Verstappen, H. Th. 1975. On palaeo climates and landform development in Malesia. In: *Modern Quaternary Research in Southeast Asia*, G.-J. Bartstra and W. A. Casparie (Ed.), Rotterdam: A. A. Balkema. 3-35.

Vuilleumier, B. S. 1971. Pleistocene changes in the fauna and flora of South America. *Science 173*. 771-780.

Whittaker, R. H. and G. E. Likens. 1975. The biosphere and man. In: *Primary Productivity of the Biosphere*, H. Lieth and R. H. Whittaker (Ed.), New York: Springer-Verlag. 305-328.

Williams, M. A. J. 1975. Late Pleistocene tropical aridity synchronous in both hemispheres? *Nature 253*. 617-618.

Sediment Mixing and Carbonate Dissolution in the Southeast Pacific Ocean

Eric Sundquist[1], D. K. Richardson[2], W. S. Broecker[2], and Tsung-Hung Peng[2]

[1]*Harvard University*, [2]*Lamont-Doherty Geological Observatory*

ABSTRACT

Core descriptions of about 280 LDGO cores from the southeast Pacific Ocean indicate that a carbonate-rich top layer, noted by Broecker and Broecker (1974) in cores from the East Pacific Rise, is widespread at or near the calcite compensation depth. Sediment mixing is indicated in these cores by mottling and burrow structures and by discordance of C^{14} and thorium sedimentation rates. A simple mathematical mixing model suggests that the high-carbonate layer may have accumulated during a low stand of the compensation level in the last glacial period. However, in many cores the carbonate layer is not repeated in down-core cycles, as would be ex-cores. Previous glacial carbonate peaks may have been largely dissolved during succeeding interglacials, or stirred into the most recent glacial layer along with small accumulations of interglacial non-carbonates. Box core studies should help to clarify the important influences of bioturbation on these problems.

INTRODUCTION

The southeast Pacific Ocean is an area of gently sloping topography and low sedimentation rates, making it a useful region for studying the sedimentary record of deep-sea carbonate dissolution. The western flank of the East Pacific Rise dips across the foraminiferal lysocline (*Parker and Berger*, 1971), and the calcite compensation depth has been mapped in the same area by *Lisitzin*

(1972) and *Berger and Winterer* (1974). Down-core fluctuations in carbonate content appear to reflect cycles of varying dissolution intensities in the past (*Broecker*, 1971).

On a local scale, *Broecker and Broecker* (1974) examined a series of closely-spaced cores from the East Pacific Rise taken at or near the compensation depth. They demonstrated an abrupt transition from water depths at which core tops show little evidence of dissolution to depths at which dissolution effects are conspicuous. Radiometric dates and gasometric carbonate analyses revealed a complicated record of carbonate dissolution over the last hundred thousand years. The deepest cores studied contained a high-carbonate top layer of late glacial age overlying carbonate-poor sediments.

This paper reports progress in determining the regional importance of these observations. It is believed that sediment mixing by bioturbation has contributed significantly to the observed features.

REGIONAL OBSERVATIONS

Figure 1 shows the carbonate content of approximately 280 piston and trigger core tops from the southeast Pacific Ocean, drawn from core descriptions at the Lamont-Doherty Geological Observatory. (A detailed listing of the cores, their locations and water depths, and their classifications is available from the authors.) Although the carbonate content classifications in these descriptions are rather subjective [1], several regional trends are clearly apparent.

As expected, the carbonate core tops are associated with relatively shallow bottom depths and the non-carbonate core tops are from deeper areas. The calcite compensation depth, defined as the depth below which core top carbonate content is nil to low, shoals from about 5000 meters near the equator to about 4000 meters at 30°S. This variation is demonstrated in Figure 2, which shows the number of carbonate vs. non-carbonate core tops in each depth and latitude range west of the crest of the East Pacific Rise. For each latitude range there is a range of transitional depths at which both carbonate and non-carbonate core tops have been described. The number of cores at these transitional depths is relatively small. However, the absence of carbonate core tops below and non-carbonate

---

[1] After megascopic and microscopic (but not smear slide) examination on ship and in the core lab, carbonate content is classified as "low", "moderate", or "high". The boundaries separating these classifications are probably near 30 and 60 percent carbonate, with large margins of error. The non-carbonate core tops in Figure 1 were described as low in carbonate; the carbonate core tops were described as moderate to high.

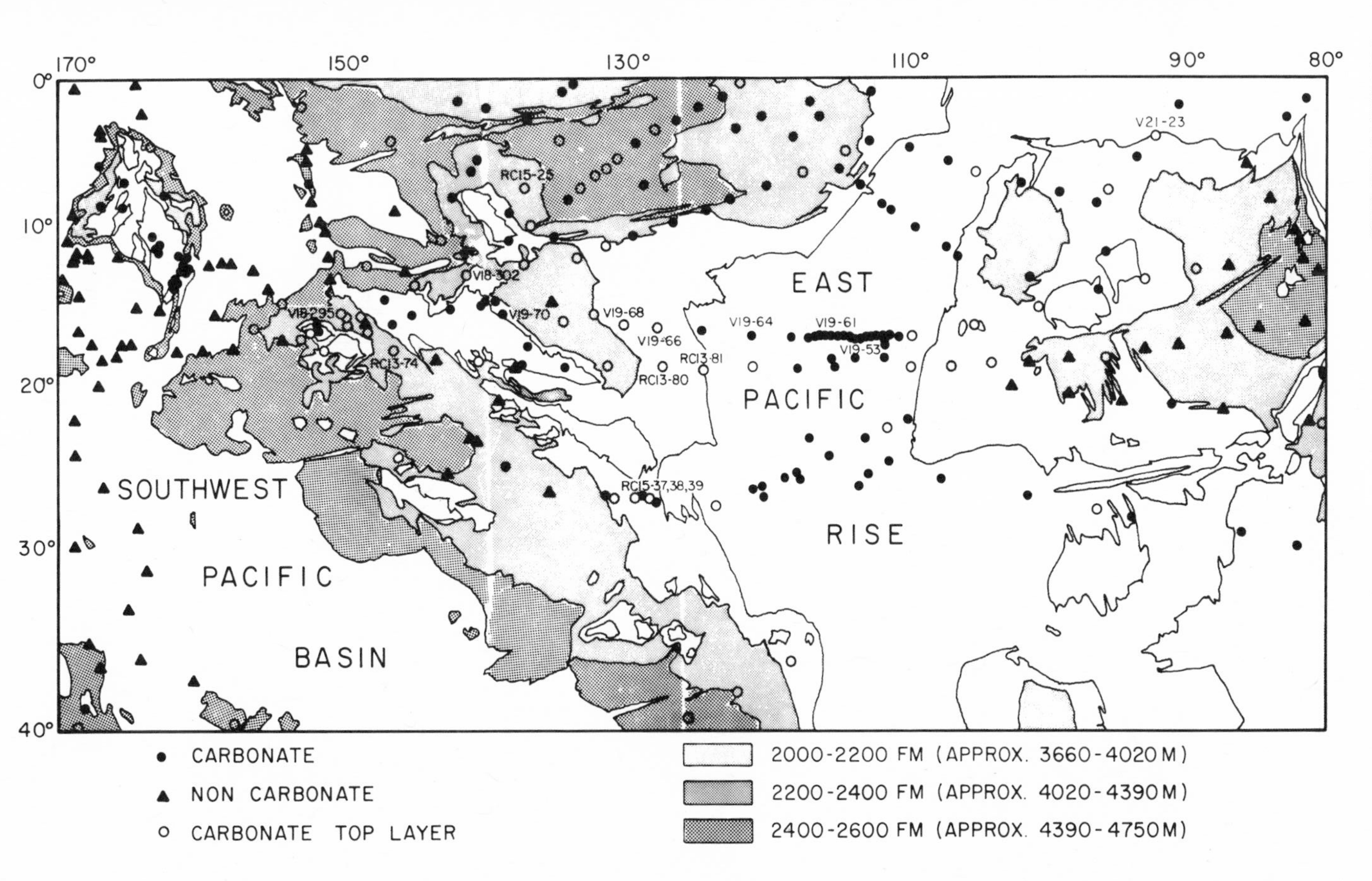

Fig. 1. Locations and carbonate contents of cores examined in this study. The shaded depth range 2000-2600 fathoms corresponds to about 3800-4900 meters. Bottom topography was taken from *Mammericx et al.*, 1973.

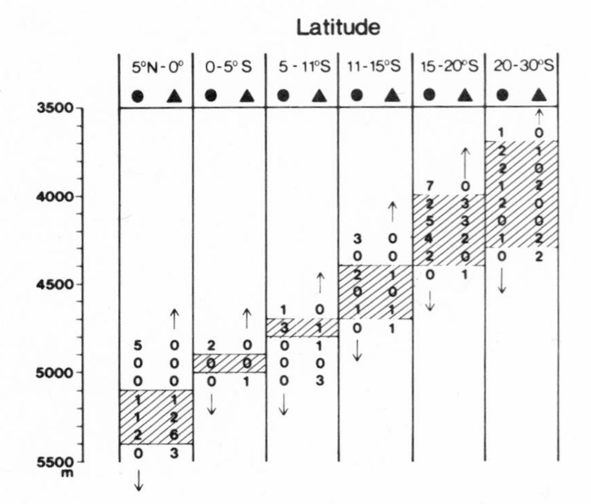

CORE-TOP CARBONATE TRANSITION DEPTHS

Fig. 2. The number of carbonate (circle columns) and non-carbonate (triangle columns) core tops observed in each 100 meter span of bottom depths in the indicated latitude ranges west of the crest of the East Pacific Rise. The transition depths from all carbonate to all non-carbonate are shaded.

core tops above restricts the compensation depth to this transitional zone.

Figure 1 also shows that the high-carbonate top layer observed by *Broecker and Broecker* (1974) is a regional characteristic among cores near the compensation depth. The carbonate layer is found uniformly in cores either from or just above the carbonate compensation depths transition depths shown in Figure 2. It is easily distinguished in core descriptions because of its lighter color and greater abundance of foraminifera, as seen in Figure 3. Figure 4 shows that these subjective criteria are supported by gasometric carbonate analyses[2] of a number of cores from various locations.

To summarize our regional observations, Figure 5 shows that carbonate-rich core tops occur at depths down to the carbonate compensation transition zone, where red clay core tops are also found. Below the transition zone carbonate is no longer present in core tops. The carbonate-rich top layer occurs within the transition zone and just above it.

---

[2] The analytical technique is similar to that of *Hülsemann*, 1966.

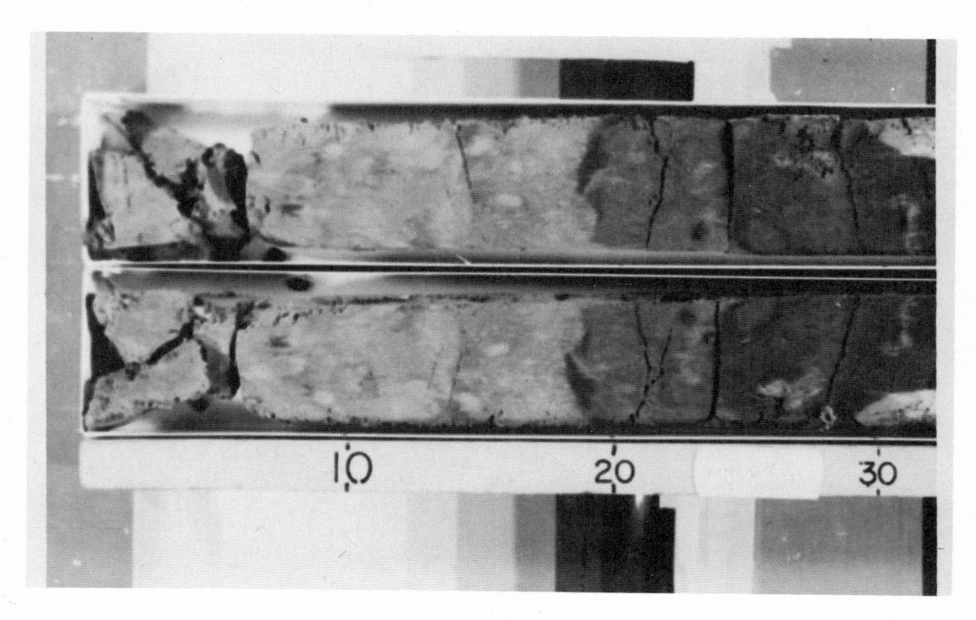

Fig. 3. Photograph of the top of core V19-67, showing the high-carbonate top layer and mottling due to bioturbation (courtesy LDGO core lab).

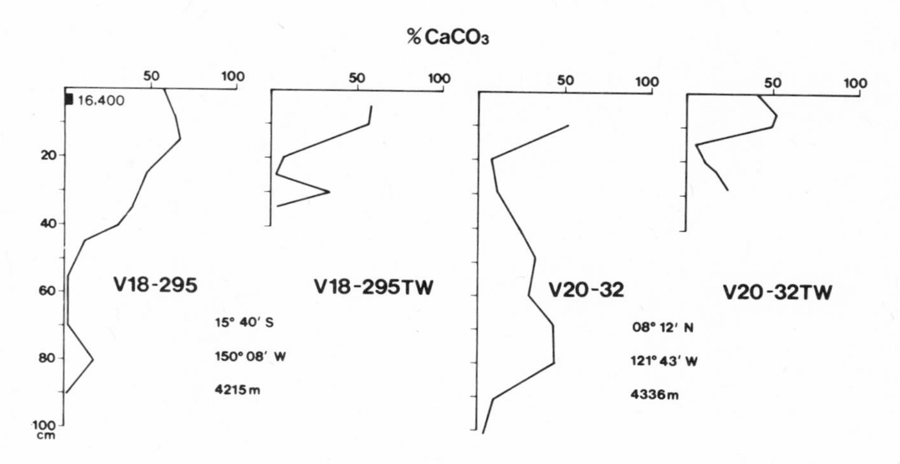

Fig. 4. Gasometric carbonate profiles for several cores with carbonate-rich top layers. Percentages refer to weight percent of bulk dry sediment. C^{14} ages are indicated on the depth axes. Ages to the right of the V19-67 depth axis are from *Broecker and Broecker*, 1974.

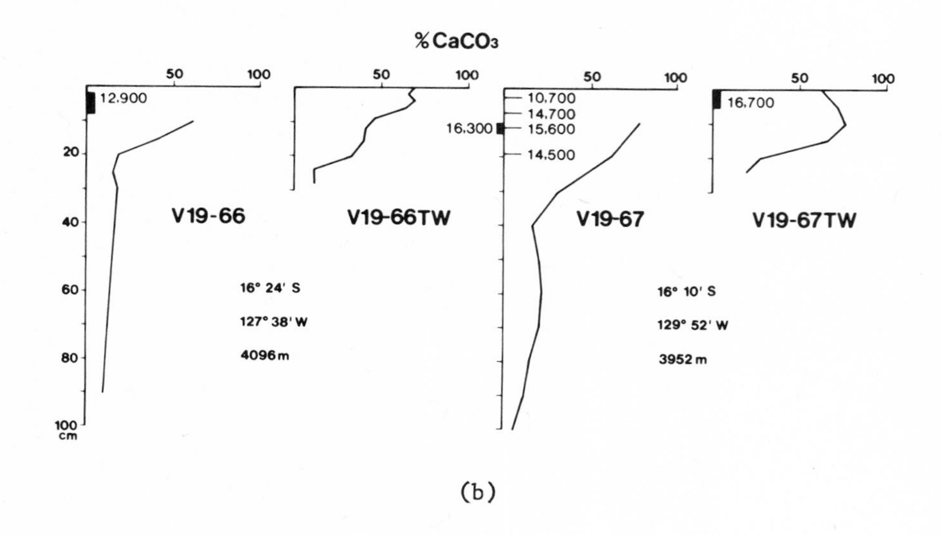
% CaCO3
50
100
12,900
20
40
60
80
100 cm
V19-66
16° 24' S
127° 38' W
4096 m
V19-66TW
10,700
14,700
16,300
15,600
14,500
V19-67
16° 10' S
129° 52' W
3952 m
16,700
V19-67TW

(b)

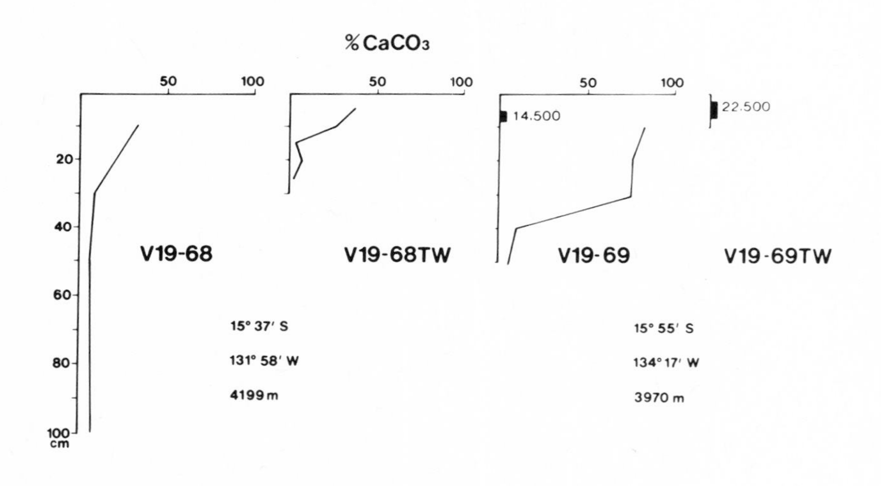
% CaCO3
50
100
20
40
60
80
100 cm
V19-68
15° 37' S
131° 58' W
4199 m
V19-68TW
14,500
V19-69
15° 55' S
134° 17' W
3970 m
22,500
V19-69TW

(c)

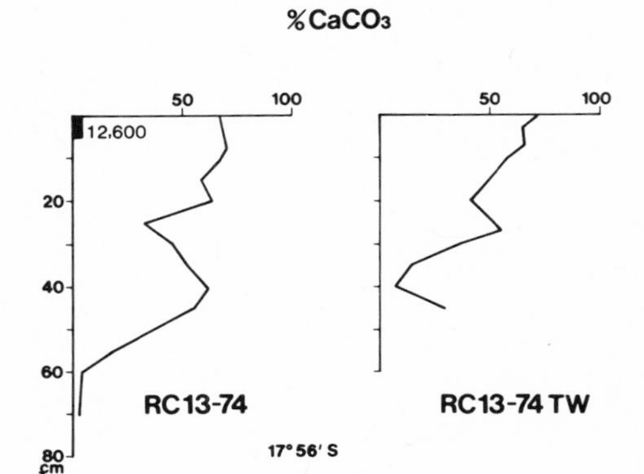
%CaCO3
50
100
12,600
20
40
60
80
cm
RC13-74
50
100
RC13-74 TW
17° 56′ S
146° 15′ W
3959m

(d)

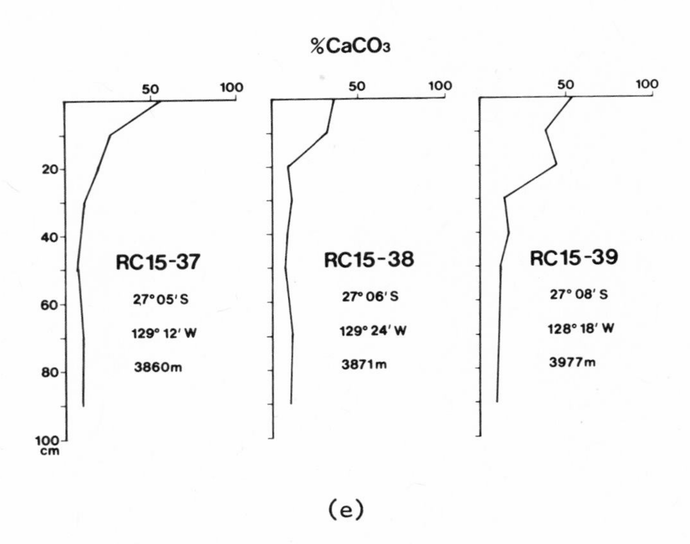
%CaCO3
50
100
20
40
60
80
100
cm
RC15-37
27° 05′ S
129° 12′ W
3860m
50
100
RC15-38
27° 06′ S
129° 24′ W
3871m
50
100
RC15-39
27° 08′ S
128° 18′ W
3977m

(e)

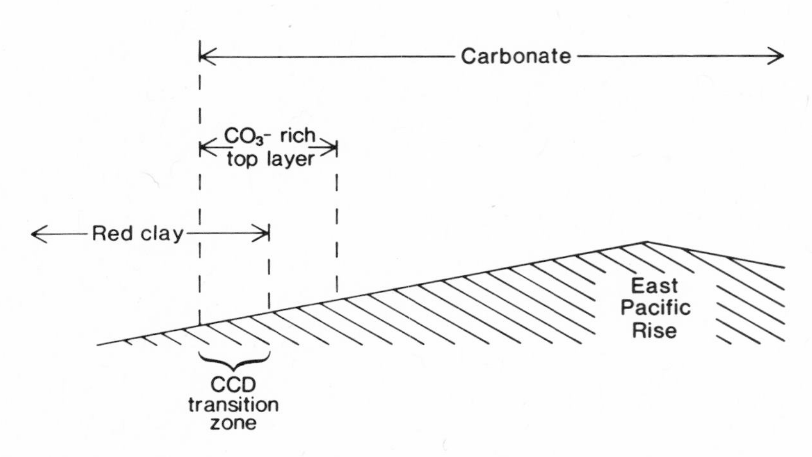

Fig. 5. Schematic diagram of the generalized distribution of carbonate in sediments on the western flank of the East Pacific Rise.

DISSOLUTION INDICATORS

The distinct association of the high-carbonate top layer with the compensation depth led us to examine indications of dissolution in these cores. *Broecker and Broecker* (1974) noted that the appearance of carbonate-rich tops in the R/V VEMA 19 traverse corresponds to a sharp increase in dissolution intensity with increasing bottom depth. This conclusion was based mainly on changes in the relative abundance of solution-resistant foraminifers first observed by *B. Luz* (pers. comm., 1973) and quantified here in Figure 6. Cores V19-53, -55, -64, and -65 are high in carbonate throughout their lengths, and their tops contain approximately the same proportion of solution-resistant foraminifers. Cores V19-67, -69TW, and -66 are high in carbonate only at their tops, and these cores define a sudden increase in resistant foraminifer content over a bottom depth range of only about 100 meters.

Variations in foraminifer dissolution indicators within the high-carbonate top layer were also examined. As shown in Figure 7, proportions of solution-resistant planktonic foraminifers, benthic foraminifers, and foraminifer fragments indicate that the down-core decrease in carbonate coincides with an increase in the extent of dissolution.

Thus, it seems clear that the carbonate-rich top layer must be explained in terms of dissolution mechanisms. *Broecker and Broecker* (1974) proposed two hypotheses: (1) the compensation level

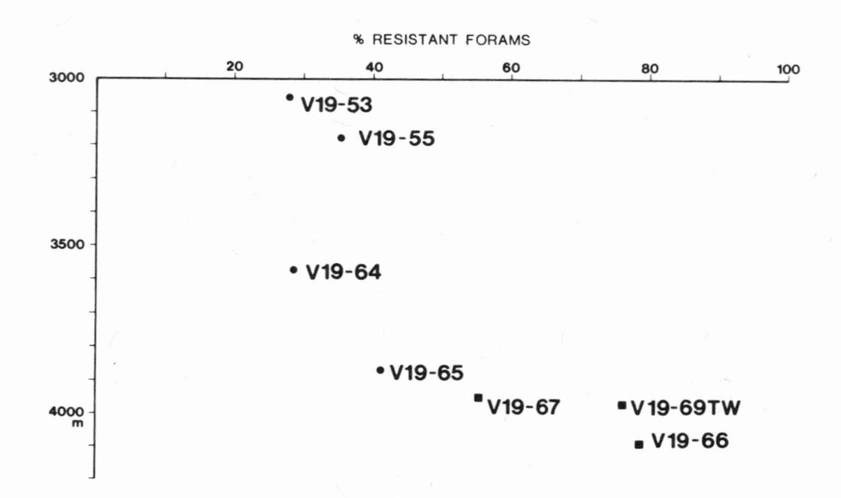

Fig. 6. Percentages of solution-resistant foraminifer species (Table 1) in the >250μ fraction of R/V VEMA 19 core tops, plotted against sea-floor depth. Core symbols are defined in Figure 1.

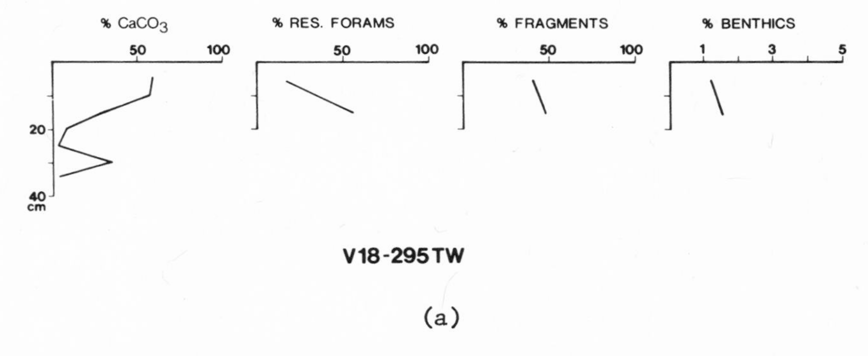

(a)

Fig. 7. Weight percentages of calcium carbonate in bulk dry sediment; and percentages of solution-resistant foraminifers, foraminifer fragments, and benthic foraminifers in the >250μ fraction of several cores with high-carbonate top layers.

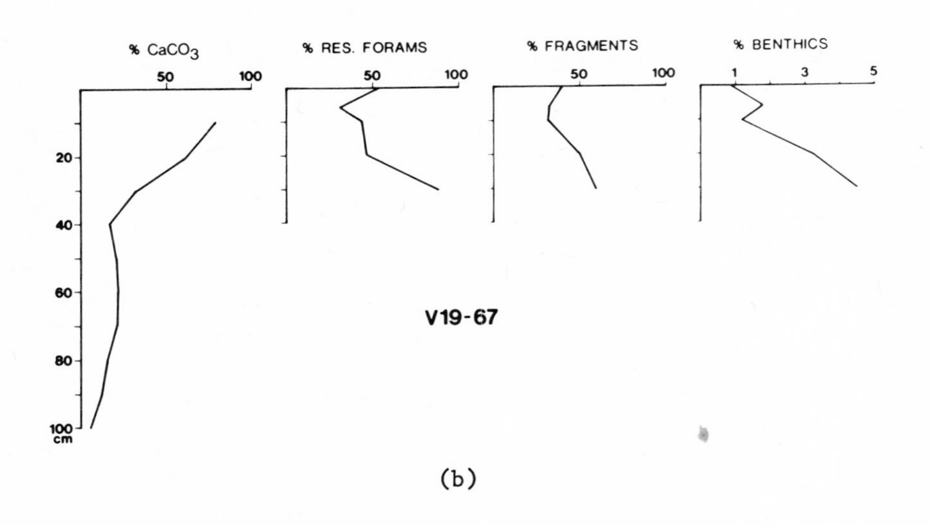
% $CaCO_3$
50
100
% RES. FORAMS
50
100
% FRAGMENTS
50
100
% BENTHICS
1
3
5
20
40
60
80
100
cm
V19-67

(b)

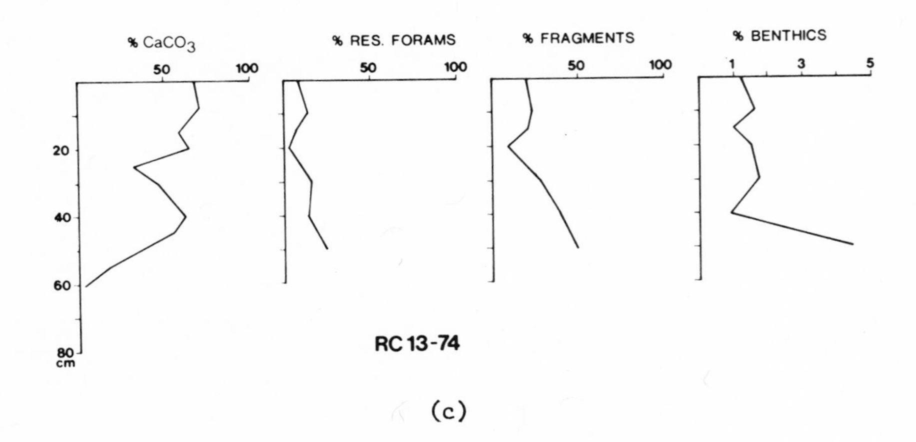
% $CaCO_3$
50
100
% RES. FORAMS
50
100
% FRAGMENTS
50
100
% BENTHICS
1
3
5
20
40
60
80
cm
RC 13-74

(c)

TABLE 1. Planktonic Foraminifer Species Indentified and Counted in this Study.

Non-resistant:

Globigerinoides ruber
Globigerinoides sacculifer
Globigerinoides conglobatus
Globigerina bulloides
Globigerinita glutinata
Globoquadrina hexagona

Resistant:

Globoquadrina conglomerata
Globorotalia hirsuta
Globorotalia truncatulinoides
Globorotalia inflata
Globorotalia crassaformis
Globorotalia cultrata
Globoquadrina dutertrei
Globigerina pachyderma
Pulleniatina obliquiloculata
Sphaeroidinella dehiscens
Globorotalia tumida
Turborotalia humilis

may have shifted downward, causing dissolution to decrease during the time the layer was deposited, or (2) dissolution in the sediments may occur so slowly that carbonate material can be buried for tens of thousands of years before it eventually dissolves.

THE SHIFTING COMPENSATION LEVEL HYPOTHESIS

If the high-carbonate top layer was caused by a shift of the compensation level, it would be expected that the layer would correlate from core to core. On the basis of C^{14} measurements in core V19-67, *Broecker and Broecker* (1974) originally estimated the base of the layer to be about 15,000 years old. However, as will be shown below, it is not now believed that this core (as well as the others) has been disturbed by sediment mixing. A more reasonable way to do justice to the shifting compensation depth hypothesis is exemplified in Figure 8 (a), taken from *Luz and Shackleton* (1975). Core V21-33, from just south of the equator,

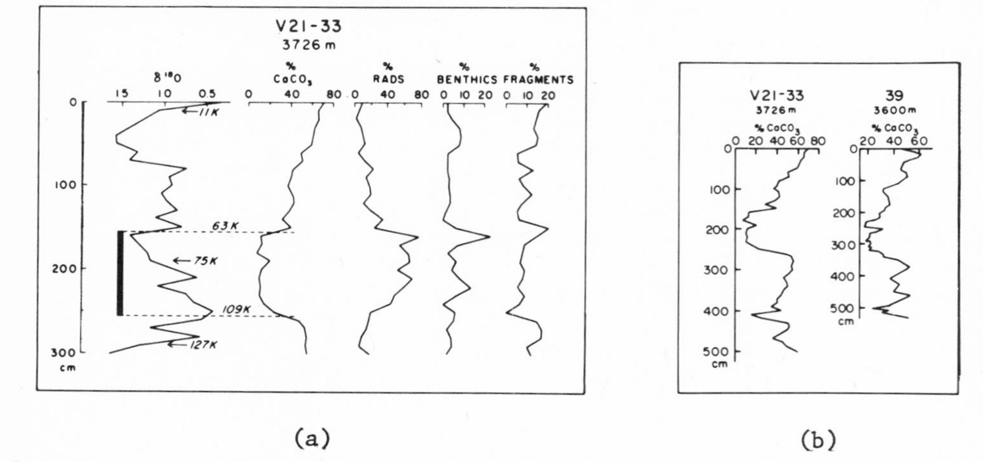

(a) (b)

Fig. 8. (a) δO^{18} and percent calcium carbonate in core V21-33. (b) Weight percent calcium carbonate for V21-33 and Arrhenius core 39. (From *Luz and Shackleton* (1975), p. 145 and 146).

clearly contains a carbonate-rich top layer[3]. Oxygen isotope measurements show that this layer must have been deposited during the last glacial period. In fact, repeated glacial carbonate peaks in equatorial Pacific Ocean sediments have been well-documented by *Arrhenius* (1952) and *Hays et al.* (1969), throughout the Brunhes magnetic interval. Thus, the top layer may be just the most recent in a series of climate-related carbonate cycles.

This argument requires that overlying low-carbonate Holocene sediments be missing from the core tops in question. As is shown in Figure 8(b), *Luz and Shackleton* (1975) showed this to be true for core V21-33 by correlation with the carbonate curve of nearby core 39 (*Arrhenius*, 1952), which does display a decrease in carbonate at the top.

Further evidence for the absence of Holocene sediments comes from a large number of C^{14} measurements reported by *Broecker and Broecker* (1974) and in this paper, shown in Table 2. Only 3 out of 25 calculated ages are clearly Holocene, in spite of efforts to sample the tops of both piston and trigger cores. These measurements should discourage efforts to seek simple relationships between

---

3 Accumulation rate estimates and carbonate content measurements in this core suggest that the upward increase in carbonate is caused by approximately equal amounts of decreasing dissolution and decreasing non-carbonate dilution (*Luz and Shackleton*, 1975).

TABLE 2. C^{14} Ages Determined in this Study and by *Broecker and Broecker* (1974).

Core	Sample depth; cm	C^{14} age; yrs.
V19-53TW	2-6	3,300
V19-54	0-5	3,500
	8.5-11.5	5,500
V19-52TW	top	32,500
V19-61TW	0-5	16,600
V19-63	6-10	9,800
V19-66	2-8	12,900
V19-67	2.5	10,700
	7	14,700
	11	15,600
	10-13	16,300
	19	14,500
V19-67TW	0-5	16,700
V19-69	5-8	14,500
	15-18	17,600
V19-70	2-6	17,700
V19-70TW	0-5	22,500
RC13-74	0-5	12,600
RC13-80	0-6	14,800
RC13-81	0-2	14,000
	14-16	>30,000
RC13-81TW	0-4	16,400
	31-34	>30,000
V18-302	5-9	12,800
RC15-25	0-2	9,400

core-top dissolution data and present bottom-water properties. However, we were encouraged to try to extend the correlation of high-carbonate top layers to cores well south of the equator.

Because many of these cores do not contain enough whole forams for oxygen isotope determinations, thorium dating techniques were employed. *Broecker and Broecker* (1974) has shown that thorium measurements in three of the R/V VEMA 19 cores yielded very similar rates of non-carbonate accumulation (40, 40, and 41 mg/cm^2-kyr in V19-61, -64, and -66, respectively). It therefore seems reasonable to calculate accumulation times for the top layer in nearby cores from the total amount of non-carbonate accumulated in the layer. Table 3 shows time spans calculated on this basis for carbonate-rich top layers in several cores. Non-carbonate content was averaged over each interval between pairs of gasometric carbonate analyses (Figure 4), and sediment dry densitites were taken from *Luz and Shackleton's* (1975) regression of *Arrhenius'* (1952)

TABLE 3. Time Spans Calculated from the Thorium-based Non-carbonate Accumulation Rate for High-carbonate Top Layers Reported by *Broecker and Broecker* (1974).

Core	Depth of layer; cm	Time span; yrs.
V19-66	0-18	97,000
V19-66TW	0-22	123,000
V19-67	0-40	191,000
V19-67TW	0-20	91,000
V19-68	0-20	121,000
V19-68TW	0-12	77,000
V19-69	0-35	176,000

data for density as a function of carbonate content in Pacific sediments. For each interval, the total sediment accumulation was multiplied by the fraction of non-carbonate and then divided by the non-carbonate accumulation rate ($40\ mg/cm^2$-kyr) to yield an approximate time span. The total time span for each high-carbonate layer is the sum of the spans for its intervals, taking the bottom of the layer to be the depth at which the carbonate curve crosses 20%.

The time span of the last glacial period was approximately 64,000 years (75,000 to 11,000 years ago). From Table 3 it is clear that the high-carbonate tops observed by *Broecker and Broecker* (1974) span significantly longer periods of non-carbonate sedimentation. Moreover, the wide variance in calculated time spans suggests the possibility that the carbonate layers do not correlate from core to core, although the discrepancies between piston cores and comparison trigger weight cores indicate that coring artifacts may be involved.

This reasoning is corroborated by more recent thorium measurements in core RC13-74, from a location 15° to the west of the cores studied by *Broecker and Broecker* (1974). Because these measurements were made down-core through a high-carbonate top layer, adjustments for the varying carbonate content were necessary before a non-carbonate accumulation rate could be determined. The unsupported Th^{230} content of the carbonate fraction was assumed to be negligible (*Ku*, 1965), so each excess Th^{230} value determined for a bulk sediment sample was normalized to the non-carbonate content of the sample by dividing by the fraction of non-carbonate (inferred from the gasometric carbonate profile of the same core). These normalized values were then plotted against the cumulative down-core mass of non-carbonate rather than depth in the core. Cumulative down-core non-carbonate values were calculated by progressive summation of non-carbonate contents averaged over intervals between gasometric carbonate analyses and adjusted for sediment dry densities, exactly as non-carbonate contents were calculated for the time span estimates in Table 3. The result of these manipulations is contained in Figure 9, a convenient plot of the non-carbonate sedimentation rate analogous to the excess-activity vs. depth-in-core plots commonly used to determine bulk sedimentation rates in less complicated cores.

The best-fit line in Figure 9 has a slope corresponding to a non-carbonate accumulation rate of 39 mg/cm^2-kyr, remarkably close to the values reported in the R/V VEMA 19 cores. This result suggests that non-carbonate accumulation times up to at least a few hundred thousand years can be calculated for cores taken from a wide area.

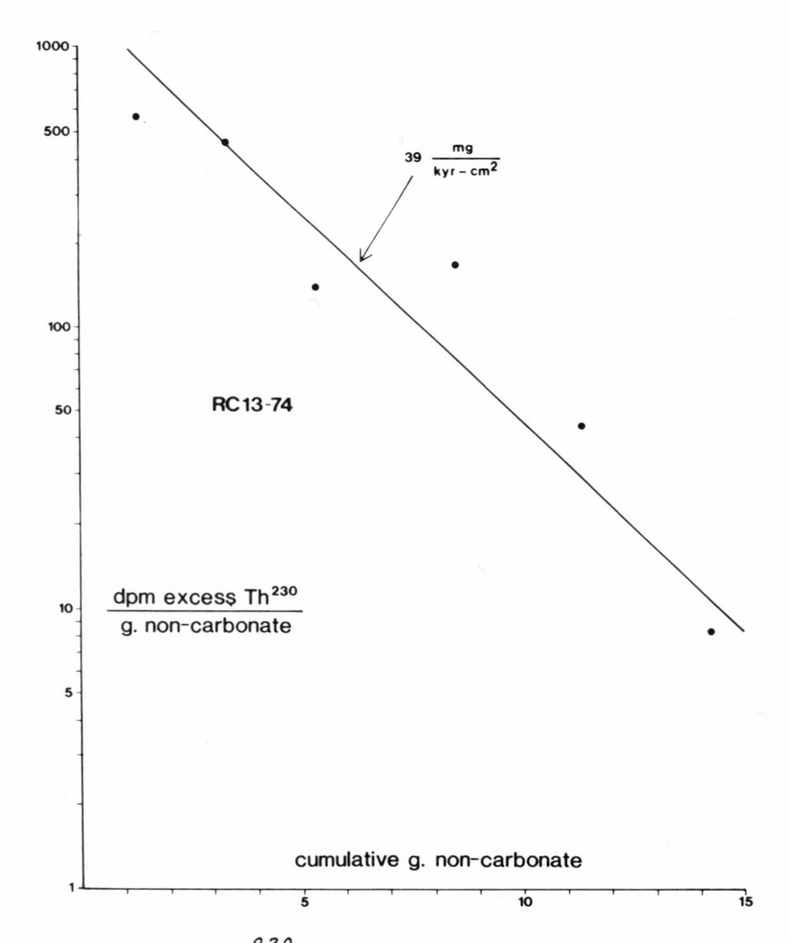

Fig. 9. Excess Th^{230} in core RC13-74, normalized to non-carbonate and plotted against cumulative down-core non-carbonate dry mass.

In core RC13-74 the time calculated for the top layer is 325,000 years, and there is a carbonate minimum in this layer at 135,000 years (see Figure 4). Like the times indicated in Table 3, these numbers are much too high to correspond only to the last glacial period. Thus, if the shifting compensation level hypothesis is to survive, it must be extensively modified. It is believed that sediment mixing by bioturbation is important enough in these cores to rectify the apparent discrepancies.

BIOTURBATION

Much of the evidence for bioturbation is obvious from an examination of the cores: burrow structures and mottling occur very frequently in the high-carbonate top layers. Figure 3 shows typical mixing structures in core V19-67, and such features are almost always mentioned in the LDGO core descriptions of carbonate-rich top layers.

In Figure 4 and Table 2 several C^{14} ages for various depths in core V19-67, mostly taken from *Broecker and Broecker* (1974), have been listed. The maximum and minimum ages in the upper 20 centimeters are 5600 years apart, yet our thorium-based non-carbonate accumulation rate suggests a sedimentation span of 80,000 years for the same interval. Core V19-69 shows the same phenomenon: two C^{14} ages only 3100 years apart bracket an interval containing the equivalent of 50,000 years of non-carbonate accumulation. Finally, both thorium (*Broecker and Broecker*, 1974, p. 48) and C^{14} bulk sedimentation rates are available for core V19-54 from the crest of the East Pacific Rise. The thorium rate is $0.6\ g/cm^2\text{-}kyr$, but two C^{14} measurements in the top 10 centimeters indicate a rate three times as high. These anomalies might be explained either by core-top sediment mixing or by large widespread core-top increases in the real rate of non-carbonate accumulation (or bulk sediment accumulation, in the case of V19-54). The abundant sedimentary structures mentioned above, as well as the monotonous character of red clay deposition throughout the Pacific Ocean basin, argue strongly that bioturbation is by far the most likely explanation.

An attempt was made, therefore, to understand the influence of sediment mixing by means of a simple mathematical model (*Sundquist*, 1976). As diagrammed in Figure 10, the model assumes homogeneous, instantaneous mixing in a layer of finite thickness at the sea-sediment interface, with no mixing below this layer. This approach was first utilized by *Goldberg and Koide* (1962) and *Berger and Heath* (1968). The principle of conservation of mass, when applied to the balance among the fluxes into and out of the mixed layer, yields differential equations for the time dependence of various components of the total mixed layer mass. The Appendix details the equations and solutions for carbonate and C^{14} in the mixed layer, assuming all other parameters to be constant.

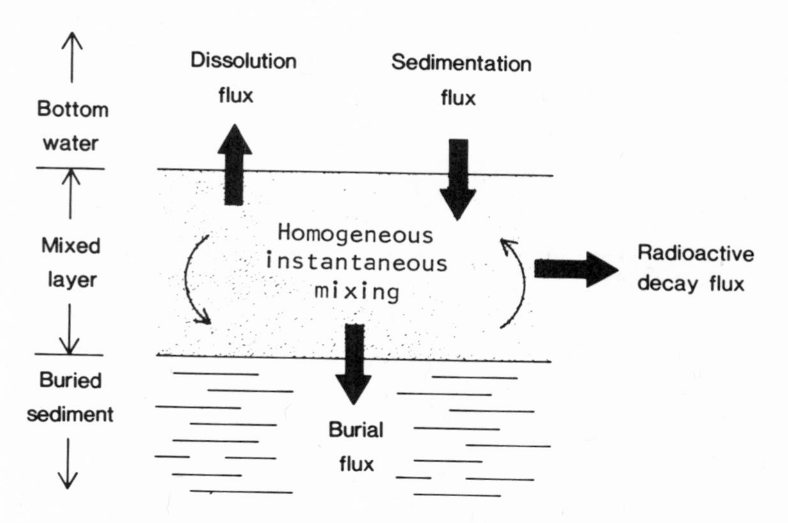

Fig. 10. Schematic diagram of the sediment mixing model detailed mathematically in the Appendix.

This model has been used to generate hypothetical sediment carbonate and C^{14} profiles for a glacial-interglacial cycle. Given the time-dependent mixed-layer solutions in the Appendix, it is easy to derive sediment profiles from sedimentation rates inherent in the constants. A glacial-interglacial cycle can be profiled by applying glacial values for the constants to interglacial initial conditions and vice versa. Implicit in this approach is the assumption that the change from one set of constants to the other occurs instantaneously.

Table 4 lists the constants and initial conditions that have been chosen to approximate the conditions of the most recent glacial-interglacial transition in the R/V VEMA 19 cores. These

TABLE 4. Constants and Initial Conditions Used to Calculate the Hypothetical Profiles in Figure 11. Refer to the Appendix for definitions of terms and units.

	F	Q	D	M	λ	δ	g_0	b_0/a
75,000-11,000 yrs.	0.4	0	0.6	10	0.12	0.9	0.10	0.83
11,000 yrs.-present	0.4	0.356	0.6	10	0.12	0.9	0.89	0.36

values are based primarily on the non-carbonate accumulation rate for cores V19-61, -64, and -66 determined by thorium dating. It has been assumed that the maximum carbonate content observed in these cores is the carbonate content of the sedimentation flux, and that the minimum carbonate content represents a steady-state balance between the sedimentation and dissolution fluxes. A glacial set of constants, with no dissolution flux, was applied to the period from 75,000 to 11,000 years ago; and an interglacial set of constants, with a dissolution flux, was applied from 11,000 years ago to the present. Interglacial conditions are assumed to have been at steady state before 75,000 years ago to provide a starting point for the calculations. The resultant hypothetical profiles are plotted in Figure 11.

Given the scarcity of relevant direct observations of mixing depths in deep sea cores, one of the most arbitrary aspects of the model is the selection of the mixing depth below the sea-sediment interface. The C^{14} measurements in core V19-67 seem to indicate that mixing may be effective to at least a depth of 20 centimeters, but evidence presented by *Peng and Broecker* (1977) in a number of cores from the Pacific and Indian Oceans suggests that 9 to 15 centimeters may be more reasonable. A mixing depth of 10 centimeters has been chosen here, but the basic conclusions drawn in this paper do not depend critically on this value.

Although the assumptions of our mixing model are much more simplistic than those of several other published models (*Guinasso and Schink*, 1975; and *Guinasso*, 1976) and C^{14} measurements indicate

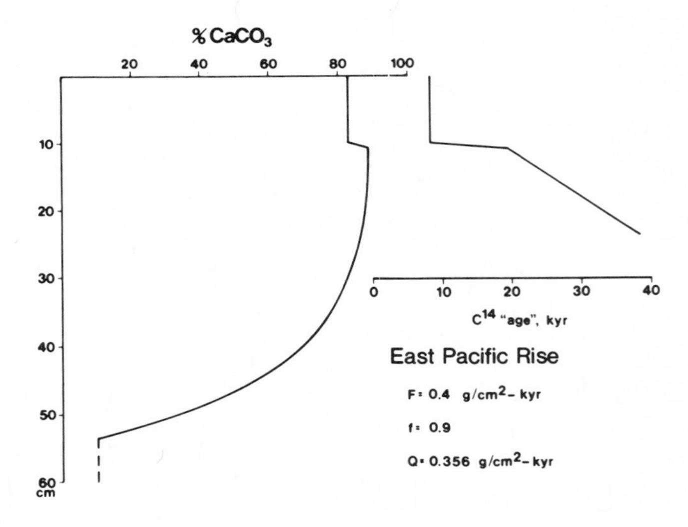

Fig. 11. Hypothetical carbonate and C^{14} profiles for the last 75,000 years in the R/V VEMA 19 cores. Parameters for the calculations are listed in Table 4.

that the mixed layer is not completely homogenized, the hypothetical profiles of Figure 11 resemble real observations in several important ways. The carbonate profiles in Figure 4 show an overall general similarity in shape to the model profile. The hypothetical Holocene carbonate decrease is so slight that it might easily go unnoticed in a series of gasometric measurements. The C^{14} age of the model mixed layer (about 8,000 years) is actually younger than most of the real core top dates. There are several reasonable factors which might explain this discrepancy: (1) mixing may be deeper than 10 centimeters, (2) some of the relatively young carbonate of the sedimentation flux may dissolve before it can be mixed downward, or (3) the true sediment tops may be missing from the tops of the cores.

The thorium-based non-carbonate accumulation rate can be applied to the hypothetical profile to yield a time span, exactly as we have done with the real profiles. One might expect the calculated time span to be 75,000 years because the hypothetical non-carbonate accumulates at the constant thorium rate through that time. However, the mixing of pre-glacial non-carbonate upward into the hypothetical profile causes the calculated time span to be about 230,000 years. Thus, sediment mixing might account for some of the difficulties we encountered in attempting to correlate the high-carbonate top layer with the last glacial period.

CONTINUING STUDIES

The demonstrated importance of sediment mixing by bioturbation requires new perspectives and approaches to problems of carbonate dissolution in the southeast Pacific Ocean. In our continuing efforts to explain the carbonate-rich top layer, several conspicuous features seem to demand particular attention.

First, the sudden downslope increase in degree of dissolution (Figure 6) seems to indicate a precise depth control on the dissolution flux, although this may be an artifact of a sensitive response to a relatively slight increase in dissolution. If this characteristic can be demonstrated on a regional scale, it may offer clues to the way bottom waters and mixed-layer pore waters interact in the sea-floor dissolution process. It is intriguing to note that a recent map of the Benthic Front by *Chung* (1975) shows this water mass boundary impinging on the western flank of the East Pacific Rise very close to the depth of the apparent dissolution increase. Also, this depth is only slightly shallower than that of a conspicuous bottom sill which channels northward bottom water flow at 127° W. by 25°S.

Another unresolved problem is the absence, in many cores, of repeated carbonate cycles below the high-carbonate top layer. As has been noted, equatorial cores show several carbonate cycles repeated through the last several hundred thousand years. However, in cores from areas of lower sedimentation rates, the carbonate top is not immediately underlain by other carbonate layers. The most obvious explanation of this feature is that the most recent carbonate peak may have been more extreme than its predecessors (*Hays et al.*, 1969). Another possibility is suggested by the fact that the dissolution flux that would be necessary to produce the observed carbonate fluctuations in equatorial cores is actually several times greater than the total sedimentation flux without dissolution farther south. Thus, it seems possible that interglacial dissolution may actually eat its way downward through underlying high-carbonate glacial sediments.

The Appendix details efforts to adapt the mixing model presented in this paper to this "negative accumulation" situation. The differential equations are basically unchanged, except that the sedimentation flux is less than the dissolution flux, so sediment is incorporated into the bottom of the mixed layer rather than buried out of it. Figure 12 shows the hypothetical C^{14} and carbonate profiles which result from applying negative accumulation to the model, R/V VEMA 19 profiles since the beginning of the Holocene. Parameters for the calculations are listed in Table 5. The carbonate decrease at the top of core V19-67 TW (Figure 4), and the rapid down-core disappearance of C^{14} in core RC13-81 (Table 2), may be indications that negative accumulation is occurring. This hypothesis can only be tested by more extensive sampling near the sea-sediment interface.

Even without downward dissolution, the total non-carbonate sedimentation during an interglacial is less than 5 centimeters. Thus, the residual carbonate from successive glacials would be stirred together by bioturbation.

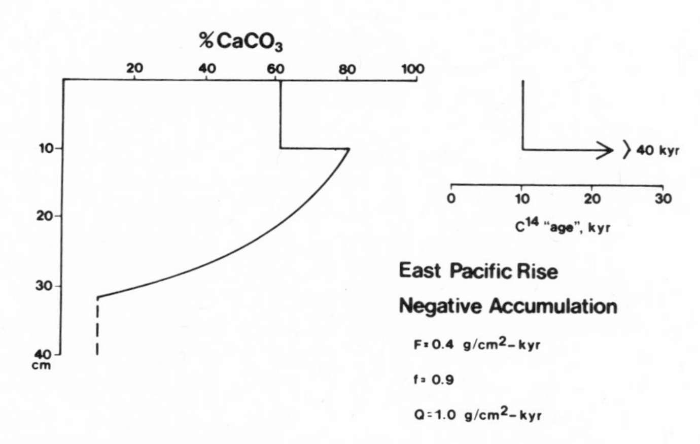

Fig. 12. Hypothetical carbonate and C^{14} profiles for the last 75,000 years in the R/V VEMA 19 cores, assuming negative accumulation during the last 11,000 years. Parameters for the calculations are listed in Table 5.

TABLE 5. Constants and Initial Conditions Used to Calculate the Hypothetical Profiles in Figure 12. Refer to the Appendix for Definitions of Terms and Units.

	F	Q	D	M	λ	δ	g_o	b_o/a
75,000-11,000 yrs.	0.4	0	0.6	10	0.12	0.9	0.10	0.83
11,000 yrs.-present	0.4	1.0	0.6	10	0.12	0.9	0.89	0.36

In the calculations for Figure 12, it was assumed that $g = g^* = 0.89$.

Some of the cores with high-carbonate tops are known to contain disconformities (*Saito*, pers. comm.; *Saito et al.*, 1974). Seismic reflection profiles and sediment structures in box cores suggest downslope mass movement in carbonate sediments on the flanks of the Ontong Java Plateau (*Berger and Johnson*, 1976). The continuous thorium profile through the carbonate top of core RC13-74 (Figure 9) seems to preclude slumping or turbidite deposition, and current-related mass transport seems unlikely as an explanation of such a widespread phenomenon. Nevertheless, the possible effects of erosion and redeposition must be considered in evaluating other hypotheses.

Finally, *Broecker and Broecker's* (1974) second hypothesis - kinetic lag of dissolution behind burial - deserves reconsideration in light of the effects of sediment mixing. As it was originally stated, *Broecker and Broecker* (1974) showed that this hypothesis implied unreasonable pore water concentrations and gradients in order to drive the dissolution flux up and out of the sediments by means of molecular diffusion. Ways in which sediment mixing might drive the dissolution flux without such unrealistic requirements are now being examined.

CONCLUSIONS

The carbonate-rich top layer appears to be a result of at least three important influences on the sediment record in the southeast Pacific Ocean: (1) variations in dissolution intensity through time; (2) benthic stirring by organisms; and (3) slow rates of deposition. The model presented illuminates these influences and suggests possibilities that should be tested by further sampling.

Because bioturbation is essentially a sea-sediment interface phenomenon, intact samples of the interface are essential. Box coring appears to be the best technique for obtaining such samples, and the large sample volume available also makes possible the unrestricted application of several techniques (e.g., C^{14} thorium, O^{18}/O^{16}, foraminifer species counts) to subsamples from the same depth below the interface. Box coring is also a standard technique for studying the biology of the benthos, which must be understood as the fundamental basis of bioturbation.

Whatever explanations finally resolve the problems that have been discussed, it is believed that bioturbation must be considered an important factor in continuing efforts to explain the features of carbonate dissolution in the southeast Pacific Ocean.

ACKNOWLEDGEMENTS

Discussions with Dr. Raymond Siever were valuable throughout the preparation of this paper. George Kipphut, Norm Guinasso, Jr., and Nicholas Shackleton provided very helpful suggestions in its early stages. Foram species counts for core V19-67 were done by Boaz Luz, and John Goddard and Ursula Middel performed the thorium analyses. Holly Fox Sundquist helped with the mapping and typing. Support was provided by NSF Grants OCE75-21210 and OCE 74-21412.

REFERENCES

Arrhenius, G. 1952. Sediment cores from the East Pacific. *Swedish Deep-Sea Exped. (1947-1948) Repts.*, *5*. 6-227.

Berger, W. H. and G. R. Heath. 1968. Vertical mixing in pelagic sediments. *Jour. Marine Res.*, *26*. 134-143.

Berger, W. H. and T. C. Johnson. 1976. Deep sea carbonates: dissolution and mass movement on Ontong-Java Plateau. *Science* (in press).

Berger, W. H. and E. L. Winterer. 1974. Plate stratigraphy and the fluctuating carbonate line. In: Hsü, K. J., and H. C. Jenkyns, Eds., Pelagic sediments: on land and under the sea, *IAS Spec. Pub. No. 1*, Blackwell. 11-48.

Broecker, W. S. 1971. Calcite accumulation rates and glacial to inter-glacial changes in oceanic mixing, In: Turekian, K. K., Ed., *The late Cenozoic glacial ages*, Yale Univ. 239-265.

Broecker, W. S. and S. Broecker. 1974. Carbonate dissolution on the western flank of the East Pacific Rise, In: Hay, W. W., Ed., Studies in paleo-oceanography, *SEPM Spec. Pub. No. 20*. 44-57.

Chung, Y. 1975. Areal extent of the benthic front and variation of the scale height in Pacific deep and bottom waters. *Jour. Geophys. Res.*, *80*. 4169-4178.

Goldberg, E. D. and M. Koide. 1962. Geochronological studies of deep-sea sediments by the ionium/thorium method. *Geochim. et Cosmochim. Acta*, *26*. 417-450.

Guinasso, N. L., Jr. 1976. Diffusion models of biological mixing in abyssal sediments as linear filters. *Abstract, EOS, 57.* 150-151.

Guinasso, N. L., Jr., and D. R. Schink. 1975. Quantitative estimates of biological mixing rates in abyssal sediments. *Jour. Geophys. Res., 80.* 3032-3043.

Hays, J. D., T. Saito, N. D. Opdyke, and L. M. Burckle. 1969. Pliocene-Pleistocene sediments of the euqatorial Pacific: their paleomagnetic, biostratigraphic, and climatic record. *GSA Bull., 80.* 1481-1514.

Hülsemann, J. 1966. On the routine analysis of carbonates in unconsolidated sediments. *Jour. Sed. Petrology, 36.* 622-625.

Ku, T. -L. 1965. An evaluation of the U^{234}/U^{238} method as a tool for dating pelagic sediments. *Jour. Geophys. Res., 70.* 3457-3474.

Lisitzin, A. P. 1972. Sedimentation in the world ocean, Ed. by K. S. Rodolfo, *SEPM Spec. Pub. No. 17,* Tulsa. 218 pp.

Luz, B. 1973. Stratigraphic and paleoclimatic analysis of late Pleistocene tropical southeast Pacific cores. *Quaternary Res., 3.* 56-72.

Luz, B. and N. J. Shackleton. 1975. $CaCO_3$ solution in the tropical east Pacific during the past 130,000 years, In: Sliter, W. V., A. W. H. Bé, and W. H. Berger, Eds., Dissolution of deep-sea carbonates, *Cushman Foundation for Foraminiferal Research Spec. Pub. No. 13.* 142-150.

Mammerickx, J., S. M. Smith, I. L. Taylor, and T. E. Chase. 1973. Bathymetry of the South Pacific. *IMR Tech. Rept. 47A.*

Parker, F. L., and W. H. Berger. 1971. Faunal and solution patterns of planktonic Foraminifera in surface sediments of the South Pacific. *Deep-Sea Res., 18.* 73-107.

Peng, T. -H. and W. S. Broecker. 1977. Benthic mixing in deep sea cores as determined by ^{14}C dating and its implications regarding climate stratigraphy and the fate of fossil fuel CO_2[1]. In: *The Fate of Fossil Fuel CO_2,* N. R. Andersen and A. Malahoff (Eds.), Plenum, N.Y., (this volume).

Saito, T., L. H. Burckle, and J. D. Hays. 1974. Implications of some pre-Quaternary sediment cores and dredgings, In: Hay, W. W., Ed., Studies in paleo-oceanography, *SEPM Spec. Pub. No. 20.* 6-43.

Sundquist, E. 1976. Carbonate dissolution events and sediment mixing in late Pleistocene and Holocene stratigraphy. *Abstract, EOS, 57.* 259.

APPENDIX

Let F = total sediment flux to the mixed layer (g/cm^2-kyr)
Q = dissolution flux from the mixed layer (g/cm^2-kyr)
D = dry density of sediment in the mixed layer (g/cm^3)
M = thickness of the mixed layer (cm)
λ = decay constant of C^{14} (kyr^{-1})
f = mass fraction of $CaCO_3$ in the sediment supply
g = mass fraction of $CaCO_3$ in the mixed layer
a = mass fraction of C^{14} in $CaCO_3$ of the sediment supply
b = mass fraction of C^{14} in $CaCO_3$ in the mixed layer
t = time (kyr)

Subscripts $(\ )_o$ refer to initial boundary conditions.

The equations are based on the assumption that mass is conserved during the delivery of fluxes to and from the mixed layer, as diagrammed in Figure 10. Other assumptions are explained in the text.

If dissolution occurs in the mixed layer, the equation for carbonate mass in the mixed layer is

$$MD\,\frac{dg}{dt} = \underbrace{fF}_{\text{Sedimentation flux}} - \underbrace{Q}_{\text{Dissolution flux}} - \underbrace{g\,(F-Q)}_{\text{Burial flux}} \tag{1}$$

If all the parameters except g are assumed to be constant, the solution is

$$g = g_o e^{-\frac{F-Q}{MD}(t-t_o)} + \frac{fF-Q}{F-Q}\left(1 - e^{-\frac{F-Q}{MD}(t-t_o)}\right) \tag{2}$$

The equation for C^{14} in the mixed layer is

$$MD\,\frac{d(gb)}{dt} = \underbrace{afF}_{\text{Sedimentation flux}} - \underbrace{bQ}_{\text{Dissolution flux}} - \underbrace{bg\,(F-Q)}_{\text{Burial flux}} - \underbrace{\lambda bgDM}_{\text{Radioactive decay flux}} \tag{3}$$

which has the solution

$$b = b_o e^{-A(t)} + e^{-A(t)} \int_{t_o}^{t} \left(\frac{afF}{gDM}\right) e^{A(t)}\,dt \tag{4}$$

where

$$A(t) = \int_{t_o}^{t} \left(\frac{fF}{gDM} + \lambda\right) dt \qquad (5)$$

assuming that all parameters except b and g are constant.

Without dissolution, the corresponding equations and their solutions are

$$MD \frac{dg}{dt} = fF - gF \qquad (6)$$

$$g = g_o e^{-\frac{F}{MD}(t-t_o)} + f\left(1 - e^{-\frac{F}{MD}(t-t_o)}\right) \qquad (7)$$

$$MD \frac{d(bg)}{dt} = afF - bgF - \lambda bgDM \qquad (8)$$

$$b = \frac{b_o g_o (f-g) e^{-\lambda(t-t_o)} + \frac{afF}{F+\lambda DM}(f-g_o)\left(1-e^{-(\lambda+\frac{F}{MD})(t-t_o)}\right)}{g\,(f-g_o)} \qquad (9)$$

Throughout our calculations we have assumed that (a) remains constant through all changes in the other parameters. This enables us to calculate C^{14} "ages" very easily according to the equation

$$\text{"Age"} = \frac{-\ln\left(\frac{b}{a}\right)}{\lambda} \qquad (10)$$

For the "negative accumulation" situation discussed in the text, three new parameters must be defined:

Let g^* = the mass fraction of carbonate immediately below the mixed layer

b^* = the mass fraction of C^{14} immediately below the mixed layer

$(F-Q)^*$ = the net rate at which the sediment immediately below the mixed layer was originally deposited (g/cm^2-kyr)

This case is defined in these models by the condition $(F-Q)<0$. The values of g^* and $(F-Q)^*$ are assumed to be constant, while b^* is defined by the equation

$$b^* = b_0 e^{\int_{t_0}^{t} \left(\frac{\lambda (F-Q)}{(F-Q)^*} - \lambda \right) dt} \tag{11}$$

The equations and solutions for mixed-layer carbonate and C^{14} are

$$MD \frac{dg}{dt} = fF - Q - g^* \; (F-Q) \tag{12}$$

$$g = g_0 + \frac{fF - Q - g^* \; (F-Q)}{MD} \; (t-t_0) \tag{13}$$

$$MD \frac{d(bg)}{dt} = afF - bQ - b^* g^* \; (F-Q) - \lambda bgDM \tag{14}$$

$$b = \left(\frac{g_0}{g} \right)^{\left(1 + \frac{Q}{fF - Q - g^*(F-Q)} \right)} e^{-\lambda (t-t_0)}$$

$$\times \left[b_0 + \frac{afF}{g_0 MD} \int_{t_0}^{t} \left(\frac{g}{g_0} \right)^{\left(\frac{Q}{fF - Q - g^* \; (F-Q)} \right)} e^{\lambda (t-t_0)} dt \right.$$

$$\left. - \frac{b_0 g^* (F-Q)}{g_0 MD} \int_{t_0}^{t} \left(\frac{g}{g_0} \right)^{\left(\frac{Q}{fF - Q - g^* \; (F-Q)} \right)} e^{\lambda \left(\frac{F-Q}{(F-Q)^*} \right) (t-t_0)} dt \right] \tag{15}$$

As before, all parameters except b and g are assumed to be constant. For all the calculations in this paper, integrals were evaluated by iteration according to Simpson's Rule.

MECHANISMS FOR CALCITE DISSOLUTION ON THE SEA FLOOR

T. Takahashi and W. S. Broecker

Lamont-Doherty Geological Observatory

ABSTRACT

Arguments in support of the hypothesis that the major dissolution of $CaCO_3$ in the deep sea takes place within a few millimeters of the water-sediment interface are presented. Three models for the dissolution of $CaCO_3$ are considered: the stagnant boundary film model, and the constant and variable resaturation time models. Assuming that the diffusive flux of $CO_3^=$ is equal to that of total CO_2 out of sediment, the validity of each model is tested using the $CaCO_3$ distribution in sediments observed in the Rio Grande Rise area. The stagnant boundary film model fails to adequately represent the data. The constant resaturation time model fits the data satisfactorily, and gives a characteristic calcite resaturation time consistent with the experimental value of 7 minutes. The variable resaturation time model when combined with the calcite solubility of Berner (1976) also fits the data satisfactorily, and predicts dissolution of calcite above the sedimentary lysocline. This prediction is consistent with the observation of Berger (1975). However, the model requires that the rate of calcite dissolution increases with decreasing $CO_3^=$ concentrations at about one-fourth of the value obtained from the data of Berner and Morse (1974).

Accepting the linear kinetics model as a working model, it has been applied to the western North Atlantic Ocean between 25° and 30°N. Using the characteristic calcite resaturation time obtained for the Rio Grande Rise area, a $CaCO_3$ rain rate of 1.3 mg/cm² yr is *obtained for the area between the Mid-Atlantic Ridge and 60°W. This is consistent with the $CaCO_3$ production rate estimated from the carbon-14 dating of the super-lysocline sediments on the Mid-Atlantic Ridge. The area east of 60°W in this latitude range gives a $CaCO_3$ production rate of 0.5 mg/cm^2 yr. Whether this value actually represents the low productivity of this area, or is due to a change in the parameters used in the model is not understood.*

INTRODUCTION

A decrease in the $CaCO_3$ concentration in deep sea sediments with increasing water depth was first observed by *Murray and Renard* (1891). Their observations has been extended to the world oceans by a series of recent studies by *Biscaye et al.* (1976) in the Atlantic Ocean, *Kolla et al.* (1976) in the Indian Ocean, and *Berger et al.* (1976) in the Pacific Ocean. The observed distribution of $CaCO_3$ in deep sea sediments is generally attributed to the dissolution. Calcite dissolution taking place within the sea can be divided into three categories:

1) Water column dissolution: This category includes all processes taking place between the time a unit of calcite is formed and the time it first reaches the sea floor. In addition to the attack of particles settling through under-saturated water, it would include attack within the guts of organisms and during encapsulation in fecal pellets.

2) Sea floor dissolution: This category includes the processes taking place after a particle reaches the ocean bottom, which involves the attack by undersaturated bottom water (as opposed to the attack by undersaturated sediment pore fluids). It includes dissolution during times of resuspension in the nepheloid layer, while lying on top of the sediment column, and while buried shallower than a few characteristic diffusion path lengths (as shown later in this paper, this length is on the order of several millimeters).

3) Pore water dissolution: This category includes all dissolution occurring in the sediment column at depths greater than a few characteristic diffusion path lengths. Although the major agent promoting such dissolution is the production of CO_2 through the oxidation of organic debris, other diagenetic reactions could be envisioned.

The previous listing is not meant to imply that these categories of dissolution occur in sequence. Since the sediment on the sea floor is stirred by benthic organisms to a depth of about ten centimeters, a given particle may move back and forth between the bottom water and pore water zones. In fact, since this stirring appears to homogenize the upper several centimeters of sediment, even core top samples will have experienced all three regimes. As no means of distinguishing the effects of these three processes are available, sediment studies will yield only an estimate of the sum of the three effects.

The approach considered here is to isolate the effects of the interface dissolution. It is assumed that the transition from the relatively high calcite content sediment found above the lysocline to the calcite-poor sediment found below the carbonate compensation depth is produced by interface dissolution alone. It is difficult to imagine a mechanism by which either water column or pore fluid dissolution could produce a feature confined to such a narrow depth range (half width from fifty to several hundred meters in water depths). This assumption is supported by the microscopic observation of *Adelseck and Berger* (1975) on calcareous skeletons in the deep net-tow and box core samples collected below the regional calcite compensation depth in the eastern tropical Pacific Ocean. They concluded that larger forams and pteropods experience little solution during settling through the water column, and dissolution takes place mainly on the sea floor.

Because of a considerable difference between the calcite solubility data in sea water obtained by *Ingle et al.* (1973) and *Berner* (1976), two internally consistent models using each of those data will be developed here. The solubility data of *Ingle et al.* (1973) are smaller than those of *Berner* (1976) by 35% at 2°C, and 16% at 25°C. This discrepancy has been extensively discussed by *Broecker and Takahashi* (in press-a). If the former data are considered, the saturation depth for calcite in the oceans appears to be consistent with the "sedimentary" lysocline and with the "foram" lysocline as shown by *Broecker and Takahashi* (in press-b). Here, the "sedimentary" lysocline is defined as the water depth at which an accelerated dissolution of $CaCO_3$ is indicated by a sharp change in a $CaCO_3$ content in sediment versus water depth plot (*Berger*, 1975). The "foram" lysocline is defined as the boundary zone between the well preserved and poorly preserved foraminiferal assemblages on the sea floor (*Berger*, 1975). On the other hand, if the solubility data of *Berner* (1976) is considered, the calcite saturation level in the oceans lies up to several kilometers above the "sedimentary" and "foram" lysoclines. *Berger* (1975) has observed dissolution of the fine fractions of skeletal carbonate at water depths above the "foram" lysocline in the central north Pacific Ocean, and *Roth and Berger* (1975) observed that coccolith assemblages start showing signs of dissolution at water depths about 1000 meters above the regional coccolith lysocline of 4000 meters. Although qualitative, these observations appear to support the calcite solubility data of *Berner* (1976). On the other hand, these observations might not represent the sea floor dissolution, for the dissolution might have taken place in the water column or pore waters.

The models presented here exclude the possibility that dissolution during resuspension in the nepheloid layer contributes significantly. The standing crop of material in the nepheloid layer is generally about $100\ \mu gm/cm^2$, and corresponds to a layer of sediment only one-thousandth of a millimeter thick. The model considered here for the sea floor dissolution suggests that a layer a few millimeters thick must be involved. It is, therefore, unlikely that dissolution of resuspended calicite can contribute a major fraction of the total interface dissolution.

FACTORS INFLUENCING DISSOLUTION OF $CaCO_3$ ON THE SEA FLOOR

Two possible limiting steps for dissolution, namely a stagnant film at the base of the water column and the calcite resaturation time in pore water by calcite dissolution, are considered for the model of calcite dissolution on the sea floor.

Diffusion Through a Film at the Base of the Water Column

A wide variety of formulations can be used to characterize an impedance against diffusion of ions, as discussed by *Danckwerts* (1970). These range from molecular diffusion through a fixed stagnant film to molecular diffusion into film which is periodically replaced with new water from the main body of fluid. Although the exact physics remains obscure, any one of the single parameter versions of the barrier model are suitable for most chemical applications. In this framework, the $CO_3^=$ flux, F, through a stagnant film of thickness, ℓ, in a steady state condition is:

$$F = D_b\ (\Delta CO_3^=)/\ell \qquad (1)$$

where D_b is the diffusion coefficient for $CO_3^=$ in the ocean bottom water (which constitutes the stagnant film layer), and $\Delta CO_3^=$ is the difference between the critical $CO_3^=$ concentration and the $CO_3^=$ concentration at the top of the stagnant film layer (i.e., the ambient bottom water value).

As little is known about either the effective thickness of the film at the base of the water column, or about resaturation times for real sediments, it is not possible to say *a priori* which limits dissolution at any given place on the sea floor. The ratio of film thickness, ℓ, to resaturation time, τ, at which these two mechanisms would impose equal resistances to transfer, is given by the following relationship (derived from Equations (1) and (14):

$$\ell = D_b/p\sqrt{D/\tau} \qquad (2)$$

where p is the porosity of surface sediment ($\sim$0.7). If the diffusion coefficient in the bottom water, D_b, is twice as large as that in sediment, D, as estimated by *Li and Gregory* (1974), then

$$\ell = 2\sqrt{\tau \cdot D}\ /p \qquad (3)$$

Thus, in order to be the dominant resistance, the film thickness is about three times the characteristic depth for dissolution within the sediments. For a resaturation time of 10 minutes, a film thickness of several millimeters would constitute significant resistance, while one of a few tenths of a millimeter would be of secondary importance.

As a geological calibration is used for the model, the question arises as to whether it makes any difference which step is assumed to be limiting. It does. For example, as long as the film is limiting, the rate of dissolution must change linearly with the difference between the critical $CO_3^=$ concentration in pore water and the bottom water $CO_3^=$ concentration (i.e., on $\Delta CO_3^=$). Also the rate of dissolution will be independent of both the calcite content of the sediment and of the nature of the calcite in the sediment. On the other hand, if the resaturation time is limiting, the dependence of the dissolution rate on $\Delta CO_3^=$ need not be linear. For example, the exponential increase in the rate of dissolution of calcite with decreasing $CO_3^=$ concentrations, which is inferred by the data of *Berner and Morse* (1974), would apply if the resaturation time were limited by the same mechanism as dissolution in their *pH* stat. Also, as will be shown later for marine calcites of the same specific reactivity, the rate of dissolution would depend on the square root of the calcite fraction in the sediment. Finally, because of grain size differences and variable degree of grain coatings, the rate of dissolution would depend on the nature of the calcite.

If a stagnant film is present, then its thickness would presumably vary from point to point on the sea floor because of differences in the dynamics of bottom water movement. Hence, were it limiting the distribution of calcite in marine sediments, the rate of $CaCO_3$ dissolution would depend strongly on the mode of bottom water movement, as suggested by *Edmond* (1974). On the other hand, if the time of resaturation is limiting, variations in rate from point to point on the sea floor would occur because of changes in the nature of the raining calcite and of differences in the bacteriological processes occurring. The productivity and ecology of the overlying water and the rain rate of non-biogenic phases might well also influence the calcite distribution on the sea floor.

The Resaturation Time for Sediment Pore Water

For calcite grains buried beneath the interface, dissolution occurs via diffusive exchange with the overlying sea water. This process is described by Fick's law:

$$D(\partial (CO_3^=)_z/\partial z^2) = (\partial (CO_3^=)_z/\partial t) \qquad (4)$$

where D is the diffusion coefficient for $CO_3^=$ in sediment, and $(CO_3^=)_z$ is the concentration of $CO_3^=$ in the pore water at a depth z below the water-sediment interface. If the rate of dissolution of $CaCO_3$ is assumed to be proportional to the $CO_3^=$ concentration in the pore water, the rate of $CO_3^=$ concentration change is given by:

$$d(CO_3^=)_z/dt = - \{(CO_3^=)_c - (CO_3^=)_z\}/\tau \quad (5)$$

where $(CO_3^=)_c$ is the critical concentration of $CO_3^=$ above which no dissolution of $CaCO_3$ takes place, and τ is the characteristics resaturation time of pore water for dissolution of $CaCO_3$. *Broecker and Takahashi* (in press-b) relate the lysocline depths to the $CO_3^=$ concentrations in the deep waters of the Pacific and Atlantic Oceans, and designate the $CO_3^=$ concentration at the calcite lysocline as the critical $CO_3^=$. The resaturation time, τ, has been determined for optical calcite grains (∿50 mesh) to be about 7 minutes, on the basis of the *pH* change recorded with the deep-sea calcite saturometer designed by *Ben-Yaakov and Kaplan* (1971). However, it should be noted that this time is a function of grain size, the nature of calcite grains, and the condition of grain surfaces.

When the rate of $CO_3^=$ loss via diffusion from pore water is balanced with the supply rate of $CO_3^=$ by dissolution, the solution of the above equations is:

$$(CO_3^=)_z = (CO_3^=)_c \, (1 - e^{-z/\sqrt{\tau D}}) + (CO_3^=)_b \, e^{-z/\sqrt{\tau D}} \quad (6)$$

where $(CO_3^=)_b$ is the $CO_3^=$ concentration in the bottom water just above the water-sediment interface, and the $CO_3^=$ concentration at the interface (i.e., where $z = 0$) is set equal to $(CO_3^=)_b$. The depth of sediment in which this diffusion-dissolution process is operative can be characterized by $Z_{1/2}$, at which the $CO_3^=$ concentration in the pore water reaches a value halfway between the bottom water concentration and the critical $CO_3^=$ concentration. From Equation (6), this depth is expressed by:

$$Z_{1/2} = \ln 2 \cdot \sqrt{\tau \cdot D} \quad (7)$$

Since the resaturation time, τ, in sediment must be a function of the $CaCO_3$ content of the sediment, it is assumed to be inversely proportional to the fraction of $CaCO_3$ in sediment, f:

$$\tau = \tau^*/f \quad (8)$$

where τ^* is the resaturation time for a sediment composed entirely of $CaCO_3$. Thus, the characteristic depth, $Z_{1/2}$, is given by:

$$Z_{1/2} = \ln 2 \cdot \sqrt{\tau^* D/f} \quad (9)$$

As will be seen later, the value of τ^* must be on the order of 10 minutes. Using a diffusivity of 2×10^{-6} cm^2/sec for $CO_3^=$ in sediments (*Li and Gregory*, 1974), this yields a characteristic depth o about one millimeter. As the temperature of ocean bottom water is quite uniform (e.g., 2±1°C) and as the porosity of carbonate-bearing sediment does not vary greatly (e.g., ±20%), the diffusion coefficient will not change much from place to place on the sea floor. The resaturation time will, of course, depend on the grain size and nature of calcite grains in the sediment, and hence, would be expected to vary from place to place.

Two aspects of the Holocene calcite distribution might prove useful in distinguishing these possibilities:

1) the depth of the transition zone between calcite-rich and calcite-poor sediment, and

2) the width and shape of this transition zone.

Broecker and Takahashi (in press-b) have examined the relationship between the depth for the top of the transition zone and the $CO_3^=$ content of the water column. Within the uncertainties of the measurements (depth definition ±200 meters and $CO_3^=$ definition ±4 μM/kg) it is concluded that the two are related by the equation

$$(CO_3^=)_C^C = 90 \exp\{0.16(Z - 4)\} \mu M/kg \qquad (10)$$

where $(CO_3^=)_C^C$ is the $CO_3^=$ content of the water in contact with the top of the calcite transition zone (i.e., sedimentary lysocline) and Z is the water depth in kilometers. The fact that this result agrees with the zero dissolution $CO_3^=$ content calculated from *in situ* saturometry, and its extrapolation to zero pressure yields the calcite solubility obtained by *Ingle et al.* (1973) leads one to suspect that it represents the saturation $CO_3^=$ content in sea water in equilibrium with calcite (*Broecker and Takahashi*, in press-a). As it is fully realized that the last word has not been spoken on this subject, it will be referred to instead as the critical $CO_3^=$ content.

With the concept of critical carbonate ion content in mind, one can consider what variations in the depth of water at the top of the transition zone might be expected in addition to the variation with $CO_3^=$ content of the deep water.

1) Phosphate retardation: *Morse* (1974) proposed that the rate of dissolution of $CaCO_3$ in seawater depends sensitively on the concentration of $PO_4^{\equiv}$ in seawater. If so, the onset of dissolution would depend on $PO_4^{\equiv}$ content of the bottom water. It has been shown by *Broecker and Takahashi* (in press-b), using the data of *Morse* (1974), that the range of deep ocean $PO_4^{\equiv}$ content (1.4 - 2.8 μM/kg) would not produce a perturbation beyond the limits of error in the lysocline depth (i.e., ± 100 to 200 meters).

2) Stagnant film thickness: Such variations would be expected to change the thickness of the transition zone, but not the depth of its top.

3) Resaturation time: Again the width of the transition zone would be expected to vary but not the depth of its top.

Thus, if valid geologic criteria for separating the various hypotheses regarding the rate limiting step do exist, they lie in the shape and absolute thickness of the transition layer and not in its depth.

THREE MODELS RELATING $\Delta CO_3^=$ TO FRACTION OF $CaCO_3$ IN SEDIMENT

Stagnant Film Model

The diffusive flux of $CO_3^=$ through a stagnant film of thickness ℓ and diffusion coefficient D_b has been given in Equation (1). Assuming that the flux of $CO_3^=$ out of sediment is equal to that of dissolved $CaCO_3$, and that a steady state exists between the diffusive loss of dissolved $CaCO_3$ (F_s), the accumulation rate of $CaCO3$ (A), and the rain rate of $CaCO_3$ (R), we obtain:

$$F_s = R - A$$

$$= R[1 - \{(f/f_L)(1-f_L)/(1-f)\}] \quad (11)$$

where f is the fraction of $CaCO_3$ in sediments below the sedimentary lysocline, and f_L is that at the lysocline. The terms $(1-f)$ and $(1-f_L)$ represent the non-carbonate fractions. Combining Equations (1) and (11), we obtain a relationship between $\Delta CO_3^=$ and the sedimentary parameters:

$$\Delta CO_3^= = (\ell \cdot R/D_b) \cdot f_1 \quad (12)$$

where $f_1 = 1 - [(f/f_L)(1-f_L)/(1-f)]$. The value of $\Delta CO_3^=$ can be computed as the difference between the critical $CO_3^=$ concentration given by Equation (10) and the $CO_3^=$ concentration measured in a near bottom water sample (corrected to the *in situ* pressure and temperature conditions). The porosity and the $CaCO_3$ fractions are readily measured in core-top sediment specimens. The calcite rain rate, R, is the Holocene accumulation rate of calcite above the sedimentary lysocline, and can be determined by the carbon-14 dating method.

Constant Resaturation Time Model

The $CO_3^=$ concentration in pore fluid at depth z below the water-sediment interface is given in Equation (6), when the rate of $CaCO_3$ dissolution is assumed to be proportional to the concentration (i.e., linear kinetics). The flux of $CO_3^=$ diffusing across the interface is thus obtained by differentiating Equation (6) and setting z equal to zero:

$$F = D\ (\partial (CO_3^=)_z/\partial z)_{z=0}$$
$$= \sqrt{D/\tau}\ [(CO_3^=)_c - (CO_3^=)_b] \qquad (13)$$
$$= \sqrt{D/\tau}\ (\Delta CO_3^=)$$

Since this equation represents the flux across a unit area of pore fluid, it should be multiplied by the sediment porosity to yield the flux across a unit area of sediment, F_s:

$$F_s = p \cdot F$$

where p is the sediment porosity and ranges between 0 and 1. Substitution of Equations (8) and (13) into this yields:

$$F_s = p\ \sqrt{D \cdot f/\tau^*}\ (\Delta CO_3^=) \qquad (14)$$

If the observed distribution of $CaCO_3$ in sediments is assumed to be in a steady state, Equation (11) describes the material balance. Thus, substituting Equation (14) into (11), we obtain:

$$\Delta CO_3^= = (\sqrt{\tau^*/D})\ (R/p)(1/\sqrt{f})\ [1 - (f/f_L)(1-f_L)/(1-f)]$$
$$= (\sqrt{\tau^*/D})\ (R/p) \cdot f_2 \qquad (15)$$

where f_2 is used to represent all the terms which include f and f_L.

Variable Resaturation Time Model

Berner and Morse (1974) determined the rate of calcite dissolution in a synthetic sea water as a function of pH, and observed that the rate increases very slowly with decreasing pH when the pH is within 0.15 pH unit of the calcite-sea water equilibrium value, and it increases rapidly with decreasing pH beyond this threshold value. *Broecker and Takahashi* (in press-b) computed the $CO_3^=$ concentration in solutions using the data of *Berner and Morse* (1974), and observed that the rate of calcite dissolution is an exponential function of carbonate ion concentration (i.e., exponential kinetics). In this case, Equation (15) which was derived for the constant resaturation time can be modified to represent the exponential kinetics by replacing the constant τ^* with a resaturation time which varies as an exponential function of carbonate ion concentration:

$$\Delta CO_3^= = (R/p)\sqrt{\tau^*/D} \cdot exp\{(-b/2)\ [(\Delta CO_3^=) - (\Delta CO_3^=)_{ref}]\} \cdot f_2 \qquad (16)$$

Here, the parameter b and $(\Delta CO_3^=)_{ref}$ can either be determined from the experimental data of *Berner and Morse* (1974), or from Holocene sediment data.

The choice of $\Delta CO_3^=$ rather than depth as the reference coordinate eliminates one oceanographic variable from the problem. Since the gradient of $CO_3^=$ content with depth varies from region to region in the ocean, the relationship between $\Delta CO_3^=$ and depth below the lysocline will also vary. It goes from a high of 4.4 μM/kg per 100 meters in the western South Atlantic Ocean to about 1.5 μM/kg per 100 meters in the equatorial Pacific Ocean. Other parameters being equal, this variation would produce a corresponding change in the width (but not the shape) of the transition zone.

Among the remaining variables, the Holocene superlysocline rain rate causes the main difficulty. Too little detailed radiocarbon dating has been conducted to permit generalizations to be drawn. Neither the average rate nor the variability in this rate has been adequately defined. An error of a factor of two in this estimate would produce a corresponding error of a factor of two in the predicted transition zone width. Such an error, however, would <u>not</u> affect the calculated <u>shape</u>. Thus, until more detailed radiocarbon dating has been carried out, the shape rather than the absolute width of the transition zone will have to be emphasized.

Figure 1 shows the difference between the shape of the calcite versus depth curves predicted by the three models. The main difference is in the character of the low calcite tail. The stagnant film hypothesis predicts a very sharp transition from cores with intermediate calcite content to cores <u>entirely</u> free of calcite. On the other hand, the constant resaturation time model predicts a long tail of low but finite calcite content. The variable resaturation time model predicts something in between these two extremes.

Another way to look at this question is to consider the weight fraction of calcite dissolved as a function of $\Delta CO_3^=$ (or depth). Predictions of this quantity are shown in Figure 2. The reason why this set of curves looks so different from those in Figure 1, is that when a sediment with 90% calcite has lost <u>half</u> its $CaCO_3$ to dissolution, it still is 80% calcite, and when it has lost nine-tenths of its $CaCO_3$ to dissolution it is still 50% calcite. This means that the water depth at which calcite starts to dissolve may not be accurately determined on the basis of the $CaCO_3$ versus water depth plot, or the preservation of the coarse foram fractions. Therefore, as observed by *Berger* (1975), calcite dissolution could have proceeded above what might be picked as the top of the transition zone (i.e., the sedimentary lysocline). Thus, it is important to develop a quantitative way of estimating the fraction of calcite lost from a sediment, so that the vital assumptions made for the formulation of the calcite dissolution models can be substantiated.

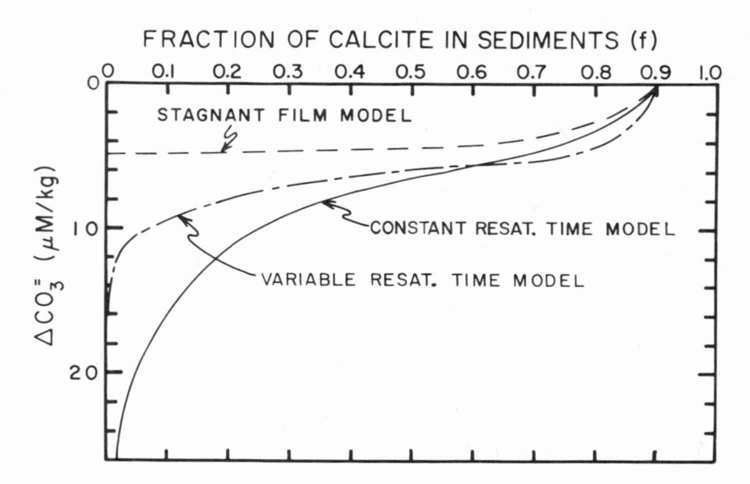

Fig. 1. The relationships between the weight fraction of $CaCO_3$ in sediments and $\Delta CO_3^=$ for three different models. The following values have been used for the computation of those curves: For the stagnant film model, $\ell = 0.14$ cm, $D = 6 \times 10^{-6}$ cm^2/sec, and $R = 2.1 \times 10^{-13}$ M/cm^2·sec; for the constant resaturation time model, $\tau^* = 1910$ seconds, $D = 3 \times 10^{-6}$ cm^2/sec, and $R = 2.1 \times 10^{-13}$ M/cm^2·sec (= 0.7 mg/cm^2 yr); and for the variable resaturation time model, $b = 3.7 \times 10^8$ cm^3/M and $(\Delta CO_3^=)_{ref} = 6$ µM/kg, and R, D and τ^* same as above.

TESTS FOR THE MODELS

The Rio Grande Rise area in the western South Atlantic Ocean has been chosen to evaluate the three models, mainly because of the availability of the data for the $CaCO_3$ concentration in sediments at various water depths and of the $CO_3^=$ concentration in the deep water.

Figure 3 shows the distribution of $CaCO_3$ in the sediments as a function of water depth (*Biscaye et al.*, 1976), and Figure 4 shows the $CO_3^=$ concentration in the water column at GEOSECS Stations 58, 59, and 60 located in the Rio Grande Rise area, and the critical $CO_3^=$ concentration for calcite. The $CO_3^=$ concentrations in the water column have been calculated from the GEOSECS alkalinity and total CO_2 data according to the scheme described by *Takahashi* (1975). Assuming that the $CaCO_3$ rain rate, R, the diffusion coefficients, D and D_b, the resaturation time constants, τ^* and b, the porosity, p, and the thickness of stagnant film, ℓ, are all constant, the adequacy of the models may be tested by observing how well the $\Delta CO_3^=$ versus f_1 or f_2 plots conform to a straight line as described by Equations (12), (15) and (16).

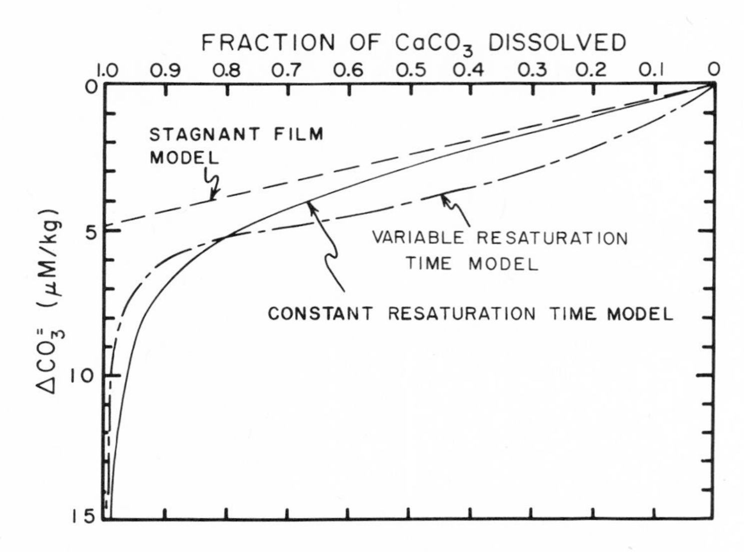

Fig. 2. The relationships between the fraction of $CaCO_3$ dissolved from sediments and $\Delta CO_3^=$ for three different models. The parameters used for the computation are the same as those for Figure 1.

Stagnant Film Model

The stagnant film model not only yields a more or less hyperbolic relationship in the $\Delta CO_3^=$ versus δ_1 plot, as shown in Figure 5, but also fails to exhibit a downward tail in the low $CaCO_3$ range in the $CaCO_3$ versus water depth plot as shown in Figure 1. Hence, this model appears to be inconsistent with the data.

Constant Resaturation Time Model

For the constant resaturation time model, the $\Delta CO_3^=$ versus δ_2 plot (Figure 6) exhibits a linear relation (with a correlation coefficient of 0.98) as postulated by Equation (15). The least squares straight line shown in Figure 6 is replotted in Figure 3 as a function of $CaCO_3$ fraction and water depth. This model appears to represent the observed $CaCO_3$ distribution satisfactorily.

As shown in Equation (15), the slope of this straight line is equal to $(R/p)\sqrt{\tau^*/D}$. When R, p, and D are known, the saturation time for calcite in sediment, τ^*, can be calculated. In the Rio Grande Rise area, the Holocene rain rate of skeletal calcite, R, has been determined to be 0.7 mg/cm^2 yr on the basis of the carbon-14 dating method (*Biscaye et al.*, 1972). The value of the diffusion

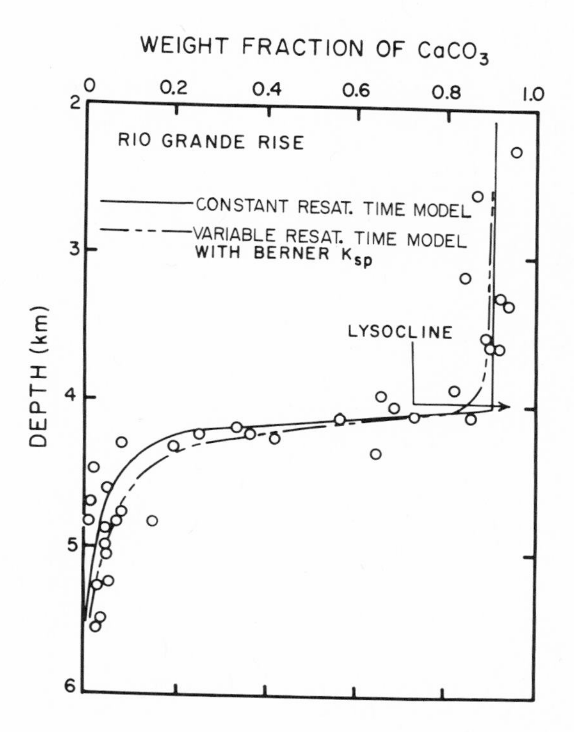

Fig. 3. Weight fraction of $CaCO_3$ in sediments in the Rio Grande Rise area versus the depth of water. The solid curve represents the constant resaturation time model (i.e., the straight line fit to the $\Delta CO_3^=$ and f_2 plot shown in Figure 6). The chain curve represents the variable resaturation time model (i.e., the straight line obtained by a least squares fit to y and f_2 shown in Figure 7). The parameter values used for the models are: $p = 0.7$, $D = 2 \times 10^{-6}$ cm^2/sec, $R = 2.2 \times 10^{-7}$ $\mu M/cm^2\ sec$, $\tau^* = 590\ sec$, $b = 0.0894\ kg/\mu M$, and $(\Delta CO_3)_{ref} = 113\ \mu M/kg$.

coefficient for $CO_3^=$ in sediment, D, has been estimated to be $2 \times 10^{-6}\ cm^2/sec$ at 2°C based on the data of *Li and Gregory* (1974). The porosity of the sediment near the water-sediment interface is assumed to be 0.7 ± 0.1. Thus, the observed slope of 5.2 μM/kg gives a characteristic resaturation time of 590 ± 160 seconds or 9.8 ± 2.7 minutes. This value is somewhat greater than the resaturation time of about 7 minutes for calcite crystals observed in the deep sea saturometer (*Ben-Yaakov*, 1970) and suggests that some additional reaction barrier such as organic coatings may be present in the sediment.

An application of this model thus calibrated will be presented in the following section.

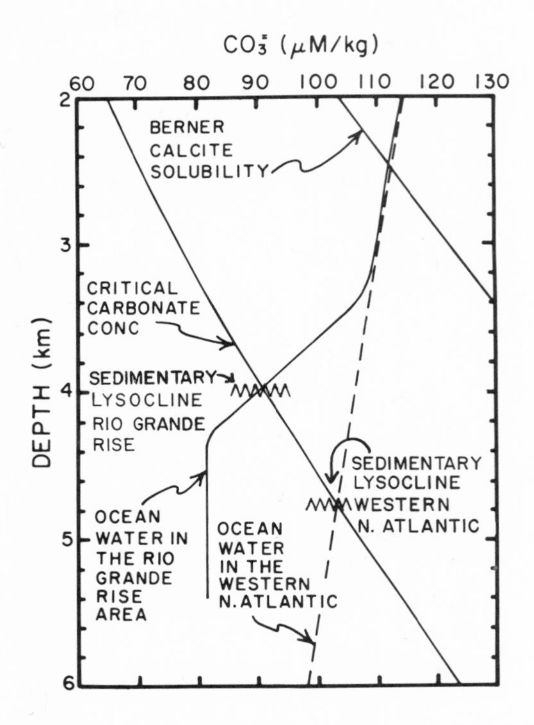

Fig. 4. Concentrations of $CO_3^=$ in the Atlantic Ocean and the critical $CO_3^=$ concentration for calcite as a function of the depth of water. The critical $CO_3^=$ concentration is expressed by:

$$(CO_3^=)_C^C = 90 \exp\{0.16(Z - 4)\}, \ \mu M/kg$$

where Z is the depth of water in Km. The calcite solubility of *Berner* (1976) at 2°C has been extrapolated to higher pressures using a $\Delta\overline{V}$ value of -35.0 cm^3/mole. The lysoclines in the Rio Grande Rise area and western North Atlantic Ocean are indicated.

Variable Resaturation Time Model

Equation (16), which describes the variable saturation time model, contains three adjustable parameters, namely b, $(\Delta CO_3^=)_{ref}$ and τ^*, which are to be determined from the data. Thus it is rearranged in a form suitable for curve fitting processes:

$$\Delta CO_3^= \cdot \exp\{(b/2)(\Delta CO_3^=)\} = E \cdot f_2 \tag{17}$$

where $E = (R/p)\sqrt{\tau^*/D} \cdot \exp\{(b/2)(\Delta CO_3^=)_{ref}\}$, and it is assumed to be constant. It must be pointed out that the exponential dependence of calcite dissolution rate predicts a finite dissolution rate, even in a supersaturated solution, in which $\Delta CO_3^=$ is negative,

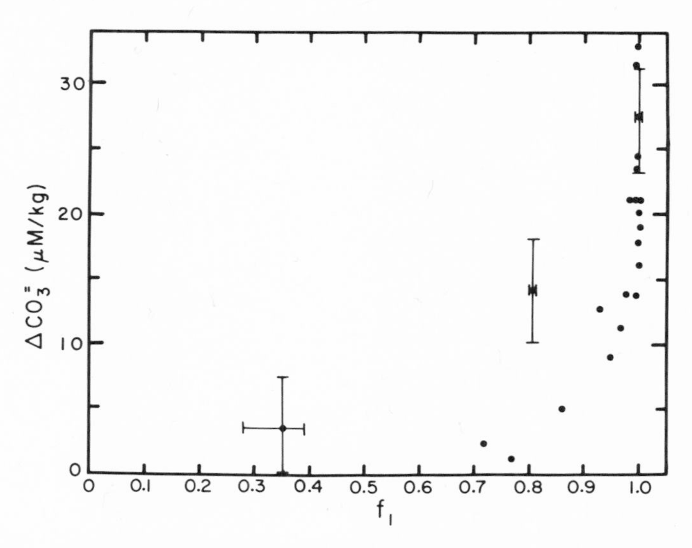

Fig. 5. A plot of $\Delta CO_3^=$ versus f_1 for testing of the stagnant film model. $\Delta CO_3^=$ is the critical $CO_3^=$ concentration minus the $CO_3^=$ concentration in sea water, and f_1 is defined by Equation (13). The errors due to ± 0.01 in $CaCO_3$ fraction and ± 4 $\mu M/kg$ in $\Delta CO_3^=$ are indicated by error bars.

whereas the critical $CO_3^=$ concept excludes calcite dissolution above the lysocline. Thus, Equation (17) is incompatible with the concept of the critical $CO_3^=$ concentration. As mentioned earlier, the calcite solubility data of *Berner* (1976) at 2^oC is as much as 35% greater than the critical $CO_3^=$ concentration, and predicts calcite dissolution a few kilometers above the lysocline. Therefore, *Berner's* (1976) calcite solubility value is used to compute $\Delta CO_3^=$ value. The value of b has been determied to be 0.37 $kg/\mu M$ for a sea water of 35°/oo salinity and 2.0 $\mu M/kg$ PO_4 concentration using the experimental dissolution rate data of *Berner and Morse* (1974). The physical significance of the parameter b is that the calcite dissolution rate increases by e times for an increase of $1/b$ $\mu M/kg$ in $\Delta CO_3^=$. A plot of $\Delta CO_3^= \cdot exp\{(0.37/2)\Delta CO_3^=\}$ versus f_2 indicates that the distribution of the data points is more or less parabolic rather than linear. Thus, Equation (17) with the Berner-Morse value for b cannot properly represent the observed $CaCO_3$ distribution. However, if b is treated as an adjustable parameter in addition to E (which includes τ^* and $(\Delta CO_3^=)_{ref}$), then these two parameters can be determined by means of a least squares fit of Equation (17) to the $CaCO_3$ data. Such a fit to the Rio Grande Rise data for $\Delta CO_3^=$ and f_2 yields:

$$b = 0.0894 \ Kg/\mu M,$$

and

$$E = 842 \quad \mu M/kg.$$

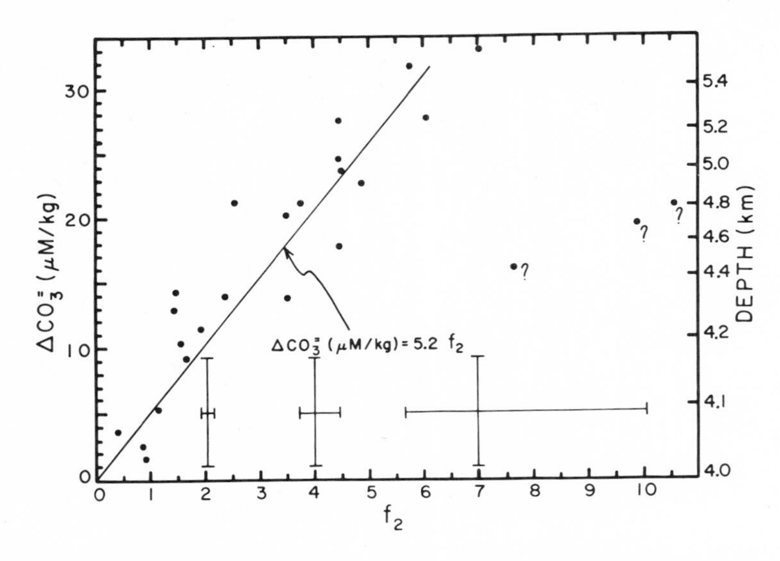

Fig. 6. A $\Delta CO_3^=$ versus f_2 plot for testing of the constant resaturation time model. f_2 is defined by Equation (16). The observed high degree of correlation, $r = 0.98$, indicates that the linear dissolution rate model represents the data satisfactorily. Errors in f_2 resulting from a ± 0.01 error in the $CaCO_3$ fraction, and from ± 4 μM/kg in $\Delta CO_3^=$ are indicated by error bars. Three data points marked with a question mark are not included in the computation.

Using this value for b, a new set of values for the left side of Equation (17) has been computed, and plotted against f_2 in Figure 7. A linear relationship is apparent, and is expressed by a least square straight line passing through the origin and having a slope of 842 μM/kg. This relationship is plotted as a function of the $CaCO_3$ fraction and water depth in Figure 3. The data are represented by the curve satisfactorily. However, it is noted that the value of b used for this successful representation of the data is about one-fourth of the value obtained from the data of *Berner and Morse* (1974). This difference might represent the difference in the surface reactivities of the calcite crystals used in their experiments and of the skeletal calcite in the sediments.

A carbonate ion flux diffusing out of the Rio Grande Rise sediment can be computed from Equation (14) and the $CO_3^=$ data shown in Figure 4, and is plotted in Figure 8 as a function of the water depth for both the constant and variable saturation time models. Assuming that a flux of $CO_3^=$ is approximately equal to that of dissolved $CaCO_3$, this represents the calcite dissolution flux out of the Rio Grande Rise sediments.

The major difference between those two models are seen in Figures 3 and 8. The variable saturation time model predicts

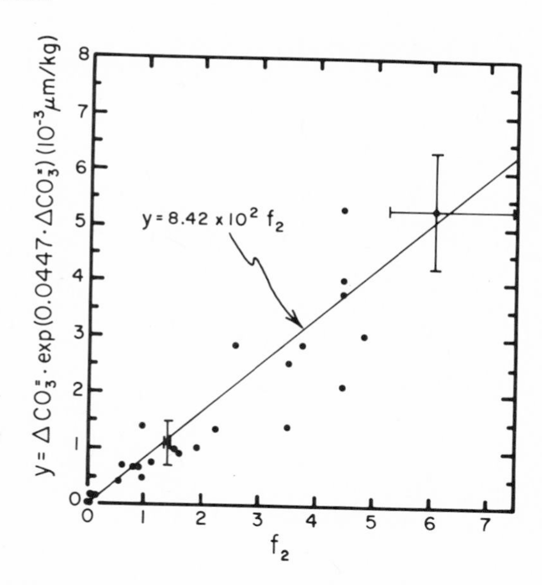

Fig. 7. A $\Delta CO_3^{=} \cdot exp\{(b/2)(\Delta CO_3^{=})\}$ versus f_2 plot for testing of the variable resaturation time model. The linear regression line, when transformed to the $CaCO_3$-depth coordinates, are shown in Figure 3 by a chained curve. Errors resulting from an error of $\pm$0.01 in $CaCO_3$ contents and ± 4 $\mu M/kg$ in $\Delta CO_3^{=}$ are indicated by the error bars. Three data points marked with a question mark in Figure 6 are not shown in this figure, and are excluded from the computation of the linear regression line.

calcite dissolution far above the sedimentary lysocline. This feature is consistent with the observation of *Berger* (1975), provided that his observation represents calcite dissolution at or near the water-sediment interface. On the other hand, the constant saturation time model predicts no calcite dissolution above the lysocline. Thus, if the dissolution of calcite observed above the lysocline actually increases with water depth, then the variable saturation time model is superior over the constant saturation time model. However, the formulation for the variable saturation time model contains three adjustable parameters, i.e., τ^*, b, and $(\Delta CO_3^{=})_{ref}$, of which the physical significances cannot be tested directly on the basis of laboratory or field experiments at present. Thus, its applications to other oceanic areas appear to be premature, and will not be discussed.

APPLICATION OF THE CONSTANT RESATURATION TIME MODEL

Figure 9 shows the distribution of $CaCO_3$ in sediments in the western Atlantic Ocean between 25° and 30°N. It is most interesting

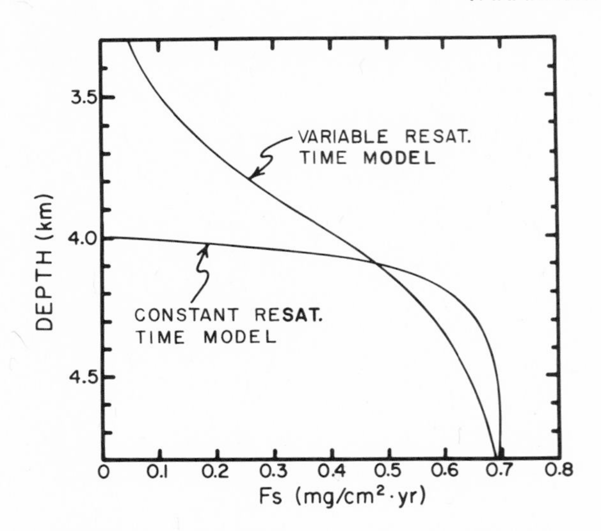

Fig. 8. The flux of $CO_3^=$ (i.e., dissolved $CaCO_3$) out of the Rio Grande Rise area sediment computed from the constant and variable resaturation time models. The parameters used are: $p = 0.7$, $D = 2 \times 10^{-6}$ cm^2/sec, $\tau^* = 590$ *seconds*, $R = 2.2 \times 10^{-7}$ $\mu M/cm^2 \cdot sec$, $b = 0.0894$ $kg/\mu M$, and $(\Delta CO_3^=)_{ref} = 113$ $\mu M/kg$. The value of $\Delta CO_3^=$ is based on the critical $CO_3^=$ concentration for calcite for the constant resaturation time model, and on the calcite solubility of *Berner* (1976) for the variable resaturation time model.

to note that the trend for the data obtained between 68°W and 60°W differs systematically from that for the area between the Mid-Atlantic Ridge and 60°W. The distribution of the $CO_3^=$ in deep water obtained at GEOSECS Stations 31 through 34 is shown in Figure 4, along with the critical $CO_3^=$ concentration. Using this information, a plot for $\Delta CO_3^=$ versus δ_2 has been constructed (Figure 10). The western area and eastern area data have been fitted to two separate straight lines: for the area between 60°W and 68°W, $\Delta CO_3^= = 3.72\ (\pm 0.14)\ \delta_2$ $(\mu M/kg)$; and for the area east of 60°W and west of the Mid-Atlantic Ridge, $\Delta CO_3^= = 9.9\ (\pm 0.9)\ \delta_2 - 4.2\ (\pm 1.7)$ $(\mu M/kg)$. The $\Delta CO_3^=$ axis intercept of -4.2 $\mu M/kg$ in the latter equation indicates either imperfection of the model, deviation from the critical $CO_3^=$ concentration concept, or geographical difference in the $CO_3^=$ distribution in the deep water. This negative intercept is considered a result of geographical difference in the $CO_3^=$ concentrations in deep water. Accordingly, two depths of the lysocline are indicated in Figure 9 to reflect the difference in the axis intercepts for the eastern and western areas.

The rain rate for $CaCO_3$ consistent with the observed calcite content versus water depth curves in those two areas have been

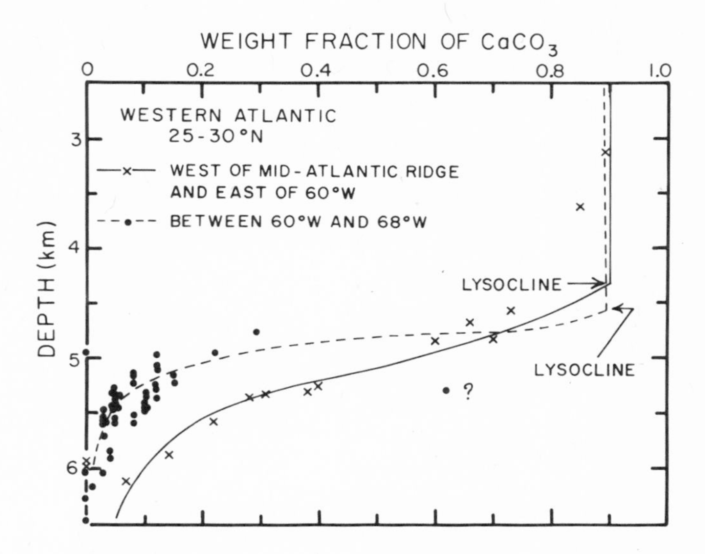

Fig. 9. Distribution of $CaCO_3$ in the sediments in the western Atlantic Ocean (North American Basin) between 25°N and 30°N. The sources of the data are fully described in *Biscaye et al.* (1976). The X's represent the sediment cores located west of the Mid-Atlantic Ridge and east of 60°W, and the circles represent those between 60°W and 68°W. The solid and dashed curves correspond to the respective linear regression lines shown on the $\Delta CO_3^=$ versus f_2 plot (Figure 10).

computed using the slope of the linear regression lines and the values of τ^* obtained from the Rio Grande Rise area. The values for the $CaCO_3$ rain rate thus obtained are: 1.3 ± 0.1 mg/cm^2 yr for the area between 60°W and the Ridge, and 0.48 ± 0.02 mg/cm^2 yr for the area between 60°W and 68°W. For both cases, a porosity of 0.7 is assumed. The former is consistent with the Holocene average $CaCO_3$ deposition rate of 1.2 mg/cm^2 yr obtained from the Mid-Atlantic Ridge sediments (*Biscaye et al.*, 1972). The latter may represent a low productivity of $CaCO_3$ in this part of the Saragasso Sea, or a change in the D, p and/or τ^* for the sediments in this area. Since the slope of the linear regression line for the $\Delta CO_3^= - f_2$ plot (Figure 10) is insensitive to the choice of the values for f_L (the slope is increased from 3.72 to 3.89 for a change in f_L from 0.9 to 0.5), the small observed slope, and hence the low $CaCO_3$ rain rate, cannot be attributed to the increased flux of terrigenous non-carbonate fraction into this area. If the $CaCO_3$ productivity in the western area is assumed to be the same as that near the Mid-Atlantic Ridge, then a factor of 3 change in the τ^* value is predicted. Since τ^* depends on the state and size of calcite grains in the

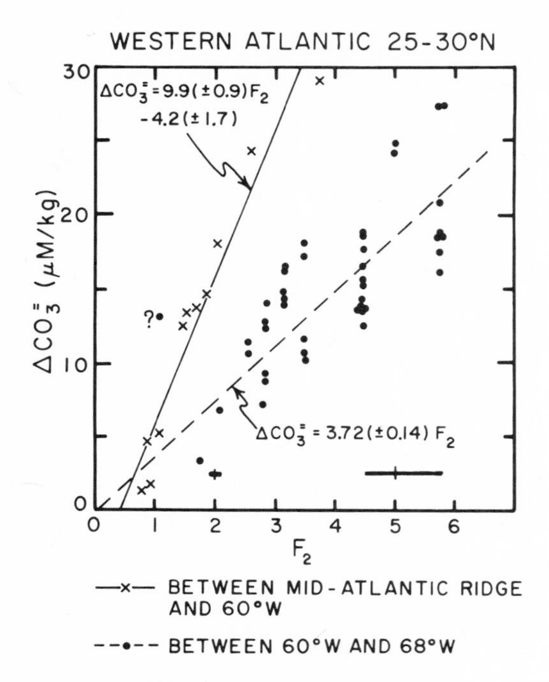

Fig. 10. A $\Delta CO_3^=$ versus f_2 plot for the data shown in Figure 8 for the western Atlantic Ocean between 25°N and 30°N. The X's are for the sediments located between the Mid-Atlantic Ridge and 60°W, and the circles are for those located between 60°W and 68°W. f_L is taken to be 0.90 for the both cases. The slopes of the linear regression lines are insensitive to the choice of the f_L values. Errors in f_2 resulting from an error of ± 0.01 in $CaCO_3$ fraction are indicated by the horizontal bars.

sediments, such a magnitude of variation in the τ^* value for deep sediments seems to be not unrealistic over the geographic areas being considered.

CONCLUSIONS

Assuming that the dissolution of skeletal $CaCO_3$ in the deep oceans takes place mainly within a short distance below the water-sediment interface, various factors which may affect the dissolution of $CaCO_3$ have been discussed. As a rate limiting step for the dissolution of $CaCO_3$, three processes have been considered: (1) diffusion of $CO_3^=$ through a stagnant film at the base of water column, (2) a dissolution rate of $CaCO_3$ proportional to the difference between the $CO_3^=$ concentration in pore fluid and the critical $CO_3^=$ concentration for calcite, and (3) a dissolution rate exponentially related to the difference between the pore fluid $CO_3^=$ concentration and the calcite solubility of *Berner* (1976). Three models describing each of those processes are designated respectively as the stagnant layer, the constant resaturation time and the variable resaturation time models. It appears that, if valid geologic criteria for separating various rate limiting steps exist, the shape and thickness of the transition layer should be the critical information rather than the depth of lysocline.

Using the $CaCO_3$ content data in the sediments and the $CO_3^=$ concentrations in deep water in the Rio Grande Rise area, the validity of each of three models has been tested. The stagnant film model not only fails to satisfy the linear relationship between $\Delta CO_3^=$ versus f_1 as described by Equation (12), but it also does not give a downward tail in the low $CaCO_3$ range in the $CaCO_3$ fraction versus water depth plot. Thus, it can be rejected. On the other hand, the constant resaturation time model in conjunction with the concept of the critical $CO_3^=$ concentration for calcite can satisfactorily account for the observed distribution of $CaCO_3$ in sediment. Using values of $0.7\ mg/cm^2\ yr$ for the rain rate of calcite, $2 \times 10^{-6}\ cm^2/sec$ for the diffusion coefficient of $CO_3^=$ in the sediment and 0.7 ± 0.1 porosity, the model yields a characteristic resaturation time of 9.8 ± 2.7 minutes for pore water by calcite dissolution. This time is in fair agreement with the field experiment value of 7 minutes obtained by the *in situ* calcite saturometry measurements. The variable resaturation time model can be fitted to the data satisfactorily if the calcite solubility of *Berner* (1976) is accepted. However, the model yields that the dissolution rate of calcite increases with increasing $\Delta CO_3^=$ value at a rate of about one-quarter of that indicated by the dissolution rate experiments of *Berner and Morse* (1974). This may be interpreted as a manifestation of the difference between the surface reactivities of the calcite crystals used in their experiments and of natural skeletal calcite in sediments.

It has been demonstrated that, in order to distinguish the mechanisms influencing the distribution of calcite in deep sea sediments, the information on the $CaCO_3$ content in sediments and $CO_3^=$ concentration in seawater is important. In particular, the shape of the $CaCO_3$-water depth (or $CO_3^=$ in deep water) curve in the high (>85%) and low (<15%) $CaCO_3$ ranges provides critical information. Since the available $CaCO_3$ data are not sufficiently precise to permit critical analyses, due perhaps to their geographical diversity of the core locations, an intense study of $CaCO_3$ contents in surface sediments in closely spaced core samples taken across lysocline is needed. In addition, more determinations of the oceanic $CaCO_3$ accumulation and rain rates which can be measured by the radiocarbon dating method are acutely needed to extend the tests for the applicability of our ideas presented here to the world oceans.

ACKNOWLEDGEMENTS

We have been benefited from the discussions with Robert A. Berner, John W. Morse, Pierre E. Biscaye, Nick Shackleton, Y-H Li, and Wolfgang Berger. This work has been supported by grants from the Office of Naval Research and from the Office of the International Decade of Ocean Exploration, National Science Foundation, to the Lamont-Doherty Geological Observatory and Queens College, City University of New York, where the first author was formerly affiliated. This is Contribution No. 2525 from Lamont-Doherty Geological Observatory. GEOSECS Publication No. 63.

REFERENCES

Adelseck, C. G. and W. H. Berger. 1975. On the dissolution of planktonic foraminifera and associated microfossils during settling and on the sea floor. In: *Dissolution of Deep-Sea Carbonates*, W. V. Sliter, A. W. H. Bé and W. H. Berger (Eds.), Spec. Publ. 13, Cushman Foundation for Foraminiferal Research. 70-81.

Ben-Yaakov, S. 1970. An oceanographic instrumentation system for *in situ* measurements, Ph.D. Thesis, U.C.L.A. 343 pp.

Ben-Yaakov, S. and I. R. Kaplan. 1971. Deep sea *in situ* calcium carbonate saturometry. *Jour. Geophys. Res.*, *76*. 722-731.

Berger, W. H. 1975. Deep-sea carbonates: dissolution profiles from foraminiferal preservation. In: *Dissolution of Deep-Sea Carbonates*, W. V. Sliter, A. W. H. Bé and W. H. Berger (Eds.), Spec. Publ. 13, Cushman Foundation for Foraminiferal Research. 82-86.

EDITOR'S NOTE: Recent unpublished data obtained by M. L. Ledbetter of the University of Rhode Island, D. A. Johnson of the Woods Hole Oceanographic Institute, and D. F. Williams of Brown University for cores from the Rio Grande Rise area show that their superlysocline carbonate rain rate is consistent with that used in this paper, whereas the sedimentation rate of non-carbonate material increases by approximately one hundred folds at a depth corresponding to the sublysocline transition region. This being the case, the sublysocline decrease in sedimentary carbonate content, which has been attributed entirely to calcite dissolution in the present paper, must be to a large extent the result of dilution by non-carbonate material. Thus, the model presented in this paper should be considered tentative. Implication of the high non-carbonate sedimentation rate on the present model is being investigated.

Berger, W. H., C. G. Adelseck and L. A. Mayer. 1976. Distribution of carbonate in surface sediments of the Pacific Ocean. *Jour. Geophys. Res., 81.* 2617-2627.

Berner, R. A. 1976. The solubility of calcite and aragonite in sea water at atmospheric pressure and 34.5 $^{o}/oo$ salinity. *Am. J. Sci., 276.* 713-730.

Berner, R. A. and J. W. Morse. 1974. Dissolution kinetics of calcium carbonate in sea water. IV. Theory of calcite dissolution. *Am. J. Sci., 274.* 108-134.

Biscaye, P. E., A. Bé, D. Ellis, J. Gardner, T. Kellog, A. McIntyre, W. Prell, M. Roche, and K. Venkatarathnarn. 1972. Holocene vs. glacial (17,000 yrs, B.P.) patterns of sedimentation in the Atlantic Ocean. *Geol. Soc. Am. 1973 Ann. Mtg. Abstracts, Vol. 5, No. 7.* 551.

Biscaye, P. E., V. Kolla, and K. K. Turekian. 1976. Distribution of calcium carbonate in surface sediments of the Atlantic Ocean. *J. Geophys. Res., 81.* 2595-2603.

Broecker, W. S. and T. Takahashi. (In press-a). The solubility of calcite in seawater. In: *Thermodynamics in Geology,* D. G. Fraser (Ed.), D. Reidel Publ. Co., Dordrecht, Holland.

Broecker, W. S. and T. Takahashi. (In press-b). The relationship between lysocline depth and *in situ* carbonate ion concentration. *Deep-Sea Research.*

Danckwerts, P. V. 1970. *Gas-Liquid Reactions.* McGraw-Hill, N.Y. 276 pp.

Edmond, J. M. 1974. On the dissolution of carbonate and silicate in the deep ocean. *Deep-Sea Res., 21.* 450-480.

Ingle, S. E., C. H. Culberson, J. E. Hawley, and R. M. Pytkowicz. 1973. The solubility of calcite in seawater at atmospheric pressure and 35 $^{o}/oo$ salinity. *Marine Chem., 1.* 295-307.

Kolla, V., A. W. H. Bé, and P. E. Biscaye. 1976. Calcium carbonate distribution in the surface sediments of the Indian Ocean. *Jour. Geophys. Res., 81.* 2605-2616.

Li, Y-H, and S. Gregory. 1974. Diffusion of ions in sea water and in deep-sea sediments. *Geoch. et Cosmoch. Acta, 38.* 703-714.

Morse, J. W. 1974. Dissolution kinetics of calcium carbonate in sea water. V. Effects of natural inhibitors and the position of the chemical lysocline. *Am. J. Sci., 274.* 638-647.

Murray, J. and A. F. Renard. 1891. Report on deep-sea deposits based on the specimens collected during the voyage of H.M.S. Challenger in the years 1872 and 1876, In: *Report of the Voyage Challenger.* Longmans, London. 525 pp.

Roth, P. H. and W. H. Berger. 1975. Distribution and dissolution of coccoliths in the south and central Pacific. In: *Dissolution of Deep-Sea Carbonates,* W. V. Sliter, A. W. H. Bé and W. H. Berger (Eds.), Spec. Publ. 13, Cushman Foundation for Foraminiferal Research. 87-113.

Takahashi, T. 1975. Carbonate chemistry of sea water and the calcite compensation depth in the oceans. In: *Dissolution of Deep-Sea Carbonates,* W. V. Sliter, A. W. H. Bé and W. H. Berger (Eds.), Spec. Publ. 13, Cushman Foundation for Foraminiferal Research. 11-26.

EQUILIBRIUM AND MECHANISM OF DISSOLUTION OF MG-CALCITES

R. Wollast and D. Reinhard-Derie

Free University of Brussels

ABSTRACT

The dissolution of Mg-calcite, the most abundant carbonate secreted by marine organisms, is discussed in terms of the thermodynamic implications concerning the dissolution of a solid solution. A method of theoretical interpretation of the experimental results is proposed. It is also shown that the composition of the equilibrium mixture of the final aqueous and solid solution depends on the initial ratio of solid to liquid used in the experiments.

The mechanism of the dissolution is further considered on the basis of kinetic data and examination of the initial and final products of the reaction. It is concluded that the experimental study of the dissolution of natural Mg-calcites is much complicated by the fact that marine organisms are generally heterogeneous solid solutions, and that reprecipitation of a series of new solid solutions on the surface of the initial Mg-calcite lead to a multiphase system which strongly prevents this system from reaching the equilibrium at the time scale used in laboratory experiments.

It is thus difficult at this time to precisely quantify the influence of the composition of the Mg-calcite on its dissolution in the marine environment and further studies are indispensable to a better comprehension of an essential step of the cycle of CO_2 in the Ocean.

INTRODUCTION

Precipitation and dissolution of $CaCO_3$ represent one of the essential steps of the carbone cycle in the Ocean. Marine organisms commonly secrete calcite containing large amounts of magnesium carbonate; up to 30 mole per cent for certain species. It is generally admitted that the stability of the *Mg*-calcite decreases as the magnesium content increases, and that a low *Mg*-calcite should be the stable phase in the environment of deposition. However, the quantitative influence of the magnesium content on the solubility of *Mg*-calcite at low temperature is poorly understood (*Chave et al.*, 1962; *Schmalz and Chave*, 1963; *Friedman*, 1964; *Lerman*, 1965; *Land*, 1967; *Winland*, 1969; *Schroeder*, 1969; *Plummer and Mackenzie*, 1974). The problem is indeed complicated by the fact that these compounds dissolve incongruently (i.e., the ratio of Mg^{++}/Ca^{++} which dissolves is different from the ratio of $MgCO_3/CaCO_3$ to be found in the initial solid solution (*Land*, 1967; *Schroeder*, 1969; *Plummer and Mackenzie*, 1974). For this reason, most of the discussions about the dissolution of $CaCO_3$ in the marine environment refer to pure calcite; its solubility is well known and its dissolution mechanism rather simple. A better understanding of this process could, however, be gained by considering the real phases involved. The thermodynamic implications of the incongruent character of the dissolution of *Mg*-calcite are, therefore, discussed first. The mechanism of the reaction and its influence on the kinetic behavior of natural *Mg*-calcites during dissolution will thus be considered.

STABILITY OF MAGNESIUM CALCITES

The dissolution of *Mg*-calcite is generally described by the following relation:

$$Mg_{x_o}\, Ca_{(1-x_o)}\, CO_3 \rightleftarrows a\, Mg_x\, Ca_{(1-x)}\, CO_3 + (x_o - ax)\, Mg^{++} + (1 - a - x_o + ax)\, Ca^{++} + (1-a)\, CO_3^{=} \qquad (1)$$

This relation actually represents the mass balance of the reaction; it has no thermodynamical significance.

The dissolution equilibrium of the solid solution is fully described by the following two conditions:

$$(CaCO_3)_{ss} \rightleftarrows Ca^{++} + CO_3^{=} \qquad K_{CaCO_3} = \frac{a_{Ca^{++}}\, a_{CO_3^{=}}}{a_{CaCO_3}} \qquad (2)$$

$$(MgCO_3)_{ss} \rightleftarrows Mg^{++} + CO_3^{=} \qquad K_{MgCO_3} = \frac{a_{Mg^{++}}\, a_{CO_3^{=}}}{a_{MgCO_3}} \qquad (3)$$

where a represents the activities of the species in the aqueous and solid solutions.

One source of confusion lies in the fact that relation (1) has been used extensively to estimate the stability of Mg-calcites, instead of relations (2) and (3). However, the mass balance equation (1) must be fulfilled at any time during a dissolution experiment, and equations (2) and (3) are satisfied only when equilibrium is reached.

The mass balance may be rewritten as

$$M_o = M + D \qquad (4)$$

where M_o and M are respectively, the amounts, in moles, of initial and final Mg-calcite, and D the amount, in moles, of $MgCO_3$ plus $CaCO_3$ dissolved in the aqueous phase.
If x_o and x represent the mole fraction of magnesium respectively in the initial and final phases, and y the molar ratio defined by

$$\frac{m_{Mg^{++}}}{m_{Ca^{++}} + m_{Mg^{++}}} \qquad (5)$$

we must also have

$$x_o\, M_o = xM + yD \qquad (6)$$

or by combining (4) and (6)

$$y = x_o \frac{M_o}{D} - \frac{M_o - D}{D}\, x \qquad (7)$$

On the other hand, the equilibrium relations (2) and (3) may be combined:

$$\frac{a_{Mg^{++}}}{a_{Ca^{++}}} = \frac{a_{MgCO_3}}{a_{CaCO_3}} \cdot \frac{K_{MgCO_3}}{K_{CaCO_3}} \qquad (8)$$

which may be rewritten as:

$$\frac{y}{1-y} = \frac{x}{1-x}\, \frac{\gamma Ca^{++}}{\gamma Mg^{++}}\, \frac{\lambda MgCO_3}{\lambda CaCO_3} \cdot \frac{K_{MgCO_3}}{K_{CaCO_3}} \qquad (9)$$

where γ and λ represent the activity coefficients respectively in the aqueous and solid phases.

Equations (7) and (9) may be usefully represented graphically in a x-y diagram (Figure 1). The mass balance equation (7) is represented in this diagram by a straight line intersecting the $x=y$

line at $x=x_0$ and with a slope equal to $-\dfrac{M_0 - D}{D}$.

This line is vertical for no dissolution, and horizontal for complete dissolution.

In the x-y diagram, the equilibrium condition represented by equation (9) is given by a curve, the curvature of which is defined by the ratios of the activity coefficients and of the equilibrium constants. The effect of $\gamma Ca^{++}/\gamma Mg^{++}$ is very close to 1 and negligible, and the ratio of K_{MgCO_3}/K_{CaCO_3} (*Robie and Waldbaum*, 1968; *Garrels and Christ*, 1965) is equal to $10^{-7.99}/10^{-8.44} = 2.82$. An estimate of the activity coefficients of magnesium and calcite phases in the solid solution obtained by *Lerman* (1965) from an extrapolation of high temperature experiments, shows that the ratio of $\lambda MgCO_3/\lambda CaCO_3$ is also much greater than 1, as for most non-ideal solid solutions. The values given by *Lerman* (1965) for the magnesium phase, seem however to be overestimated and do not fit the actual solubility measurements carried out at low temperature. The shape of the curve represented schematically in Figure 1, indicates that

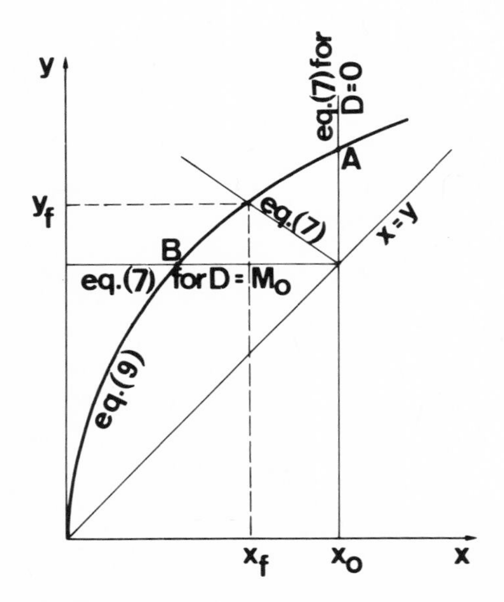

Fig. 1. Graphical solution for the determination of the final composition of a system containing initially pure water and Mg-calcite. $x = MgCO_3/MgCO_3 + CaCO_3$; $y = Mg^{++}/Mg^{++} + Ca^{++}$. The curve represents the equilibrium condition given by equation (9). Different cases corresponding to various amounts of solid dissolved are represented according to the mass balance equation (7).

according to the above discussion, the aqueous solution is enriched in Mg^{++} with respect to the solid solution ($y > x$). Due to the uncertainties regarding the activity coefficients of the solid solution, it is felt unreasonable at this time to construct a more quantitative diagram.

At equilibrium, both equations (7) and (9) must be satisfied, and the composition of the aqueous and solid phase (represented by y_f, x_f) is given by the intersection of the line representing the mass balance equation with the theoretical equilibrium curve. It appears, therefore, that the equilibrium composition of the aqueous phase and of the final solid solution may vary from A to B according to the amount of the initial solid phase dissolved. This effect is well demonstrated in Table 1, which reports the results of a dissolution experiment of the rich Mg-calcite alga *Goniolithon*(77.2M per cent $MgCO_3$). Similar results were obtained with a rather pure synthetic calcite (0, 62 per cent $MgCO_3$) which gives a ratio of Mg^{++}/Ca^{++} in solution equal to 0,6 per cent for 100 mg of initial calcite per 100 $m\ell$ of solution, and 1,80 per cent for an initial suspension of $5\ g/100\ m\ell$.

TABLE 1. Dissolution of *Goniolithon* in NH_4Ac-HAc Aqueous Solutions After One Week.

Solid $(gr/1)$	pH	$Ca^{++}(10^{-2}M/1)$	$Mg^{++}(10^{-2}M/1)$	ratio Mg^{++}/CA^{++}
4	6.45	3.43	2.80	0.82
6	6.54	4.53	4.65	1.03
8	6.60	4.13	6.25	1.51
10	6.61	4.35	8.44	1.94
100	7.98	6.00	10.70	1.78

This effect has already been pointed out by *Schroeder* (1969), but never previously taken into consideration in the estimation of the stability of Mg-calcites. It does account for the discrepant apparent solubilities reported by various authors studying the same organisms.

Figure 1 also shows that it is necessary to realize successive dissolution - reprecipitation reactions in pure water, in order to convert Mg-calcite into pure calcite. In the presence of seawater with a fixed $Mg^{++}/(Ca^{++}+Mg^{++})$ ratio, the diagenis of natural Mg-calcite does not necessarily lead to purer calcite, and explains the occurrence of high Mg-calcite in marine cements versus low Mg-calcite in meteoric cements, now well documented (*Bricker*, 1971).

It is also easy to demonstrate that the solubility of a given *Mg*-calcite, expressed as the total amount of cations released at a fixed $a_{CO_3^=}$:

$$\frac{a_{Ca^{++}} + a_{Mg^{++}}}{a^*_{Ca^{++}}} = (1-x)\ \lambda_C + x\ \lambda_M \frac{KM}{Kc} \qquad (10)$$

where $a^*_{Ca^{++}}$ refers to the activity of Ca^{++} for the pure calcite.

Even if the influence of λ_C and λ_M is neglected, which, according to *Lerman* (1971), are both greater than 1, one already obtains an increase of solubility of 20 per cent for a 10 per cent molar magnesium calcite. Such large variations justify an increased interest for the thermodynamic properties of these phases in order to ameliorate our comprehension of the behavior of carbonates in the ocean and in marine sediments.

MECHANISM OF DISSOLUTION OF MAGNESIUM-CALCITES

The above thermodynamic considerations are valid regardless of the reaction path leading to equilibrium. They do not offer, however, any information regarding the mechanism. The incongruency of the reaction has been attributed to a faster dissolution rate of $MgCO_3$ than $CaCO_3$ (*Land,* 1967). In these conditions, a residual reaction layer, less rich in *Mg*, remains on the surface of the initial *Mg*-calcite. The kinetics of the reaction is then controlled by the diffusion of its products through the residual layer. *Plummer and Mackenzie* (1974) have, however, postulated a congruent dissolution reaction followed by the precipitation reaction of a new *Mg*-calcite less rich in *Mg*. This hypothesis is based on the evolution in time of the concentrations of Ca^{++} and Mg^{++} obtained by the dissolution of *Amphiroa* in distilled water saturated by CO_2 (P_{CO_2} = 1 atm.) .

During a first phase, *Plummer and Mackenzie* (1974) observed (Figure 2) that the ionic ratio Mg^{++}/Ca^{++} in solution remained constant, which led them to suggest a congruent dissolution of the most soluble solid phase. In a second phase, they noted a decrease in the concentration of Ca^{++}, which they attributed to the precipitation of a purer calcite than the initial *Mg*-calcite.

Our examinations of the surface of *Mg*-calcites during dissolution, by means of a scanning electron microscope, have clearly shown the phenomenon. The dissolution experiments were performed on *Mg*-calcites during 14 days; 5 grams of each sample sized between 10 and 150 μ were suspended in 100 *ml* of distilled water at 25°C, under a pressure of CO_2 equal to *1 atm*. The electron pictures of *Amphiroa fragilissima* before and after dissolution (Figure 3) demonstrate the formation of small crystals of about 2 μ at the surface of the initial solid phase. Furthermore, redissolution experiments on

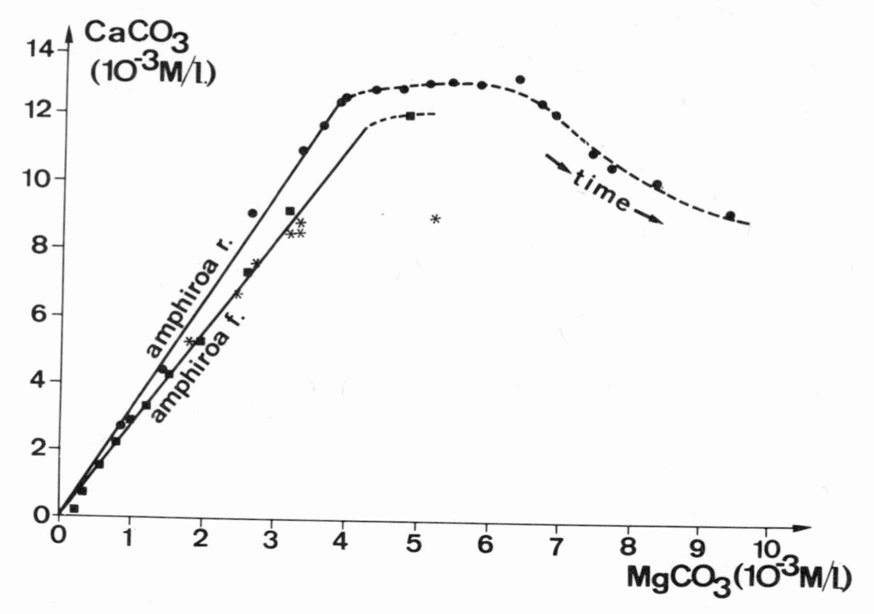

Fig. 2. Dissolution curves Ca^{++} versus Mg^{++}, of *Amphiroa fragilissima* and *Amphiroa rigida* in distilled water under P_{CO_2} = 1 *atm.* reported from *Plummer and Mackenzie* (1974); in these experi-
between 44 and 74 µ. * denotes our results on *Amphiroa fragilissima* sized between 10 and 44 µ; the ratio solid /solution is *5 gr/100 ml.*
In both experiments, the aqueous suspensions were shaked continuously.
both experiments, the aqueous suspensions were shaked continuously.
Our results on *Amphiroa fragilissima* are in perfect agreement with the results of *Plummer and Mackenzie*, for the initial congruent dissolution step (up to 80 minutes). The discrepancy for larger dissolution times is due to differences in the rate of reprecipitation related to differences of the solid/solution ratio.

Mg-calcites previously submitted to a prolonged dissolution under the conditions mentioned above have been performed. These calcites then underwent a series of controlled flash dissolutions by *HCl*, in order to cause a layer by layer peeling of the carbonate. The results for this experiment on *Lytechinus verigatus* (tests) are represented in Figure 4. The successive etchings by *HCl* release increasing ratios of Mg^{++}/Ca^{++} until the value of the ratio for the initial solid solution is reached. These experiments confirm the hypothesis of the formation of a purer calcite on the surface of the initial *Mg*-calcite. Information on the composition of the superficial layer could be gained by extrapolating the dissolution curve (Figure 4) to zero. However, the outline of the experimental

Fig. 3. Scanning electron microscopy of *Amphiroa fragilissima* sized between 10 et 150 μ before dissolution (a) and after a dissolution run of 14 days (b). Fig. (b) shows small crystals of 2 μ or less, growing at the surface of the initial material.

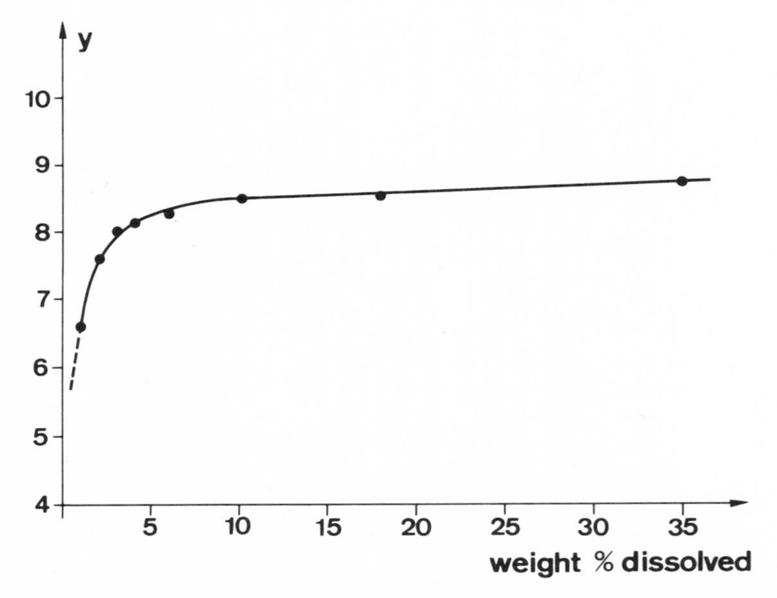

Fig. 4. Successive dissolution by *HCl* of previously aged tests of *Lytechinus verigatus*.

$y = \frac{MgCO_3}{CaCO_3 + MgCO_3}$ *per cent* reaches rapidly the value of 8,75, close to the mean value in the original (initial) material.

curve does not give a faithfull extrapolation. One obtains similar results by slowly redissolving *Mg*-calcites previously submitted to a first dissolution under the same conditions in distilled water, under a pressure of CO_2 equal to 1 atmosphere, during 14 days. Also, relatively pure calcites behave in a similar way: a suspension of *Aequipecten Gibbus* containing 0,84 mole per cent of $MgCO_3$ is stabilized at a concentration of $5.5\ 10^{-4}$ moles Mg^{++}/ℓ after a first run in distilled water saturated with $P_{CO_2} = 1$ atmosphere, and at a concentration of $1.6\ 10^{-4}$ moles Mg^{++}/ℓ after a second dissolution. In both cases, the concentration of Ca^{++} is practically constant (1.05 and $1.10\ 10^{-2}$moles $Ca^{++}/1$ respectively).

The degree of supersaturation of Ca^{++} observed during the experiments of *Plummer and Mackenzie* (1974) give complementary proof of the reprecipitation of a new phase on the surface of the initial *Mg*-calcite. If one accepts this mechanism, the supersaturation corresponds in fact to the excess of free energy necessary to overcome the activation potential of the heterogenous nucleation. The energy required is a function of the interfacial energy between the initial calcite and the new phase (*Wollast*, 1971).

In the case of the experiments of *Plummer and Mackenzie* (1974), the degree of supersaturation equals 2.2, which according to the theory, corresponds to a difference of surface energy between the two phases of the order of 10 *ergs*/cm^2. This difference is relatively small and reflects a slight change in composition between the initial phase and the phase of reprecipitation. It is comparable to the difference between the surface energies of calcite and aragonite which equals approximately 20 *ergs*/cm^2 (*Wollast*, 1971).

The mechanism of dissolution and reprecipitation on the surface of the initial *Mg*-calcite accounts for the preservation of the structure of marine organisms during diagenesis, and does not necessarily imply a preferential dissolution of $MgCO_3$. Also, from a kinetic point of view, the rate of the dissolution reaction is controlled, after the initial period of nucleation, by the diffusion of the products of the reaction through the layer formed by the new phase, as in the case for the preferential dissolution of $MgCO_3$, leaving an altered residual layer. The experimental determination of the dissolution equilbrium of *Mg*-calcites is impeded by the process of reprecipitation. The formation of the new layer progressively inhibits the dissolution of the initial phase, which becomes isolated from the aqueous solution. The initial phase could therefore exist indefinitely, on the scale of laboratory experiments, due to the extremely small rate of diffusion of the species concerned through the new phase.

In considering Figure 1, on the other hand, one can show that the composition of the solid phase which precipitates, varies with time. In reference to a congruent dissolution, the first phase to precipitate is in equilibrium with a solution corresponding to $y_0=x_0$; as the solution becomes richer in Mg^{++} during the reprecipitation stage, y progressively increases until $y=y_f$, corresponding to the final composition of the aqueous solution. At the same time, the composition of precipitating solid solutions varies between the limits of x_i and x_f. A mass balance, according to equation (1), based on the final composition of the aqueous solution does not, therefore, give any indication of the composition of the final solid phase in equilibrium with the aqueous phase.

MICROSTRUCTURE OF NATURAL *Mg*-CALCITES

The study of thermodynamic and kinetic properties implies a sufficient preliminary knowledge of the quality of the samples utilized. This condition is particularly imperative in the case of samples of biological origin, which often present heterogeneities related to the biological structure. In the case of organisms with carbonate squeletons, *Lowenstam* (1964), *Schmalz* (1965), *Schroeder et al.* (1969), *Milliman et al.* (1971) have suggested the existence of such heterogeneities at a macroscopic level. *Moberly* (1970) showed by means of the electron microprobe the existence of different *Mg*-calcite phases in the same organism.

We have quantitatively confirmed the existence of important heterogeneities, in particular, in the case of algae such as *Amphiroa* and *Goniolithon*, which are used abundantly as experimental material for studies of dissolution. Figure 5 shows, for example, the micrography of two zones of a specimen of *Goniolithon*. In the

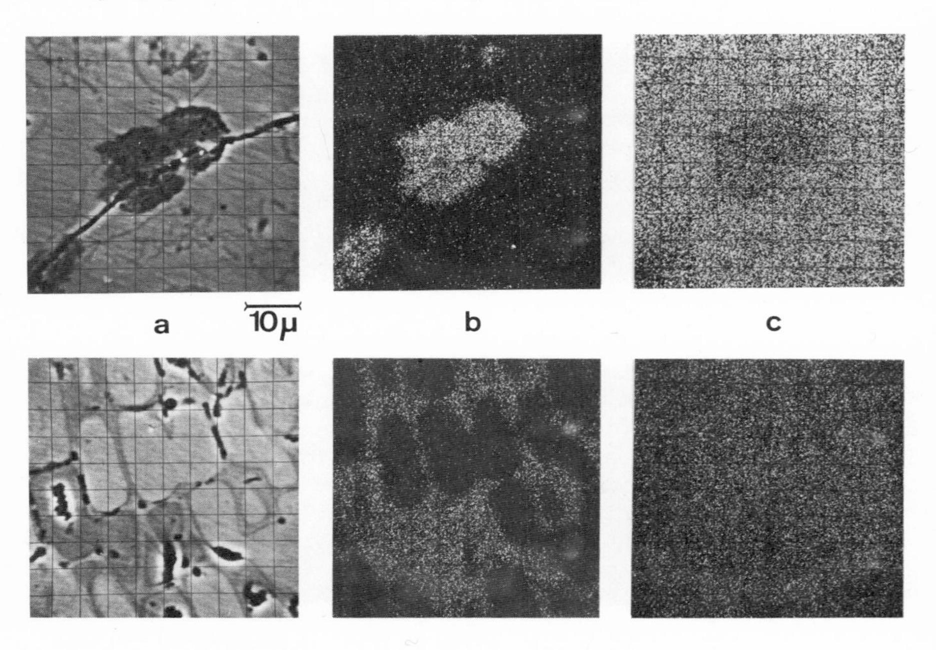

Fig. 5. Electronic (a) X-ray pictures {MgK_α: (b); CaK_α: (c)} of *Goniolithon* showing the heterogenous distribution of Magnesium.

first zone, one sees inclusions of high magnesium content embedded in a cement of relatively pure calcite. The second zone shows heterogeneities in the distribution of magnesium governed by the biological structure of the organism. Figure 6 represents the intensities of the *Mg* and *Ca* signals recorded during a scanning of the inclusion seen in Figure 5. Examination of several other zones of *Goniolithon* allows us to say that the *Mg* content varies from about 2 to 35 moles per cent. Similarly, in the case of *Amphiroa*, zones exempt of magnesium and others containing up to 30 *Mg* mole per cent were observed. On the other hand, echinoderms are much more homogeneous. The tests of *Lytechinus verigatus* present variations of $MgCO_3$ content lower than 2-3 mole per cent, whereas

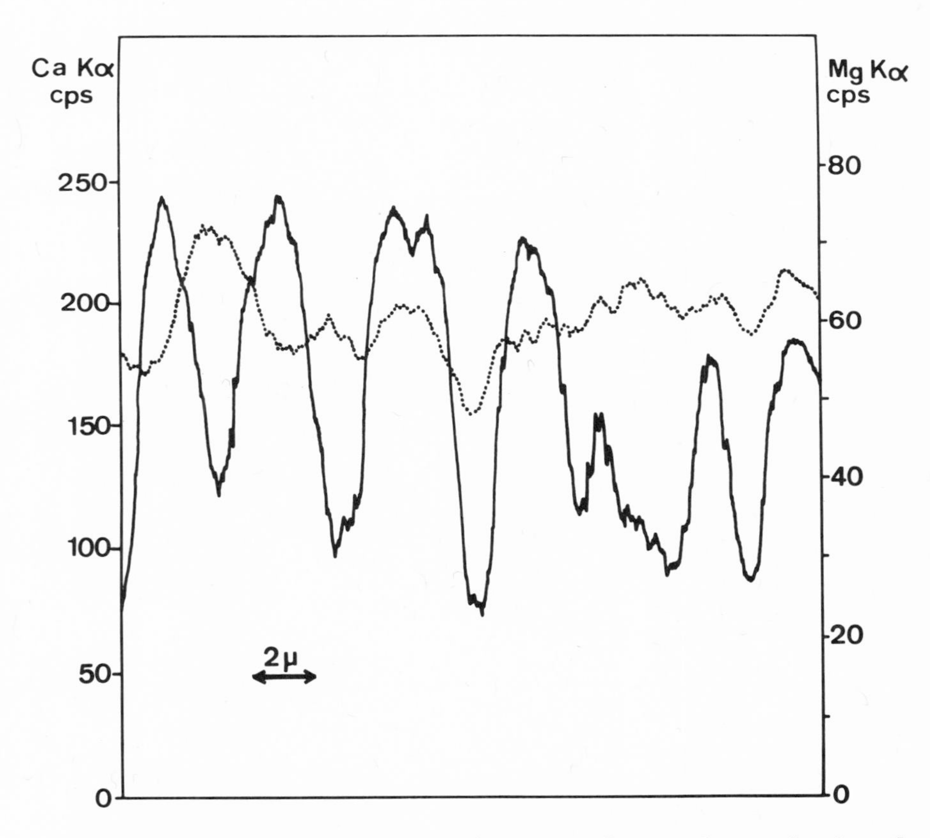

Fig. 6. Semi-quantitative electron microprobe analysis of *Goniolithon*: intensities of the signals for calcium and magnesium during the scanning of a rich *Mg* portion of the sample. Under the same conditions, the MgK_α signal obtained for a pure dolomite is about 100 c.p.s.; $MgCO_3$ content in this portion of *Goniolithon* varies roughly between 15 and 35 moles per cent.

tests of *Melita* are perfectly homogeneous on the scale of the electron microprobe.

These heterogeneities account for the anomalies observed in the shift of the X-ray diffraction peak in studies of natural calcites as a function of their *Mg* content (*Chave*, 1952; *Chave*, 1954; *Jones and Jenkins*, 1970; *Milliman et al.*, 1971; *Plummer and Mackenzie*, 1974). The analysis of the *Mg* content by X-ray diffraction is applicable only to homogeneous solid solutions as defined by *Goldschmitt et al.* (1961). Otherwise, this method is useless and should be abandoned in the case of natural magnesium calcites.

The dissolution experiments reflect precisely these heterogeneities. Figure 7 shows, for example, a series of successive dissolutions performed on *1 g* of *Amphiroa* in *100 ml* of distilled water saturated at $P_{CO_2} = 1$ atmosphere. These dissolutions were interrupted after 20 minutes; in other words during the initial congruent dissolution stage and before the reprecipitation of a new calcite. The first phase dissolved corresponds to the less stable solid solution corresponding to a Mg^{++}/Ca^{++} ratio in solution equal to 0,30, whereas this value is lowered to 0,17 after the fifth dissolution run. The results are in good agreement with the microprobe analyses which showed, for the richest phase of this organism, a content of 30 mole per cent of *Mg*. The mean concentration of *Mg* in the sample was equal to 18 moles per cent, close to the value obtained after the fifth dissolution.

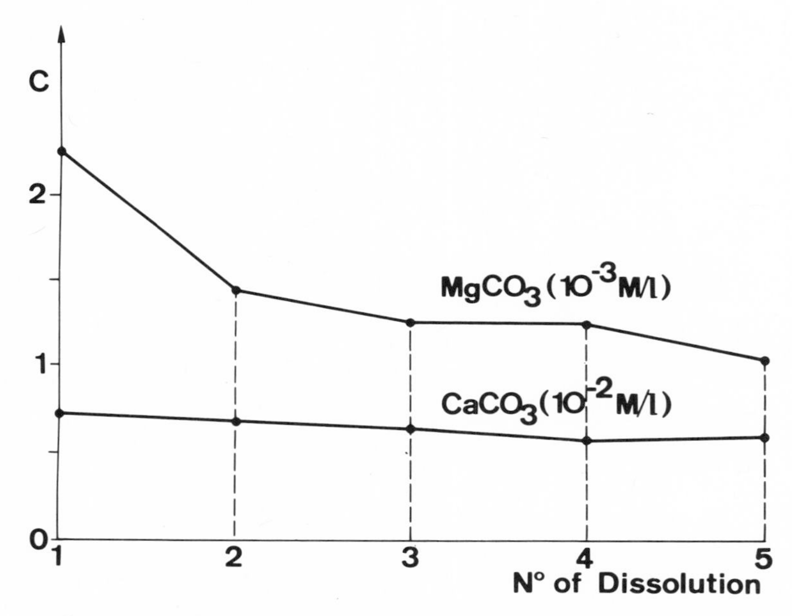

Fig. 7. Five successive dissolutions of 1 gram of *Amphiroa fragilissima* each in *100 ml* distilled water under $P_{CO_2} = 1$ *atm.*, during 20 minutes.

In the case of samples submitted to prolonged dissolution during 14 days in distilled water under pressure $P_{CO_2} = 1$ atmosphere, one observes that the activity of $MgCO_3$ in the solid phase defined by

$$a_{MgCO_3} = \frac{a_{Mg^{++}} \cdot a_{CO_3^=}}{K_{MgCO_3}} \qquad (11)$$

is in the range of 1 (0.95 to 1.28) for several magnesium rich specimens (*Lytechinus verigatus*, mouth and tests; and *Amphiroa fragilissima*). This value is not imputable to the activity coefficient of $MgCO_3$ in these solid solutions, because the same experiments show that this coefficient rapidly approaches 1 when the amount of $MgCO_3$ exceeds 7 mole per cent. It is thought that the activity close to 1 may be attributed to the presence of inclusions of pure magnesite as the microprobe observations of Moberly (1970) previously led him to suspect. From our investigations with the microprobe, it would seem that the inclusions of magnesite have a size less than 1 micron.

CONCLUSIONS

We have voluntarily abstained from presenting more quantitative data regarding the equilibrium and the kinetics of the dissolution of natural Mg-calcites. These compounds present actually a marked heterogeneity which seriously complicates the interpretation of the experimental results. We may add that certain organisms which present a good homogeneity, as for example *Melita*, do not manifest reprecipitation during experiments of dissolution, and behave as if they were dissolving congruently. The absence of adequate nucleation sites could explain the difference of behavior in these organisms.

We envision a new approach to the problems of equilibrium and of the kinetics of Mg-calcite dissolution from synthetic and homogeneous solid solutions, on which one could perform separate experiments of dissolution and precipitation. This approach is indispensable to a better comprehension of fundamental and varied phenomena like the diagenesis of Mg-calcites, the formation of dolomite, the depth of the lysocline and the interpretation *in situ* saturometry, all essential steps to an improved knowledge of the cycle of CO_2 in the ocean.

REFERENCES

Bricker, O. P. 1971. In: *Carbonate Cements, Introduction to the Chapter on chemistry*. Johns Hopkins Press, Baltimore. p. 235.

Chave, K. E. 1952. A solid solution between calcite and dolomite. *Journ. Geology 60*. 190-2.

Chave, K. E. 1954. Aspects of the biochemistry of *Mg*. *Journ. Geology 62*. 366-83, 587-99.

Chave, K. E., K. S. Deffeys, P. K. Weyl, R. M. Garrels, and M. E. Thompson. 1962. Observations on the solubility of skeletal carbonates in aqueous solutions. *Science 137*. 33-34.

Friedman, G. M. 1964. Early diagenesis and lithification in carbonate sediments. *J. Sediment. Petrol. 34*. 777-813.

Garrels, R. M. and C. L. Christ. 1965. *Solutions, Minerals and Equilibria*. Harper and Row.

Goldschmith, J. R., D. L. Graf, and H. C. Heard. 1961. Lattice constants of the *Ca-Mg* carbonates. *Am. Mineralogist 46*. 453-457.

Jones, W. C., and D. A. Jenkins. 1970. Calcareous sponge spicules: A study of *Mg*-calcites. *Calc. Tiss. Res. 4*. 314-329.

Land, L. S. 1967. Diagenesis of skeletal carbonates. *J. Sediment. Petrol. 37*. 914-930.

Lerman, A. 1965. Palaeoecological problems of *Mg* and *Sr* in biogenic calcites in light of recent thermodynamic data. *Geochim. Cosmochim. Acta 29*. 977-1002.

Lowenstam, H. A. 1964. Coexisting calcites and aragonites from skeletal carbonates of marine organisms and their *Sr* and *Mg* contents. In: *Recent Research in the Fields of Hydrosphere, Atmosphere and Geochemistry*. Editorial Committee of Sugawara Festival, Vol. 373-404.

Milliman, J. D., M. Gastner, and J. Muller. 1971. Utilization of *Mg* in coralline algae. *Geol. Soc. America Bull. 82*. 573-580.

Moberly, R. J. 1970. Microprobe study of diagenesis in calcareous algae. *Sedimentology 14*. 113-123.

Plummer, N. L. and F. T. Mackenzie. 1974. Predicting mineral solubility from rate data: Application to the dissolution of magnesian calcites. *Am. J. of Sci. 274*. 61-83.

Robie, R. A. and D. R. Waldbaum. 1968. Thermodynamic properties of minerals and related substances at 298.15^{o}K and one atmosphere. *Geological Survey Bulletin 1259*.

Schmalz, R. F. and K. E. Chave. 1963. $CaCO_3$: Factors affecting saturation in ocean water of Bermuda. *Science 139*. 1206-7.

Schmalz, R. F. 1965. Brucite in carbonate secreted by red *Alga Goniolithon* sp. *Science 149*. 993-6.

Schroeder, J. H. 1969. Experimental dissolution of calcium, magnesium and strontium from recent biogenic carbonates: A model of diagenesis. *J. Sediment. Petrol. 39*. 1057-73.

Schroeder, J. H., T. J. Dvornik, and J. J. Papike. 1969. Primary protodolomite in echinoids skeletons. *Geol. Soc. Am. Bull. 80*. 1616-13.

Winland, H. D. 1969. Stability of calcium carbonate polymorphs in warm, shallow seawater. *J. Sediment. Petrol.* 39. 1579-87.

Wollast R. 1971. Kinetic aspects of nucleation and growth of calcite from aqueous solutions. In: *Carbonate Cements*, O. P. Bricker (Ed.), Johns Hopkins Press, Baltimore. 264-273.

III

SEDIMENTS

Recommendations of the Working Group on Fate of Fossil Fuel CO_2 and the Sedimentary Record

W. H. Berger, Chairman

Scripps Institution of Oceanography

In discussing the problems of the fate of fossil fuel CO_2, it is not immediately clear which role the geologists might play in attaining solutions. The main challenge for the investigator is how to realistically model the path of the CO_2 between the atmosphere, the biosphere, the ocean, and the sediment. Study of the sedimentary record is necessary for setting up the models. In particular, the record bears on three major questions.

1. Are the fundamental assumptions of a given model reasonable; that is, are the relevant reservoirs all considered and are they properly partitioned according to their response times?

For example, the traditional reservoir labeled "carbonate sediments" is probably quite unrealistic. As Hay and Southam emphasized, much carbonate is being precipitated on the shelves. The net accumulation of this carbonate is of the order of deep-sea sedimentation, and is one of the primary factors governing the alkalinity of the ocean. From the papers by Milliman and the comments by Rezak concerning carbonate cementation, it becomes apparent that the shallow-water carbonates are highly reactive. In the present context, inhibition of cementation by precipitation can be considered equivalent to a dissolution response to increased pCO_2. Contrary to the opinion that supersaturation prevents a dissolution response in shallow waters to the entry of fossil fuel CO_2, the large amount of biological micritization and the diurnal changes in pCO_2 of microenvironments should make such a response possible. As Chave and his collaborators have emphasized for some years, precipitation of calcium carbonate on the world's shelves is significant, and much of this carbonate finds its way into slope sediments. If this flux were stopped, it would decrease the pCO_2 from the regular background. A realistic budget of production, destruction, and net accumulation of shallow-water carbonate is badly needed. The upper slope environment, with its rapid precipitation and re-dissolution of mollusc aragonite, also needs to be studied from this aspect, in addition to a closer look at the aragonite story of the deep sea, which Berner has called for.

Besides the carbonate reservoir, there is the marine organic carbon reservoir to consider. Heath presented much-needed information on this topic. By far, the bulk of the organic carbon sedimented ends up in continental slope deposits. The exponential diagenetic reduction suggested by Heath and co-workers introduces a damping function into the carbon transfer from the atmosphere or the biosphere to the ocean. This effect also allows slope sediments to act as a quick, temporary sink for carbon, regardless of the long-term stability of the carbonate-carbon ratio indicated by $\delta^{13}C$ contents of ancient sediments. The efficiency of such a sink is greatly increased by increased productivity in coastal waters, which enhances anaerobic sedimentation. Such anaerobic enhancement may now be taking place in certain basins off Southern California, and perhaps elsewhere, due to eutrophication. The organic-rich slope sediments need to be studied more closely in order to assess their role in sequestering carbon, and to determine the natural background fluctuations occurring under anaerobic conditions.

On a longer time-scale, the carbon decay rate suggested by Heath would provide a delay mechanism for converting terrigenous carbon deposited on slopes during glacials into CO_2. Much of the slope carbon is of such terrigenous origin, as shown for example, by C-13 studies by Sackett in the Gulf of Mexico. If Heath's decay mechanism does indeed exist, the continental slopes are right now releasing subfossil CO_2.

In going from the slopes to the deep sea, it is the rate at which the abyssal carbonates respond to an increase in pCO_2 that becomes important. Broecker urges that we find out whether the response is linear or exponential with increased departure from a critical level of saturation. It appears that as near as can be determined, the depth gradient of dissolution rate is linear, when averaged over the Quaternary Period or the post-Eocene Tertiary. Heath and co-workers present evidence that this gradient has been linear through much of the Cenozoic, although it varied by more than an order of magnitude. Hence, if the ocean responds to a pCO_2 increase by simply raising uniformly the levels of equal saturation, a linear response would seem to be expected. The observed linearity of the depth gradient could be due to a non-linear response, as seen by Berner and Morse in the laboratory, combined with the enrichment of increasingly resistant particles as dissolution proceeds at depth.

According to Berger, Johnson, and Hamilton, an increase in dissolution might be expected to result in increased mass wasting by removing down-slope support from carbonate layers, and by changing size distributions leading to a decrease in shear strength. Such a process would accelerate the alkalinity increase in response to the pCO_2 pulse. More needs to be known about these processes related to mass wasting. Changes in grain size in response to dissolution, as reported by Thiede, are important in this context.

2. Are there any major excursions in pCO_2 that have occurred in the past that may indicate the response of the system to disturbance, or may serve as analogues to the present CO_2 pulse?

Surely the deglaciation event some 10,000 years ago is the most important phenomenon to be studied for this purpose. The response of the biosphere to this event is of great importance. Shackleton has made the very interesting suggestion that the growth and decay of tropical rain forests can explain an overall negative correlation between $\delta^{13}C$ and $\delta^{18}O$ in Quaternary benthic foraminifera. He is satisfied that the amplitude of the $\delta^{13}C$ signal corresponds to the envisaged change in the size of the terrestrial biosphere. Perhaps the possibility also exists that organic carbon, including terrigenous carbon in slope sediments during the glacial, would release isotopically light carbon to the bottom-living foraminifera. In planktonic foraminifera, a fertility signal may be the overriding factor in producing the $\delta^{13}C$ fluctuations, as suggested by Broecker. Shackleton's rain forest effect and the increased fertility of such forests may combine with high glacial sedimentation rates to sequester significant amounts of carbon in slope sediments, perhaps many times more than one terrestrial biomass. This might result in a drawdown of pCO_2, which would in part explain the better overall preservation of calcareous shells during glacials.

An intriguing phenomenon is the preservation pulse near the end of the last glacial, as evident in the box cores from the western equatorial Pacific Ocean, and also evident from data given by Diester-Haass and by Thiede for the Atlantic Ocean. There is a possibility that the growth of the biosphere at deglaciation might draw down the pCO_2 sufficiently to provide for a short time interval of increased preservation of carbonate. When first considering this possibility, it appeared that the biosphere growth was insufficient because the mass of carbon involved would be less than 1% of the total ocean carbon, and if spread out over the duration of the preservation pulse of a thousand years or more, the rate of the disturbance would appear rather small and could conceivably be buffered in other ways. However, Shackleton suggests the biosphere growth effect might indeed be sufficient to explain the preservation pulse. This is an important question, and in order to resolve it, it will be necessary to map the actual extent of the preservation pulse on the seafloor in some detail as well as do a mass balance à la Broecker. It is noteworthy that the preservation spike occurs well before the deglaciation event. Perhaps a warming of the ocean is also involved, or perhaps the overriding influence of a Worthington effect is seen; that is, a meltwater lid on the deep ocean, below which the pCO_2 can rise.

From Broecker's $\delta^{13}C$ data (see Figure 19 in Berger's paper in this volume), there may be an indication for such a CO_2 buildup in the ocean during deglaciation proper, perhaps from Heath-type decay of organic carbon and from the Worthington effect. To solve this problem, the deglaciation event needs to be studied in great detail, and the record unscrambled using the best mixing models that can be constructed. Obviously, the crucial point is whether or not the sequence and duration of the phenomena can be exactly identified; that is, the "leads and lags" which Moore has identified with high precision for the entire Quaternary Record. One major problem in ascribing such phase shifts to the workings of the carbon system is differential mixing. The typical mixing lengths of the sediment particles differ. This can produce leads and lags if one signal resides in the coarse fraction and the other in the fine fraction. This particular problem can be attacked by looking at the isotopic values in coccoliths, as is being done by Kroopnick and Margolis.

In the present context of the fate of fossil fuel CO_2, the deglaciation event is especially important as an analogue to the mini-termination of the Little Ice Age through which our grandparents lived. It would seem that the change in vegetation that goes with the Little Ice Age needs to be assessed to establish a pre-industrial CO_2 base-line. The marine record of anaerobic coastal basins will be very important in this respect. It may turn out that the pCO_2 of the last century is anomalously low to begin with, and a level more nearly average with respect to the last few thousand years is being approached.

3. <u>What is the natural long-term frequency and amplitude of climatic disturbance and the reaction of the biosphere to such disturbance</u>?

If a point should ever be reached where climatic effects can be predicted, it may be desirable to know the ecological consequences of such predictions. Although the effects on human ecology and economy may seem to be a more pressing problem, in the long view, the effects on the biosphere, as a whole, are more important, because ultimately it is the biosphere that provides the substrate on which mankind can exist. In order to answer this question, it is necessary to take the long-range view and to study the carbon system over millions of years, as exemplified by the presentations of Heath, Moore, Winterer, Hay, Margolis, and Kroopnick. Winterer's carbonate monitors, the island chains and plateaus, would seem to deserve much closer attention than they have received so far. To extract the maximum information out of such monitors, the paleodepths of deposition must be found. It is encouraging to see that Winterer thinks this problem may be tractable. As emphasized by Wise, the history of early diagenesis will play an important part in the interpretation of these records.

In summary, the sedimentary record does indeed contribute significantly to the solution of the CO_2 problem. Important problems remaining are: (1) a definition of carbon budgets and cycles of the coastal ocean; (2) the poor definition of major climatic excursions, including the Little Ice Age; (3) ignorance about the response of the biosphere to changes in the overall chemistry of ocean and atmosphere.

RECOMMENDATIONS

Recommendations for future activities with regard to the fate of fossil fuel CO_2, by the sediment panel, relate to 4 principal topics:

1. Immediate response of the environment to increased pCO_2.
2. Long-term response of the environment.
3. Climatic effects of pCO_2 increase.
4. Possible remedial action to prevent excessive pCO_2.

Specific recommendations for research priorities were made by the following session participants: W. H. Berger, W. W. Hay, G. R. Heath, C. D. Hollister, T. C. Johnson, S. V. Margolis, J. Milliman, T. C. Moore, R. Rezak, N. Shackleton, E. L. Winterer.

1. Immediate response problem:

Dissolution of accessible carbonate and deposition of organic carbon.

(a) The sedimentation processes of shallow-water carbonates need to be studied, in some detail, on all major shelves and reefs of the world. The rates and modes of precipitation of the various carbonate phases need to be determined, as well as the rates and modes of destruction and the rates of net accumulation. The factors affecting these various rates need to be assessed.

(b) Similar studies are necessary for deeper carbonates in contact with actively sinking waters; that is, carbonates that are likely to respond quickly to increased pCO_2.

(c) Experiments need to be made in the field and in the laboratory on the reactions of shallow water carbonate producers to increased pCO_2.

(d) The effect of temperature variation, expected from climatic change, on the shallow-water carbonate system needs to be assessed.

(e) Recent effects on reef carbonates by pollution and other effects have to be assessed, to find the background signal.

(f) In addition, the effects of eutrophication on organic carbon sedimentation in coastal oceans need to be assessed.

2. Long-term response:
Sedimentation and dissolution of deep-sea carbonates.

(a) The degree of undersaturation in ocean waters needs to be more closely tied to the actual dissolution rates of deep-sea sediments on the seafloor, at all depth levels and in different oceanographic settings.

(b) The particle flux of carbon has to be assessed for an estimate of carbon removal. Direct measurement by filtration and trapping are necessary.

3. Climatic effects of changing pCO_2:

(a) The possible changes of pCO_2 over the last centuries and millenia have to be investigated in some detail.

(b) A good place to do this is in areas of high rates of sedimentation. Examples of favorable conditions for this purpose are the Cariaco Trench, anaerobic shallow basins on ocean islands, alpine-type lakes, and glaciers. The $\delta^{13}C$ signal of these stratigraphic sequences should be analyzed.

(c) The record should be scanned for evidence of the strongest climatic events conceivable, such as deglaciations and volcanic episodes and Mediterranean salinity crises, and the subsequent second order perturbations following such events should be identified.

(d) A close monitoring of the effects of the climatic optimum and the Little Ice Age is important. The pteropod preservation in slope sediments and the cementation history in shallow water carbonates, if datable, might contain information on the correlations between climate and pCO_2 in the Holocene.

4. Remedial action:

There were several suggestions for remedial action in case an increased CO_2 level should prove to be highly undersirable for the environment. Fundamentally, there are two ways to soak up CO_2 quickly: (a) increase the rate of dissolution of shallow-water carbonates, and (b) deposit carbon in an out-of-the-way place.

The first approach, for example, could be implemented by dumping organic matter (grass, kelp) onto exposed or slightly submerged carbonate flats. The scale of this operation in order for it to be effective would be prohibitive.

The second approach is more realistic, but nevertheless hardly feasible. We assume that an annual input of 10^{16}gC has to be neutralized by increasing the dead carbon part of the biosphere (the traphosphere, as it were). This could be done, by fertilizing certain coastal waters by artificial upwelling and increasing the supply of organic matter to sediments or even deep waters. To make such a scheme viable, the effluent from 10,000 power plants would have to provide upwelling plumes, each responsible for sequestering 10^{12}gC per year. Assuming an upwelling area of $10^8 m^2$ for each plant, 10 kg of carbon would have to come from each square meter.

This is 2 orders of magnitude higher than possible. It appears that such a scheme would not be feasible.

One could also envisage a scheme whereby kelp would be harvested continuously and dumped into deep waters (one problem being that kelp tends to float, at least initially). This process goes on in nature through transport of a mixture of beach sand and kelp down submarine canyons. The process sequesters carbon very efficiently. Kelp grows very quickly, and 10 kg $C/m^2/yr$ would be acceptable as a harvestable amount, according to a calculation of Lawrence Small. To obtain $10^{16}gC$, $10^{12}m^2$ or 10^6km^2 have to be harvested. This corresponds to a coastal strip 20 km wide and 50,000 km long. Even if all that were required would be to cut the kelp and let it drift, it would be difficult to find enough kelp beds to neutralize even a moderate fraction of the CO_2 input.

The magnitude of the problem is readily appreciated if one calculates the rate of carbonate shelf erosion necessary to provide a sufficient sink for the added CO_2. For a carbonate shelf area of the order of 10^7km^2, each m^2 has to neutralize 1 kgC annually to account for an input of $10^{16}gC$. Dissolution of 100 g $CaCO_3$ takes up 12 g of C. Hence, the rate of erosion required to balance CO_2 input is of the order of $1g/cm^2/yr$. How high a pCO_2 is required to produce a rate of erosion approaching this magnitude?

Clearly, this problem is a very difficult one and undoubtedly requires further investigation.

CARBON DIOXIDE EXCURSIONS AND THE DEEP-SEA RECORD: ASPECTS OF THE PROBLEM

W. H. Berger

Scripps Institution of Oceanography

ABSTRACT

Deep-sea sediments bear on the CO_2 problem in three major ways: (1) The sediment-ocean system is a powerful buffer for a geologically short-term CO_2 pulse like the present industrial input; (2) deep-sea sediments provide a detailed record of the natural background fluctuation in climate against which a CO_2 induced change has to be measured; (3) substantial CO_2 excursions may have occurred in the past, leaving a record in deep-sea sediment.

The buffer action is mainly due to dissolution of calcareous sediments. Response is fastest in an ocean where dissolution effects are already widespread, such as the present one. The great extent of dissolution on the sea floor is evident from the shallowness of the R_0-level, the level where foraminiferal species lists first change noticeably. The lysocline (the level of maximum change) is about halfway between R_o-level and CCD.

Climatic fluctuations are thought to be tied to a large extent to astronomical causes. To verify this hypothesis, the fluctuations of sedimentation rates in sediments unaffected by dissolution have to be determined. Excursions of the saturation level during the Pleistocene (taken as change in the compensation depth of Globigerina rubescens) exceed 1 km and interfere with the signal produced by productivity changes alone.

Excursions of the pCO_2 are expected, for example, from volcanic episodes, from warming and cooling of the ocean, from changes in stratification, from changes in the size of the ocean or in the size of the biosphere, and from changes in the transport of terrigenous carbon to the ocean and its subsequent combustion on the continental slope. The excursions should be recognizable in *preservation and dissolution pulses or reversals, and in $\delta^{13}C$ changes. Such signals do exist in the deep sea record, but their unambiguous interpretation awaits further study.*

INTRODUCTION

The industrial revolution started a global experiment on what will happen if the carbon dioxide content of the atmosphere is suddenly greatly increased. The resources of carbon fuel are such that a peaking out of the input rate is to be expected sometime during the next century (*Hubbert*, 1969; see Figure 1).

Some authors have predicted more or less dire consequences for the biosphere or for mankind, or both, from the expected CO_2 increase (e.g., *Pytkowicz*, 1972; *Fairhall*, 1973; *Broecker*, 1975).

How does the deep-sea record bear on this problem? What aspects of the record should we study in order to gain perspective and to find grounds for hope, despair, or even -- if there is indeed Intelligent Life on Earth, as *Cloud* (1973) speculates there might be -- for wise decisions in energy management?

Deep-sea sediments bear on the CO_2 problem in three major ways:

(1) The sediment-ocean system is a powerful buffer for a geologically short-term CO_2 spike like the one we are witnessing.

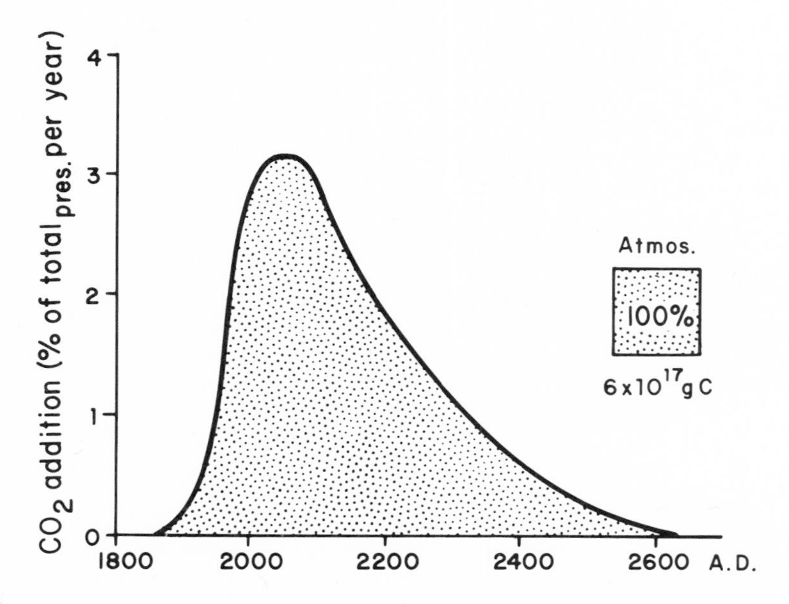

Fig. 1. Sketch of expected CO_2 input to atmosphere, from fuel combustion. Annual input may reach 3% of present atmospheric content in the 21st Century. Based on data in *Hubbert*, 1969.

Without dissolution of carbonate sediments, an increase of 10 percent in the partial pressure of CO_2 is balanced by only 1 percent increase in total CO_2 in the ocean since without alkalinity increase, the *pH* drop in the ocean quickly raises the pCO_2 and prevents further entry (*Deffeyes*, 1967). Upon dissolution of carbonate, the *pH* rises and allows additional CO_2 into the ocean reservoir, until roughly the present partitioning (50:1) is reached (*Keeling*, 1973; see Figure 2).

(2) Deep-sea sediments provide the most detailed record of climatic fluctuations and their effects on life in the sea. A change in CO_2 level presumably causes a change in climate. The sign and the magnitude of such a change can only be predicted by realistic modeling, taking into account the background of natural climatic fluctuation. Long and detailed time series are necessary to obtain the frequencies and amplitudes of this natural signal. Ice cores (*Dansgaard et al.*, 1971) and the deep-sea record (*Emiliani*, 1955, 1966; *Imbrie and Kipp*, 1971; *Shackleton and Opdyke*, 1973) have great promise for providing useful series. However, the systematic variation of sedimentation rates with climatic fluctuations, are a major obstacle to spectrum analysis, so far (see Figure 3).

(3) Deep-sea sediments should contain a record of substantial -- perhaps even drastic -- CO_2 changes in the past. The atmospheric reservoir is highly sensitive to seasonal flux reversals, beautifully illustrated by the Mauna Loa records (*Ekdahl and Keeling*, 1973).

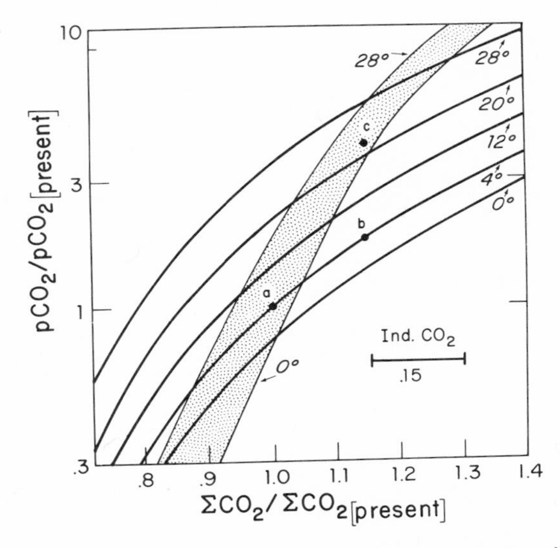

Fig. 2. Response of pCO_2 to industrial input (15% of Σ CO_2 in ocean and atmosphere, from Figure 1), neglecting sedimentation of organic carbon. a) present pCO_2; b) pCO_2 after equilibration with solid $CaCO_3$ by dissolution; c) pCO_2 without dissolution of $CaCO_3$. Based on a diagram by *Plass* (1972).

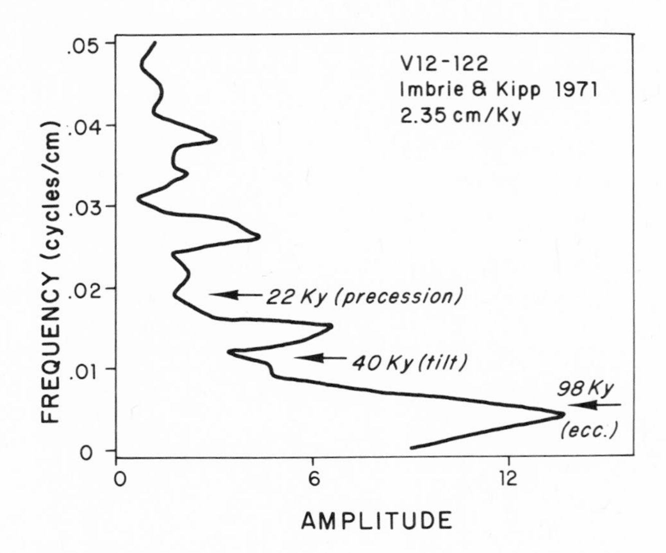

Fig. 3. Spectrum of fluctuations (winter temperature index) in a Caribbean core, as found by *Imbrie and Kipp* (1971). Apparently systematically varying sedimentation rate fluctuations distort the time signal. Of the expected astronomical cycles only the long-term 98 ky cycle is observed.

Thus, it seems highly unlikely that CO_2 concentrations in the atmosphere stayed constant even over relatively short geologic time spans. Climate changed significantly and suddenly many times in the past, on various time scales (Figure 4). Incidentally, in view of the very recent pronounced climatic change due to the termination of the "Little Ice Age" (Figure 4a), "pre-industrial" CO_2 concentrations cannot be implied to represent the normal background. The growth of forests and peat deposits following retreat of the "Little Ice Age" glaciers should have led to anomalously low pCO_2 in the last century.

Even a simple warming or cooling of the ocean can greatly change pCO_2, if the rate of temperature change exceeds the rate of dissipation to the sediment reservoir (Figure 2). Other geologic processes effecting pCO_2 excursions can be readily envisaged, such as large scale changes in volcanic activity (*Damon*, 1971; *Baksi and Watkins*, 1973; *Kennett and Thunell*, 1975). However, it is obviously highly desirable to generate some ideas of how such postulated CO_2 excursions might be recognized in deep-sea sediments.

In the following, I shall outline what I think are the salient problems in the study of the deep-sea record with a view to CO_2 excursions, beginning with the present system and moving through the Quaternary and the Cenozoic.

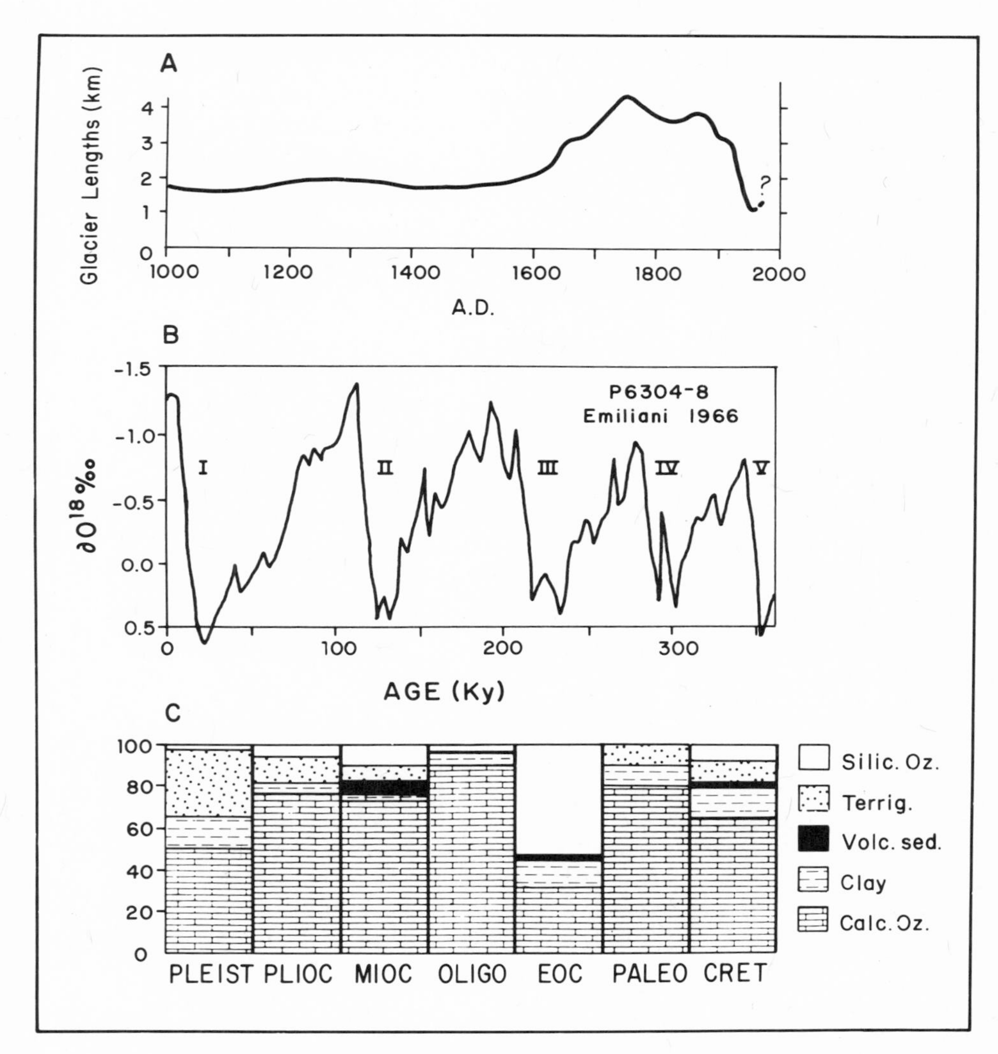

Fig. 4. Evidence for large climatic variations on three different time scales. (A) Growth and retreat of Norwegian glaciers during the last 1000 years (Ahlmann, in *Worthington*, 1968). The Little Ice Age (A.D. 1650 to 1850) terminates with a mini-deglaciation event in the years after A.D. 1900. (B) Fluctuations in oxygen isotopic composition of planktonic foraminifera, Caribbean Sea (data of Emiliani as given in *Broecker and van Donk*, 1970). (C) Facies composition of Cenozoic sediments recovered by the Deep Sea Drilling Project (*Davies and Supko*, 1973). Note the drastic differences in carbonate content and in volcanic material between adjacent epochs.

SEDIMENTATION OF CARBON: THE PRESENT SYSTEM

If we wish to predict how the ocean-sediment system deals with a CO_2- pulse, we have to first understand how it sediments carbon. It does so in two ways, as calcium (-magnesium)-carbonate, and as organic carbon, in the following represented as $CaCO_3$ and CH_2O respectively:

$$CO_2 + CaSiO_3 = CaCO_3 + SiO_2 \quad (1)$$

$$CO_2 + CaSO_4 + H_2O = CaCO_3 + H_2SO_4 \quad (2)$$

$$CO_2 + H_2O = CH_2O + O_2 \quad (3)$$

The first reaction (*Urey*, 1952) describes the weathering of calcium-bearing silicates and the subsequent precipitation of carbonate and silica. It illustrates that in order to get rid of CO_2 as carbonate, new calcium has to be found. The weathering of pre-existing carbonate and its subsequent precipitation merely shifts the place of deposition, without lasting effect on pCO_2. Calcium may also be obtained from the dissolution of gypsum. For reaction (2) to proceed, the sulfate must be removed by reduction and precipitation as iron sulfide. In this process, as well as in the photosynthetic reaction (Reaction 3), oxygen is released to the atmosphere. To the extent that such oxygen is subsequently used in the combustion of sedimentary carbon, new CO_2 is produced, and there is no decrease of pCO_2. For reactions (2) and (3) to be effective, therefore, the oxygen has to be used up for oxidizing crustal iron. These matters are treated elsewhere in some detail (e.g., *Garrels and Perry*, 1974).

The point is that sedimentation of carbon as carbonate depends on the availability of calcium, and sedimentation as organic carbon on that of an oxygen acceptor someplace in the system.

The entry of new calcium proceeds at a slow rate compared with the time scales of CO_2- pulses from fossil fuel burning or volcanic episodes.

Thus, a sustained reduction of a strong CO_2- pulse on a relatively short time scale (millenia) is only possible through sedimentation of CH_2O. Anaerobic conditions, of course, are favorable for deposition of organic carbon. Such conditions, incidentally, may be enhanced at present by eutrophication of the coastal zone through input of agricultural nitrogen and phosphorous fertilizers.

A short-term reduction of a sudden CO_2 increase is provided by the dissolution of carbonate:

$$CO_2 + H_2O + CaCO_3 = Ca^{++} + 2HCO_3^- \quad (4)$$

The capacity of this sink is limited, however.

Fundamentally, these rules apply whether the ocean is modelled as being in equilibrium with its sediments (*Sillén*, 1961, 1967) or whether the chemistry of the ocean is considered to be intimately tied to its fertility (*Redfield*, 1942; *Harvey*, 1957; *Redfield et al.*, 1963; *Broecker*, 1971b). However, there is an important difference. The "Harvey Ocean" is greatly undersaturated with all nutrient substances and is therefore able to react much more vigorously with its biogenous sediments than the "Sillén Ocean", whose kinetics are slow by definition.

In the present context, we are interested in the pronounced undersaturation of the Harvey Ocean (the real ocean) with calcium carbonate throughout much of its depth (*Berner*, 1965; *Pytkowicz*, 1965; *Peterson*, 1966), because the alkalinity in this ocean will rise much more quickly in response to increased pCO_2 than it would in the Sillén Ocean, whose dissolving waters are restricted to the greatest depths and therefore have maximum response times (Figure 5).

In the Sillén Ocean, the saturation level and the calcite compensation depth (CCD) coincide, as proposed by *Broecker and Broecker* (1974). Fortunately for the buffering response of the ocean-sediment system, this is not the case in the actual ocean (*Berner*, 1974; *Edmond*, 1974).

The major features of dissolution processes in the pelagic realm are classically displayed in the Atlantic Ocean, which collects carbonate because its deep waters are, on the whole, young and have not had time to acquire additional CO_2 from combustion of organic matter at depth (see Figure 6). A preservation level, the

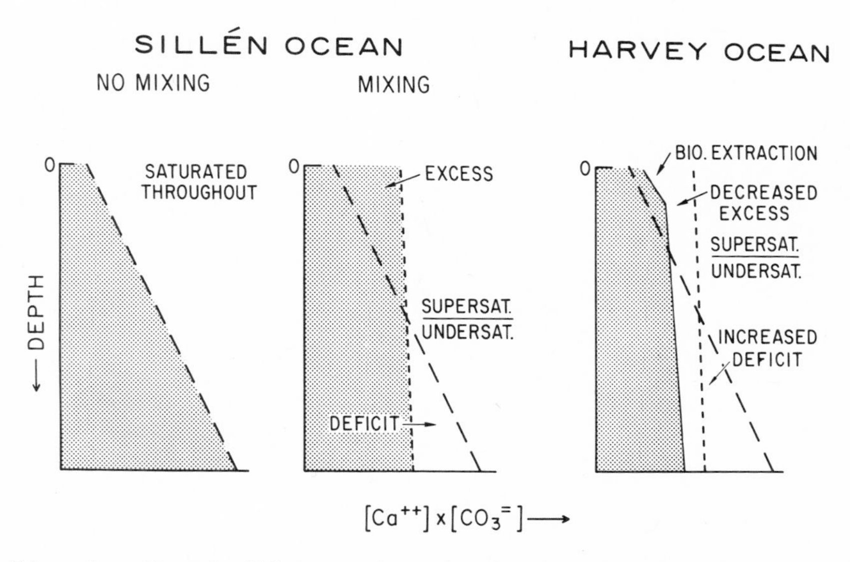

Fig. 5. Sketch illustrating the fundamental difference between a sterile ocean, whose composition is controlled by equilibria (Sillén model) and a fertile ocean, which is undersaturated with biogenic precipitates whose rates of dissolution are crucial in controlling composition ("kinetic" models). The Sillén Ocean has high alkalinity, the Harvey Ocean has low alkalinity.

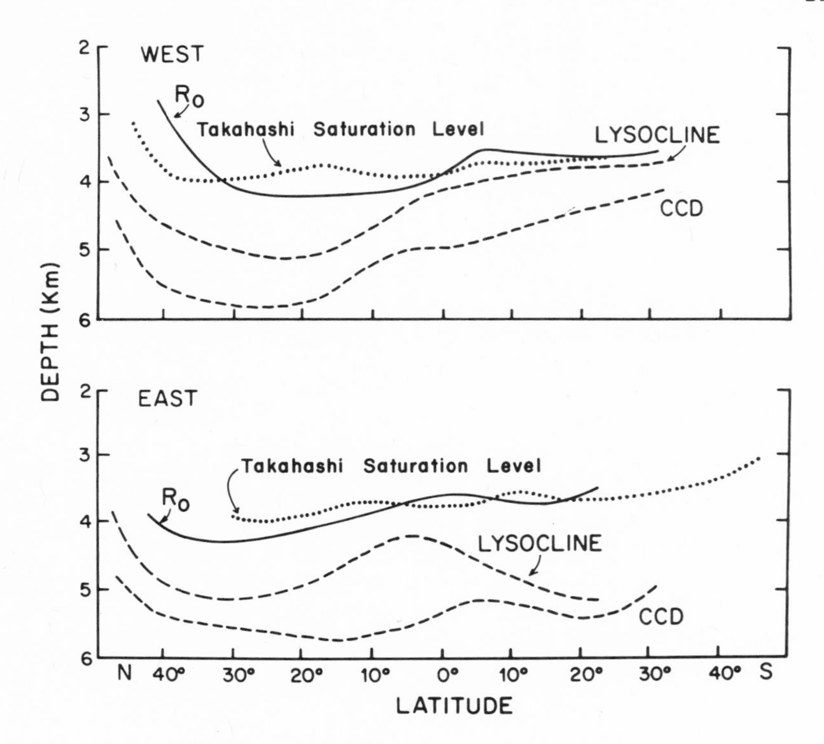

Fig. 6. Foram preservation levels in the Atlantic, compared with saturation level of *Takahashi* (1975). R_o-level: level below which the percentage of the most resistant species first increases detectably. Lysocline: level below which the foram assemblages are strongly affected by dissolution (abundances of resistant species triple near this level). CCD: level at which carbonate percentages fall to values of a few percent or less. No whole tests of forams are present at this level. The changes in vertical distance between the levels imply changes in the depth gradient of dissolution rates. Data from *Berger* (1974b).

R_o-level, has been identified, above which no dissolution effects can be detected from foraminiferal species lists (*Berger*, 1974b). Its bathymetry is the same as *Takahashi's* (1975) Saturation Level. The lysocline, where foram assemblages undergo maximum change in response to dissolution, is about halfway between the R_o-level and the CCD. For a linear increase with depth, therefore, the dissolution rate at the lysocline would be (in loss per time).

$$L/t = 1/2 \ (\text{sedim. rate at } R_o) \cong 0.5 g/cmKy = 0.05 \ moles/m^2yr,$$

and the rate gradient is 0.07 to 0.10 moles/m^2yr per km depth increase This is similar to the rate gradient in the equatorial Pacific, which can be estimated from CCD topography (Figure 7) using a simple supply-preservation model (Figure 8). Quaternary rate gradients and average post-Eocene ones are comparable (Figure 9).

Once the R_o-level, the lysocline, and the CCD are mapped, the rate of alkalinity input to the deep ocean from pelagic sediments, at various depths, can be readily estimated. However, large scale mass wasting (*Berger and Johnson*, 1976) and dissolution on the continental slope adds significant but unknown amounts. Pronounced dissolution on the continental slope is documented in the eastern Atlantic (Figure 10) and in the eastern Pacific (Figure 11). The mass of $CaCO_3$ deposited annually in slope sediments is roughly equal to that in deep sea carbonates, and alkalinity input, therefore, would seem to be of similar magnitude also. The slope input, in fact, may exceed the abyssal input because of high supply of aragonitic mollusc shells (pteropods and larvae of pelecypods) in coastal waters (for pteropods see *Berger and Soutar*, 1970, Table 1; Size classes 1, 2, 3, 4, correspond to > 250, 250-177, 177-125, and > 62 μ, not as printed in caption. For pelecypod larvae see *Thiede*, 1974, 1975). An assessment of rates of $CaCO_3$ supply and dissolution as a function of fertility and depth needs to be made

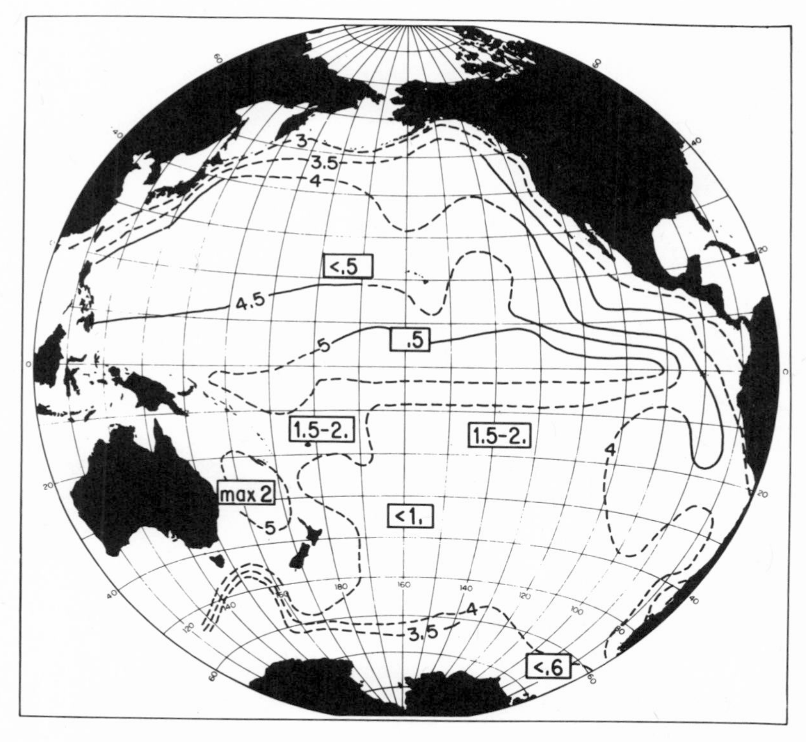

Fig. 7. Topography of the CCD in the Pacific Ocean. In boxes: depth of the pteropod limit (ACD), in km. From *Berger* (1974b).

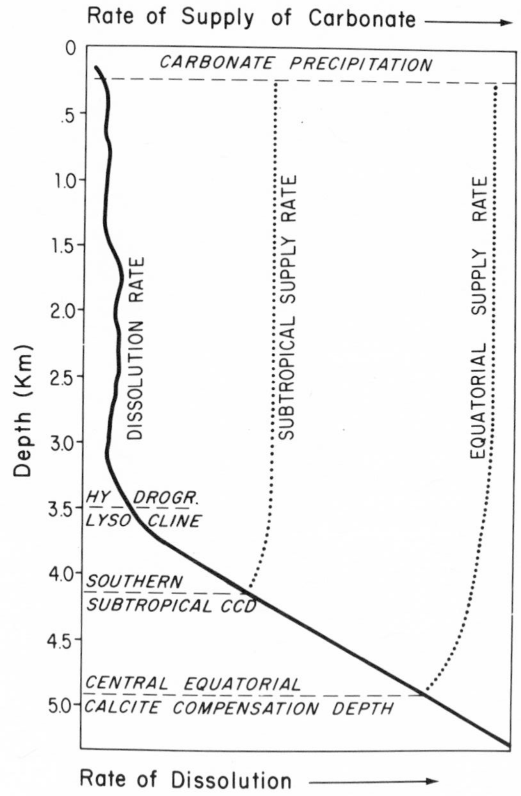

Fig. 8. Simple model accounting for the depression of the CCD in the equatorial Pacific. From *Berger, Adelseck, Mayer* (1976).

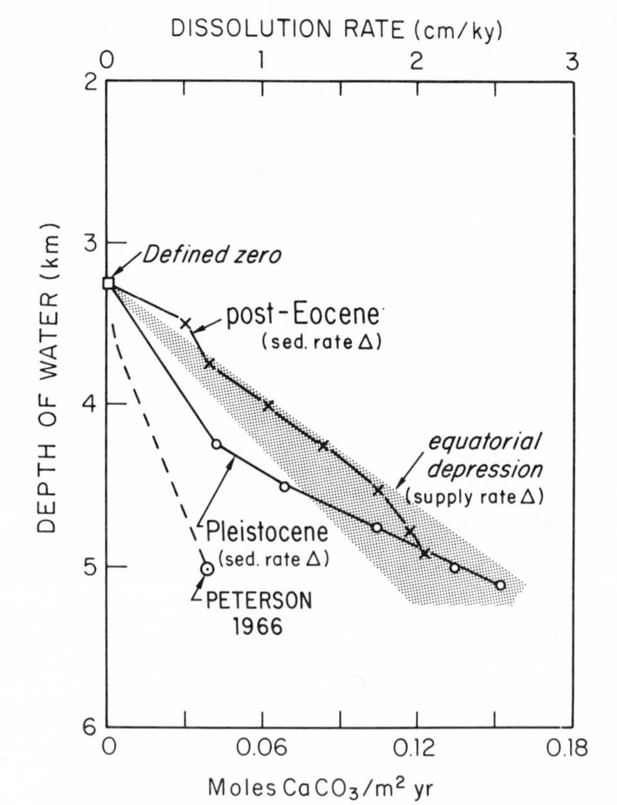

Fig. 9. Depth gradients of carbonate dissolution rates in the equatorial Pacific. Estimates based on equatorial CCD depression (model in Fig. 8), and sedimentation rate distribution in the Quaternary (*Berger*, 1974b) and in the post-Eocene (*Berger*, 1973a). Note the comparatively low rates of dissolution of suspended calcite spheres (*Peterson*, 1966).

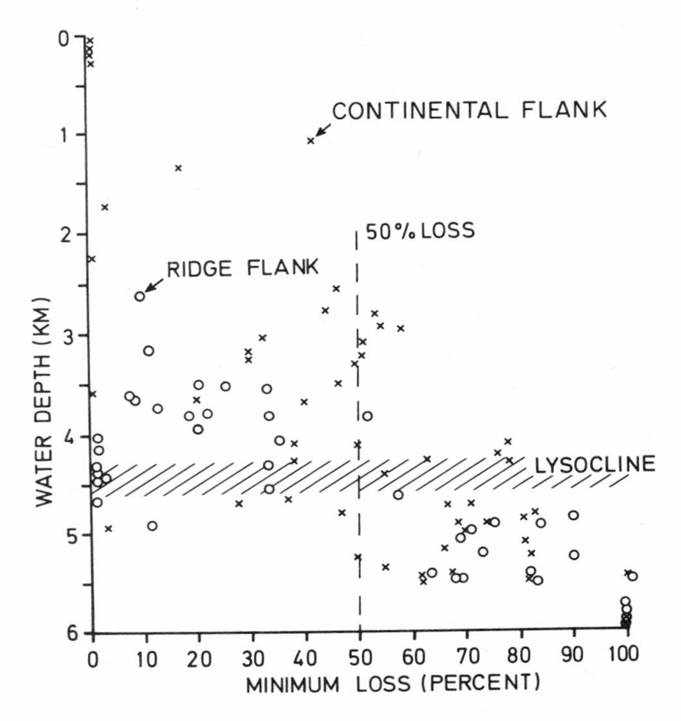

Fig. 10. Evidence for increased dissolution of planktonic forams in slope sediments, off NW Africa. Minimum loss calculated from enrichment of resistant specimens in the residual faunal assemblages, assuming resistant forms are insoluble. Tropical North Atlantic, eastern trough. Data from *Schott* (1935), *Phleger, Parker, Peirson* (1953), and *Ruddiman and Heezen* (1967).

(Figure 12), not only for the deep sea, but especially for the coastal areas where production rates are high (*Chave et al.*, 1972; *Smith*, 1972).

THE QUATERNARY RECORD: RESPONSE TO GLACIAL OSCILLATION

The waxing and waning of vast continental ice sheets dominates our own geologic era. A hundred years ago, *Croll* (1875, p. 54) suggested that "There are two causes affecting the position of the earth in relation to the sun, which must, to a very large extent influence the earth's climate; viz., the precession of the equinoxes and the change in the eccentricity of the earth's orbit." His calculations implied a long succession of cold and warm epochs, the last cold one he took as the "glacial epoch (the time of the till and boulder clay)" between 240,000 and 80,000 years. The problems

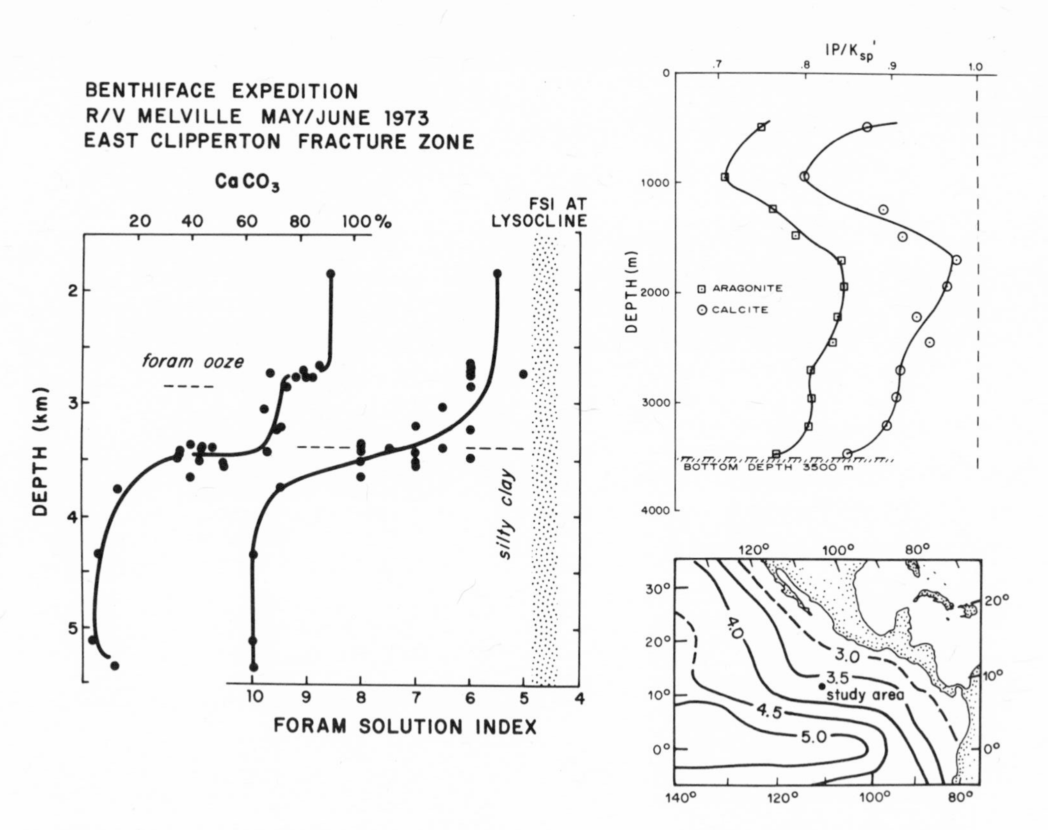

Fig. 11. Evidence for increased carbonate dissolution in a fertile, continent-near area, eastern tropical Pacific. Highly undersaturated waters of the oxygen minimum intersect the continental slope (but not the deep sea floor). Foram Solution Index (FSI): semiquantitative measure of preservation aspect, based on proportions of resistant species and on abundance of fragments. Determination of $CaCO_3$ and FSI by *Valenzuela* (1975). Saturometer results by Ben-Yaakov, Kaplan, and Ruth (In: *Adelseck and Berger*, 1975).

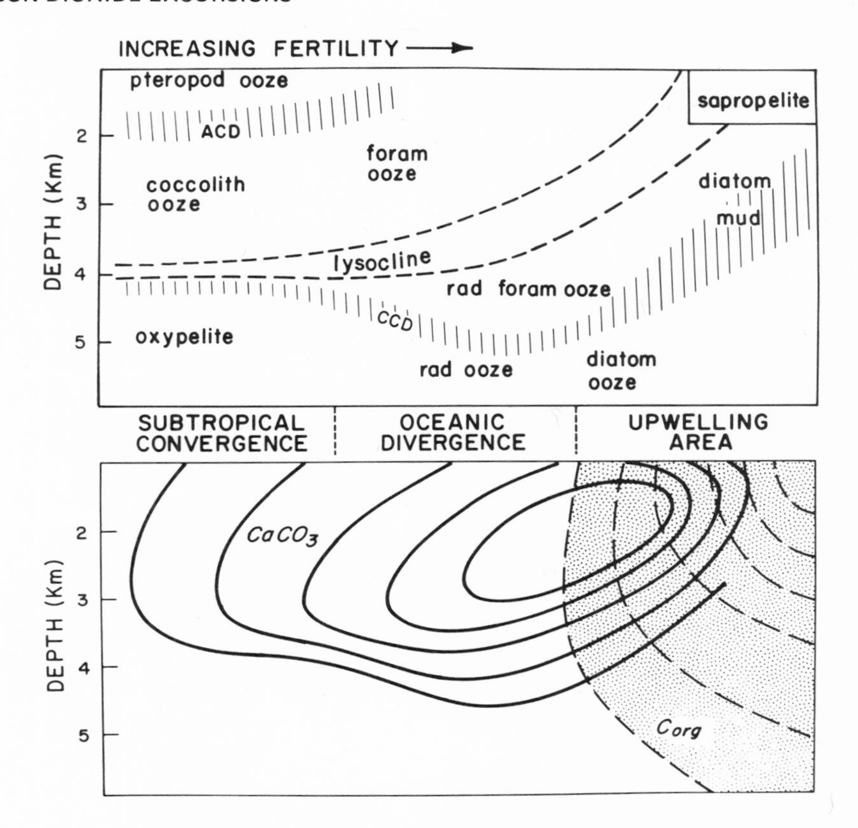

Fig. 12. Sketch of the type of matrix needed to determine the budgets for deposition and redissolution of carbonate and organic carbon on the sea floor. Too few numbers are as yet available to construct satisfactory matrices for these budgets in the major oceanic regions.

plaguing Croll - the relevance of the astronomical signals to the climatic signals, and the lack of exact dates for the geologic events to be compared with the astronomical ones - are still with us.

If these problems are not solved, the background climatic effect from astronomical causes, if any, cannot be computed and any superimposed effects from CO_2 excursions cannot be assessed, therefore. The most important development in this connection, as far as the deep sea record, started with *Emiliani's* (1955) pioneering work on oxygen isotope fluctuations in calcareous sediments.

Shackleton and Opdyke's (1973) oxygen isotope profile of a deep sea core from Ontong Java Plateau represents the state of the art (Figure 13). Using an average sedimentation rate of 1.71×10^{-3} cm per year, as given by these authors, one can date the peaks and

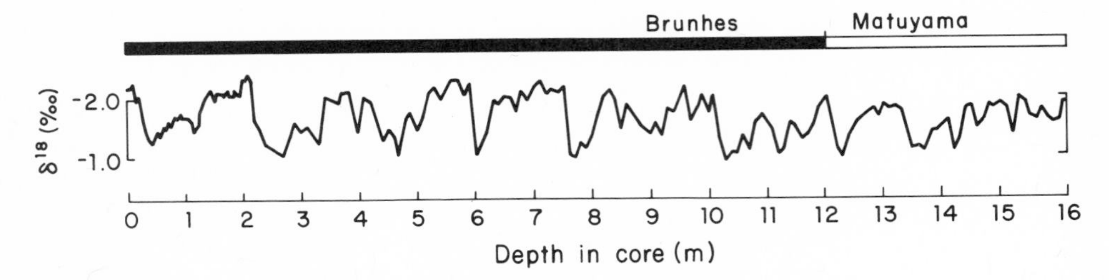

Fig. 13. Fluctuations of oxygen isotopic composition of foraminiferal tests in a core from Ontong-Java Plateau (*Shackleton and Opdyke*, 1973). Age estimates of Pleistocene climatic events based on this core are among the best, although the fluctuations do not constitute a time-series. Systematic variations in sedimentation rates are due to increased production and decreased dissolution during glacials.

valleys of $\delta\ O^{18}$ which are presumably directly related to warm and cold epochs. However, this procedure does not deliver dates with which astronomical theory can be tested rigorously, unless sedimentation rates are indeed constant. The date of 123,000 yrs. for the last prominent warm peak, found by interpolation from the Brunhes/ Matuyama boundary, corresponds nicely to "Barbados II" of *Broecker et al.* (1968). However, it does not indicate a uniform average rate of sedimentation. Deviations from the mean are highly probable, and are likely tied systematically to glacial and interglacial conditions (*Turekian*, 1965; *Broecker*, 1971a). The fluctuations are especially severe at depths near and below the lysocline. A special problem is posed by the loss of surface material and the assumption of a zero age for core-top sediment. Box cores are the answer to this particular problem.

The first step is to establish a reliable time-scale; recent efforts are summarized in Table 1. The bench-mark date of Barbados III has been thrown into some doubt by *Emiliani and Rona* (1969; see also *Emiliani and Shackleton*, 1974). However, properly plotted, the data given by *Emiliani and Rona* (1969) support Broecker's time scale (Figure 14).

The second step is to sort out the nature of gross variations in sedimentation rates, which are due to fluctuation of rates of supply and dissolution of carbonate, as well as dilution effects. All of these rate fluctuations have to be assessed separately in the various regions. In the equatorial Atlantic, for example, there is evidence for a change in the level of the lysocline from pre-Holocene to Holocene, presumably in response to a fall of the level of Antarctic Bottom Water (Figure 15). The setting is complex, however, and a number of factors may combine to produce the observed phenomenon (*Gardner*, 1975). In the equatorial Pacific, carbonate cycles

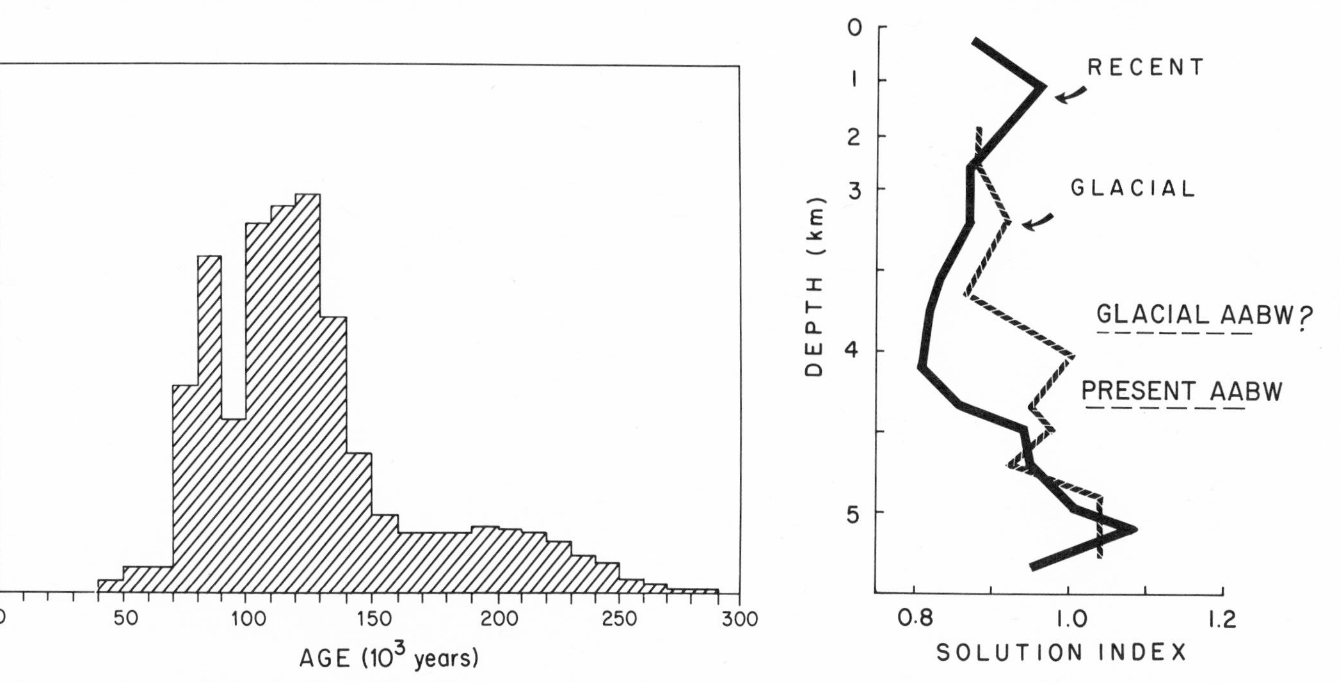

Fig. 14. Histogram of coral age determinations as compiled by *Emiliani and Rona* (1969). Each sample has equal area. Width of each sample area is appropriate to the confidence limit given in op. cit. Note the peaks at 80-90 ky and at 100 to 130 ky.

Fig. 15. Evidence for vertical excursion of the lysocline in the equatorial Atlantic, presumably in response to a higher level of AABW during the last glacial or increased fertility, or both (*Berger*, 1968). The solution index is based on relative abundance of resistant species, a high index means poor preservation.

TABLE 1. Climatic Events in Late Quaternary (in ky)

KUKLA[1]	B & vD[2]	S & O[3]	BLOOM[4]	EVENT
W 6				OPTIMUM
	--11			--DEGLACIATION
C 17	18	20		MAX. GLAC.
W 28			28	
C 36				
W 42			41	
C 51				
W 55	55	50^+	61	
C 67	75	73		
W 77	82	∿ 80	87	BARBADOS I
C 89				
W 101	103	∿ 100	106	BARBADOS II
C 111				
W 122	124	123	124	BARBADOS III
C 134				

1) "Milankovitch" mechanism plus Kukla hypothesis. Adjustable by ± a few ky, depending on "sensitive" season. *Kukla*, 1975.
2) Coral (*Broecker et al.*, 1968) and deep sea cores. *Broecker and van Donk*, 1970.
3) Core V28-238, Ontong Java Plateau. *Shackleton and Opdyke*, 1973.
4) Coral. *Bloom et al.*, 1974.

---

were produced by dissolution cycles (Figure 16). Fluctuations of productivity, although less important, reenforce the effect of dissolution on sedimentation rates, as shown by evidence from box cores from Ontong Java Plateau in the west equatorial Pacific (Figure 17).

The box cores are ideally suited for the study of the deglaciation event, the most dramatic global change in climate that may still be in human memmory (*Emiliani et al.*, 1975). In the deep sea record, the event is associated with a marked change in ecologic and preservational aspects of foraminiferans and coccoliths, and with drastic variations in isotopic composition. In the North Pacific, the the CCD rose well over 1000 m from the last glacial to the Holocene and dissolution increased markedly even on the upper continental slope (*Berger*, 1970; see also data in *Barnard and McManus*, 1973). Dissolution effects increased in the equatorial Pacific (*Olausson*, 1965; *Broecker*, 1971a), and on the East Pacific Rise the CCD apparently shallowed by some 200 m (*Broecker and Broecker*, 1974). Preliminary study of the box cores from Ontong Java Plateau indicates that the same increase in carbonate dissolution took place here also,

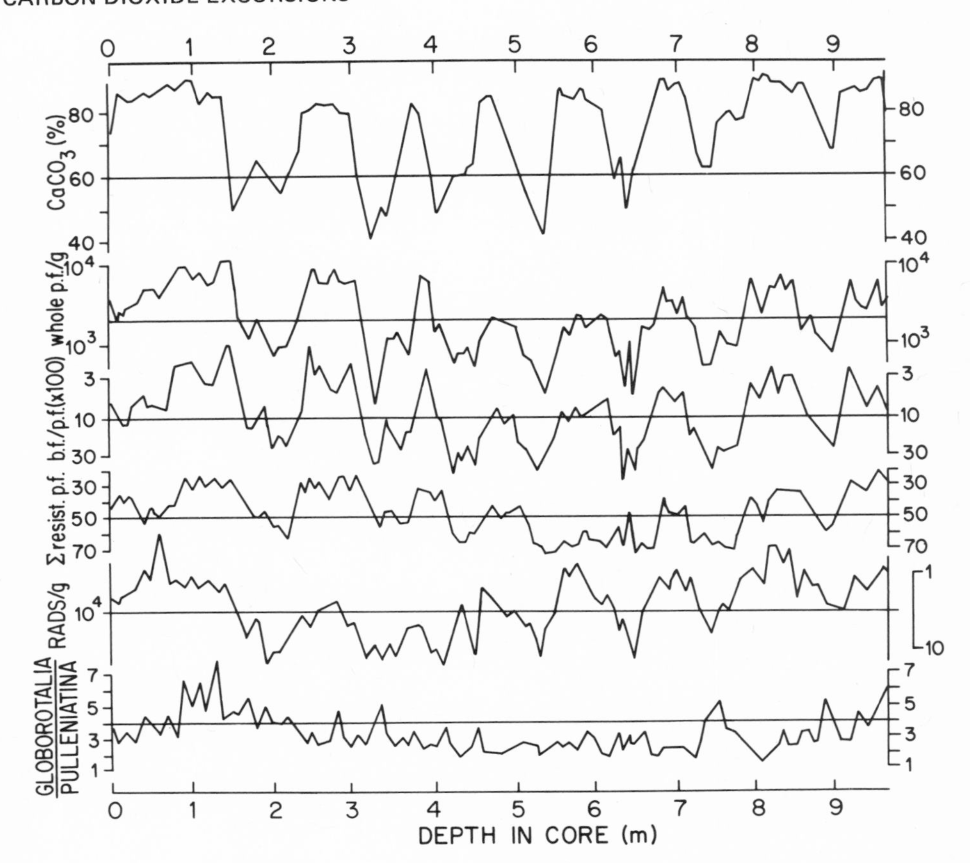

Fig. 16. Evidence for carbonate cycles in the eastern equatorial Pacific being dissolution cycles. Note the correspondence of low carbonate values with low numbers of whole planktonic forams per gram (dissolution destroying the tests), high ratios of benthic to planktonic forams (benthics being concentrated during dissolution), high percentages of resistant planktonic forams, and increased numbers of siliceous radiolarians (a resistant dilutant). The correspondence is nearly quantitative. The ratio Globorotalia/Pulleniatina is a fertility index. It shows no correlation with the cycles, except for the last one, suggesting that productivity variations are of secondary importance (*Berger*, 1973b. Data from *Arrhenius*, 1952).

and that the phenomenon essentially involved the entire depth range in this area (see Figure 18). The aragonite compensation depth rose from 1.8 km to its present depth somewhere near 1 km (see Figure 7). The calcite saturation level (or a level of slight undersaturation), rose by about 1 km, from 4 km to 3 km, as shown by the rise in the

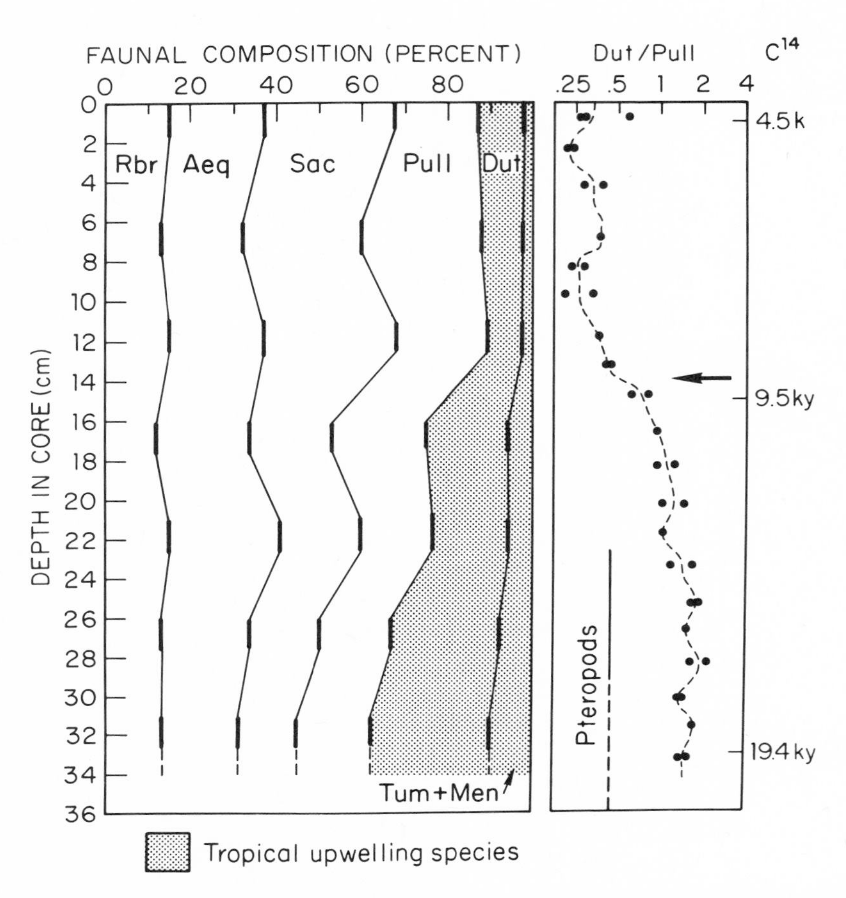

Fig. 17. Evidence for changes in productivity in the western equatorial Pacific, in a box core unaffected by dissolution of calcite (core ERDC BX 92, Ontong Java Plateau, depth 1598 m; 2° 13.5'S; 156° 59.9E). Diagram shows changes in foram assemblage >295μ during the last 20,000 years. Rbr, *Globigerinoides ruber*; Aeq, *Globigerinella aequilateralis*; Sac, *Globigerinoides sacculifer*; Pull, *Pulleniatina obliquiloculata*; Dut, *Globoquadrina dutertrei*; Tum, *Globorotalia tumida*; Men, *G. menardii*. Dut/Pull: upwelling index. C^{14} dates supplied by W. S. Broecker, L.D.G.O. Note that topmost age is too old by 4000 years, mid-point age too young by 2000 years (arrow shows transition, taken as 11 ky). Discrepancies result from upmixing of C^{14}-depleted sediment and down-mixing of young material. Mixing is completed below 10 cm, thus, lowermost C^{14} age probably also is too young by 2000 years. Glacial sedimentation rate: 1.8 cm/ky; post-glacial rate: 1.26 cm/ky. Winnowing may account for relatively low average rates at this elevation.

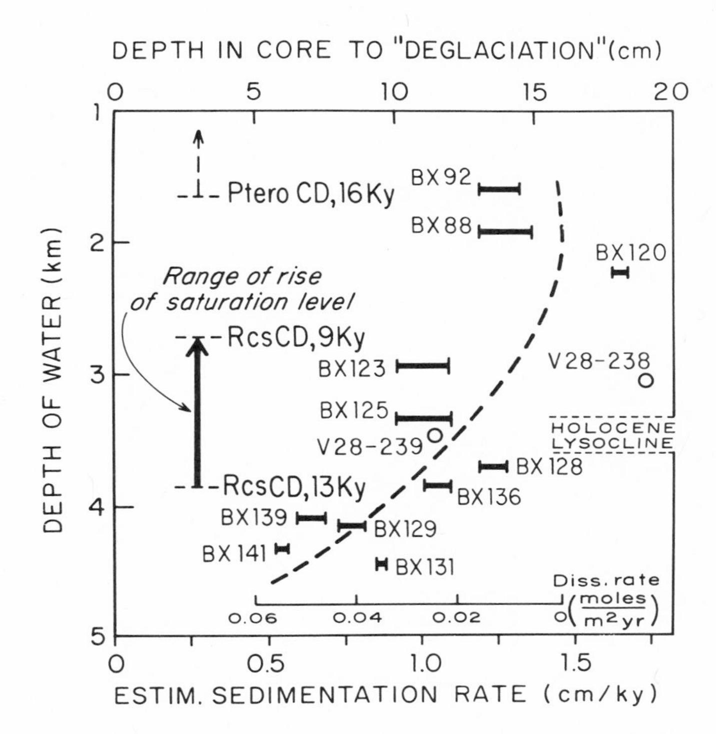

Fig. 18. Thickness of sediment above the maximal faunal change in box cores from Ontong Java Plateau, as a function of water depth. The faunal change is interpreted as marking the 11 ky transition (scale at bottom) and the general decrease in sedimentation rates with depth is ascribed to dissolution effects (inset scale). The compensation depth of pink *Globigerina rubescens* (RcsCD), the most solution susceptible common tropical foram, is taken as marking the calcite saturation level (or a level close to it). This level rose by about 1 km during the glacial-interglacial transition. The pteropod compensation depth (Ptero CD) rose through a similar depth range. Pteropods are aragonitic.

compensation depth of *Globigerina rubescens*, the most soluble common foram (RcsCD in Figure 18).

The maximum change in preservation may be slightly earlier than the maximum change in ecologic aspect. Also, the maximum preservation apparently occurs between maximum glaciation and deglaciation proper (19 to 16 K years for pteropods, Figure 17). Such phase shifts between climatic change and preservation (*Luz*, 1973; *Luz and Shackleton*, 1975) or "leads and lags" (*Moore et al.*, 1977) are intriguing and invite speculation about response times of the ocean - biosphere - atmosphere system (*Pisias et al.*, 1975). Extremely accurate dating is necessary to establish such details, because the ecologic and preservational signals may be time-transgressive as a function of depth.

Assuming that maximum change in the foram assemblages ("deglaciation" event in Figure 18) occurred 11,000 years ago at all water depths (bottom x-axis, Figure 18), the average rate for V 28-239 (*Shackleton and Opdyke*, 1973) falls near the visual-fit sediment rate line drawn without regard to this point (Figure 18). V 28-238, at 1.71 cm/ky, is distinctly high-rate. A Holocene rate of 1.44 would bring this core back into the general trend, a difference of 16% from the average presumably due to lowered production at the present time. It is not clear to me why V 28-239 shows no such pattern, perhaps it is incomplete. In any case in Core Bx 92 (Figure 17) the sedimentation rates for Holocene and last glacial (Wurm) differ by a factor of 1.4, supporting the notion of glacial/interglacial productivity change.

In connection with the deglaciation event proper, *Diester-Haass* (1975) reports pteropods preserved at the end of the Wurm down to 3.5 km off Africa, where none are now below 600 meters. *Thiede* (1973) detected a marked preservation event just at the Wurm-Holocene boundary, in a core from Tagus Abyssal Plain.

From these somewhat scattered data it appears that the ocean reaches a high state of saturation at the end of the Wurm, with a low ACD in the North Atlantic, a low lysocline in the equatorial Pacific (and probably Indian Ocean) and a low CCD in the North Pacific. These levels rose drastically in response to the deglaciation event, suggesting a decrease in alkalinity or an increase in pCO_2, or both.

What may be some of the factors providing for such a change?

Fundamentally, deglaciation exposes uplands but submerges lowlands by putting water in the ocean. Mixing is inhibited while the salinity stratification from the melt water (shown as $\delta^{18}O$ excursion, *Kennett and Shackleton*, 1975) is being dissipated. The exposure of land is followed by a rapid growth of temperate forests and peat-producing bogs. If the melting of the glaciers is preceded by wide-spread reduction of snow cover and thawing of permafrost, such growth could start reducing the pCO_2 in anticipation of the deglaciation proper, and thus help produce the late glacial preservation pulse (see *Shackleton*, 1977). About one half of the mas of atmospheric carbon could be involved in this, on the scale of a few centuries, reducing the ΣCO_2 of the ocean by perhaps 0.5% under favorable circumstances. The sheer increase in ocean water is much to the point. The additional CO_2 that can be held in the ocean after deglaciation exceeds the amount in the atmosphere.

As a "nutrient", alkalinity is affected by overall fertility (Figure 5) and as a "salt", by precipitation on shelves. The decrease in fertility at the end of the glacial would presumably lead to an increase in alkalinity. However, increased production at high latitudes might nullify such an effect. On the other hand, due to the high productivity of carbonate fixers on the shelf, alkalinity will drop in the open ocean during a transgression. As we shall see, this is a very general phenomenon. In addition, pCO_2 is likely to rise in intermediate waters, as carbon on the slopes is no longer quickly buried by terrigenous sedimentation but stays available for recycling. While mixing is being inhibited, pCO_2 also should rise in deep waters, providing a dissolution pulse during deglaciation (*Worthington*, 1968). Evidence for such a pulse should be looked for, if it exists it implies a subsequent large-scale release of CO_2 to the atmosphere, when the ocean finally turns over.

This kind of speculation can go on and on. The task, obviously, is to first collect systematically the pertinent observations on the deglaciation phenomenon, as it relates to CO_2, and then to model the system with some rigour. This has not been attempted. An initial effort has been made in detecting the $\delta\,^{13}C$ signal from forams across this event (*Broecker*, 1973). As a general trend, low $\delta\,^{13}C$ go with warm times, and vice versa. The indication of a low $\delta\,^{13}C$ spike at the deglaciation, at least in the Caribbean (Figure 19), is intriguing. We hope to have further information on this matter very shortly, from analysis of our box cores. In order to understand the system fully, of course, the $\delta\,^{13}C$ signal and the associated production and preservation of $CaCO_3$ and CH_2O have to be defined as a function of the entire glaciation cycle, not just the deglaciation event.

THE CENOZOIC RECORD: RESPONSE TO SHELF SEA OSCILLATION

In pre-Quaternary time the rise and fall of sea level and the associated transgressions and regressions continue a pattern that is in analogy to the Quaternary one, irrespective of the cause of change of sea level. A general trend is superimposed on the pulsations of the last 70 million years: an overall withdrawal of the ocean into its deep basins, from the super-transgression of the Late Cretaceous to the super-regression of our epoch. A general drop in pCO_2 is likely to go parallel, because of the increased weathering and the draw-down of alkalinity by increased fertility, as illustrated by the expansion of the Urey reaction:

$$\begin{aligned} CaSiO_3 + CO_2 \xrightarrow{a} & \; Ca^{++}\,CO_3^{=} + SiO_{2(aq)} \\ \xrightarrow{b} & \; CaCO_3 + SiO_2 \end{aligned} \qquad (1)$$

Reaction $\underline{a}$ is favored by increased exposure of igneous rocks, reaction $\underline{b}$ by increased production of carbonate, especially of a kind that is hard to redissolve (dense tests of pure calcite, increasingly common in late Cenozoic). One might speculate that such a decrease in pCO_2 is at least partially responsible for the general cooling trend from late Cretaceous to the present. Apparently,

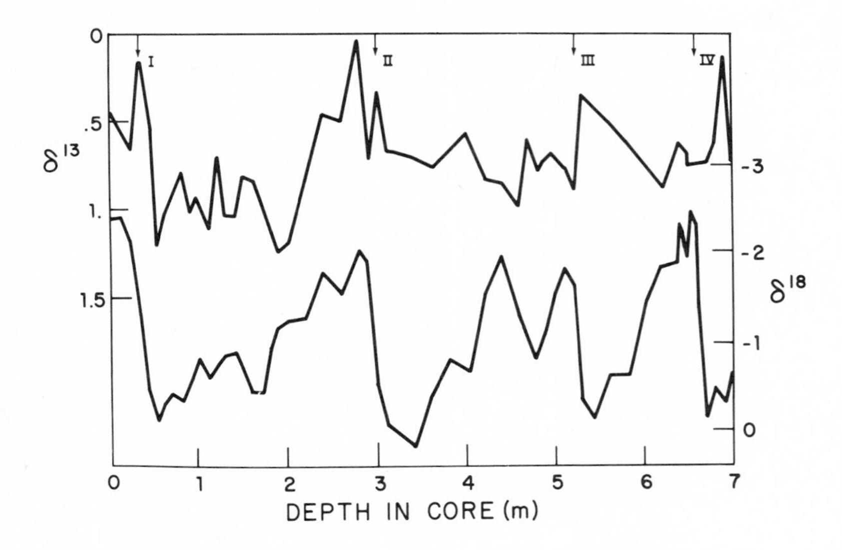

Fig. 19. Fluctuations of $\delta^{18}O$ and $\delta^{13}C$ in a core from the Caribbean (data in *Broecker*, 1973). Note the δ excursions at deglaciation. The $\delta^{13}C$ variations contain signals from several processes: changes in ocean fertility, fluctuations in burial and release of terrestrial carbon, changes in anaerobism, growth and decay of biosphere and associated dead matter, and carbonate dissolution cycles.

we live in an unusually cold time with an unusually low CO_2 content. The biosphere may actually remember a "golden age" of higher CO_2 (*Rubey*, 1951, p. 1132): many terrestrial plants photosynthesize at optimum rates at CO_2 levels that are up to five times higher than the present one (*Leopold*, 1964).

The response of the ocean to changes affecting the carbonate system are evident in CCD fluctuations (Figure 20) and sedimentation rate fluctuations (Figures 21 and 22). The original depth of deposition is found by backtracking along subsidence paths (*Berger and Winterer*, 1974). The global response of deep sea carbonate deposition to the various causes, notably transgression and regression is most closely monitored by the Pacific equatorial system. The sudden drop of the CCD at the end of the Eocene is especially noteworthy. This event may record the great regression at the end of the Eocene with a concomitant transport of shelf carbonates to the deep sea. The central Atlantic experienced a remarkably shallow CCD in the middle Miocene (Figures 20, 21) presumably in response to the unopposed invasion of Antarctic Bottom Water, which was allowed to age and rise in the entire basin. Carbonate thus prevented from

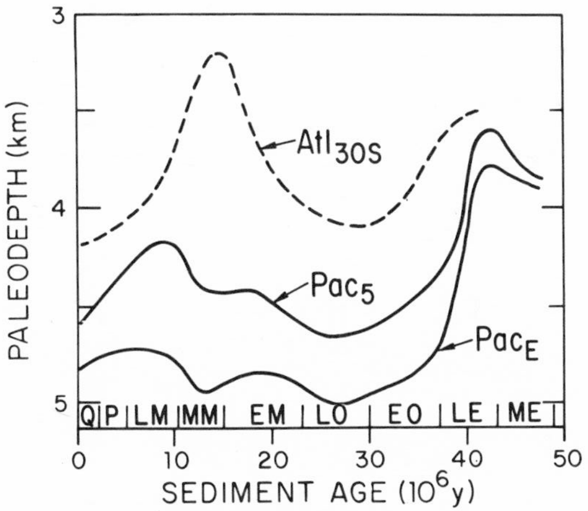

Fig. 20. Fluctuations of the CCD in the Atlantic (DSDP Leg 3) and in the east equatorial Pacific (DSDP Legs 8, 9, 16) found by backtracking of ocean crust along an average ridge flank subsidence curve. Pac_5, average elevation of CCD for 5°N and 5°S. Pac_E, elevation of CCD at equator. (Data from *Berger*, 1972, 1973a).

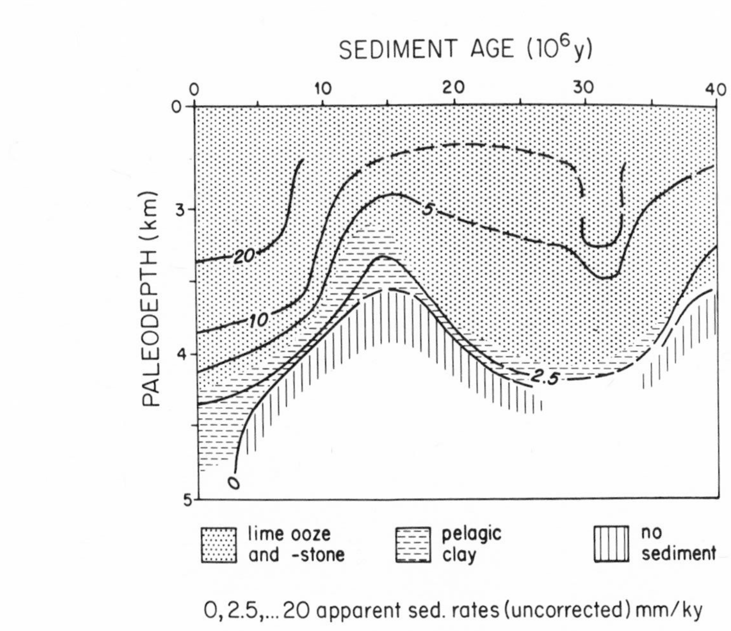

Fig. 21. Fluctuations of sedimentation rates as a function of time and depth, Atlantic, 30°S (DSDP Leg 3). Note the weak dissolution depth gradient in the Oligocene and the sharpening of the gradient in the Miocene, presumably due to AABW production. (*Berger*, 1972, redrawn).

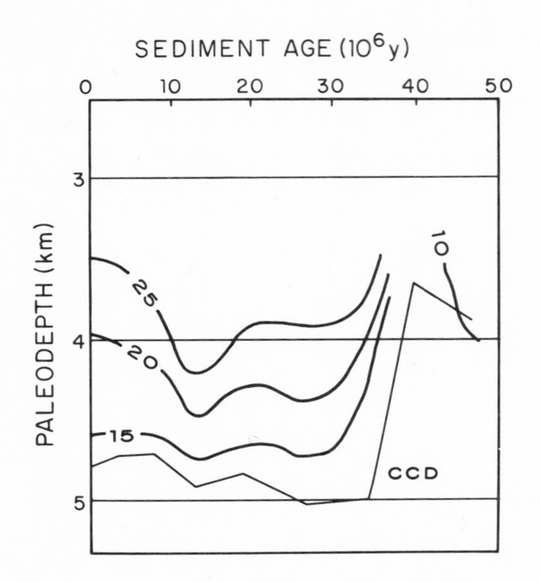

Fig. 22. Sketch of fluctuations of maximum sedimentation rates as a function of time and depth, in the central equatorial Pacific (DSDP Legs 8, 9, 16). Note the increase of accumulation rates in the Miocene, during a general trend of shallowing CCD: opposing effects of fertility and abyssal dissolution. From facies matrices (*Berger*, 1973a) with rates uncorrected for compaction.

accumulation in an Atlantic basin with estuarine circulation and CO_2-rich deep water, accumulated elsewhere. Consequently, sedimentation rates increased in the Pacific (Figure 22), partially counteracting a general rise of the Miocene CCD (Figure 20, Pac_5, Pac_E).

In both oceans, the Oligocene period of depressed CCD position is characterized by widely spaced accumulation rate isolines (Figures 21, 22). This wide spacing implies widely separated solution levels (R_o, lysocline, CCD) and associated weak depth gradient of dissolution. The gradients sharpen considerably during the early Miocene, presumably reflecting increased influx and activity of Antarctic Bottom Water.

The dissolution rate gradient also reflects the overall fertility of the ocean. A low fertility ocean resembles the Sillén model (Figure 5): there is no need to achieve high rates of dissolution at depth in order to balance any high rates of accumulation on shallow parts of the seafloor. In contrast, a high fertility Harvey Ocean is greatly undersaturated and vigorously removes carbonate from the deep sea floor to make up for the "excess supply" by organisms. (The "excess supply" is that part of carbonate supply to the seafloor which exceeds the input from weathering and other primary sources.) Accumulation rates in shallow areas are high. Thus, in the Harvey Ocean the zero level of carbonate, that is the CCD, rises as sedimentation rates increase above it, providing for a pronounced depth gradient of dissolution rates. It is evident that the analysis of fluctuations of dissolution rate gradients promises to become a major tool in paleoceanography, as exemplified in the comprehensive study of Pacific equatorial sedimentation by *van Andel, Heath, and Moore* (1975).

The degree to which fluctuations in dissolution patterns reflect changes in the global chemistry and fertility is readily apparent when comparing the changes in CCD and carbonate scatter with changes in isotopic composition of calcareous tests and with changes in productivity (Figure 23, 24). Sea level appears as a prime candidate for a basic cause of many of the changes. A comparison of the fluctuations of parameters of carbonate deposition shows (Figure 23): a general co-occurrence of shallow CCD, abundance of low carbonate percentages, high $\delta^{13}C$ values, low $\delta^{18}O$ values and high sea level, and vice versa.

Possible explanations for these correspondences can be readily enumerated, and this has been done for some of the present correlations as well as for others (e.g., paleotemperature and plankton diversity) in the literature (see *Berger and Roth*, 1975, for review). The testing of such hypotheses is entirely a different matter, and there is a remarkable lack of suggestions of how to go about it. The following propositions, unfortunately, are on this very same level of speculation, not by choice but from a lack of comprehensive geochemical-stratigraphic-ecologic models. Although such models are being attempted, at least partially (*Berger and Roth*, 1975; *Berger and Winterer*, 1974; *Berggren and Hollister*, 1974; *Broecker*, 1971a,b; *Codispoti and Piper*, 1975; *Lipps*, 1970; *Ramsey*, 1974; *Tappan*, 1968, 1971; *Tappan and Loeblich*, 1973; *van Andel*, 1975; *van Andel, Heath, Moore*, 1975) their merits need more critical discussion than can be given here.

The interpretations given in Figure 24 are based on the following assumptions, in addition to the notion that the data are meaningful and representative:

1. The abundance-distribution of carbonate percentages between 20 and 100% (Figure 23B) reflects type and stability of dissolution rate gradients and dilution effects, notably from silica supply.

2. The difference between nannofossil $\delta^{13}C$ and planktonic foram $\delta^{13}C$ reflects the gradient of $\delta^{13}C$ in surface waters, in post-Eocene sediments (Figure 23C). The gradient is strong when the oxygen minimum is strong and/or has a shallow upper boundary.

3. The difference between $\delta^{13}C$ of planktonic and of benthic forams reflects the degree to which surface waters are impoverished in C^{12}, relative to deep waters (Figure 23C). This is primarily a matter of stratification and stability of the water column.

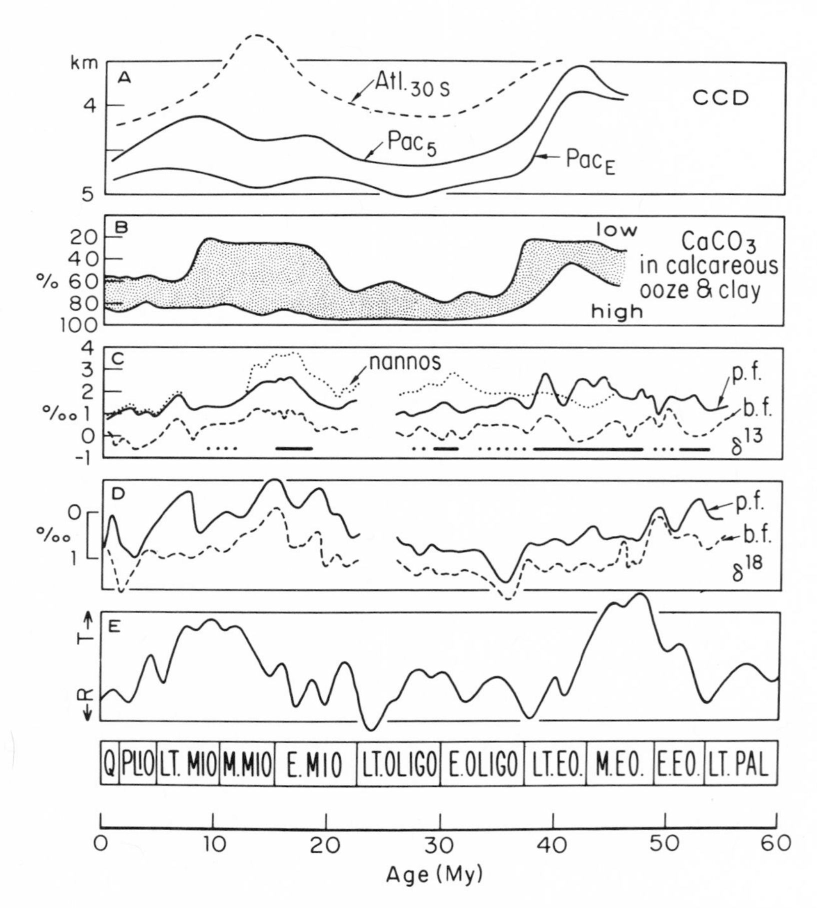

Fig. 23. Comparison of fluctuations in various parameters of the carbon system, through the Cenozoic. Presumed driving function at bottom (E): Transgression - Regression cycles, highly generalized from a compilation by *Flemming and Roberts* (1973), modified by comparison with *Holmes* (1965) and verbal descriptions by *Hallam* (1963). Other curves: (A), CCD fluctuations (Figure 20); (B) carbonate scatter (20% to 100%) of drill sites in central equatorial Pacific (*van Andel, Heath, Moore*, 1975); (C) $\delta\ ^{13}C$ in nannoplankton (*Margolis et al.*, 1975) and in planktonic and benthic foraminifera (*Shackleton and Kennett*, 1975) in DSDP Sites 277, 279, and 281; bottom solid line: $\Delta\delta^{13}$ is large; stippled line: $\Delta\delta^{13}$ is small; (D) $\delta\ ^{18}O$ in planktonic and benthic forams, DSDP Sites 277, 279, 281 (*Shackleton and Kennett*, 1975); plot shows the difference of the signals to the general trend to heavier oxygen isotopes ($0.8‰\ \delta\ ^{18}/10^6 y$ for both p.f. and b. f. for the last 40 my). Note the correspondence between transgressions, shallow CCD (sequestration of $CaCO_3$ on shelves), abundance of low $CaCO_3$ values (increased dilutants), high δ^{13} values (anaerobic C_{org} deposition), low $\delta\ ^{18}$ values (warming of ocean).

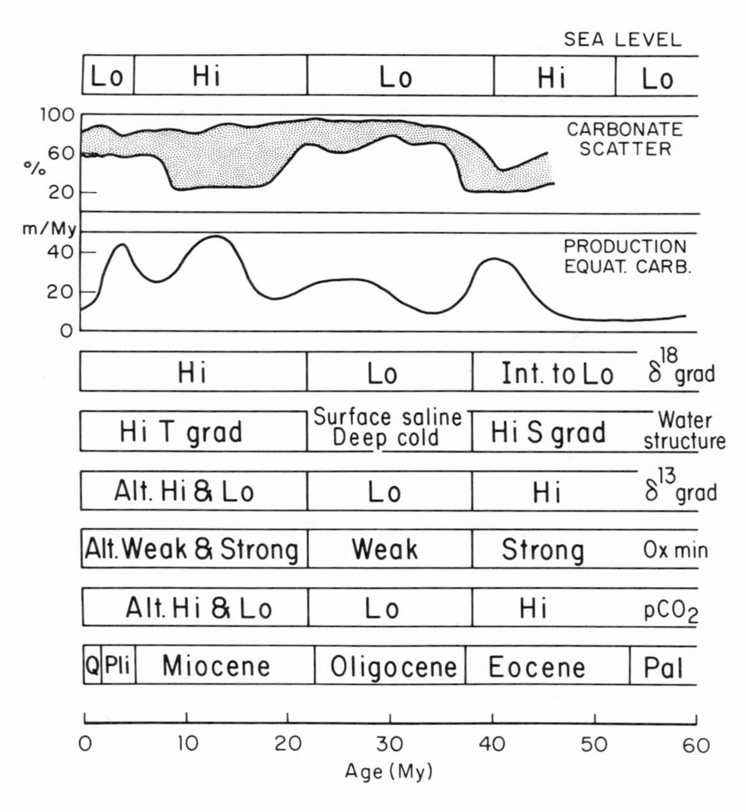

Fig. 24. Interpretation of fluctuations of carbon-related parameters through Cenozoic time. Carbonate scatter (same as in Figure 23) corresponds to productivity variation (uncorrected accumulation rates of DSDP Site 289, Ontong Java Plateau), mainly due to changes in dilution by siliceous material. The early Cenozoic is characterized by intermediate to low overall δO^{18} depth gradient in upper waters and a high δC^{13} gradient, reflecting a strong oxygen minimum and a high overall pCO_2. The Oligocene is characterized by low productivity, low δ^{18} gradient, low stability of the water structure (opposing effects of saline warm waters and cold, less saline cold waters), weak oxygen minima, and relatively low pCO_2. The late Cenozoic is typified by alternating conditions. On the whole, the establishment of a modern cold deep-water sphere dominates. The Miocene had a strong oxygen minimum during much of the time, with associated high pCO_2.

4. The difference in δO^{18} of planktonic and of benthic foraminifera reflects the difference in temperature as well as in salinity of shallow and deep waters (Figure 23D).

5. The oxygen minimum reflects mainly stratification and stability (that is, the range of ages of subsurface waters). It is largely produced at the continental margins, rather than in the open ocean as generally assumed. Therefore, productivity changes of the open ocean, by themselves, have little influence on its extent, whereas the interaction between thermocline and shelf edge becomes crucial. This interaction produces leaks of nutrients through the thermocline, which generates the organic matter whose combustion produces the oxygen minimum which spreads out by lateral eddy diffusion.

6. Deep haline circulation is favored during transgressions, when evaporative shelf seas are established, and deep thermal circulation is favored by regression, which produces cold climate and allows restricted high latitude basins to act as major sinks.

Inspection of the data shown, together with the above assumptions, produces the following scenario for the development of the Cenozoic (compare also refs, cited in Figure captions, and *Savin, Douglas, Stehli*; 1975).

The early Paleogene is characterized by relatively warm oceans with stable salinity gradients and a strong oxygen minimum. The pCO_2 is believed to have been high. The Oligocene starts with a rapid (perhaps catastrophic) change to an unstable stratification by major invasions of cold deep water interfering with salinity stratification. The oxygen minimum is weak. (Interestingly, Oligocene deep sea sediments are quite generally low in organic matter.) Productivity appears to have been low and alkalinity was high. Low pCO_2 is suggested.

The Miocene starts with a drop in carbonate saturation and a rise in fertility. It is characterized by variability, heralding the cold-warm oscillations typical of the Late Cenozoic, and which culminate in the Quaternary glacial-interglacial cycles.

Each of the transitions from one state to the other are thought to be associated with important pCO_2 changes. This is especially true for the deglaciation events and their earlier geologic analogues: first there is a draw-down of pCO_2 in response to (a) continued delivery of terrigenous carbon to the slopes, but (b) without efficient oxidation as stability rises in response to warming and to (c) the meltwater effect putting a lid on the deep waters. In addition, (d) the biosphere expands, that is forests, peat bogs, swamps and marshes and their associated dead-carbon reservoirs. Subsequently there is an over-supply of pCO_2, as the meltwater lid is removed and as the biosphere and associated taphospere carbon reservoirs are saturated, while previously deposited terrigenous carbon leaks out from the slope sediments. After a few thousand years, the system forgets the deglaciation event.

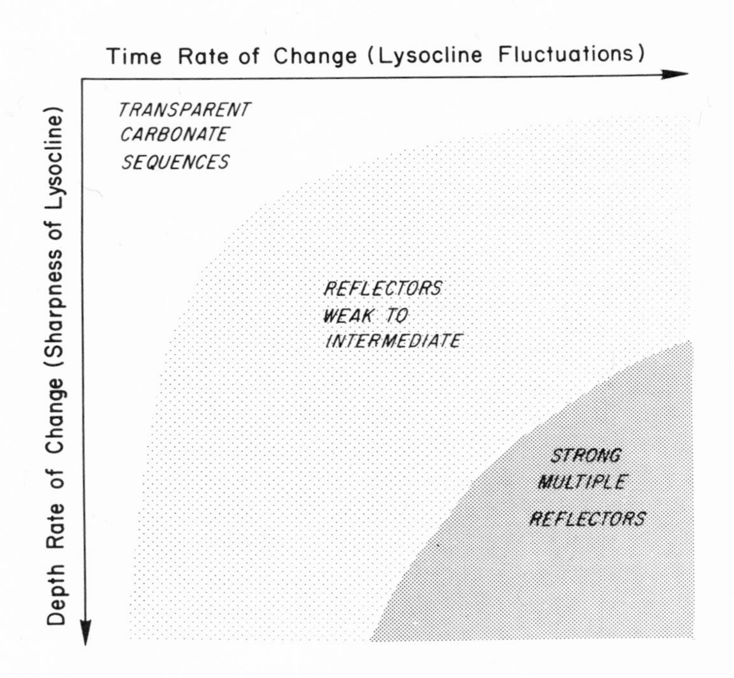

Fig. 25. Model of the origin of acoustic reflectors in pelagic carbonate sequences. The "echo potential" (defined as the ability of a sediment to produce marked changes in acoustic impedance after burial and diagenesis) increases to the right and downward. It is at a maximum when depth gradients and time gradients of dissolution are strong.

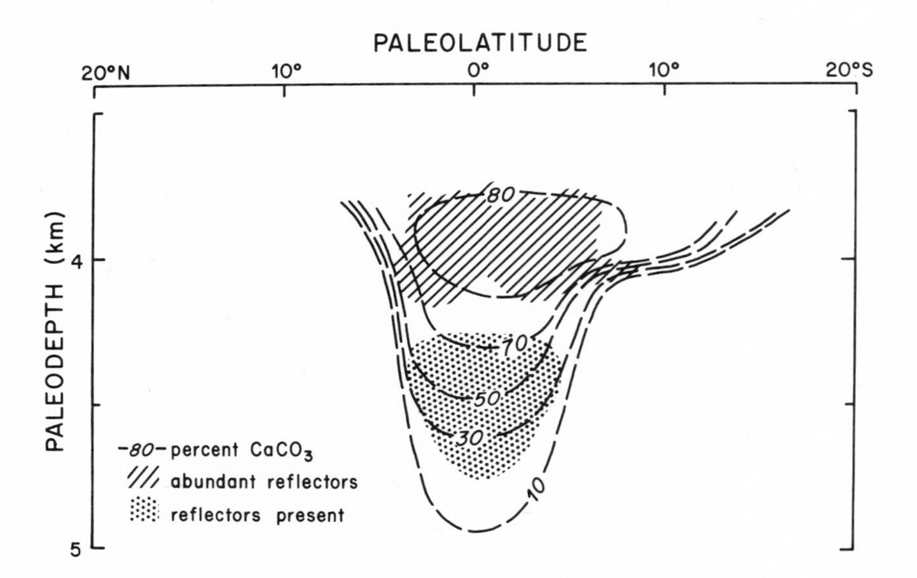

Fig. 26. Miocene carbonate sediments with strong multiple reflectors in the central equatorial Pacific (DSDP Legs 8, 9, 16) backtracked into a paleo depth- paleo latitude matrix. $CaCO_3$ percentage matrix also shown. The echo potential is greatest in shallow (reactive) carbonates, where diagenetic transformation is fast and highly dependent on original preservation state.

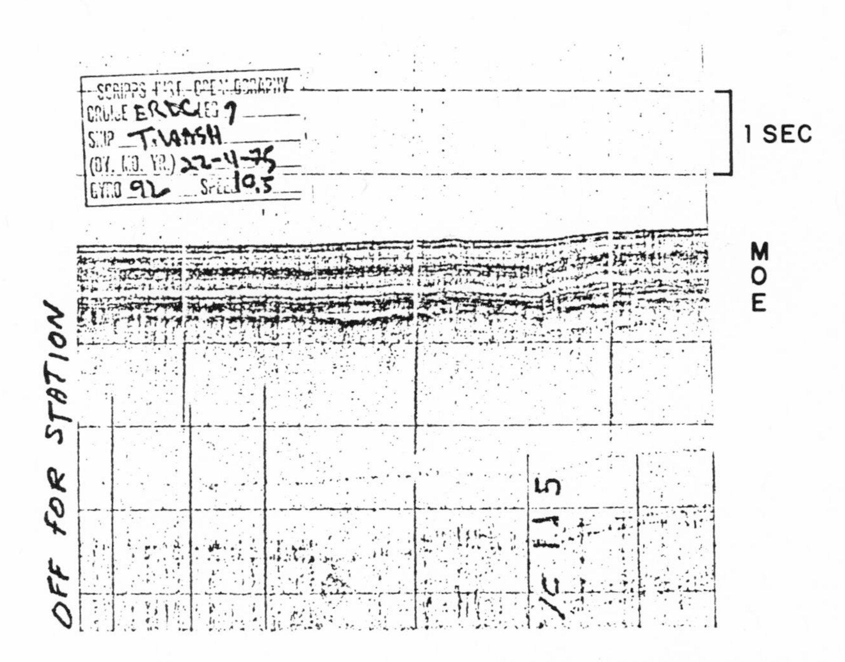

Fig. 27. Seismic profiling record from the Ontong-Java Plateau, Eurydice Expedition, 1975. Transparent Oligocene carbonates are sandwiched between highly reflective Miocene and Eocene sediments. According to the model, Oligocene echo potential should be low, because of a low depth gradient of dissolution rates (Figure 22) and relatively high uniformity of preservation states (carbonate scatter, Figure 23). The Miocene band of reflectors presumably is a result of both conditioning from intermittent dissolution pulses due to fluctuating AABW activity and of a favorable depth and time of burial, allowing diagenesis to amplify the effects of the dissolution pulses.

These Pleistocene events may serve, at least conceptually, as analogues to the much larger transition events, classically major sea level changes which separate the Cenozoic epochs. The boundaries between Eocene and Oligocene and between Oligocene and Miocene are times of great changes in the pelagic biota paralleling global changes in evolutionary pace. Of course, that is how the boundaries were defined in the first place. Marked changes in pCO_2 in the ocean atmosphere system should be expected at these critical times.

As noted earlier, any such pCO_2 excursions should be reflected in the preservation state of fossil calcareous assemblages. This notion has interesting implications for acoustic stratigraphy, because the course of diagenetic history of buried carbonates closely follows initial preservation states. *Schlanger and Douglas* (1974) have coined the term "diagenetic potential" to emphasize this relationship.

The diversity of the "diagenetic potential", that is, of the state of preservation of calcareous fossils, goes parallel to the carbonate scatter (Figure 23) and has maxima in the Eocene and in the Miocene. Depth gradients of dissolution are believed to be high during this time (Figures 21 and 22), and also highly variable through time. The combination of sharp depth gradients and rapid change juxtaposes carbonates of greatly differing diagenetic transformation rates, resulting in fluctuations of diagenetic states in the stratigraphic column. Such fluctuations produce alternating sonic impedance values. Impedance discontinuities and/or oscillations produce acoustic reflectors. Thus, Miocene sediments, through their high diagenetic diversity, acquire a high "echo potential", defined as the ability to reflect sound after burial and diagenesis (Figure 25). The hypothesis is being tested (Figure 26). If correct, the changing saturation state of the ocean, with respect to carbonate, is directly responsible for the multiple closely spaced reflectors seen in pelagic carbonates (Figure 27).

ACKNOWLEDGEMENTS

I am indebted to Brian Funnell, for lively discussions on the topic of CO_2 fluctuations through geologic time, and to Joris Gieskes and C. D. Keeling for useful advice. Research supported by the Office of Naval Research, USN N00014-69-A-0200-6049 and by the National Science Foundation (Oceanography), DES 75-04335.

Adelseck, C. G., and W. H. Berger. 1975. On the dissolution of planktonic foraminifera and associated microfossils during settling and on the sea floor. *Cushman Found. Foram. Res., Spec. Publ. 13.* 70-81.

Arrhenius, G. 1952. Sediment cores from the East Pacific. *Rept. Swedish Deep-Sea Exped. 1947-1948, 5.* 1-228.

Baksi, A. K., and N. D. Watkins. 1973. Volcanic production rates: comparison of oceanic ridges, islands, and the Columbia Plateau Basalts. *Science, 180.* 493-496.

Barnard, W. D., and D. A. McManus. 1973. Planktonic foraminifera-radiolarian stratigraphy and the Pleistocene-Holocene boundary in the northeast Pacific. *Geol. Soc. Amer. Bull., 84.* 2097-2100.

Berger, W. H. 1968. Planktonic Foraminifera: selective solution and paleoclimatic interpretation. *Deep-Sea Res., 15.* 31-43.

Berger, W. H. 1970. Planktonic foraminifera: selective solution and the lysocline. *Marine Geol., 8.* 111-138.

Berger, W. H. 1972. Deep-sea carbonates: dissolution facies and age-depth constancy. *Nature, 236.* 392-395.

Berger, W. H. 1973a. Cenozoic sedimentation in the eastern tropical Pacific. *Geol. Soc. Amer. Bull., 84.* 1941-1954.

Berger, W. H. 1973b. Deep-sea carbonates: Pleistocene dissolution cycles. *Jour. Foram. Res., 3.* 187-195.

Berger, W. H. 1974a. Deep-sea sedimentation. In: *The Geology of Continental Margins*, C. A. Burk and C. L. Drake (Eds.), Springer-Verlag, N. Y. 213-241.

Berger, W. H. 1974b. Sedimentation of deep-sea carbonate: maps and models of variations and fluctuations, Presented in Kiel, Germany, Sept. 1974, Conference on Marine Plankton and Sediments, E. Seibold, Convener, to be published by Micropaleontology Press, Washington, D. C.

Berger, W. H., and T. C. Johnson. 1976. Deep-sea carbonates: dissolution and mass wasting on Ontong Java Plateau. *Science, 192.* 785-787.

Berger, W. H., and P. H. Roth. 1975. Oceanic micropaleontology: progress and prospects. *Rev. Geophys. Space Phys., 13.* 561-585 and 624-635.

Berger, W. H., and A. Soutar. 1970. Preservation of plankton shells in an anaerobic basin off California. *Geol. Soc. Amer. Bull., 81.* 275-282.

Berger, W. H., and E. L. Winterer. 1974. Plate stratigraphy and the fluctuating carbonate line. In: *Pelagic Sediments on Land and Under the Sea,* K. J. Hsü and H. C. Jenkyns (Eds.), Spec. Publ. Internat. Assoc. Sedimentologists, 1. 11-48.

Berger, W. H., C. G. Adelseck, and L. A. Mayer. 1976. Distribution of carbonate in surface sediments of the Pacific Ocean. *J. Geophys. Res., 81.* 2617-2627.

Berggren, W. A., and C. D. Hollister. 1974. Paleogeography, paleobiogeography, and the history of circulation in the Atlantic Ocean. In: *Studies in paleo-oceanography*, W. W. Hay (Ed.), *Soc. Econ. Paleont. Mineral. Spec. Publ. 20.* 126-186.

Berner, R. A. 1965. Activity coefficients of bicarbonate, carbonate, and calcium ions in sea water. *Geochim. et Cosmochim. Acta, 29, (8).* 947-965.

Berner, R. A. 1974. Physical chemistry of carbonates in the oceans. In: *Studies in paleo-oceanography,* W. W. Hay (Ed.), *Soc. Econ. Paleont. Mineral. Spec. Pub. 20.* 37-43.

Bloom, A. L., W. S. Broecker, J. M. A. Chappell, R. K. Matthews, and K. J. Mesolella. 1974. Quaternary sea level fluctuations on a tectonic coast: new $^{230}Th/^{234}U$ dates from the Huon Peninsula, New Guinea. *Quaternary Res., 4.* 185-205.

Broecker, W. S. 1971a. Calcite accumulation rates and glacial to interglacial changes in oceanic mixing. In: *The Late Cenozoic Glacial Ages,* K. K. Turekian (Ed.), Yale University Press, New Haven. 239-265.

Broecker, W. S. 1971b. A kinetic model for the chemical composition of sea water. *Quat. Res., 1, (2).* 188-207.

Broecker, W. S. 1973. Factors controlling CO_2 content in the oceans and atmosphere. In: *Carbon and the Biosphere,* G. M. Woodwell and E. V. Pecan (Eds.), AEC Symposium, vol. 30. 32-50.

Broecker, W. S. 1975. Climatic change: are we on the brink of a pronounced global warming? *Science, 189.* 460-463.

Broecker, W. S., and S. Broecker. 1974. Carbonate dissolution on the western flank of the East Pacific Rise. *Soc. Econ. Geol. and Paleont. Spec. Publ. No. 20.* 44-57.

Broecker, W. S., and J. van Donk. 1970. Insolation changes, ice volumes, and the O^{18} record in deep-sea cores. *Rev. Geophys. and Space Phys., 8, (1).* 169-198.

Broecker, W. S., D. L. Thurber, J. Goddard, T. L. Ku, R. K. Matthews, and K. J. Mesolella. 1968. Milankovitch hypothesis supported by precise dating of coral reefs and deep-sea sediments. *Science, 159.* 297-300.

Chave, K. E., S. V. Smith, and K. J. Roy. 1972. Carbonate production by coral reefs. *Marine Geol., 12.* 123-140.

Cloud, P. 1973. Is there intelligent life on Earth? In: *Carbon and the Biosphere,* G. M. Woodwell and E. V. Pecan (Eds.), AEC Symposium No. 30, Natl. Techn. Info. Service, Springfield, VA. 264-280.

Codispoti, L. A., and D. Z. Piper. 1975. Marine phosphorite deposits and the nitrogen cycle. *Science, 188. (4183).* 15-18.

Croll, J. 1875. *Climate and Time,* Daldy, Isbister, & Co., London. 577 pp.

Damon, P. E. 1971. The relationship between Late Cenozoic volcanism and tectonism and orogenic-epeirogenic periodicity. In: *The Late Cenozoic Glacial Ages,* K. K. Turekian (Ed.), Yale Univ. Press, New Haven. 15-35.

Dansgaard, W., S. J. Johnsen, H. B. Clausen, and C. C. Langway. 1971. Climatic record revealed by Camp Century ice core. In: *The Late Cenozoic Glacial Ages*, K. K. Turekian (Ed.), Yale Univ. Press, New Haven. 37-56.

Davies, T. A., and P. R. Supko. 1973. Oceanic sediments and their diagenesis: some examples from deep sea drilling. *J. Sediment. Petrol.*, *3*. 381-390.

Deffeyes, K. S. 1967. Carbonate equilibria: a graphic and algebraic approach. *Limnol. Oceanogr.*, *10*, (3). 412-426.

Diester-Haass, L. 1975. Sedimentation and climate in the Late Quaternary between Senegal and the Cape Verde Islands. *"Meteor"-Forsch.-Ergebn. Reihe C No. 20*. 1-32.

Edmond, J. M. 1974. On the dissolution of carbonate and silicate in the deep ocean. *Deep-Sea Res.*, *21*. 455-480.

Ekdahl, C. A., and C. D. Keeling. 1973. Atmospheric carbon dioxide and radiocarbon in the natural carbon cycle: I. Quantitative deductions from records at Mauna Loa Observatory and at the South Pole. In: *Carbon and the Biosphere*, G. M. Woodwell and E. V. Pecan (Eds.), Techn. Info. Center, United States Atomic Energy Commission, Washington, D. C. 51-85.

Emiliani, C. 1955. Pleistocene temperatures. *J. Geol.*, *63* (6). 538-578.

Emiliani, C. 1966. Paleotemperature analysis of Caribbean Cores P6304-8 and P6304-9 and a generalized temperature curve for the past 425,000 years. *J. Geol.*, *74*, (2). 109-126.

Emiliani, C., and E. Rona. 1969. Caribbean Cores P6304-8 and P6304-9: new analysis of absolute chronology. A reply, *Science*, *166*. 1551-1552.

Emiliani, C., and N. Shackleton. 1974. The Brunhes Epoch: isotopic paleotemperatures and geochronology. *Science*, *183*. 511-514.

Emiliani, C., S. Gartner, B. Lidz, K. Eldridge, D. E. Elvey, T. C. Huang, J. J. Stipp, M. F. Swanson. 1975. Paleoclimatological analysis of Late Quaternary cores from the Northeastern Gulf of Mexico. *Science*, *189*. 1083-1088.

Fairhall, A. W. 1973. Accumulation of fossil CO_2 in the atmosphere and the sea. *Nature*, *245*. 20-23.

Flemming, N. C., and D. G. Roberts. 1973. Tectono-eustatic changes in sea level and seafloor spreading. *Nature*, *243*. 19-22.

Gardner, J. V. 1975. Late Pleistocene carbonate dissolution cycles in the eastern equatorial Atlantic. *Cushman Found. Foram. Res. Spec. Publ. 13*. 129-141.

Garrels, R. M., and E. A. Perry. 1974. Cycling of carbon, sulfur, and oxygen through geologic time. In: *The Sea, Vol. 5*, E. D. Goldberg (Ed.), Wiley-Interscience, N. Y. 303-336.

Hallam, A. 1963. Major epeirogenic and eustatic changes since the Cretaceous, and their possible relationship to crustal structure. *Amer. J. Sci.*, *261*. 397-423.

Harvey, H. W. 1957. *The Chemistry and Fertility of Sea Waters.* Cambridge Univ. Press, U. K. 240 pp.

Holmes, A. 1965. *Principles of Physical Geology.* Ronald Press, N. Y. 1288 pp.

Hubbert, M. K. 1969. Energy resources. In: *Resources and Man,* a study and recommendations by the Committee on Resources and Man of the Division of Earth Sciences, National Academy of Sciences No. 1703 National Research Council. W. H. Freeman, San Francisco. 157-242.

Imbrie, J., and N. G. Kipp. 1971. A new micropaleontological method for quantitative paleoclimatology: application to a Late Pleistocene Caribbean core. In: *The Late Cenozoic Glacial Ages,* K. K. Turekian (Ed.), Yale Univ. Press, New Haven. 71-181.

Keeling, C. D. 1973. The carbon dioxide cycle: reservoir models to depict the exchange of atmospheric carbon dioxide with the oceans and land plants. In: *Chemistry of the Lower Atmosphere,* S. I. Rasool (Ed.), Plenum Publ. Corp. 251-329.

Kennett, J. P., and N. J. Shackleton. 1975. Laurentide ice sheet meltwater recorded in Gulf of Mexico deep-sea cores. *Science, 188, (4184).* 147-150.

Kennett, J. P., and R. C. Thunell. 1975. Global increase in explosive volcanism. *Science, 187.* 497-503.

Kukla, G. J. 1975. Missing link between Milankovitch and climate. *Nature, 253 (5493).* 600-603.

Leopold, A. C. 1964. *Plant Growth and Development,* McGraw-Hill, N. Y. 466 pp.

Lipps, J. H. 1970. Plankton evolution. *Evolution, 24 (1).* 1-22.

Luz, B. 1973. Stratigraphic and paleoclimatic analysis of Late Pleistocene tropical Southeast Pacific cores (with an Appendix by N. J. Shackleton). *Quaternary Res., 3.* 56-72.

Luz, B., and N. J. Shackleton. 1975. $CaCO_3$ solution in the tropical east Pacific during the past 130,000 years. In: *Dissolution of Deep-Sea Carbonate,* W. V. Sliter, A. W. H. Bé, and W. H. Berger (Eds.), Cushman Found. Foram. Res., Vol. 13. 142-150.

Margolis, S. V., P. M. Kroopnick, D. E. Goodney, W. C. Dudley, and M. E. Mahoney. 1975. Oxygen and carbon isotopes from calcareous nannofossils as paleoceanographic indicators. *Science, 189.* 555-557.

Moore, T. C., Jr., N. G. Pisias, and G. R. Heath. 1977. Climate changes and lags in Pacific carbonate preservation, sea surface temperature and global ice volume. In: *The Fate of Fossil Fuel* CO_2, N. R. Andersen and A. Malahoff (Eds.), Plenum, N.Y., (this volume).

Olausson, E. 1965. Evidence of climatic changes in North Atlantic deep-sea cores, with remarks on isotopic paleotemperature analysis. In: *Progress in Oceanography, Vol. 3,* M. Sears (Ed.). 221-252.

Peterson, M. N. A. 1966. Calcite: rates of dissolution in a vertical profile in the Central Pacific. *Science, 154, (3756).* 1542-1544.

Phleger, F. B., F. L. Parker, and J. F. Peirson. 1953. North Atlantic foraminifera. *Rept. Swedish Deep-Sea Exped., 7.* 1-122.

Pisias, N. G., G. R. Heath, T. C. Moore. 1975. Lag times for oceanic responses to climatic change. *Nature, 256 (5520).* 716-717.

Plass, G. N. 1972. Relationship between atmospheric carbon dioxide amount and properties of the sea. *Environmental Science and Technology, 6, (8).* 736-740.

Pytkowicz, R. M. 1965. Calcium carbonate saturation in the ocean. *Limnol. and Oceanol., 10, (2).* 220-225.

Pytkowicz, R. M. 1972. Fossil fuel burning and carbon dioxide: a pessimistic view, Comments on Earth Sciences. *Geophysics, 3.* 15-22.

Ramsey, A. T. S. 1974. The distribution of calcium carbonate in deep-sea sediments. In: *Studies in Paleo-oceanography*, W. W. Hay (Ed.), Soc. Econ. Paleont. Mineralog. Spec. Publ. 20. 58-76.

Redfield, A. C. 1942. The processes determining the concentration of oxygen, phosphate, and other organic derivatives within the depths of the Atlantic Ocean. *Papers Phys. Oceanog. Met., Mass. Inst. Tech. & Woods Hole Oceanog. Inst., 9, (2).* 1-22.

Redfield, A. C., B. H. Ketchum, and F. A. Richards. 1963. The influence of organisms on the composition of sea-water. In: *The Sea, Vol. 2*, M. N. Hill (Ed.), Interscience, N. Y. 26-77.

Rubey, W. W. 1951. Geologic history of sea water, an attempt to state the problem. *Geol. Soc. Amer. Bull., 62.* 1111-1148.

Ruddiman, W. F., and B. C. Heezen. 1967. Differential solution of planktonic foraminifera. *Deep-Sea Res., 14.* 801-808.

Savin, S. M., R. G. Douglas, and F. G. Stehli. 1975. Tertiary marine paleotemperatures. *Geol. Soc. Amer. Bull., 86.* 1499-1510.

Schlanger, S. O., and R. G. Douglas. 1974. The pelagic ooze-chalk-limestone transition and its implication for marine stratigraphy. *Spec. Publs. Int. Ass. Sediment, 1.* 117-148.

Schott, W. 1935. Die Foraminiferen in dem äquatorialen Teil des Atlantischen Ozeans. *Deutsch. Atl. Exped. Meteor 1925-1927, 3, (3).* 43-134.

Shackleton, N. J. 1977. Carbon-13 in *Uvigerina*: Tropical rainforest history and the equatorial Pacific carbonate dissolution cycles, In: *The Fate of Fossil Fuel CO_2*, N. R. Andersen and A. Malahoff (Eds.), Plenum, N.Y., (this volume).

Shackleton, N. J., and J. P. Kennett. 1975. Paleotemperature history of the Cenozoic and the initiation of Antarctic glaciation: oxygen and carbon isotope analyses in DSDP Sites 277, 279, and 281. In: *Init. Reports Deep Sea Drilling Project, 29*, U. S. Govt. Printing Office, Washington, D. C. 743-755.

Shackleton, N. J., and N. D. Opdyke. 1973. Oxygen isotope and paleomagnetic stratigraphy of equatorial Pacific core V28-238: oxygen isotope temperatures and ice volumes on a 10^5 year and 10^6 year scale. *Quaternary Res., 3.* 39-55.

Sillén, L. G. 1961. The physical chemistry of sea water. In: *Oceanography*, M. Sears (Ed.), Am. A. Sci. Pub. No. 67. 549-582.

Sillén, L. G. 1967. The ocean as a chemical system. *Science, 156*. 1189-1197.

Smith, S. V. 1972. Production of calcium carbonate on the mainland shelf of southern California. *Limnol. and Oceanog., 17*, (1). 28-41.

Takahashi, T. 1975. Carbonate chemistry of sea water and the calcite compensation depth in the oceans. In: *Dissolution of Deep-Sea Carbonates*, W. V. Sliter, A. W. H. Bé, W. H. Berger (Eds.), *Cushman Found. Foram. Res., Vol. 13*. 11-26.

Tappan, H. 1968. Primary production, isotopes, extinctions and the atmosphere. *Paleogeogr. Paleoclimatol. Paleoecol., 4*. 187-210.

Tappan, H. 1971. Microplankton, ecological succession and evolution. *North Am. Paleont. Convention Proc., Pt. H*. 1058-1103.

Tappan, H., and A. R. Loeblich. 1973. Evolution of the oceanic plankton. *Earth-Science Reviews, 9*. 207-240.

Thiede, J. 1973. Planktonic foraminifera in hemipelagic sediments: shell preservation off Portugal and Morocco. *Geol. Soc. Amer. Bull., 84*. 2749-2754.

Thiede, J. 1974. Marine bivalves: distribution of meroplanktonic larval shells in eastern North Atlantic surface waters. *Palaeogeogr., Palaeoecol., Palaeoclimat., 15*. 267-290.

Thiede, J. 1975. Shell- and skeleton-producing plankton and nekton in the eastern North Atlantic Ocean. *Meteor-Forsch. Ergeb. Reihe C, 20*. 33-76.

Turekian, K. K. 1965. Some aspects of the geochemistry of marine sediments. In: *Chemical Oceanography, Vol. 2*, J. P. Riley and G. Skirrow (Eds.), Academic Press, London. 81-126.

Urey, H. C. 1952. *The Planets, Their Origin and Development*. Yale Univ. Press, New Haven. 245 pp.

Valenzuela, E. A. 1975. Deep-sea sediments and topography of an area near Clipperton Island, eastern tropical Pacific, M.S. Thesis, Univ. of California, San Diego. 160 pp.

van Andel, Tj. H. 1975. Mesozoic/Cenozoic calcite compensation depth and the global distribution of calcareous sediments. *Earth and Planetary Science Letters, 26*. 187-194.

van Andel, Tj. H., G. R. Heath, T. C. Moore. 1975. *Cenozoic History and Paleooceanography of the Central Equatorial Pacific Ocean*, Geol. Soc. Am. Memoir 143. 1-134.

Worthington, L. V. 1968. Genesis and evolution of water masses. *Meteorological Monographs, 8*, (30). 63-67.

SEDIMENTATION ON ONTONG JAVA PLATEAU: OBSERVATIONS ON A CLASSIC "CARBONATE MONITOR"

W. H. Berger[1], T. C. Johnson[2], and E. L. Hamilton[3]

[1]Scripps Institution of Oceanography, [2]University of Minnesota, [3]Naval Undersea Center

ABSTRACT

Ontong-Java Plateau in the western equatorial Pacific is the classic example for a low-latitude deep-sea carbonate monitor. Box cores and conventional cores were taken, and seismic profiles were obtained on the Plateau for the study of the sedimentation processes governing the operation of the monitor.

Properties of the sediments are greatly influenced by increasing dissolution of carbonate with depth of deposition and by winnowing processes above the lysocline. Burrowing patterns seen in the box cores depend on depth and suggest intermittent slow gravity flow in sub-lysoclinal calcareous ooze. Seismic profiles indicate large-scale slumps whose morphology also depends on depth, and hence presumably on dissolution processes. It is proposed that much of the slumping arises from the removal of downslope support, at the lower flanks of the Plateau, by dissolution. It appears likely that both ooze flow and slumps are triggered by earthquakes generated at the nearby plate boundary.

There is a general decrease of sedimentation rates with depth, but the scatter is considerable. Carbonate percentages are relatively inefficient predictors of changes in sedimentation rates, and hence of the amount of dissolution experienced.

The origin of acoustic reflectors within carbonate sequences is obscure, but should be, at least in part, tied to cycles of dissolution, winnowing, and productivity, as well as to erosional surfaces. A close relationship of reflectors to depth-dependent zones of dissolution fluctuations could explain the striking differences between reflection profiles above the lysocline and the profiles from greater depth. Such a relationship also would reconcile our *observations that individual reflectors cannot be correlated over long distances, with previous suggestions that long-distance correlations do exist.*

INTRODUCTION

The best places to study the history contained in carbonate sequences are the relatively shallow large plateaus, which contain thick sections of pelagic sediment. *Winterer* (1977) has drawn attention to these plateaus as "carbonate monitors", because their shallowness and constant buildup insures a long period of carbonate accumulation well above the levels of severe dissolution. Ontong-Java Plateau is the classic example for a low-latitude carbonate monitor (Figure 1). It is less rugged than Manihiki Plateau (*Winterer et al.*, 1974), considerably shallower than Magellan Rise (*Winterer et al.*, 1973), and much less exposed to corrosive waters than, for example, Shatsky Rise (*Fischer et al.*, 1971; *Larson et al.*, 1975).

During the spring of 1975, two of us (W. H. Berger and T. C. Johnson) led an expedition across Ontong-Java Plateau, to collect box cores and long cores in a fashion allowing maximum resolution of the influence of water depth and distance from the equator on carbonate sedimentation. Samples were taken for physical properties and were subsequently analyzed at the Naval Undersea Center, San Diego, (E. L. Hamilton). We also obtained seismic profiles along the steamed track in order to see whether the types of profiles recovered would change with depth and with distance from the equator.

Here we present a first overview of the results obtained so far. Further studies are in progress. Details of certain aspects of the present topic have been presented elsewhere: on mass movement by *Berger and Johnson* (1976), and on physical properties of surficial sediments by *Johnson et al.* (in press). The Hawaii Institute of Geophysics has done considerable work in the area (*Kroenke*, 1972; *Valencia*, 1972, 1973; *Resig et al.*, 1976, and references therein). Access to unpublished data was generously provided by the Hawaii Institute of Geophysics and greatly facilitated the planning of our cruise. Also, two legs of the Deep Sea Drilling Project crossed the area (Legs 7 and 30), and three holes were drilled (*Winterer et al.*, 1971; *Andrews et al.*, 1975). Work by Lamont-Doherty Geological Observatory and Japanese scientists on the Ontong-Java Plateau was reported by *Ewing et al.* (1969), *Shackleton and Opdyke* (1973), and *Murauchi et al.* (1973).

PRESENT PATTERNS OF SEDIMENTATION: THE BOX CORES

One of the problems in determining sedimentation patterns for the present ocean (the only one for which we have good control on

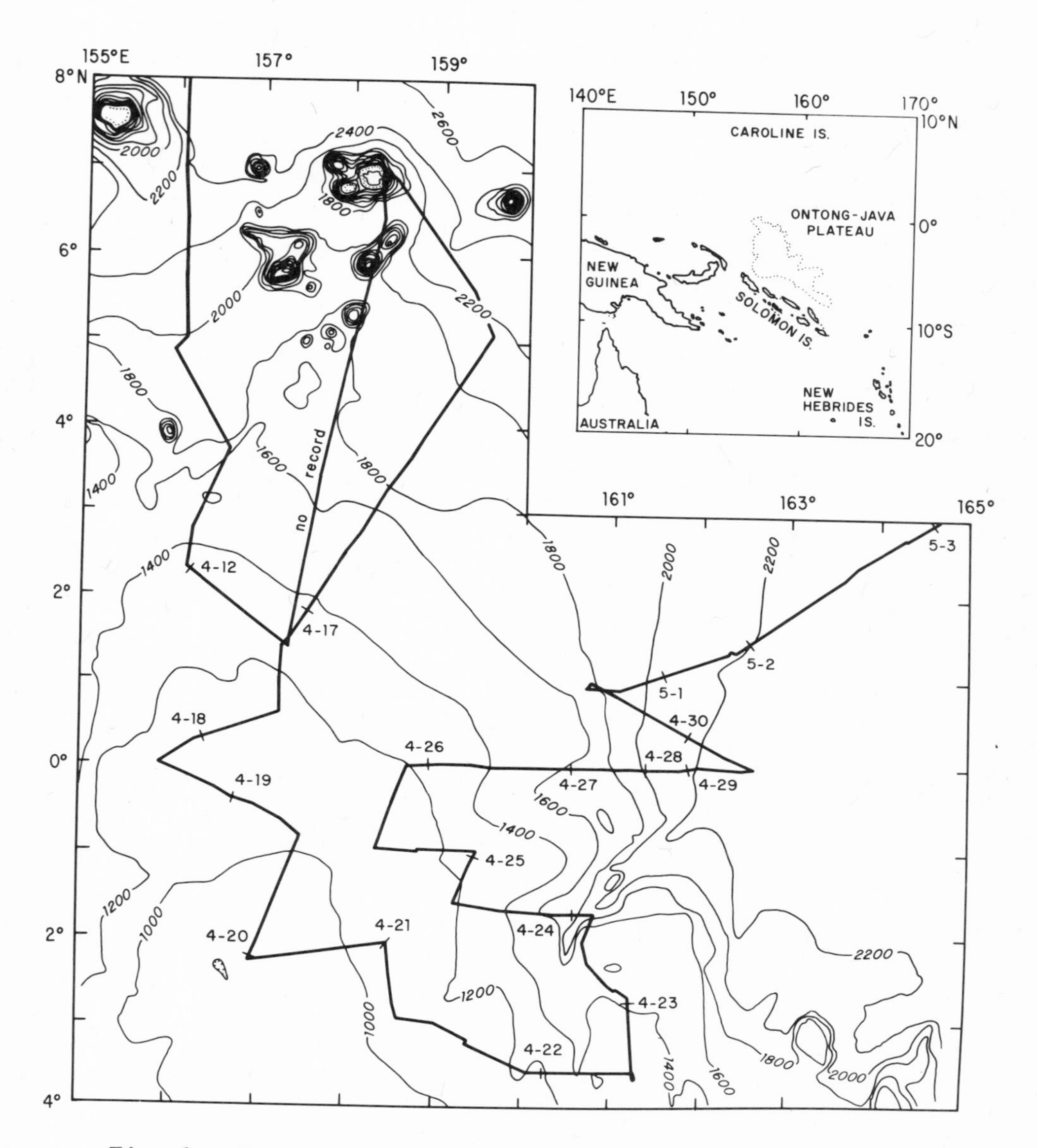

Fig. 1. Topographic map of the northern Ontong-Java Plateau, based on bathymetric data collected by Lamont-Doherty Geological Observatory, Hawaii Institute of Geophysics, and Scripps Institution of Oceanography, through June 1975. Ship track enters in NW corner. Days are marked as Mo-day, 0 hours Z.

the causative agents) is the doubtful consistency of core-tops of conventional gravity and piston cores. Core noses with finger catchers, for example, strain the soft surficial sediment through the fingers before it becomes hard enough to push the fingers back against the corer wall. On board ship, the relatively thin cylinder of core-top sediment settles to fill the greater width of the barrel. Sedimentation rates found for such sediment are low. Other complications also exist, resulting from the pressure wave in front of the core nose hitting the sea floor. In some cases, a considerable amount of surface-near sediment may be lost outright.

Box cores largely overcome these shortcomings. In addition, they make available a large amount of material for comparative studies of many different properties. On board, we soon learned to take advantage of the fact that even after extensive sampling, much material was available to cut in various planes to study burrow structures.

Simple cutting and scraping of a vertical face showed the "mottles" often referred to in the literature (Figure 2A). Gentle washing with spray from a water hose gradually revealed a rich network of burrows (Figure 2, B and C). Some of the burrows were seen to line up horizontally over some considerable distance, due to intersection of a single meandering horizontal track. A multitude of small vertical burrows randomize stratigraphic information over a 1 to 2 cm interval: there appear to be no grains left undisturbed. Longer burrows (∿ 5 cm) represent more sporadic disturbances, and in some cores, accentuate a remarkable grid pattern of predominantly vertical and horizontal burrows. The sediment-water interface is characterized by low mounds and lumps, a few centimeters in diameter. Some of these are white and appear to be fresh, while others look older and are covered with greyish matter.

The sedimentary structures in the box cores change in characteristic fashion, depending on the water depth (Figure 3). The shallow cores (Figure 2 and Figure 3A) show the typical orthogonal "grid" pattern, while deeper ones show an oblique grain, indicative of strain (Figure 3B, C). The deepest ones have considerable deformation and loss of vertical burrows (Figure 3D, E). This variation in burrow structure presumably is due to downhill creep of the sediment (*Berger and Johnson*, 1976), perhaps initiated by intermittent fluidization of the sediment through earthquakes.

During sampling, we noted that sediments near the top of the plateau are stiff, while those on the lower flanks are soft. This subjective observation is supported by shear strength measurements (Figure 4A), showing the weakening of the sediment as a function of depth. The change in shear strength may be largely a function of grain size, which has similar trends (Figure 4B). In turn, the grain size distribution presumably is a result of dissolution and winnowing processes. Sound velocities in these sediments are also affected by grain size (*Johnson et al.*, in press) and, accordingly are higher in supralysoclinal sediments than in sublysoclinal ones (Figure 4C).

Fig. 2. Box Core ERDC Bx 83 (1°24.1'N, 157°52.1'E, 2343 m). (A) after cutting vertically with an aluminum sheet and cleaning of surface; (B) after gentle washing with a spray of water; (C) after prolonged washing. Differences in texture between coarse burrows and finely burrowed matrix allow for differential removal by spraying. Photos T. Walsh, S.I.O.

Fig. 3. Evidence for deformation of uppermost sediment strata by "ooze flows", as a function of depth. (A) box core face, cut and washed, showing "grid pattern" due to horizontal and vertical burrowing and typical for cores taken above 3,000 m; (B) and (C) strain pattern superimposed on slightly deformed "grid"; (D) and (E) elimination of vertical burrows and deformation of strata by shear and mass flow. Photos W. H. B.

The most important controlling factor on the distributional patterns of these properties (shear strength, grain size, sound velocity) is the depth gradient of carbonate dissolution. This gradient can be estimated from the increase of insoluble residue

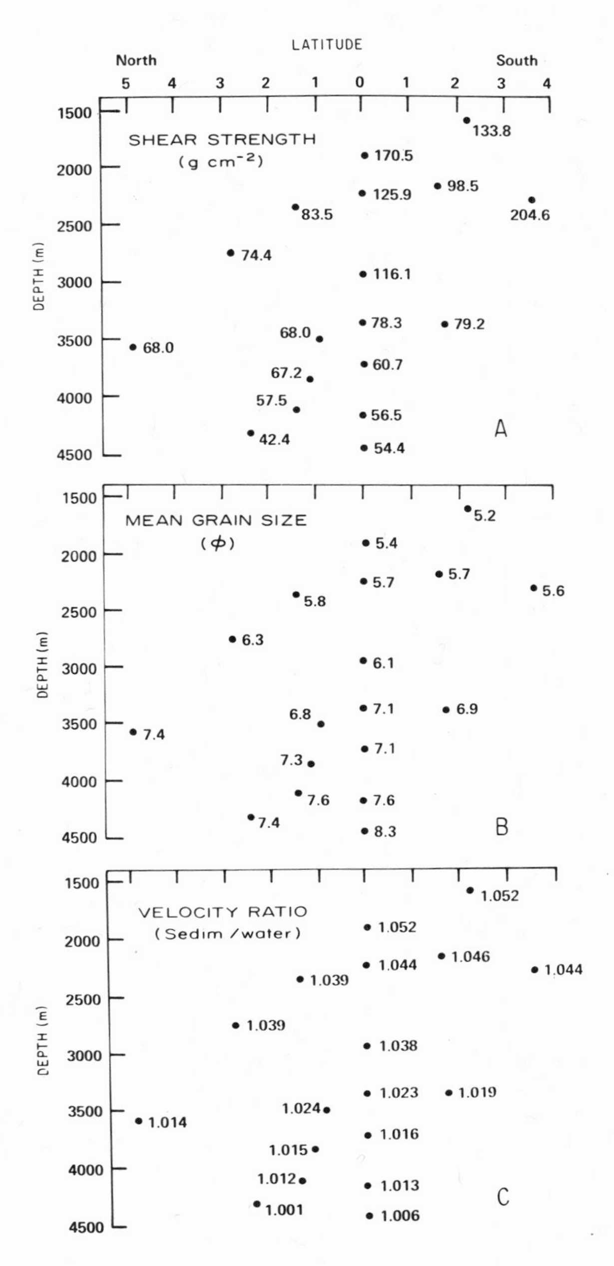

Fig. 4. Physical properties of sediments, as measured in box core samples. (A) shear strength decreases with water depth; (B) grain size decreases with water depth; (C) sediment velocity decreases with water depth.

with depth, assuming that such increase is entirely a function of carbonate dissolution, and also from estimates of Holocene sedimentation rates (*Berger*, 1977). The two methods give comparable results (Figure 5). A direct comparison of sedimentation rate estimates shows a large scatter (Figure 6). The particular comparison shown assumes a carbonate content of 85% at 1.4 cm/ky. Changing this condition will change the plot somewhat, but will not reduce the scatter. It appears that an "initial" carbonate content cannot be specified for a given sample, with the desirable degree of reliability. Without a reliable estimate of "expected" carbonate, the limits on the amount of dissolution that a sample may have experienced according to its remaining carbonate content become very large. This puts large error limits on depth gradients of dissolution deduced from $CaCO_3$ percentages (*Heath and Culberson*, 1970). Processes other than dissolution, such as subtraction or addition of fine material with a different carbonate content, apparently play a considerable role in producing "noise" in the carbonate percentage patterns. Winnowing in conjunction with burrowing, settling of fines in small depressions, and diagenetic processes can provide this noise (cf. *Moore et al.*, 1973).

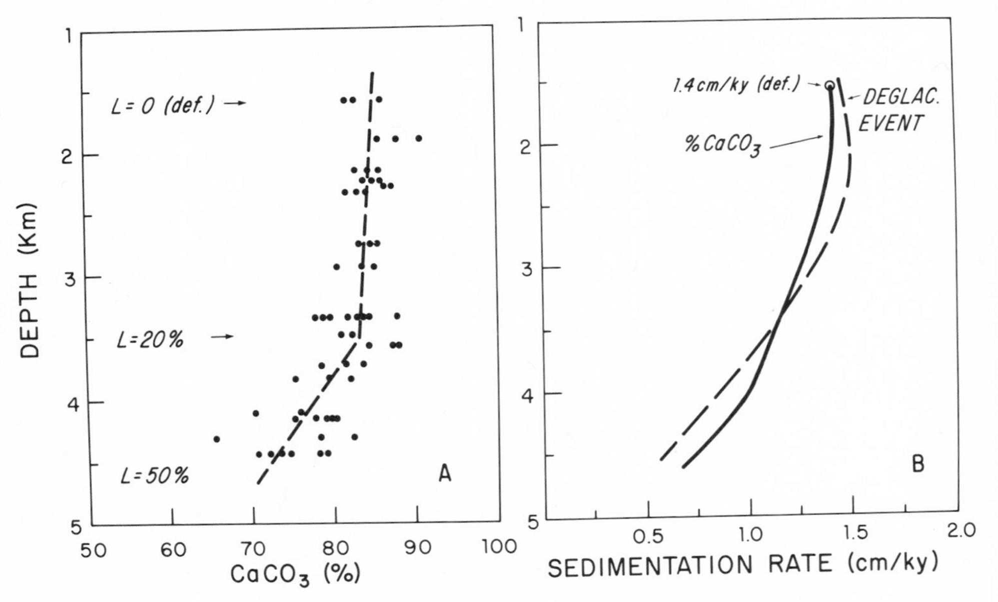

Fig. 5. Depth gradient of $CaCO_3$ content in box core samples (A) and associated calculated sedimentation rate, assuming 1.4 cm/10^3 years corresponds to 85% $CaCO_3$, (B). The dashed line is a generalized sedimentation rate profile, based on faunal transitions in the box cores.

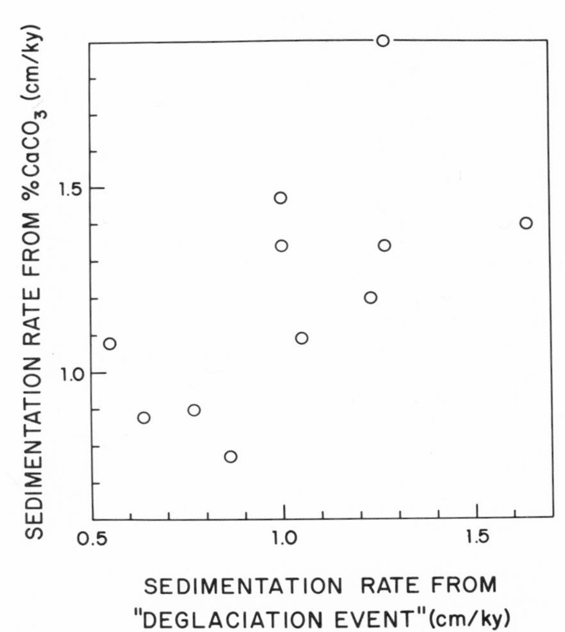

Fig. 6. Comparison of calculated and measured sedimentation rates, based on assumptions stated in text.

Prominent color banding was seen in most of the cores, a dark layer overlying a light one in the lower one-half of most of the box cores. The sediment in the upper half is a rather uniform light grey buff color (2.5y 7/2). Apparently there is some upward mobilization of iron and manganese from below about 30 cm, with reprecipitation producing a dark layer. Conceivably, this is due to a lowering of the Eh within the older sediment. This older sediment was deposited during times of increased fertility (*Berger*, 1977) and is expected to be relatively enriched in organic matter. The phenomenon is being studied. It is somewhat reminiscent of the iron-rich layers in hemipelagic sediments observed elsewhere below surface sediments (*Lynn and Bonatti*, 1965; *McGeary and Damuth*, 1973).

Stable isotope determinations were made on foraminifera from a number of box cores (*Berger and Killingley*, in press; *Berger* et al., in press). Approximately 200 samples were analyzed. The ranges in δO^{18}, between Holocene and Würm, for *G. sacculifer* and *P. obliquiloculata* are both about 1.5°/oo. δO^{18} values for different sub-samples of speciments (about 20 each) from the same sample can vary routinely by as much as 0.1 to 0.3°/oo. The transition from glacial to interglacial is quite noisy, presumably mostly due to burrowing, but perhaps also due to considerable oscillation in the retreat of the glaciers.

The δC^{13} values show a slight trend from light to heavy in going from Würm to Holocene. The change is about 0.3 - 0.4°/oo. The scatter between samples of the same inferred age is of the same magnitude. Occasional samples deviate by as much as 1°/oo from the trend average, to either side. We are investigating the possibility that such deviations cluster and contain messages about cessation or intensification of upwelling on the scale of centuries. The general trend toward heavier δC^{13} values is expected from a decrease in upwelling intensity across the Würm-Holocene transition (*Berger*, 1977). Shackleton's rain forest effect (*Shackleton*, 1977) also may play a role. Caution must be exercised in the interpretation of δC^{13} fluctuation of planktonic foraminifera, since differential dissolution removes the shallower living variants of a species, thus decreasing the C^{13}/C^{12} ratio of the species. An analogous effect exists for δO^{18}, as pointed out previously (*Berger*, 1971; *Savin and Douglas*, 1973; *Shackleton and Opdyke*, 1976).

A number of piston cores and gravity cores also were taken during the Ontong-Java Cruise and will be reported on after further study. *Olausson* (1960), *Valencia* (1972), *Shackleton and Opdyke* (1973) (1976), and *Thompson and Saito* (1974) provide information on Pleistocene patterns of sedimentation. In the present context, it is interesting to note how the box core results compare with these patterns and how they bear on the interpretation of the records from long cores.

The Quaternary sedimentation rates obtained from paleomagnetic dating of piston cores (HIG: *Valencia*, 1972; Lamont: *Shackleton and Opdyke*, 1973) compare favorably with the Holocene rates of our box cores (*Berger*, 1977). However, there is no clear depth trend in the piston core rates, and the scatter is considerable (Figure 7). Erosion and redeposition processes apparently are responsible for much of this scatter: Core 68 PC 15, for one, has a distinct hiatus (*Valencia*, 1972).

Sedimentation rate patterns are rather irregular geographically, hence they are probably equally irregular through time, in any one core. Support for this suggestion may be found in a paper by *Thompson and Saito* (1974), who correlated Cores V28-238 and V28-239 at several levels on the basis of the abundances of *Pulleniatina obliquiloculata* and *Globigerinoides ruber*. Their correlations imply that the ratio between the sedimentation rates of the two cores varied from 1.2 to 1.8, from one cycle to another (∿ 100,000 year interval). The corresponding overall ratio, based on *Shackleton and Opdyke* (1973), is 1.6, with V28-238 having the higher rate of the two (1.71 cm/ky *versus* 1.06 cm/ky).

The inference that sedimentation rates fluctuated *between* cycles (and not just *within*, as a function of glacial-interglacial

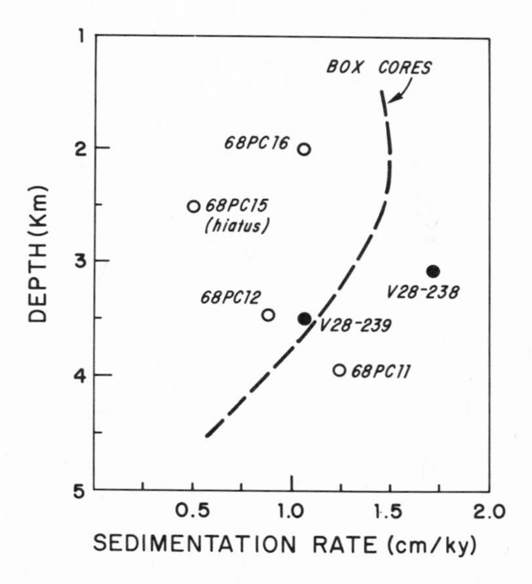

Fig. 7. Pleistocene sedimentation rates based on paleomagnetism (Brunhes-Matuyama boundary) compared with box core sedimentation rates, based on faunal changes at deglaciation time. Open circles: data from *Valencia* (1972), Hawaii Institute of Geophysics; solid circles: data from *Shackleton and Opdyke* (1973), Lamont-Doherty Geological Observatory.

differences in fertility and dissolution) depends on the correctness of the correlations by *Thompson and Saito* (1974). The abundance profiles they use contain fertility (and temperature) signals as well as the dissolution signal that they mention. It is not entirely clear, therefore, what oceanographic events are being correlated. However, all the signal inputs may be nearly synchronous, so the correlations cannot be dismissed on this basis. Although much of the variation in rate ratios between the two cores may be due to sedimentation rate excursions in V28-239, it is highly probable that considerable distortion remains also in V28-238, due to both productivity and dissolution fluctuations. The corresponding error in the Pleistocene age-scale given by *Shackleton and Opdyke* (1973) cannot be assessed without further study. In any case, their scale systematically assigns too young an age to odd-over-even boundaries and too old an age to even-over-odd boundaries, because the warm periods (odd) need to be expanded to correct for decreased fertility and increased dissolution. The correction is thought to be non-trivial (factor of 1.5 or more) on the basis of the box-core results.

CENOZOIC SEDIMENTATION AND MASS-WASTING: THE SEISMIC RECORDS

Seismic profiling records were taken along most of the track shown in Figure 1, using 40-cubic-inch airguns. In agreement with earlier work (*Winterer et al.*, 1971; *Hussong*, 1972; *Kroenke*, 1972; *Maynard*, 1973; *Murauchi et al.*, 1973; *Andrews et al.*, 1975), our records show that the shallow part of the plateau is characterized by great expanses of flat-lying sediment containing about a dozen reflectors, generally almost perfectly parallel to acoustic basement and to each other (Figures 8a, b). Many of these reflectors can be correlated over considerable distances.

The origin of acoustic reflectors in deep-sea carbonate sequences is still rather a mystery. Changes in lithology which are presumed to provide the necessary change in sonic impedance (velocity times bulk density) turn out to be rather subtle in many cases, where checked by drilling. According to *Winterer et al.*, (1971, p. 491), "reflectors are associated with changes in the degree of induration of the calcareous sediments, and the more abrupt and important the change, the more sharp and strong the reflection". They suggest a time-stratigraphic control on induration, contingent upon depositional units of wide extent, to explain the occurrence of widespread sub-parallel reflectors such as those found on Ontong-Java Plateau. In addition, the transitions from ooze to chalk and from chalk to limestone are thought to be important in providing velocity changes, and hence reflectors (ibid., p. 481; see also *Packham and van der Lingen*, 1973).

Changes in velocity can also be caused by changes in grain size without induration. As noted by *Johnson et al.* (in press; see also Figure 4C), coarse-grained foraminiferal sediments on top of the Ontong-Java Plateau have velocities about 5% greater than sediments in deeper water. Any layer of coarser-grained sediments within a fine-grained sediment would be apt to form a reflector; this is the usual situation in turbidites.

Schlanger and Douglas (1974) examined in some detail the mechanics of the transition from ooze to chalk to limestone. They suggested the term "diagenetic potential" to emphasize the dependency of diagenetic reaction rates on the original composition of skeletal carbonate deposits. Original composition can vary widely due to differential supply and differential removal or destruction of the various types of calcareous shells, whether on the shelf (*Chave*, 1964) or in the deep sea (*Schott*, 1935; *McIntyre and McIntyre*, 1971; *Sliter et al.*, 1975). The presence of skeletal particles of aragonite or high Mg-calcite due to excellent conditions for preservation, appears to favor early diagenesis and cementation (*Friedman*, 1964; *Gevirtz and Friedman*, 1966; *Milliman*, 1966; *Fischer and Garrison*, 1967). In deep-sea oozes, the more easily dissolved coccoliths and foraminifera with open structures and relatively high Mg content (*Savin and Douglas*, 1973) can play a similar role of providing potential cementing material. Alternating conditions of good and poor preservation, therefore, could produce alternations of unlithified and lithified calcareous sediment (e.g., *Bartlett and Greggs*, 1970; *Wise and Kelts*, 1972; *Berger and von Rad*, 1972; *Milliman and Müller*, 1973).

It may be reasonably argued that if the initial preservation states of carbonate sediments are determined by the oceanographic conditions at the time of deposition, then the subsequent cementation patterns likewise reflect these conditions and the associated acoustic properties are, therefore, synchronous over the areas over which these conditions prevailed. In support of this hypothesis, *Schlanger and Douglas* (1974) presented evidence for long-distance correlation of certain strong reflectors, whose ages they identified as 3, 5 to 6, 13 to 14, 21 to 26, and about 40 to 44 million years (ibid, p. 135-139).

In the present context, an extensive discussion of the question of long-distance correlation of carbonate reflectors would lead too far afield. It may be noted, however, that one way to produce an oceanographic event of the type envisaged by *Schlanger and Douglas* (1974) would be to drastically change the CO_2 content of the ocean-atmosphere system. The current industrial CO_2-pulse, in this view, could eventually show as an acoustic reflector in the carbonate monitors of the deep ocean.

The inverse problem of which reflectors might be due to CO_2 pulses and which to other causes has not been attempted (or even formulated). Preservation or dissolution pulses or alternations, such as described earlier, while ocean-wide or even global in nature, act in different ways on the surficial carbonate on the sea floor, depending on geographic location and depth. Thus, the "diagenetic potential" resulting from such an event or events also will depend on circumstances, and so will the resulting "echo potential", that is, the potential reaction to insonification. In support of this cautious approach, it should be pointed out that there are subtle changes in strength and spacing of reflectors as well as outright disappearances and appearances which accumulate sufficiently on the scale of 200 km or so to make it hazardous to assert that individual reflectors correlate over distances greater than this.

We prefer to view a sequence of carbonate reflectors as a manifestation of a series of oceanographic events such as dissolution or winnowing cycles, or productivity cycles, or combinations of these. These events precondition the sediment depending on local circumstances, especially depth, which is itself a function of subsidence through time and of upbuilding of the sediment stack. Subsequently the signal is amplified and later destroyed by diagenesis. It appears likely, in view of the limited extent of individual reflectors here proposed, that long-distance correlations in carbonates are due to zones of dissolution fluctuations, rather than single events, as well as to the drastic changes in sedimentation regime (silica deposition, erosional surfaces) which have been recognized for some time as causative agents.

In our records from Ontong-Java Plateau, we find that there are two striking differences between the shallow profiles above the lysocline (Figure 8a, b) and the profiles from greater depth (Figure 8c, d): (1) a loss of distinct parallel reflectors, and (2) a decrease in sediment thickness, as also noted by several previous investigators cited in the Introduction (e.g., *Ewing et al.*, 1969, p. 2484-2486).

The difference in the nature of reflectors is in part primary, that is, caused by initial sediment conditioning and subsequent diagenesis. The range for possible conditioning, it may be surmised, decreased with depth because of the degrease of the diversity of particles whose partial removal constitutes conditioning. However, to a large degree, the deterioration of the reflectors would seem to be due to internal disturbance, such as sliding and flow.

The decrease in thickness, on the whole, may be ascribed to increasing dissolution at depth (thicknesses are derived from sonobuoy refraction experiments; report in preparation). On the average, there is a decrease of roughly 100 m of sediment thickness for each 200 m increase in depth of water between 2.5 and 4.5 km depth (Figure 9). Assuming 90% initial carbonate content, and assuming that the entire decrease is due to dissolution, an additional 10% of initial carbonate would have to be lost for each 200 m depth increase, until all carbonate would be thus removed at 4,500 m, with 100 m of residual sediment remaining.

As shown earlier (Figure 5), the actual decrease of carbonate with depth for present-day sediments is much less drastic than suggested by this simple calculation. The present pattern is much more favorable to carbonate deposition in this area than were the last 100 million years, on the whole. Judging from uncorrected sedimentation rate distributions in Site 289 (*Andrews et al.*, 1975), the times unfavorable for carbonate deposition are mainly pre-Late Oligocene, and especially pre-Eocene. In fact, more than 80% of the stack accumulated in less than the last 50% of the time represented. It is apparent that the present carbonate sedimentation pattern, if allowed to prevail for the last 40 to 50 million years (without redistribution of sediment), would produce a different cover from the one seen, that is, one in which thicknesses stay virtually constant down to 3 km depth, and drop off suddenly below 3.5 km depth.

The relationship between depth and sediment thickness (Figure 9) shows several other noteworthy features besides the trend described. Above the lysocline, slumping provides for points far to the left of the "normal" trend. Also, on the uppermost plateau, thicknesses appear to decrease. Several possible causes can be envisaged: (1) a latitudinal gradient which introduces asymmetry because the shallowest portions are off the equator by 2½ degrees (although this is only true for the present), (2) a difference in diagenetic reactions which favors the development of false basement in the shallowest parts of the plateau, for example by enhanced limestone

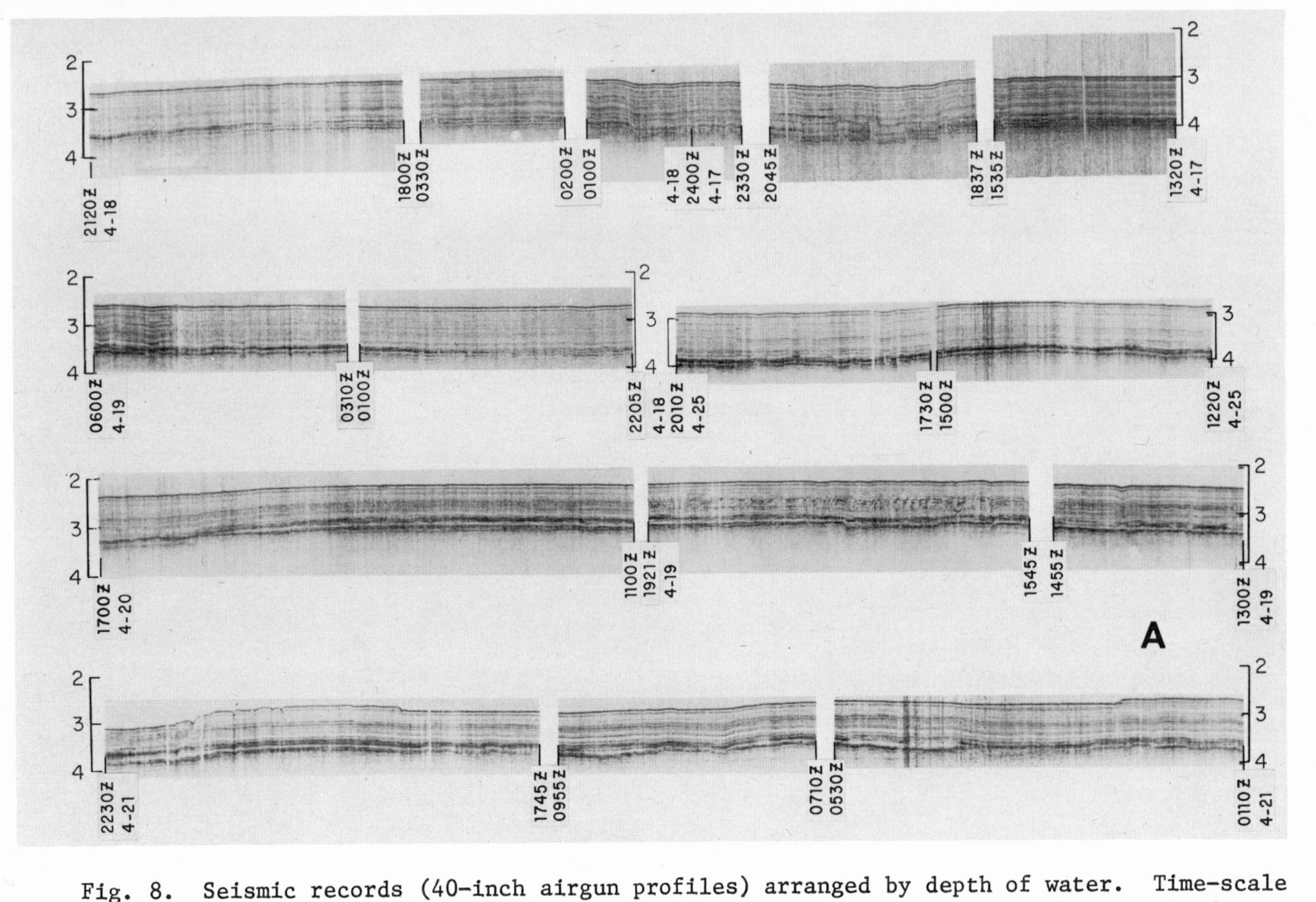

Fig. 8. Seismic records (40-inch airgun profiles) arranged by depth of water. Time-scale in two way travel time. (A). Shallow plateau top, subparallel multiple reflectors. Evidence for deformation (Tension?) at 4-19, 1545Z-1900Z.

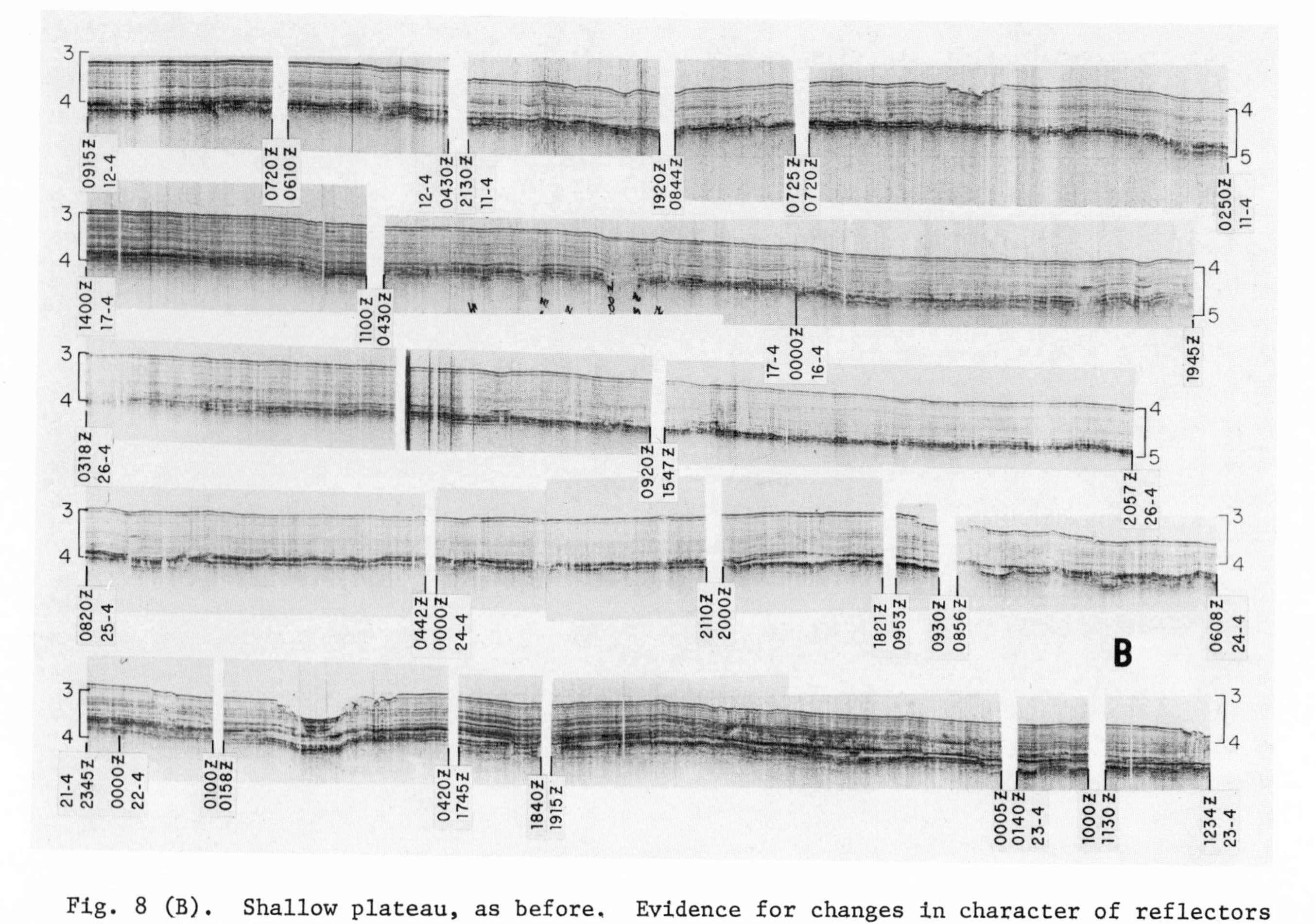

Fig. 8 (B). Shallow plateau, as before. Evidence for changes in character of reflectors downslope, incipient slumping.

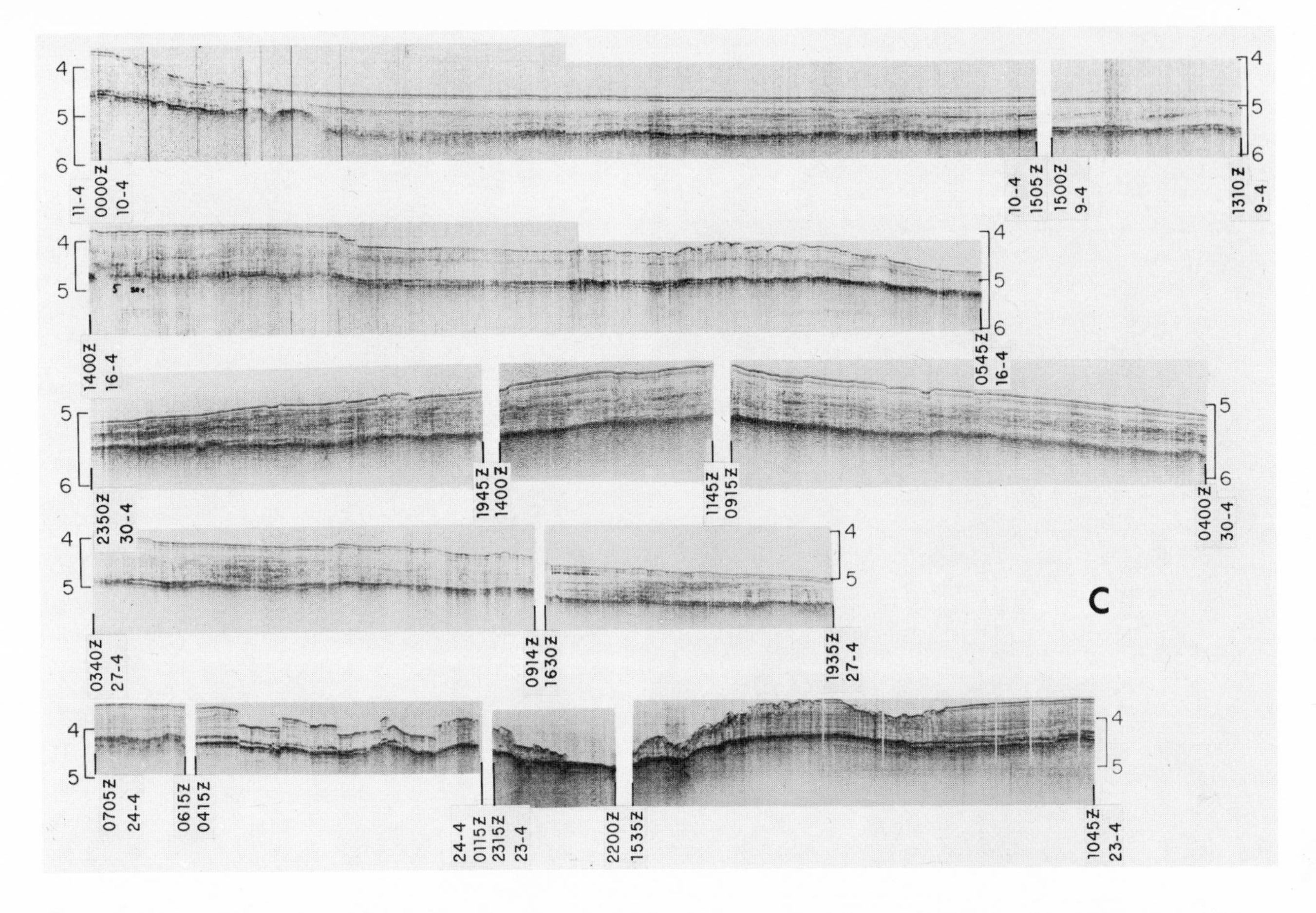

Fig. 8 (C). Plateau rim, reflectors change character, abundant evidence for slumping, flow.

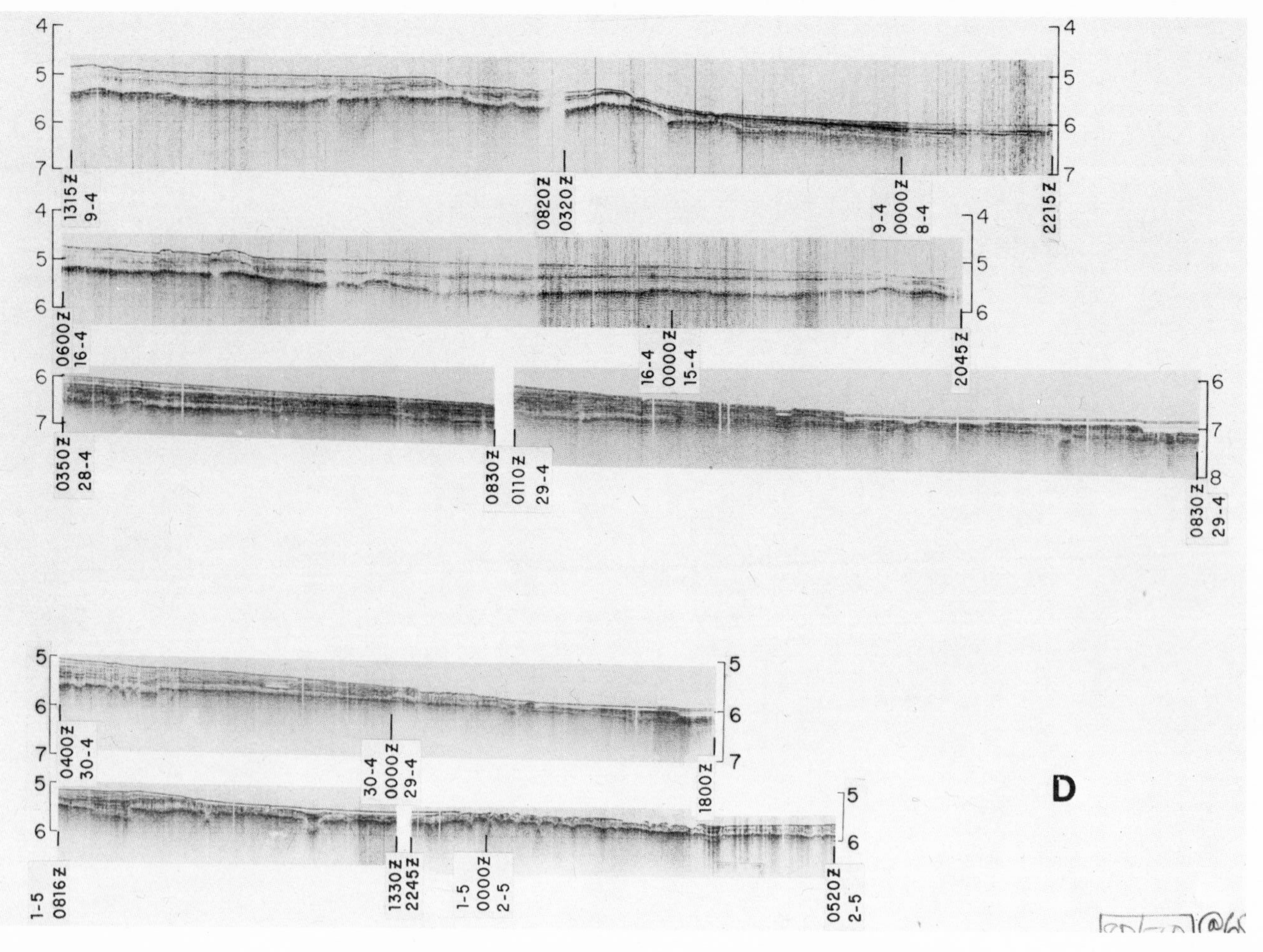

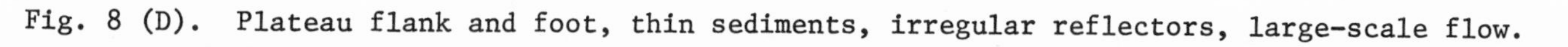

Fig. 8 (D). Plateau flank and foot, thin sediments, irregular reflectors, large-scale flow.

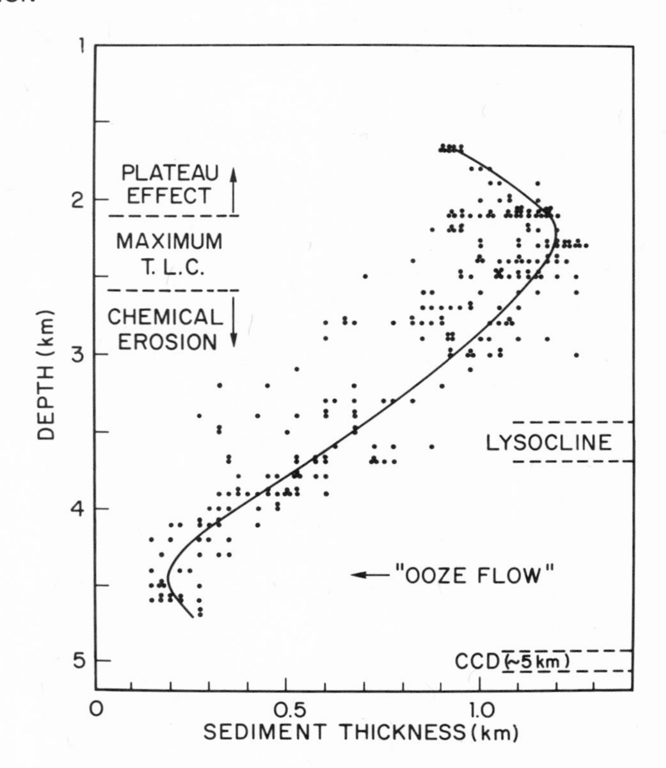

Fig. 9. Thickness distribution of sediments as a function of depth, northern Ontong-Java Plateau. Plateau Effect: thinning of sediments above 2 km water depth, reason unknown (winnowing?); Maximum T. L. C.: maximum thickness of transparent carbonate; Chemical Erosion: arrow marks average position of start of thinning, presumably due to dissolution of carbonate, through Tertiary time; Lysocline: present position of marked change in sediment character, due to dissolution; "ooze flow" addition of material, opposing the dissolution effect, by earthflow type movement, probably due to fluidization of ooze during earthquakes.

and chert formation in the initially less dissolved and hence more reactive sediments at shallow depths, and (3) differential winnowing of material, especially removal of fines at the shallowest depths. The relative importance of these factors is as yet undetermined.

Another deviation from the overall depth gradient of sediment thickness is the failure of the profiles to continue thinning as the CCD is being approached. The effect is greater than that expected from the fact that much of the soluble part of the sediment has been removed. Also at 4,500 m, our cores still show near 70% carbonate. We suggest that piling-up of flowing sediment helps produce this phenomenon. This interpretation is supported by the sediment structures and the physical properties of box core sediments, as presented above (Figures 3, 4).

Evidence for mass wasting on an even larger scale is abundant in the seismic records themselves, as pointed out elsewhere (*Berger and Johnson*, 1976). Briefly, sliding and slumping of carbonate into the deep N-S valley intersecting the plateau imparts a rugged topography to the area of mass waste (Figure 8c, bottom). Sediments are stiff and support steep cliffs, even in the upper part of the stack where cementation is still far from complete (Figure 10). The apparent softness of uppermost oozes reported from drilling results may largely result from disturbance during coring. In contrast, sublysoclinal sediments are seen to smooth the underlying rougher topography and give the impression of flowage (Figure 8d). Such flowage also would result in acoustically transparent sediments, because of homogenization of the material.

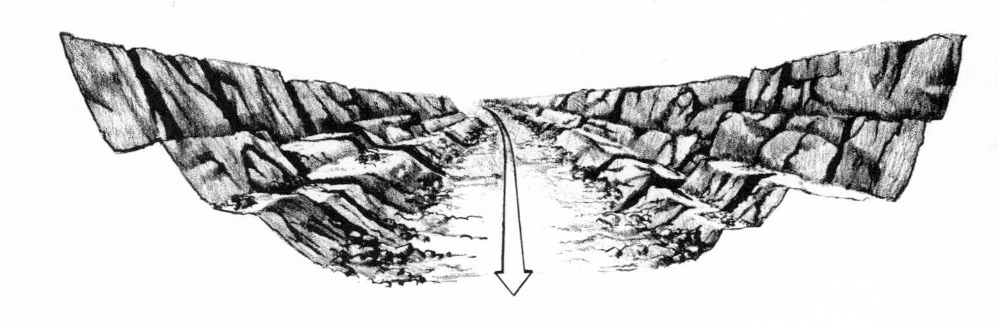

Fig. 10. Artist's concept of nature of upper part of the major north-south valley intersecting Ontong Java Plateau. Slumping calcareous sediment moves down on Eocene chert as on a glide. Drawn by Judy Lackmund Clinton according to instructions by W. H. B.

It appears that mass-wasting occurs on two scales on Ontong-Java Plateau: (1) ooze flow and creep near the surface, and (2) large scale slumping and sliding. Both phenomena owe much to dissolution processes, the first showing a depth dependence of preserved burrow patterns and of shear strength, the second showing a depth dependence of topographic morphology. Debris flows and turbidity currents may be generated from slumping and sliding (see *Hampton*, 1972; *Klein*, 1975a, b).

Both processes are geologically young and may depend on earthquake activity for a triggering mechanism. The collision of the Ontong-Java Plateau on the Pacific Plate with the Australian Plate, which presumably resulted in uplift of deep-sea carbonates in the Solomon Islands (*Kroenke*, 1972) could have produced the necessary earthquake activity. Intensification of bottom water circulation during the late Cenozoic (*Watkins and Kennett*, 1971) and associated changes in depth gradients of carbonate dissolution (*Berger*, 1972; *van Andel et al.*, 1975) could have provided for removal of support on the flanks of the plateau. Presumably, ooze flow also would be affected by such variations in dissolution gradients, in as yet unknown ways.

Concerning carbonate chemistry, any large-scale flow of calcareous ooze at depths approaching the CCD would result in a depression of the CCD. Thus, the CCD would no longer directly relate to observed or calculated parameters of water chemistry, such as a calcite saturation level (*Broecker and Broecker*, 1974), or certain levels of undersaturation (*Takahashi*, 1975), or to physical levels such as the top of the Antarctic Bottom water (*Berger*, 1968; *Edmond*, 1974). Also, such flow would constitute additional input of alkalinity to deep waters, beyond that which is derived from dissolution of particles falling to or resting on the abyssal sea floor. Mass wasting, and especially surficial sediment flow processes, would also affect the apparent diffusion constants for flux of matter and heat through the sediment water interface.

In order to assess the importance of such mass wasting processes in the context of carbonate chemistry, we need to find out (1) to what extent are these movements episodic over large areas of the sea floor, in response to bottom water action and to what extent are they regional and result from tectonic events, and (2) what are the conditions (slope, type of ooze, depth range) under which carbonate transfer by mass wasting occurs and what are the applicable rates of such transfer to abyssal depths.

If increased CO_2-levels in the atmosphere-ocean system prove to be highly undesirable, it might be worthwhile to contemplate artificially setting off large-scale carbonate slides, which would decrease free CO_2 in deep waters and hence the delivery of CO_2 to the atmosphere by these waters when they reach the surface. Any benefits of such an action would accrue to future generations only.

ACKNOWLEDGEMENTS

This study was supported primarily by the Office of Naval Research, Contract USN-N-00014-69-A-0200-6049 (W. H. Berger and E. L. Winterer) and by the National Science Foundation (Oceanography) DES 75-04335 (W. H. Berger, post-doctoral support for T. C. Johnson). Supplementary support was obtained from NSF Grant OCE 76-02255 (T. C. Johnson) and from the Naval Sea Systems Command (Code 06H14), the Naval Electronic Systems Command (Code 320), and the Office of Naval Research (Code 480) (E. L. Hamilton).

REFERENCES

Andrews, J. E., G. Packham, J. V. Eade, B. K. Holdsworth, D. L. Jones, G. deV. Klein, L. W. Kroenke, T. Saito, S. Shafik, D. B. Stoeser, G. J. v.d. Lingen. 1975. Initial Repts. *Deep Sea Drilling Project, Vol. XXX,* U. S. Government Printing Office, Washington, D. C. 1-753.

Bartlett, G. A. and R. G. Greggs. 1970. The Mid-Atlantic Ridge near 45°00' north. VIII. Carbonate lithification on oceanic ridges and seamounts. *Can. J. Earth Sci., 7, (2-1).* 257-267.

Berger, W. H. 1968. Planktonic Foraminifera: selective solution and paleoclimatic interpretation. *Deep-Sea Res., 15, (1).* 31-43.

Berger, W. H. 1971. Sedimentation of planktonic Foraminifera. *Marine Geol., 11.* 325-358.

Berger, W. H. 1972. Deep-sea carbonates: dissolution facies and age-depth constancy. *Nature 236.* 392-395.

Berger, W. H. 1977. Carbon dioxide excursions and the deep-sea record: aspects of the problem. In: *Fate of Fossil Fuel CO_2,* N. R. Andersen and A. Malahoff (Eds.), Plenum, N.Y., (this volume)

Berger, W. H. and T. C. Johnson. 1976. Deep-sea carbonates: dissolution and mass wasting on Ontong-Java Plateau. *Science, 192.* 785-787.

Berger, W. H. and J. Killingley. (In press). Glacial-Holocene transition in deep-sea carbonates: selective dissolution in the stable isotope signal. *Science.*

Berger, W. H. and U. von Rad. 1972. Cretaceous and Cenozoic sediments from the Atlantic Ocean. Initial Repts. *Deep Sea Drilling Project, Vol. XIV,* U. S. Government Printing Office, Washington, D. C. 787-954.

Berger, W. H., R. F. Johnson, and J. S. Killingley. (In prep.). "Unmixing" of the deep-sea record: an experiment in the glacial-Holocene transition.

Broecker, W. S. and S. Broecker. 1974. Carbonate dissolution on the western flank of the East Pacific Rise. In: *Studies in Paleo-oceanography,* W. W. Hay (Ed.), *Soc. Econ. Paleont. Mineral. Spec. Pub. 20.* 44-57.

Chave, K. E. 1964. Skeletal durability and preservation. In: *Approaches to Paleoecology*, J. Imbrie and N. Newell (Eds.), John Wiley, N. Y. 377-387.

Edmond, J. M. 1974. On the dissolution of carbonate and silicate in the deep ocean. *Deep-Sea Res.*, *21*. 455-480.

Ewing, M., R. Houtz, and J. Ewing. 1969. South Pacific sediment distribution. *J. Geophys. Res.*, *74*. 2477-2493.

Fischer, A. G. and R. E. Garrison. 1967. Carbonate lithification on the sea floor. *J. Geol.*, *75*. 488-496.

Fischer, A. G., B. C. Heezen, R. E. Boyce, D. Bukry, R. G. Douglas, R. E. Garrison, S. A. Kling, V. Krasheninnikov, A. P. Lisitzin, and A. C. Pimm. 1971. Initial Repts. *Deep Sea Drilling Project, Vol. VI*, U. S. Government Printing Office, Washington, D. C. 1-1329.

Friedman, G. M. 1964. Early diagenesis and lithification in carbonate sediments. *J. Sediment. Petrol.*, *34*. 777-813.

Gevirtz, J. L. and G. M. Friedman. 1966. Deep-sea carbonate sediments of the Red Sea and implications on marine lithification. *J. Sedim. Petrol.*, *36*, *(1)*. 143-151.

Hampton, M. A. 1972. The role of subaqueous debris flow in generating turbidity currents. *J. Sed. Pet.*, *42*, *(4)*. 775-793.

Heath, G. R., and C. Culberson. 1970. Calcite: Degree of saturation, rate of dissolution, and the compensation depth in the deep oceans. *Geol. Soc. Amer. Bull.*, *81*. 3157-3160.

Hussong, D. M. 1972. Detailed structural interpretations of the Pacific Oceanic crust using ASPER and ocean-bottom seismometer methods. Ph.D. Dissertation, University of Hawaii. 165 pp.

Johnson, T. C., E. L. Hamilton, and W. H. Berger. (In press). Physical properties of calcareous ooze on the Ontong-Java Plateau, western equatorial Pacific. *Marine Geology*.

Klein, G. deV. 1975a. Resedimented pelagic carbonate and volcaniclastic sediments and sedimentary structures in Leg 30 DSDP cores from the western equatorial Pacific. *Geology*, *3*, *(1)*. 39-42.

Klein, G. deV. 1975b. Sedimentary tectonics in Southwest Pacific marginal basins based on Leg 30 Deep Sea Drilling Project cores from the South Fiji, Hebrides, and Coral Sea Basins. *Bull. Geol. Soc. Amer.*, *86*. 1012-1018.

Kroenke, L. W. 1972. Geology of the Ontong-Java Plateau. Hawaii Institute of Geophysics, HIG-72-5. 1-119.

Larson, R. L., R. Moberly, D. Bukry, H. P. Foreman, J. V. Gardner, J. B. Keene, Y. Lancelot, H. Luterbacher, M. C. Marshall, A. Matter. 1975. Initial Repts. *Deep Sea Drilling Project, Vol. XXXII*, U. S. Government Printing Office, Washington, D. C. 1-980.

Lynn, D. C. and E. Bonatti. 1965. Mobility of manganese in the diagenesis of deep-sea sediments. *Marine Geol.*, *3*. 457-474.

Maynard, G. L. 1973. Seismic wide-angle reflection and refraction investigation of the sediments on the Ontong-Java Plateau. Ph.D. dissertation, University of Hawaii. 156 pp.

McGeary, D. F. R. and J. E. Damuth. 1973. Postglacial iron-rich crusts in hemipelagic deep-sea sediment. *Geol. Soc. Amer. Bull.*, *84*. 1201-1212.

McIntyre, A. and R. McIntyre. 1971. Coccolith concentrations and differential solution in oceanic sediments. In: *The Micropalaeontology of Oceans*, B. M. Funnell and W. R. Riedel (Eds.), Cambridge Univ. Press, Cambridge, U. K. 253-261.

Milliman, J. D. 1966. Submarine lithification of carbonate sediments. *Science, 153*. 994-997.

Milliman, J. D. and J. Müller. 1973. Precipitation and lithification of magnesian calcite in the deep-sea sediments of the eastern Mediterranean Sea. *Sedimentology, 20*. 29-46.

Moore, T. C., G. R. Heath, and O. Kowsmann. 1973. Biogenic sediments of the Panama Basin. *Jour. Geol., 81, (4)*. 458-472.

Murauchi, S., W. J. Ludwig, N. Den, H. Hotla, T. Asanuma, T. Yoshii, A. Kubotera, and K. Hagiwara. 1973. Seismic refraction measurements on the Ontong-Java Plateau northeast of New Ireland. *J. Geophys. Res., 78*. 8653-8663.

Olausson, E. 1960. Sediment cores from the West Pacific. Repts. *Swed. Deep-Sea Exped. 1947-1948, 6, (5)*. 161-214.

Packham, G. H. and G. J. van der Lingen. 1973. Progressive carbonate diagenesis at Deep Sea Drilling Sites 206, 207, 208, and 210 in the Southwest Pacific and its relationship to sediment properties and seismic reflectors. Initial Repts. *Deep Sea Drilling Project, Vol. XXI*, U. S. Government Printing Office, Washington, D. C. 495-507.

Resig, J. M., V. Buyannanonth, and K. J. Roy. 1976. Foraminiferal stratigraphy and depositional history in the area of the Ontong-Java Plateau. *Deep-Sea Res., 23*. 441-456.

Savin, S. M., and R. G. Douglas. 1973. Stable isotope and magnesium geochemistry of recent planktonic foraminifera from the South Pacific. *Geol. Soc. Amer. Bull., 84*. 2327-2342.

Schlanger, S. O., and R. G. Douglas. 1974. The pelagic ooze-chalk-limestone transition and its implication for marine stratigraphy. *Spec. Publs. Int. Ass. Sediment., 1*. 117-148.

Schott, W. 1935. Die Foraminiferen in dem äquatorialen Teil des Atl. Ozeans. *Wiss. Ergebn. d. deutsch. Atl. Exped. "Meteor" 1925-1927. Bd. III (3), Sect. B*. 43-134.

Shackleton, N. J. and N. D. Opdyke. 1973. Oxygen isotope and paleomagnetic stratigraphy of equatorial Pacific Core V28-238: Oxygen isotope temperatures and ice volumes on a 10^5 year and 10^6 year scale. *Quaternary Research, 3*. 39-55.

Shackleton, N. J. and N. D. Opdyke. 1976. Oxygen-isotope and paleomagnetic stratigraphy of Pacific Core V28-239 late Pliocene to latest Pliocene. *Geol. Soc. Amer.. Memoir 145*. 449-464.

Shackleton, N. J. 1977. Carbon-13 in Uvigerina: Tropical rainforest history and the equatorial Pacific carbonate dissolution cycles. In: *The Fate of Fossil Fuel* CO_2, N. R. Andersen and A. Malahoff (Eds.), Plenum, N. Y., (this volume).

Sliter, W. V., A. W. H. Bé, and W. H. Berger (Eds.). 1975. Dissolution of deep-sea carbonates. *Cushman Found. Foram. Res., Spec. Publ. 13.* 159 pp.

Takahashi, T. 1975. Carbonate chemistry of seawater and the calcium carbonate compensation depth in the oceans. *Cushman Found. Foram. Res. Spec. Publ. 13.* 11-26.

Thompson, P. R. and T. Saito. 1974. Pacific Pleistocene Sediments: Planktonic Foraminifera dissolution cycles and geochronology. *Geology, 2, (7).* 333-335.

Valencia, M. J. 1972. Tertiary and Quaternary sediments of the Ontong-Java Plateau area. Hawaii Institute of Geophysics, Data Rep. No. 21. HIG-72-17. 52 pp.

Valencia, M. J. 1973. Calcium carbonate and gross-size analysis of surface sediments, western equatorial Pacific. *Pacific Science, 27, (3).* 290-303.

van Andel, Tj. H., G. R. Heath, T. C. Moore. 1974. Cenozoic history and paleooceanography of the central equatorial Pacific Ocean. *Geol. Soc. Amer. Memoir 143.* 1-134.

Watkins, N. D. and J. P. Kennett. 1971. Antarctic bottom water: major change in velocity during the late Cenozoic between Australia and Antarctica. *Science, 173.* 813-817.

Winterer, E. L., W. R. Riedel, R. M. Moberly, J. M. Resig, L. W. Kroenke, E. L. Gealy, G. R. Heath, P. Brönnimann, E. Martini, T. R. Worsley. 1971. Site 64, Initial Repts. *Deep Sea Drilling Project, Vol. VII (1)*, U. S. Govt. Printing Office, Washington, D. C. 473-606.

Winterer, E. L., J. I. Ewing, R. G. Douglas, R. D. Jarrard, Y. Lancelot, R. M. Moberly, T. C. Moore, P. H. Roth, S. O. Schlanger. 1973. Intial Repts. *Deep Sea Drilling Project, Vol. XVII*, U. S. Govt. Printing Office, Washington, D. C. 930 pp.

Winterer, E. L., P. F. Lonsdale, J. L. Matthews, and B. R. Rosendahl. 1974. Structure and acoustic stratigraphy of the Manihiki Plateau. *Deep-Sea Res., 21.* 793-814.

Winterer, E. L. 1977. Carbonate sediments on the flanks of seamount chains. In: *The Fate of Fossil Fuel CO_2*, N. R. Andersen and A. Malahoff (Eds.), Plenum, N.Y., (this volume).

Wise, S. W. and K. R. Kelts. 1972. Inferred diagenetic history of a weakly silicified deep-sea chalk. *Gulf Coast Assoc. Geol. Soc. Trans. 22.* 177-203.

Modulation of Marine Sedimentation By the Continental Shelves

William W. Hay and John R. Southam

University of Miami

ABSTRACT

Estimates of the magnitudes of fluctuations in the mechanical and chemical loads of rivers through geological time are obtained by considering the relation between river loads and area, elevation, hypography and climate of the present continental configuration. On long time scale it is shown that the continental shelves bypass material to deeper oceanic accumulation sites. However, for periods of rapidly rising sea level following glaciations, the sedimentation regime operates in such a way that continental margins consume a flux of $CaCO_3$ equal to or several times larger than the input from rivers. The initial effect on the ocean's sedimentation system due to the introduction of CO_2 by the burning of fossil fuels will be the dissolution of aragonite from shallow water tropical shelves.

INTRODUCTION

Many investigators have followed *Broecker* (1971, 1974), considering oceanic pelagic sedimentation to be a kinetically dominated system consisting of two reservoirs, a surface ocean and a deep ocean. The interaction between these two reservoirs takes place through vertical mixing resulting from the downwelling of surface water and upwelling of deep water, and through the sinking of particulate matter produced by organisms in the surface ocean.

The input flux to this system is the river flow which constitutes run-off from land entering the surface ocean. The output is the particulate matter which is fixed by biological processes in the

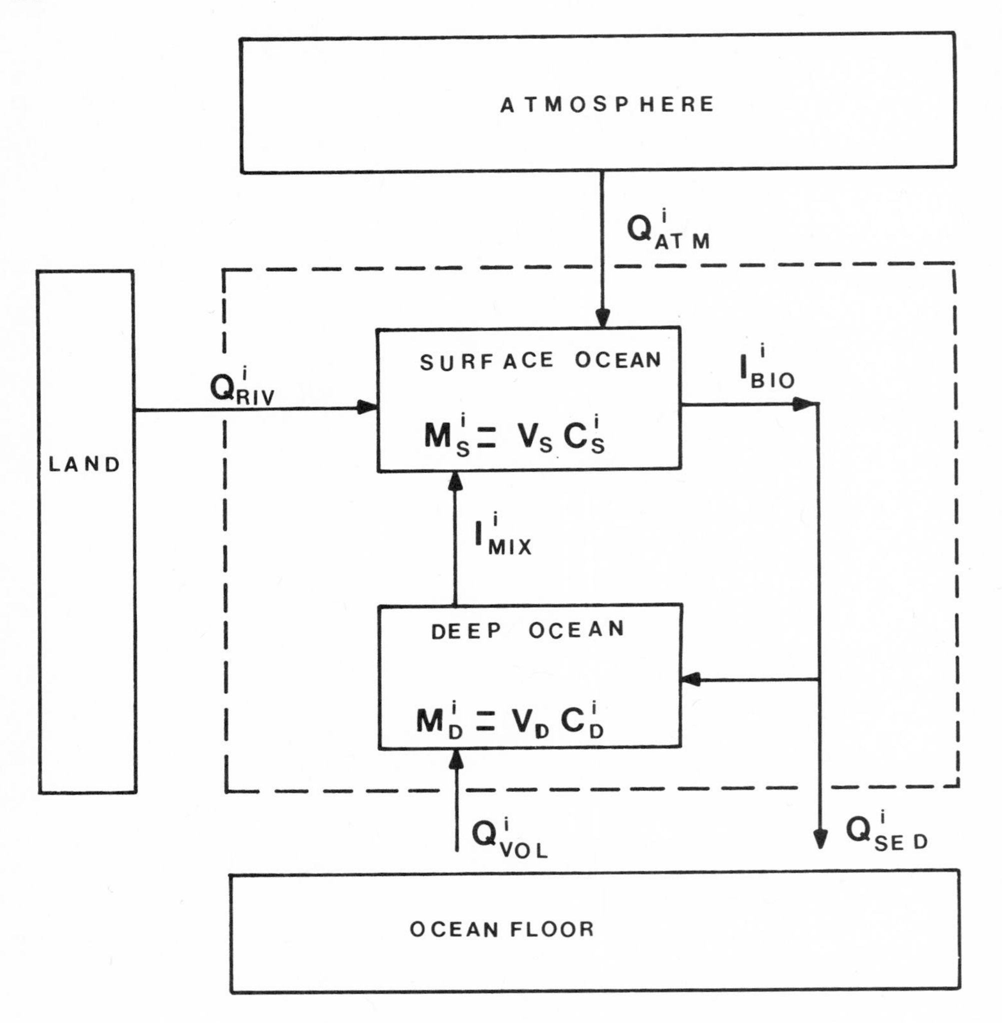

Fig. 1. A two-box model of the biogenous pelagic sedimentation system (after *Southam and Hay*, 1975). V_S and V_D are the volumes of the surface and deep ocean respectively. $C_S{}^i$ and $C_D{}^i$ are the concentrations of chemical species i in the surface and deep oceans. Q^i_{RIV}, Q^i_{ATM}, Q^i_{SED}, and Q^i_{VOL} are fluxes describing the exchange of material into and out of the system as a result of river runoff, atmospheric exchange, loss of particulates to the sediment reservoir and exchange occurring as a result of the interaction between sea water and basalts, respectively. I^i_{MIX} and I^i_{BIO} are internal fluxes representing transport of material within the system as a consequence of vertical mixing and biological processes respectively.

surface ocean and settles out to become sediment deposited on the floor of the deep ocean. A simple representation of this system is presented in Figure 1.

The dynamic equations governing the behavior of the system are obtained by balancing all contributions to the rate of increase in each chemical species in both the surface and deep oceans. The result is two coupled equations for each chemical species under consideration. When coupling exists between chemical species it is necessary to solve the sets of equations for these species simultaneously (for formulation, solution, and discussion of Broecker's model in dynamic form, see *Southam and Hay*, 1975, 1976).

Figure 2 shows a more complex model of the world ocean which accounts for all the major processes contributing to the biogenous component to pelagic sedimentation. The world ocean is represented as a system of boxes which exchanges material with the exterior world, i.e. the land, atmosphere and sea floor. The ocean system consists of two reservoirs - a surface ocean and a deep ocean - separated by the main thermocline. Each of these reservoirs is further divided into boxes representing the forms in which each chemical species may exist, namely, dissolved, non-living particulate, or incorporated in living organisms, and fluxes between boxes are indicated by arrows.

Although these models attempt to describe realistically the internal processes of the ocean system, they are naive in terms of the inputs to the system. Most authors have considered only the dissolved load of rivers as the input to the system, although some have also included atmospheric exchange. In this paper we will consider first the inputs from land into the general marine system, then the shelf seas as a special part of the general system, and finally the dynamics of fluxes across the shelf sea-ocean interface.

INPUTS FROM LAND TO SEA

Garrels and Mackenzie (1971, pp. 96-131) have presented an excellent review and discussion of the sediment input, including river, groundwater, glacial, and atmospheric dust contributions to the general marine system. They estimated the present annual fluxes to be, in terms of importance: 1) mechanical load of rivers, 183×10^{14} g; 2) dissolved load of rivers, 39×10^{14} g; 3) mechanical load of glaciers, 20×10^{14} g; 4) dissolved load of groundwater, 4×10^{14} g; 5) erosion on coasts, 2.5×10^{14} g of solids; and 6) atmospheric dust load, $.6 \times 10^{14}$ g.

In their discussion of the rates of input and the controlling factors they pointed out that the inputs are dependent on the area of land, precipitation (climate), slope, original material, and type of weathering or soil development. They then concluded that for the most part these factors tend to average out for all continents so that the two most important fluxes, the mechanical and dissolved

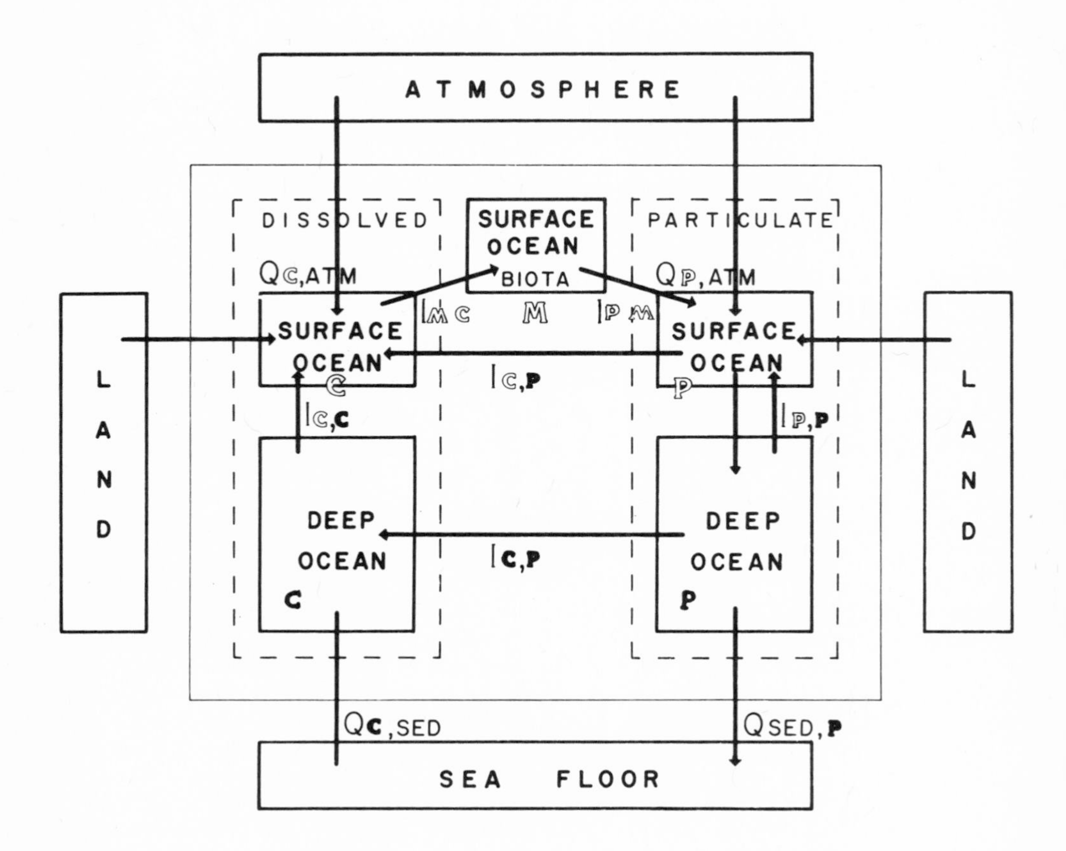

Fig. 2. A five-box model of the biogenous pelagic sedimentation system C, M, P are concentrations of chemical species associated with the three forms: dissolved, incorporated in living organisms, particulate. The Q's are fluxes describing the exchange of material with reservoirs external to the ocean system. The I's are internal fluxes describing the transport of material interior to the system due to the physical and biogeochemical processes, vertical mixing, biological productivity, dissolution, oxidation, settling of particulates.

load of rivers, appear to obey simple rules: 1) the dissolved load of rivers is proportional to the area of the continent and independent of elevation of the continent, and 2) the mechanical load of rivers is proportional to the area of the continent but also increases exponentially with the average elevation of the continent. They noted, too, that average distribution of elements in the dissolved load is very similar for all continents and would appear to be independent of climate weathering, and the relatively slight variations in kinds of original material available for weathering.

In examining other data, it becomes evident that there are additional general relationships which might also be considered. First, there is a distinct relationship between continental elevation and size; second, there is also a relationship between continental size and the area of the continent with interior drainage; third, it appears from studies of individual drainage basins that mechanical erosion is related to local relief rather than average elevation.

Examining the continent-area-average elevation relationship first, consider the data presented in Figure 3. Plots of the land area, land plus shelf areas (continental platform) and land plus shelf plus slope areas (continental block) all show a close exponential relation between area and average elevation for the six non-glaciated continents. The differences in elevation reflect different total thickness of the continental blocks, which, using a density of 2.8 for continental material and 3.3 for displaced mantle material should be an average thickness of 30 km for Asia and 26.7 km for Europe and Australia. Antarctica has been plotted in its isostatically adjusted position, assuming a present average elevation of 2200 meters, an ice thickness of 1880 meters (*Bardin and Suyetova*, 1967), and an ice density of .92 (*Lawson*, 1940). It does not obey the rule that the other continents seem to follow, but is unusually high for its area. As will be discussed below, this may be related to its peculiar erosional history.

Figure 4 shows the relation between size of continent and area with exterior drainage, i.e., draining to the ocean, based on data from *Livingstone* (1963). It is evident that large continents are more likely to have large areas of interior drainage for two reasons, 1) the larger the continent, the further the center from the oceanic moisture supply, and 2) the larger the continent, the more likely that some part of it will lie in a zone of net evaporation. In the case of Asia, about one-third of the continent has interior drainage. Because data on the mechanical load of rivers is based solely on rivers which enter the sea, it is important to consider whether their loads are better related to the area with exterior drainage than to the total area of the continent.

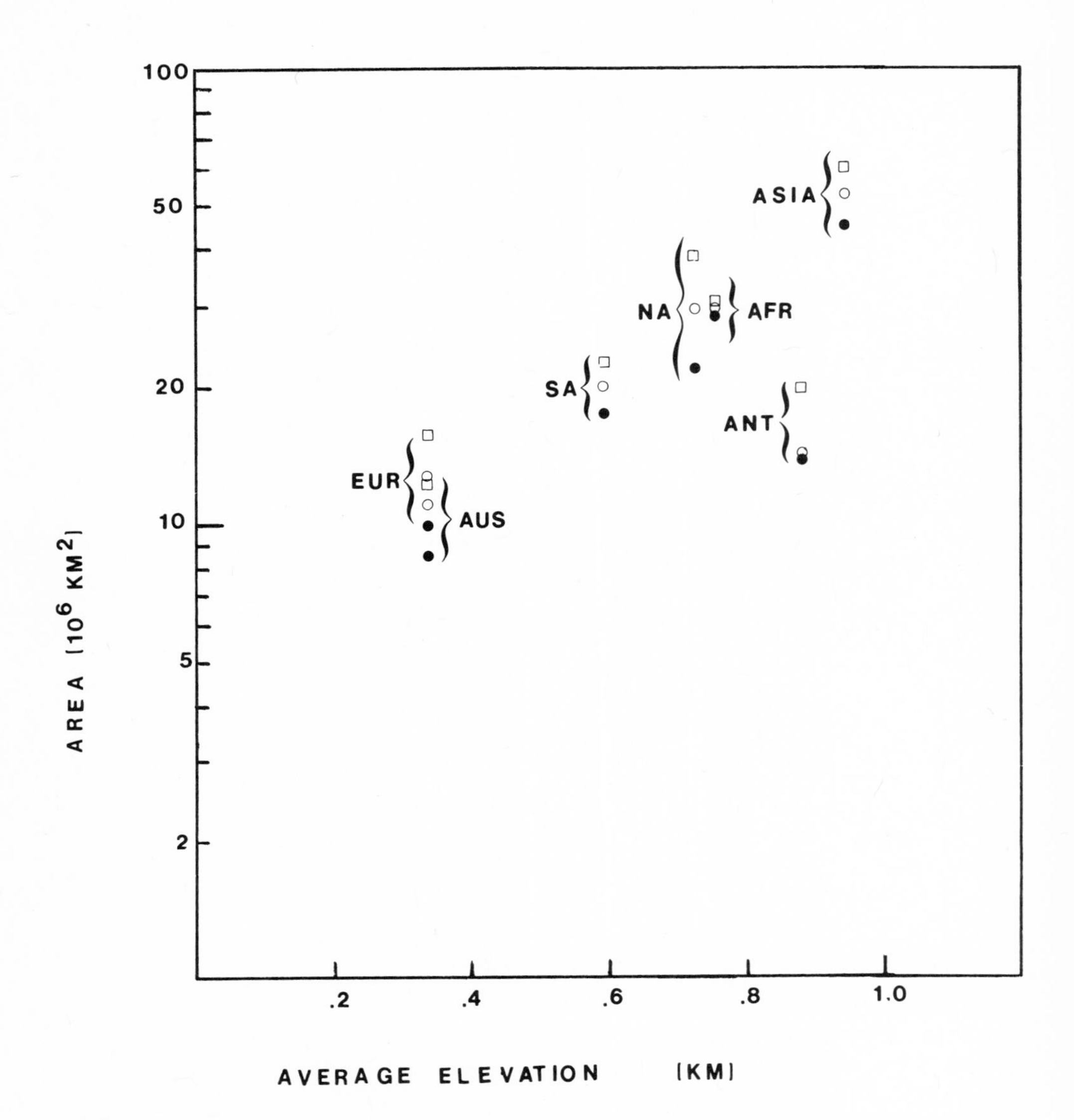

Fig. 3. Continental area plotted against average elevation. The solid circles, open circles and open boxes correspond to land areas, land plus shelf areas (continental platform) and land plus shelf plus slope areas (continental block). Data after Kossina, Meinárdus and Thiel (in *Fairbridge*, 1968). Position of ice-free Antarctica is plotted using data discussed in text. AFR = Africa; ANT = Antarctica; AUS - Australia; EUR = Europe, NA = North America; SA = South America.

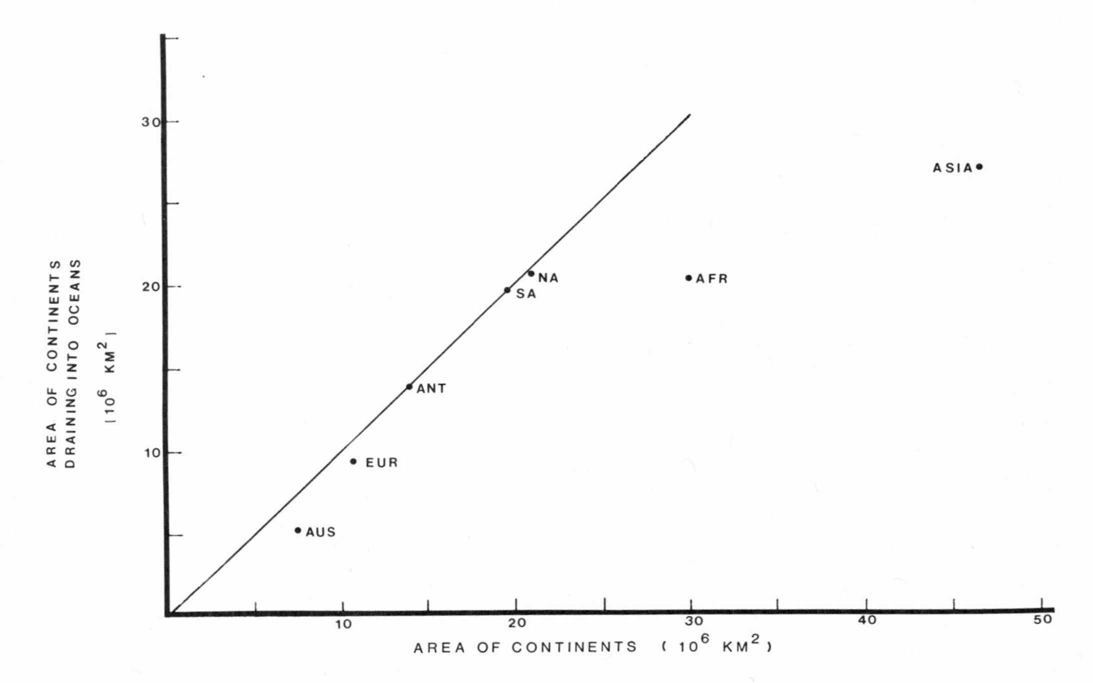

Fig. 4. Total continental area plotted against area of the continent having exterior drainage, i.e., draining to the sea. For abbreviations see Fig. 3.

Figure 5 shows runoff per unit area plotted against average elevation. The effect of climate is obvious. Australia, which is located almost wholly within a region of net evaporation, receives almost an order of magnitude less precipitation than the other continents. South America, which lies largely in the tropics, receives the most precipitation. If total continental area is used to calculate runoff per unit area, the runoff of all other continents average out at about 20 g/cm^2 yr. If runoff per unit area is calculated using the area of continent having exterior drainage, Asia has the same rate as South America, this being a reflection of the importance of its tropical areas. For the parts of the continent which drain to the exterior, the runoff could be generalized as follows: for a continent in the desert region 5-6 g/cm^2 yr; for a continent in the temperate zone, 25 g/cm^2 yr; for a continent in the tropics 40 g/cm^2 yr.

What would the effect of glaciation and deglaciation be on these runoff rates? If Pleistocene ice caps were formed exclusively by water which would otherwise be river runoff, they would reduce the rate of river discharge by about 10% below average during the period of melting. Because this would not be evenly distributed among

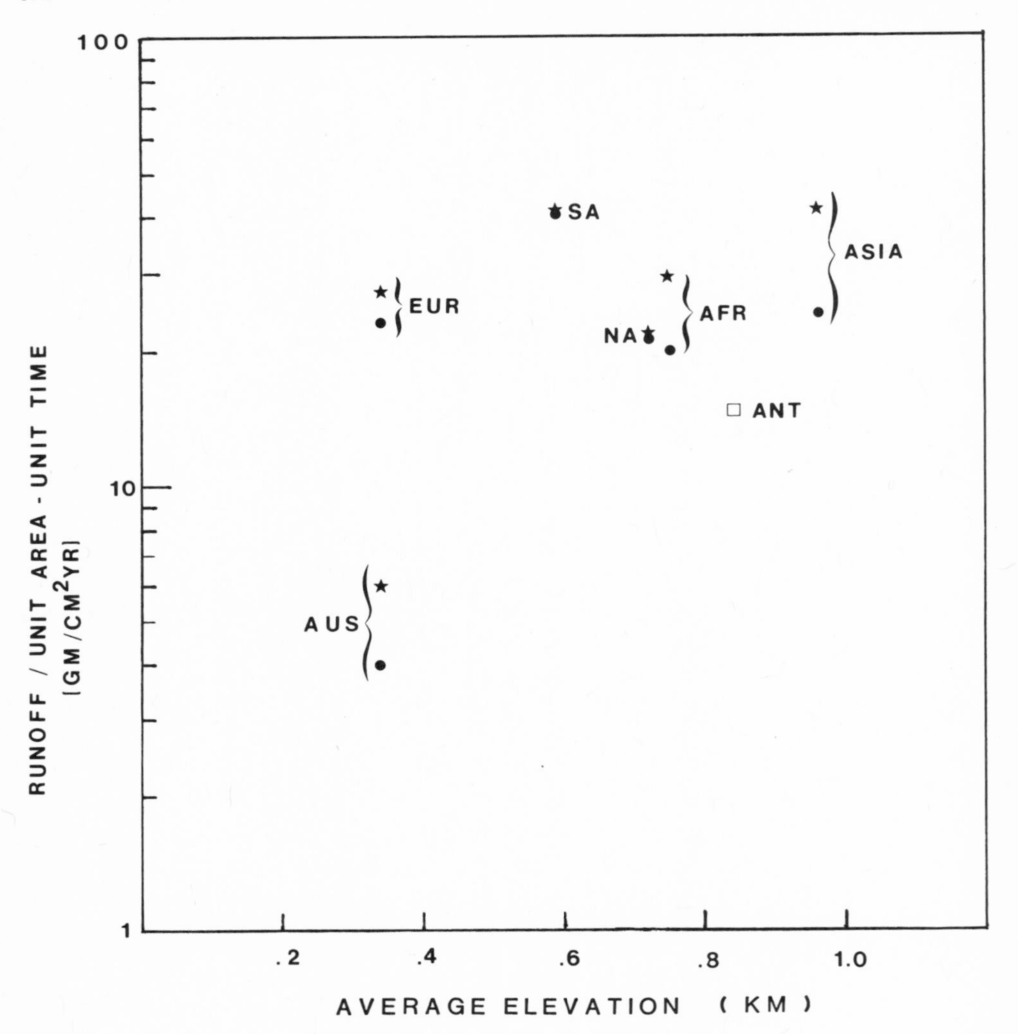

Fig. 5. Semi-log plot of run off per unit area-time plotted against average elevation of the continents. Runoff data are those of *Livingstone* (1963). Circles are runoff/yr divided by total continental area. Stars are runoff/yr divided by area of continent with exterior drainage. The square for Antarctica is to denote that its runoff is in the form of ice. For abbreviations see Fig. 3.

continents, but would be largely limited to Europe and North America, it would more markedly affect the river discharges for those continents. For North America, the present rate of river discharge is

21 g/cm^2 yr. Assuming that the Laurentide Ice Sheet had a volume equal to 27 x 10^{3}km^3 of water (*Flint*, 1971), and melted in 10,000 years, the additional runoff would be equivalent to 13 g/cm^2 yr. If the supply rate for runoff water to the continent were constant through time, the growth and melting of the Laurentide Ice Sheet would modulate stream flow so that the runoff rate would vary from 8 g/cm^2 yr during time of glacial growth to 34 g/cm^2 yr, or by a factor of 4. For Europe the results are similar: the present runoff rate is 23 g/cm^2 yr; assuming a volume of the Scandinavian Ice Sheet equivalent to 12.11 km^3 of water (*Flint*, 1971), and the growth and melting of the ice sheet each lasted about 10,000 years, the decrease or increase in runoff would be 11 g/cm^2 yr. The variation would be from 12 to 34 g cm yr, again about a factor of 3. For the difference in runoff rate to be any less than this, it is necessary that the net supply of moisture to the continent increase during the period of formation of glaciers and decrease during the period of melting of glaciers, i.e., that the changes in moisture supply are out of phase with and precede glacials and interglacials. If the changes in moisture supply were in phase with glacials and interglacials, the amplitude of the runoff rate variation would be larger than the factor of 3 or 4.

Figure 6 derived from the data of *Holeman* (1968), shows the relation between runoff rate and fluvial mechanical erosion rate; increase of the runoff rate does not affect fluvial mechanical erosion rate, which varies over a wide range with runoff rates of 20-30 g/cm^2 yr. The runoff rates of Africa and Asia differ by only 50%, but the fluvial mechanical erosion rate varies by more than an order of magnitude. From this, it would appear that changing the runoff of rivers per se would have little effect on the mechanical load delivered to the ocean.

Figure 7 derived from data of *Livingstone* (1963), shows the relation between runoff rate and fluvial dissolved erosion rate. For all continents except Australia, dissolved load per unit area-time is clearly independent of runoff rate. Australia plots in an anomalous position, suggesting that it may not receive enough water to enable the normal chemical weathering processes, which would be the immediate control on the dissolved load, to proceed at the same rate as on other continents. If this is true, it indicates that there is a threshold level below which the dissolved load of rivers decreases sharply with decreased runoff.

Figure 8 examines the relation between mechanical load and average elevation of the continents, noted by *Garrels and Mackenzie* (1971). If the mechanical load of rivers is divided by the total area of the continent to obtain the rate, there appears to be a very close approximation to an exponential relationship between runoff rate and average elevation. If the mechanical load is divided by the area of the continent having exterior drainage to obtain the rate, the increase is greater than that of a simple exponential function.

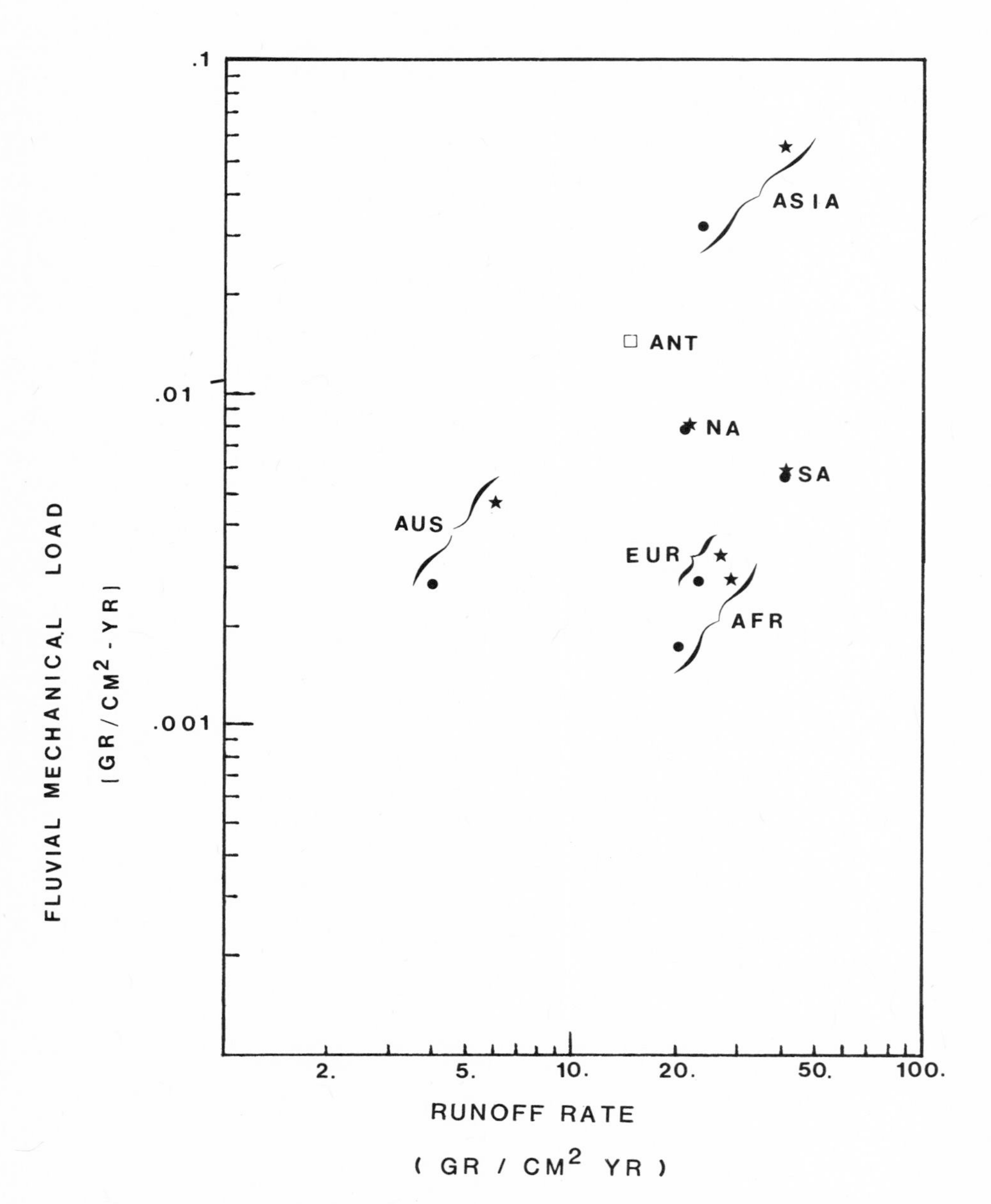

Fig. 6. Log-log plot of runoff per unit area-time against fluvial mechanical erosion load per unit area-time. Circles and stars as in Fig. 5; abbreviations as in Fig. 3.

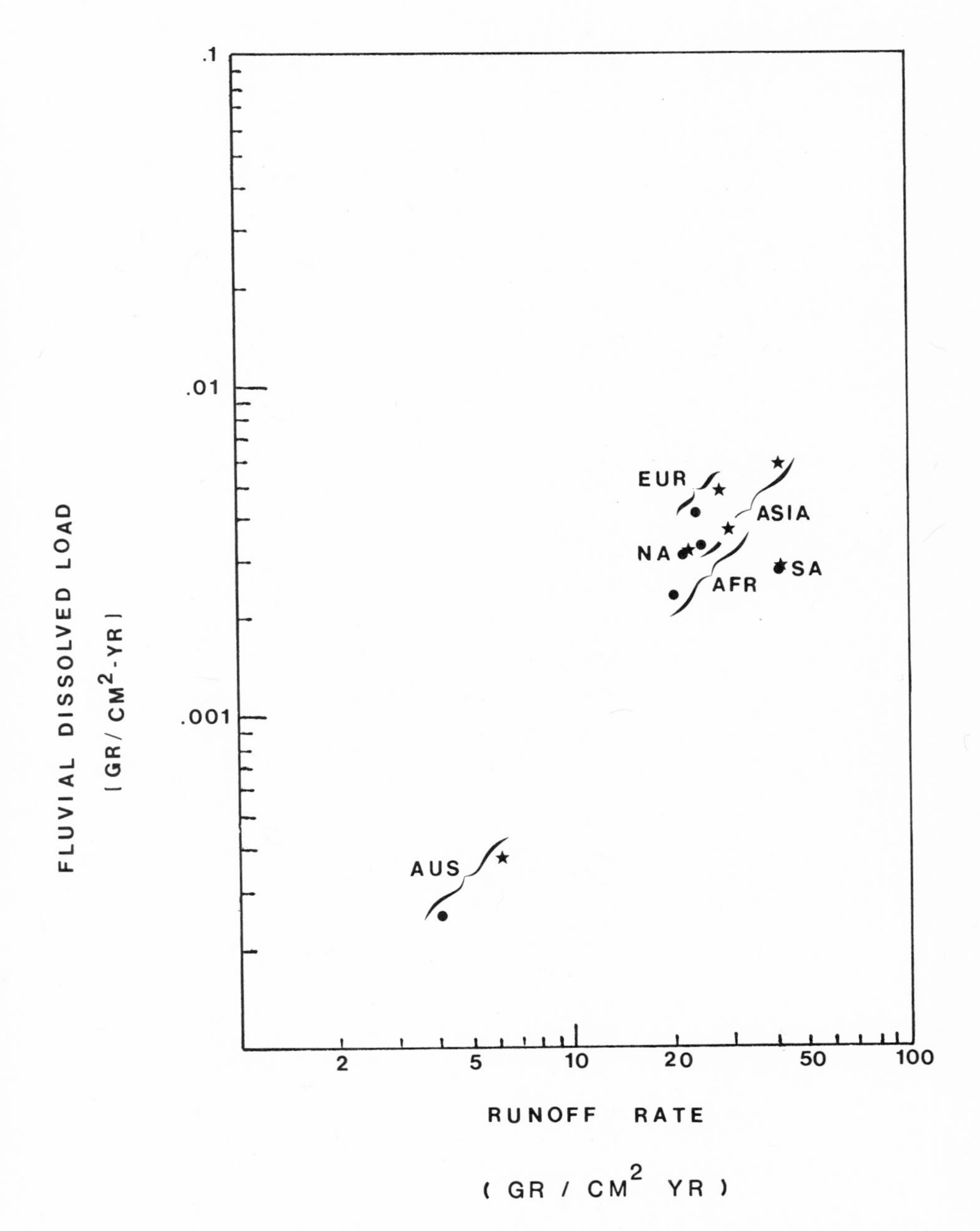

Fig. 7. Log-log plot of runoff per unit area-time against fluvial dissolved load per unit area-time. Circles and stars as in Fig. 5; abbreviations as in Fig. 3.

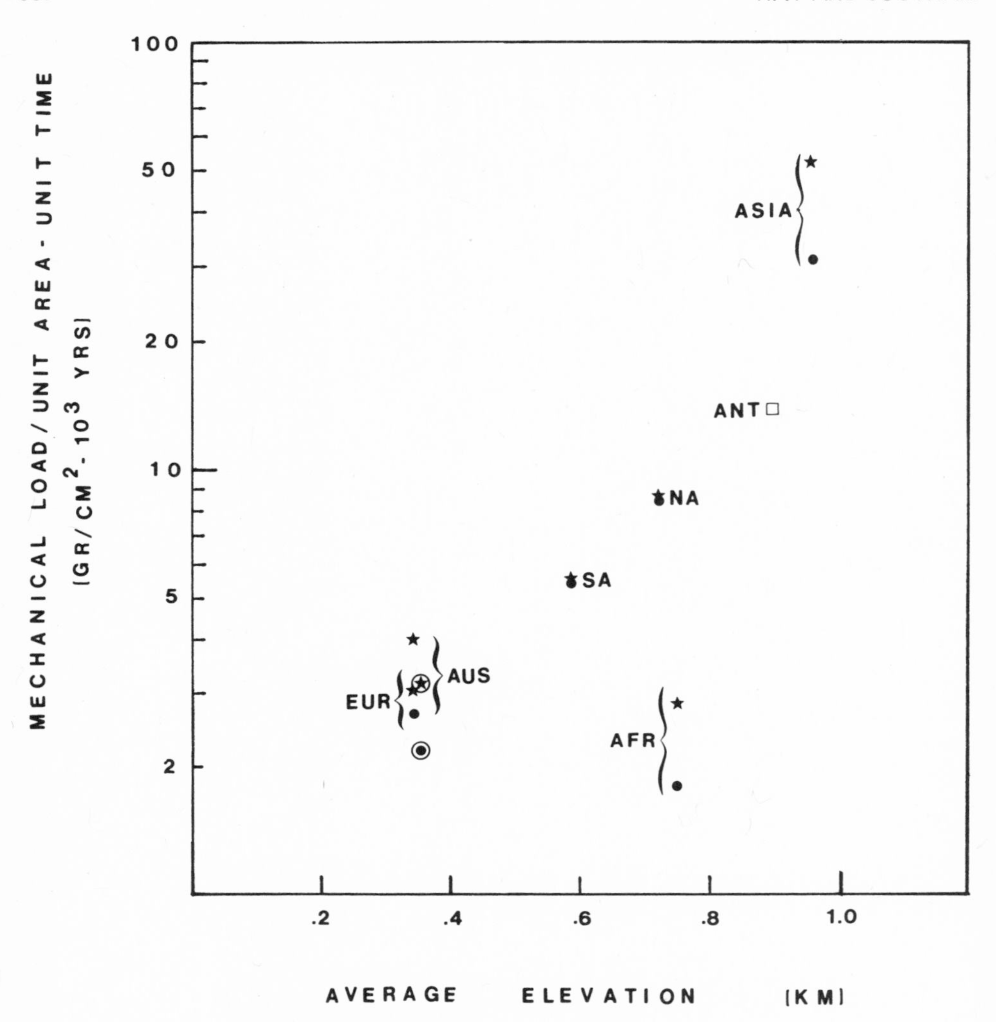

Fig. 8. Semi-log plot of mechanical load per unit-time against average elevation of continent. Circles and stars as in Fig. 5; abbreviations as in Fig. 3.

Africa plots anomalously low, indicating that the relation must have some important exception. That the other continents should plot out so well is surprising, because from studies of individual drainage basins it is known that erosion rates depend on local relief (*Ruxton and McDougall*, 1967; *Ahnert*, 1970). Although it seems intuitively obvious that the highest continent should have the greatest local relief and that local relief may be a simple function of continental elevation, we could find no reliable data on the local relief of entire continents; we have examined continental hypsographic curves instead.

Figure 9 is a set of hypsographic curves for the continents, with cumulative area plotted on a logarithmic scale against elevation. This plot reveals a number of interesting features not readily apparent on conventional representations of hypsographic curves. Curves for Asia, South America, and North America show distinctive high "plateaus". On the Asian curve, the high "plateau" of the hypsographic curve corresponds closely to the area of the Tibetian plateau, the part of the continent which has been underridden by continental crust of the Indian Plate. The curve for South America shows a much smaller "plateau", corresponding to 5-10% of the continent, i.e., the Andean Cordillera. The North American curve shows a broad flexure probably due to the western part of the continent overriding young (warm) oceanic crust. Four of the curves, Australia, Europe, Antarctica (rock surface) and Africa are smooth slopes. The European and Australian curves closely approximate each other, which is not surprising because both continents have the same average elevation. Africa displays a distinct peculiarity, shared only, perhaps, by Anarctica: its "break in slope" or "edge of the continental shelf" occurs at 200m above sea level, rather than 200 meters below sea level. If one assumes that the mechanical erosion of continents is related to local relief within the continent, and that the local relief is related to average elevation above the break in slope, then it would be appropriate to consider Africa's average elevation of 750 meters to be 400 meters too high. If this correction is made on the plot of mechanical load per unit area, Africa closely approximates Europe and Australia (see Figure 8).

As has been noted above, Antarctica is anomalously high for its size, indicating an unusual thickness of continental crust. Antarctica delivers no appreciable chemical load to the oceans, and its rate of glacial mechanical erosion has been estimated by *Yevteyev* (1959) to be about .69 km^3/yr, which Garrels and Mackenzie have calculated is about 19 x 10^{14} g/yr or only half that which would be expected from a river-drained continent of equal size and similar average elevation. Its unusually low erosion rate, which are peculiar to its glacial cover and history, is probably related to its apparent thickness, as is its anomalous hypsography.

Figure 10 is a plot of the fluvial dissolved load per unit area-time against average elevation. With the exception of Australia, which may be due to the very low level of moisture available to that continent, chemical denudation appears to be independent of average elevation. It would at first appear that chemical load is independent of climate, because South America, North America, Africa and Asia have very similar rates, but this is more probably a relfection of the fact that each of these continents spans several climatic zones.

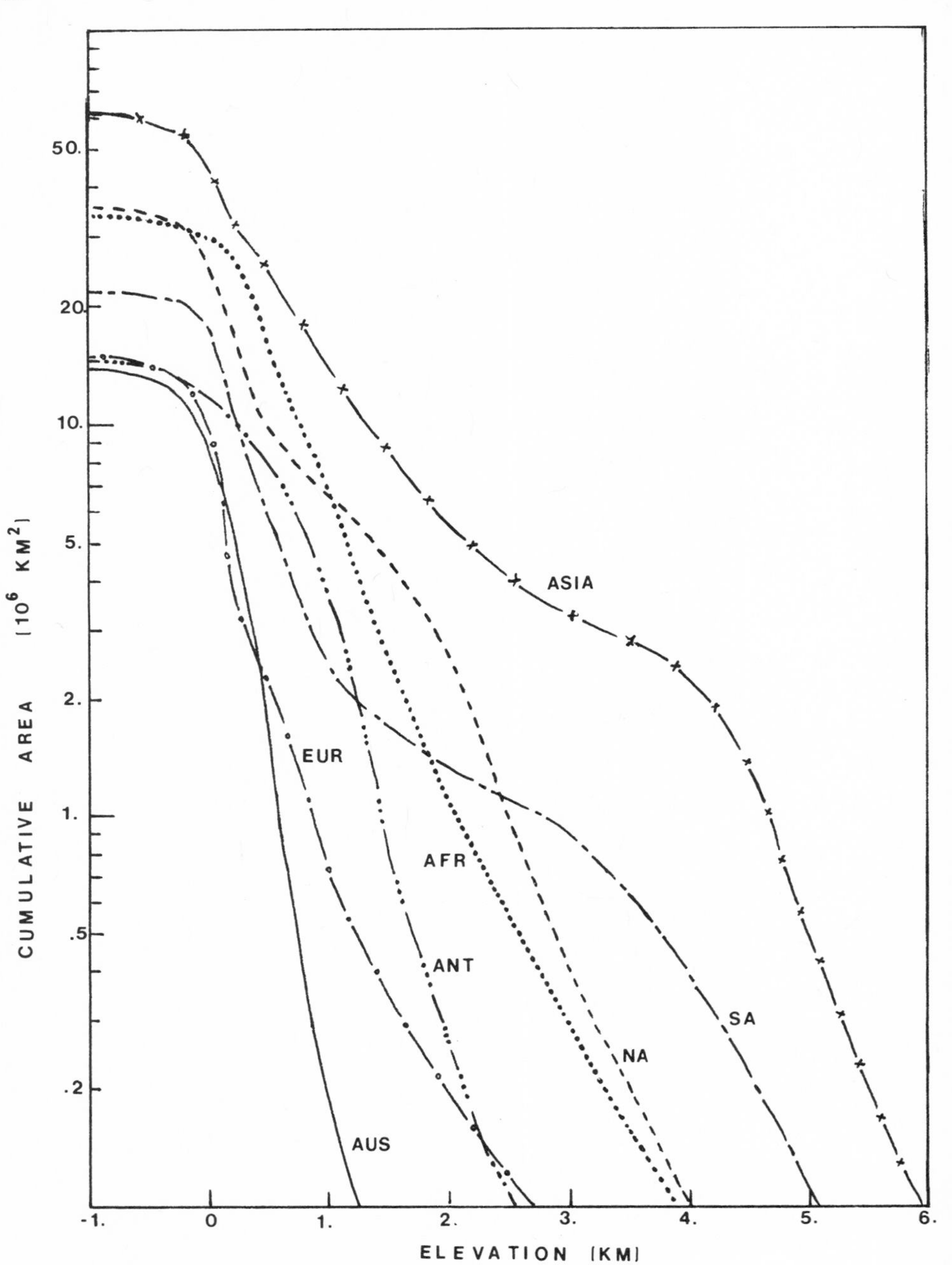

Fig. 9. Semi-log plot of continental hypsographic curves. Data after Kossina in *Fairbridge* (1968). Antarctica plotted from measurement of bedrock topographic map by *Bentley* (1972) adjusted for isostatic compensation after removal of ice.

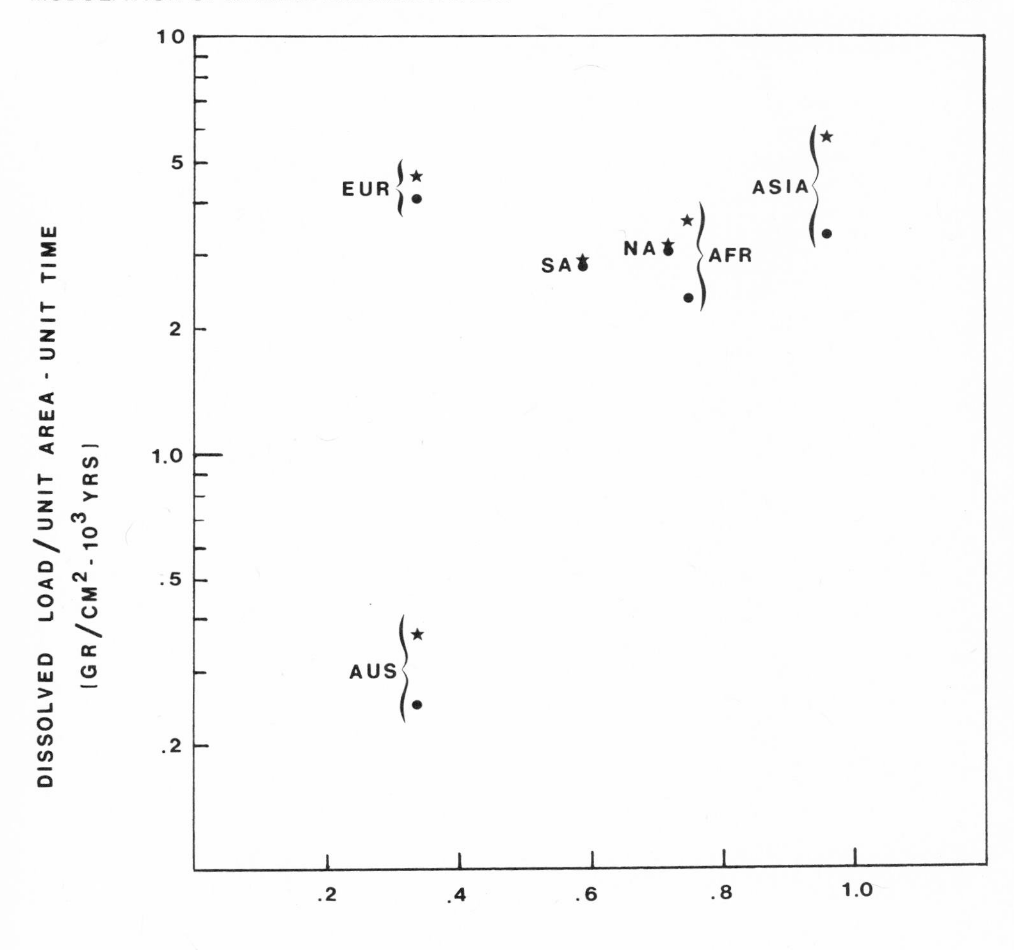

Fig. 10. Semi-log plot of dissolved load of rivers per unit area-time against average elevation of continent. Circles and stars as in Fig. 5; abbreviations as in Fig. 3.

The extreme values for Europe and Australia are a reflection of continents which lie almost wholly within one climatic regime. According to *Livingstone* (1963) they have the extreme values for average river water concentrations 52 ppm for dry Australia and 182 ppm for humid Europe, i.e., a factor of 3.5.

Table 1 shows the chemical composition of the dissolved load of rivers expressed as percent of the total dissolved load, calculated from the data of *Livingstone* (1963). Examination of Livingstone's computations of the fluvial chemical flux of different continents indicates that his value for the composition of African river waters is strongly dominated by the Congo and Niger, both tropical rivers. Thus for purposes of comparison, South America and Africa are both "tropical" supplies to the ocean. They differ from the other continental supplies chiefly in being enriched in silica by a factor of 2, depleted by 40% in calcium and depleted by 20% in HCO_3^-. Tropical rivers also have recycled salts as a more significant component, with Na and Cl slightly increased. Thus if all weathering and river supply on earth was "tropical" the major change noted would be an increase in the supply of silica by a factor of almost 2, but other fluxes would change only slightly.

TABLE 1. The earth's major river types

	Percent of total dissolved load in rivers		
Species	Temperate	Tropical	Volcanic Arc
HCO_3	53.4	41.2	36.7
SO_4	9.0	8.7	18.4
Cl	5.7	8.3	12.1
Ca	14.0	10.0	11.8
Mg	3.9	2.4	3.3
Na	5.6	7.3	6.9
K	1.3	2.5	1.6
Fe	0	1.6	0.8
SiO_2	6.9	17.9	22.1
Avg. dissolved load (ppm)	131.0	95.0	100.0
Estimated Discharge (10^{20} g/yr)	1.67	1.25	0.32

Glacial, groundwater, marine erosion and atmospheric dust loads are too inadequately known to speculate on changes in their importance except as included in discussions below.

From consideration of the relationships described above, it is possible to speculate on the possible excursions of the land-sea input system.

For the present, there are seven continents, with an average size of about 22 x 10^6km^2; these have an average exterior drainage area of 17 x 10^6km^2 (75%) and an internal drainage area of 6 10^6 km^2 (25%). The total runoff is about 3.24 x 10^9 g/yr. The total dissolved load is 4.0 x 10^{15} g/yr. If all these continents had the runoff rate of Australia, the runoff would total only .85 x 10^{19} g/yr; if all had the runoff rate of South America, the runoff would total 5.6 x 10^{19} g/yr (the average of these two values is 3.2 x 10^{19} g/yr, equal to the present total runoff estimate). Applying the extreme dissolved load concentrations represented by Australia and Europe, the minimum dissolved load would be .5 x 10^{14} g/yr, the maximum dissolved load would be 7.9 x 10^{14} g/yr (these two values also average out to closely approximate the present dissolved load total).

If all the continents were the size of Australia (7.5 x 10^6km), there would be 20 continents; the total area of exterior drainage might be 67-100%. The maximum total runoff would be 6.2 x 10^{19} g/yr if all these continents had runoff rates similar to South America, or .62 x 10^{19} g/yr if the runoff rates were similar to Australia. The total dissolved load extreme values are 8.76 x 10^{14}g and .37 x 10^{14}g respectively.

If all the continents were assembled as a single large block, the area with exterior drainage might be only 43 x 10^6km^2 (29%). The maximum runoff might be 6.2 x 10^{19} g/yr although if there were the extensive interior drainage suggested it would be 1.8 x 10^{19} g/yr; the minimum runoff .27 x 10^{19} g/yr. The maximum dissolved load might be 8.76-2.51 x 10^{14} g/yr; depending on the area with interior drainage, the minimum dissolved load would be .10 x 10^{14} g/yr.

The changes that any large portion of the continental mass could have extreme values of Europe, South America or Australia are very small. With the configurations of continents postulated for the Mesozoic and Cenozoic, it is very unlikely that the chemical load of rivers has changed by more than 30%, both in terms of mass and elemental composition, due to the break-up and dirft of the continents alone.

During the Mesozoic and Cenozoic, the maximum flooding of the continents occurred during the Late Creteceous when 33 x 10^6km^2 of present land surface were covered by water (*Hallam*, 1971). This reduced the total land area from the present 148 x 10^6km^2 to 125 x 10^6km^2, and could be expected to modulate the chemical load to the oceans, reducing it by 26% to 3.4 x 10^{14} g/yr. On the other extreme, it is possible that very low stands of sea level during the Pleistocene might have exposed the 26 x 10^6km^2 of continental shelf presently flooded, which would increase the chemical load to the oceans, if it were not for the fact that ice sheets built up on land and removed 29 x 10^6km^2 of land surface from the chemical weathering process. During glacials, it is likely that 50% or more of the continental shelves was exposed, suggesting that the net loss of continental area exposed to chemical weathering was probably less than 16 x 10^6km^2, or 12% of the land surface exclusive of Antarctica. If this were the case, glacial-interglacial cycles would be expected to change the chemical load to the ocean by only about 10%, with the lesser load coming during times of glaciation.

All of this discussion leads to a simple conclusion: the chemical input from land into the sea does not appear to be subject to significant fluctuations, even on a long-term time scale. Even if the removal of *Ca* from the continents by calcareous plankton postulated by *Hay and Southam* (1975) is real and has been operating for 100 million years, it still has not altered the composition of the continents sufficiently to affect appreciably the chemical composition of rivers. For the fluvial mechanical erosion rate, significant changes with time would be implied by changes in the average elevation of continents. If the relationship *Garrels and Mackenzie* (1971) suggested, that there is a logarithmic relation between average continental elevation and mechanical load per unit area-time, were correct, glacial lowering of sea level would be expected to approximately double the load of rivers and the mechanical erosion rate. If the relationship depends on average elevation of the break in slope as we suggest, glacial lowering of sea level produces changes of the same type as recognized for chemical weathering. The area of continent available for fluvial mechanical erosion is somewhat reduced, but glaciers would supply additional material from the areas they occupy. It may be that glaciers erode less effectively than rivers, so that the mechanical load of rivers would be increased only slightly by the additional detritus supplied by glaciers. The overall net effect for the world would probably be minor, but the continents of Europe and North America might be significantly affected.

SEDIMENTATION OF THE CONTINENTAL SHELF

The total area of the continental shelves to 200 meters is about 26 x 10^6km^2. Of this, about two-thirds is in the form of passive continental margins which originated as the continents broke away from each other. Most of these margins appear to follow the simple rules of subsidence with distance from a spreading center postulated by *Sleep* (1971) and *Kinsman* (1975). *Poldervaart's* (1955) estimate of the volume of shelf sediments (120 x 10^3km) suggests an average thickness of 4.6km, which seems reasonable. Assuming that the average age of these sediments is 100 million years, this gives an average sedimentation rate on the shelves of the world at 4.62 cm/

10^3 yrs which is about the same as that calculated as an average rate for Florida from information from wells. Assuming an average porosity of about 22% for this sediment column (*Atwater and Miller*, 1965), there would be a 2.1 g of sediment per cc. The average sedimentation rate would be 9.7 g/cm^2-10^3 yrs. In other words, the long-term consumption rate of this sink for sediments is only 25.2 x 10^{14} g/yr. This is 10% of the present total flux from land to sea, which is estimated by Garrels and Mackenzie to be 250 x 10^{14} g/yr. If one considers Garrels and Mackenzie's estimate of the long-term land-sea flux, 50 x 10^{14} g/yr, the present shelves would still consume only half of the supply. The remainder would be lost off the shelves to be returned to the continents in other forms through obduction, subduction and volcanism or through mountain building associated with continental collisions. On a long-term basis, it seems evident that the continental shelves bypass material to deeper oceanic accumulation sites.

The long-term rates of sedimentation may be proper averages, but it is possible that short-term fluctuations of significance can occur. Sedimentation rates in many areas of carbonate deposition are presently an order of magnitude higher than the average rate (*Milliman*, 1974). This is surprising, because consideration of sea floor spreading arguments suggests that shelf subsistence rates should decrease with time and young sediments should be deposited at a rate much below average.

Figure 11 is a schematic hypsographic diagram for continental shelves (the continental slope is shown only diagramatically). Present shelves slope from the shore out to an average depth of 200 m and occupy 26 x $10^6 km^2$. Most of their isostatic readjustment due to water loading after the post-glacial rise in sea level has already occurred (*Chappell*, 1974). If one considers the shape of this part of the hypsographic curve during a glacial epoch when sea level was 100 meters lower than at present (with reference to the continent, which is equivalent to about 140 meters with respect to an oceanic island "dipstick" as described by *Bloom*, 1967), the inshore part of the curve will readjust isostatically upward as the loading occur due to storage of water as ice on land. If isostatic adjustment goes to completion, the final shape of the curve would be that shown as "glacial shelf". Of the former continental shelf, 75% would be land if erosion did not modify the pattern. Rise in sea level back to the present level would transform the "glacial shelf" back into the present shelf through intermediate stages such as that marked "deglaciating shelf" by isostatic readjustment to water loading.

Very little is known about marine erosion, but many geologists think that the present position of the "break in slope" at about 200 meters represents adjustment to Pleistocene low sea level stands. The position of this break in slope is 50 m below sea level for the model suggested. If the shelf break were at the same level (-50 m)

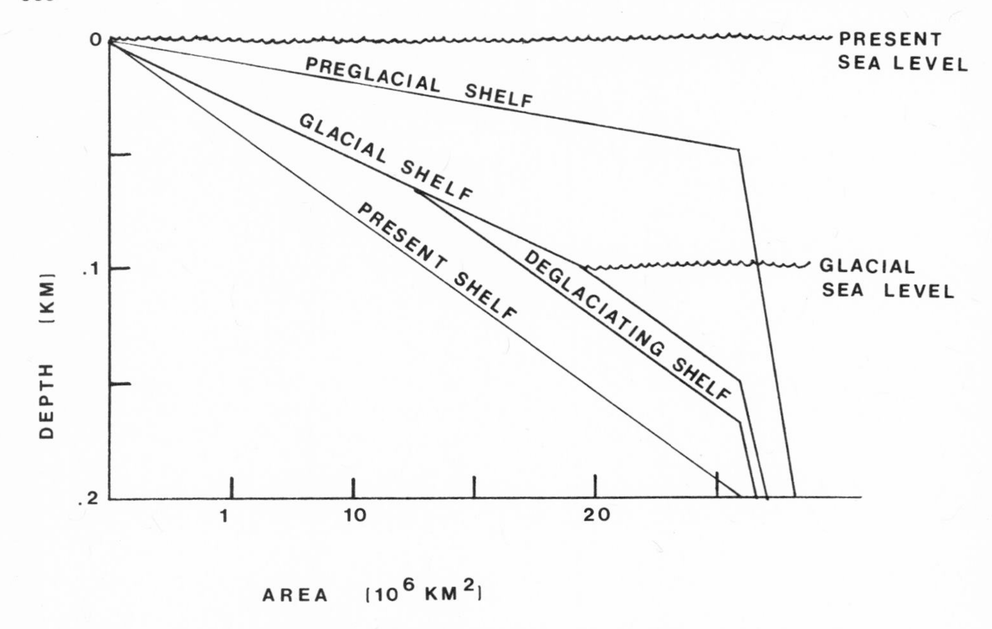

Fig. 11. Schematic hypsographic diagram for continental shelves. For discussion, see text.

prior to the onset of glacial-interglacial sea level fluctuations, a large volume of sediment would need to be removed in order to adjust the shelf break to the new lower sea level stand. Assuming the parameters of the model described above, the volume of material which would be removed would be $1.14 \times 10^{21} cm^3$, not counting the effects of isostatic readjustment as the sediment is removed for shallow buried. Using an estimated porosity of 30% for shallow buried shelf sediments, and allowing for isostatic readjustment due to removal of the sediment load, the mass of sediment would be 5.10×10^{21} g. If this were accomplished during 350,000 years of lowered sea levels during the Pleistocene, it means a flux of 146×10^{14} g/yr over the shelf break during the low stands of sea level. This is almost the same as the total mechanical load of rivers (183×10^{14} g/yr), which would also be discharged over the shelf edge during low stands of sea level.

When sea level rises, these processes reverse. The shelf should be able to consume as much sediment during a high stand of sea level as it lost during a low stand, plus a small increment due to subsidence with distance from a spreading center. There is one important difference, however; at low stands of sea level, marine erosion and young streams working into the recently deposited but now exposed sediment would rapidly and more or less evenly erode the shelf area below and above water. After a sea level rise the shelf can accumulate sediment, but the supply is not evenly distributed. The supply consists of a series of point sources, the river mouths, which can fill in a local area of the shelf while most of the shelf remains sediment starved. Once the local area is filled in, the river may discharge over the shelf break (as in the case of the Mississippi).

From these considerations it seems evident that the reason submarine canyons are spectacular features is that they would ordinarily carry 90% of the mechanical load of rivers to the deep sea; they are empty today because the loads of the rivers which would ordinarily supply them are consumed by filling of the estuaries created by the recent postglacial rise in sea level. It is intuitively obvious that because of the different nature of the supply and removal mechanisms operating on the detrital sediments of the shelf, the shelf will become more readily adjusted to low sea level, and would require a long period of stable conditions to achieve equilibrium with high sea level stands.

The situation with chemical sedimentation on the shelves is strikingly different. Carbonates tend to become cemented upon exposure to fresh water, and are much more difficult to erode than detrital sediments. The shelf breaks in areas of carbonate deposition tend to be very shallow. Further, when the opportunity for deposition returns with rising sea level, the supply of carbonate is diffuse and large, being in effect the ocean itself. The process we describe here would tend to favor deposition of carbonate on shelves during time of fluctuating sea level for the following reasons: 1) most shelf carbonate is in the form of skeletal particles produced by benthic organisms; 2) the grain size of these skeletal particles may be quite large, and they may form in areas where currents cannot transport detrital particles of such size; and 3) the biological productivity of the area may be initially limited only by the availability of nutrients and the shelf has the advantage of being able to draw on the nutrient supply of both rivers and the world ocean.

High deposition rates of carbonate cannot be indefinitely sustained, because the shelf would soon approach sea level. The availability of space becomes the ultimate factor limiting further accumulation. Sea level fluctuations on the time scale of those of the Pleistocene cannot alter the long-term sedimentation rates on the continental shelves, but they may bias the shelf as a deposition site favoring carbonate over terrigenous detrital sediment. Although their long-term effect may be slight, the processes postulated here have a very significant effect on short-term rates and could produce a stratigraphic record in which the beds of the sequence were formed during a very small fraction of the geologic time represented by the whole section.

CARBONATE SEDIMENTATION RATES ON THE SHELVES

Milliman (1974) has reviewed the literature and calculated long-term and Holocene rates of sedimentation for areas of carbonate deposition. Based on studies of bore holes on Bikini, Eniwetok, Midway, Florida, the Bahama Platform, Kita-Daito-Jima, and Funafuti, he claculated the average long-term (post-Miocene) sedimentation rate to be 3 cm/10^3 yr or 2.5 g $CaCO_3$/$cm^2 10^3$ yr. Based on studies of sedimentation in various parts of Florida, the Bahamas, Bermuda, Jamaica, Belize, Western Australia, and Hawaii, he calculated the average Holocene sedimentation rate to be 40 cm/10^3 yr or 35 g $CaCO_3$/cm^2-10^3 yr. He noted that the long-term rates, based on post-Miocene deposition, range between 1.7 and 6 cm/10^3 years, with the exception of the Sunda Shelf which reportedly has a post-Miocene rate of 10 cm/10^3 yr. Holocene rates range between 12 and 500 cm/10^3 yr.

This striking order-of-magnitude difference between long-term and short-term rates is possible only if unusual amounts of shelf space are made available for carbonate deposition on the short term. Normal subsidence on aging sea floor is in close agreement with the long-term sedimentation rates calculated by *Milliman* (1974), but does not provide sufficient space to account for the Holocene rates. Subsidence related to the thermal history of the sea floor beneath the Bahamas would only account for subsidence of about 50 cm during a glacial epoch, including the effects of isostatic adjustment to the subsequently added sediment load. This is less than the average rate of 70 cm/10^3 years, and far less than the several meters of sediment which have accumulated over most of the Bahama Platform in the past 7,000 years. Reduction of the elevation of the platform by dissolution and erosion during the low stands of sea level must have created the additional space made available for deposition during the Holocene. The Holocene carbonate sedimentation rates, although surprising when compared with long-term averages, are well below the maximum rates that can be achieved by carbonate secreting organisms under optimum conditions. *Neumann and Land* (1968) and *Marszalek* (1975) have shown that the rates of carbonate production for calcareous organisms exceed the rates of Holocene sedimentation so that a considerable part of the biogenic carbonate fixed by benthic shelf benthos must be redissolved or lost to the deep sea.

CARBONATE FLUX ONTO THE SHELF

To obtain an idea of the importance of the flux of carbonate onto the shelves, Milliman calculated the mass of $CaCO_3$ deposited in different sinks during the post-Miocene to obtain the long-term rate, and for the Holocene to determine the short-term rate. For the purposes of comparison it is useful to know the annual dissolved load of $CaCO_3$ carried by rivers. The Ca^{2+} flux is 4.9 x 10^{14} g/yr, and of this a negligible amount is due to atmospheric recycling from the ocean (*Garrels and Mackenzie*, 1971). If all of this calcium were removed from the sea as carbonate, the $CaCO_3$ flux would be 12.2 x 10^{14} g/yr. The carbon flux associated with this process would be 1.46 x 10^{14} g/yr. This leaves a net annual input of carbon of 5.5 x 10^{14} g/yr which is available for recycling through the atmosphere or for sedimentation as organic carbon.

Milliman's figures have been recalculated here to correct some typographical errors which appeared in his tables. The post-Miocene fluxes for the shelf sinks are: reefs, .45 x 10^{14} g/yr; all shelves except the Sunda Shelf, .17 to .34 x 10^{14} g/yr; Sunda Shelf, 1.85 x 10^{14} g/yr. Milliman postulates that during this period the Sunda Shelf consumed more carbonate than all other shelves and reef areas combined. If the anomalous Sunda Shelf were omitted, the consumption rate of the shelves would be .63-.79 x 10^{14} g/yr, or 5-7% of the river supply; with the Sunda Shelf, the rate of consumption is about 21% of the river supply. Other post-Miocene rates for carbonate sinks were determined to be: Red Sea .22 x 10^{14} /g; Black Sea .15 x 10^{14} g/yr; Mediterranean Sea .61 x 10^{14} g/yr; continental slopes, 4.6 x 10^{14} g/yr; and the remainder (33%) is available to the deep sea.

For the Holocene, the fluxes were calculated to be: reefs, 5.04 x 10^{14} g, all shelves, 2.37 x 10^{14} g/yr. Milliman's estimate of the total of these two alone would be 61% of the river input. Other Holocene fluxes into sinks were estimated to be: Red Sea, 23 x 10^{14} g/yr; Black Sea, .06 x 10^{14} g/yr; Mediterranean Sea, .82 x 10^{14} g/yr; and continental slopes, 4.6 x 10^{14} g/yr. The total flux to all of these sinks is 13.1 x 10^{14} g/yr, which exceeds the river supply by .9 x 10^{14} g/yr and means that the ocean reservoir should supply the deficit.

We have prepared a new estimate of the Holocene flux onto the shelves. By planimetering the shelves of the world by ocean and latitude increment, the areas of shelf in each increment of 10° have been determined. The results of these measurements are presented in Figures 12, 13 and 14. Using the map of distribution of sediments in the Physical Geographic Atlas edited by *Gerasimov* (1964) we have noted the proportion of shelf in each increment covered by sediment containing more than 30% carbonate. The total area of shelf with more than 30% carbonate is estimated to be 10.3 x $10^{6}km^{2}$. Using Milliman's average Holocene sedimentation rate of 40 cm, porosity of 60% and a $CaCO_3$ content of 30%, the annual flux is calculated to be 13 x 10^{14} g/yr, or slightly more than the present river input. This calculation would appear to be extremely conservative, because it assumes no $CaCO_3$ removal on 15.7 x $10^{6}km^{2}$ of shelf, although $CaCO_3$ contents of 5-15% and higher appear to be common in these areas. It also assumes that the carbonate content of many areas of carbonate deposition is only 30% although it is more nearly 80-90% in many of these areas. Thus, our estimate may well be low by a factor of 3 and it may be that this flux greatly exceeds the river input.

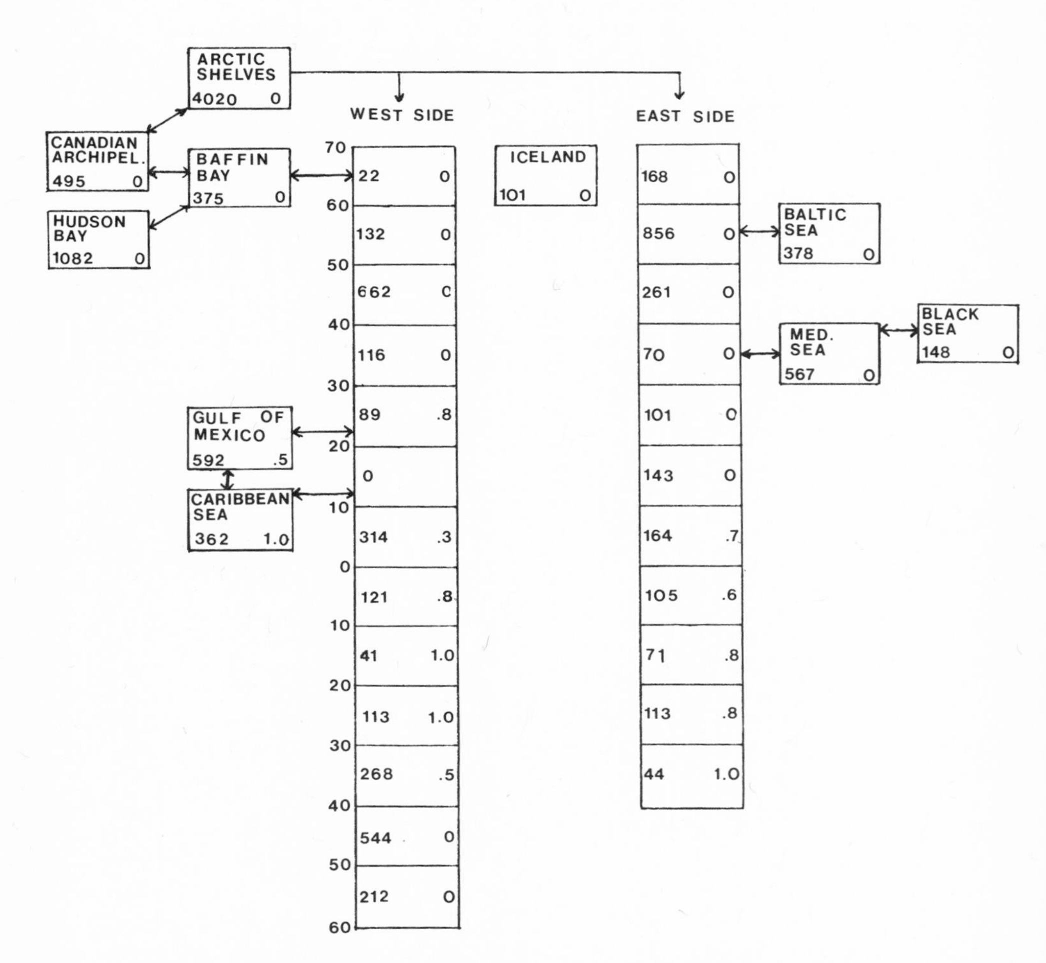

Fig. 12. Shelf areas and carbonate proportion of the Atlantic and Arctic Oceans and their tributary seas; number in the left of each box is area in $10^3 km^2$, number on the right is proportion of the shelf area covered by sediment having more than 30% $CaCO_3$.

Using any of these estimates, it would appear that the present sedimentation regime operates in such a way that the continental margins consume a flux of $CaCO_3$ equal to or several times larger than the input from rivers. This means that for the short term, shelf carbonate sedimentation must draw on the ocean reservoir.

PACIFIC OCEAN

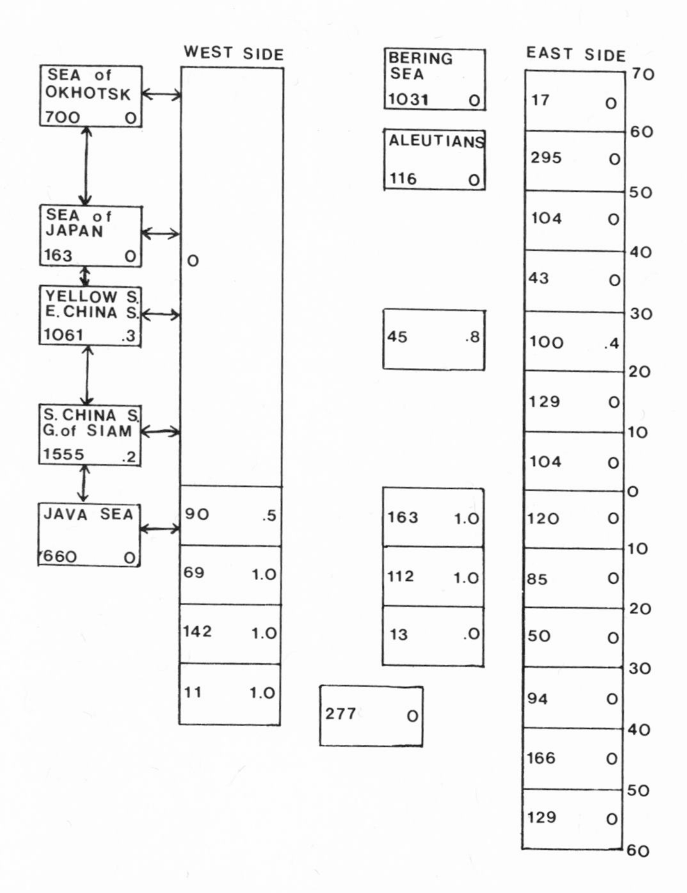

Fig. 13. Shelf areas and carbonate proportion of the Pacific Ocean and their turbidity seas; number in the left of each box is area in $10^3 km^2$, number on the right is proportion of the shelf area covered by sediment having more than 30% $CaCO_3$.

At the same time, the oceanic calcareous plankton continue to deposit $CaCO_3$ on the deep sea floor. The long-term rate for this process has been estimated by Milliman to be 11.8 x 10^{14} g/yr; which, added, to the long-term output on shelves, slopes and Mediterranean Sea prior to the formation of the Black Sea and Red Sea, makes a total flux of 17.8 x 10^{14} g/yr. Broecker's estimate of the flux out of the ocean can be calculated to be 18.0 x 10^{14} g. Both of these totals

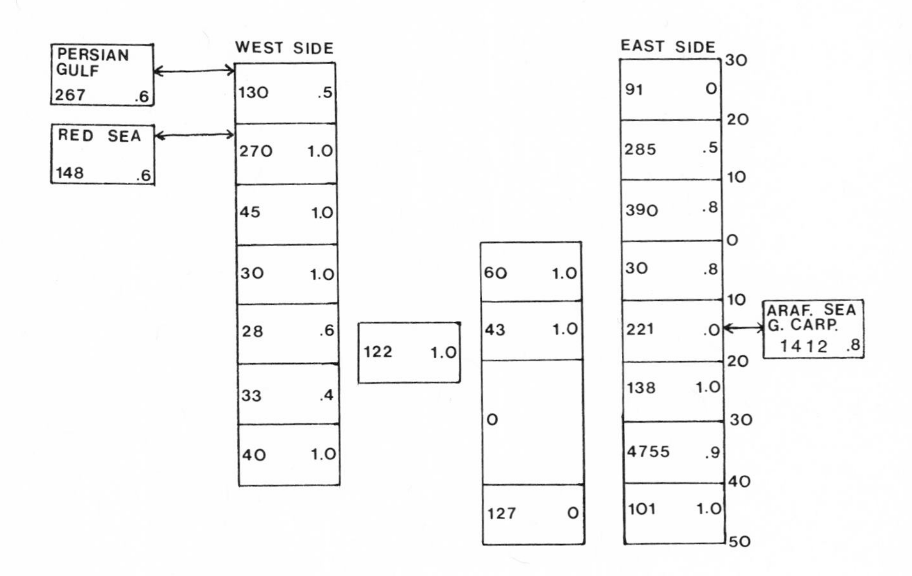

Fig. 14. Shelf areas and carbonate proportion of the Indian and Antarctic Oceans and their tributary seas; number in the left of each box is area in $10^3 km^2$, number on the right is proportion of the shelf area covered by sediment having more than 30% $CaCO_3$.

are 50% higher than the present river input and suggest that either there is a source which has not been taken into account, or, more likely, that the flux of carbonate to the deep sea has been overestimated. The overestimation of the flux to the deep sea is likely because most calculations were based on studies of piston cores of Pleistocene sediments, and the flux to the deep sea must have been higher than average during the Pleistocene because the $CaCO_3$ compensation level was lower than average. Our independent estimates of the long-term flux to the deep sea, based on DSDP cores, suggest that it was about 8 x 10^{14} g/yr.

The dissolved $CaCO_3$ reservoir of the ocean is 10 x 10^{20} g, so that if shelf sedimentation of carbonate consumed the entire river input for an indefinite period and the flux into the deep sea sink did not change, the oceanic reservoir would be exhausted in 847,500 years. But the present rate of sedimentation on shelves cannot be sustained for more than a few tens of thousands of years and must already be sharply declining as carbonate platforms grow to equilibrate to the present sea level.

A MODEL OF THE MARINE SEDIMENTATION SYSTEM

Milliman (1974) noted that sedimentation of carbonate on shelves and oceanic atolls would affect the carbonate supply to the deep sea and alter the position of the carbonate supply to the deep sea and alter the position of the carbonate compensation depth in the deep sea. The world hypsographic curve indicates that most of the deep sea and the lower parts of continents have almost the same slope (see Fig. 15). If carbonate were evenly sedimented onto shelves at the same rate that it is sedimented in the deep sea, then a transgression of the sea would cause the compensation level in the deep sea to move with sea level to remove an equivalent amount of the deep sea floor from carbonate sedimentation. If the rate of sedimentation on the shelves were double that of the average for the ocean, a rise of sea level on the shelves would cause the compensation level in the deep sea to rise twice as far. Most studies of the history of calcium carbonate compensation in the deep sea have referred to movement of the clacium carbonate compensation level without reference to the past history of sea level, but it works out that average long-term sedimentation rates in the deep sea are nearly the same as the carbonate sedimentation rate average for all shelves. This means that for the long term, the fluctuations of the carbonate compensation level in the deep sea should closely approximate fluctuations of sea level. Milliman was correct in his description of how the system should operate, but only for the long-term case. For the short-term case, the carbonate sedimentation is a highly dynamic system dependent upon the availability of nutrients and sediment storage space on the continental shelves.

Figure 16 is a new model of the marine sedimentation system, in which the shelves are taken into account. The inputs from land must remain positive and must pass through a shelf system to reach the oceanic system, represented by the box in the lower right. The oceanic system represented by a single box here is presented in more complete form in Figure 2. Fluxes from land to sea are always positive in this model, because return of oceanic sediments to the continents due to obduction, subduction and volcanism, and mountain building resulting from continental collision have been neglected, although the atmosphere may interchange material between the land,

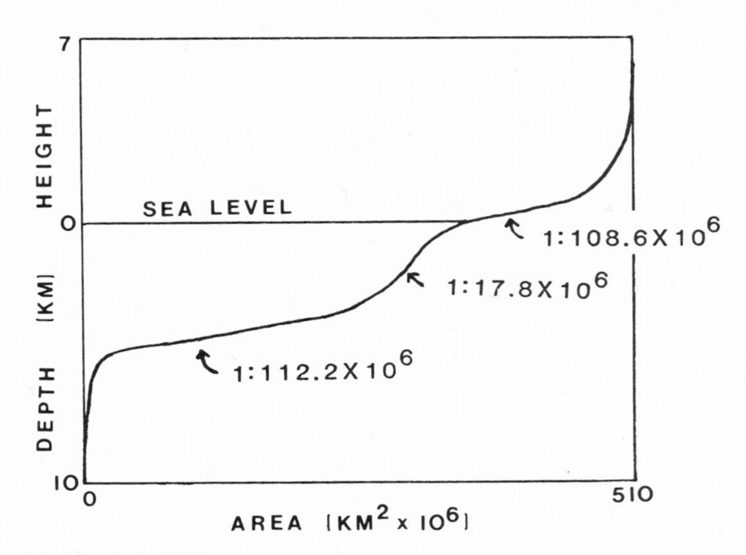

Fig. 15. World hypsographic curve. Note that most of the deep sea and the lower part of the land area have almost equal slope.

shelf and ocean. Atmospheric fluxes need not be positive but may reverse in short-term dynamic situations. Fluxes from the shelf to the ocean may be either positive or negative on a short-term base, but must be positive on a long-term scale. The shelf acts to modulate the amplitude of inputs to the ocean subsystem and may change their sign.

In terms of dynamical modeling of biologically influenced systems, it is most important to follow the flow of nutrients through the system. In most models, phosphorus is considered the limiting nutrient although it is not clear whether this is correct or whether the true limiting nutrients are fixed nitrogen or even iron (*Steward*, 1971). Unfortunately for any attempt to follow nutrients through the system, none have been measured with precision in sediments. Many oceanographic data exist for phosphorus, but its abundance is usually given as .1 or .2% in all rocks except phosphates.

The response time of phosphorus is about 170,000 years (*Southam and Hay*, 1975, 1976). The ocean reservoir of phosphorus is sufficiently large to damp out all but very large variations in phosphorus input, and it has been shown by Broecker that changes in vertical mixing rate in the ocean would greatly affect the phosphorus supply on a short-term base. There are no useful data on the average abundance of phosphorus in post-Miocene or Holocene continental shelf sediments, so it becomes necessary to examine the nutrient flow indirectly, in terms of biogenic carbonate output.

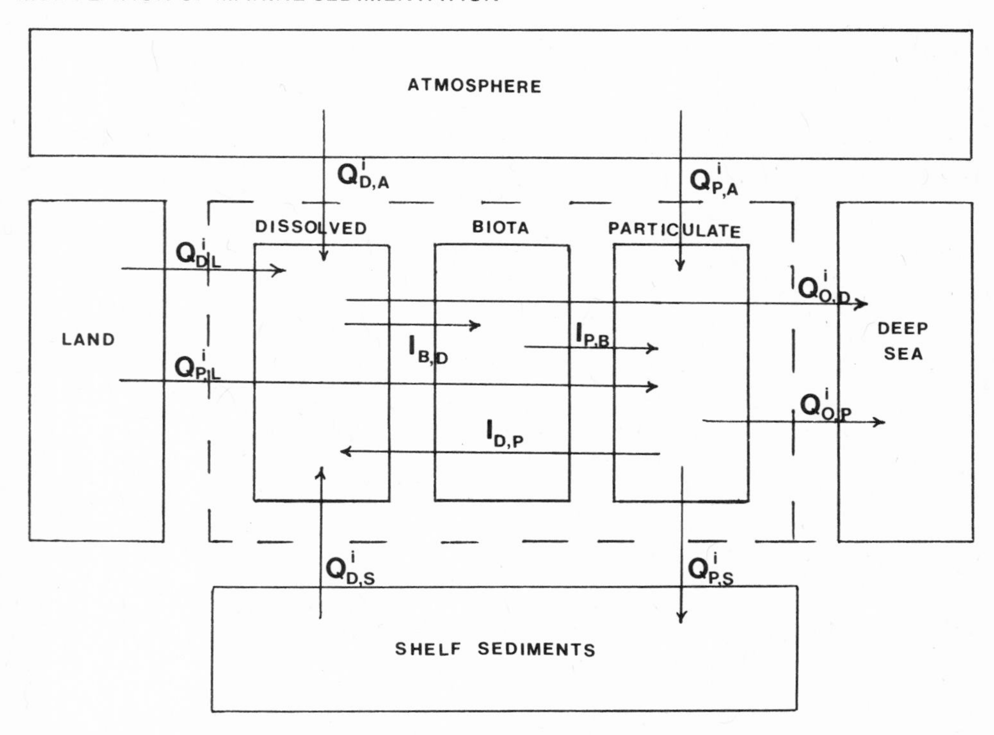

Fig. 16. A model of the shelf sedimentation system. The Q's are fluxes describing the exchange of material with reservoirs external to the shelf system. The I's are internal fluxes, describing the transport of material interior to the system due to such processes as biological productivity, physical circulation, precipitation, dissolution, oxidation, etc.

Ca is supplied to the oceans by rivers and groundwater; the amount derived from submarine weathering of basalts is probably small compared with the river flux. Because of the modulation effect of the shelf, it is possible for the oceanic input of calcium to become negative on the short term, immediately following a sudden rise in sea level; this would draw down the oceanic reservoir. Because *Ca* is responding to the availability of phosphorus as a nutrient, the response time associated with this change is not the response time of calcium, 920,000 years, but that of phosphorus, 170,000 years.

It is not clear what happens to phosphorus in this process; with unusually high sedimentation rates it might be expected that phosphorus would be less effectively recycled from the sediment back into the overlying waters, but the magnitude of this effect cannot be judged from available data.

Carbon has a high affinity for calcium; at least two thirds of the earth's carbon is locked up as carbonate. If the rate of calcium output on the shelves were as large as or larger than the river supply, then the rate of carbon removal would also be high. It would not exceed the river input, but most of the river input comes from CO_2 refluxed back from ocean to land. The oceanic reservoir would also become depleted in carbon, and rate of supply of CO_2 to the atmosphere would decrease with time, the response time of this process also being that of phosphorus.

Figure 17 shows the long-term flux of carbon through the system, using data derived from *Garrels and Mackenzie* (1971). One part of the carbon cycle is from sedimentary rocks on land to sediment in the sea. If it is assumed that all of the *Ca* flux of rivers is due to dissolution of carbonates on land, then the flux of carbon from the land rock reservoir to the ocean sediment reservoir is about 1.46×10^{14} g/yr. This must be a maximum value because some of the *CA* is supplied from weathering of other rocks. Assuming that the average sediment contains about .3% organic carbon, there would be an organic carbon flux from the land rock reservoir to the marine sediment reservoir of $.7 \times 10^{14}$ g/yr. To move the carbonate from the land to marine reservoir requires that additional carbon be used as a transport agent, because the mobile form is bicarbonate. Thus, 1.46×10^{14} g/yr must be recycled from the ocean to the atmosphere and back onto land to maintain the transport process for *Ca* in steady state. Rivers would then carry 3.62×10^{14} g/yr directly associated with the carbonate and organic carbon sedimentation. The carbon load of rivers exceeds this amount, being estimated at 3.74×10^{14} g/yr of carbon as bicarbonate ion and 3.2×10^{14} g/yr of carbon as organic carbon. These total 6.94×10^{14} g/yr, so that the excess is 3.32×10^{14} g/yr; this amount is available for recycling between ocean and land and is not immediately related to the sedimentation system.

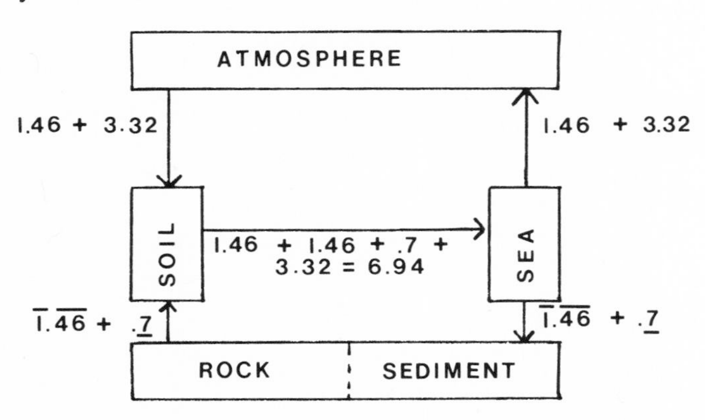

Fig. 17. Long-term flux of carbon in units of 10^{14} g/yr. Numbers underlined are organic carbon in rock or sediment numbers overlined are carbon in carbonate.

These would be the long-term fluxes for the system, but we do not believe them to be applicable to the present situation because of the high rates of shelf carbonate sedimentation which flood the Holocene rise in sea level.

Figures 18 and 19 are our estimates of the extreme values for the short-term Holocene fluxes. The maximum estimate assumes that the shelves consume three times as much carbonate as the rivers supply and that the ocean pelagic sedimentation subsystem keeps on sedimenting out carbonate at its average rate. The marine sedimentation system would then consume 5.84 x 10^{14} g/yr of carbon as carbonate, and 2.8 x 10^{14} g/yr of carbon as organic carbon disseminated in all sediments. This provides a total sedimentary output of carbon of 8.64 x 10^{14} g/yr, exceeding the river supply by 1.7 x 10^{14} g/yr. Flux to the atmosphere must draw on the oceanic reservoir of carbonate, bicarbonate and carbon dioxide. If the flux to the atmosphere from the ocean were shut off completely, the atmosphere could supply CO_2 for the calcium transport mechanism for over 4000 years. In reality, however, the flux into the atmosphere from the ocean would be only slightly diminished because the ocean contains about 400,000 x 10^{14} g of dissolved carbon as organic carbon, carbonate and bicarbonate. Running the maximum output short-term sedimentation system for 10,000 years would only draw down the ocean carbon reservoir by 20% and this would probably be the magnitude of the change in the flux into the atmosphere.

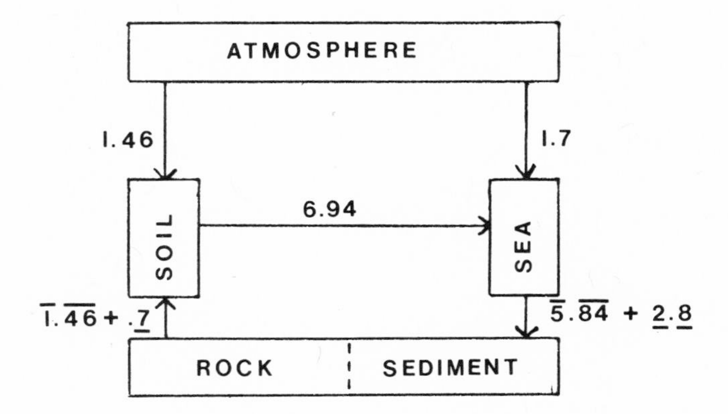

Fig. 18. Short-term flux of carbon in units of 10^{14} g/yr, assuming maximum reasonable Holocene sedimentation rate estimate. Numbers underlined are organic carbon in rock or sediment, numbers overlined are carbon in carbonate.

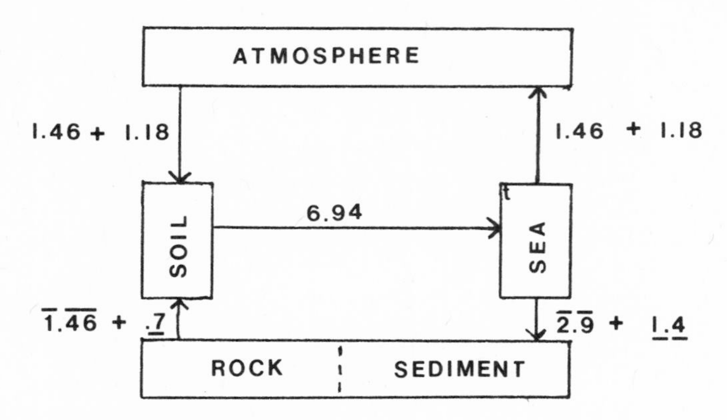

Fig. 19. Short-term flux of carbon in units of 10^{14} g/yr, assuming minimum reasonable Holocene sedimentation rate estimate. Numbers underlined are organic carbon in rock or sediment, numbers overlined are carbon in carbonate.

Figure 19 presents the minimum short-term fluxes which might apply to the Holocene. These assume that shelf sedimentation is equal to the river input of calcium carbonate and that the oceanic pelagic system has slowed somewhat below average. Although this system would draw more calcium out of the ocean system than rivers supply, the atmospheric recycling is allowed an excess of 1.18 x 10^{14} g/yr of carbon over that needed to maintain the calcium carbonate transport system alone.

WHAT IS THE EFFECT OF THE BURNING OF FOSSIL FUELS?

The system envisaged here is sufficiently complex that it is impossible to predict the dynamical behavior induced by any change in the system without formulating and solving the required sets of differential equations. Before solutions to such complex models are undertaken, it is very useful to have conferences of this sort which can elaborate and evaluate the fluxes and reservoirs and estimate the time scales on which different processes operate. The present rate of supply of C to the atmosphere from the burning of fossil fuels is estimated to be 46.5 x 10^{14} g/yr, a very large flux when compared with those in Figures 18 and 19. Several scenarios can be offered for the effects of this perturbation.

This flux would nearly double the C content of the atmosphere in 150 years if interaction with the ocean were not allowed. But it is doubtful that doubling the CO_2 level of the atmosphere would greatly increase weathering processes because the concentrations of CO_2 in soils are many times that of the atmosphere. Soils act as a buffer in the weathering system, resisting change. It is unreasonable to consider the change in the C content of the atmosphere without considering exchange with the ocean, where CO_2 can interact with the very large reservoirs of dissolved carbonate and bicarbonate. According to *Broecker and Peng* (1974) the equilibration time for atmospheric CO_2 with the ocean is only about 1.5 years, so that the surface mixed layer of the ocean instantly feels the addition of carbon by the burning of fossil fuels. The surface ocean is mixed in about 10^2 years, so that the initial effect on the ocean will be to lower the *pH* of the surface ocean by addition of CO_2.

According to our model of Holocene sedimentation, this process of lowering the *pH* of the surface ocean should already be going on, so that the effect of man's influence is to accelerate sharply an existing rate change. This would probably be felt as a narrowing of the band of shelf carbonate sedimentation. Because the surface ocean contains about $21,000 \times 10^{14}$ g of dissolved carbon in one form or another, the burning of fossil fuels should supply an additional quantity of about 30% of this amount in 100-150 years.

The problem becomes more complex if charge balance in the ocean is considered. Excessive removal of *Ca* on the shelves decreases the alkalinity of the sea, which must be equilibrated by readjustment of the carbonate-bicarbonate balance. The sense would be to enrich the ocean in bicarbonate. In shelf areas, where the fluctuations of calcium flux depart to the greatest degree from the average, the reduction in alkalinity would be most pronounced. The effect would probably be to restrict carbonate sedimentation to lower latitudes than would be expected from a consideration of the surface mixed layer of the sea as a unit.

Scenarios become more complex if the nutrient supply is taken into account. A modern account of the phosphorus cycle has been presented by *Lerman, Garrels and Mackenzie* (1971) and *Garrels, Mackenzie and Hunt* (1975). Phosphorus is being mined and supplied to land areas for cultivation at a rate of approximately $.12 \times 10^{14}$ g/yr; its rate of loss to the sea through rivers is only about $.017 \times 10^{12}$ g/yr. It would appear that for the present, the phosphorus used for cultivation is being incorporated into solid and has not yet markedly affected the world river supply to the sea. If the supply to the sea were increased to steady state with the rate at which phosphorus is being mined, and if the other nutrients did not become limiting, organic productivity and carbon sedimentation in the shelf regions and ocean might increase by a factor of 7. Multiplying the sedimentary output carbon flux rates suggested in Figures 18 and 19 by 7 yields consumption rates of 30 to 60 x 10^{14} g/yr. Such increased fertility of the sea and concomitant sedimentation would consume the carbon made available by the burning of fossil fuels. Most of this increased fertility would be concentrated around river mouths and in estuaries.

From this it would appear that a critical bit of information is the response time of soils to these changes. Specific questions are: does the plowing of soils alter their organic carbon content, and if so, how long must this process go on before a new steady state is achieved? Is phosphorus presently being accumulated in cultivated soils, and if so, how long can this process be continued before a new steady state is reached? Does the increase in area covered by soil since the melting of the Continental Ice Sheets play a significant role?

We guess the response time of soils to these changes to be in the order of 10^3 years. If this is correct, then the increased supply of phosphorus to the sea will become important only after the surface sea has been affected by the increase in C due to burning of fossil fuels. If the soil response time approximates the mixing time of the ocean, then all of these effects may be damped out by interaction in the surface sea, particularly on the shelves, and the ocean as a whole would be only slightly perturbed. This scenario is outlined schematically in Figure 20.

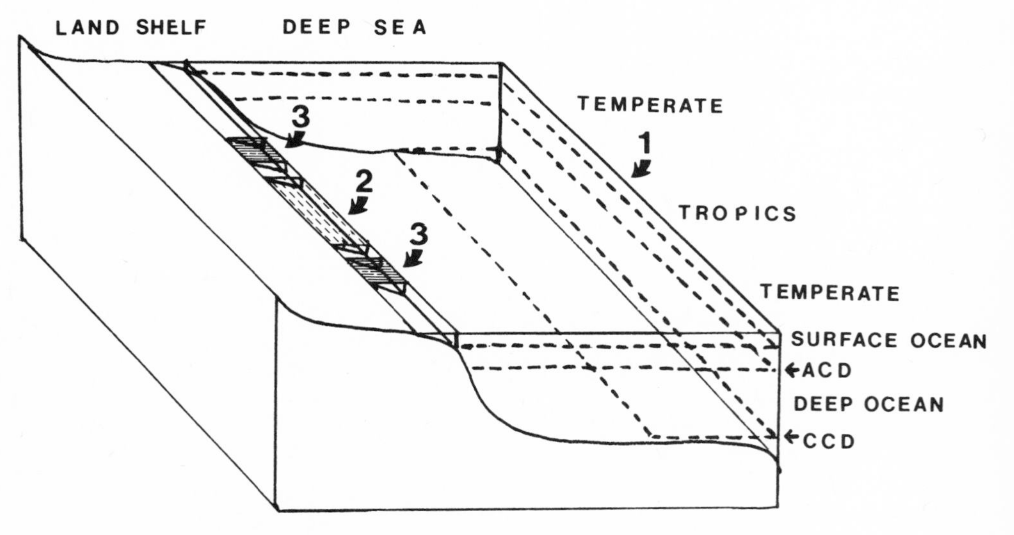

Fig. 20. Schema of the postulated effects of burning fossil fuel and mining phosphorus. The environmental stress would initially be felt (10^2 yrs) as CO_2 enrichment in the surface ocean. (1) The effects of this enrichment would be concentrated in the shallow water tropical shelves where aragonite would be selectively dissolved to compensate the inbalance, (2) this would most likely be a restriction of the shallow water aragonite deposition areas to lower latitudes. Subsequently (10^3 yrs) increase in the supply of phosphorus to the ocean as the through put of soils increases. (3) This would

Fig. 20 (Continued) be concentrated in mid-latitude estuarine coasts. The deep sea would be affected last and least. ACD = Aragonite compensation depth; CCD = Calcite compensation depth.

ACKNOWLEDGEMENTS

Financial support was provided by the National Science Foundation (grant DES74-02825).

REFERENCES

Ahnert, F., 1970. Functional relationship between denudation, relief and uplift in large mid-latitude drainage basins. *Amer. Jour. Sci., V. 268.* 243-263.

Atwater, G. I., E. E. Miller, 1965. The effect of decrease in porosity with depth on the future development of oil and gas reserves in south Louisiana: *Amer. Assoc. Petrol. Geol. Bull., V. 49.* p 334.

Bardin, V. I., and I. A. Suyetova, 1967. Basic morphometric characteristics for Antarctica and budget of the Antarctic ice cover: *Tokyo Nat. Science Mus., Jare Scientific Repts., Spec. Issue 1.* 92-100.

Bently, C. R., 1972. Subglacial rock surface topography: In: B. C. Heezen, M. Tharp, and C. R. Bentley (eds.) Morphology of the earth in the Antarctic and Subantarctic: *Amer. Geog. Soc. Antarctic Map Series, Folio 16.* 14-15 pl. 7.

Bloom, A. L., 1967. Pleistocene shorelines: a new test of isostasy: *Geol. Soc. Amer. Bull., V. 78.* 1477-1494.

Broecker, W. S., 1971. A kinetic model for the chemical composition of sea water: *Quaternary Res., V. 1.* 188-207.

Broecker, W. S., 1974. Chemical Oceanography: Harcourt Brace Jovanovich, Inc., New York, x + 214 pp.

Broecker, W. S., and T.-H. Peng, 1974. Gas exchange rates between air and sea. *Tellus, V. 26.* 21-35.

Chappell, J., 1974. Late Quaternary glacio- and hydro-isostasy on a layered Earth. *Quanternary Res., 4.* 405-428.

Fairbridge, R. W., 1968. The Encyclopedia of Geomorphology. Reinhold Book Corp., New York, XVI + 1249 pp.

Flint, R. F., 1971. Glacial and Quanternary Geology: John Wiley and Sons, Inc., New York xii + 892 pp.

Gerasimov, I. P., et al., 1964. Fiziko-geograficheski Atlas Mira. Akad. Nuak SSSRi Plavnoe Upravlenis Geodezzi i Kartografii GGK SSSR, Moscow, 298 pp.

Garrels, R. M. and F. T. Mackenzie, 1971. Evolution of Sedimentary rocks. W. W. Norton & Co., New York, xvi + 397 pp.

Garrels, R. M., F. T. MacKenzie and C. Hunt, 1975. Chemical cycles and the global environment. William Kaufmann, Inc., Los Altos, Calif., xv + 206 pp.

Hallam, A., 1971. Re-evaluation of the paleographic argument for an expanding earth. *Nature, V. 232.* 180-182.

Hay, W. W. and J. R. Southam, 1975. Calcareous plankton and loss of CaO from the continents. *Geol. Soc. Amer., Abstracts with Programs, V. 7.* p. 1105.

Holeman, J. N., 1968. The sediment yield of major rivers of the world. *Water Resources Res., V. 4.* 737-747.

Kinsman, D. J. J., 1975. Rift valley basins and sedimentary history of trailing continental margins. In: *A. G. Fischer and S. Judson (eds.) Petroleum and Global Tectonics.* Princeton Univ. Press, Princeton, N. J. 83-126.

Lawson, A. C., 1940. Isostatic control of fluctuations of sea level. *Science, V. 29.* 162-164.

Lerman, A., F. T. Mackenzie and R. M. Garrels, 1975. Modeling of geochemical cycles: phosphorus as an example, In: E. H. T. Whitten (ed.). *Geol. Soc. Amer., Memoir 142.* 205-218.

Livingstone, D. A., 1963. Chemical composition of rivers and lakes. In: *M. Fleischer (ed.) Data of geochemistry 6th ed., U. S. Geol. Surv. Prof. Pap. 440G.* 64 pp.

Marszalek. D. S., 1975. Calcisphere ultrastructure and skeletal aragonite from the alga *Acetabularia antillana: Journal Sed. Pet. V. 45.* 266-271.

Milliman, J. D., 1974. Marine Carbonates, In: *J. D. Milliman, G. Mueller and U. Foerstuer (eds.): Recent Sedimentary Carbonates. Springer Verlag, New York - Heidelberg - Berlin, pt. 1.* xv + 375 pp.

Neumann, A. C., and L. S. Land, 1968. Algal production and lime mud deposition in the bright of Abaco, Bahamas: a budget: *Geol. Soc. Amer., Abstracts for 1968, Spec. Paper 121.* p. 219.

Poldervaart, A., 1955. Chemistry of the earth's crust: In: *A. Poldervaart (ed.). Crust of the Earth, Geol. Soc. Amer., Spec. Paper 62.* 119-144.

Ruxton, B. P. and I. McDougall, 1967. Denudation rates in northeast Papua from Potassium-Argon dating of lavas. *Amer. Jour. Sci. V. 265.* 545-561.

Sleep, N. H., 1971. Thermal effects of the formation of Atlantic continental margins by continental break up. *Geophys. Jour. Royal Astronom. Soc., V. 24.* 325-350.

Southam, J. R., and W. W. Hay 1975. A dynamical model of biogenous pelagic sedimentation. *IX Internat. Sedimentological Congr. Theme 6.* 177-184.

Southam, J. R., and W. W. Hay 1976. Dynamical formulation of Broecker's model for marine cycles of biologically incorporated elements. *Internat. Jour. Math. Geol, V. 8, No. 5.* 511-527.

Steward, T. D. P., 1971. Nitrogen fixation in the sea. In: *J. D. Costlow, Jr. (ed.) Fertility of the Sea, Gordon and Breach Science Publ., V. 2.* 537-564.

Yevetyev, S. W., 1959. Determination of the amount of morainal material carried by glaciers to the east Antarctic coast. *Inform. Bull. Society Antarctic Expedition, V. 11.* 14-16.

Organic Carbon in Deep-Sea Sediments

G. Ross Heath, Ted C. Moore, Jr. and J. Paul Dauphin

University of Rhode Island

ABSTRACT

The rate of accumulation of organic carbon in marine sediments is closely related to the bulk sediment accumulation rate and can be described by the equation: (Organic Carbon Accumulation Rate) = 0.01 (Sedimentation Rate)$^{1.4}$. Diagenetic reduction of the organic carbon content of older sediments appears to be exponential in both reduced and oxidized sediments, with half lives in the 15-55,000 year range. Pleistocene climatic fluctuations have resulted in marked variations in the supply of organic carbon to the deep-sea, in part due to oceanographic changes, and in part due to the destruction of nearshore depositional environments by lowered sea level. Carbonate-rich deposits contain organic matter that is intimately associated with test calcite. The climate and carbonate effects can mask the diagenetic pattern evident in uniformly deposited sediments, but they do not alter the basic relation between organic-carbon and total-sediment accumulation rates.

INTRODUCTION

Remarkably little is known of the geochemistry of organic carbon in a broad spectrum of deep-sea sediments. Previous studies (*Bordovskiy* 1965a, b; *Müller*, 1975; *Stevenson and Cheng*, 1972) have concentrated their efforts on relatively carbon-rich reduced sediments. Not only are data on the phases containing the carbon rare, but even the source or sources of the carbon, its distribution and rate of deposition in the ocean basins, and the degree of post-depositional alteration are not well known. This study addresses the last two questions: where and how rapidly is organic carbon being deposited in the ocean basins, and how does early diagenesis modify the depositional pattern?

Within the past five years, large numbers of deep-sea sediment samples have been analyzed for organic carbon in the course of studies of organic matter in general (e.g. *Müller*, 1975), or as a byproduct of analyses for carbonate using combustion techniques (e.g. *Heath et al.*, 1976). At the same time, accumulation rates for large numbers of cores have been established by radiometric dating and by paleomagnetic and oxygen isotope stratigraphic determinations (e.g. *Ku et al.*, 1968; *Opdyke and Foster*, 1970; *Climap*, 1976). As a result, it is now possible to map not only the gross distribution of organic carbon in deep-sea sediments, but also its rate of accumulation and diagenesis in different depositional environments.

The data discussed in this paper are largely derived from samples analyzed under the CLIMAP Program. Cores studied (Table 1, Figure 1) are from the collections of the Lamont-Doherty Geological Observatory, Oregon State University, Scripps Institution of Oceanography, University of Southern California and the New Zealand Oceanographic Institute. Additional analyses from *Müller* (1975) were made on cores collected by the R/V's Meteor and Valdivia of the German Research Society. *Stevenson and Cheng* (1972) have also analyzed Lamont-Doherty Geological Observatory cores collected from the Argentine Basin.

METHODOLOGY

All organic carbon analyses in our laboratory have been made by Leco 714 carbon analyzer. Samples are ground, split, dried at 110°C and weighed. One split is analyzed in duplicate for total carbon. The other is heated to 500°C for two hours to oxidize organic carbon, then analyzed in duplicate for carbonate carbon (expressed as a percentage of 110°C weight). The difference between the two analyses is defined as organic carbon. Comparison with other techniques that use strong oxidizers to titrate organic carbon (e.g. *Müller*, 1975) or acid removal of carbonate before Leco analysis (e.g. *Boyce and Bode*, 1972) suggest good agreement for terrigenous and pelagic clays. For carbonate-rich deposits, however, our organic carbon numbers are consistently higher than those involving wet chemical treatments of the samples. It has been determined experimentally that the problem lies in the acid treatment used to remove carbonate prior to the chemical or combustive determination of organic carbon. Apparently some of the organic carbon is intimately associated with the calcite of foraminiferal or coccolith tests, either as surface or intercrystalline films. During acid treatment,

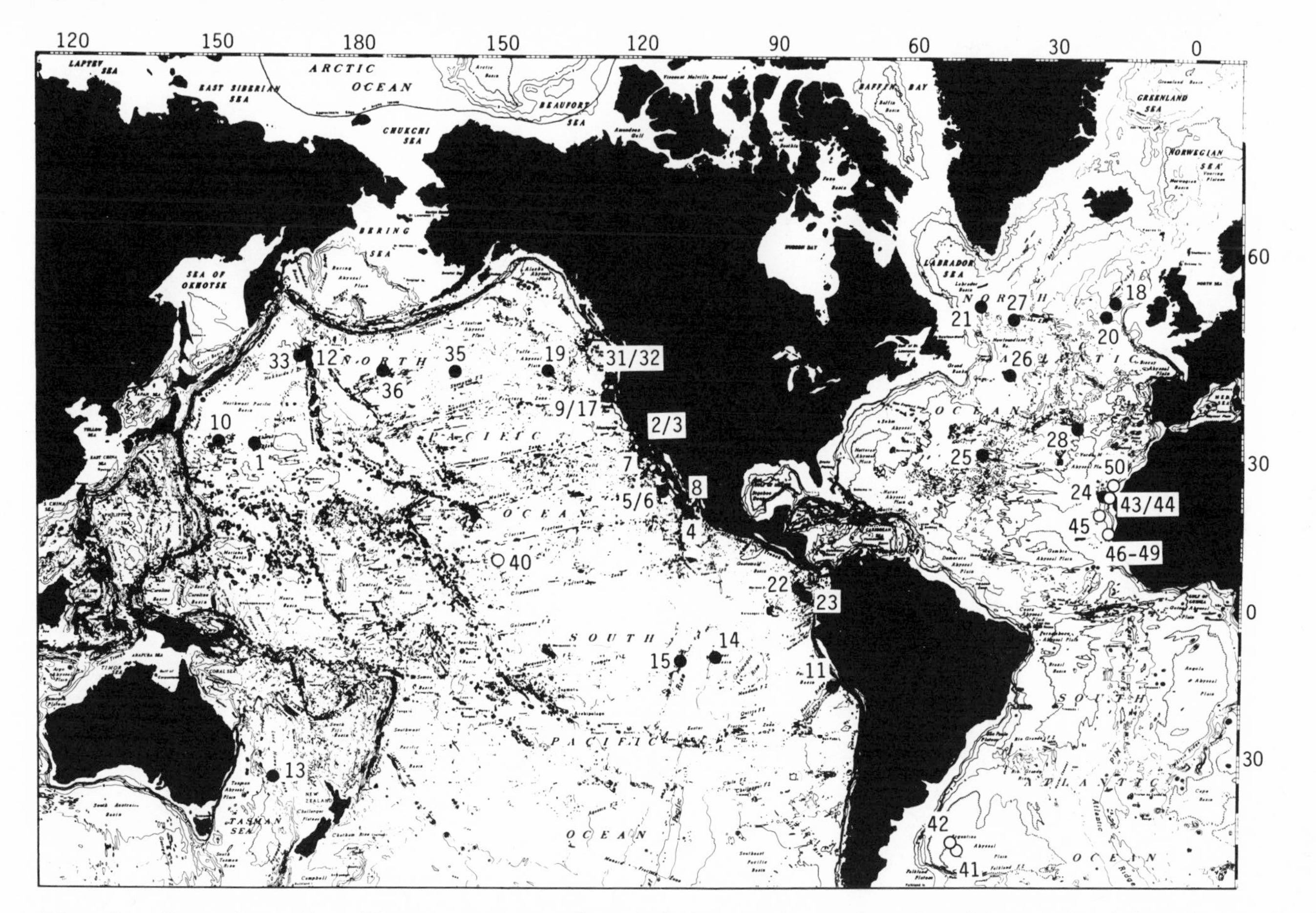

Fig. 1. Location of sediment cores analyzed for organic carbon in this study (closed circles), and by *Stevenson and Cheng* (1972), *Hartmann et al.* (1973) and *Müller* (1975). Identification numbers refer to Table 1. Base map from *Chase* (1975).

TABLE 1. Identification of Sediment Cores

Map Location	Core Identification	Latitude	Longitude	Water Depth (m)	Ref.*
1	RC10-161	33°55′N	158°00′E	3587	
3	Y71-10-117P	34°16′N	120° 3′W	570	1
4	BNFC-43PG3	20°10′N	109° 1′W	2720	
5	Y73-2-4P	24°36′N	114°35′W	4138	
6	Y73-2-3P	24°35′N	114°35′W	4138	
7	AHF-10614	32°52′N	119°49′W	1275	
8	LAPD-1G	22°53′N	110°59′W	2322	
9	Y6910-2	41°16′N	127° 1′W	2743	2
10	RC10-167	33°24′N	150°23′E	6092	3
11	Y71-6-12P	16°26′S	77°33′W	2734	
12	V20-119	47°57′N	168°47′E	2736	3
13	Z2108	33°23′S	161°37′E	1448	
14	Y71-7-36P	10° 3′S	102°51′W	4541	4
15	Y71-7-45P	11° 5′S	110° 6′W	3096	4
17	Y6910-4	41°19′N	128° 9′W	3130	
18	V23-81	54°15′N	16°50′W	2393	
19	Y70-1-5P	45°00′N	140°00′W	4374	
20	RC9-228	52°33′N	18°45′W	3981	
21	V27-20	54°00′N	46°12′W	3510	
22	Y69-108P	4° 9′N	85° 2′W	3390	
23	Y69-109P	2° 5′N	82°59′W	3721	
24	V23-98	23° 7′N	19°18′W	3506	
25	V19-309	31°10′N	45° 8′W	4544	
26	V27-16	44° 9′N	39°52′W	4797	
27	V27-19	52° 6′N	38°48′W	3466	
28	V27-141	35°29′N	24°56′W	4506	
31	Y6604-10	43°16′N	126°24′W	3002	2
32	Y6609-5	43°34′N	126°28′W	2978	2
33	V20-120	47°24′N	167°45′E	6216	3
35	Y70-1-13P	44°58′N	159°58′W	5492	
36	Y70-1-19P	45°00′N	175° 7′W	5973	
40	VALDIVIA	~10°N	~150°W	~5000	5
41	V15-141	45°44′S	50°45′W	5934	6
42	V15-142	44°54′S	51°32′W	5885	6
43	M12310	23°30′N	18°43′W	3076	7
44	M12327	23° 8′N	17°44′W	2037	7
45	M12329	19°22′N	19°56′W	3314	7
46	M12337	15°58′N	18° 7′W	3085	7
47	M12344	15°26′N	17°21′W	711	7
48	M12345	15°30′N	17°22′W	966	7
49	M12347	15°50′N	17°51′W	2710	7
50	M12392	25°10′N	16°51′W	2575	7

*1. *Koide et al.* (1972); *Sholkovitz* (1973); 2. *Heath et al.* (1976); 3. *Robertson* (1975); 4. *Kendrick* (1974); 5. *Hartmann et al.* (1973); 6. *Stevenson and Cheng* (1972); 7. *Müller* (1975).

this organic component is peptized and filtered off with the spent acid. As a result, organic carbon analyses on acid-treated carbonate-rich sediments tend to be uniformly low.

ORGANIC CARBON IN RECENTLY DEPOSITED SEDIMENTS

The organic carbon content of open-ocean deposits at the sea floor ranges from less than 0.1 percent in the most oxidized red clays to slightly more than 5 percent in reduced hemipelagic deposits close to continental margins - a range of about two orders of magnitude. The total-sediment accumulation rates of these same deposits range from about 0.03 to 100 gm/cm^2/1000 yr. - a span of 3½ orders of magnitude. The two parameters, carbon content and accumulation rate, co-vary to some extent (Figure 2), so that the organic carbon accumulation rate ranges from about 4 x 10^{-5} to about 6 gm/cm^2/1000 yr, a span of more than five orders of magnitude. Despite the diverse lithologies of the cores studied, the total sediment and organic carbon accumulation rates are highly correlated on a log-log plot (Figure 3). The regression line through the data of Figure 3 is given by:

$$\text{(Organic carbon accumulation rate)} = 0.01\ \text{(sedimentation rate)}^{1.4} \tag{1}$$

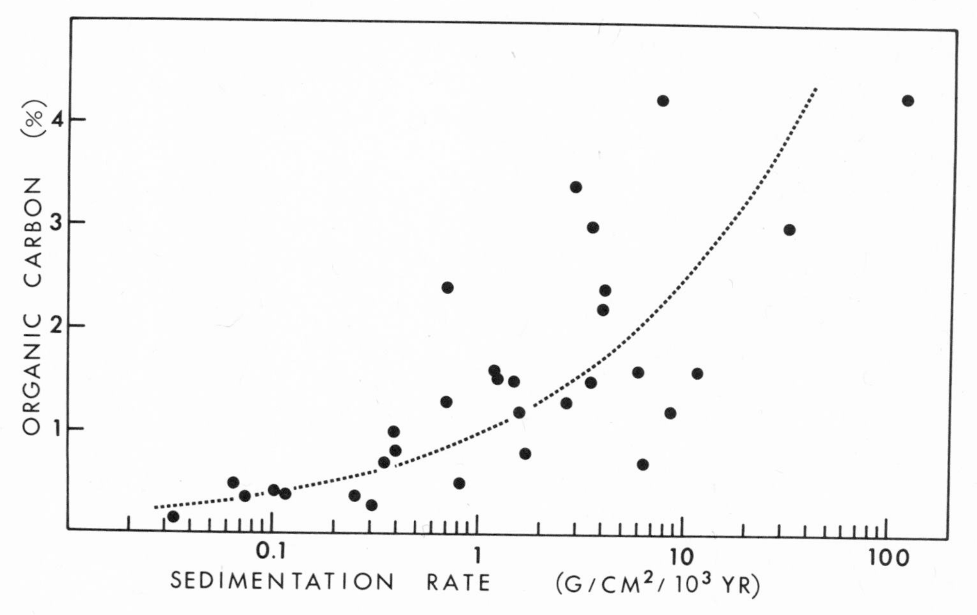

Fig. 2. Organic carbon content of marine sediments as a function of bulk-sediment accumulation rate (log scale). Dotted curve is the projected regression line of Figure 3.

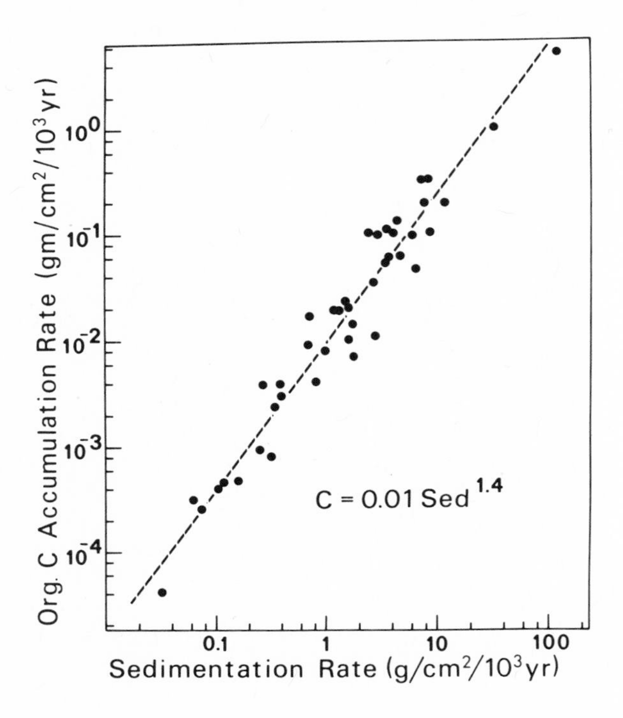

Fig. 3. Accumulation rate of organic carbon versus bulk-sediment accumulation rate (both log scales) in the cores of Figure 1. The empirical regression line suggests that the organic carbon accumulation rate is 1 percent of (sedimentation rate)$^{1.4}$.

This relation predicts organic carbon accumulation rates from total sediment accumulation rates to within a factor of about 2.5 - a degree of accuracy that is certainly adequate for first-order distribution and geochemical balance studies.

The correlation of Figure 3 is surprising, given the diverse provenance of the original sediments, ranging from carbonate-rich biogenic ooze to hemipelagic muds rich in terrigenous plant debris (*Emery*, 1960). The correlation could be fortuitous, in which case the validity of the regression equation will have to be assessed for each new depositional environment. An alternative hypothesis, though, is that Figure 3 is a reflection of the "efficiency" of benthic organisms. In areas of very slow deposition, the benthos utilizies virtually all the organic material available, whereas in rapidly accumulating deposits luxury feeding takes place and the organic carbon content is never depleted. The regression equation suggests that the "efficiency" of the benthos is highly correlated to the accumulation rate, and may be independent of the organic content of incoming detritus. Organic carbon analyses of sediment-trap collections from areas with contrasting accumulation rates and sea-floor lithologies are needed to test this suggestion and to distinguish between the two hypotheses.

Figure 4 shows the distribution of organic carbon in surface sediments of the Equatorial Pacific and North Pacific Oceans. The influence of both terrigenous input (along the continental margins) and surface biological productivity (particularly along the equator) are very apparent. Areas with high organic carbon concentrations also have the highest total sediment accumulation rates, in accord with the relations contained in Figures 2 and 3. These trends are also evident in *Bezrukov et als.'* (1961) compilation, although their map and Figure 4 differ considerably in detail, presumably due to our greater sample coverage.

PROFILES OF ORGANIC CARBON IN DEEP-SEA CORES

The organic carbon content of deep-sea cores changes with depth below the sea floor in virtually all cases. Such changes can be attributed to three dominant effects: 1) variations in the content of biogenic carbonate; 2) glacial-interglacial variations in the supply of organic carbon to the sea floor; and 3) post-depositional oxidation or diagensis of organic components of the sediment.

Effect of Carbonate Fluctuations

Variations in the carbonate content of pelagic sediments deposited at a given location usually reflect either climatically induced variations in dissolution at the sea floor (*Berger*, 1973; *Thompson and Saito*, 1974) or migration of the depositional site into deeper and deeper water due to sea floor spreading. Figures 5 and 6 show examples of these two situations. Core V20-119, from the northern Emperor Seamounts in the North Pacific Ocean, shows a series of pronounced carbonate maxima at roughly 100,000 year intervals during the Brunhes normal polarity epoch. Such fluctuations are typical of Quaternary carbonate-bearing deposits from the Pacific Ocean, and reflect differences in the corrosiveness of bottom waters between interglacial and glacial periods, (*Berger*, 1973; *Thompson and Saito*, 1974).

In contrast, core Y71-7-36P (Figure 6) from the Bauer Deep in the southeast Pacific Ocean, is from a site that was formed some 16 million years ago at the crest of the Galapagos Rise (*Hart, Yeats et al.*, 1976) and subsided as the underlying crust cooled and moved away from the spreading center. Some 4 million years ago *Kendrick* (1974) the site sank below the calcite compensation level, to produce the sharp boundary evident in Figure 6.

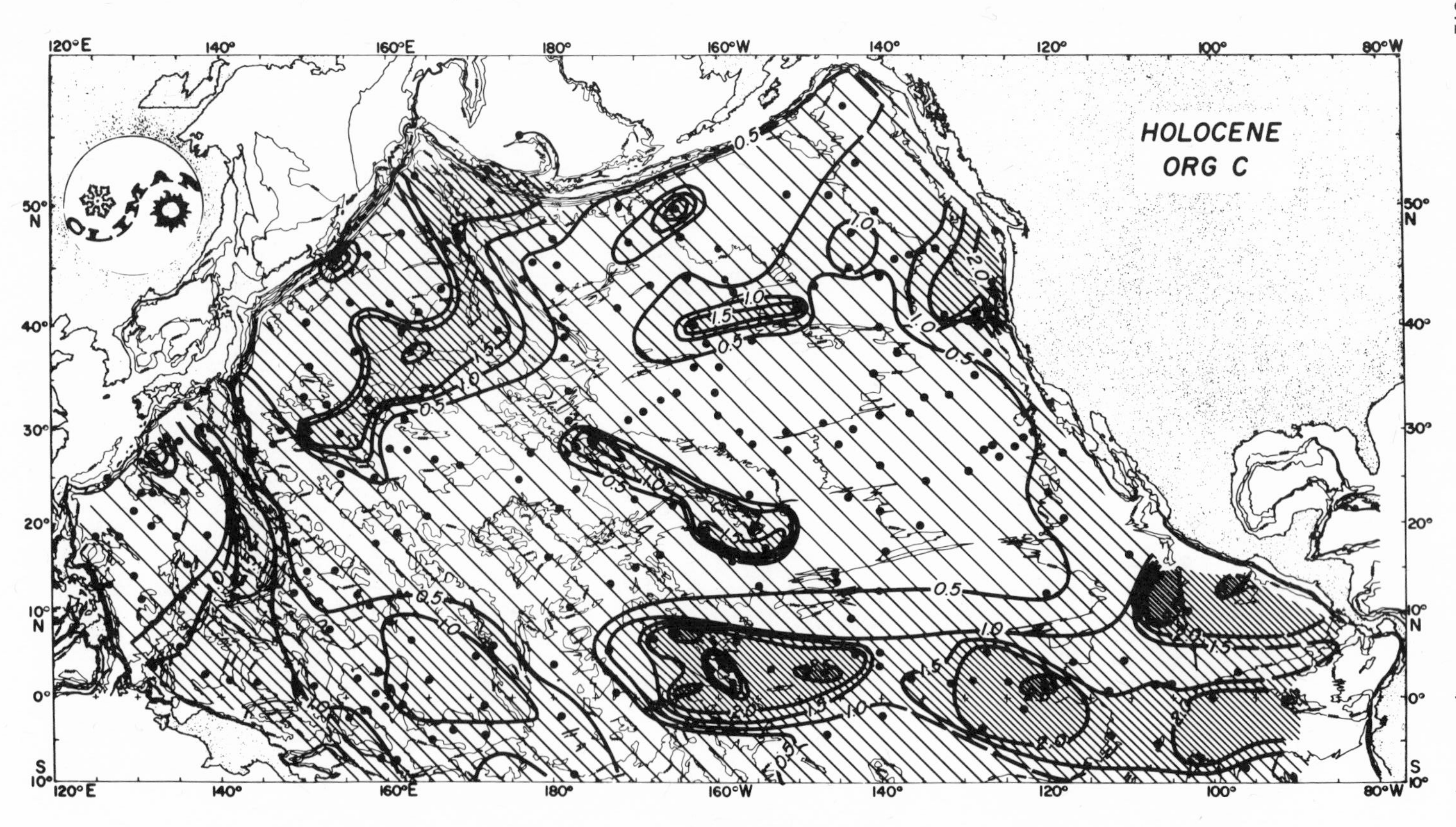

Fig. 4. Distribution of organic carbon in surface sediments from the equatorial and North Pacific Ocean, shows the influence of high surface productivity and high accumulation rates on the pattern of organic carbon deposition. Isopleths in weight percent organic carbon.

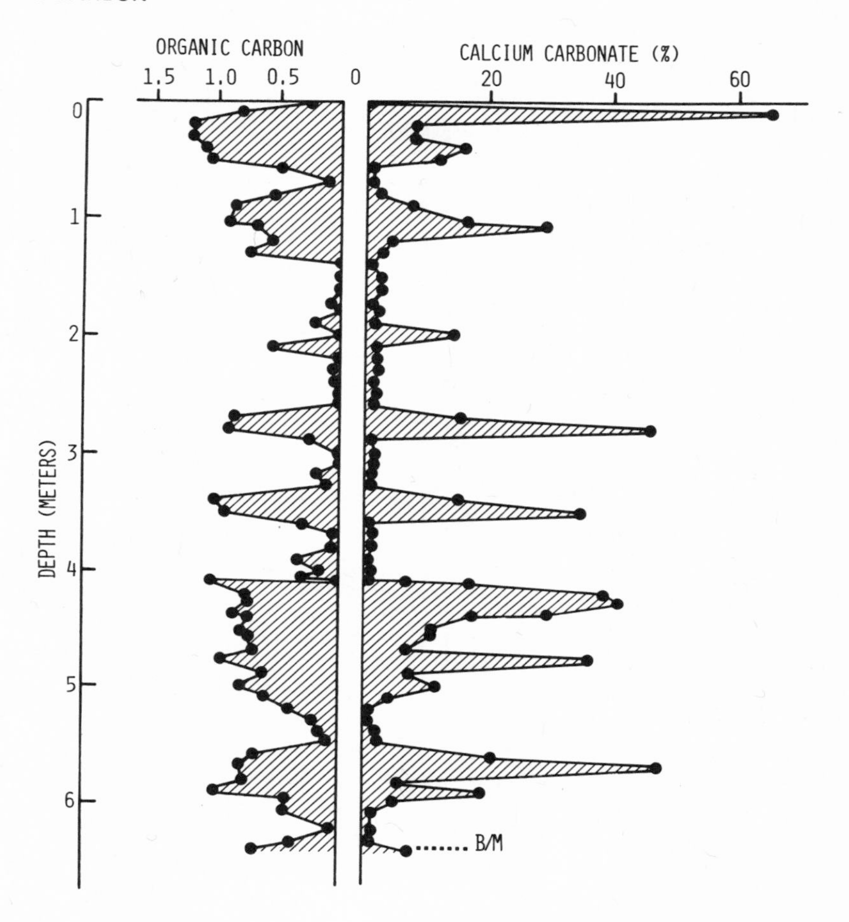

Fig. 5. Covariation of organic carbon and calcium carbonate in core V20-119 from the North Pacific Ocean. B/M is the boundary between the Brunhes and Matuyama magnetic epochs (700,000 years b.p.).

In both of these cores, the carbonate fluctuations are accompanied by conforming variations in organic carbon. Such variations could in part reflect simply the faster accumulation rate of the carbonate-rich sediments, as discussed previously. In the case of core V71-7-36P, however, the nannofossil ooze in the lower part of the core apparently accumulated at less than $0.3 gm/cm^2/1000$ yr *Dymond et al.*, 1976), rather than at about $5 gm/cm^2/1000 yr$ as would be implied from Figure 2. This anomalously high organic carbon content, in conjunction with *Lipps'* (1972 pers. comm.) observation of a thin organic layer covering the calcite plates of coccoliths,

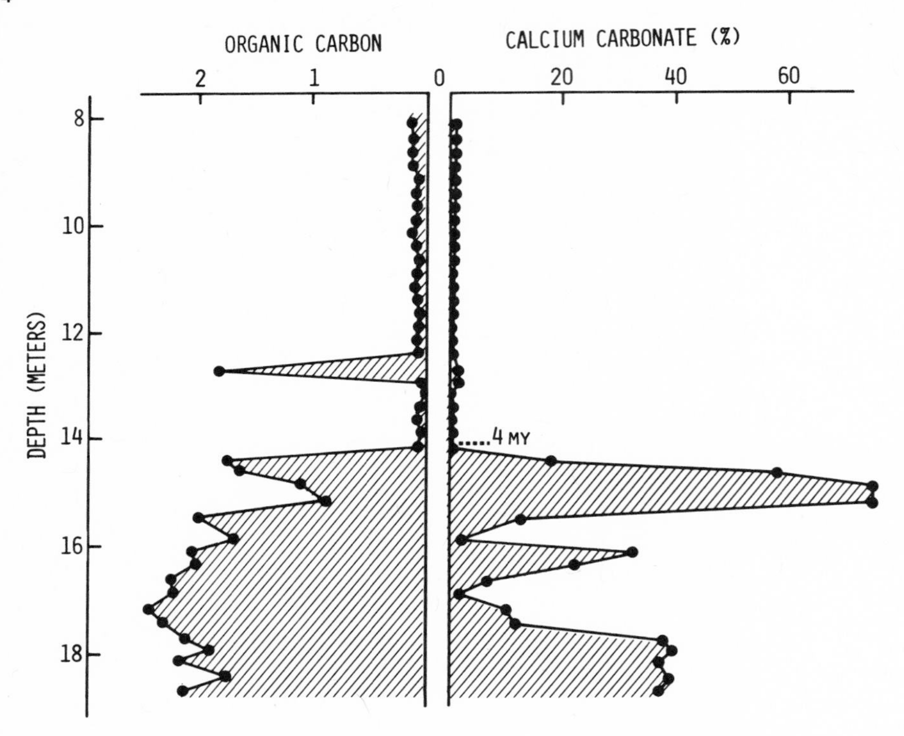

Fig. 6. Covariation of organic carbon and calcium carbonate in core Y71-7-36P from the southeast Pacific Ocean. The decrease in calcium carbonate at 5 million years b.p. marks the migration of the core site through the calcite compensation depth due to seafloor spreading.

suggests that the organic carbon is structurally associated with the carbonate debris, rather than separate particles incidentally buried by the calcareous ooze.

Effects of Paleoclimatic Fluctuations

In addition to producing the carbonate and concomitant organic carbon fluctuations discussed above, Quaternary climatic and paleoceanographic changes produced systematic changes in the supply of organic carbon to deep-sea sediments. The changes are most striking in the North Atlantic Ocean, where the equatorward shift of surface isotherms during the last glacial period was most pronounced (*Climap*, 1976).

In cores that were very close to the glacial ice margin (e.g. core V27-20 in the Labrador Sea; Figure 7) the glacial deposits are depleted in organic carbon, presumably due to influx of organic-poor glacial rock flour combined with low surface-water biologic productivity. The change from the last glacial period to the present interglacial period is marked by a rapid increase in the organic carbon content of the sediment.

Cores north of the glacial polar front, but further from the ice margin (e.g. core V23-81, west of Ireland; Figure 7) again have relatively low organic carbon values in glacial deposits. During the transition from glacial to interglacial conditions, organic carbon values peaked as rejuvenated North Atlantic Deep Water swept accumulated hemipelagic sediment to the south (*McIntyre*, 1976). Subsequently, a slower rate of deposition has resulted in lower organic carbon values again.

Finally, cores from areas where neither the surface water characteristics nor the provenance of the sediments varied markedly (e.g. core V27-141, south of the Azores; Figure 7) have relatively uniform organic carbon contents throughout.

A global comparison of glacial and interglacial rates of organic carbon deposition is not yet possible, because most well-dated glacial sections come from the North Atlantic Ocean. The existing data suggest, however, that for a given sedimentation rate, glacial deposits are enriched in organic carbon relative to Holocene deposits. Since glacial deposition rates in the deep oceans were generally greater that interglacial rates (*Broecker et al.*, 1958; *Ericson et al.*, 1961; *Heath et al.*, 1976) we infer that during glacial times,

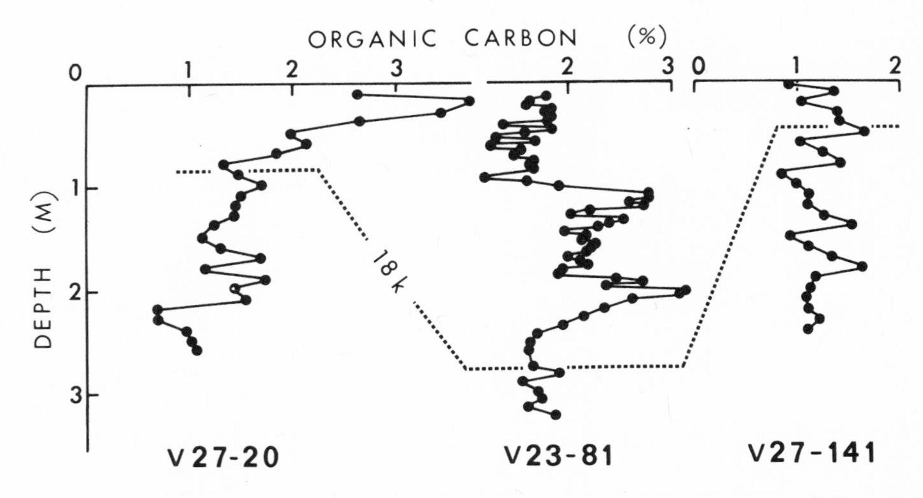

Fig. 7. Variations in the organic carbon content of North Atlantic Ocean sediments from the last glacial maximum 18,000 years ago (dotted line marked "18K") to the present. See text for a discussion of the different patterns.

organic carbon was supplied to deep-sea deposits at perhaps two to four times the present rate. It remains to be determined whether this increase in deep-sea organic carbon deposition is sufficient to balance the loss of depositional sites in paralic environments during periods of lowered sea level.

Effects of Post-Depositional Alteration.

The assessment of the degree of post-depositional alteration or destruction of organic carbon is severely impeded by the geologically recent climatic events discussed above. Fortunately, however, we have data for several cores which have sufficiently uniform lithologies that variations in organic carbon content attributable to diagenesis can be distinguished from changes in provenance.

Santa Barbara Basin

Deposition in the Santa Barbara Basin is so rapid (400 cm/1000 yr at the surface; *Koide et al.*, 1972) that several meters of sediment have accumulated during the relatively uniform climatic conditions of the Holocene. Analyses of organic carbon as a function of depth in core Y71-10-117P from the Basin (Figure 8) reflect the two sediment sources - carbon-poor detritus from the Los Angeles Basin deposited episodically after periods of high rainfall (*Drake*, 1971), and carbon-rich diatomaceous detritus deposited in the well-known annual varves *Hulsemann and Emery* (1961). The organic carbon content of both of these components decreases with depth of burial, apparently due to diagenesis. The existence of varves in the core allows the depths of Figure 8 to be converted to post-depositional ages. If the organic carbon values are plotted on a log scale versus age (Figure 9), both the end member high- and low-carbon samples can be fitted well by straight lines. The general equation for such lines is:

$$C = C_0 e^{-\lambda t} \qquad (2)$$

where C is the carbon content at time t, C_0 is the initial carbon content, and λ is a decay constant. For the lines of Figure 9, $\lambda = 4.7 \times 10^{-5} yr^{-1}$ (half-life about 15,000 years) for the carbon-rich material, and $\lambda = 1.3 \times 10^{-5} yr^{-1}$ (half-life about 53,000 years) for the carbon-poor terrigenous component. The sense of these two values is reasonable. Presumably the terrigenous phase has been much more intensely degraded in the source area and in transit to the basin than the diatomaceous component. Thus, its residual carbon is more refractory and breaks down more slowly than the varved material.

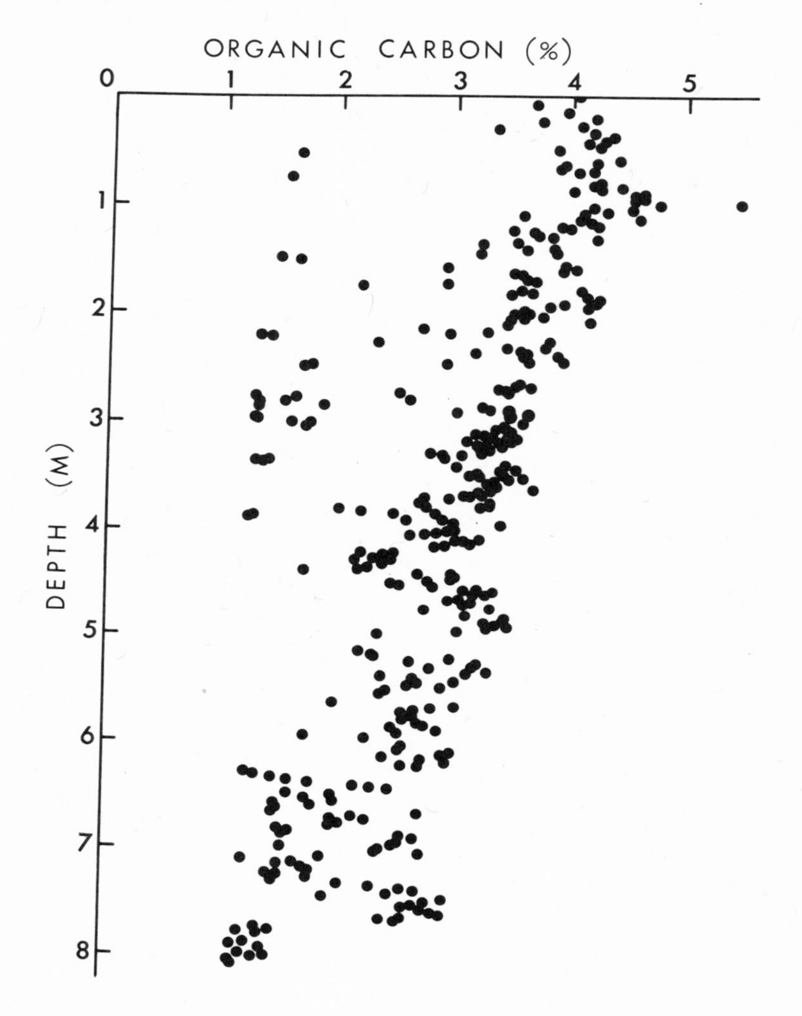

Fig. 8. Organic carbon content as a function of subbottom depth in core Y71-10-117P from the Santa Barbara Basin, off southern California.

The Santa Barbara Basin deposits are anoxic. *Sholkovitz* (1973) has determined that sulfate reduction occurs in the top meter or so, with methane formation at greater depths. The rate-controlling step in the breakdown of the organic matter is unknown, although Figure 9 makes clear that the amount of remaining organic carbon (or a component proportional to this value) is a key factor.

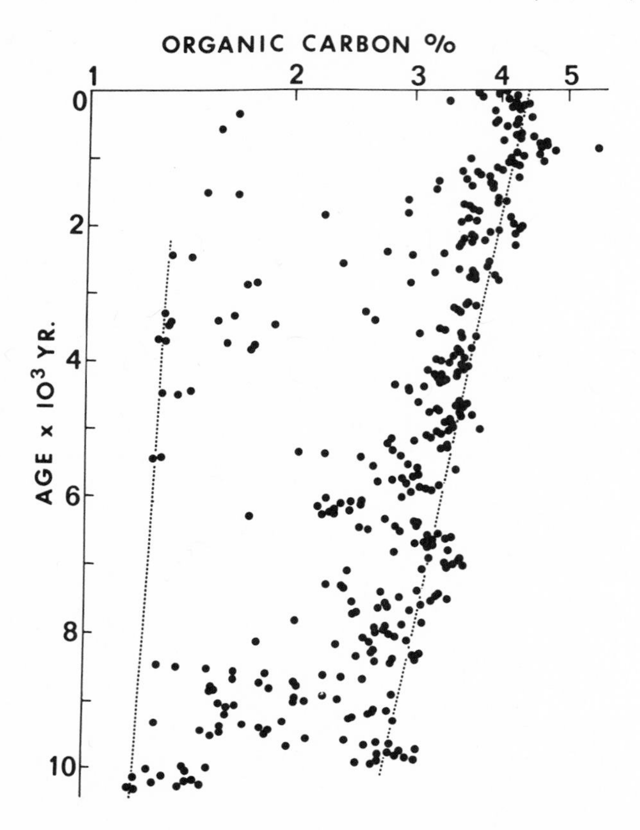

Fig. 9. Organic carbon content (log scale) as a function of time of deposition in core Y73-10-117P from the Santa Barbara Basin. Dotted lines are suggested decay curves for the turbidite (low carbon) and pelagic (high carbon) organic components.

Deviations for the straight lines of Figure 9 presumably reflect mixing of the two main sedimentary components, or injections of excess organic carbon due to minor oceanographic fluctuations over the site. The latter phenomenon appears to occur about every 2500 years, a periodicity reminiscent of the spacing of "little ice ages" *Denton and Karlén* (1973).

North Pacific Oceanic Deposits

In contrast to Santa Barbara Basin sediments, some North Pacific Ocean deposits appear to accumulate at a very slow rate that has been little affected by global climatic changes. Despite their highly oxidized state and slow accumulation rates (fractions of 1cm per 1000 yr) these deposits show exponential decreases in organic carbon with depth that are reminiscent of the Santa Barbara Basin core.

Figure 10a shows examples of two such North Pacific Ocean cores. One core, from the vicinity of 10°N;150°W is described by *Hartmann et al.* (1973). No time scale is available for this section, but if a range of 3-6mm/1000yr (typical of this area of the Pacific Ocean) is used, the decay constant is 2-4 x $10^{-5}yr^{-1}$ (half-life 16-32,000 yr). The other core, RC10-167, from the northwest Pacific Ocean *Robertson* (1975) has a decay curve defined by only 4 points (Figure 10b). The decay constant in this case, 2.2 x 10^{-4} yr^{-1} (half-life 3000 yr), is obviously less reliable than the others.

The effect of post-depositional loss of organic carbon is well displayed by maps of the 600,000 year (in a glacial interval) and 700,000 year (Brunhes-Matuyama boundary) levels in North Pacific Ocean cores (Figures 11 and 12). Much of the oceanographic info-mation evident in the surface map (Figure 4) has been lost. Only the equatorial zone of carbonate-rich deposits is marked by high organic carbon values. Such maps, together with the profiles discussed previously, illustrate why attempts to extract paleo-climatic and paleoceanographic information from organic carbon analyses of North Pacific cores have not been successful.

DISCUSSION AND CONCLUSIONS

When we began this study, we anticipated that the accumulation of organic carbon would be markedly influenced by oceanographic factors - biologic productivity of surface waters, relation of the depositional site to the oxygen minimum, etc. Such factors obviously have some influence, but are dominated by bulk sedi-mentation rate and post-depositional breakdown of the organic matter. Inasmuch as the latter effects have long time constants relative to the overturn time of the oceans, it is difficult to see how they can react quickly to changes in the feedback mechanisms that maintain the present balance between oxygen and carbon dioxide on the one hand, and recycled and mineralized organic carbon on the other. In particular, our first impression is that benthic activity will rapidly oxidize any additional organic carbon deposited as a result of the input of fossil-fuel carbon dioxide to surface ocean waters. If, on the other hand, the fossil-fuel carbon spike displaces the regression line of Figure 3 towards a higher organic carbon accumulation rate for a given bulk-sediment accumulation rate, the long time constant for post-depositional oxidation will allow the dissolution of pelagic carbonates to absorb the spike with minimal disturbance of existing depositional patterns. Our know-

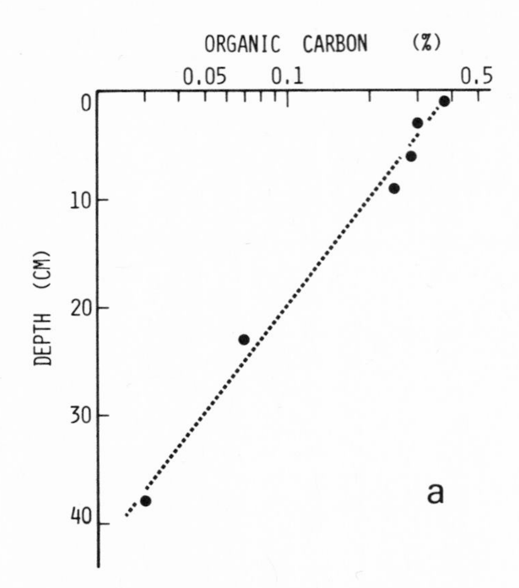

Fig. 10a. Organic carbon content (log scale) as a function of depth of burial in a North Pacific Ocean box core from about 10°N; 140°W (*Hartmann et al.*, 1973), showing suggested exponential decay of organic matter.

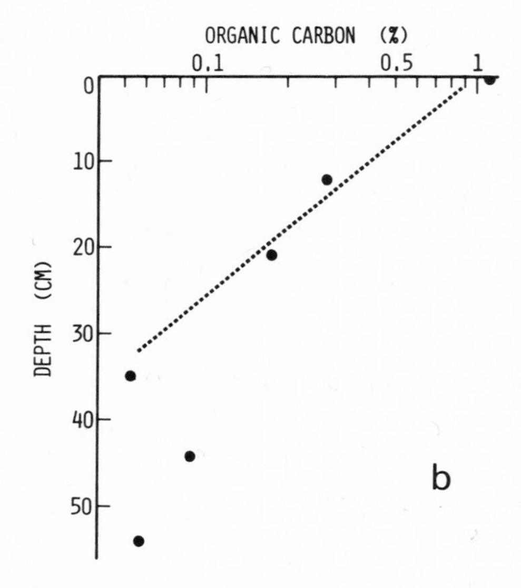

Fig. 10b. Organic carbon content (log scale) as a function of depth of burial in core RC10-167 from the northwest Pacific Ocean, showing suggested exponential decay of organic matter.

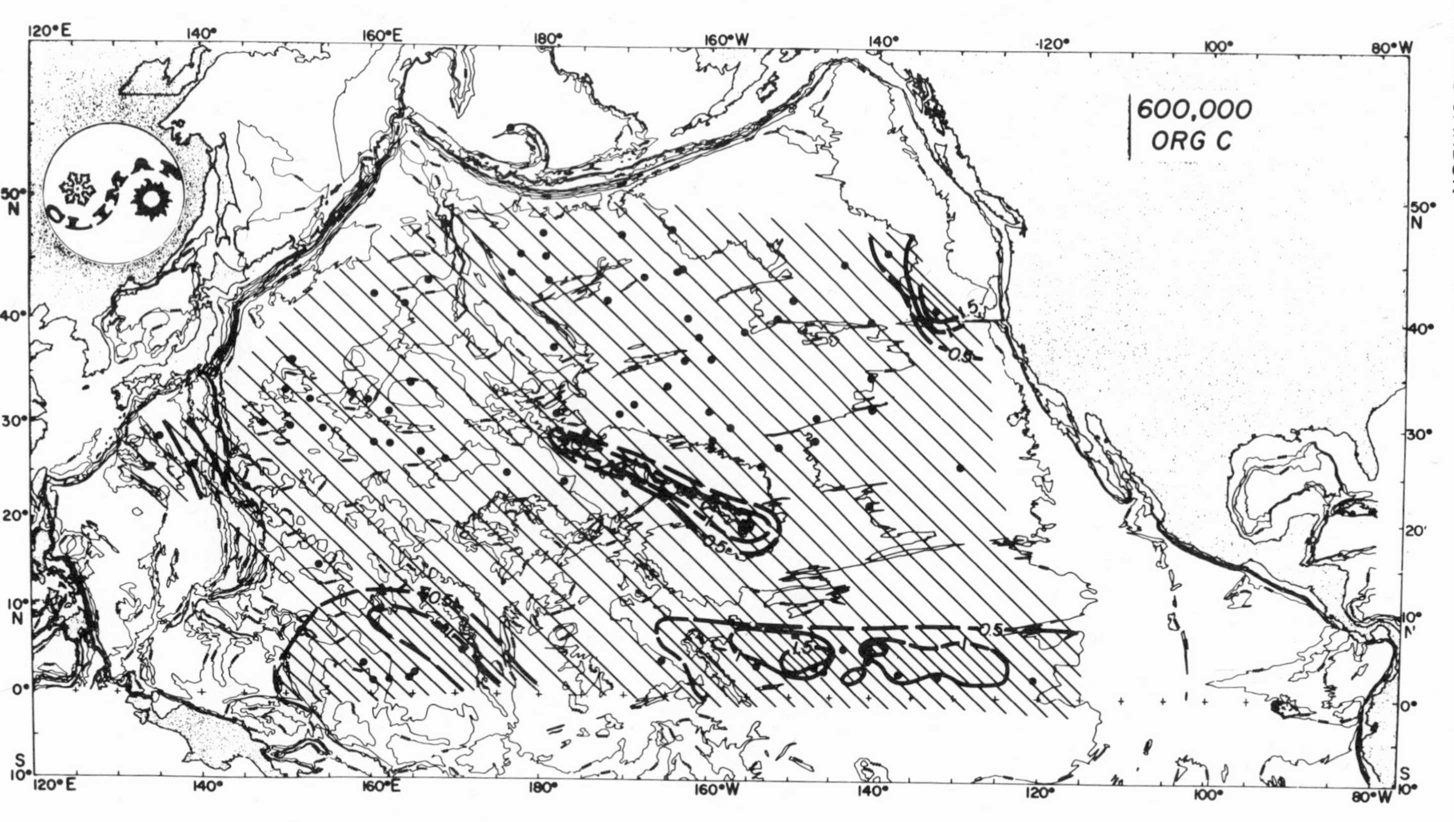

Fig. 11. Organic carbon content of equatorial and North Pacific Ocean sediments deposited during a glacial interval about 600,000 years ago. Contours in weight percent carbon.

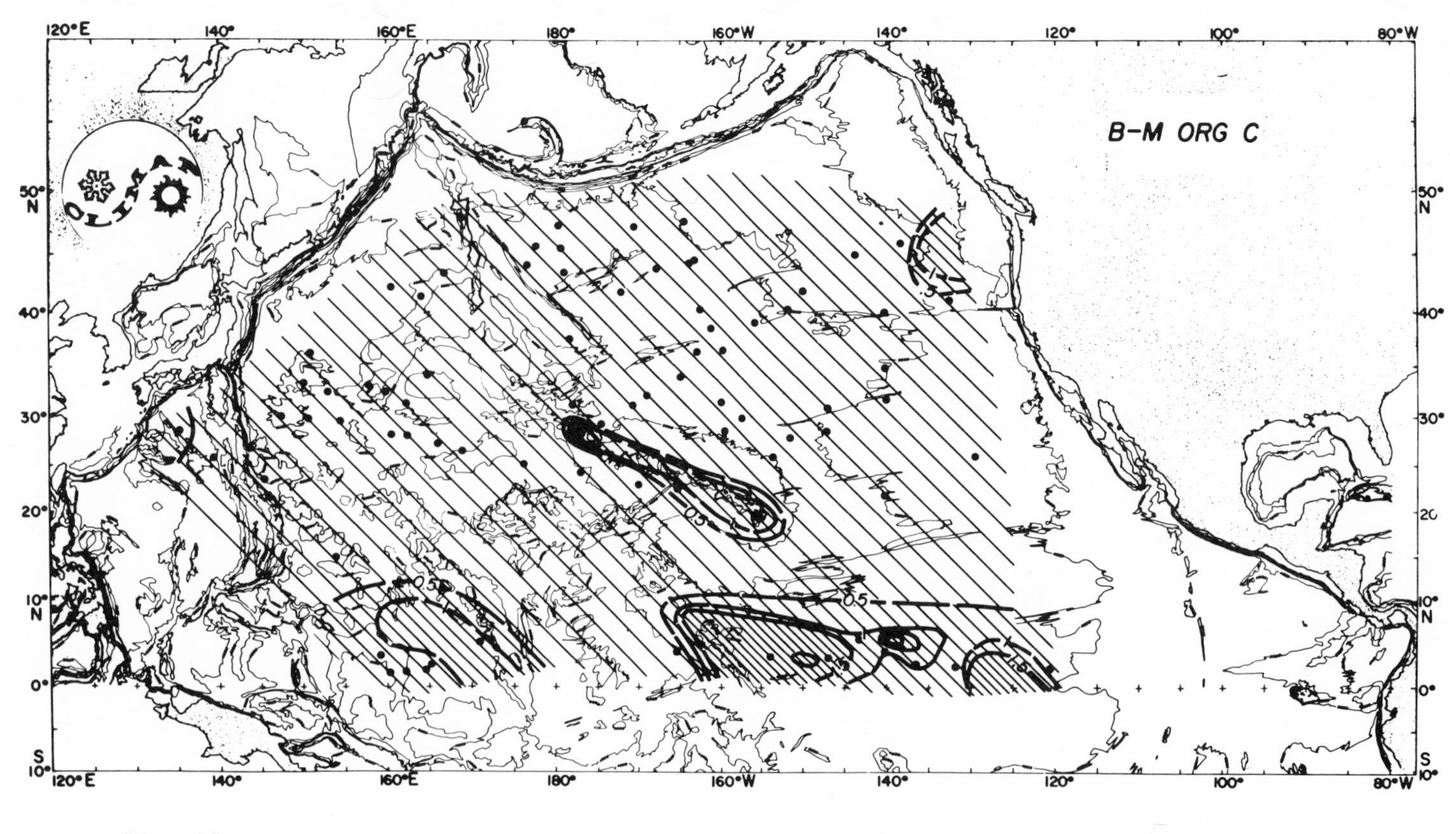

Fig. 12. Organic carbon content of equatorial and North Pacific Ocean sediments deposited during the transition from the Matuyama to the Brunhes magnetic epoch (700,000 years ago). Climatic conditions were similar to those of the present. Contours in weight percent carbon.

ledge of the rate of degradation of organic carbon from the time it reaches the sea floor until it is beyond the reach of the benthos is inadequate to choose between these alternatives at the present time.

The ultimate form of mineralized organic matter has been studied in detail by *Müller* (1975). In general, the refractory component is enriched in nitrogen relative to carbon. In the extreme, in highly oxidized pelagic clays, the C/N ratio decreases to as little as 1.3. The only organic matter to survive oxidation in such sediments are amino acids and derivatives that are tightly bound to clay minerals.

In conclusion, the rate of accumulation of organic carbon in marine sediments is primarily determined by the bulk accumulation rate of the sediment. Labile organic matter is destroyed during diagenesis, with half lives of tens of thousands of years. The ultimate concentration of mineralized or fossilized organic carbon is undetermined in organic-rich deposits. In the case of organic-poor pelagic deposits, however, it appears to be controlled by the association of organic matter with calcium carbonate or of amino acid groups with clay minerals.

ACKNOWLEDGEMENTS

This study would not have been possible without the excellent analyses made in the sedimentology laboratories of the School of Oceanography at Oregon State University, largely by C. Rathbun. We are also grateful to our CLIMAP colleagues for helpful discussions, and to the core curators, particularly those at Oregon State University and the Lamont-Doherty Geological Observatory (supported by NSF Grant GA-29460 and ONR Contract N00014-67A-0108-0004) who supplied us with samples. Portions of this research were supported by NSF Grant IDO75-20358 (CLIMAP) and ONR Contract N00014-76-C-0226.

REFERENCES

Berger, W. H., 1973. Deep-sea carbonates: Pleistocene dissolution cycles, *J. Foraminiferal Research*, 3, 187-195.

Bezrukov, P. L., A. P. Lisitzin, Ye.A. Romankevich, and N. S. Skornyakova, 1961. Contemporary sedimentation in the northern Pacific. Contemporary marine and oceanic sediments, Izd. Akad. Nauk, SSSR, Moscow.

Bordovskiy, O. K., 1965a. Accumulation of organic matter in bottom sediments, *Marine Geology*, 3, 33-82.

Bordovskiy, O. K., 1965b. Transformation of organic matter in bottom sediments and its early diagenesis, *Marine Geology*, 3, 83-114.

Boyce, R. E., and G. W. Bode, 1972. Carbon and carbonate analyses, Leg 9, Deep Sea Drilling Project, *Initial Reports of the Deep Sea Drilling Project, IX*, 797-816.

Broecker, W. S., K. K. Turekian, and B. C. Heezen, 1958. The relation of deep sea sedimentation rates to variations in climate, *American J. Sci. 256*, 503-517.

Chase, T. E., 1975. Topography of the oceans, Scripps Institution of Oceanography, *IMR Tech. Rept. Series TR57.*

Climap, 1976. The surface of the ice-age, *Science, 191.* 1131-1137.

Denton, G. H., and Karlen, W., 1973. Holocene climatic variations - their pattern and possible cause, *Quaternary Research, 3,* 155-205.

Drake, D. E., 1971. Transport and deposition of flood sediment, Santa Barbara Channel, In: *Kolpack, R. L.,* ed., Biological and Oceanographical Survey of the Santa Barbara Channel Oil Spill 1969-1970, Vol. 11. 181-217, Allan Hancock Found., Los Angeles.

Dymond, J., J. B. Corliss, and R. Stillinger, 1976. Chemical composition and metal accumulation rates of metalliferous sediments from sites 319, 320B and 321, *Initial Reports of the Deep Sea Drilling Project, XXXIV.* 575-588.

Emery, K. O., 1960. The sea off southern California, 366 pp., John Wiley, New York.

Ericson, D. B., M. Ewing, G. Wollin, and B. C. Heezen, 1961. Atlanti deep-sea sediment cores, *Geol. Soc. America Bull., 72,* 193-286.

Hart, S. R., R. S. Yeats, et al., 1976. Initial Reports of the Deep Sea Drilling Project, XXXIV. 814 pp.

Hartmann, M., F. -D. Kögler, P. Müller, and E. Suess, 1973. Preliminary results of geochemical and soil mechanical investigations on Pacific Ocean sediments, *Papers on the Origin and Distribution of Manganese Nodules in the Pacific and Prospects for Exploration,* Univ. of Hawaii, Honolulu, 71-76.

Heath, G. R., T. C. Moore, Jr., and J. P. Dauphin, 1976. Late Quaternary accumulation rates of opal, quartz, organic carbon, and calcium carbonate in the Cascadia Basin area, northeast Pacific, *Geol. Soc. America Memoir 145.* 393-409.

Hülsemann, J., and K. O. Emery, 1961. Stratification in recent sediments of Santa Barbara Basin as controlled by organisms and water character, *J. Geology, 69,* 279-290.

Kendrick, J. W., 1974. Trace element studies of metalliferous sediments in cores from the East Pacific Rise and Bauer Deep, 10°S, M.S. dissertation, Oregon State Univ., Corvallis.

Koide, M., A. Soutar, and E. D. Goldberg, 1972. Marine geochronology with ^{210}Pb, *Earth and Planetary Science Letters, 14,* 442-446.

Ku, T.-L., W. S. Broecker, and N. Opdyke, 1968. Comparison of sedimentation rates measured by paleomagnetic and the ionium methods of age determination, *Earth and Planetary Science Letters, 4,* 1-16.

Müller, P., 1975. Zur diagenese stickstoffhaltiger Substanzen in marinen Sedimenten unter oxydierenden und reduzierenden Bedingungen, Doctoral Dissertation, Christian-Albrechts Univ. Kiel

Opdyke, N. D., and J. H. Foster, 1970. Paleomagnetism of cores from the North Pacific, *Geol. Soc. America Memoir 126*, 83-119.

Robertson, J. H., 1975. Glacial to interglacial oceanographic changes in the northwest Pacific, including a continuous record of the last 400,000 years, Ph.D. dissertation, Columbia University, New York.

Sholkovitz, E. R., 1973. Interstitial water chemistry of the Santa Barbara Basin sediments, *Geochimica Cosmochimica Acta*, *37*, 2043-2073.

Stevenson, F. J., and C.-N. Cheng, 1972. Organic geochemistry of the Argentine Basin sediments: carbon-nitrogen relationships and Quaternary correlations, *Geochimica Cosmochimica Acta*, *36*, 653-671.

Thompson, P. R., and T. Saito, 1974. Pacific Pleistocene sediments: planktonic Foraminifera dissolution cycles and geochronology, *Geology*, *2*, 333-335.

Carbonate Accumulation and Dissolution in the Equatorial Pacific During the Past 45 Million Years

G. R. Heath[1], T. C. Moore, Jr.[1], and T. H. van Andel*

[1]*University of Rhode Island*
[2]*Oregon State University*

ABSTRACT

Cenozoic carbonate-bearing sediments deposited beneath both the biologically productive equatorial zone and the less productive waters more than 4° *from the equator in the central Pacific show decreasing accumulation rates with increasing water depth due to dissolution of carbonate at the sea floor. Below 3.5 km, the rate of dissolution increases linearly with depth. The vertical gradient of the dissolution rate has varied by a factor of thirty during the Cenozoic, from 0.6gm/*cm^2*/1000 yr/km during the Early Oligocene to 20gm/*cm^2*/1000 yr/km during the Late Miocene. These variations probably reflect changes in bottom-water circulation due to tectonic and climatic changes during the past 45 million years. The calcite compensation depth beneath the equator dropped almost 2 km from the Late Eocene to the Early Oligocene, and rose about 500 m from the Middle to Late Miocene. Because of the higher vertical gradient of the Miocene dissolution rate, however, the Miocene rise represents about twice the reduction in carbonate supply as does the Eocene rise.*

INTRODUCTION

Knowledge of the rate of dissolution of calcite as a function of depth in the open oceans is derived almost entirely from field experiments (*Peterson*, 1966; *Berger*, 1967; *Milliman and Laine*, 1975). The results of these experiments, which involved measuring the weight loss of calcite samples attached to vertical moorings, are difficult to translate to the sea floor because of the markedly different water-flow regime at the boundary relative to that in the unimpeded water column.

*Present address: Stanford University

Relatively little use has been made of the sediments themselves to determine rates of dissolution. *Ellis and Moore* (1973), *Kowsmann* (1973) and *Moore et al.* (1973) have attempted to estimate relative dissolution rates as a function of water depth by assuming a constant flux of the non-carbonate components to the sea floor. These estimates all suggest relatively rapid dissolution in the upper 1-3 km with slower rates at depth - a result in direct conflict with the mooring experiments.

In general, the use of modern sediments to assess the contemporary supply and dissolution of calcite at the sea floor is complicated by late Cenozoic climatic changes which led to marked variations in the calcite content of deep-sea deposits (*Arrhenius*, 1952; *Hays et al.*, 1969), and by technical problems of recovering material from the sediment-water interface, and of determining the rate of deposition of such material (*Broecker*, 1971).

Such problems are less severe for older Cenozoic deposits recovered by the Deep Sea drilling Project (DSDP). The carbonate content and bulk density of these sediments are measured routinely. In addition, the recent development of high-resolution biostratigraphic zonation schemes for several groups of microfossils (*Bukry*, 1973; *Riedel and Sanfilippo*, 1971) allows precise correlation between drill sites and facilitates the assignment of absolute ages to levels of interest in the sediment sections (*Berggren*, 1972; *Berggren and van Couvering*, 1974). With few exceptions, Tertiary changes in oceanographic conditions were slow enough (*Shackleton and Kennett*, 1975) that miscorrelations of 10^5 years do not affect the picture of calcite accumulation and dissolution - a situation that does not hold for the Quaternary.

On the other hand, plate tectonics introduces complications to the Tertiary paleoceanographer which do not affect his alter ego working with contemporary sediments. Both the depth and geographic position of a point on the sea floor change over geologic time due to subsidence of the oceanic crust as it cools and migrates away from a mid-ocean ridge, and to rotation of each lithospheric plate relative to the earth's spin axis. Each sample must be rotated back to its location of deposition and backtracked to its depositional depth (*Berger*, 1972) before it can be incorporated into a paleoceanographic reconstruction.

CENTRAL EQUATORIAL PACIFIC

For this study, we have chosen a suite of DSDP cores from the central Pacific, an area of relative tectonic simplicity that has been intensively sampled by DSDP. Samples representing 1 million-year intervals have been selected on the basis of calcareous nannofossil (*Bukry*, 1973) and radiolarian (*Riedel and Sanfilippo*, 1971) zonations converted to absolute ages (*Berggren*, 1972). Samples were rotated back to their depositional locations according to

Minster et al's (1974) scheme for the absolute motion of the Pacific lithospheric plate (*van Andel et al.*, 1975) and have been backtracked to their depositional depths using *Sclater et al's* (1971) subsidence curve as modified by *van Andel et al.* (1975).

Figure 1 shows the paleopositions of 1-million-year samples from fourteen DSDP sites used in this study. The samples have been divided into two suites. Samples within 3° latitude of the paleo-equator are grouped as an equatorial ("E") suite which lies beneath presently highly productive surface waters. Samples more than 4° of latitude from the paleoequator underlie less productive waters and are grouped as a Pacific ("P") suite. The poor sample coverage in the northwestern P area reflects hiatuses in the DSDP sections due to large-scale dissolution and erosion of Tertiary sediments southeast of Hawaii (*Moore et al.*, 1976).

Details of the tectonic reconstruction, selection of sample intervals, and data reduction, as well as a listing of the data used in the present study, are given by *van Andel et al.* (1975).

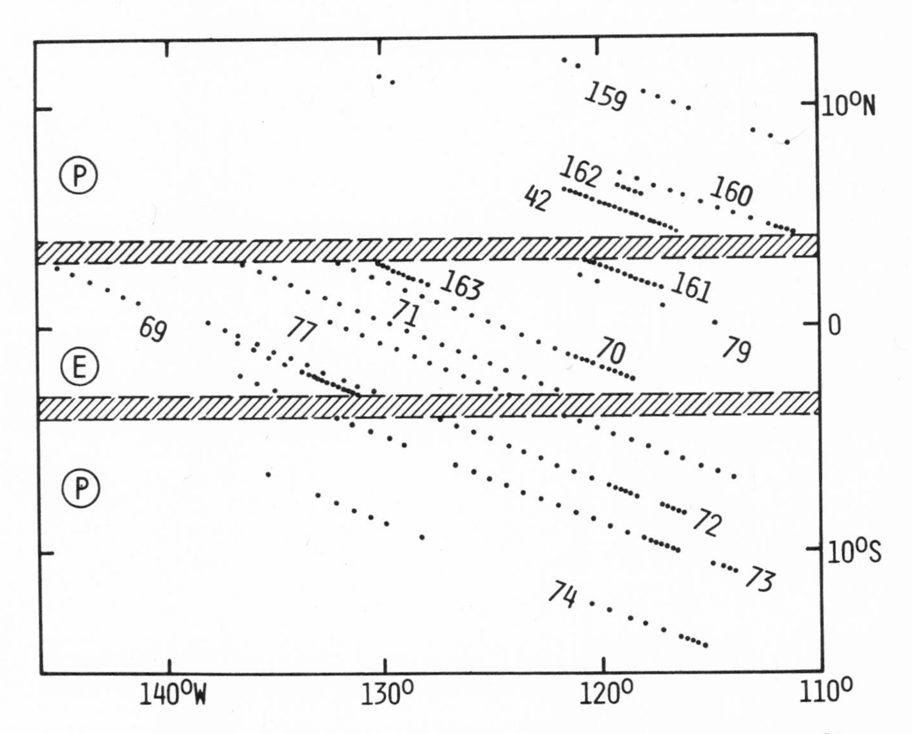

Fig. 1. Data points from the central equatorial Pacific used to construct equatorial (E) and non-equatorial Pacific (P) profiles of calcium carbonate accumulation rates (Figs. 2 and 3 respectively). Numbers are DSDP sites. Dots are backtracked site locations at 1 million year intervals (see *van Andel et al.* (1975) for a full description of the backtracking protocol).

RATE OF CARBONATE DEPOSITION

Figures 2 and 3 show isopleths of calcite accumulation rate as a function of paleodepth and depositional age for equatorial and Pacific samples.

Since the early Oligocene, the present pattern of carbonate-rich deposition along the equator and carbonate-poor deposition away from the equator has clearly prevailed. As a consequence, the calcite compensation depth (CCD), here defined as the depth where the carbonate accumulation rate is zero, has been 600m deeper beneath

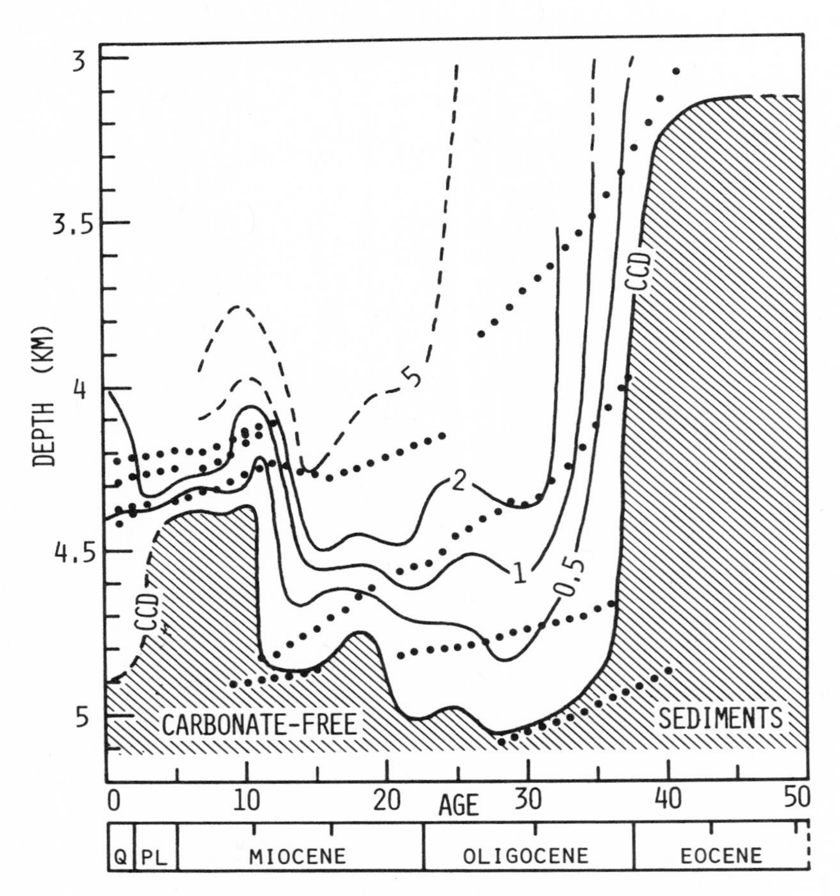

Fig. 2. Isopleths of calcium carbonate accumulation rate (gm/cm^2/10^3 yr) in a plaeodepth versus age of deposition reference system for DSDP cores within 3° latitude of the paleoequator (E region of Figure 1). CCD is the calcite compensation depth (accumulation rate = 0 gm/cm^2/10^3 yr). Dots are site positions at 1 million year intervals (Figure 1), using the backtracking protocol of *van Andel et al.* (1975)

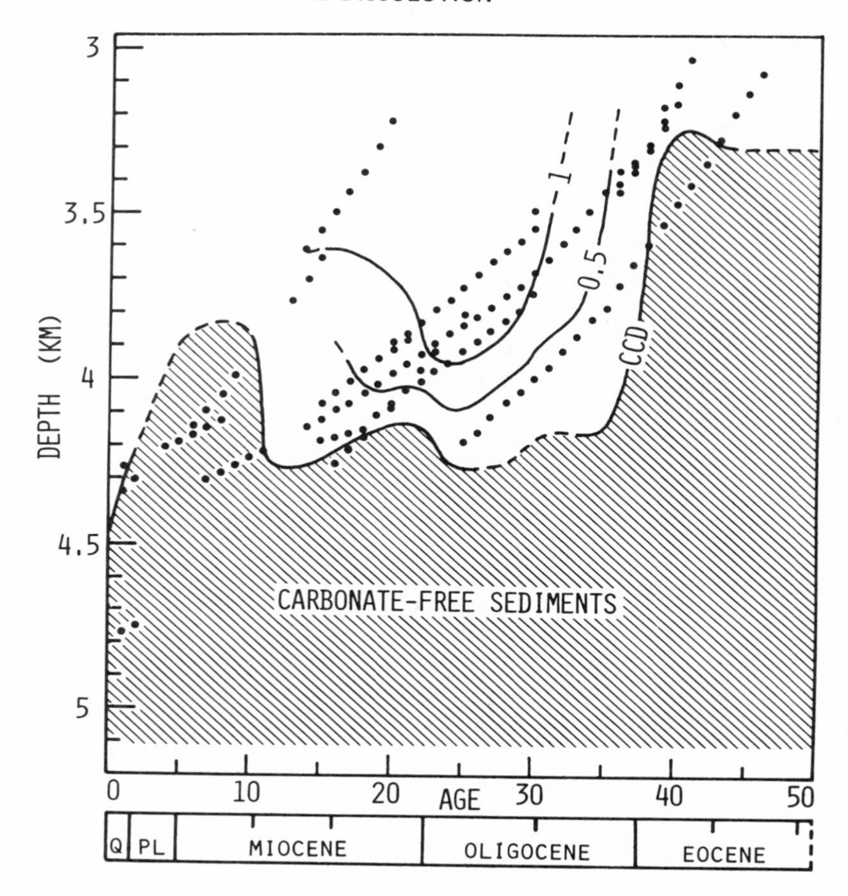

Fig. 3. Isopleths of calcium carbonate accumulation rate (gm/$cm^2/10^3$ yr) in a paleodepth versus age of deposition reference system for central Pacific DSDP cores more than 4° from the paleoequator (P regions of Figure 1). CCD is the calcite compensation depth (accumulation rate = 0 $gm/cm^2/10^3$ yr). Dots are site positions at 1 million year intervals (Figure 1), using the backtracking protocol of *van Andel et al.* (1975).

the E area than the P area for the past 35 million years (cf *Berger*, 1973). Apart from this depth offset, the E and P patterns are similar, with a very abrupt drop in the accumulation-rate isopleths during the Early Oligocene and an equally abrupt rise in the Middle to Upper Miocene. Because of the paleodepth distribution of samples, the isopleths for both P and E regions are poorly defined for the most recent 5 million years.

CALCITE DISSOLUTION GRADIENTS

If the carbonate accumulation rates for a particular interval for either P or E areas are plotted against paleodepth, the points lie on a straight line. Figure 4 shows such plots for three sets of E samples. Since the study area was restricted so as to minimize productivity variations within the P and E regions, the lines of Figure 4 represent dissolution curves. The slope of these lines is the calcite dissolution gradient (mass per unit area per unit time per unit depth increment).

Such linear gradients are not presently predictable from a knowledge of biogenic calcite supply and carbonate-system geochemistry in deep and bottom waters, but are consistent with *Peterson's* (1966) buoy experiment and with *Heath and Culberson's* (1970) interpretation of modern carbonate deposition in the equatorial Pacific. Unfortunately, the paleodepth range of our samples is only about 3.5 to 5 km. Thus, even though the linear form of the dissolution gradient is well established below 3.5 to 4 km, we cannot determine the gradient or even the functionality of the gradient at shallower depths. Thus, the paradox of the experimental

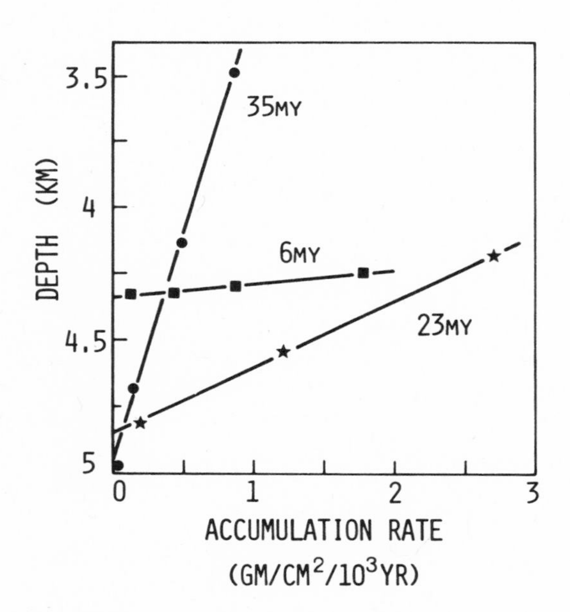

Fig. 4. Variation of the accumulation rate of calcium carbonate in equatorial Pacific (E) cores as a function of paleodepth of water for 1 million-year intervals starting at 6, 23 and 35 million years b.p. The slopes of the lines give the dissolution gradients (20, 4 and 0.6 $gm/cm^2/10^3$ yr/km respectively) for each interval. The intercept of each line is the CCD for that interval.

and sediment-derived curves cited in the introduction (see also *Berger*, 1971) remains unresolved.

At the present time, the degree of calcite undersaturation of deep waters in the equatorial Pacific increases roughly linearly with depth below 3 km (*Edmond*, 1974). If this linear form of the increase is typical of the past 40 million years, our results imply that the rate of dissolution of calcite at the sea floor is directly proportional to the degree of undersaturation. This conclusion does not agree with the experimental work of *Morse* (1977) and *Berner* (1977) which points to an exponential increase in dissolution rate with increasing degree of undersaturation. The reasons for this discrepancy remain to be determined, although *Berger* (1977) suggests that the rapid dissolution of fragile tests at low levels of undersaturation followed by slower and slower dissolution of the residual tests as the level of undersaturation increases could modulate the experimental curve sufficiently to produce the observed field relations.

A plot of the slopes of dissolution curves like those of Figure 4 shows the variation of the dissolution gradient as a function of time (Figure 5a). Both P and E curves drop from the Eocene to an Early Oligocene minimum of less than 1 gm/cm^2/1000 yr/km then increase irregularly to a maximum in the Late Miocene. Apart from an interval during the Late Oligocene, there is little difference between the two curves, suggesting that the rate-limiting step in the dissolution reaction is rather insensitive to differences in calcite supply.

Possible errors in the gradient values are difficult to assign. Any error in the duration of a 1 million-year interval will be reflected directly in the gradient for that interval. The ordinate scale on Figure 5a is logarithmic. This keeps the error bars due to such age uncertainties constant throughout the figure. Inasmuch as the assignment of absolute ages to biostratigraphic boundaries is still in a state of flux (compare *Berggren* (1972) and *Berggren and van Couvering* (1974); also *Moore* (1972)) it is clear that twofold errors in individual values of Figure 5a are possible. It is equally clear, however, that the thirty-fold increase in gradient from the Early Oligocene to the Late Miocene is much greater than any conceivable errors in the data. The absence of single point "spikes" (other than at 6 my) in Figure 5a lends further support to the reality of the overall trend.

The similarity of the dissolution gradients in the E and P areas allows an estimate of the extra flux of carbonate required to depress the equatorial CCD about 600 m below its Pacific counterpart (Figures 2 and 3). If the depth difference of the CCD (ΔCCD) is given by:

$$\Delta CCD = CCD_E - CCD_P$$

then the excess equatorial supply for any time interval will be given by:

$$\text{"E-EXCESS"} = D \times \Delta CCD$$

where $\mathcal{D}$ is the average dissolution gradient.

Figure 5b is a plot of the "excess" supply at the equator relative to the extra-equatorial (P) regions. Again, errors in the duration of our "1 million-year" intervals may affect individual points, but will not modify the overall trend. Apparently the present-day marked difference in carbonate deposition between the E and P areas is not a fundamental and constant consequence of the surface circulation pattern in the central Pacific, but rather has increased fairly uniformly since the Early Oligocene.

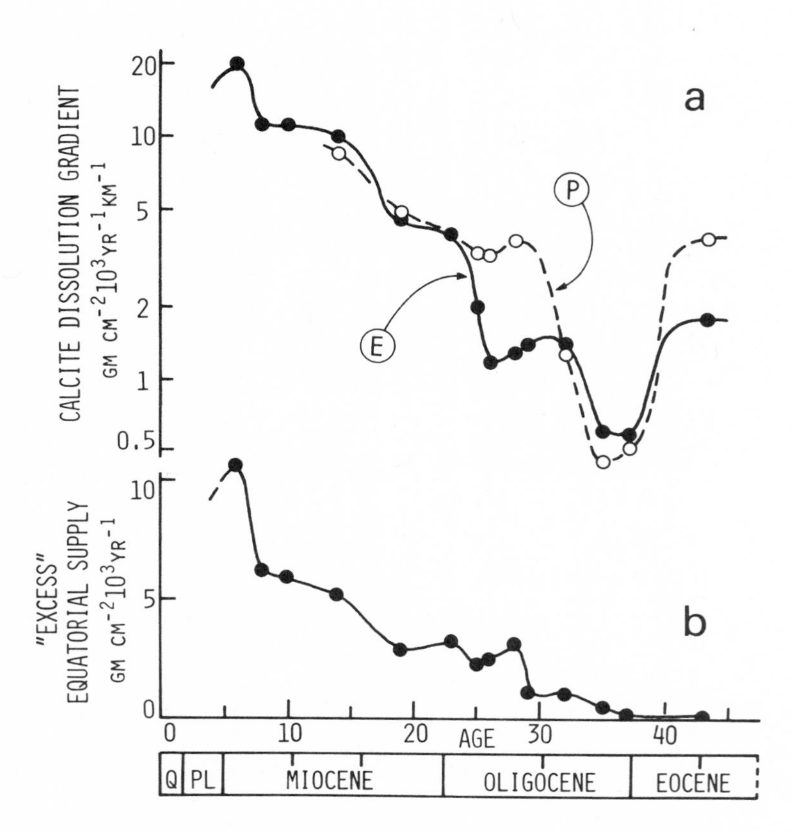

Fig. 5. a) Vertical gradient of the calcium carbonate dissolution rate (corresponding to the slopes of the lines in Figure 4) as a function of geologic age for equatorial ("E" curve, closed circles) and Pacific ("P" curve, open circles) samples (see Figure 1 for locations).

b) Additional calcium carbonate flux required to depress the CCD in the E area below the CCD in the P area by the observed depth difference (Figures 2 and 3), assuming identical vertical dissolution-rate gradients in the two areas.

Given the present uncertainty in the quantitative relation of upwelling to biologic productivity to calcite fixation, it is not yet possible to provide an unequivocal explanation for the trend of Figure 5b. One possibility, however, is that the high equatorial productivity largely results from the decrease of vertical stability (and, hence, accelerated upwelling of nutrients) due to the equatorial undercurrent. *Van Andel et al.* (1975) have suggested that the undercurrent developed only when northward migration of Australia interrupted the early Cenozoic circum-global equatorial current system and produced a piling up of surface water in the western equatorial Pacific. If these speculations are correct, Figure 5b reflects the increase in surface biological productivity along the equator as the undercurrent grew in intensity from the Early Oligocene to the Late Miocene.

DISCUSSION

The striking drop in the CCD at the beginning of the Oligocene (Figures 2 and 3) has been the subject of considerable discussion during the past decade (*Heath*, 1969; *Berger*, 1973; *van Andel and Moore*, 1974). More recently, the Miocene rise has received more attention, as it appears to be present in all the major ocean basins (*Berger and von Rad*, 1972; *van Andel*, 1975). We have attempted to assess the flux change represented by these two CCD shifts in the equatorial Pacific. The mean CCD for the past 35 million years appears to have been about 4.7 km for the E area and 4.1 km for the P areas. Deviations of the actual CCD from these values can be expressed in terms of calcite fluxes (ΔF), such that for the E area, for example:

$$\Delta F_E = (CCD_E - 4.7) \times D_E$$

where D, as before, is the dissolution gradient. CCDs below the reference depths result from an excess flux; shallower CCDs from a flux deficit. Figure 6 shows that from about 37 to 13 million years ago, ΔF was stable, slightly positive and virtually identical for the E and P areas. More importantly, however, Figure 6 shows that the flux deficit responsible for the shallow Miocene CCD was significantly greater than the deficit responsible for the shallow Eocene CCD, even though in the latter case the CCD shoaled by almost 2 km compared to about 0.5 km in the Miocene. Only when the dissolution gradient is considered in conjunction with the depth shift of the CCD does the importance of the Miocene CCD change in the equatorial Pacific become evident. This may be seen by comparing Figures 4 and 6. At 35 million years ago, only a small change in accumulation rate was required to produce a large change in the CCD, whereas at 6 million years ago, a very large change in accumulation rate would have produced but a slight change in CCD.

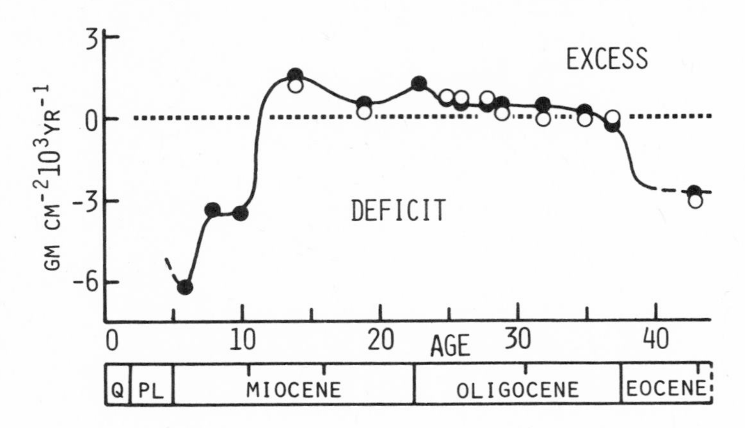

Fig. 6. Increase ("excess") or reduction ("deficit") in carbonate flux required to produce the observed differences between the CCD and 4.7 km for E sites (closed circles; Figure 2), and between the CCD and 4.1 km for P sites (open circles; Figure 3).

The factors responsible for the variations in carbonate accumulation and dissolution summarized in Figures 2 to 6 are still unknown. *Van Andel* (1975) has emphasized the role of low-latitude, shallow-water carbonate deposition in modifying the global CCD. During times of transgression such as the middle Miocene (*Vail et al.*, 1974), much of the annual river flux of carbonate to the oceans is deposited on the continental margins. To maintain a mass balance, less pelagic carbonate is preserved and the CCD rises. A related argument is that periods of continental peneplanation, such as the Eocene, are characterized by a reduced river flux of carbonate to the oceans. Again, either through reduced supply or enhanced dissolution, the maintenance of a geochemical steady state leads to a shoaling of the CCD. A third mass-balance argument attributes CCD changes to oceanographic factors that favor deposition in one ocean basin relative to another (*Berger's* (1970) euphonic "basin-basin fractionation"). Such an explanation is not tenable if CCD variations in all the ocean basins are sympathetic, as they were during the Miocene (*van Andel*, 1975), but may account for antipathetic differences (*Moore*, 1972).

Mass balance and steady state arguments may be applicable to the global distribution of the CCD (*Berger and Winterer*, 1974) but they cannot explain the dramatic changes in dissolution gradient and "excess" equatorial supply of Figure 5. These changes are much more likely to be related to the circulation of the oceans, and particularly to changes in the rate of formation of bottom water and in its "age" (as measured by its CO_2 load) when it reaches the equatorial Pacific.

Several events may have changed the deep circulation during the Cenozoic. In their isotopic analyses of benthic foraminifera from a site located at 65°S, *Kennett and Shackleton* (1976; also *Shackleton and Kennett*, 1975) record an abrupt decrease 38 million years ago in the temperature of water at 1200 m. This decrease of 4-5°C took only a few tens of thousands of years. They attribute it to the development of Antarctic alpine glaciers reaching to the sea, leading to the formation of large quantities of bottom water near 0°C. Sinking of large volumes of such water would reduce the residence time of deep and bottom waters. The lowered CO_2 content of the younger bottom water could well be responsible for the lowered calcite dissolution gradient and depressed CCD observed in Oligocene equatorial Pacific sediments. It could also explain the poor silica preservation in middle and late Oligocene deposits in the area.

Another event recorded by *Shackleton and Kennett* (1975) is the development of the Antarctic ice cap which began about 12 million years ago. This is roughly coincident with the appearance of modern North Atlantic deep water (*Vogt*, 1972). The combination of these events could have produced a thin layer of more corrosive (older) bottom water, comparable to that now found east of Hawaii. Such bottom water would explain the high dissolution gradients and shoal CCD in the equatorial Pacific during the Upper Miocene.

We have no simple explanation for the gradual change in the calcite dissolution gradient and the "excess" equatorial supply from the middle Oligocene to the Late Miocene. It is easy to hypothesize a gradual evolution of deep and surface circulation as the ocean basins developed their present physiography and as modern climatic and oceanographic zonations evolved. Unfortunately, however, nothing in our data allows us to test such hypotheses, so, regretfully, we will leave the embellishments of such speculations to our successors.

ACKNOWLEDGEMENTS

We are grateful to our colleagues M. Bender, P. Dauphin, M. Leinen, and N. Pisias for helpful discussions. Portions of this research were supported by ONR Contract N00014-76-C-0226 and NSF Grant GA-31478.

REFERENCES

Arrhenius, G. O. S. 1952. Sediment cores from the east Pacific. *Rept. Swedish Deep-Sea Exped.*, *5*. 189-201.

Berger, W. H. 1967. Foraminiferal ooze: solution at depths. *Science*, *156*. 383-385.

Berger, W. H. 1970. Biogenous deep-sea sediments: fractionation by deep-sea circulation. *Geol. Soc. America Bull.*, *81*. 1385-1402.

Berger, W. H. 1971. Sedimentation of planktonic foraminifera. *Marine Geol.*, *11*. 325-358.

Berger, W. H. 1972. Deep-sea carbonates: dissolution facies and age-depth constancy. *Nature*, *236*. 392-395.

Berger, W. H. 1973. Cenozoic sedimentation in the eastern equatorial Pacific. *Geol. Soc. America Bull.*, *84*. 1941-1954.

Berger, W. H. and U. von Rad. 1972. Cretaceous and Cenozoic sediments from the Atlantic Ocean. *Init. Repts. Deep Sea Drilling Project*, *14*. 787-954.

Berger, W. H. and E. L. Winterer. 1974. Plate stratigraphy and the fluctuating carbonate line. *Internat. Assoc. Sedimentol. Spec. Publ.*, *1*. 11-48.

Berger, W. H. 1977. Carbon dioxide excursions and the deep-sea record: aspects of the problem, In: *The Fate of Fossil Fuel* CO_2, N. R. Andersen and A. Malahoff (Eds.), Plenum, N.Y., (this volume).

Berggren, W. A. 1972. A Cenozoic time scale: some implications for geology and paleobiogeography. *Lethaia*, 5. 195-215.

Berggren, W. A. and J. A. van Couvering. 1974. The late Neogene, Paleogeography, Paleoclimatology, Paleoecology, 16. 1-216.

Berner, R. A. 1977. Sedimentation and dissolution of pteropods in the ocean, In: *The Fate of Fossil Fuel* CO_2, N. R. Andersen and A. Malahoff (Eds.), Plenum, N.Y., (this volume).

Broecker, W. S. 1971. Calcite accumulation rates and glacial to interglacial changes in oceanic mixing, In: *The Late Cenozoic Glacial Ages*, K. K. Turekian (Ed.), Yale Univ. Press, New Haven. 239-269.

Bukry, D. 1973. Coccolith stratigraphy, eastern equatorial Pacific. *Init. Repts. Deep Sea Drilling Project*, *16*. 653-711.

Edmond, J. M. 1974. On the dissolution of carbonate and silicate in the deep ocean. *Deep-Sea Research* *21*. 455-480.

Ellis, D. B. and T. C. Moore, Jr. 1973. Calcium carbonate, opal and quartz in Holocene pelagic sediments and the calcite compensation level in the South Atlantic Ocean. *Jour. Marine Res.*, *31*. 210-227.

Hays, J. D., T. Saito, N. D. Opdyke, and L. H. Burckle. 1969. Pliocene-Pleistocene sediments of the eastern equatorial Pacific: their paleomagnetic, biostratigraphic and climatic record. *Geol. Soc. America Bull.*, *80*. 1481-1515.

Heath, G. R. 1969. Carbonate sedimentation in the abyssal equatorial Pacific during the past 50 million years. *Geol. Soc. America Bull.*, *80*. 689-694.

Heath, G. R. and C. Culberson. 1970. Calcite: degree of saturation, rate of dissolution and the compensation depth in the deep ocean. *Geol. Soc. America Bull.*, *81*. 3157-3160.

Kennett, J. P. and N. J. Shackleton. 1976 (in press). Development of the psychrosphere 38 m.y. ago: oxygen isotopic evidence from subantarctic Paleogene sediments. *Nature*.

Kowsmann, R. O. 1973. Coarse components in surface sediments of the Panama Basin, eastern equatorial Pacific. *Jour. Geology, 81.* 473-494.

Milliman, J. D. and E. P. Laine. 1975. Dissolution of calcium carbonate in the North Atlantic. *Geol. Soc. America Abstracts with Programs, 7.* 1201-1202.

Minster, J. B., T. H. Jordan, P. Molnar, and E. Haines. 1974. Numerical modeling of instantaneous plate tectonics. *Geophys. Jour. Roy. Astron. Soc., 36.* 541-576.

Moore, T. C., Jr. 1972. DSDP: successes, failures, proposals. *Geotimes, 17.* 27-31.

Moore, T. C., Jr., G. R. Heath, and R. O. Kowsmann. 1973. Biogenic sediments of the Panama Basin. *Jour. Geology, 81.* 458-474.

Moore, T. C., Jr., T. H. van Andel, C. Sancetta, and N. Pisias. 1976 (in press). Cenozoic hiatuses in marine sediments, In: *Marine Plankton and Sediments,* W. R. Riedel and T. Saito (Eds.), Micropaleo. Press.

Morse, J. W. 1977. The carbonate chemistry of North Atlantic Ocean deep-sea sediment pore water, In: *The Fate of Fossil Fuel CO_2,* N. R. Andersen and A. Malahoff (Eds.), Plenum, N.Y., (this volume).

Peterson, M. N. A. 1966. Calcite: rates of dissolution in a vertical profile in the central Pacific. *Science, 154.* 1542-1544.

Riedel, W. R. and A. Sanfilippo. 1971. Cenozoic Radiolaria from the western tropical Pacific, Leg 7. *Init. Repts. Deep Sea Drilling Project, 7.* 1529-1672.

Sclater, J. G., R. N. Anderson, and M. L. Bell. 1971. Elevation of ridges and evolution of the central eastern Pacific. *Jour. Geophys. Res., 76.* 7888-7915.

Shackleton, N. J. and J. P. Kennett. 1975. Paleotemperature history of the Cenozoic and the initiation of Antarctic glaciation: oxygen and carbon isotope analyses in DSDP sites 277, 279 and 281. *Init. Repts Deep Sea Drilling Project, 29.* 743-755.

Vail, P. R., R. M. Mitchum, Jr., and S. Thompson. 1974. Eustatic cycles based on sequences with coastal onlap. *Geol. Soc. America Abstr. with Programs, 6.* 993.

van Andel, T. H. 1975. Mesozoic/Cenozoic calcite compensation depth and the global distribution of calcareous sediments. *Earth Planet. Sci. Letters, 26.* 187-194.

van Andel, T. H., G. R. Heath, and T. C. Moore, Jr. 1975. Cenozoic history and paleo-oceanography of the central equatorial Pacific. *Geol. Soc. America Memo. 143.* 134 pp.

van Andel, T. H and T. C. Moore, Jr. 1974. Cenozoic calcium carbonate distribution and calcite compensation depth in the central equatorial Pacific. *Geology, 2.* 87-92.

Vogt, P. R. 1972. The Faeroe-Iceland-Greenland aseismic ridge and the western boundary undercurrent. *Nature, 239.*

Dissolution of Calcium Carbonate in the Sargasso Sea (Northwest Atlantic)

John D. Milliman

Woods Hole Oceanographic Institution

ABSTRACT

Results from four dissolution experiments in the Sargasso Sea show that: 1) The lysoclines for aragonite, magnesian calcite and calcite occur at critical levels of undersaturation, about 50 percent in each instance; 2) The fact that the aragonitic compensation depth is shallower than the aragonitic lysocline reflects the high degree of dissolution (and thus, undersaturation) above the lysocline; 3) A shallow zone of dissolution, between 2000 and 2500 m, may be related to concentration of organic matter and its subsequent recycling; 4) Dissolution of carbonate particles is not uniform, but rather occurs at points and zones of weakness; 5) Until effective surface areas can be measured, experimental dissolution data cannot be quantified.

INTRODUCTION

Insights into the rate and manner in which calcium carbonate dissolves in the deep sea have numerous geochemical, sedimentological and paleontological implications (*Broecker*, 1971; *Milliman*, 1974; *Berger and Roth*, 1975). To date, however, the only measurement of *in situ* dissolution rates was by *Peterson* (1966), in which he exposed finely milled calcite balls (several cm in diameter) to various depths of sea water for one year. The results of this elegant experiment (Figure 1) have been quoted widely, and yet while the data are qualitatively valuable (see *Berger*, 1967), a number of potentially troublesome assumptions seriously restrict their quantitative value:

1) Perhaps the most serious problem was Peterson's implicit assumption that ground Iceland spar dissolves in a manner similar to biogenic calcite. Data by *Honjo and Berner* (pers. comm.) indicate a wide discrepancy in the dissolution characteristics of various calcites, particularly Iceland spar (*Morse*, pers. comm.).

2) Peterson calculated the surface area of the calcite balls by assuming them to be smooth spheres; weight loss then could be equated to unit surface areas. In fact, these balls contain numerous pittings, fractures and surface films that severely alter the surface character of the "spheres". Preliminary analyses indicate that the actual surface area of the spheres is more than 5 times that assumed by Peterson (*Honjo*, pers. comm.)

3) The low ratio of surface area to weight (due to the relatively large diameter of the balls) necessitated that Peterson measure extremely small weight losses relative to the original weight.

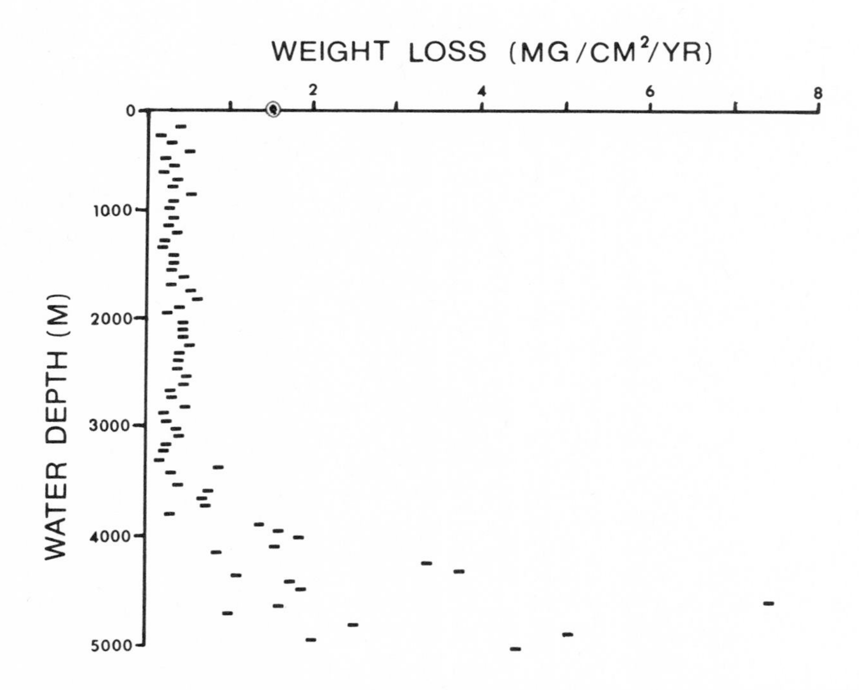

Fig. 1. Rate of calcite dissolution in the central Pacific as measured by *Peterson* (1966).

This paper presents the results of 4 experiments on the dissolution of calcite, aragonite, and magnesian calcite in the Sargasso Sea, in which some of the problems in Peterson's experiment were at least partially corrected. To alleviate the surface area/weight problem and at the same time use "natural" marine carbonate, aragonite and magnesian calcite (12 mole percent $MgCO_3$) ooids and "spherical" planktonic foraminifera (mainly *Orbulina universa* and *Globigerinoides sacculifer*) were used as carbonate specimens. Primary and secondary organic coatings which retard dissolution (*Chave and Suess*, 1970) were left intact, so that the particles would react more naturally with sea water. Particles between 250 and 500 microns in size were separated by sieving, and spheroidal grains concentrated by rolling down a slightly inclined plane. The resulting surface area/weight ratios of these particles were about 4 to 5 orders of magnitude greater than those in Peterson's experiment. At first, the writer mistakenly assumed that ooids had effective surface areas of spheres (and calcitic forams, 4 times the surface area of the ooids); thus the early results (which reported as weight loss relative to surface area; *Milliman*, 1975) were as erroneous as those of *Peterson* (see below).

Pre-weighed samples (generally 150 to 400 *mg*; 300 to 1000 grains of each carbonate species) were placed in porous cloth sacks and suspended within containers designed to allow maximum contact with ambient sea water but protect the samples from agitation and abrasion (Figure 2)[1]. To ensure consistent results, replicate samples of the carbonate species were placed in each container. The containers were then attached at fixed intervals to MODE and POLYMODE moorings in the Sargasso Sea (Figure 3) and left for 4 month (Station 1) and 9 month (Stations, 543, 545, 548) intervals. The nature of the MODE and POLYMODE experiments did not

---

[1]Nylon pantyhose (with mesh openings of approximately 20 to 40 μ), cut into 10 *cm* square patches, proved ideal, low-cost cloth. To measure the degree to which the cloth hampered circulation of sea water, pre-weighed specimens of coralline algae were suspended within various containers, some in cloth and others hanging freely. Variations in dissolution between the exposed and covered fragments were minimal, indicating that the cloth was permeable to sea water (*Milliman*, 1975). To measure the degree of abrasion by vertical and horizontal motion of the mooring, preweighed specimens were placed within polyethylene sacks and placed in the container; no abrasion was noted. To minimize the corrosion of the galvanized bolt which connected the two parts of the container (Figure 2), the bolt was coated with several layers of Scotch-Coat(r) and fitted with polyester "shrink-tubing"; very little corrosion was noted, even after emersion for 9 months.

Fig. 2. Preweighed carbonate samples were placed in cloth sacks and hung from the top of a porous container. A protective outer cylinder of the container was then placed over the sacks and secured with a metallic bolt. The container had 30 to 40 0.6 *cm* holes to allow complete communication with sea water. All metallic parts were thoroughly coated to prevent corrosion (see text).

permit the close vertical control achieved by Peterson (Figure 1), and only 8 to 10 containers were placed on any single mooring. On the other hand, mooring these containers was essentially cost-free to the writer, thus allowing a number of experiments to be placed at various sites for a minimal price.

Upon recovery and return to the laboratory, the samples were washed with buffered (*pH* 8.0 to 8.5) distilled water, dried and weighed. The percent weight loss (per year) was then calculated.

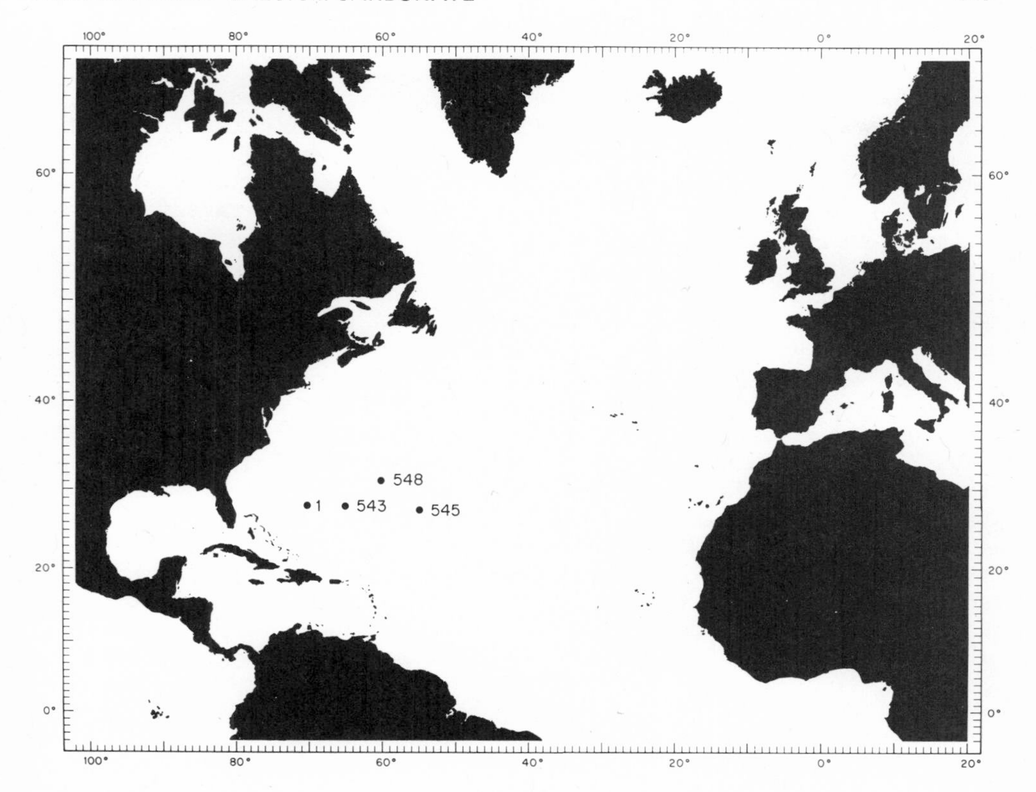

Fig. 3. Location of 4 dissolution experiments in the Sargasso Sea. Station 1 was emplaced in mid-December, 1973, and recovered in mid-April, 1974. Stations 543, 545, 548 were emplaced in mid-April, 1974, and recovered in early January, 1975. Deepest moored containers were generally 50 meters above the bottom.

OBSERVED RATES OF DISSOLUTION

Dissolution patterns measured were roughly similar in all four experiments (Figures 4-7): Solution in the upper 1000 meters was small or not detected. Dissolution of aragonite became noticeable at about 2000 *m* and the aragonitic lysocline (the depth at which rate of dissolution accelerates - *Berger*, 1968) occurred between 3000 and 3500 *m*. The magnesian calcite lysocline was between 3500 and 4000 *m* and the calcitic lysocline between 4500 and 5000 *m*.[2]

The depth of the chemical calcitic lysocline measured in this experiment agrees closely with the observed foraminiferal lysocline in the north Atlantic (*Berger*, 1975) and is about 500 to 1000 *m* shallower than the calcitic CCD (carbonate compensation depth) (*Lisitzin*, 1972; *Berger and Winterer*, 1974). The aragonitic lysocline, however, falls below the observed compensation depth of aragonite (*Chen*, 1964; *Pilkey and Blackwelder*, 1968). This apparent anomaly is explained by the fact that even above the lysocline, aragonite dissolution is sufficiently rapid (see footnote) to remove any aragonitic tests; thus, the aragonitic lysocline measured in these experiments is strictly a chemical parameter and has no paleontological or sedimentological implication.

One unexpected anomaly was noted in the first experiment (Station 1): dissolution of all three carbonate species at 2500 *m* was markedly higher than in waters immediately above or below. In fact, calcite dissolution at 2500 *m* was greater than it was at 4500 *m* (Figure 4). Without further data, the writer (*Milliman*, 1975) could not dismiss the possibility of an analytical or observational error (although replicates agreed closely) or perhaps an abnormal microenvironment within the container. However, when 2 of the next 3 experiments also showed similar dissolution patterns (and rates) at 2000 *m* (Figures 5-7), it became apparent that dissolution at the 2000-2500 level was real. It seems unlikely that calcite (or aragonite or magnesian calcite) is sufficiently undersaturated to effect a marked dissolution at this depth. Perhaps dissolution results from local changes related to the concentration and recycling or organic matter (*Morse*, pers. comm.); the exact process, however, is not clear. The extent to which dissolution is reflected in carbonate sediments at the 2000-2500 *m* level is probably small, due to the relatively high carbonate productivity in most continental margin areas.

MODES OF DISSOLUTION AND PROBLEMS IN QUANTIFYING RATES

Dissolution of carbonate particles appears to occur primarily in zones of weakness (Figure 8). As these surface flaws deepen

---

[2] Because the aragonite and magnesian calcite ooids are solid spheres and because they have relatively few surface irregularities, their effective surface areas (per unit weight) are estimated to be 1 to 2 orders of magnitude smaller than those of the calcitic foraminifera. Thus, equal weight losses of ooids and forams indicate 10 to 100 times greater dissolution (per unit surface area) for the ooids.

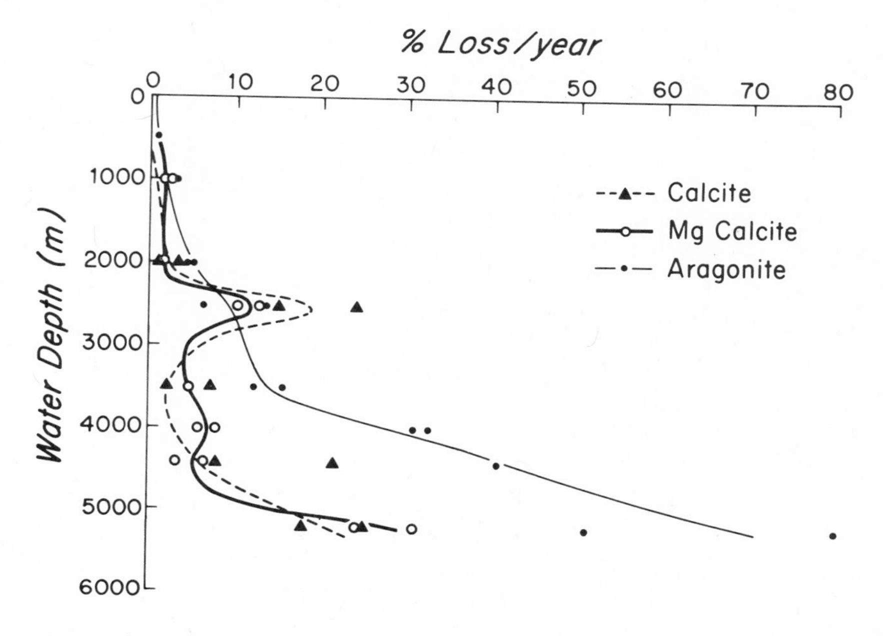

Fig. 4. Calculated weight loss (per year) of various carbonate species at Station 1.

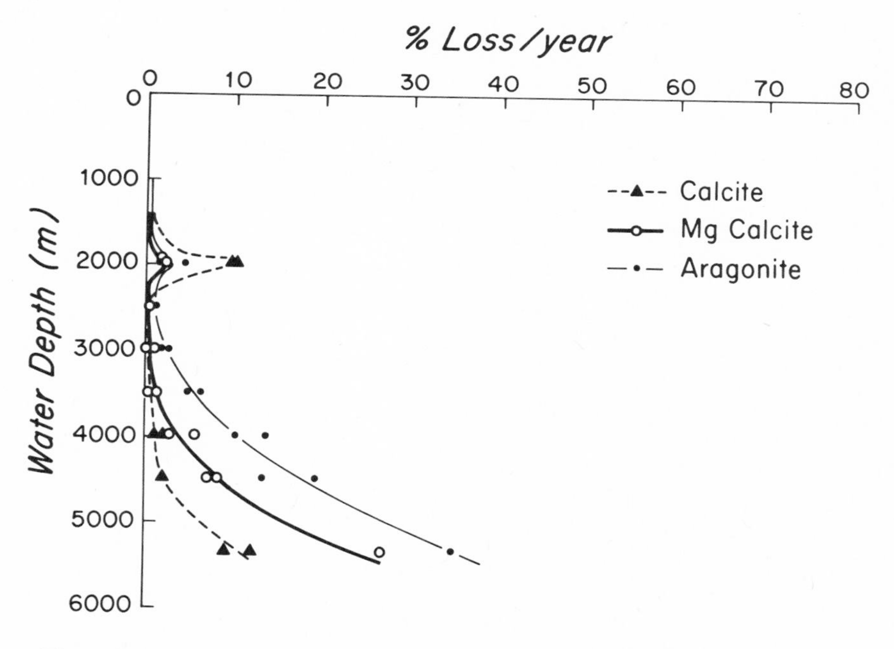

Fig. 5. Calculated weight loss (per year) for various carbonate species at Station 543.

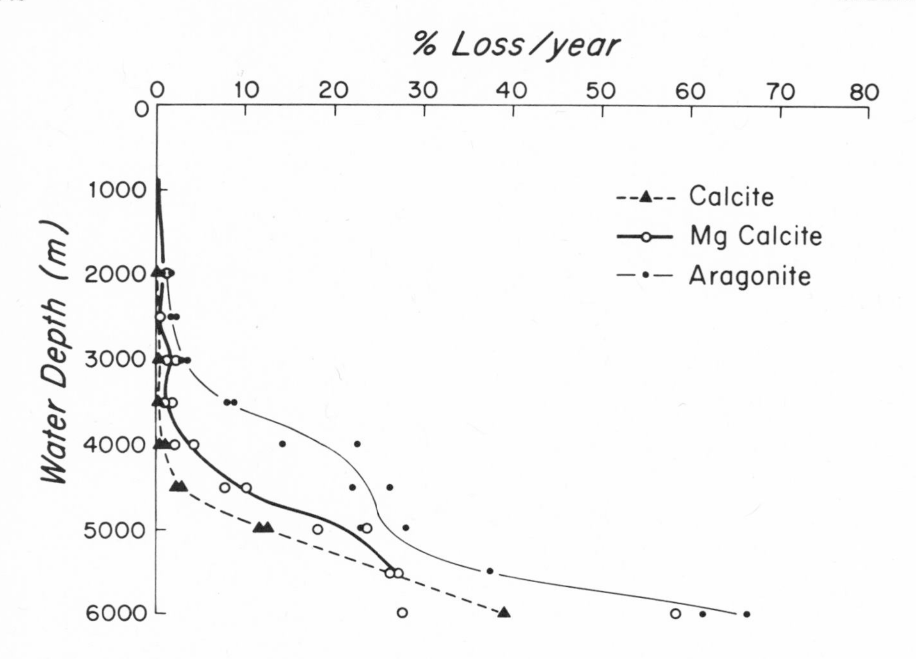

Fig. 6. Calculated weight loss (per year) for various carbonate species at Station 545.

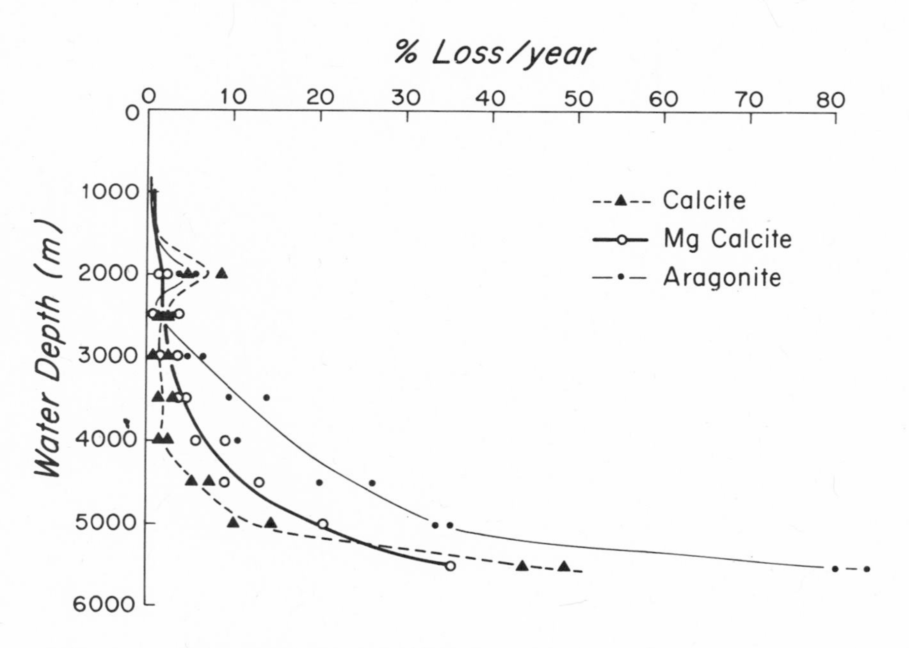

Fig. 7. Calculated weight loss (per year) for various carbonate species at Station 548.

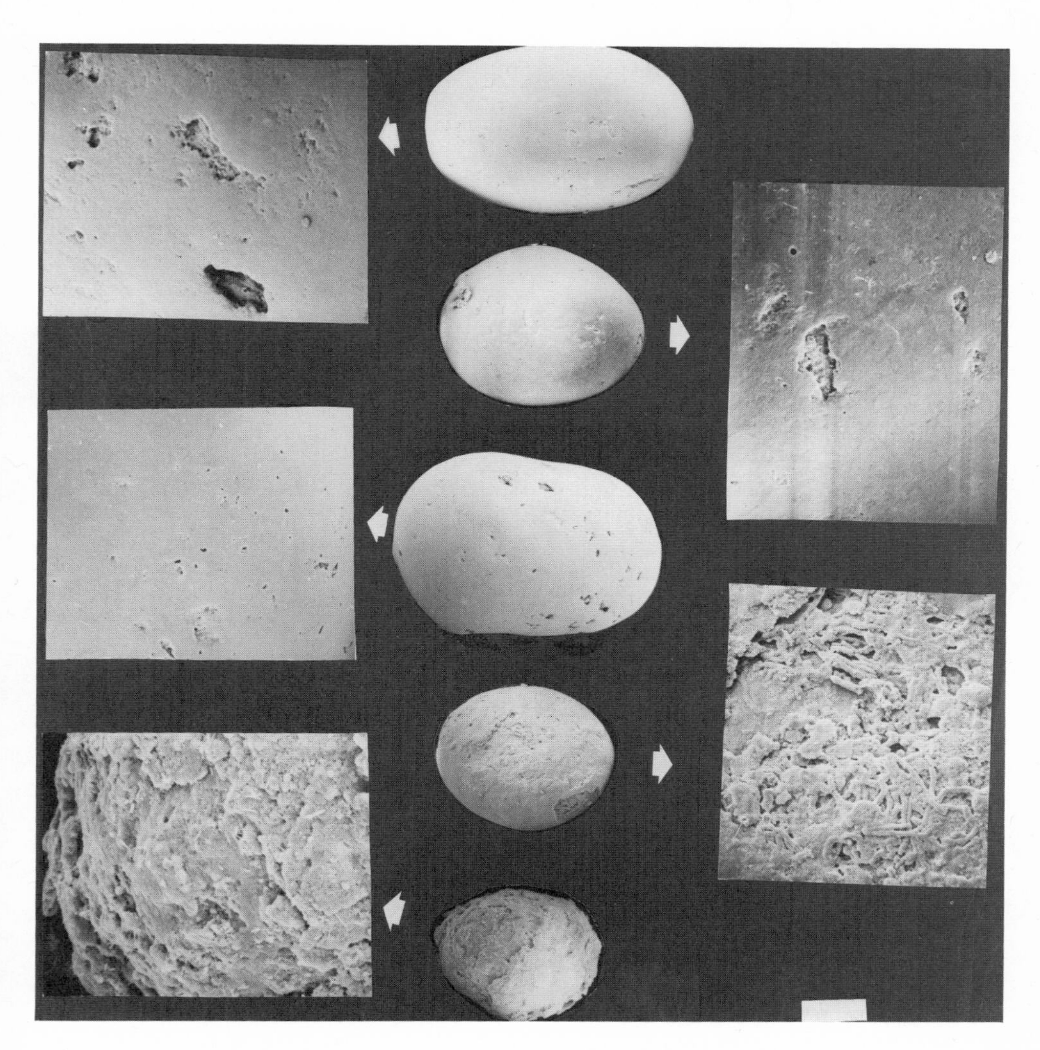

Fig. 8. SEM photomicrographs showing dissolution of aragonitic ooids at Station 548. These ooids were exposed to ambient seawater for approximately 9 months, at depths of 500, 2500, 3000, 4500 and 5500 *m*, (top to bottom, respectively). The surface blemishes of the ooid at 500 *m* are algal borings; it is in these local points of "weakness" where dissolution occurs. Central photos show entire ooids; photos on either side show higher magnifications of ooid surfaces. Scale (lower right) is 50 microns for central photos and 250 microns for the side photos.

and widen, surface layers begin to flake off[3], surface area of the particle increases (compare photomicrographs at 4500 to 5000 *m*), and dissolution accelerates. Thus, quantifying dissolution rates becomes increasingly difficult since the effective surface area of any particle changes markedly as it dissolves. Even if the surface area of the original particles has been correctly estimated, the quantitative values of *Peterson* (1966) and *Milliman* (1975) would have been misleading in that they did not take into account the constantly changing surface area with increasing dissolution. The rates shown by both Peterson and Milliman below the lysocline are certainly too great, but whether the acceleration of dissolution is linear (*Broecker and Broecker*, 1973) or exponential (*Morse and Berner*, 1972) cannot be determined until surface areas can be accurately measured.

One final point should be noted. In Peterson's experiment, the calcitic lysocline occurred where the water was approximately 50 percent saturated relative to calcite (*Edmond*, 1974). Similar calculations show that the aragonite and calcite lysoclines in the Sargasso Sea also occur at approximately 50 percent saturation levels (Figure 9). These data indicate that dissolution of both calcite and aragonite accelerates at a critical undersaturation; the absolute level of saturation, of course, depends upon the thermodynamic constants used in computing saturation levels. However, in the case of aragonite, dissolution begins at depths far shallower than the lysocline (see above), thus indicating that these waters also are undersaturated. This infers that the lysocline does not occur at the transition from saturated to undersaturated conditions (*Ben-Yaakov et al.*, 1974), but rather at some critical level of <u>under</u>saturation.

FUTURE WORK

Although the data from these 4 experiments delineate the various carbonate lysoclines within the Sargasso Sea and illustrate the mode of dissolution, quantifying these data requires two additional measurements. First, it must be proved that the pantyhose sacks do not limit access of ambient ocean water nor restrict "interstitial" circulation between the particles themselves. This latter situation could result in retention of relatively saturated "pore waters", thus reducing the level of dissolution of spheres not directly in contact with sea water. The fact that the pantyhose appear permeable to sea water, that replicate samples show strong

---

[3]In some sacks, very fine (less than 20 microns in diameter) powder was noted, presumably derived by the flaking off of surface carbonate during "dissolution".

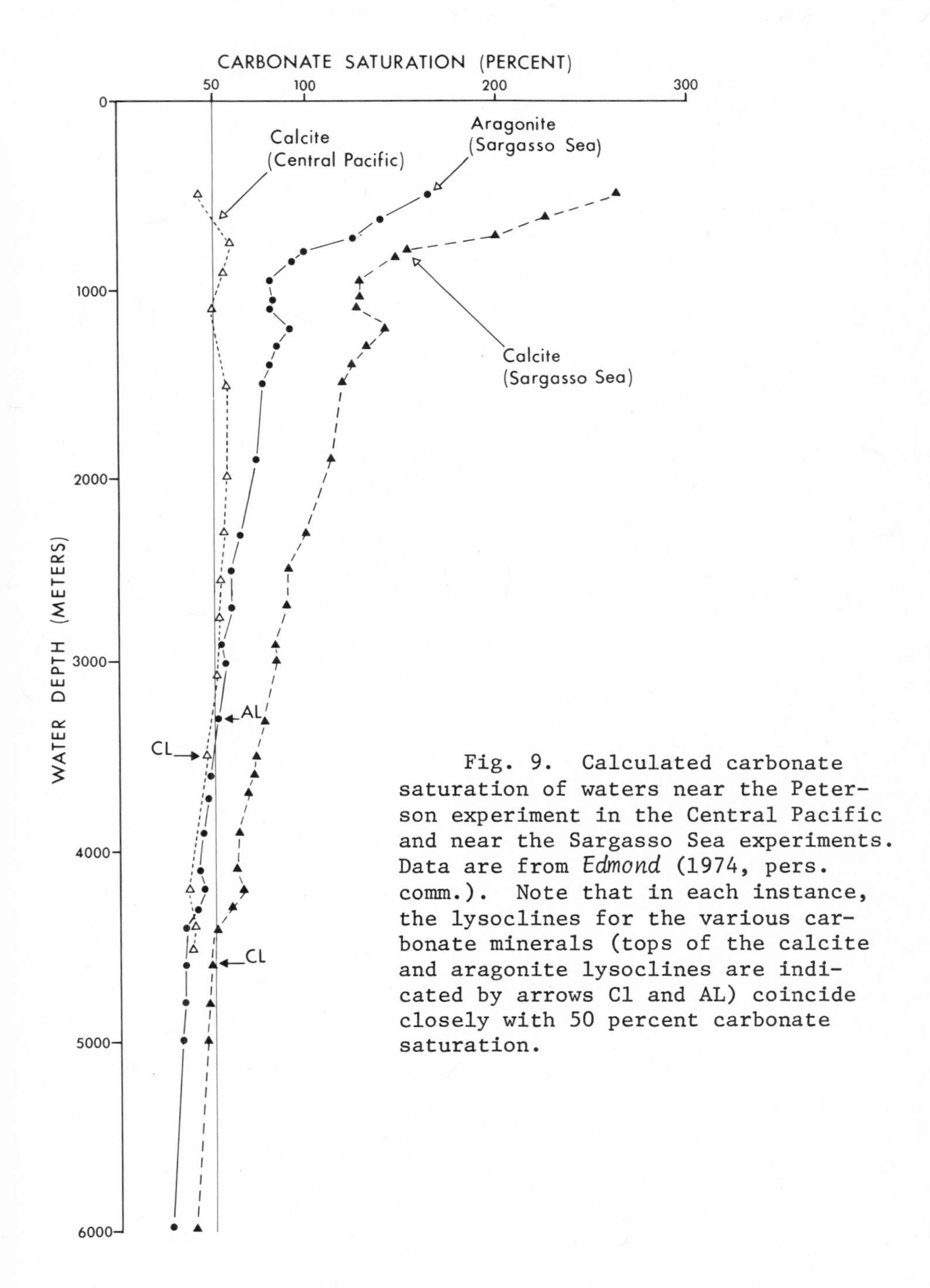

Fig. 9. Calculated carbonate saturation of waters near the Peterson experiment in the Central Pacific and near the Sargasso Sea experiments. Data are from *Edmond* (1974, pers. comm.). Note that in each instance, the lysoclines for the various carbonate minerals (tops of the calcite and aragonite lysoclines are indicated by arrows Cl and AL) coincide closely with 50 percent carbonate saturation.

agreement, and that the carbonate particles are sand-size should favor optimal permeability of both the cloth sacks and the particles. On the other hand, particles within any one sample tended to show a wide range of dissolution; whether this reflects a natural situation among carbonate particles (*Berger*, 1967) or indicates poor circulation within the cloth sacks will be tested when the dissolution rates are compared with data derived from a more sophisticated experiment by Honjo (in which plates coated with carbonate grains are exposed to constant currents at various depths).

Second, the effective surface areas of the grains must be accurately measured, not only the original samples, but also those that have undergone dissolution. The latter measurements are particularly important, since these data may help delineate whether the acceleration of dissolution below the lysocline is linear or exponential (see above). However, measured surface areas may not accurately reflect the effective surface exposed to dissolution. Organic coatings retard dissolution and some pits and crevices may be too small to allow sufficient "circulation" (and therefore optimal dissolution); moreover, dissolution appears to begin at points of weakness (Figure 8) and not over the entire particle. Thus, analytically measured surface area may be far greater than the "effective" surface exposed to dissolution. Resolving this problem awaits further experimental data.

ACKNOWLEDGEMENTS

This study was supported by the Office of Naval Research (Contract N00014-74-0262 and NR 083-004); MODE and POLYMODE support came from the International Decade of Ocean Exploration (NSF contracts 10071-04215 and 10075-04215). I thank C. Offinger, L. Toner, D. Allison and A. Milliman for contributing their pantyhose, thereby braving cold legs during New England winters. Special thanks go to Christopher Milliman for unknowingly supplying nylon fishing line which was used to tie and attach the cloth to their containers.

J. Ellis helped in the original design of the containers, and subsequently manufactured nearly 100 of them. D. Moller, R. Heinmiller and G. Tupper set and recovered the containers from the MODE and POLYMODE moorings. C. Offinger and M. Goreau aided greatly in the subsequent analyses. I thank particularly Susumu Honjo for help with the scanning electron microscopy and for many fruitful discussions concerning this experiment in particular and dissolution of calcium carbonate in general. Drs. R. A. Berner and W. H. Berger offered helpful criticisms of the manuscript. Contribution No. 3777 from the Woods Hole Oceanographic Institution and 55 of Polymode.

REFERENCES

Ben-Yaakov, S., E. Ruth, and I. R. Kaplan. 1974. Carbonate compensation depth: relation of carbonate solubility in ocean waters. *Science, 184.* 982-984.

Berger, W. H. 1967. Foraminiferal ooze: solution at depth. *Science, 156.* 383-385.

Berger, W. H. 1968. Planktonic foraminifera: selective solution and paleoclimatic interpretation. *Deep-Sea Res., 15.* 31-43.

Berger, W. H. 1975. Sedimentation of deep-sea carbonate: maps and models of variations and fluctuations. In: W. R. Riedel and T. Saito (Eds.), *Marine Plankton and Sediments,* Micropaleo. Press, N. Y., in press.

Berger, W. H. and P. M. Roth. 1975. Oceanic micropaleontology: progress and prospect. *Rev. Geopys. Space Phys., 13.* 561-635.

Berger, W. H. and E. L. Winterer. 1974. Plate stratigraphy and the fluctuating carbonate line. In: K. J. Hsü and M. C. Jenkyns (Eds.), *Int. Assoc. Sedimentol. Spec. Publ. 1.* 11-48.

Broecker, W. S. 1971. Calcite accumulation rates and glacial to inter-glacial changes in ocean mixing. In: K. K. Turekian (Ed.), *The Late Cenozoic Glacial Ages,* Yale Univ. Press. 239-265.

Broecker, W. S. and S. Broecker. 1974. Carbonate dissolution on the flank of the East Pacific Rise. *Soc. Econ. Paleont. Mineal. Spec. Publ. 20.* 44-57.

Chen, C. 1964. Pteropod ooze from Bermuda Pedestal. *Science, 144.* 60-62.

Chave, K. E. and E. Suess. 1970. Calcium carbonate saturation in seawater: effects of dissolved organic matter. *Limnol. Oceanogr., 15.* 633-637.

Edmond, J. M. 1974. On the dissolution of silicate and carbonate in the deep sea. *Deep-Sea Res., 21.* 455-480.

Lisitzin, A. P. 1972. Sedimentation in the world ocean. *Soc. Econ. Paleont. Mineral. Spec. Publ. 17.* 218 pp.

Milliman, J. D. 1974. *Marine Carbonates.* Springer-Verlag, Heidelberg. 375 pp.

Milliman, J. D. 1975. Dissolution of aragonite, Mg-calcite, and calcite in the North Atlantic Ocean. *Geology 3.* 461-462.

Morse, J. W. and R. A. Berner. 1972. Dissolution kinetics of calcium carbonate in sea water: II. A kinetic origin for the lysocline. *Amer. J. Sci., 272.* 840-851.

Peterson, M. N. A. 1966. Calcite: rates of dissolution in a vertical profile in the Central Pacific. *Science, 154.* 1542-1544.

Pilkey, O. H. and B. W. Blackwelder. 1968. Mineralogy of the sand size carbonate fraction of some recent marine terrigenous and carbonate sediments. *J. Sed. Petrology, 38.* 799-810.

CHARACTERISTICS AND GENESIS OF SHALLOW-WATER AND DEEP-SEA LIMESTONES

John D. Milliman[1] and Jens Müller[2]

Woods Hole Oceanographic Institution[1] and Technische Universität, West Germany[2]

ABSTRACT

The cements in shallow-water limestones and beachrocks, micritic crusts and non-skeletal carbonates (ooids, grapestone and cryptocrystalline lumps) display many petrographic and mineralogic similarities and particularly close isotopic compositions. Genesis of these various shallow-water cements and grains appears to be biologically controlled, but whether by direct biologic precipitation or biologic catalysis is not known; blue-green algae seem a likely agent.

In contrast, the cements of many deep-sea limestones appear to have precipitated inorganically. Chalks cement through the reprecipitation of carbonate released by dissolution of more soluble grains. Other deep-sea limestones require igneous activity to facilitate lithification. Particularly interesting are those limestones and lutites occurring in non-depositional areas (plateaus and guyots) and hypersaline basins; isotopic data of these carbonates suggest precipitation from ambient water at or near the sediment-water interface.

The apparent lack of inorganic precipitation in shallow tropical waters probably reflects the strong dominance of biologic calcification, particularly by algae. Physiochemical processes can dominate in deeper aphotic waters, but only in those few environments which are supersaturated with respect to calcium carbonate.

---

[1]Woods Hole Oceanographic Institution, Woods Hole, Mass. 02543
[2]Lehrstuhl für Geologie; Technische Universität, D-8 München 2, West Germany

INTRODUCTION

The past ten years have seen some major revisions in concepts about the occurrence and mode of precipitation and cementation of calcium carbonate in the marine environment. For example, the "chemical precipitation" of shallow-water aragonitic muds (e.g., *Cloud*, 1962) may actually represent biological mineralization by codiacean algae (e.g., *Lowenstam and Epstein*, 1957; *Stockman et al.*, 1967; *Neumann and Land*, 1975). Even the sacred concept of "inorganic" precipitation of ooids and grapestones may require reconsideration in light of the possible biological role suggested by recent studies (see below). On the other hand, recent studies also have uncovered numerous examples of submarine carbonate lithification at both shallow and great depths (e.g., *Friedman*, 1964; *Gevirtz and Friedman*, 1966; *Milliman*, 1966; *Fischer and Garrison*, 1967; *Milliman et al.*, 1969; *Ginsburg et al.*, 1971; *Shinn*, 1969; *Milliman and Müller*, 1973; *Müller and Fabricius*, 1974).

In this paper we try to place these various concepts into perspective with one another, with particular emphasis on isotopic composition, which seems to differentiate various groups of precipitates.

SHALLOW-WATER LIMESTONES, NON-SKELETAL FRAGMENTS AND BEACHROCK

Shallow water tropical and subtropical environments offer ideal sites for the precipitation of calcium carbonate. In warm and hypersaline water, sea water may be several times supersaturated with respect to aragonite. Moreover, it is in these shallow water areas where photosynthesis can affect great diurnal fluctuations in *pH*, thus offering an ideal triggering mechanism for the precipitation (*Cloud*, 1962; *Schmalz and Swanson*, 1969; *Friedman*, 1975). Such arguments have been used to explain the chemical formation of shallow-water carbonate muds, ooids and grapestone deposits, as well as intertidal beachrocks, reef cavity cements and subtidal limestones (see discussions and references in *Bathurst*, 1971; *Milliman*, 1974; and *Friedman*, 1975). All these cements, as well as those found within micritic rinds have compositional and petrographic similarities, particularly with respect to isotopic composition; only methane-derived cements (*Russell et al.*, 1967; *Garrison et al.*, 1969; *Hathaway and Degens*, 1969; *Allen et al.*, 1969; *Roberts and Whelan*, 1975) are distinctly different.

1) Intragranular cements and grains within non-skeletal fragments (ooids, pelletoids and grapestones) and many micritic envelopes are composed mainly of microcrystalline aragonite needles, many crystals being shorter than 2 microns. Apparently magnesian

calcite cement does not occur within grapestone (although it may precipitate during original stages of formation and occasionally continue during later growth; *Winland and Matthews*, 1974; *Friedman et al.*, 1973) and only rarely within individual layers of modern ooids (*Frishman and Behrens*, 1969). Relict magnesian calcitic ooids, have been found in continental shelf deposits, suggesting that under certain conditions magnesian calcite might form (*Milliman and Barretto*, 1975; *Marshall and Davies*, 1975).

In contrast, intergranular cements in shallow-water limestones, reef fillings and intertidal beachrocks are often exclusively composed of magnesian calcite scalenohedra, with $MgCO_3$ content often exceeding 15 mole per cent (e.g., *Ginsburg et al.*, 1971; *Shinn*, 1969; *Moore*, 1973; *Alexandersson*, 1972a; *Ginsburg and Schroeder*, 1973; *Friedman et al.*, 1974). Because of differential growth of individual crystals, the magnesian calcitic cement often forms in "micro-nodules", 25 to 50 microns in diameter (*Alexandersson*, 1972b; *Macintyre and Milliman*, 1970). While aragonite also acts as a cementing agent, its role tends to decrease with increasing water depth (*Milliman*, 1974); it is interesting to note, however, that aragonitic needles are often more than an order of magnitude longer than those within intragranular cements.

2) In most shallow-water cements studied, calcified blue-green algal filaments and tubes are obvious components (*Schroeder*, 1972a; *Moore*, 1973). Many workers have tended to regard these tubes as the remains of boring (endolithic) algae (e.g., *Newell et al.*,1960). However, similar calcified tubes also occur in micritic rinds (*Alexandersson*, 1972b), suggesting that the algae that formed these tubes and filaments played an important role in the formation of the inter- and intragranular cements (*Moore*, 1973; *Schroeder*, 1972b; *Fabricius*, in press). Moreover, the amino-acid composition of ooids suggests that organic matter plays an important role in ooid formation (*Mitterer*, 1968, 1972). Whether precipitation occurs after death of boring algae (*Bathurst*, 1966) or whether epilithic algae catalyze or actively precipitate calcium carbonate (*Fabricius*, in press) is not known. However, it is interesting to note that in higher latitudes, algally bored carbonates seldom contain calcified filaments (*Alexandersson*, 1976a) suggesting that the supersaturation of $CaCO_3$ from the surrounding waters also is an important factor (*Fabricius*, in press).

3) Perhaps the most intriguing similarity between the cements within beachrocks, shallow-water limestones, ooids, grapestones, etc. is their stable isotope composition. Most of these sediments have δO^{18} values between *+1* and *-1 o/oo*, suggesting equilibrium with ambient tropical waters. In contrast, the δC^{13} values cluster between *+3* and *+5 o/oo* (Figure 1), which is clearly greater than what one would expect for carbonate precipitated from tropical

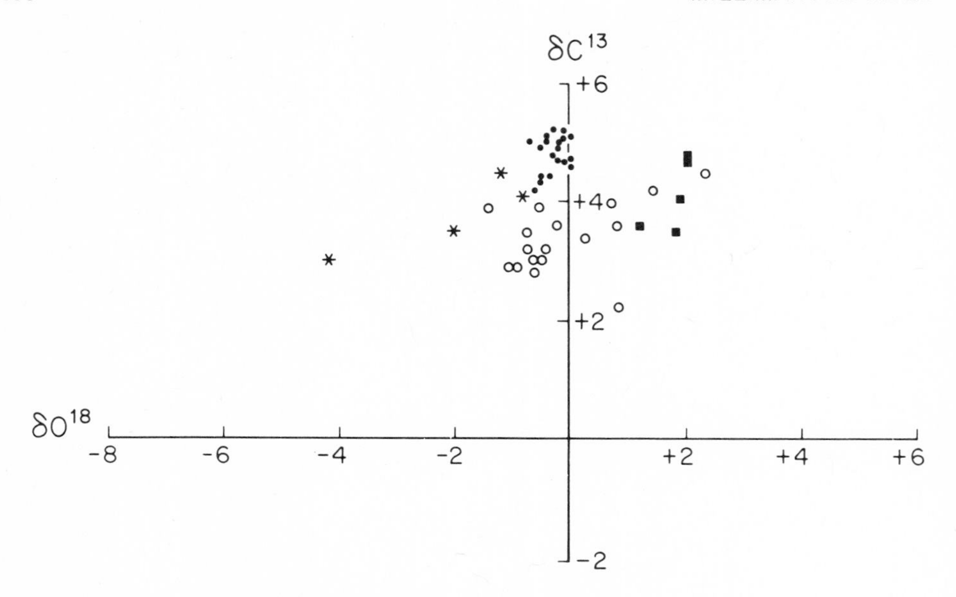

Fig. 1. Stable isotope variation in shallow-water limestones and ooids. Data from *Lowenstam and Epstein* (1957), *Shinn* (1969), *Deuser and Degens* (1969), *Land and Goreau* (1970), *Land* (1973), *Moore* (1973), *Milliman and Barretto* (1975).

marine waters (*Lloyd*, 1971). The close similarity between these cements suggests a unique mode of formation[1] (see below).

---

[1]Much of this spread in isotopic values is due to the difficulty of isolating sufficient quantities of cement (particularly intergranular cements). For example, *Buddemeier et al.* (1975) report *Mg*-calcite cements within a cemented reef rock to have δC^{13} content of *+5.0 o/oo*, whereas the entire rock shows a value of *+2.6 o/oo*. If similar care had been taken in all the analyses shown in Figure 1, the values undoubtedly would have fallen nearer those of ooids and micritic rinds.

DEEP-SEA LIMESTONES

Limestones occurring in water depths greater than 200 *m* can be classified as deep-sea limestones in that cementation occurred at essentially aphotic depths, thus excluding any influence of photosynthic algae; this, of course, excludes limestones formed during low stands of sea level or displaced by gravitational slides or tectonic subsidence. Based on their character and occurrence, deep-sea limestones can be divided into three general types; 1) limestones, crusts and nodules from hypersaline basins (Mediterranean and Red Sea, in particular) and from sites of non-deposition (such as guyots, slopes and plateaus); 2) limestones and carbonate inclusions associated with igneous activity (both intrusive and extrusive); 3) nannofossil chalks. Each limestone type will be discussed in the following paragraphs.

1) Hypersaline and Non-Depositional Limestones - Limestones recovered from these environments generally have rim cements and/or micritic matrices. To some extent the amount of matrix depends upon whether the limestone formed in muddy sediment (such as the Mediterranean and Red Sea) or in a coarse sediment (on a guyot or plateau). The primary cement appears to be magnesian calcite (9-13 mole per cent $MgCO_3$); crystals are 5 to 30 microns long, but generally not nearly as well formed as those in shallow-water cements (see *Milliman*, 1974). Some aragonitic cements occur, notably in Red Sea limestones, formed during the last lower stand of sea level (*Milliman et al.*, 1969); stable isotope studies idicate that temperatures and salinities during this interval were sufficiently high to favor aragonite precipitation (*Deuser and Degens*, 1969; see also *Murray and Irvine*, 1891; *Kinsman and Holland*, 1969). With time, the *Mg*-calcite and aragonite cements invert to calcite; most all pre-Pliocene limestones are entirely calcitic, although maintaining similar stable isotope compositions (*Milliman*, 1966).

One interesting aspect of hypersaline basin sediments is the abundance of magnesian calcite within the lutite fraction. In both the Mediterranean Sea and Red Sea, lutite often contains more than 50 per cent *Mg*-calcite (9 to 13 mole per cent $MgCO_3$) and the clay-size fraction can contain more than 80 per cent (*Milliman et al.*, 1969; *Müller and Fabricius*, 1974; *Milliman and Müller*, 1973; *Sartori*, 1974). The close petrographic and compositional similarity of these deep-sea limestone cements and the lack of any apparent biogenic source (either planktonic or benthic) suggest that these lutites precipitate in a similar way to the ambient limestone cements (*Milliman et al.*, 1969).

A plot of the stable isotopes within these deep-sea cements and lutites follows a generally linear trend passing through the origin (Figure 2). In many of these limestones and fragments, the oxygen isotope values have been shown to reflect ambient conditions at the time of cementation (*Milliman*, 1966, 1971; *Deuser and Degens*,

1969). The fact that the δO - δC relationship is linear, therefore, would suggest that the carbon also is in equilibrium with ambient waters.

2) Limestones associated with igneous activity - A number of limestones occur on volcanic seamounts or near contact with igneous bodies. While non-depositional and basin limestones tend to occur in water depths less than 3000 *m*, igneous-associated limestones apparently can form in water depths greater than 4000 *m* and under

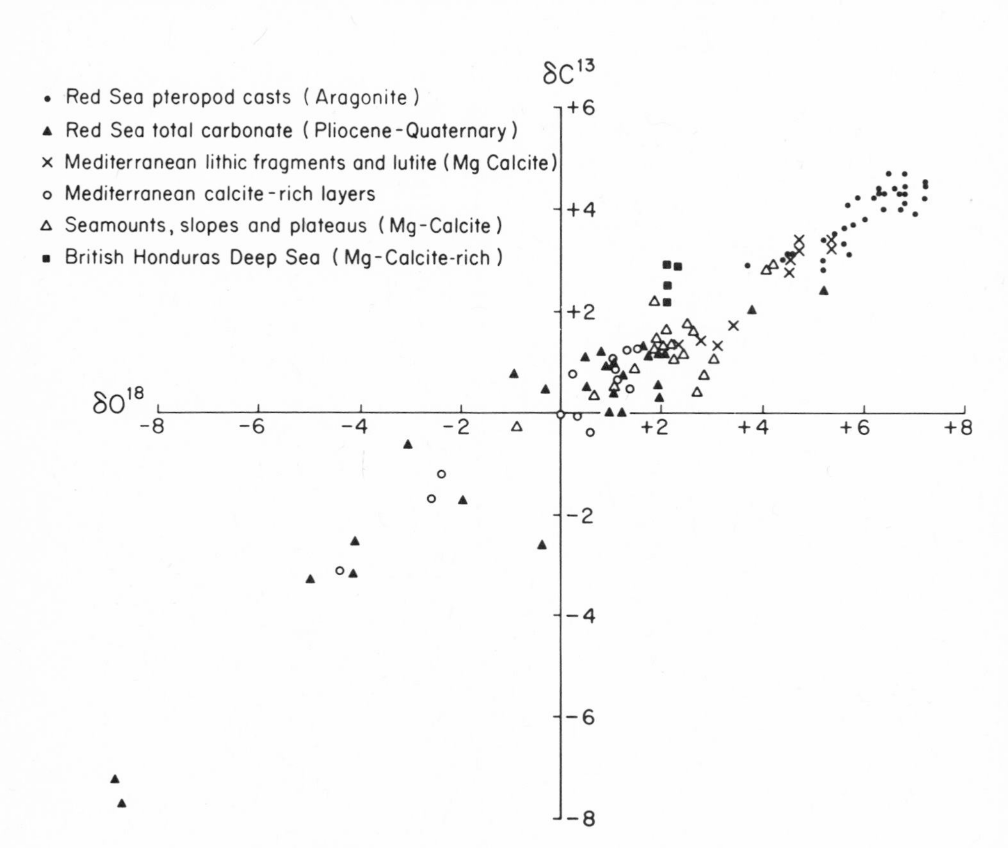

Fig. 2. Variation of stable isotopes in deep-sea limestones from warm, hypersaline waters and areas of non-deposition. The British Honduras lutites come from depths of 150 to 2000 *m* and are characterized by *Mg*-calcite contents commonly greater than 50 percent of the carbonate fraction; W. J. Koch and R. M. Lloyd have suggested that this *Mg*-calcite was precipitated *in situ*, and the relatively close correlation between these stable isotope values and others within this diagram would support such a supposition. Data from *Milliman* (1966, 1971), *Deuser and Degens* (1969), *Lawrence* (1973), *Supko et al.*, (1974), *Fontes et al.*, (1975), *Neumann* (pers. comm.), *Koch and Lloyd* (in preparation), *Fontes and Desforges* (1975), and *Deuser* (pers. comm.).

relatively great sediment thickness (e.g., *Anderson and Schneidermann*, 1973). Petrographically, they range from solid, well-indurated limestones (*Thompson*, 1972) to fragile "baked" carbonate sediments (*Saito et al.*, 1966) to single large crystal inclusions within volcanic tuffs (*Milliman*, 1971). These limestones and inclusions are almost entirely low-*Mg* calcite, although aragonite has been reported and usually in association with serpentinized ultramafic intrusions (*Thompson*, 1972; *Honnorez et al.*, 1975; *Kharin*, 1974). The matrix and rim cements, so common in hypersaline and non-depositional limestones, are mostly lacking.

The isotopic composition of igneous-associated limestones also differs from those found in hypersaline and non-depositional types (Figure 3). Although a slightly positive relation between δC and δO is seen, the trend passes substantially above the origin and the scatter is great. Clearly, either these limestones formed in a

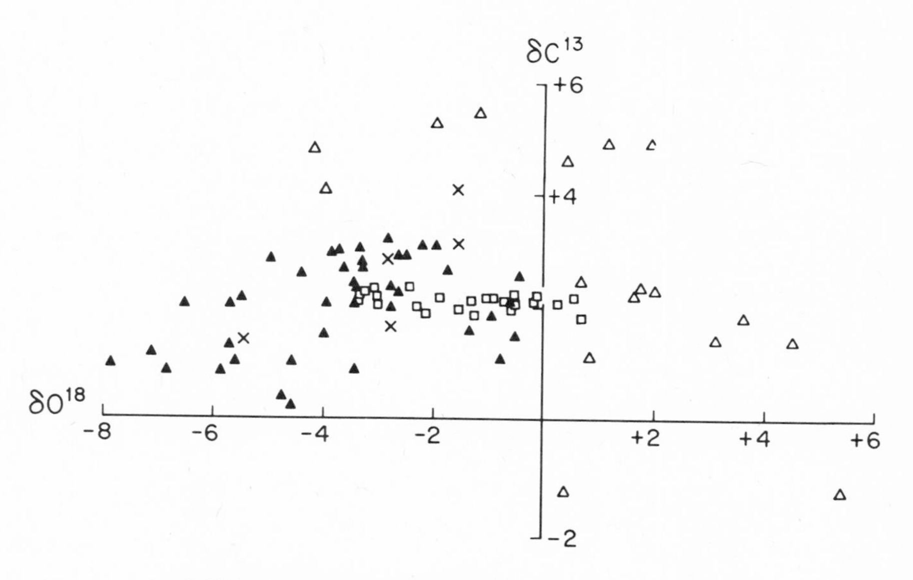

Fig. 3. Isotopic contents of various deep-water limestones associated with igneous activity, either intrusive or extrusive. Data from *Thompson et al.* (1968), *Milliman* (1971), *Thompson* (1972), *Anderson* (1973), *Garrison et al.* (1973), *Coplen and Schlanger* (1973), *Anderson and Schneidermann* (1973), *Bonatti and Honnorez* (1971).

manner different from other limestones or they formed in equilibrium with an environment no longer present. Perhaps the truth lies somewhere between.

3) Deep-Sea Chalks - The occurrence of deep-sea chalks and nannofossil limestones has been noted in numerous cores recovered by recent DSDP drilling (e.g., *Moberly and Heath*, 1971; *Roth and Thierstein*, 1972; *Wise*, 1973; *Davies and Supko*, 1973; *Matter et al.*, 1975), and these discoveries have sparked a number of summary papers on the formation of deep-sea chalks (*Wise and Hsü*, 1971; *Schlanger and Douglas*, 1974; *Wise*, 1977). Apparently these chalks are the prominent examples of carbonate lithification in the deep sea.

Chalks can be defined as relatively poorly cemented sediments, representing a transition from unconsolidated nannofossil oozes to well lithified limestones. Although the ooze-chalk-limestone transformation generally proceeds with increasing depth of burial (*Schlanger and Douglas*, 1974), chalks can form at or near the sediment surface during periods of low sedimentation (*Wise and Hsü*, 1971; *Wise*, 1977). Lithification generally results from compaction of the sediment and cementation by overgrowths of small (.5 to 1.5 micron) calcite crystals upon the coccoliths and discoasters (e.g., *Schlanger and Douglas*, 1974; *Matter et al.*, 1975). Dissolution of planktonic foraminifera and less stable nannofossils apparently increases carbonate saturation to the point where cementation can occur. Available stable isotope data suggest that chalks are not in equilibrium with ambient pore waters, but rather reflect compositions between those of the nannofossils and the dissolved carbonate components (*Lawrence*, 1973; *Matter et al.*, 1975). A more detailed description of chalks and their genesis has been given by Wise in this volume.

GENESIS OF MARINE CARBONATE CEMENTS

1) Shallow-water Cements - Shallow-marine limestones and non-skeletal fragments have a number of similar compositional and petrographic properties that are quite distinct from deep-sea cements (Table 1). In discussing the origin of shallow-water cements, a number of factors must be considered: a) although aragonite is common in intertidal beachrocks and dominant in non-skeletal fragments, it decreases markedly with increasing water depth; b) the aragonite contains more *Sr* (generally 1.0 per cent and the magnesian calcite considerably more *Mg* (15 to 29 mole per cent $MgCO_3$) (*Milliman*, 1974) than theoretical and experimental data predict for inorganically precipitated carbonates (*Kinsman*, 1969; *Weyl*, 1967; *Berner*, 1975); c) δC^{13} contents of shallow-marine cements appear to be anomalously high to be in equilibrium with ambient sea water (compare Figures 1 and 2).

The fact that shallow-water cements do not appear to be in either isotopic or compositional equilbrium suggests that they did

TABLE 1. Synthesis of Petrographic and Compositional Properties of Marine Carbonate Cements.

		Cement Type	Mineralogy	Stable Isotopes	Origin
Shallow-water	Non-skeletal fragments micritic rinds, beachrock, cemented reef rock, limestone	Intra- and inter-granular (generally rim) cements	Intrag.-Arag, Mg calcite Interg.-increasing Mg calcite with increasing water depth Mg calcite—15-25 mole percent $MgCO_3$	δO^{18} = +1 to -1 o/oo δC^{13} = +3 to +5 o/oo (Fig. 2) Apparent disequilibrium with ambient ocean waters	Algal-direct precipitation, catalysis, post-mortum alteration??
Deep-sea	Hypersaline and non-depositional limestones, crusts and nodules	Intergranular rim and matrix cements. Individual components can have intragranular cement	Mg calcite (9-13 mole percent), altering to calcite with time	Direct relation between δO^{18} and δC^{13} (Fig. 3) Equilibrium with ambient oceanic waters	Inorganic precipitation??
	Igneous-associated limestones	Wide range of cement types, often poorly developed	Mostly calcite; occasionally aragonite	Wide scatter of δO^{18} values; δC^{13} generally between +2 and +4 o/oo	Processes related to igneous activity and/or alteration of igneous rocks
	Nannofossil chalks	Calcitic overgrowths on coccoliths; often poorly cemented	Calcite	Composition of dissolved planktonic carbonates which provided the source of overgrowth cement	Dissolution and re-precipitation of calcite from less stable planktonic foraminifera and nannofossils

not form strictly by inorganic precipitation. One alternative explanation would be the influence of algae; this would explain why these carbonates appear to form only in shallow photic depths, the occurrence of calcified algal filaments in many cements, and the marked C^{13}/C^{12} fractionation within the cements (*Calder and Parker*, 1973). Exhaustive studies by *Fabricius* (in press) indicate that many of the filaments in shallow-water cement are depositional, not erosional. Moreover, amino acid content of the organic matrix (in ooids) is similar to those in organically-produced carbonates, suggesting a biological influence in precipitation (*Mitterer*, 1968, 1972). Two other observations also indicate the potential importance of algae in shallow-marine cementation: a) Micritic rinds are thin crytocrystalline crusts which often "coat" shallow marine carbonate grains. These rinds are formed by blue-green algae, which bore the carbonate substrate and eventually die; carbonate (both aragonite and magnesian calcite) precipitates within the excavated cavities, although whether it happens during algal activity or subsequently is not known (*Bathurst*, 1966; *Winland*, 1968). *Lloyd* (1971) found that the δC^{13} contents of these algally-caused micritic rinds average *4.3 o/oo*, which considering the possible analytical contamination with the host substrate, is very close to the values of other shallow marine cements (Figure 1).

b) Although marine blue-green algae are not considered to precipitate calcium carbonate (*Gebelein*, 1969; *Monty*, 1967), *Friedman et al.* (1973) and *Horodyski and Vondar Haar* (1975) found thin cemented laminae within modern algal stromatolies; cementation apparently was by blue-green algae. The δC^{13} contents of the Red Sea stromatolites cements ranged as high as *4.7 o/oo* (*Friedman*, pers. comm.), which is remarkably close to the values shown in Figure 1, and further suggests an algal influence in shallow-marine cementation.

If algae are important in the cementation of shallow-water carbonates, what type(s) of algae (species, genus, phylum?) is (are) involved? Does cementation occur as a result of direct precipitation (*Fabricius*, in press), through photosynthetic activity that can elevate and depress *pH* levels (*Schmalz and Swanson*, 1969; *Friedman*, 1975), through post-mortum catalysis (*Bathurst*, 1966) or does the organic matrix act as a template on to which carbonate can precipitate (*Shearman et al.*, 1970)? At present these questions cannot be answered. *Friedman et al.* (1974) have suggested that excessively high *pH's* (as great as or greater than 9.5) are critical in reef cementation, and offer as proof the presence of etched and corroded quartz grains within some rocks. Presumably these high *pH* levels, which also have been noted in beachrock environments (*Revelle and Emery*, 1957; *Davies and Kinsey*, 1973), result from (blue-green?) algal photosynthesis (*Helsinger and Friedman*, 1973). The presence of magnesian clacite or aragonite may depend upon the mineralogy of the substrate, the type of organic matter present or the *pH* of the precipitating medium (*Friedman et al.*, 1974). The tendency of increasing magnesian calcite cementation with increasing water depth may simply indicate the greater solubility of aragonite (*Friedman*, 1965).

Another problem involves the wide divergency of cement morphologies within shallow-marine carbonates. The fact that ooids and grapestones form as individual grains or clusters and contain short (1-2 microns), stubby aragonitic crystals may reflect the shallow depths (and high supersaturations of calcium carbonate) and high energy conditions that favor rapid precipitation (and thus small crystal size; *Füchtbauer and Müller*, 1970) and laminar cements. The longer (10 to 50 microns) needles within shallow-marine limestones may reflect relatively more stable conditions of cementation. Even here, however, cementation can be rapid: micritic fillings, for instance, can form within 10's of years after death of the host shell (*Moberly*, 1973), and in coralline algae, cementation can occur during plant growth (*Alexandersson*, 1976b).

2) Deep-Sea Limestones - The marked divergency of character of hypersaline and non-depositional limestones, chalks, and igneous limestones is difficult to ignore. One type (hypersaline and non-depositional) appears to lithify in equilibrium with normal, ambient sea water. Chalk forms through pressure-dissolution of less stable carbonate tests and subsequent reprecipitation of overgrowth cement. Igneous limestones appear to be out of equilibrium with ambient sea waters and probably form as a direct or indirect result of igneous activity.

While it is difficult to imagine the specific way in which hypersaline and non-depositional limestones form, it is interesting to note that they occur above the lysocline of magnesian calcite (less than 3000 *m*; *Milliman*, 1975; 1977). The source of cement, however, is not known. Perhaps the Mediterranean and Red Sea supersaturated with respect to magnesium calcite, facilitates precipitation directly from surrounding sea water; whether precipitation occurs within the water column (*Emelyanov and Shimkus*, 1972) or at the sediment-water interface (*Milliman and Müller*, 1973) is not known. In non-depositional evnironments, such as the North Atlantic, waters deeper than about 1000 *m* are undersaturated with respect to magnesian calcite. Thus, lithification must involve dissolution of more soluble species and subsequent reprecipitation of magnesian cement. The exact process, however, is not understood.

At the other end of the spectrum are igneous-associated limestones, which are markedly out of equilibrium with present-day ambient deep-sea environment (Figure 3). These anomalous compositions, together with their association with volcanic fragments (either as fillings or overgrowths), strongly suggest the influence of igneous activity. However, whether it occurs by exposure by juvenile water (*Kharin*, 1974), alteration of volcanic rocks (*Milliman*, 1971) or thermal metamorphism (*Anderson and Schneidermann*, 1973) is not known. Within any given situation, any or all of these mechanisms might trigger cementation.

CONCLUSION

It is well known that shallow tropical waters are supersaturated with respect to calcium carbonate and that most of the deep sea is markedly undersaturated. Thus, the fact that most examples of inorganically precipitated carbonates (both cements and muds) occur in the deep sea at first seems incongruous. This apparent discrepancy is undoubtedly related to the ability of shallow-water biota, particularly algae, to calcify at high rates and also effect large *pH* fluctuations through photosynthetic activities; purely physicochemical precipitation appears to be severely hindered. Only in deeper aphotic environments can physicochemical processes take over. Precipitation, however, can occur only in those areas in which the overlying water is supersaturated with respect to calcium carbonate (hypersaline basins, such as the Red Sea and Mediterranean Sea) or in which dissolution of less stable grains (aragonite, soluble nannofossils?) enables local supersaturation and subsequent precipitation. The exact mechanism (s) of precipitation awaits further study.

ACKNOWLEDGEMENTS

We are particularly grateful to Dr. W. Stahl (BGR, Hanover) and Dr. W. D. Deuser (WHOI) for their isotopic analyses of eastern Mediterranean *Mg*-calcite nodules and lutites and to Dr. Frank Fabricius for allowing us access to his most interesting manuscript concerning ooid formation. This study was sponsored by the Office of Naval Research (ONR 262.64) and the Deutsche Foreschungsgemeinschaft. We thank Dr. E. T. Alexandersson, G. M. Friedman, and G. Thompson for their comments on the manuscript. Woods Hole Contribution Number 3775.

REFERENCES

Alexandersson, T. 1972a. Intergranular growth of marine aragonite and *Mg*-calcite: evidence of precipitation from supersaturated sea water. *J. Sediment. Petrol.*, *42*. 441-460.

Alexandersson, T. 1972b. Micritization of carbonate particles: processes of precipitation and dissolution in modern shallow-water sediments. *Bull. Geol. Inst. Univ. Uppsala, N.S.*, *3*. 201-236.

Alexandersson, T. 1976a. Marine carbonate diagenesis in cold waters in the Skagerrak, North Sea. In: R. W. Fairbridge (Ed.), *Paleoclimatic Indicators in Sediments, SEPM 50th Anniversary Symposium*. In press.

Alexandersson, T. 1976b. Carbonate cementation in recent coralline algal constructions. In: E. Flügel (Ed.), *First Int. Symp. on Fossil Algae*, Springer-Verlag Heidelberg. In press.

Allen, R. C., E. Gavish, G. M. Friedman, and J. E. Sanders. 1969. Aragonite-cemented sandstone from outer continental shelf off Delaware Bay: submarine lithification mechanism yields product resembling beachrock. *J. Sediment. Petrol.*, *39*. 136-149.

Anderson, T. F. 1973. Oxygen and carbon isotope composition of altered carbonates from the western Pacific, core 53.0, Deep Sea Drilling Project. *Marine Geology*, *15*. 169-180.

Anderson, T. F. and N. Schneidermann. 1973. Stable isotope relationships in pelagic limestones from the Central Caribbean: Leg 15, Deep Sea Drilling Project. In: *Initial Repts, Deep Sea Drilling Project 15*. 795-803.

Bathurst, R. G. C. 1966. Boring algae, micritic envelopes and lithification of molluscan biosparites. *Geol. J.*, *5*. 15-32.

Bathurst, R. G. C. 1971. Carbonate sediments and their diagenesis. Developments in Sedimentology, 12. *Elsevier Publ., N.Y.* 620 p

Berner, R. A. 1975. The role of magnesium in the crystal growth of calcite and aragonite from sea water. *Geochem. Cosmochim. Acta 39*. 489-504.

Bonatti, E. and J. Honnorez. 1971. Non-spreading crustal blocks at the mid-Atlantic Ridge. *Science*, *174*. 1329-1331.

Buddemeier, R. W., S. V. Smith, and R. A. Kinzie. 1975. Holocene windward reef-flat history. *Enewetak Atoll. Geol. Soc. Amer. Bull.*, *86*. 1581-1584.

Calder, J. A. and P. L. Parker. 1973. Geochemical implications of induced changes in fractionation by blue-green algae. *Geochim. Cosmochim. Acta*, *37*. 133-140.

Cloud, P. E., Jr. 1962. Environment of calcium carbonate deposition west of Andros Island, Bahamas. *U.S. Geol. Survey Prof. Paper 350*. 138 p.

Coplen, T. B. and S. O. Schlanger. 1973. Oxygen and carbon isotope studies of carbonate sediments from site 167, Magellan Rise, Leg 17. *Initial Repts, Deep-Sea Drilling Project*, *17*. 505-509.

Davies, P. J. and D. W. Kinsey. 1973. Organic and inorganic factors in recent beachrock formation, Heron Island, Great Barrier Reef. *J. Sediment. Petrol.*, *43*. 59-81.

Davies, T. A. and P. R. Supko. Oceanooic sediments and their diagenesis: some examples from deep-sea drilling. *J. Sediment. Petrol.*, *43*. 381-390.

Deuser, W. G. and E. T. Degens. 1969. O^{18}/O^{16} and C^{13}/C^{12} ratios of fossils from the Hot Brine deep area of the central Red Sea. In: E. T. Degens and D. A. Ross (Eds.), *Hot Brines and Recent Heavy Metal Deposits in the Red Sea*, Springer-Verlag, Heidelberg. 336-347.

Emelyanov, E. M. and K. M. Shimkus. 1972. Suspended matter in the Mediterranean Sea. In: D. J. Stanley (Ed.), *The Mediterranean*

Sea, a natural sedimentation laboratory. Dowden, Hutchinson and Ross, Stroudsburg (Pa.). 417-439.

Fabricius, F. In press. Origin of marine ooids and grapestones. Contrib. *Sedimentology*.

Fischer, A. G. and R. E. Garrison. 1967. Carbonate lithification of the sea floor. *J. Geol., 75*. 488-497.

Fontes, F. G. H. and G. Desforges. 1975. Oxygen 18, carbon 13 and radiocarbon as indicators of a Würmian cold and deep diagenesis in western Mediterranean carbonate sediments IX Congres Internationale de Sedimentologie, Nice.

Friedman, G. M. 1964. Early diagenesis and lithification of carbonate sediments. *J. Sediment. Petrol., 34*. 777-813.

Friedman, G. M. 1965. Occurrence and stability relationships of aragonite, high-magnesium calcite and low-magnesium calcite under deep-sea conditions. *Geol. Soc. Amer. Bull., 76*. 1191-1196.

Friedman, G. M. 1975. The making and unmaking of limestones, or the down and ups of porosity. *J. Sediment. Petrol., 45*. 379-398.

Friedman, G. M., A. J. Amiel, M. Braun, and D. S. Miller. 1973. Generation of carbonate particles and laminites in algal mats: example from sea-marginal hypersaline pool, Gulf of Aqaba, Red Sea. *Amer. Assoc. Petroleum Geol. Bull., 57*. 541-557.

Friedman, G. M., A. J. Amiel, and N. Schneidermann. 1974. Submarine cementation in reefs: example from the Red Sea. *J. Sediment. Petrol., 44*. 816-825.

Frishman, S. A. and E. W. Behrens. 1969. Geochemistry of oolites, Baffin Bay, Texas (abst.), *Program, Ann. Meeting Geol. Soc. Amer.* p. 71.

Fuchtbauer, H. and G. Müller. 1970. Sedimente und Sedimentgesteine. E. Schweizerbartsche-Verlag, Stuttgart. 726 p.

Garrison, R. E., J. R. Hein, and T. F. Anderson. 1973. Lithified carbonate sediment and zeolitic tuff in basalts, Mid-Atlantic Ridge. *Sedimentology, 20*. 399-410.

Garrison, R. E., J. L. Luternauer, E. V. Grell, R. D. MacDonald, and J. W. Murry. 1969. Early diagenetic cementation of recent sands, Fraser River Delta, British Columbia. *Sedimentology, 12*. 27-46.

Gebelein, C. D. 1969. Distribution, morphology, and accretion rate of recent subtidal algal stromatolites, Bermuda. *J. Sediment. Petrol., 39*. 49-69.

Gevirtz, J. L. and G. M. Friedman. 1966. Deep-sea carbonate sediments of the Red Sea and their implications on marine lithification. *J. Sediment. Petrol., 36*. 143-157.

Ginsburg, R. N. and J. H. Schroeder. 1973. Growth and submarine fossilization of algal cup reefs, Bermuda. *Sedimentology, 20*. 575-614.

Ginsburg, R. N., J. H. Schroeder, and E. A. Shinn. 1971. Recent synsedimentary cementation in subtidal Bermuda reefs. In: O. P.

Bricker (Ed.), *Carbonate Cements, The Johns Hopkins Univ. Studies in Geology, 19.* 54-58.

Hathaway, J. C. and E. T. Degens. 1969. Methane-derived marine carbonates of Pleistocene age. *Science, 165.* 690-692.

Helsinger, Mitt. and G. M. Friedman. 1973. Diagenetic modification and cementation in carbonate sediments at intertidal levels (abs.), *Geol. Soc. Amer. Abs. with Programs, 5(4).* 177.

Honnorez, J., E. Bonatti, C. Emiliani, P. Brönniman, M. A. Furrer, and A. A. Meyerhoff. 1975. Mesozoic limestone from the Vema offset zone, Mid-Atlantic Ridge? *Earth Planet. Sci. Letters. 26.* 8-12.

Horodyski, R. J. and S. P. Vonder Harr. 1975. Recent calcareous stromatolites from Laguna Mormona (Baja California). *J. Sediment. Petrol., 45.* 894-906.

Kharin, G. 1974. Chemogenic aragonite on the seafloor in some abyssal areas of the Atlantic Ocean. *Oceanology, 14.* 232-237.

Kinsman, D. J. J. 1969. Interpretation of Sr^{+2} concentrations in carbonate minerals and rocks. *J. Sediment. Petrol., 39.* 486-508.

Kinsman, D. J. J. and H. D. Holland. 1969. The co-precipitation of cations with $CaCO_3$. IV. The co-precipitation of Sr^{+2} with aragonite between 16 and 96°C. *Geochim. Cosmochim. Acta, 33.* 1-17.

Land, L. S. 1973. Contemporaneous dolomitization of middle Pleistocene reefs by meteoric water, North Jamaica. *Bull. Marine Sci., 23.* 64-92.

Land, L. S. and T. F. Goreau. 1970. Submarine lithification of Jamaican reefs. *J. Sediment. Petrology, 40.* 457-462.

Lawrence, J. R. 1973. Interstitial water studies, Leg 15 - Stable oxygen and carbon isotope variations in water, carbonates, and silicates from the Venezuela Basin (Site 149) and the Aves Rise (Site 148). In: *Initial Reports, Deep-Sea Drilling Project 20.* 891-899.

Lloyd, R. M. 1971. Some observations on recent sediment alteration ("micritization") and the possible role of algae in submarine lithification. In: O. P. Bricker (Ed.), *Carbonate Cements, The Johns Hopkins Univ. Studies in Geology, 19.* 72-79.

Lowenstam, H. A. and S. Epstein. 1957. On the origin of sedimentary aragonite needles of the Great Bahama Bank. *J. Geol., 65.* 364-375.

MacIntyre, I. G. and J. D. Milliman. 1970. Physiographic features on the outer shelf and upper slope, Atlantic continental margin, southeastern United States. *Geol. Soc. Amer. Bull., 81.* 2577-2598.

Marshall, J. F. and P. J. Davies. 1975. High-magnesium calcite ooids from the Great Barrier Reef. *J. Sediment. Petrol., 45.* 285-291.

Matter, A., R. G. Douglas, and K. Perch-Nielsen. 1975. Fossil preservation, geochemistry, and diagenesis of pelagic carbonates

from Shatsky Rise, northwest Pacific. *Initial Repts, Deep-Sea Drilling Project, 32.* 891-921.
Milliman, J. D. 1966. Submarine lithification of carbonate sediments. *Science, 153.* 994-997.
Milliman, J. D. 1971. Examples of submarine lithification, In: O. P. Bricker (Ed.), *Carbonate Cements, The Johns Hopkins Univ., Studies in Geology, 19.* 95-102.
Milliman, 1974. Marine Carbonates. Springer-Verlag, Heidelberg, 375 p.
Milliman, J. D. 1975. Dissolution of aragonite, Mg-calcite, and calcite in the North Atlantic Ocean. *Geology, 3.* 461-462.
Milliman, J. D. 1977. Dissolution of calcium carbonate in the Sargasso Sea (Northwest Atlantic), In: *The Fate of Fossil Fuel CO_2,* N. R. Andersen and A. Malahoff (Eds.), Plenum, N.Y., (this volume).
Milliman, J. D. and H. T. Barretto. 1975. Relict magnesian calcite oolite and subsidence of the Amazon shelf. *Sedimentology, 22.* 137-145.
Milliman, J. D. and J. Müller. 1973. Precipitation and lithification of magnesian calcite in the deep-sea sediments of the eastern Mediterranean Sea Sediment., 20. 29-45.
Milliman, J. D., D. A. Ross, and T. L. Ku. 1969. Precipitation and lithification of deep-sea carbonates in the Red Sea. *J. Sediment. Petrol. 39.* 724-736.
Mitterer, R. M. 1968. Amino acid composition of organic matrix in calcareous oolites. *Science, 162.* 1498-1499.
Mitterer, R. M. 1972. Biogeochemistry of aragonite mud and oolites. *Geochim. Cosmochim. Acta, 36.* 1407-1422.
Moberly, R. 1973. Rapid chamber-filling growth of marine aragonite and Mg-calcite. *J. Sediment. Petrol., 43.* 634-635.
Moberly, R., Jr. and G. R. Heath. 1971. Carbonate sedimentary rocks from the western Pacific Leg 7, Deep Sea Drilling Project. In: *Initial Repts., Deep Sea Drilling Project, 7.* 977-986.
Monty, C. L. V. 1967. Distribution and structure of recent stromatolitic algal mats, eastern Andros Island, Bahamas. *Bull. Ann. Soc. Geol. Belg., 90.* 55-100.
Moore, C. H., Jr. 1973. Intertidal carbonate cementation, Grand Cayman, West Indies. *J. Sediment. Petrol., 43.* 591-602.
Müller, J. and F. Fabricius. 1974. Magnesian calcite nodules in the Ionian deep sea. An actualistic model for the formation of some nodular limestones. *Spec. Publ. Int. Assoc. Sediment., 1.* 235-248.
Murray, J. and R. Irvine. 1891. On coral reefs and other carbonate of lime formations in modern seas. *Roy. Soc. Edinburgh Proc., 17.* 79-109.
Neumann, A. C. and L. S. Land. 1975. Lime mud deposition and calcareous algae in the Bight of Abaco, Bahamas: *A budget. J. Sediment. Petrol., 45.* 763-786.
Newell, N. D., E. G. Purdy, and J. Imbrie. 1960. Bahamian oolitic sand. *J. Geol., 68.* 481-497.

Revelle, R. and K. O. Emery. 1957. Chemical erosion of beachrock and exposed reef rock. *U.S. Geol. Survey Prof. Paper 260-T.* 699-709.

Roberts, H. H. and T. Whelan, III. 1975. Methane derived carbonate cements in barrier and beach sands of a subtropical delta complex. *Geochim. Cosmochim. Acta, 39.* 1085-1089.

Roth, P. H. and H. Thierstein. 1972. Calcareous nannoplankton. In: *Initial Repts, Deep-sea Drilling Project, 14.* 421-486.

Russell, K. J., K. S. Deffeyes, G. A. Fowler, and R. M. Lloyd. 1967. Marine dolomite of unusual isotopic composition. *Science, 155.* 189-191.

Saito, T., M. Ewing, and L. H. Burckle. 1966. Tertiary sediment from the Mid-Atlantic Ridge. *Science, 151.* 1075-1079.

Sartori, R. 1974. Modern deep-sea magnesian calcite in the central Tyrrhenian Sea. *J. Sediment. Petrol., 44.* 1313-1322.

Schlanger, S. O. and R. G. Douglas. 1974. The pelagic ooze-chalk-limestone transition and its implications for marine stratigraphy. *Spec. Publ., Int. Assoc. Sediment., 1.* 117-148.

Schmalz, R. F. and F. J. Swanson. 1969. Diurnal variations in the carbonate saturation of sea water. *J. Sediment. Petrol., 39.* 255-267.

Schroeder, J. H. 1972a. Calcified filaments of an endolithic alga in Recent Bermuda reefs. *Neues Jb. Geol. Paläont. Mn., 1.* 16-33.

Schroeder, J. H. 1972b. Fabrics and sequences of submarine carbonate cements in Holocene Bermuda cup reefs. *Geol. Rundschau, 61.* 708-730.

Shearman, D. J., J. Tyman, and M. Zand Karimi. 1970. The genesis and diagenesis of oolites. *Proc. Geol. Assoc., 81.* 561-575.

Shinn, E. A. 1969. Submarine lithification of Holocene carbonate sediments in the Persian Gulf. *Sedimentology 12.* 109-144.

Stockman, K. W., R. N. Ginsburg, and E. A. Shinn. 1967. The production of lime mud by algae in south Florida. *J. Sediment. Petrol., 37.* 633-648.

Supko, P. R., P. Stoffers, and T. B. Coplen. 1974. Petrography and geochemistry of Red Sea dolomite, In: *Initial Repts., Deep Sea Drilling Project 23.* 867-878.

Thompson, G. 1972. A geochemical study of some lithified carbonate sediments from the deep sea. *Geochim. Cosmochim. Acta, 36.* 1237-1253.

Thompson, G., V. T. Bowen, W. G. Melson, and R. Cifelli. 1968. Lithified carbonates from the deep sea. *J. Sediment. Petrology, 38.* 1305-1312.

Weyl, P. K. 1967. The solution behavior of carbonate minerals in sea water. *Studies in Tropical Oceanogr., Univ. Miami, 5.* 178-228.

Winland, H. D. 1968. The role of high Mg-calcite in the preservation of micrite envelopes and textural features of aragonite sediments. *J. Sediment. Petrol., 38.* 1320-1325.

Winland, H. D. and R. K. Matthews. 1974. Origin and significance of grapestone, Bahama Islands. *J. Sediment. Petrol.*, *44*. 921-927.

Wise, S. W., Jr. 1973. Calcareous nannofossils from cores recovered during Leg 18, Deep Sea Drilling Project: biostratigraphy and observations of diagenesis. In: *Initial Repts., Deep-Sea Drilling Project 18*. 565-615.

Wise, S. W., Jr. and K. J. Hsü. 1971. Genesis and lithification of a deep sea chalk. *Ecolog. Geol. Helv.*, *64*. 273-278.

Wise, S. W., Jr. 1977. Chalk formation: Early diagenesis. In: *The Fate of Fossil Fuel* CO_2, N. R. Andersen and A. Malahoff (Eds.), Plenum, N.Y., (this volume).

TEXTURAL VARIATIONS OF CALCAREOUS COARSE FRACTIONS IN THE PANAMA BASIN (EASTERN EQUATORIAL PACIFIC OCEAN)

Jörn Thiede[1]

Oregon State University

ABSTRACT

Panama Basin sediment surface coarse fractions are dominantly composed of planktonic foraminiferal remains. Textural studies of these coarse fractions by means of a large diameter settling tube system reveal characteristic grain size spectra with important modes at 2.0-2.25 phi, 2.3-2.45 phi, 2.5-2.75 phi, 3.0-3.3 phi, and 3.4-3.75 phi. The coarser modes consist of large Globoquadrina dutertrei and Globorotalia menardii shells, the finer ones of small planktonic foraminiferal species and of shell fragments of the larger species. Analyses of samples from the Carnegie Gap provide sufficient information such that the extent of the high energy environment close to the sill depth can be mapped; the textural analyses also seem to indicate south and northward flowing components of the bottom currents which transport particle assemblages with distinct textural characteristics. The samples bear evidence for large scale removal of calcareous fines from the crests of structural highs; the fines are then dumped on the flanks of these elevations.

INTRODUCTION

Sorting of Biogenous Deep-Sea Sediment Components

Deep-sea sediment components are frequently used to reconstruct the paleo-environment of former oceans. However, the sediment particle assemblages have passed through a number of filters which may have considerably altered their characteristics or their composition. In the case of biogenous grains of pelagic sediments, the

Present address: [1]Universitetet i Oslo, Post Boks 1047, Blindern, Oslo 3, Norway production of shell and skeletal material by living faunas and floras determines the input of particles into the depositional system. At the same time, processes are occurring which can alter the composition of these assemblages, as well as the shape and the nature of the single particles. These processes include dissolution in the water column, at the water-sediment interface and within the sediment, mechanical sorting due to gravity, tidal currents and thermohaline circulation and, finally, diagenesis in the sediment column.

The dissolution of calcareous components such as shells of planktonic foraminifers (*Berger*, 1970), pelagic gastropods (*Melguen and Thiede*, 1974), coccoliths (*Berger*, 1973), siliceous skeletons of radiolarians (*Berger*, 1968) and diatoms (*Schrader*, 1971), and the diagenesis of these biogenous particles (*Schlanger and Douglas*, 1974; *Calvert*, 1974; *Hurd and Theyer*, 1975) are processes which are believed to be understood qualitatively and quantitatively. Aspects of the alteration of biogenous deep-sea sediment particle assemblages due to mechanical sorting have attracted much less attention in the scientific community. This is probably for two reasons: this problem was not recognized until recently, and, many institutions lack the instrumentation (hardware as well as software) necessary to carry out the tedious measurements with sufficient accuracy. The development of efficient settling tubes, accessories to subsample specific size fractions for compositional studies, and computer software suitable to produce and process large numbers of samples and data (*Thiede et al.*, 1976) is opening a new field which until now has been largely disregarded.

Planktonic foraminiferal shells and their fragments are particularly well suited to study the effects of mechanical sorting because these components are easily identified to the species or at least to the genus level. Previous studies are essentially restricted to the study of the settling behavior of foraminiferal shells (*Berthois and Le Calvez*, 1960; *Berger and Piper*, 1972) and to observations of foraminiferal distributions in turbidite deposits. *Diester-Haass*, 1975 tried to evaluate textural characteristics of calcareous coarse fractions from the Vema Channel (southwestern Atlantic Ocean) which consists dominantly of planktonic foraminiferal remains (*Melguen and Thiede*, 1974; *Thiede*, 1976), by subdividing these coarse fractions into five artificial size classes through dry sieving.

In this study I intend to discuss textural characteristics of calcareous deep-sea sediment coarse fractions which have been investigated using a newly-developed and fully automated large diameter settling tube (see below). A collection of samples from the Panama Basin seems to be well suited to elucidate the above mentioned complex problems for a number of reasons: First, because this basin is particularly well known for large scale reworking and displacement of its deep-sea sediments (*van Andel*, 1973; *Lonsdale and Malfait*, 1974); second, because of the ample evidence available to evaluate the influence of dissolution of calcareous components (*Moore et al.*, 1973; *Yamashiro*, 1975); third, because of the impact of bottom water transport on the distribution of fines in the deep basin (*Heath et al.*, 1974; *Laird*, 1971); and fourth, because of our understanding of Holocene accumulation rates of various sediment components within this region (*Swift*, 1976 a and b)

Panama Basin: Distribution and Dissolution of Calcareous Sediments

Calcareous sediments dominate the bottom of the southwestern Panama Basin in the eastern equatorial Pacific Ocean because of the high accumulation rates of biogenous calcareous particles and because of the elevation of the basin floor above the calcite compensation depth (Fig. 1). Towards the northeastern edge of the basin the calcite compensation depth rises rapidly, from water depths of > 4000 m close to the Galapagos Islands to < 3000 m. Since the northeastern corner of the basin is deep and close to the continental margin, the $CaCO_3$ content of the surface sediments drops to 15% (*Moore et al.*, 1973). The distribution of fine grained (< 0.063 mm) and total calcium carbonate and the texture of the surface sediments suggest that large quantities of fine grained calcareous material are winnowed off the ridge crests to be dumped on the slopes of these elevations (mainly Carnegie and Cocos Ridges), see Fig. 2.

Fragments of planktonic foraminiferal shells have been found to contribute most significantly to sediments on the slopes of structural highs in the Panama Basin: *Kowsmann* (1973) detected the peculiar distribution pattern of whole and fragmented foraminiferal shells in the Panama Basin surface sediment samples shown in Fig. 3. He suggested the distribution was primarily due to winnowing processes that removed hydrodynamically lighter material from the ridge crest, although it was impossible for him to ignore the fact that the dissolution of calcium carbonate had a major impact on the distribution of foraminiferal fragments (compare *Moore et al.*, 1973). Thus, even after recent extensive marine geological and sedimentological studies in this area the question remains: does mechanical sorting, chemical dissolution or both processes control the distribution of calcareous sediment components in a small ocean basin, which seems to receive a fairly homogenous input of shell material from the surface water masses (*Bradshaw*, 1959). I will attempt to approach this problem using textural and compositional data from Panama Basin surface sediment coarse fractions. The dominant portion of these samples has been taken from a small region on the Carnegie Ridge (*Malfait*, 1974).

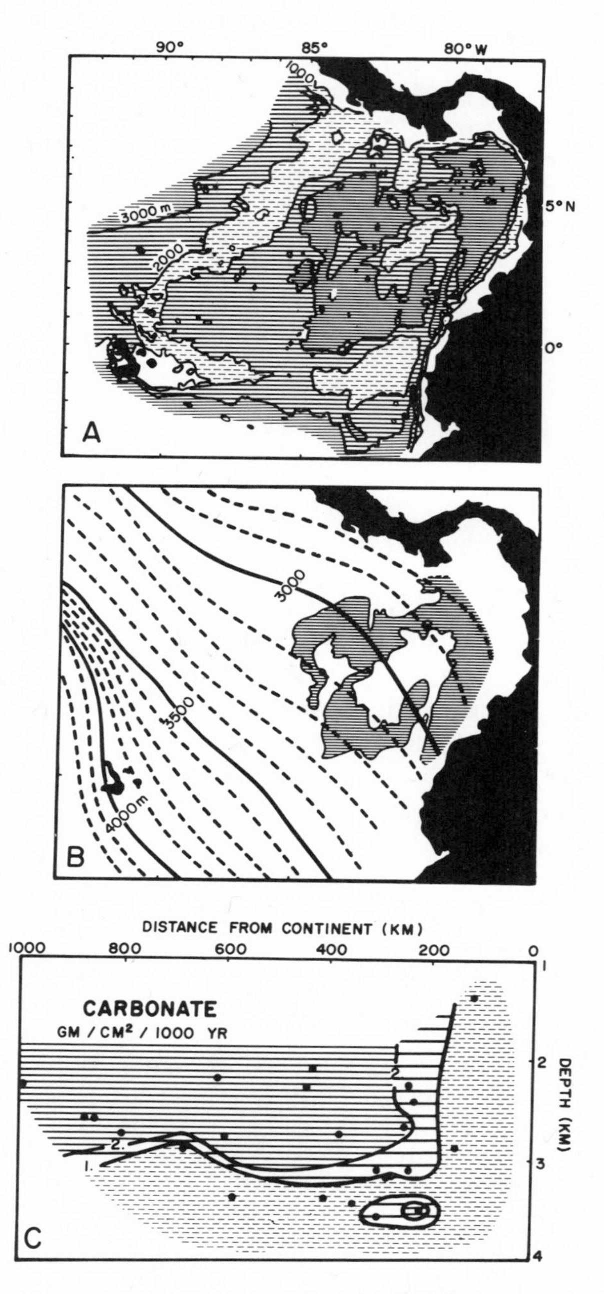

Fig. 1. a) Panama Basin morphology (after *van Andel et al.*, 1971), b) calcite compensation depth (after *Berger and Winterer*, 1974; the shaded part of the basin is situated below the calcite compensation depth), and c) Holocene calcium carbonate accumulation rates in the Panama Basin (from *Swift*, 1976 a).

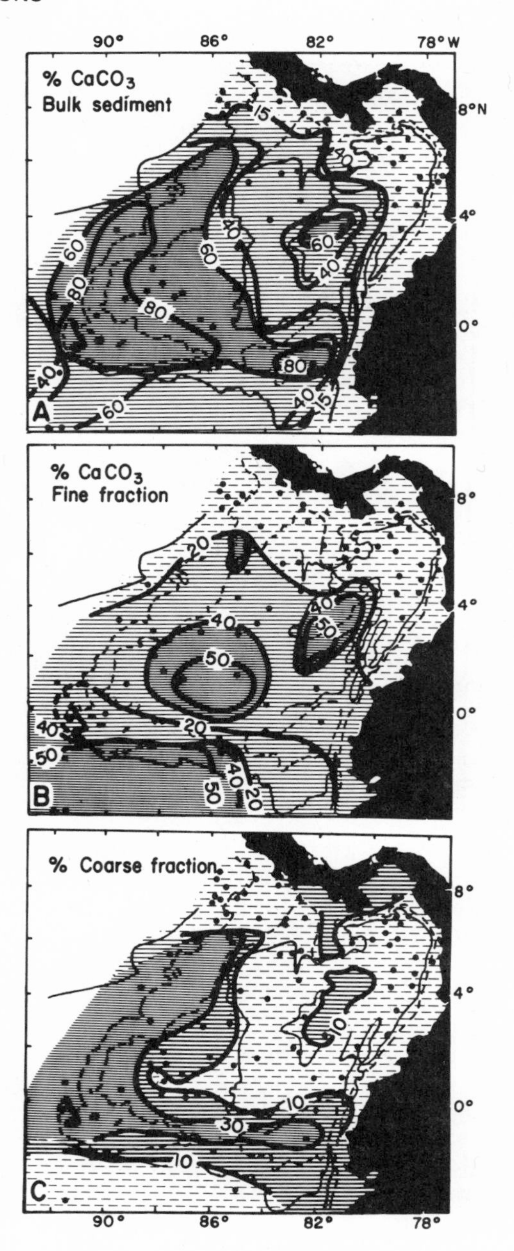

Fig. 2. Distribution of calcium carbonate and proportion of coarse material in surface sediment samples from the Panama Basin. a) $CaCO_3$ contents of the bulk sediment; b) $CaCO_3$ contents of the size fraction < 0.063 mm; c) proportion of coarse fraction (> 0.063 mm) of the bulk sediment. a) and b) after *Moore et al.*,1973 and *Yamashiro*, 1975, c) after *van Andel*, 1973.

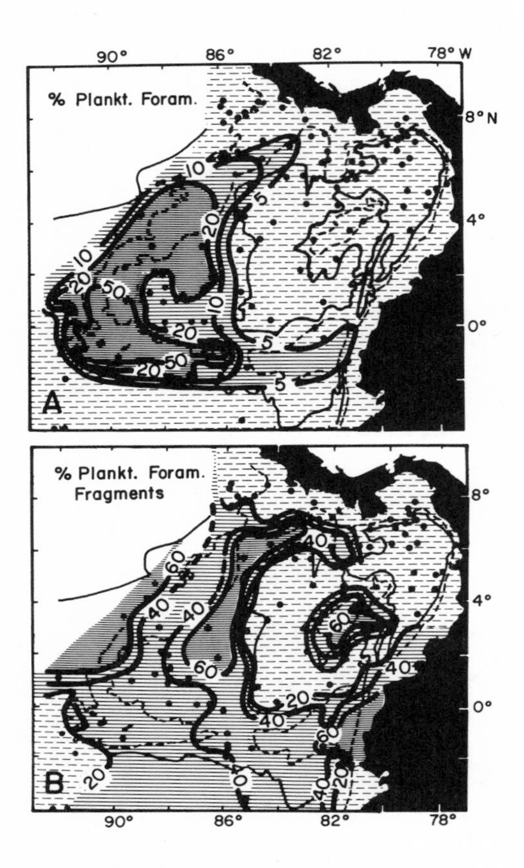

Fig. 3. Distribution of whole planktonic foraminifers (a) and of planktonic foraminiferal fragments (b), both in percent of the total coarse fraction (after *Kowsmann*, 1973).

METHODS

Sample locations for this study have been listed in Table 1. The best regional coverage has been obtained in the area of the Carnegie Gap (*Malfait*, 1974) while samples are sparsely distributed throughout the remainder of the Panama Basin. The reason for this uneven distribution is due to occasional insufficient volumes of bulk sediment obtained from the surfaces of the old, small diameter piston and pilot gravity cores. As shown in Fig. 2, the proportion of coarse fraction is too small to yield sufficient sample for settling tube analyses in all samples from the central, and north-eastern Panama Basin (except from the Coiba Ridge), since a minimum of 300-400 mg of material is needed for each size analysis (see

TABLE 1. Listing of surface sample locations from the Panama Basin. PC = piston core, FF = free fall core, MC = multi-gravity core, PG = pilot gravity core.

Core no.	Type of Corer	Latitude	Longitude	Water depth (corr. m)	Sample interval(cm)	OSU Acc. no.
V15-29	PC	06° 21'N	083° 16'W	1889	11-14	00013
V19-24	PC	03° 12'N	080° 08'W	1712	9-12	00045
V19-25	PC	02° 28'N	081° 42'W	2404	10-13	00047
V19-27	PC	00° 28'S	082° 04'W	1373	10-13	00049
V21-25	PC	05° 43'N	081° 03'W	1359	15-18	00055
V21-29	PC	00° 57'S	089° 21'W	712	12-14	00062
V21-30	PC	01° 13'S	089° 41'W	617	10-13	00064
V21-211	PC	03° 00'N	088° 23'W	1443	7-10	00066
V21-214	PC	03° 50'N	080° 38'W	2246	13-16	00072
V24-36	PC	06° 30'N	085° 13'W	1878	0-3	00086
RC8-102	PC	01° 25'S	086° 51'W	2180	7-10	00088
RC11-238	PC	01° 31'S	085° 49'W	2573	16-19	00117
Y69-71	PC	00° 06'N	086° 29'W	2740	14-20	00153
Y71-3-14 FF-1	FF	01° 40'S	085° 43'W	2518	0-2	07937
Y71-3-15 MC-1	MC	01° 28'S	085° 42'W	2660	0-2	07939
Y71-3-16	PC	01° 12'S	085° 36'W	2227	18-20	07940
Y71-3-18 MC-2	MC	01° 38'S	085° 21'W	2537	0-2	07941
Y71-3-19 MC-4	MC	01° 34'S	085° 15'W	2404	0-2	07942
Y71-3-22 MC-3	MC	01° 08'S	085° 25'W	2338	10-13	07945
Y71-3-23 MC-1	MC	01° 05'S	085° 11'W	2334	0-2	07946
Y71-3-24 MC-1	MC	01° 06'S	085° 01'W	2099	0-2	07947
Y71-3-25 MC-2	MC	00° 59'S	085° 07'W	2395	0-2	07949
Y71-3-29	PG	00° 37'S	085° 34'W	2779	0-2	07951
Y71-3-31	PG	00° 32'S	085° 42'W	2840	0-3	07953
Y71-3-32 FF-2	FF	00° 44'S	085° 25'W	2641	0-2	07954
Y71-3-32 FF-4	FF	00° 44'S	085° 25'W	2574	0-2	07955
Y71-3-32 FF-5	FF	00° 44'S	085° 25'W	2522	0-2	07956
Y71-3-32 FF-7	FF	00° 44'S	085° 25'W	2444	0-2	07957
Y69-102	PC	01° 04'S	085° 51'W	2220	1-3	07958

Clauson in *Thiede et al.*, 1976). The samples from the Carnegie Gap also encompass a water depth interval, which corresponds to the region where large quantities of foraminiferal fragments have been observed on the flanks of Carnegie and Cocos Ridges (*Kowsmann*, 1973). It should therefore be possible to separate effects of dissolution from those of mechanical sorting. The region of the Carnegie Gap has been studied in great detail by *Malfait*, 1974 and *Lonsdale and Malfait*, 1974, by evaluating deep-tow data. Thus, the relationship of the sample locations to the morphologic features in this area is much better known than in the remainder of the Panama Basin; this relationship is important to understand the textural characteristics of these size distributions. The samples used were available at Oregon State University from previous studies where further details can be found (*van Andel*, 1973; *Heath et al.*, 1974; *Kowsmann*, 1973; *Moore et al.*, 1973; *Dinkelman*, 1974; *Dowding*, 1975; *and Yamashiro*, 1975), see also Table 1.

The coarse fractions of these samples have been separated from the fines by wet sieving through a 0.063 mm sieve. Textural characteristics of those samples which contained sufficient material to determine size distribution have been analyzed with a newly developed fully automated settling tube system (*Thiede et al.*, 1976). Approximately 400-800 mg of dry sample are needed to obtain reliable and reproducible size distributions. Before carrying out these textural measurements the samples have to be processed to clean them, and to re-suspend them in an aqueous solution; the latter is quite difficult since small air bubbles are trapped in the single chambers of many foraminiferal shells. If these air bubbles are not removed prior to the analysis, settling velocities representative for these particles cannot be measured. The settling velocity data are then processed to plot cumulative and frequency curves of the size distributions of the single coarse fraction samples. The polymodal frequency curves are next resolved using a DuPont 310 curve resolver, following a technique which has been discussed at length by *van Andel*, 1973. Since the version of the curve resolver which is available to this institution at present contains only five channels, only the main modes have been analyzed to date. The curve resolver specifies the height, width and position of the single modes, which can also be summed to match the original size frequency distribution. The computer program developed for the large diameter settling tube data calculates grain sizes from settling velocities assuming that the particles correspond to spheres with a density of 2.65 g/cm^3. This introduces an error into the size distributions, since the coarse fractions analyzed consist dominantly of calcitic components with a specific gravity of 2.72 g/cm^3. This error might be counterbalanced by the fact that we are not dealing with spheres in these samples, but instead with dominantly planktonic foraminiferal shells which have a relatively rough surface, or with platy foraminiferal fragments.

Once the position of the single modes within the size spectrum of the total coarse fractions is known, the bottom part of the large diameter settling tube can be reassembled to subsample up to six different size intervals which are usually chosen to correspond to the most important modes. These subsamples can then be studied for their compositional characteristics. It is obvious from visual frequency estimates of the single particle categories that the assemblages are quite different in the single modes. This fact has already been determined by *Malfait* (1974).

COARSE FRACTION GRAIN SIZE SPECTRA FROM PANAMA BASIN SURFACE SEDIMENT SAMPLES

Position of Main Modes

The positions of the main modes found in the dominantly calcareous fractions (*Kowsmann*, 1973) of Panama Basin surface samples are listed in Table 2. As can be seen from the few selected samples illustrated in Fig. 4 the main modes which can be traced in the bulk of the samples (Table 2) are not local phenomena, but a basin-wide feature typical of the depositional environment in this part of the eastern tropical Pacific Ocean. The four samples shown in Fig. 4 come from such diverse locations as the Carnegie Gap, the northeastern extension of the Cocos Ridge and from the eastern flank of the Malpelo Ridge. The dominant modes in these samples have been found at approximately 2.3, 2.65, 3.15 and 3.5 *phi*, although the positions of these modes vary somewhat from sample to sample (Table 2). Table 1 shows that the four samples cover from 1712-2573 m water depth which is well above the calcite compensation depth throughout the basin (Fig. 1).

Since the four modes found in these samples have been observed at most locations throughout the Panama Basin (Table 2) they are thought to represent typical size distributions of calcareous coarse fractions in the region of the eastern tropical Pacific Ocean. The dominant portions of these coarse fractions consist of the shells of planktonic foraminifers. Therefore, it was to be expected that species compositions throughout the basin would be relatively constant; this observation seems to be confirmed by studies of planktonic foraminiferal faunas living in this area (*Bradshaw*, 1959). Coarser modes around 2.0-2.35 *phi* and in very few samples between 1.45 and 1.95 *phi* have been observed on most samples from the Carnegie Gap, which is known to represent a high energy environment (*Malfait*, 1974; *Lonsdale and Malfait*, 1974); positions of samples containing these coarse modes outside the Carnegie Gap area have been found to be situated on top or on the shoulders of local elevations, primarily in relatively shallow water depths (compare

TABLE 2. Positions of main modes (in *phi*) of Panama Basin sediment surface samples (compare Table 1).

Core no.	1.45 1.90 *phi*	2.0- 2.25	2.3- 2.45	2.5- 2.75	3.0- 3.3	3.4- 3.75	3.95- 4.15	4.25- 4.55 *phi*
V15-29			2.3	2.65	3.1	3.6		
V19-24			2.35	2.65	3.15	3.65		
V19-25			2.45	2.65	3.1	3.6		
V19-27		2.15		2.5	3.3			4.55
V21-25				2.75	3.3	3.6	4.0	
V21-29		2.15		2.7	3.25		4.05	
V21-30		2.05		2.75			3.95	4.55
V21-211			2.3	2.65	3.1	3.6		
V21-214		2.2	2.45	2.6			4.05	
V24-36				2.55	3.15	3.65		4.25
RC8-102		2.2		2.65	3.15	3.55		
RC11-238		2.25		2.65	3.15	3.4		
V69-71	1.65		2.4			3.45	4.1	
Y71-3-14 FF-1			2.3	2.65	3.1	3.6		
Y71-3-15 MC-1			2.35	2.6	3.1	3.6		
Y71-3-16				2.65		3.75	4.05	4.45
Y71-3-18 MC-2			2.3	2.65	3.1	3.5		
Y71-3-19 MC-4			2.3	2.6	3.1	3.45		
Y71-3-22 MC-3		2.25		2.65	3.0	3.5		
Y71-3-23 MC-1		2.0		2.65	3.05	3.7		
Y71-3-24 MC-1	1.9	2.15		2.55			4.0	
Y71-3-25 MC-2	1.45	2.25	2.4			3.65		
Y71-3-29		2.0		2.6		3.55	3.95	
Y71-3-31		2.2		2.65		3.7		4.3
Y71-3-32 FF-3		2.15		2.6		3.55	4.15	
Y71-3-32 FF-4		2.1		2.6		3.6	4.05	
Y71-3-32 FF-5		2.15		2.65		3.75		4.35
Y71-3-32 FF-1		2.15		2.6		3.7	4.05	
Y69-102		2.0		2.6	3.25	3.7		

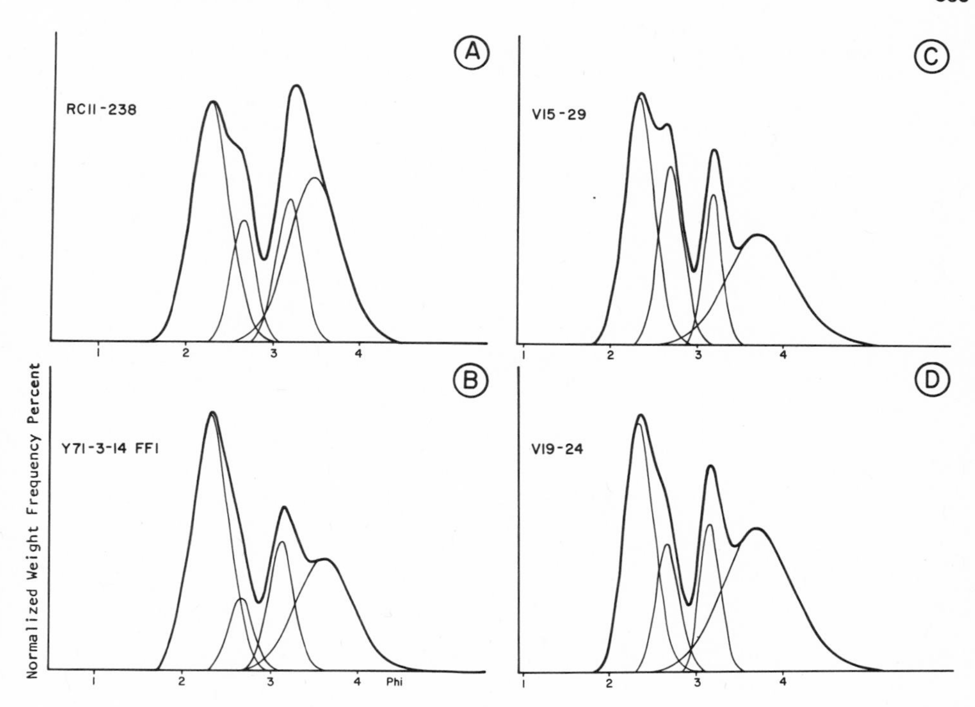

Fig. 4. Selected size distributions of coarse fractions from the Carnegie Gap (a and b), the northeastern extension of Cocos Ridge (c), and from the eastern flank of Malpelo Ridge (d).

Table 1 and 2). It is also interesting to note that grain size spectra in many Pleistocene Panama Basin samples investigated in the course of this study, were similar to the corresponding surface samples.

Although the main modes in these samples can be traced throughout the basin (Table 2) and although many samples from distinct settings reveal almost identical size distribution, the large variability of samples from adjacent localities and the sparse sample coverage of the whole basin prevented us from tracing these modes regionally. Plots against water depth revealed an undecipherable scatter. However, excluding the samples from the Carnegie Gap, where the coarsest sediments are found in the saddle itself (see below), the finer modes were observed to be more prominent in the deeper samples, as suggested by the overall grain size distribution of the Panama Basin surface sediments (Fig. 2). A major obstacle to the better understanding of these size distributions is the lack of knowledge of the local setting of each sample, beyond such general information as water depths and echo sounding records from surface ships. It seems a prerequisite that information about the local setting of each sample is available either from extensive deep-tow surveys, as has been illustrated by *Lonsdale and Malfait*, (1974) for the Carnegie Gap, or from bottom observations from submersibles. This may be why the set of samples obtained from the Carnegie Gap region is better understood than those samples from the remainder of the basin (see section on Regional Distribution of Modes).

The range of local differences in the size distributions can be better understood by comparing size distributions of the samples illustrated in Fig. 5. Although most samples are found to have polymodal size distributions (Fig. 4), the variability from sample to sample ranges from samples which consist almost entirely of one specific mode, to samples which contain all major modes observed in this region, even though their positions may be adjacent (Fig. 5, Table 1). The variability of the size spectra ranges from the dominance of one single mode in one sample to the almost complete absence of the same mode in the next sample with the result that - as mentioned above - it was impossible to trace these changes regionally, except in the rather well-covered and well-surveyed region of the Carnegie Gap.

Regional Distribution of Modes

Modes are not only characterized by the position of their peak within the size spectrum of a particle assemblage, but also by their height and width. Since height and width of the modes are highly variable (compare Table 3), the importance of the single modes is better expressed by their frequency percentage of the total coarse fractions (compare *van Andel*, 1973), which is determined while processing the data on the curve resolver, and which is a function of their height and width. Although it was impossible to understand the frequency distributions of samples from the entire Panama Basin, as has been successfully done for silt-sized material (*van Andel*, 1973), the samples from the Carnegie Gap reveal a very interesting size distribution. This distribution can be related to previous observations of the sediment cover in this region (*Lonsdale and Malfait*, 1974; *Malfait*, 1974) and to the inferred hydrography within the gap (*Lonsdale*, 1976). The contours (Fig. 6) on the eastern side of the Carnegie Gap are only poorly controlled. However, these contours are supported by size distributions from samples taken on the Carnegie Ridge further to the east (V19-27), and by samples to the north of the Carnegie Gap (Y69-71).

Modes > 2.45 *phi* (mainly those between 2.0 and 2.25 *phi*, see Table 2) make up an important portion of all samples stretching in a N-S belt over the saddle of Carnegie Gap close to the area, where *Malfait*, 1974 observed a stretch of relatively coarse grained material

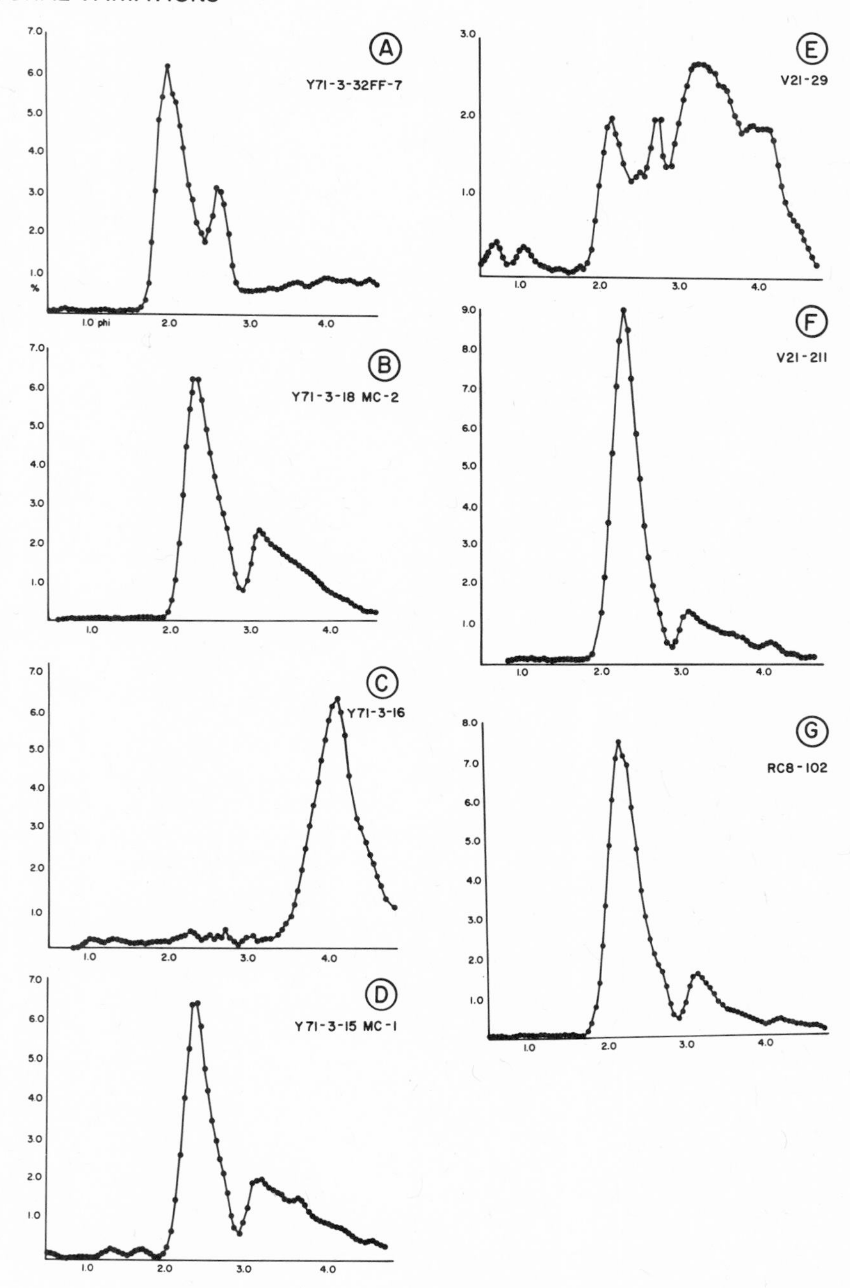

Fig. 5. Size distributions of samples from the Carnegie Gap (a, b, c, d), from the Carnegie Ridge just east of the Galapagos Islands (e), from the southwestern Cocos Ridge (f), and from the southwestern Panama Basin just north of the Carnegie Ridge (g).

TABLE 3. Position, height, width and area of two dominant modes in the Panama Basin in surface sediment samples.

Core no.	Position of mode (*phi*)	Height (Freq.%)	Width (x0.5, in *phi*)	Area (freq.%)	Position of mode (*phi*)	Height (freq.%)	Width (x0.5, in *phi*)	Area (freq.%)
V15-29	2.65	41.8	6.7	21	3.6	26	18.4	33
V19-24	2.65	29.5	6.5	13	3.65	33.3	18.8	42
V19-25	2.65	8.2	3.5	2	3.6	60	21.3	75
V19-27	2.5	57.8	5.5	15				
V21-25	2.75	2.5	5	1	3.6	43	9.5	36
V21-29	2.7	29.8	5	7				
V21-30	2.75	56	8.7	29				
V21-211	2.65	8.5	6.3	9	3.6	5	2	9
V21-214	2.6	20.5	7	13				
V24-36	2.55	33	13	22	3.65	53	10	28
RC8-102	2.65	11.8	6.5	11	3.55	5	17	10
RC11-238	2.65	28.2	6.5	11	3.4	38.5	14	34
R69-71					3.45	33.5	7.2	13
Y71-3-14 FF-1	2.65	16.8	5	8	3.6	26.2	15.5	30
Y71-3-15 MC-1	2.6	21.4	7	19	3.6	14	15	25
Y71-3-16	2.65	4.2	7	3	3.75	24.6	8.3	23
Y71-3-18 MC-2	2.65	21	6.5	15	3.5	21	6.5	15
Y71-3-19 MC-4	2.6	23	7	18	3.45	19	15.5	31
Y71-3-22 MC-3	2.65	46	5	18	3.5	31	11.7	26
Y71-3-23 MC-1	2.65	26.3	5.8	17	3.7	9	15.5	15
Y71-3-24 MC-1	2.55	35	4	13				
Y71-3-25 MC-2					3.65	8.7	9.3	5
Y71-3-29	2.6	28	5.3	22	3.55	3.3	6.7	3
Y71-3-21	2.65	60	5	30	3.7	12.4	6.7	6
Y71-3-32 FF-3	2.6	39	5.5	22	3.55	6.8	8.5	5
Y71-3-32 FF-4	2.6	26	5	18	3.6	2	4	1
Y71-3-32 FF-5	2.65	44.6	4.7	20	3.75	7.2	8.7	6
Y71-3-32 FF-7	2.6	47.5	4.6	24	3.7	5.7	6	3
Y69-102	2.6	61	5	37	3.7	9.3	7.6	8

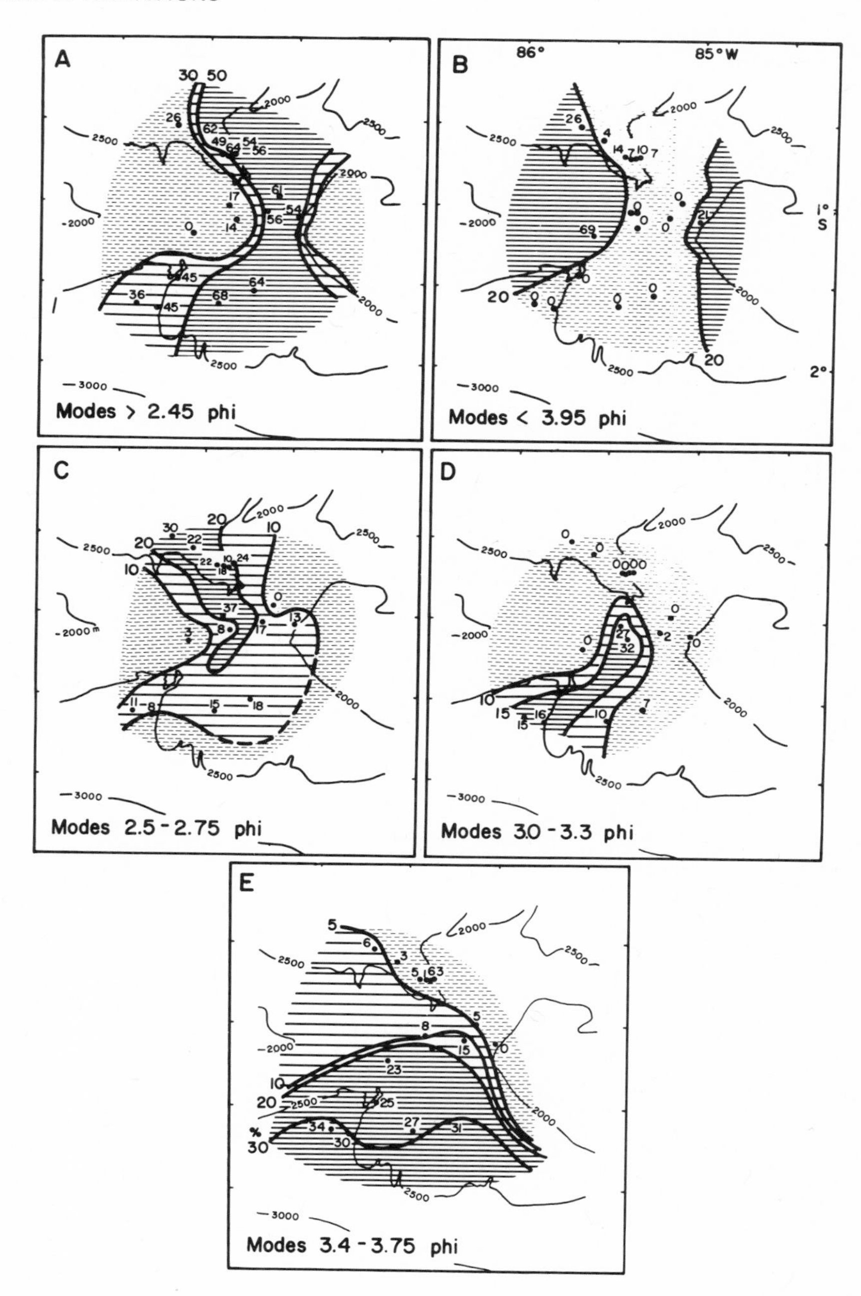

Fig. 6. Distribution of main modes in the Carnegie Gap area (compare Table 2). Contours are in frequency percent of coarse fraction. a) Summed modes > 2.45 *phi*, b) summed modes < 3.95 *phi*, c) mode at 2.5-2.75 *phi*, d) mode at 3.0-3.3 *phi*, and e) mode at 3.4-3.75 *phi*.

(see his Fig. 37). The highest proportions (> 60%) of these coarse modes are confined to the central part of the Gap (Fig. 6a), where submarine dunes built up by foraminiferal sands have been observed. These modes can also be traced well to the north and to the south of the actual sill and suggest N-S directed bottom water currents on both sides of the sill. Such currents have actually been observed recently (*Lonsdale*, 1975, 1976) though the evidence is not overwhelmingly convincing.

The finest modes in these samples (< 3.95 *phi*) reveal a distribution (Fig. 6b) opposite to the coarse modes. They are missing in the areas where coarse modes are concentrated and have probably been removed from the floor of the Carnegie Gap. It is interesting to note that their frequency increases immediately to the west and to the east of the sill region proper.

The regional distribution of the remaining mappable modes (Fig. 6c, d, and e) is considerably more difficult to understand. Two of them follow essentially the Carnegie Gap, the frequency of the 2.5-2.75 *phi*-modes increasing to the north, the frequency of the 3.0-3.3 *phi*-modes increasing to the south. We know that thick deposits of probably reworked sediments are found to the north as well as to the south of the Carnegie Gap (*Malfait*, 1974). These modes might be explained by bottom water masses moving back and forth through the gap (as has been observed by *Lonsdale*, 1975), but working at different strength.

The increase of the 3.4-3.75 *phi*-modes on the southern flank of the Carnegie Ridge (Fig. 6e) is difficult to explain, since a similar distribution should be expected on its northern flank. The poor sample coverage on the northern flank prevented us from detecting this mode; however, it is represented at Y69-71 (13% of the total coarse fraction) to the north of the region, thereby confirming this expectation.

COMPOSITION OF PANAMA BASIN SURFACE SEDIMENT COARSE FRACTIONS

The main modes have been separated from each other in the large diameter settling tube (see above) to study their composition. The consistent position of the main coarse fraction modes within the grain size spectra throughout the basin (Table 2) suggests that their composition may be homogenous, which has been verified when visually screening the separated subsamples (compare also *Malfait*, 1974; and *Lonsdale and Malfait*, 1974). The coarse modes (> 2.45 *phi*) consist dominantly of intact shells of *Globoquadrina dutertrei* with relatively few *Globorotalia menardii* and *G. inflata*, while foraminiferal fragments are almost entirely absent from these modes. The 2.5-2.75 *phi*-modes contain small *G. dutertrei* and *G. menardii* shells, broken specimens of the above species and up to 60% foraminiferal fragments (mainly of the above mentioned species). The small modes consist almost entirely of planktonic foraminiferal fragments largely derived from *G. dutertrei* and *G. menardii*; small species such as *Globigerinita glutinata*, *Globoquadrina hexagona* and *Globigerina quinqueloba* are also concentrated in these fine modes.

It is a deficiency of this study that this set of data does not cover any samples from an area close to the calcite compensation depth. Therefore, it is almost impossible to evaluate the influence of $CaCO_3$ dissolution on the textural characteristics of these calcareous particle assemblages. Since *G. dutertrei* and *G. menardii* are both known to represent solution resistant planktonic foraminifers (*Berger*, 1970), and since most samples have been taken well above the calcite compensation depth (*Berger and Winterer*, 1974) we have to assume that the fragmentation of these shells occurs largely due to mechanical stress. The inhomogeneities from sample to sample are therefore thought to be controlled by the basin-wide input of planktonic foraminiferal shells which is locally altered according to the prevailing bottom water currents.

CONCLUSIONS

1. Panama Basin surface sediment coarse fractions are composed of remains of planktonic foraminifers. The texture of these coarse fractions reveals well-defined modes which dominate the grain size spectra throughout the basin.
2. The position of the modes and their composition is controlled by the production of planktonic foraminiferal shells in the surface water masses of the Panama Basin (largely within the water masses of the eastern equatorial current system) and by the alteration of the shell assemblages by local bottom currents.
3. These alterations include separation of various shell assemblages due to their distinct hydraulic behavior, as well as the fragmentation of larger specimens.
4. While the proportion of the coarse grain sizes diminishes downslope of the ridges bordering the Panama Basin to the south and to the west (Fig. 2), this sequence is reversed in the Carnegie Gap because bottom currents remove all fine particles from its bottom sediments.
5. Water movement over the bottom of the Carnegie Gap seems to consist of north- and southward oriented components of different strength, each transporting their own particle assemblage through the gap.
6. These Panama Basin surface sediment samples bear evidence for large scale removal of fine-grained calcareous material from the crest of morphologic highs. This process might conceptually be responsible for building thick accumulations of deep-sea sediments along the flanks of mid-ocean ridges; this observation has been traced back in time in the South Atlantic by *Berger*, 1972.

ACKNOWLEDGEMENTS

These investigations have been supported by the Office of Naval Research under ONR contract no. N00014-76-C-0067. I have benefitted from discussions with S. A. Swift (Corvallis), Nancy A.Brewster (Corvallis) and various other members of the marine geology group at the School of Oceanography of Oregon State University. This study would have been impossible without the assistance of the team involved in the development of the settling tube system for grain size analysis of deep-sea sediments (T. Chriss, M. Clauson and S. A. Swift) and of K. Klaffke who ran most of the samples described in this paper through the large diameter settling tube. I am indebted to all colleagues who have supported these activities.

REFERENCES

Berger, W. H., 1968. Radiolarian skeletons: Solution at depths.- *Science, 159*. 1237-1238.

Berger, W. H., 1970. Planktonic foraminifera: Selective solution and the lysocline. *Marine Geol., 8*. 111-138.

Berger, W. H., 1972. Deep sea carbonates: Dissolution facies and age-depth constancy. *Nature, 236*. 392-395.

Berger, W. H., 1973. Deep sea carbonates: Evidence for a coccolith lysocline. *Deep Sea Res., 20*. 917-921.

Berger, W. H. and D. J. W. Piper, 1972. Planktonic foraminifera: Differential settling, dissolution and redeposition. *Limnol. Oceangr., 17*. 275-287.

Berger, W. H. and E. L. Winterer, 1974. Plate stratigraphy and the fluctuating carbonate line. *Internat. Assoc. Sediment., Spec. Publ. 1*. 11-48.

Berthois, L. and Y. Le Calvez, 1960. Étude de la vitesse de chute des coquilles des foraminifères planctoniques dans une fluide comparativement a celle des grains de quartz. *Inst. Pêches Marit., Rev. trav., 24*. 293-301.

Bradshaw, J. S., 1959. Ecology of living planktonic foraminifera in the North and Equatorial Pacific Ocean. *Cushman Found. Foram; Res. Contr., 10*. 25-64.

Calvert, S. E., 1974. Deposition and diagenesis of silica in marine sediments. *Internat. Assoc. Sediment., Spec. Pub. 1*. 273-299.

Diester-Haass, L., 1975. Influence of deep-oceanic currents on calcareous sands off Brazil. *9. Congr. Internat. Sediment. 8*. 25-28.

Dinkelman, M. G., 1974. Late Quaternary radiolarian paleo-oceanography of the Panama Basin, eastern equatorial Pacific. PhD. Thesis, 123 pp., (School Oceanogr., Oregon State Univ.) Corvallis.

Dowding, L. B., 1975. Sedimentation within the Cocos Gap, Panama Basin. MS-Thesis, 69 pp., (School Oceanogr., Oregon State Univ.) Corvallis.

Heath, G. R., T. C. Moore and G. L. Roberts, 1974. Mineralogy of surface sediments from the Panama Basin, eastern equatorial Pacific. *J. Geol.*, *82*. 145-160.

Hurd, D. C. and F. Theyer, 1975. Changes in the physical and chemical properties of biogenic silica from the central equatorial Pacific. *Adv. Chem. Ser. (Anal. Meth. Oceanogr.) 147*. 211-230.

Kowsmann, R. O., 1973. Coarse components in surface sediments of the Panama Basin, eastern equatorial Pacific. *J. Geol.*, *81*. 473-494.

Laird, N. P., 1971. Panama Basin deep water - properties and circulation. *J. Marine Sci.*, *29*. 226-234.

Lonsdale, P., 1975. Detailed abyssal sedimentation studies in the Panama Basin: Cruise report of expedition Cocotow legs 2b and 3. *SIO-Ref.*, *75-4*. 17 pp.

Lonsdale, P., 1976. Inflow of bottom water to the Panama Basin: I. Hydrography. *Deep Sea Res.* (in press).

Lonsdale, P. and B. Malfait, 1974. Abyssal dunes of foraminiferal sand on the Carnegie Ridge. *Geol. Soc. Amer. Bull.*, *85*. 1697-1712.

Malfait, B. T., 1974. The Carnegie Ridge near 86° W: Structure, sedimentation and near bottom observations. Ph.D. Thesis, 131 pp., (School Oceangr., Oregon State Univ.) Corvallis.

Melguen, M. and J. Thiede, 1974. Facies distribution and dissolution depths of surface sediment components from Vema Channel and the Rio Grande Rise (southwest Atlantic Ocean). *Marine Geol.*, *17*. 341-353.

Moore, T. C., G. R. Heath and R. O. Kowsmann, 1973. Biogenic sediments of the Panama Basin. *J. Geol.*, *81*. 458-472.

Schlanger, S. O. and R. G. Douglas, 1974. Pelagic ooze-chalk-limestone transition and its implications for marine stratigraphy. *Internat. Assoc. Sediment.*, *Spec. Publ.*, *1*. 117-148.

Schrader, H.-J., 1971. Ursache and Ergebnis der Auflösung von Kieselskeletten in den oberen Sedimentbereichen am Beispiel zweier Kern-Profile vor Marokko und Portugal. *Proc. 2. Plankt. Conf.*, *2*. 1149-1155.

Swift, S. A., 1976 a. Holocene accumulation rates of pelagic sediment components in the Panama Basin, eastern equatorial Pacific. MS Thesis, 91 pp., (School Oceangr., Oregon State Univ.) Corvallis.

Swift, S. A., 1976 b. Holocene rates of sediment accumulation in the Panama Basin, eastern equatorial Pacific: Pelagic sedimentation and lateral transport. *J. Geol.* (in press).

Thiede, J., 1976. Calcareous oozes on Rio Grande Rise (SW Atlantic Ocean). Dissolution and Texture of coarse fraction components. *EOS, Transact. Amer. Geophys. Union*, *57*. 257.

Thiede, J., T. Chriss, M. Clauson and S. A. Swift, 1976. Settling tubes for size analysis of fine and coarse fractions of oceanic sediments. *Tech. Rep., 76-8.* 87 pp. (School Oceangr., Oregon State Univ.) Corvallis.

Van Andel, T. H., 1973. Texture and dispersal of sediments in the Panama Basin. *J. Geol., 81.* 434-457.

Van Andel, T. H., G. R. Heath, B. T. Malfait, D. F. Heinrichs and J. I. Ewing, 1971. Tectonics of the Panama Basin, eastern equatorial Pacific. *Geol. Soc. Amer. Bull., 82.* 1489-1580.

Yamashiro, C., 1975. Differentiating dissolution and transport effects in foraminiferal sediments from the Panama Basin. *Cushman Found. Foram. Res., Spec. Publ., 13.* 151-159.

CARBONATE SEDIMENTS ON THE FLANKS OF SEAMOUNT CHAINS

Edward L. Winterer

Scripps Institution of Oceanography

ABSTRACT

The summits and flanks of seamount chains, aseismic ridges, and oceanic plateaus are good storage areas for carbonate sediments of a wide range of depth facies. Drill data suggest that except for seamounts in their very youthful stage, which have special rates and special sedimentary facies, subsidence of these regions may be similar to that of normal oceanic lithosphere, and that the methods of plate stratigraphy could be applied to their sedimentary record.

INTRODUCTION

The normal orderly hypsometry of the ocean basins, where new hot crust is created at rise crests and then cools and subsides along a regular path as a function of the age of the crest (*Sclater et al.*, 1971), is disturbed by seamount chains, aseismic ridges, and plateaus. Even though altogether these abnormal highland areas constitute only a few percent of the sea floor, they nevertheless influence sedimentation patterns over far more extensive areas, both by perturbing normal oceanic sedimentation and by generating special highland facies. Indeed, it is one of these special facies -- the coral atoll -- that from *Darwin's* (1842) time to our own has challenged the imagination of every marine geologist. Because pelagic carbonate sediments are better preserved on shallow than deep sea floors, the highland areas should have a particularly strong influence on carbonate facies, and it is thus appropriate to take up these special regions in the context of this conference.

From a purely morphologic point of view, one can see a virtual continuum of forms ranging from simple isolated volcanic cones, through seamount chains and aseismic ridges, to plateaus. With some minor exceptions, e.g., Rockall Bank and perhaps the Seychelles Bank and Mascarene Plateaus and a few others, the oceanic highland areas appear to be oceanic volcanic features rather than continental fragments. Some of these highlands -- notably the plateaus -- appear to be generated at rise crests, perhaps close to triple junctions (*Winterer et al.*, 1974) and are made of abyssal tholeiite (*Jackson et al.*, 1976). The seamount chains on the other hand, can appear almost anywhere, on young crust or old. What little is known of aseismic ridges, e.g., Ninetyeast Ridge, suggests that they may be formed close to active rise crests.

PLATEAUS AND ASEISMIC RIDGES

The summit regions of plateaus and of many aseismic ridges are broad and smooth enough to provide accumulation sites for pelagic sediments, uninterrupted by influxes of redeposited sediments from higher ridges, and thus these features have been a favorite target for deep-sea drilling efforts. It would be useful to have a model to reconstruct the history of paleodepth changes on plateaus and aseismic ridges, and since thermal contraction models for subsidence of normal crust produced at spreading ridges appear to accord well with observed age-depth relations for normal ocean crust (*Tréhu*, 1975), a model of that type, where depth is equal to a constant times the square root of the age of the crust, can be tried.

The drilling results from Manihiki Plateau (*Schlanger, Jackson, et al.*, 1976). (Figure 1), where shallow water sediments rest on abyssal-type tholeiite, give a mean rate of subsidence of about 25 m/m.y. (after removing the isostatic loading of the sediments), which is consistent with a whole range of models, including a linear model (depth $\approx$ 30 x age) or an exponential model somewhat like the familiar curve of *Sclater et al.*, (1971) (depth $\approx 300 \times \text{age}^{1/2}$). At other plateaus in the Pacific (Ontong Java and Magellan), the drill cores show pelagic sediments on basement, and thus the points plotted for these sites on Figure 1 show the maximum amounts of subsidence possible. When these and other plateaus where basement was not recorded (Shatsky and Hess) are combined with the aseismic ridges (Ninetyeast, Cocos, and Carnegie), a general increase in depth with increasing age is seen. It is clear that we need far more data points before asserting any law of subsidence that could then be used as a firm basis for plate stratigraphy. The curve Depth $\approx 320 \times \text{Age}^{1/2}$ is drawn in to show the fit to the type of subsidence curve common for normal oceanic crust (*Tréhu*, 1975).

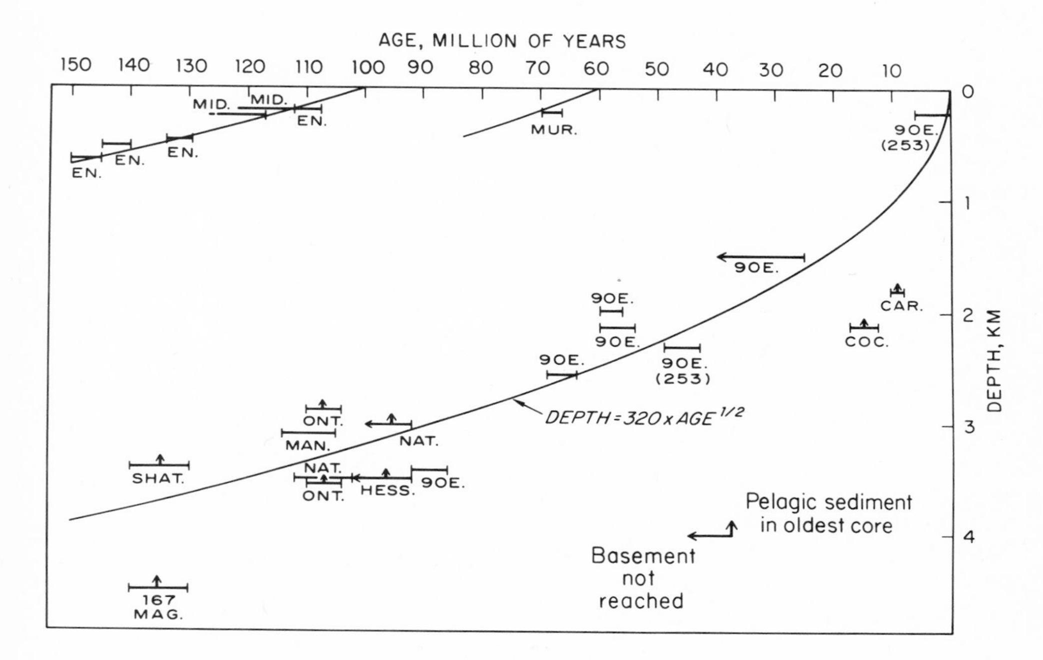

Fig. 1. Age vs. depth for basement (or oldest sediments reached) on drilled plateaus, aseismic ridges, and atolls. Depths are corrected for isostatic effect of sediment cover. Atoll data plotted by offsetting to estimated age of underlying lithosphere. Abbreviations: Shat., Shatsky Rise; Mag., Magellan Plateau; Man., Manihiki Plateau; Ont., Ontong Java Plateau; Nat., Naturaliste Plateau; 90E, Ninetyeast Ridge; Coc., Cocos Ridge; Car., Carnegie Ridge; En., Eniwetok Atoll; Mid., Midway Atoll; Mur., Mururoa Atoll. Drill data from *van Andel, Heath, et al.*, 1973; *Winterer, Ewing, et al.*, 1973; *von der Borch, Sclater, et al.*, 1974; *Davies, Luyendyk, et al.*, 1974; *Veevers, Heirtzler, et al.*, 1974; *Andrews, Packham, et al.*, in press; *Larson, Moberly, et al.*, 1975; and *Schlanger, Jackson, et al.*, 1976.

A special element of the history of these features is that they may be the only monitors of conditions in the upper 2500 meters of the water column, where there may be important dissolution of aragonite, and where an oxygen minimum may be developed. Mid-Cretaceous black shales occur on Manihiki (*Schlanger, Jackson, et al.*, 1976). Hess,and Shatsky (*Larson, Moberly, et al.*, 1975) plateaus at paleodepths (using the curve of Fig. 1) of from about 500 to 1500 m, whereas contemporaneous sediments in the Pacific at "normal oceanic" paleodepths are well oxidized, e.g., at Site 164 (*Winterer, Ewing, et al.*, 1973). A suggestion of shallow dissolution of aragontie in early Cretaceous times is given by the occurrence of *aptychi* in basal sediments on Magellan Plateau (*Renz*, 1973) at an estimated paleodepth of about 500 m. A possible way to get a more direct answer to the question of how plateaus or aseismic ridges subside is to drill holes on atolls on such features, e.g., at Ontong Java or Manihiki Atolls.

SEAMOUNT CHAINS

Subsidence

The subsidence histories of a few seamounts can be reconstructed in more detail, because capping coral reefs provide us with sea-level benchmarks. Drilling on atolls on four Pacific seamounts suggests subsidence rates during coral growth that are nearly linear, and at the rate expected for a volcano riding passively on the underlying crust that is subsiding at the rate appropriate to its age (Fig. 1). But as *Menard* (1973) and others have pointed out, reversals in the general subsidence curve are recorded on some seamounts. Some of these are doubtless due to eustatic changes in sea level, as evidenced by their simultaneous occurrence on widely scattered seamounts where an uplift and subaerial exposure of reefs is recorded, as for example the Mid-Miocene emergence recorded on Eniwetok (*Schlanger*, 1963), Bikini (*Emery, Tracey, and Ladd*, 1954), and Midway (*Ladd, Tracey, and Gross*, 1970). Others are clearly tectonic, as for example at Makatea, an uplifted atoll with foundations probably about 50 million years old in the Tuamotu Chain, which is the only atoll in the group to show any evidence of uplift. It may be close enough to be caught up in the bulge of the archipelagic apron around the very young volcanic pile at Tahiti (Fig. 5). Recent volcanism along older parts of chains, for example the post-erosional alkalic volcanism that so commonly trails some 3-20 million years behind the original main cone-building stage, may be accompanied by renewed uplift. *Menard* (1973) suggested that volcanoes may be elevated as the lithosphere on which they ride passes over bumps on the asthenosphere.

The very early part of the subsidence history of seamounts is probably somewhat different than the later part, when reefs may become established over a major part of the area of the seamount. The slow pick-a-back descent of seamounts as recorded by atoll drilling is of course augmented by isostatic sinking under the load of the growing coral or sediment cap, so the actual rate of sinking during this reef-building stage may be about twice the underlying lithospheric rate. But this rate of sinking is clearly much slower than the rates recorded at very young volcanic piles, as for example in the Hawaiian Islands, where rates of more than a hundred meters per million are estimated from water-well data (*Stearns and Chamberlain*, 1967). The common -- but not universal -- occurrence of moats next to seamounts is suggestive of isostatic sinking under the load of new volcanics. A plausible way to reconcile this contrast in very early and later subsidence rates is to hypothesize a lithosphere weak enough so that isostatic response is fast: Very soon after the end of volcanism, most of the load is compensated.

Archipelagic Aprons

When first built, seamount chains are flanked by an archipelagic apron (often indented by a moat) extending away from the chain several hundred km. The upper end of the apron, where it meets the steeper slopes of the volcano above, commonly stands about 800 m higher than the regional sea floor, and slopes gradually away almost across a relatively smooth topography to merge almost imperceptibly with the normal surroundings. As *Menard* (1956) pointed out, the aprons -- even young ones -- are fairly smooth. Profiler records show that part of the smoothing is due to turbidites derived from the island chain, but part is probably due to deep-sea volcanism, as evidenced by the abnormal smoothness of the pre-sediment basement surface as seen on some profiler records close to seamount chains. If the apron were built up by deep-sea extrusive volcanism contemporaneous with adjacent seamount volcanism, one would expect the pattern of sea floor spreading magnetic anomalies to be obliterated on aprons, but this does not seem to be the case, at least not for the Hawaiian and Marshall chains, where Mesozoic anomalies can be recognized well up onto the aprons (*Pitman, Larson, and Herron*, 1974).

Profiles drawn across archipelagic aprons of increasing age (Figure 2) show that the original slope tends gradually to flatten out. Part of this is due to the isostatic load of sediments, as for example along the Marshall and Gilbert chains, but the total thickness of sediments appears inadequate to explain all the apron subsidence. Perhaps the aprons are partly of thermal origin, a regional bulge associated with the new heat manifest in the seamount volcanism. Reliable and systematic heat flow measurements across relatively young volcanic chains have yet to be made.

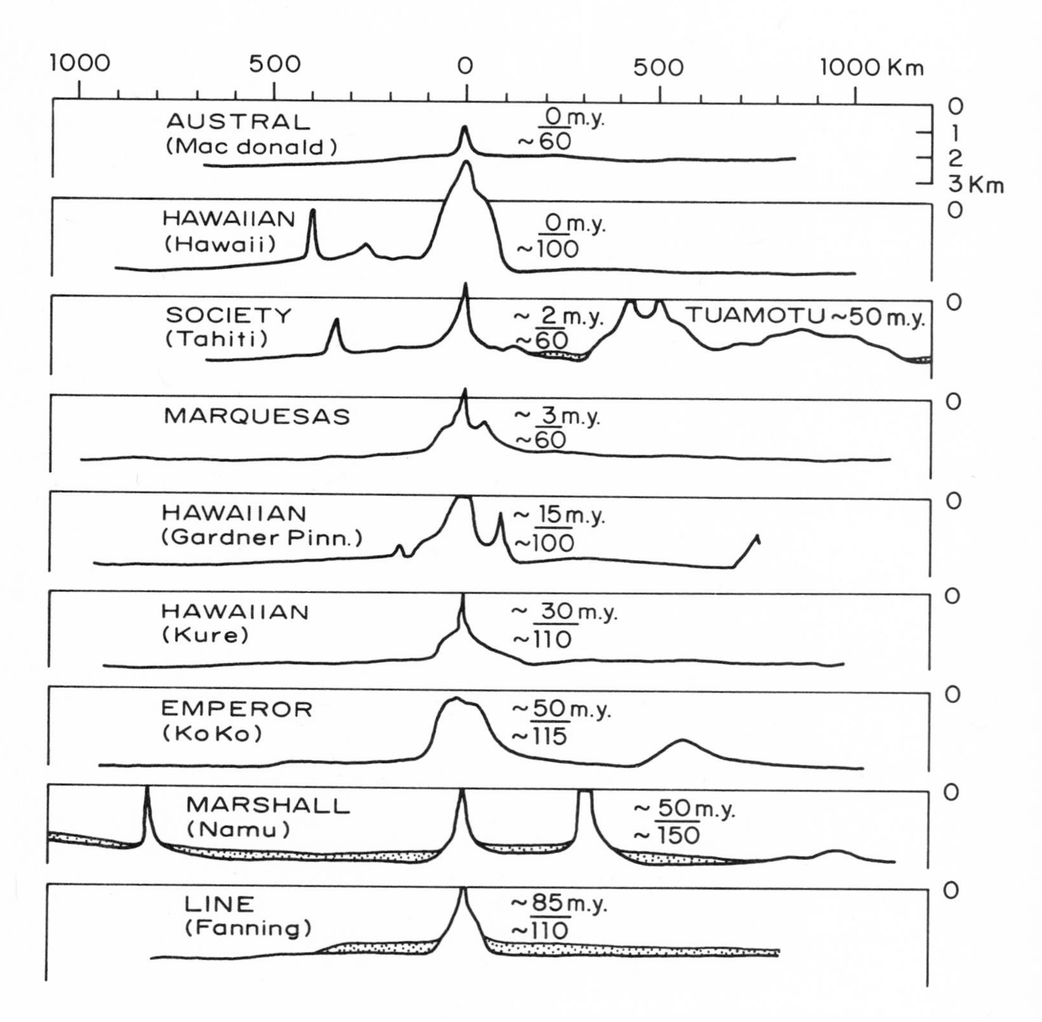

Fig. 2. Profiles of seamount chains. Estimated age of seamount is numerator, estimated age of lithosphere is denominator of fraction. Sediments stippled. The archipelagic apron tends to flatten with age.

Sediments

Three different realms of sedimentation occur on and near seamount chains: (1) the summit area, (2) the archipelagic apron area, and (3) upland areas, such as benches and intermontane valleys. Slopes steeper than about 5 degrees are generally virtually sediment-free. Carbonates can occur in each of the realms, and their rate of accumulation, as elsewhere in the world ocean is mainly controlled by latitude and depth. The special geography and subsidence history of seamount chains gives rise to distinctive facies.

Summit Area: As *Darwin* (1842) hypothesized, subsidence of a seamount permits upward growth of reefs, which evolve from fringing to barrier to atoll. Reef-building corals thrive in today's ocean only in warm tropical waters; seamounts farther than about 25 to 30 degrees from the equator generally have no reefs, and any carbonates in the shallow water sediments on these seamounts consist largely of the remains of bryozoans, molluscs, and benthonic foraminifera. But because seamounts move with the underlying lithospheric plate they change latitude, with the result that flourishing tropical reefs can be carried north or south into less favorable climes where the reefs die, while the seamount continues to subside. The reefs and guyots of the Hawaiian-Emperor seamount chain show this sequence: The seamounts are probably born near latitude 19° and migrate northwest along the path marked out by the traces of the chains; reefs grow on the subsiding foundations up to about latitude 28, but beyond only guyots remain with sunken fossil coral caps.

The extinction of reefs is of course not simply a matter of moving into poor latitudes; ocean climates can also change with time. But a perhaps even more devastating event is a rapid rise in sea level, as suggested, for example, by *Matthews et al.*, (1974) for the extinction of Mid-Cretaceous rudisted reefs on the Mid-Pacific Mountains, leaving guyots that since have collected only a thin veneer of pelagic calcareous sediments as they sank a further 1800 meters.

Intermontane Troughs: Some seamount chains are broad and complex in structure, and include at intermediate depths large intermontane valleys or flattish areas where sediments can accumulate. The deep terraces off the Hawaiian Islands are well known, and intermontane troughs are common in the Tuamotu, Marshall, and Line Islands chains (Figures 3, 4, and 5). These intrachain flattish areas are commonly blanketed by calcareous sediments which can attain considerable thicknesses. In the Line Islands, intermontane valleys with more than 1 km of sediment thickness can be seen on reflection profiler records (Figure 4). A drill site in a similar intermontane setting in the Tuamotu chain (Figure 5), and located about 60 km from a large atoll (*Schlanger, Jackson, et al.*, 1976; Site 318 penetrated 745 m of sediments (basement was not

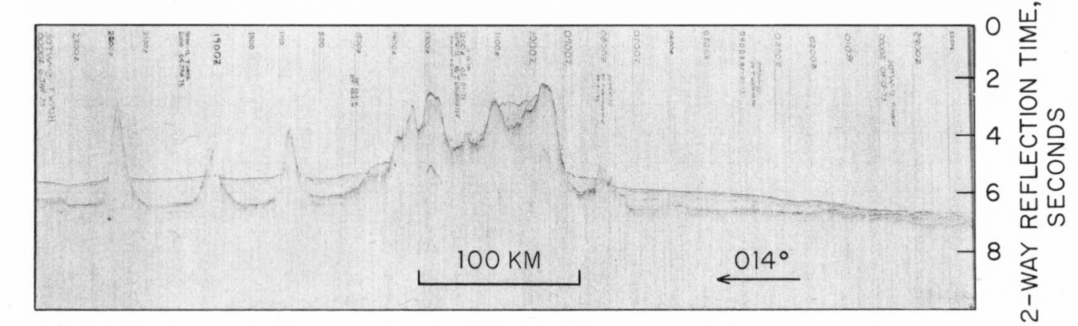

Fig. 3. Seismic reflection profiler record across the Line Islands Chain in the vicinity of Washington Island showing intermontane valley within chain partly filled with sediment. Record taken by Scripps Institution of Oceanography from R/V THOMAS WASHINGTON, Expedition Southtow, Leg 12.

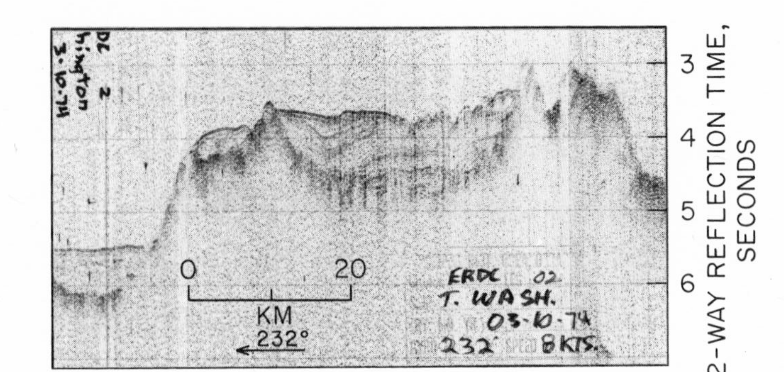

Fig. 4. Seismic reflection profiler record across the Line Islands Chain in the vicinity of Flanning Island, showing an intermontane valley containing about 1 km of sediments. The sediments have been partly eroded away. Record taken by Scripps Institution of Oceanography from R/V THOMAS WASHINGTON, Expedition Eurydice, Leg. 2.

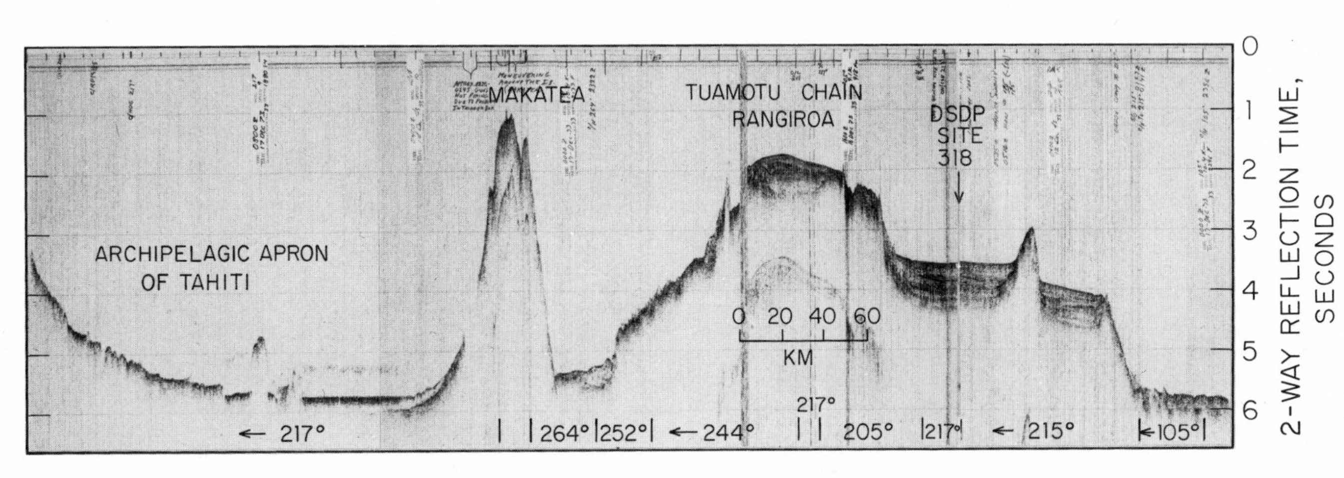

Fig. 5. Seismic reflection profiler record across the northern part of the Tuamotu seamount chain, showing intermontane sediment trough to JOIDES Site 318. Rangiora Atoll rises only a few km away from the line of the profile at the place indicated. The island of Tahiti is about 20 km farther along from the left edge of the profile. Record taken from D/V GLOMAR CHALLENGER on Leg 33.

reached), consisting of highly calcareous pelagic oozes at the top, grading downward into more and more volcanoclastic (Eocene) sediments near the base (Figure 6). The volcanic debris has been mainly highly altered to montmorillonitic clays. Graded beds, carrying redeposited shallow-water sketetal debris and volcanic rock fragments (Figure 7A) are intercalated in the pelagic sediments, but many of the graded layers consist very largely of planktonic sketetal debris (Figure 7B), which is of the same age as fossils in the adjacent normal pelagic layers. These relations suggest that pelagic calcareous sediments falling on the side slopes of the sediment troughs do not accumulate permanently there, but that downslope processes carry the sediments, commonly by means of turbidity currents, farther down onto the floor of the trough. The change from thick graded beds rich in volcanoclastic material near the base of the section to highly clacareous beds with only rare volcanic grains and thin graded layers above suggests that the newly formed volcanic slopes are mantled with great quantities of altered loose volcanic debris that is quickly swept away by mass wasting. Gradually the slopes are denuded of nearly all loose material, except for that supplied by the rain of pelagic skeletons from the waters above,by the drifting downslope of reef debris from the summit area, and by the weathering out of occasional grains of much altered volcanic bedrock. The highly volcanoclastic thick graded beds in the lower part of the sedimentary column include conspicuous amounts of shallow-water calcareous debris: large benthonic forams, calcareous algae, molluscs, echinoid debris, etc. (Figure 7A), possibly because only very large-volume turbidity currents are energetic enough to transport coarse reef debris far from its source.

In some places, the intermontane sediments themselves may constitute a source area for sediments at still deeper levels. In the Line Islands, for example, reflection profiler records (Figure 4) show erosional surfaces developed across intermontane sediments.

Archipelagic Apron: The sedimentary section lying at the foot of the volcanic chain has been explored at three Pacific JOIDES sites close to the Line Islands in the Pacific (Site 165: *Winterer, Ewing, et al.*, 1973; Sites 315 and 316: *Schlanger, Jackson, et al.*, 1976) and at two sites close to the New England seamount chain in the Atlantic (Sites 382 and 385, *Joides*,1975). The basal part of the section is of nearly the same age (Campanian) and the same lithology in both areas, and is remarkably like that just described for the lower sediments in the Tuamotus, commonly in thick graded units, becoming finer grained and thinner bedded upward (Figures 8 and 9). The volcanogenic Line Islands sediments also contain redeposited shallow-water skeletal remains beginning in sediments 3-15 m.y. younger than the oldest apron sediments, but no redoposited shallow-water fossils are reported from the apron sediments of the New England seamount, possibly because that chain was generated too

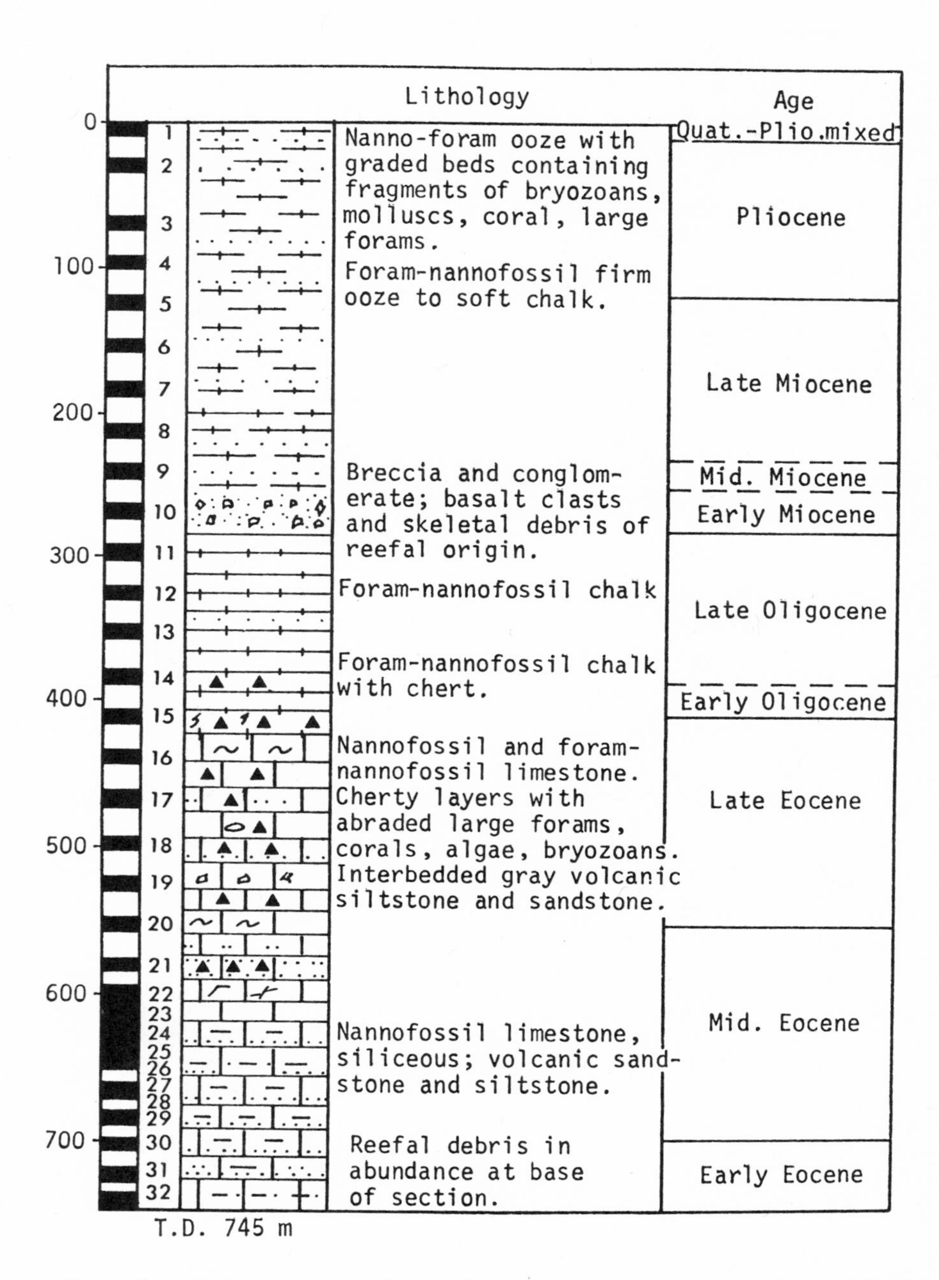

Fig. 6. Columnar section of sediments encountered at JOIDES Site 318 *Schlanger, Jackson, et al.*, (1976), Chapter 6.

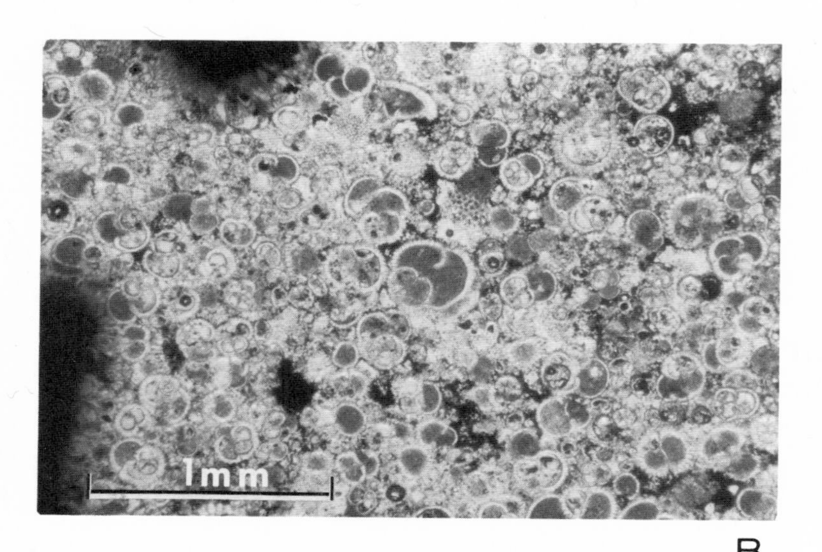

Fig. 7. Photomicrographs of sediments near seamounts.

A. Volcanic sandstone with resedimented shallow-water calcareous fossils (benthonic foraminifera, molluscs, algae). From a graded bed. JOIDES Site 318, Core 32-4, 744 m below sea floor. Age: Early Eocene.

B. Packstone of planktonic foraminifera from a graded bed. JOIDES Site 318, Core 2CC, 28 m below sea floor. Age: Late Pliocene.

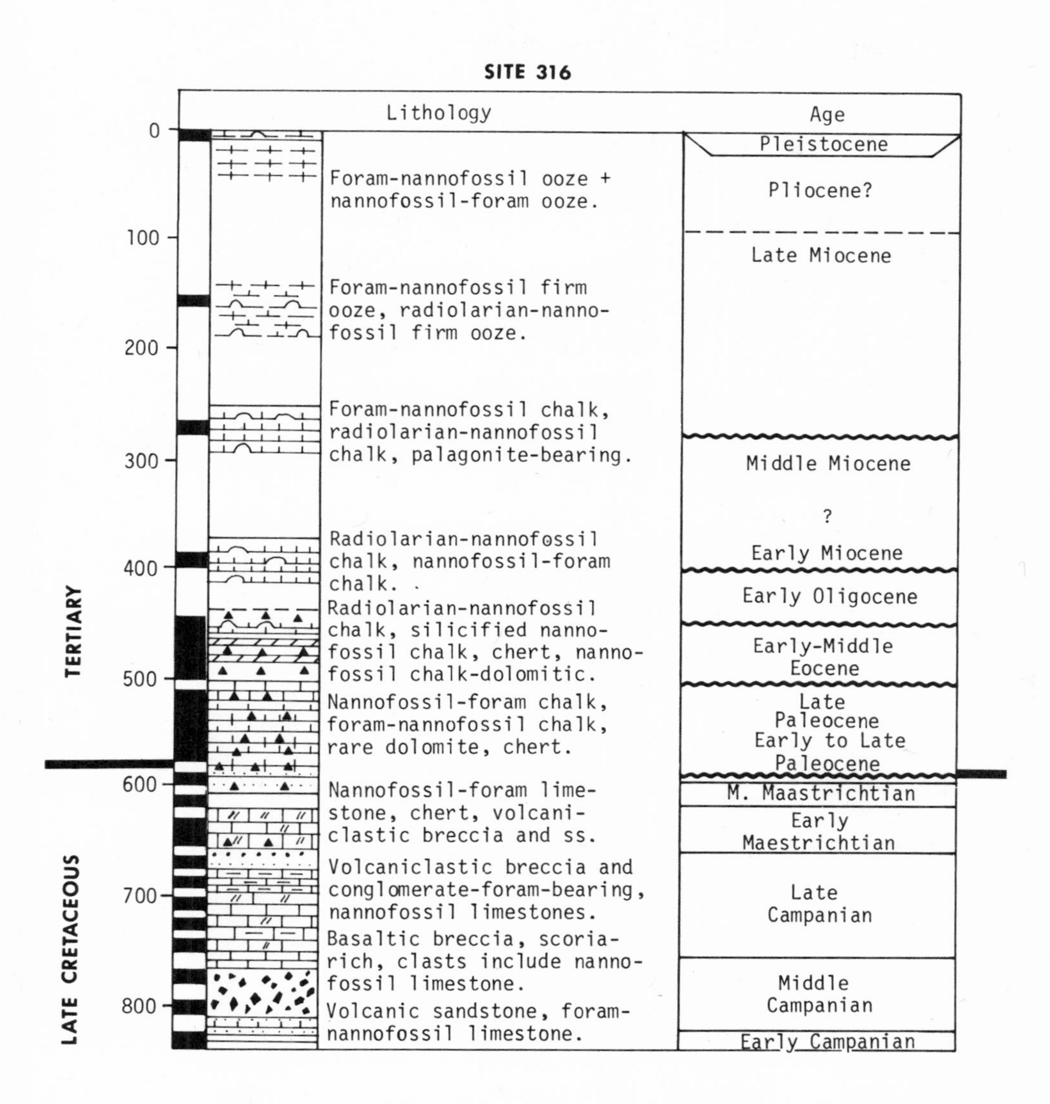

Fig. 8. Part 1. Columnar section of sediments encountered at JOIDES Sites 165 (*Winterer, Ewing, et al.*, 1973, Chapter 3) 315, and 316 (*Schlanger, Jackson, et al.*, 1976, Chapters 3 and 4).

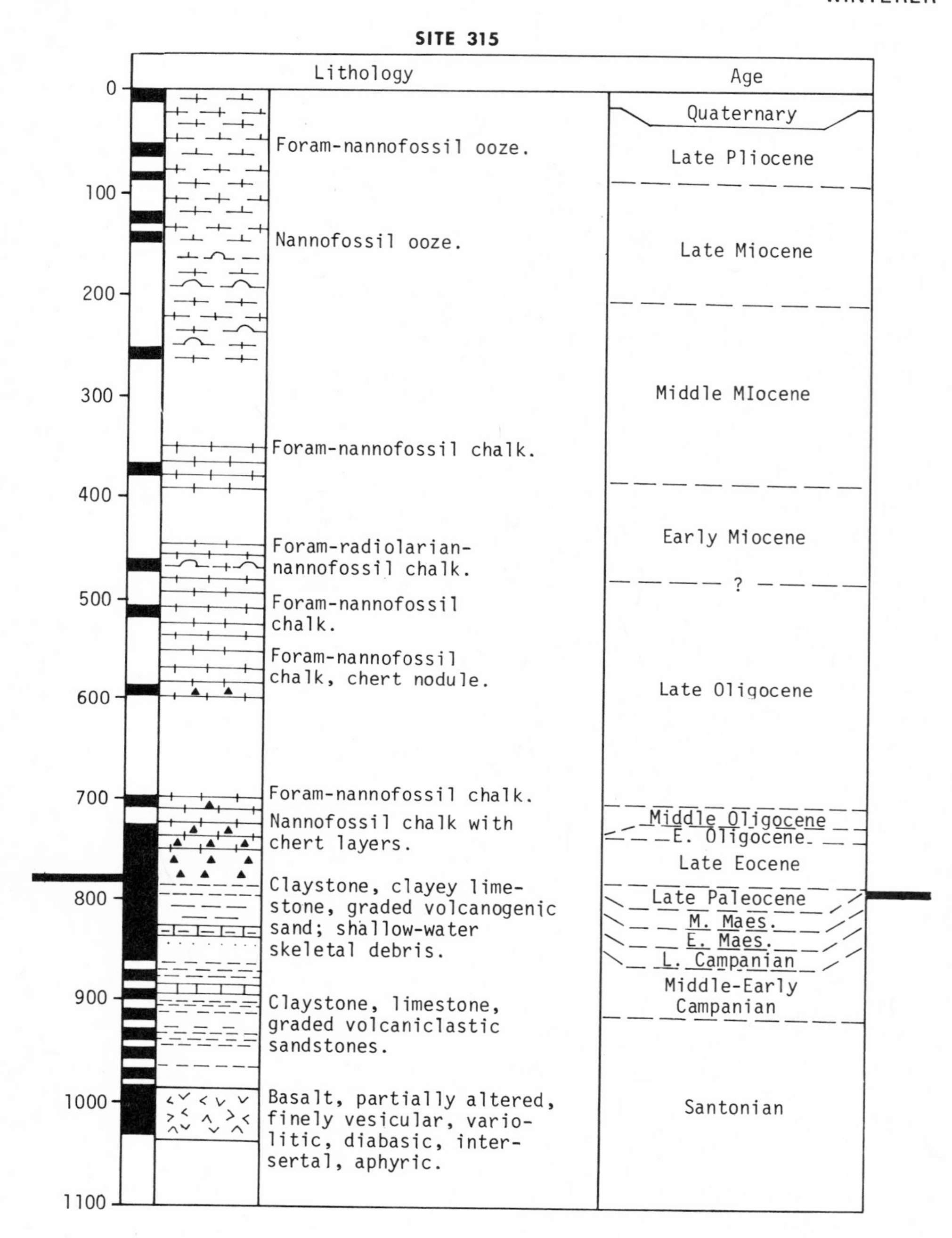

Fig. 8. Part 2. Columnar section of sediments encountered at JOIDES Sites 165 (*Winterer, Ewing, et al.*, 1973, Chapter 3), 315, and 316 (*Schlanger, Jackson, et al.*, 1976, Chapters 3 and 4).

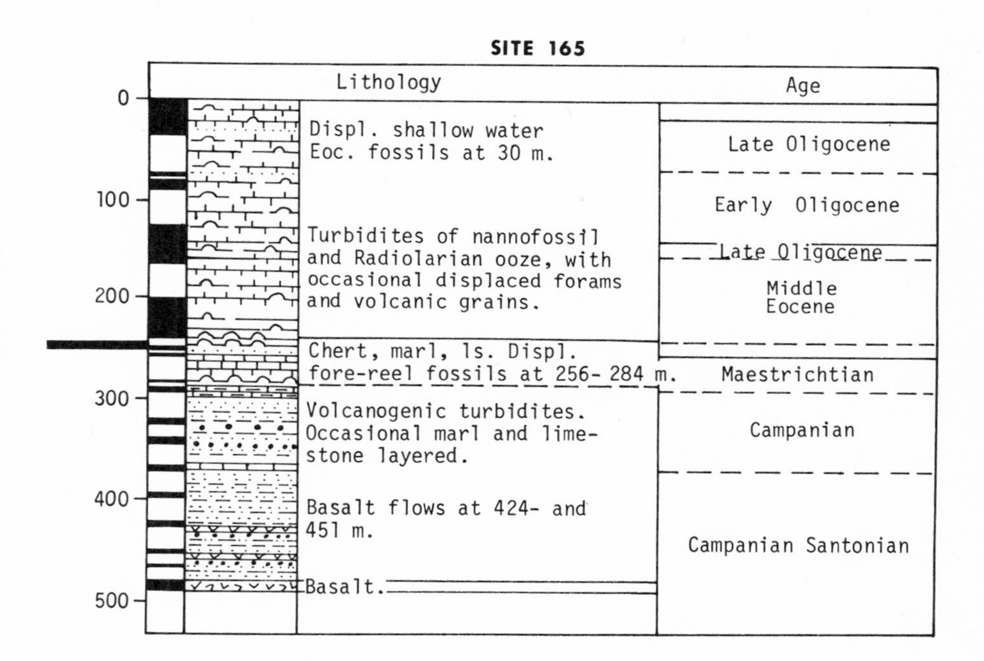

Fig. 8. Part 3. Columnar section of sediments encountered at JOIDES Sites 165 (*Winterer, Ewing, et al.*, 1973, Chapter 3) 315, and 316 (*Schlanger, Jackson, et al.*, 1976, Chapters 3 and 4).

far north to have supported vigorous reefs or because the summit area close to sea level was too small or short-lived to generate appreciable quantities of reefal debris. The delayed appearance of redeposited reefal or shallow-water materials in the Line Islands probably reflects the time required for the seamount subsidence that permits reefs to establish themselves widely around the seamount summit area.

At one of the drill sites (165), thin basalt flows were penetrated 30 and 56 m up in the sediment section lying on the thicker (9 m +) alkalic basalt flow in which drilling ceased (*Winterer, Ewing, et al.*, 1973). It would seem likely that basal sediments and flows may be commonly interlayered at the heads of the aprons.

The lithology of the sediments overlying the basal volcanoclastic unit depends very much on the paleodepth-paleolatitude history at the site. Around the Line Islands, because the seamounts were built on crust only about 25 m.y. old (*Winterer*, in press), the early paleodepths were relatively shallow (~ 3500 m) over much of the apron area, and even pelagic carbonates could be preserved. On the other hand, those parts of the chain more than a few degrees from the equator probably had a relatively low pelagic supply rate. Most of the Late Cretaceous and Early Tertiary carbonate on the apron is redeposited from turbidity currents, probably originating on the middle and upper slopes of the seamount chain. As parts of the chain moved close to the equator, the supply rate of pelagic carbonate increased; post-Cretaceous pelagic carbonates have dominated over turbidites in the paleolatitudes within a few degrees of the equator. Furthermore, the supply rate of redeposited pelagic sediments probably also increased close to the equator, since these are derived from the slopes of the volcanic highlands. A check on this is given by the Late Eocene sediments at Site 165, which show in their accumulation rates the equatorial crossing time of the site, even though the Upper Eocene at this site consists almost entirely of turbidites. The site was within a few degrees of the equator at the time, but because of a shallow equatorial CCD (*van Andel*, 1975), pelagic sediments dissolved at the paleodepth of the site.

The Hawaiian chain, on the other hand, has very little carbonate in the apron sediments, partly because of the low pelagic supply rate in mid-latitudes, and partly because the chain is built on old (~ 100 m.y.) crust and hence has deep (mainly greater than 5000 m) aprons. Turbidity currents bring a little shelf carbonate to the moat (*Fan and Grunewald*, 1971), but mainly volcanic sediments or brown clays prevail. The direction of motion of seamounts in the chain means that the chances for carbonate apron sediments diminish rather than improve with time.

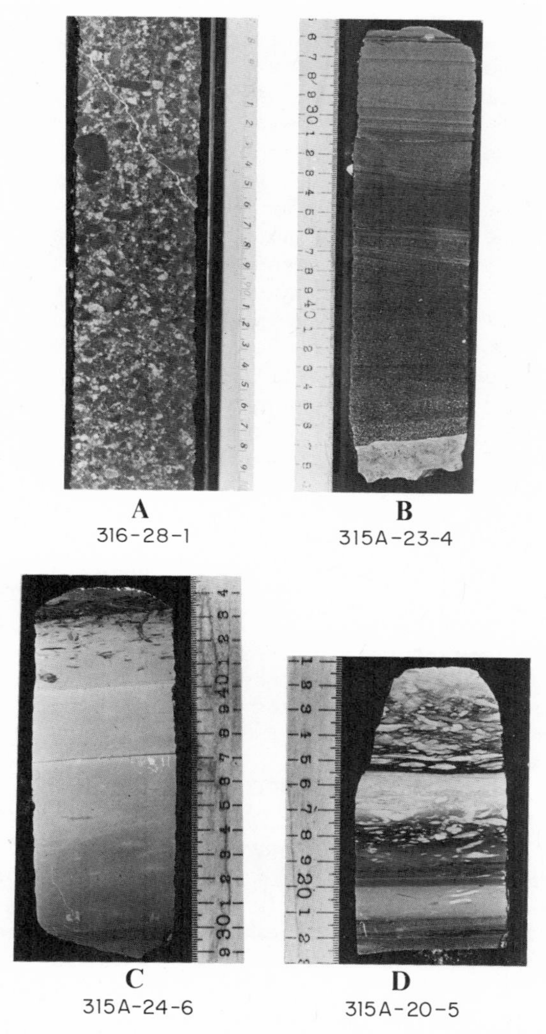

Fig. 9. Cores from lower parts of sediment column at JOIDES drill sites on archipelagic apron of Line Islands Chain. Scales in cm.

A. Volcanic breccia. Part of thick graded bed.

B. Volcanoclastic sandstone (part of a graded bed) showing grading, current bedding, and lamination.

C. Graded bed, volcanoclastic at base and calcareous at top. Burrow mottling of darker volcanoclastic silt from overlying bed.

D. Graded bed, volcanoclastic at base, calcareous near top. Burrow mottling of light-colored highly calcareous, probably from calcareous pelagic top of bed, later eroded off.

In the Line Islands, most parts of the apron have long since drifted north past the equatorial zone and, since water depths are too great for pelagic carbonates, radiolarian ooze or brown clay sediments overlie older apron carbonates. Indeed, in many places erosion prevails, and older apron sediments are channeled. Calcareous apron sediments are now being redeposited, partly by turbidity currents, into still greater depths of water. This channeling appears on reflection profiler records (Figure 10) all to be of late Cenozoic age: no old buried channels are seen in the records. Similar late channeling is seen cutting into sediments within and close to deep passageways through the seamount chain (*Schlanger and Winterer*, 1976, p. 679, 680, 687), and it seems plausible to ascribe much of this erosion to vigorous Antarctic Bottom Water flow in the very late Cenozoic. It is probably also the acceleration of these currents close to the topography of the seamount chain that accounts for much of the erosion (by solution, sapping, and subsequent mass wasting) of sediments in the intermontane valleys.

Evolution of seamount chain sediments: An idealized sequence of events on and near a typical seamount chain, synthesized from many chains at various stages of development and on various ages of crust in a wide range of altitudes, is given in Table 1. The special case of a seamount chain that begins to grow on a relatively young part of a lithospheric plate, and which remains largely in warm tropical seas, and hence should be the richest in carbonate sediments, is depicted diagrammatically in Figure 11.

CONCLUSIONS

1. The assumption that oceanic plateaus and aseismic ridges subside along a (crustal age)$^{\frac{1}{2}}$ law appears reasonable, and thus the methods of plate stratigraphy (e.g., backtracking for paleodepths) may be applicable to the sedimentary sections preserved on them.

2. Available data from drilling on atolls suggests that after an initial period of very rapid subsidence, mainly during the period of active volcanism, oceanic seamounts generally subside at the rates appropriate to the crustal age of the underlying lithospheric plate.

3. The sedimentary record on the summit is strongly influenced by the interplay of subsidence history, paleolatitude changes (paleoclimate), and eustatic sea level changes, which are the major determinants in the establishment and survival of coral reefs -- the only sediments that can accumulate fast enough to keep up with the normal subsidence rates of the volcanic foundation.

4. Archipelagic aprons flanking very young seamount chains commonly stand 500-1000 m above the surrounding sea floor, thus enhancing the environment for preservation of pelagic carbonate sediments.

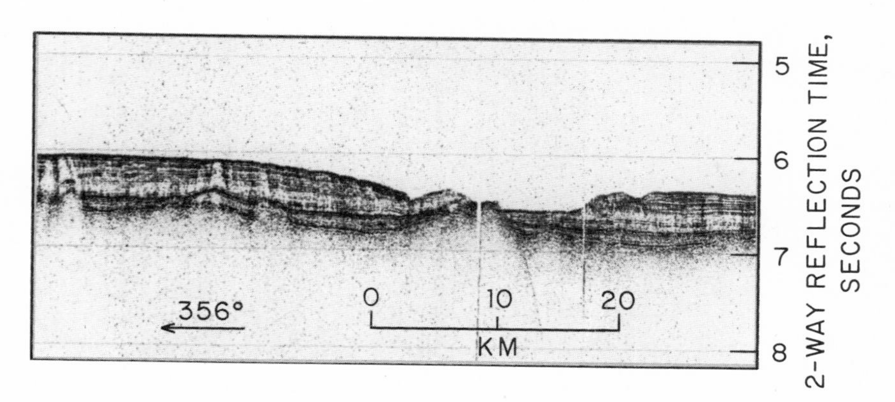

Fig. 10. Seismic reflection profiler record across a part of the archipelagic apron sediments west of the Line Islands, at about 3°N, 161°W, showing a channel cut into the sediments. Record taken by Scripps Institution of Oceanography from R/V THOMAS WASHINGTON, Expedition Eurydice, Leg 2.

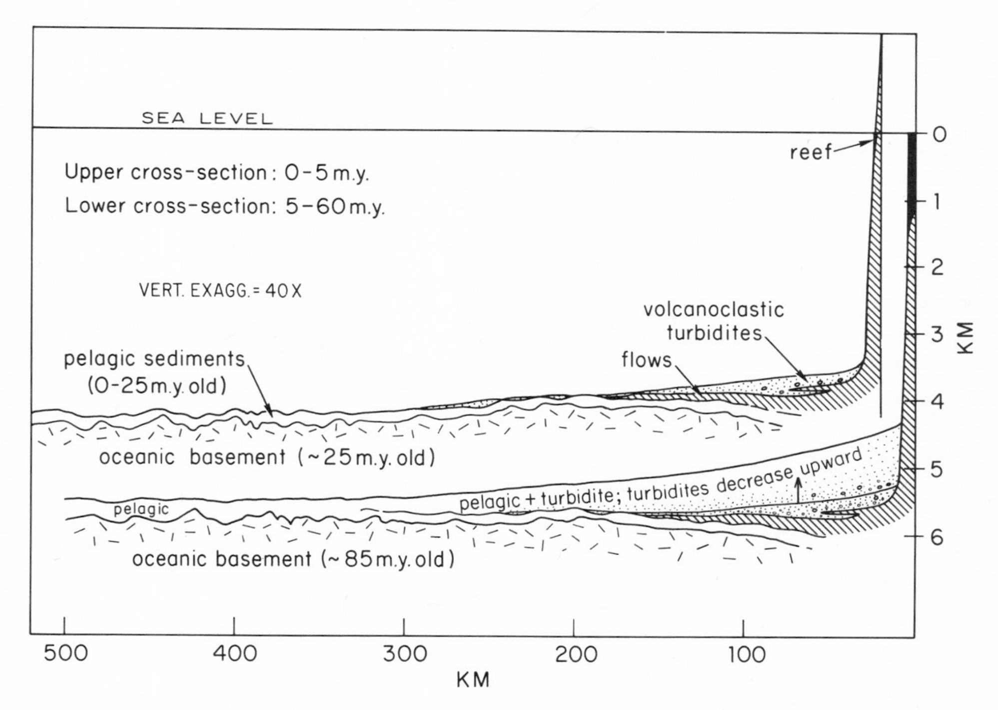

Fig. 11. Idealized sequence of events on and next to a seamount chain that remains in warm tropical seas, where carbonate sediments are best developed.

TABLE 1. Events on and Near Seamount Chains.

	VOLCANISM, SUBSIDENCE	SEDIMENTATION AND EROSION	
		Summit Area	Apron
0–1 m.y.	Rapid upbuilding of volcanic pile, commonly to a height a km or two above sea level. Isostatic sinking goes on almost apace with upbuilding. Creation of a broad apron at the foot of the new part of the seamount, standing from 500 to 1000 m above the surrounding older sea floor. The load of the volcanic pile commonly results in a moat at the head of the apron slope, next to the new seamounts.	Narrow fringing reefs (in tropical waters). Subareal volcanic slopes eroded. Narrow wave-cut shelves may develop during subsidence, and the shelves may be wide where the volcanic rocks are especially weak, e.g., where volcanism close to sea level has led to production of much tuff and agglomerate. Normal calcareous skeletal debris of calcareous organisms accumulates here, but is diluted with volcanic materials.	Very rapid accumulation, often in the moat, of volcanoclastic breccia, sand, and mud, swept off the volcanic slopes by slides, debris flows, mudflows, and turbidity currents. Minor amounts of shallow-water calcareous debris, derived from shelves and/or fringing reefs, carried down onto apron by turbidity currents.
1–5 m.y.	Major cone-building volcanism ceased, though there may be alkalic eruptions that cover parts of the summit regions.	Fringing reefs broaden and evolve to barrier reefs and trap muddy swamp and lagoonal deposits derived from eroding subareal volcanic terranes. Shelves broaden as subsidence slows.	Turbidity currents carry shallow-water reef and shelf skeletal debris to apron, along with volcanogenic sediments from shallow water and from the debris still mantling the submarine volcanic slopes. The volcanogenic supply rate falls off steadily through this period.
			Pelagic carbonate sediments accumulate on the apron (if the apron is shallow enough for these sediments to escape dissolution), and turbidity currents sweep pelagic carbonate sediments off the volcanic slopes onto the apron, where the carbonate fraction commonly constitutes the upper parts of graded beds with volcanogenic debris in the lower parts.
5+ m.y.	Continued subsidence of the volcanic pile, but at close to the normal rate of the underlying older lithospheric plate, augmented by isostatic sinking under the load of any summit reefs. Generally the subareal volcanic terrane is virtually gone by 10 m.y. Later volcanism or tectonism at or near the cone may result in minor uplifts during the long normal history of subsidence.	Reefs continue to grow upward on the subsiding foundations, most likely in the form of atolls. Where reefs have failed to start, or where they have been killed, pelagic sediments accumulate on the seamount (guyot), but these commonly show strong effects of winnowing and shaping by currents.	Pelagic sediments on the apron accumulate as elsewhere on the deep sea floor, taking into account the depth and latitude of the site, its nearness to continents, etc. The continuing influence of the seamount chain is felt in two ways: First, pelagic carbonate sediments accumulate temporarily on the volcanic slopes, and are swept onto apron by turbidity currents, along with minor amounts of volcanic detritus weathered from the slopes and of reef debris. The supply rate of the redeposited pelagic calcareous sediment to the apron reflects pelagic supply rates, which are controlled by surface-water fertility patterns. Second, the topography of the seamount chain perturbs the flow of bottom waters, and probably of internal tides, resulting in local acceleration of flows and the concomitant effects on sedimentation, e.g., erosion of already deposited sediments.

5. The initial sediments along the flanks of the seamount chain -- on the archipelagic apron, in moats, and in intermontane valleys -- generally consist very largely of graded beds of volcanic debris brought down the volcanic slopes by mass wasting and turbidity currents.

6. These sediments generally grade upward into sediments normal for the depth and latitude, augmented by pelagic sediments redeposited from the seamount slopes. The redeposition generally takes place very quickly, as judged by the rarity of reworked older fossils in apron turbidites. Carbonate sediment in rapidly deposited turbidites generally escapes dissolution even below the regional CCD.

7. The combination of shallow depths and rapid redeposition processes makes seamount chains efficient storage areas for carbonate sediments representing a wide range of depth facies.

ACKNOWLEDGEMENT

This study was supported by the Office of Naval Research, Contract USN-N-0014-69-A-0200-6049.

REFERENCES

Andrews, J. E., G. Packham, L. W. Kroenke, G. J. van der Lingen, G. d. Klein, D. L. Jones, J. V. Eade, T. Saito, S. Shafik, B. K. Holdsworth, and D. G. Stoeser, in press, *Initial Reports of the Deep Sea Drilling Project*, v. *30*, Washington (U. S. Government Printing Office), 754 p.

Darwin, C., 1842. The structure and distribution of coral reefs, London.

Davies, T. A., B. P. Luyendyk, K. S. Rodolfo, D. R. C. Kempe, B. C. McKelvey, R. Leidy, G. Horvath, R. Hyndman, H. R. Thierstein, R. Herb, E. Boltovskoy, and P. Doyle, 1974. *Initial Reports of the Deep Sea Drilling Project*, v. *26*, Washington (U. S. Government Printing Office), 1129 p.

Emery, K. O., J. I. Tracey, Jr., and H. S. Ladd, 1954. Geology of Bikini and nearby atolls, *U. S. Geol. Surv. Prof. Paper 260-A*, 1-255.

Fan, Pow-foong and R. R. Grunwald, 1971. Sediment distribution in the Hawaiian Archipelago, *Pacific Sci.*, *V*. 25, 484-487.

Jackson, E. D., K. E. Bargar, B. F. Fabbi, and C. Heropoulos, 1976. Petrology of the basaltic rocks drilled on DSDP Leg 33, In: *Schlanger, S. O., E. D. Jackson, et al., Initial Reports of the Deep Sea Drilling Project, Washington (U. S. Government Printing Office), V*. 33, 571-630.

JOIDES Scientific Staff, 1975. Glomar Challenger drills in the North Atlantic, *Geotimes, V. 20*, 18-21.

Ladd, H. S., J. I. Tracey, Jr., and G. Gross, 1970. Deep drilling on Midway Atoll, *U. S. Geol. Surv. Prof. Paper 680-A*, A1-A21.

Larson, R. L., R. Moberly, D. Bukry, H. P. Foreman, J. V. Gardner, J. B. Keene, Y. Lancelot, H. Luterbacher, M. L. Marshall, and A. Matter, 1975. *Initial Reports of the Deep Sea Drilling Project, V. 32, Washington (U. S. Government Printing Office)*, 980 pp.

Matthews, J. L., B. C. Heezen, et al., 1974. Cretaceous drowning of reefs in Mid-Pacific and Japanese guyots, *Science, V. 184*, 462-464.

Menard, H. W., 1956. Archipelagic aprons, *Bull. Amer. Assoc. Petroleum Geologist, V. 40*, 2195-2210.

Menard, H. W., 1973. Depth anomalies and the bobbing motion of drifting islands, *Jour. Geophys. Res., V. 78*, 5128-5137.

Pitman, W. C., III, R. L. Larson, and E. M. Herron, 1974. Magnetic lineations of the oceans, *Geol. Soc. Amer. Chart.*

Renz, O., 1973. Two Lamellaptychi (Ammonoidea) from the Magellan Rise in the Central Pacific, In: *Winterer, E. L., J. Ewing, et al., Initial Reports of the Deep Sea Drilling Project, V. 17, Washington (U. S. Government Printing Office)*, 895-898.

Schlanger, S. O., 1963. Subsurface geology of Eniwetok Atoll, *U. S. Geol. Surv. Prof. Paper 260-BB*, 901-1066.

Schlanger, S. O., E. D. Jackson, E. L. Winterer, H. E. Cook, H. C. Jenkyns, K. R. Kelts, R. E. Boyce, C. L. McNulty, Jr., A. G. Kaneps, E. Martini, and D. A. Johnson, 1976. *Initial Reports of the Deep Sea Drilling Project, V. 33, Washington (U. S. Government Printing Office)*, 973 pp.

Schlanger, S. O., and E. L. Winterer, 1976. Underway geophysical data: navigation, bathymetry, magnetics, and seismic profiles, In: *Schlanger, S. O., E. D. Jackson, et al., Initial Reports of the Deep Sea Drilling Project, V. 33, Washington (U. S. Government Printing Office)*, 655-693.

Sclater, J. G., R. N. Anderson and M. L. Bell, 1971. The elevation of ridges and the evolution of the central eastern Pacific, *Jour. Geophys. Res., V. 76*, 7888-7915.

Stearns, H. T. and T. K. Chamberlain, 1967. Deep cores at Oahu, Hawaii, and their bearing on the geologic history of the central Pacific Basin, *Pacific Sci., V. 21*, 153-165.

Tréhu, A. M., 1975. Depth versus $(\text{age})^{1/2}$: a perspective on mid-ocean rises, *Earth Planetary Sci. Lett., V. 27*, 287-304.

van Andel, T. H., 1975. Mesozoic/Cenozoic calcite compensation depth and global distribution of calcareous sediments, *Earth Planet. Sci. Lett., V. 26*, 187-194.

van Andel, T. H., G. R. Heath, R. H. Bennett, J. D. Bukry, S. Charleston, D. S. Cronan, M. G. Dinkelman, A. G. Kaneps, K. S. Rodolfo and R. S. Yeats, 1973. *Initial Reports of the Deep Sea Drilling Project, V. 16, Washington (U. S. Government Printing Office)*, 949 pp.

Veevers, J. J., J. R. Heirtzler, H. M. Bolli, A. N. Carter, P. J. Cook, V. Krashenimikov, B. K. McKnight, F. Proto Decima, G. W. Renz, P. T. Robinson, K. Rocker, Jr., and P. A. Thayer, 1974. *Initial Reports of the Deep Sea Drilling Project, V. 27, Washington (U. S. Government Printing Office)*, 1060 pp.

von der Borch, C. C., J. G. Sclater, S. Gartner, Jr., R. Hekinian, D. A. Johnson, B. McGowran, A. C. Pimm, R. W. Thompson, J. J. Veevers, and L. S. Waterman, 1974. *Initial Reports of the Deep Sea Drilling Project V. 22, Washington (U. S. Government Printing Office)* 890 pp.

Winterer, E. L., J. I. Ewing, R. G. Douglas, R. D. Jarrard, Y. Lancelot, R. M. Moberly, Jr., T. C. Moore, Jr., P. H. Roth, and S. O. Schlanger, 1973. *Initial Reports of the Deep Sea Drilling Project, V. 17, Washington (U. S. Government Printing Office)*, 930 pp.

Winterer, E. L., P. F. Lonsdale, J. L. Matthews, and B. R. Rosendahl, 1974. Structure and acoustic stratigraphy of the Manihiki Plateau, *Deep-Sea Research, V. 21*, 793-814.

CHALK FORMATION: EARLY DIAGENESIS

S. W. Wise, Jr.

Florida State University

ABSTRACT

Induration of chalk layers during the early diagenesis of deep sea carbonate oozes is effected primarily by processes of cementation superimposed on the normal depth dependent effects of gravitational compaction. Units having a high "diagenetic potential" due to the deposition of greater amounts of metastable particles are likely to become lithified earliest so that the induration of the sediment layers does not proceed uniformly. Calcium carbonate cement is derived from the in situ dissolution of metastable particles and is precipitated on more stable particles as low magnesium calcite overgrowths. This process may be inhibited by the presence of non-carbonate impurities such as clay minerals, volcanic ash or siliceous microfossils, but where effective, it allows lithification to proceed in a manner which conserves calcium carbonate within the sediment pile, thus allowing little carbonate or CO_2 to be recycled into the hydrosphere or atmosphere during diagenesis.

INTRODUCTION

Chalks consist primarily of coccolith/foraminiferal oozes which have been lithified through post depositional diagenetic processes such as cementation, gravitational compaction, pressure solution or recrystallization. *Wolf* (1968) recognized that chalk formation could occur during early or late diagenesis and that the processes involved and the end results in each case are somewhat different. In the oceanic environment, induration of chalk layers during early diagenesis is effected primarily by processes of cementation superimposed on the normal depth dependent effects of gravitational compaction. Early cementation supports an open framework of carbonate grains which may resist intergrain penetration during subsequent late diagenesis. Late diagenesis begins after sufficient burial (600 to 1000 meters) allows compaction, pressure solution and recrystallization to consolidate the sediment. Chalks formed during late diagenesis are characterized by grain interpenetration, welding and "ameboid mosaics" (*Fischer et al.*, 1967).

This paper is a brief review of the processes of chalk formation during early diagenesis. Consideration is given to the question of whether or not processes of early diagenesis allow recycling of significant amounts of CO_2 from the sediment back into the hydrosphere or atmosphere. The scope of this paper does not allow a comprehensive review of chalk formation. Recent articles by *Schlanger and Douglas* (1974), *Neugebauer* (1974), *Hakasson et al.* (1974), *Matter and Perch-Nielsen* (1975), and *Mapstone* (1975) provide in depth studies of the subject.

Schlanger and Douglas (1974) list a succession of diagenetic realms (Table 1) through which pelagic carbonate skeletal material passes in the oceanic realm from its initial production to its lithification as chalk, thence ultimately to a recrystallized limestone. The productivity phase (1 to 200 m) can be quite important in determining the initial composition and, therefore, the ultimate diagenetic history of the ooze available for lithification. Most carbonate oozes are hetrogeneous mixtures of skeletal components that are variable in their stability in the marine environment. The stability of skeletal components varies within and between major taxonomic groups of shell producing organisms according to minerology, trace element and organic content, ultrastructure, and morphology of tests. Metastable components, if buried within the sediment, may become important donors of calcium carbonate for cementation during lithification.

Highly metastable aragonitic components such as pteropod and cephalopod tests are only locally abundant in deep sea oozes, and will not be considered further in this paper although where they do occur in sediment they should be considered as important potential donors of carbonate for cementation. Instead, more subtle variations will be considered within the major groups contributing to pelagic calcitic carbonates, the calcareous nannoplankton (coccoliths) and the foraminifera. These must be examined closely if the variation in component stability is to be appreciated. For purposes of this paper, calcareous nannofossils will be used chiefly to illustrate these differences since they are often the principal constituents of calcareous microfossil oozes and chalks.

TABLE 1. Diagenetic Realms (from *Schlanger and Douglas*, 1974).

Depth	Realm	Residence Time	Petrography	Porosity (ϕ)% Velocity(V_C)km/s	Diagenetic potential 0 ⟷ ∞
0-200 m (surface water)	I Initial production	Weeks	Highly dispersed calcite-sea water system; 10-10^2 forams/m^3, 10^4-10^6 nannoplankton/m^3		
200 m to sea floor	II Settling	Days to weeks for forams; months to years for coccoliths depending on pelletization (Smayda, 1971)	Pelletized coccoliths, ratio of broken to whole nannoplankton increases downward, ratio of living to empty foram tests decreasing during settling		
3000-5000 m (see Diagenetic Potential)	III Deposition	Inversely proportional to sedimentation rate and dissolution rate	'Honeycombed' structure. Large foram tests supported by chains of coccolith discs. This surface is actually part of Realm IV	$\phi \approx 80\%+$ $V_C \approx 1{\cdot}45$-$1{\cdot}50$	←slope at 3000 m ←Slope at 5000 m
0-1 m sub-bottom)	IV Bioturbation	50,000 years (at 20 m/10^6 years sedimentation rate)	Remoulded 'honeycomb', slight compaction, burrowing, destruction by ingestion and solution	$\phi \approx 75$-80% $V_C \approx 1{\cdot}45$-$1{\cdot}6$	
1-200 m (sub-bottom)	V Shallow-burial	10 X 10^6 years (at 20 m/10^6 years sedimentation rate)	Ooze affected by gravitational composition, establishment of firm grain contacts; dissolution of fossils and initiation of overgrowths	$\phi \approx 75$-60% $V_C \approx 1{\cdot}6$-$1{\cdot}8$	
200-1000 m + (sub-bottom)	VI Deep-burial	Up to ≈120 X 10^6 years (by then either subducted or uplifted)	Chalk with strong development of interstitial cement and overgrowths; transition down to limestone with dissolution of forams, pervasion by cement and overgrowths--grain interpenetration, welding and 'ameboid mosaics' (Fischer et al., 1967)	$\phi \approx 60\%$ down to 35-40% $V_C \approx 1{\cdot}8$ increasing to 3·3 km/s	
1-10 km sub-surface	VII Metamorphic	10^6-10^7 years	Recrystallization trending to 'pavement mosaic' (Fisher et al. 1967) of completely interlocking crystals	$\phi \approx 40\%$ down to <5% $V_C \approx 3$ + up to 6 km/s	

COCCOLITH ASSEMBLAGES: COMPOSITION AND DISSOLUTION CHARACTERISTICS

As with most calcareous planktonic organisms, coccoliths attain their greatest diversity in the tropics, with the number of species decreasing sharply in the higher latitudes. This is well illustrated by numerous studies of the distribution of coccoliths in living plankton and in modern sediment (see, *McIntyre and Be*, 1967; *Honjo*, 1975; *Roth and Berger*, 1975; *Berger and Roth*, 1975). Certain fossil groups such as the discoasters are considered nearly exclusively warm water forms. A second major variable in coccolith distribution is their dramatic increase in diversity within continental margin sediments. This is thought to be partly a function of primary productivity (some forms are believed to be linked to a benthic phase in their life cycle, and therefore are found only in nearshore habitats). The diversity is also a function of enhanced preservation in shallower waters where deposition well above the carbonate compensation depth is followed by rapid burial, often along fine clastics. Thus the record of delicately constructed forms or those that are not resistant to dissolution such as the holococcoliths and members of the braarudosphaerid group is far better preserved in continental margin sediments than in those of the deep sea.

This leads to a second factor that is of over-riding importance in determining coccolith ooze composition: dissolution in the water column and at the sediment/water interface (diagenetic realms II and III, Table 1). Dissolution results in the etching, fragmentation, and differential removal of coccolith species. The most fragile coccoliths are eliminated in the upper reaches of the water column through fragmentation by plankton feeders or by chemical dissolution.

The dissolution characteristics of those coccoliths which accumulate as sediment have been studied by a number of authors (*McIntyre and McIntyre*, 1971; *Schneidermann*, 1973; *Roth and Berger*, 1975). *Roth and Berger* (1975) determined the degree of dissolution of high and low latitude Recent coccoliths from the South Pacific by statistical data treatment of assemblage compositions using pairing analysis. They showed that dissolution removes only restricted species at depths as shallow as 3000 m, but dissolution increases rapidly at 4000 m. Below 5000 m near the base of the calcite compensation zone, only three species are left. There is a level of maximum change in diversity at about 4000 m, which is close to the lysocline. It is interesting to note that the dissolution characteristics of planktonic foraminifera and coccoliths are quite different since well preserved foraminiferal assemblages are dominated by delicate species, whereas well preserved coccolith assemblages contain a fairly high percentage of dissolution resistant forms.

Obviously, any skeletal carbonate dissolved in the water column or near the sediment water interface is recycled back into the hydrosphere. Most workers assume that between these two environments, most dissolution occurs at the sediment/water interface where residence time is much greater. Secondly, the fact that much coccolith material in the photic zone is "packaged" by plankton feeders for rapid transport to the bottom (*Smayda*, 1971; *Honjo*, 1975, 1976; *Roth, Mullin, and Berger*, 1975) reduces significantly the opportunity for dissolution to be effective in the water column. Calculations by *Morse* (1977), however, suggest that dissolution rates are far greater within the water column than at the sediment water interface. The relative effectiveness of dissolution in the two zones, therefore, is still open to question.

Within the top meter of sediment, bioturbation further subjects coccoliths to disaggregation by mechanical action and chemical dissolution as well as to possible re-exposure to undersaturated bottom waters. Residence time at the sediment water interface and within the first meter of sediment varies as a function of sedimentation rate; *Schlanger and Douglas* (1974) estimate 50,000 years for a sedimentation rate of 20 m/10^6 years. It should be noted that exposure to this type of destructive activity could be considerably increased during depositional hiatuses.

Dissolution characteristics for more deeply buried fossil coccolith assemblages can be determined best where the environments of deposition can be reconstructed accurately, particularly in terms of paleodepth and proximity to the CCD and lysocline. Absolute paleodepths can be estimated by allowing for compaction and by correcting for tectonic subsidence along oceanic ridges by backtracking (for example, see *Berger*, 1973). Figures 1 and 2[1] show Pliocene assemblages deposited slightly above the CCD. Compared to assemblages of equivalent age deposited well above the CCD, those shown here exhibit little diversity and consist of a residue of heavily etched, highly dissolution resistant forms, notably discoasters and placolith fragments of the species *Coccolithus pelagicus*. Placoliths are collar button-shaped and consist of two shields composed of cycles of small lath-shaped elements (Figure 3). In life, placoliths are articulated to form a coccosphere (bottom left, Fig. 4) around the algal cell. Upon death or excystment of the cell, coccospheres usually become disarticulated, and are seldom found intact in the sediment. In assemblages which have undergone extensive dissolution, such as those in Figures 1 and 2, the shields are frequently separated and the central elements missing. Also note that the numbers of placoliths are greatly reduced in comparison with the more dissolution resistant discoasters. Although our knowledge of the processes of dissolution of fossil coccolith assemblages is limited and is largely qualitative, the work of *Bukry* (1971, 1973), *Roth* (1973), *Wise* (1973), and *Thierstein* (1974), indicates a dissolution ranking for Cenozoic and Mesozoic nannofossil genera (see summary by *Roth et al.*, 1975). Those placoliths with overlapping elements such as *Coccolithus* show

[1]All micrographs in this report are scanning electron micrographs.

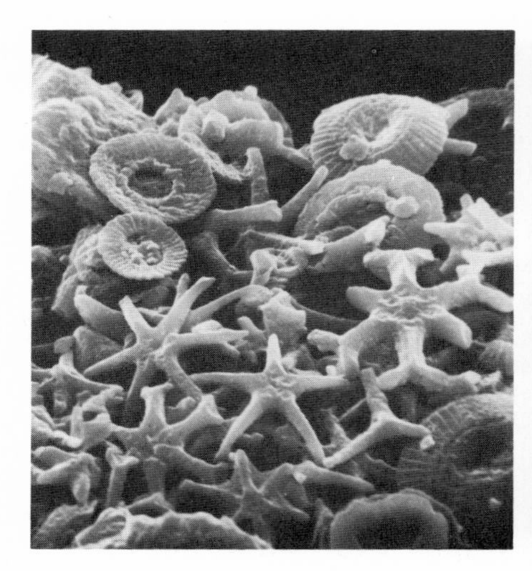

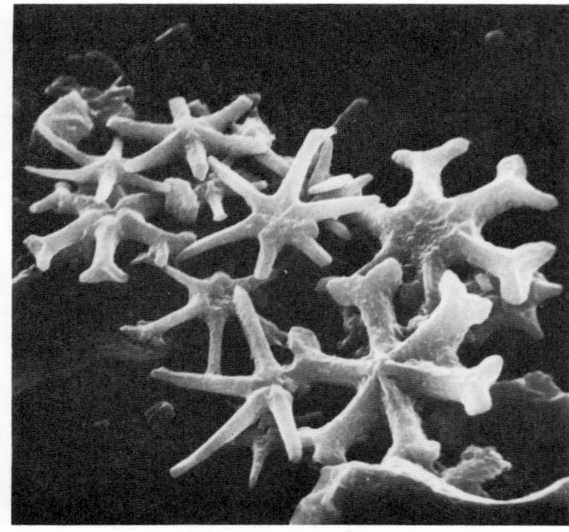

Figs. 1, 2. Scanning electron micrographs of a heavily etched Pliocene ooze (DSDP Sample 14-141-7-1, 50 cm). Note high concentration of discoasters. Dissolution has destroyed the proximal shields and central areas of most of the placoliths which remain (they are all specimens of the dissolution resistant *Coccolithus pelagicus* Wallich). Fig. 1, 2,500X; Fig. 2, 2,700X.

Fig. 3. Placolith (*Coccolithus* sp.) showing normal construction of a proximal shield (smaller) attached at the central area to the distal shield (larger) (DSDP Sample 8-71-34-4, 48 cm.); 7,000X.

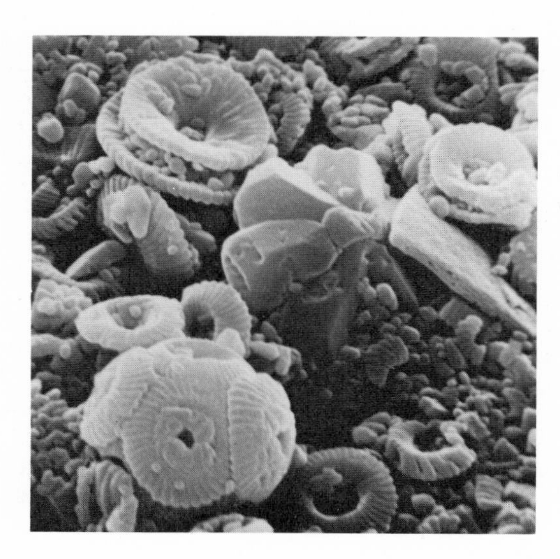

Fig. 4. Miocene coccolith ooze from the central Equatorial Pacific (water depth 4,419 m, sub-bottom depth 304 m); bottom left, coccosphere of placoliths; center, large discoaster which has accreted considerable calcium carbonate in the form of secondary calcite overgrowths. Note particularly the large amount of skeletal rubble composed mostly of unidentifiable fine particles in the ooze (DSDP Sample 8-71-34-4, 148 cm); 2,200X.

greater resistance as a group than those with non-overlapping elements. Heterococcoliths (composed of elements of different sizes and shapes) are generally more resistant than holococcoliths (elements all of equal size and shape). Dissolution is also correlated with the thickness of the elements although the length and widths of the elements show no correlation (*Adelseck and Roth*, 1975). It is not surprising, therefore, that the discoasters, which are composed of very large elements, are more resistant than all the other groups.

CARBONATE DISSOLUTION AND REPRECIPITATION IN SEDIMENT

Coccoliths may not only be affected by dissolution at the sediment/water interface, but also within the sediment. Following burial, calcareous fossils dissolve to the extent that surrounding pore waters become saturated with respect to carbonate and equilibrium is established with the skeletal material. Within the top 200 meters of sediment (realm IV of *Schlanger and Douglas*, 1974), gravitational compaction reduces porosity to about 75-80%, squeezing out some pore water which will carry some dissolved carbonate back to the hydrosphere. If this were the total extent of the process, dissolution of the *in situ* material would be negligible with comparatively little skeletal material dissolved to establish equilibrium. Evidence from cores recovered by the Deep Sea Drilling Project, however, shows that such is not always the case. Instead, large stable skeletal particles which exhibit optical and crystallographic continuity such as discoasters act as seed crystals on which dissolved carbonate in pore waters precipitates as secondary overgrowths. As this tends to reduce concentrations of carbonate in pore waters, less stable particles such as placolith elements begin to dissolve in an attempt to bring pore waters back up to equilibrium. The donor-acceptor relationships in this dissolution-diffusion-reprecipitation phenomenon may be quite complex due to the considerable variability in the size and shape of the heterococcoliths which constitute the vast majority of forms found in deep sea sediments. Dissolution and overgrowth are not only species selective, but different ultrastructural parts of individual coccoliths may vary in their susceptibilities to these processes. For instance, the central areas of placoliths frequently dissolve whereas their shields remain intact (Figure 1). Conversely, secondary overgrowths are frequently seen on the undersides (proximal side) of the distal shields of placoliths of the genus *Coccolithus* (*Wise and Kelts*, 1972, Pl. 3, Figures 1, 2); whereas, the proximal shield and central areas may show no overgrowths whatsoever. In some environments, the dissolution-diffusion-reprecipitation reaction may begin shortly after burial. *Roth and Berger* (1975, Pl. 3, Figures 3, 4) illustrate placoliths from a core top sample which shows significant secondary overgrowth. Field evidence of this transfer process in the upper 200 meters of sediment was obtained by a study of the core from Deep Sea Drilling Hole 172 (eastern Pacific) where a 15 cm thick Oligocene coccolith ooze was recovered near the base of a 23 meter red clay section underlain by basalt. Illustrations of the coccoliths assemblage (*Wise*, 1973, Plates 1-7) show considerable overgrowths on the more stable and resistant taxa such as Discoaster and heavy etching of the placoliths. Since no carbonate was present anywhere else in the entire sediment sequence and the carbonate ooze is considered autochtonous, calcite cement for the overgrowths must have come from within the carbonate layer itself.

Such observations have been confirmed by the laboratory experiments of *Adelseck et al.* (1973) which utilized elevated temperatures and pressures to accelerate the rate of the dissolution-diffusion-reprecipitation reaction in an initially well preserved Pliocene coccolith ooze. In this way they reproduced quite faithfully etch-overgrowth features observed in nature.

Thus it is possible to have in a sediment simultaneous dissolution and reprecipitation of skeletal calcite, a process which can eventually lead to lithification of the sediment. An excellent example of a chalk formed by this process under conditions of shallow burial (13 to 133 m) within the otherwise unconsolidated ooze sequence is the Oligocene *Braarudosphaera* chalk of the South Atlantic (*Wise and Hsu*, 1971; *Wise and Kelts*, 1972). This chalk is composed primarily of braarudasphaerid pentalith segments. These are rather large skeletal elements which exhibit optical continuity, and thus are ideal "seed crystals". Lithification of the chalk was effected when secondary calcite overgrowth on the *Braarudosphaera* particles became sufficiently well developed to bind the particles together. The chalk is quite solid with porosity of about 40%.

Braarudosphaera, however, is relatively non-resistant to dissolution and is normally found only along continental margins. The atypical proliferation of *Braarudosphaera* in the South Atlantic at selected intervals during the Oligocene created a condition in which the same taxa provided both donor and acceptor material to fuel the dissolution-reprecipitation process. This rather contradictory situation can only be explained by considering the size, shape, and orientation in the sediment of each particle individually as well as assuming some differences in the dissolution histories of the particles. It is quite evident from other studies that all speciments of a taxon in a given sedimentary deposit do not show the same diagenetic effects. This may be due in part to differences in dissolution histories of the individual specimens which result from variations in resident time in the water column and at the sediment/water interface. This would also depend on whether the speciments were ingested by grazers, were covered quickly upon deposition by other sediment particles, or were re-exposed by erosion or burrowing activity, etc. *Wise* (1973) illustrates numerous dissolution series showing well to poorly preserved specimens of a given taxon in the etched ooze at DSDP Site 172. Experimental etching produces similar results (*Bé and Morse*, 1975; *Morse et al.*, 1976). Thus such diagenetic effects are best characterized statistically.

Perhaps the most dramatic examples of early diagenetic dissolution-reprecipitation phenomena are found mid way down in diagenetic realm IV (shallow-burial zone of *Schlanger and Douglas*, 1974). Although oozes in this zone are strongly affected by gravitational compaction, lithification is not uniform throughout the zone. As had been observed by others during studies of *Braarudosphaera* chalk, Schlanger and his colleagues noted in their pelagic sequences of the equatorial Pacific, abnormally well consolidated layers within otherwise unconsolidated ooze sequences. Diagenetic effects within these layers and in the adjacent coccolith oozes are intense, as illustrated by photographs in Figures 3 through 12. One is immediately impressed by the high percentage of skeletal rubble

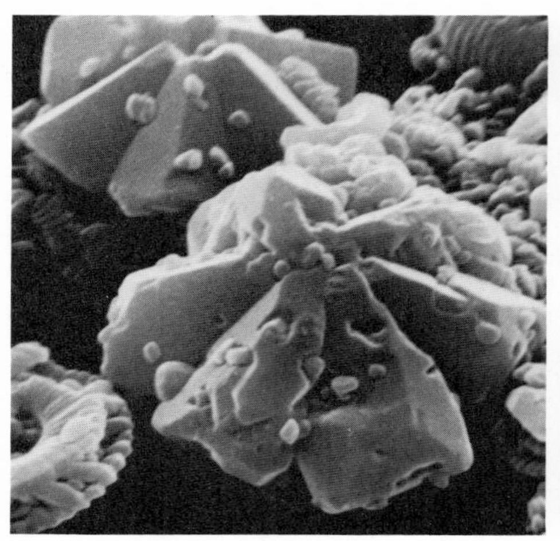

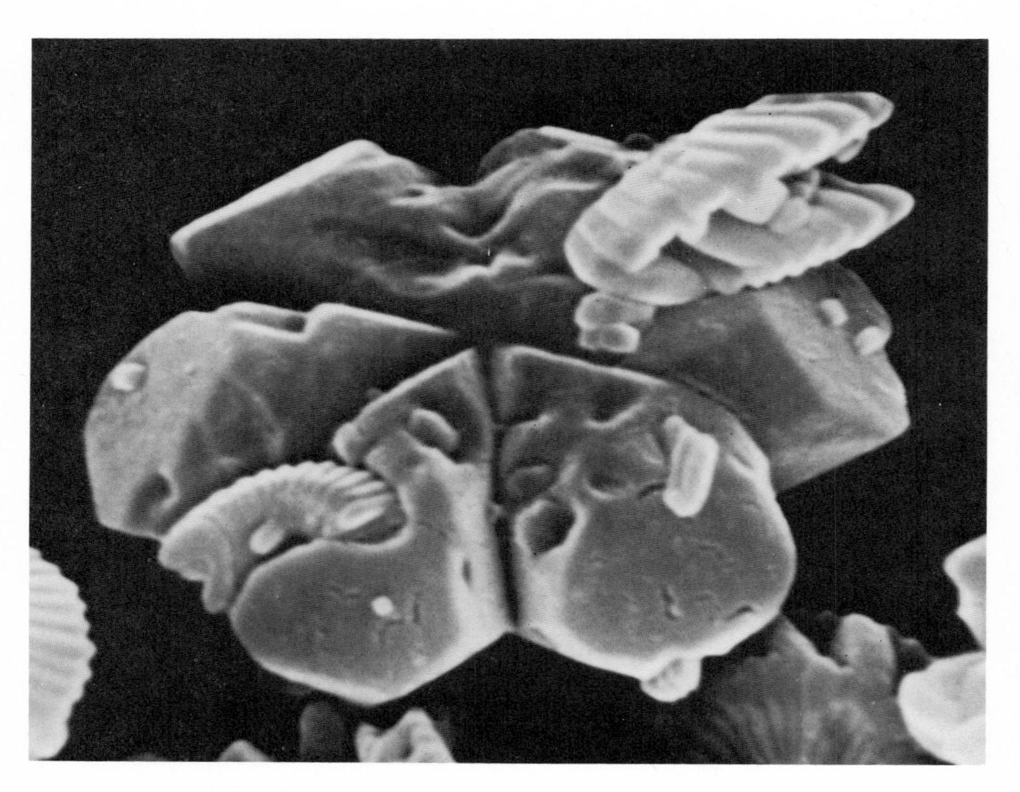

Figs. 5, 6, 7. Heavily overgrown discoasters from DSDP Sample 8-71-34-4, 48 cm. Tell-tale signs of overgrowth include two to three fold increases in size, euhedral calcite crystal faces developed along rays, obfuscation of bifurcations which are normally visible at ray tips, and the puckering of the surface of the astro-

and fine particles in such a zone (Figure 4). Equally impressive are the discoasters (center, Figure 4; Figures 5-7), most of which have expanded to two or three times their normal size through the accretion of secondary calcite. The ray tip bifurcations normally seen on Oligocene discoasters are completely obscured (compare with bifurcate forms in Figure 2), and distinct crystal faces are developed along the arms, whereas well preserved discoasters are more smoothly sculptured. The curiously dimpled surfaces form where particles of debris have been incorporated into the overgrowths; in fact, entire placoliths appear to have been engulfed or swallowed up in a pseudocanabilistic fashion by the growing calcite masses (Figures 6, 7).

Other calcareous nannofossils in the assemblage show unmistakable signs of calcite accretion. Corresponding elements of adjacent shields of the dissolution resistant *Cyclicargolithus* are joined by overgrowths which mimic the shield elements in size and shape (*Wise*, 1972). These overgrowth features cannot be natural products of biological calcification because their presence in life would have precluded the articulation of the coccoliths into coccospheres. The overgrowths appear to originate between the shields of the placoliths (Figures 8-10). They are then guided in their growth outwards along corresponding elements so that when fully developed, they form parallel inclined units along the outer margin of the coccoliths (Figure 9). Overgrowths on other taxa are identified by the dimples and well developed crystal faces on robust elements such as those on the sphenolith in Figure 11 and on the thoracosphaerid in Figure 12.

There are many implications of the diagenetic processes discussed above, the most obvious being taxonomic problems which arise from post-depositional alterations. For instance, before diagenetic phenomena in chalk oozes were fully appreciated, considerable confusion arose among nannofossil specialists over the status of a number of nannofossil species, particularly discoasters described from equatorial regions (example, *Discoaster cubensis*, *D. extensus*, *D. lidzi*, *D. woodringii*, and *D. aster*). Although several of these forms were considered to have stratigraphic utility, these species could not be correlated with any degree of certainty or consistency outside of the local areas in which they were originally described. It was concluded (*Wise*, 1972; *Roth and Thierstein*, 1972; *Bukry*, 1973; *Wise*, 1973; *Roth*, 1973) that these are not true biological species at all, but merely highly altered forms of discoasters so heavily overgrown with inorganically

---

liths where minute particles have been enveloped by secondary calcite. Entire placoliths have become incorporated into the specimens in Figures 6 and 7. Compare with discoasters shown in Figures 1 and 2 which show no calcite overgrowths. Figure 5, 5,000X; Figure 6, 4,500X; Figure 7, 12,500X.

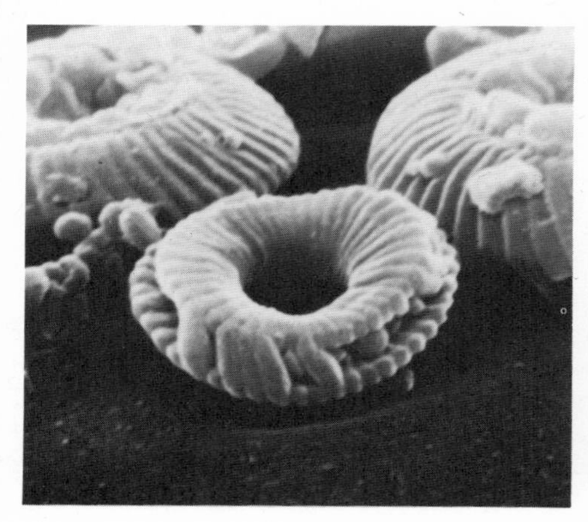

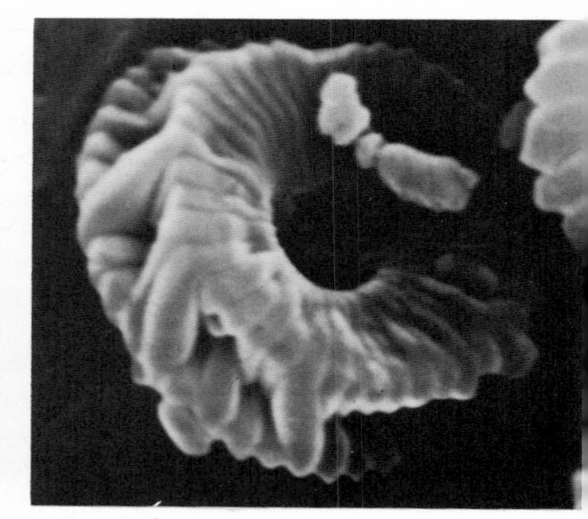

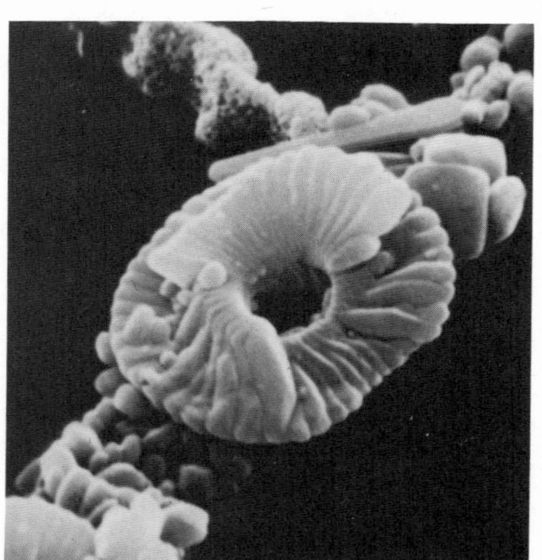

Figs. 8, 9, 10. Calcite overgrowths on placoliths of *Cyclicargolithus* sp. mimic in size and shape the shield elements; however, the overgrowths are clearly diagenetic features since in life their presence would prevent articulation of the placoliths into coccospheres (DSDP Sample 8-71-34-4, 48 cm). Figure 8, 8,000X; Figure 9, 11,500X; Figure 10, 12,000X .

Fig. 11. Specimen of *Sphenolithus moriformis* (Brönniman and Stradner) showing crystal faces and dimpled surfaces (arrow) characteristic of calcite overgrowths (DSDP Sample 8-71-34-4, 48 cm); 11,500X.

precipitated secondary calcite that it is impossible to ascertain their true specific identities. After study by scanning electron microscopy, it appeared that a significant number of the discoasters described in previous years by light microscopy could be derived from certain common generalized forms or one from another as suggested by the sketch in Figure 13. Correlations based on such species were actually correlations of diagenetic facies rather than biostratigraphic correlations. These faces, however, may be widespread and time equivalent within a given region.

The problem of calcite overgrowth on discoasters and other forms first became apparent when the GLOMAR CHALLENGER began exploratory activities in the Equatorial Pacific. *Bukry* (1973) and *Wise* (1973) illustrate several dissolution-overgrowth sequences in discoasters and placoliths of Tertiary age. *Roth* (1973) catalogs diagenetic susceptibilities of many Mesozoic species, and *Thierstein* (1974) illustrates overgrowth sequences of Cretaceous forms. Overall, diagenetic alterations produce a false picture of the true taxonomic diversity of the preserved fossil assemblages. The alterations obscure features necessary for taxonomic discrimination in some groups (thus producing artificially an apparently lower taxonomic diversity; see *Berger and Roth*, 1975). On the other hand, in groups such as the

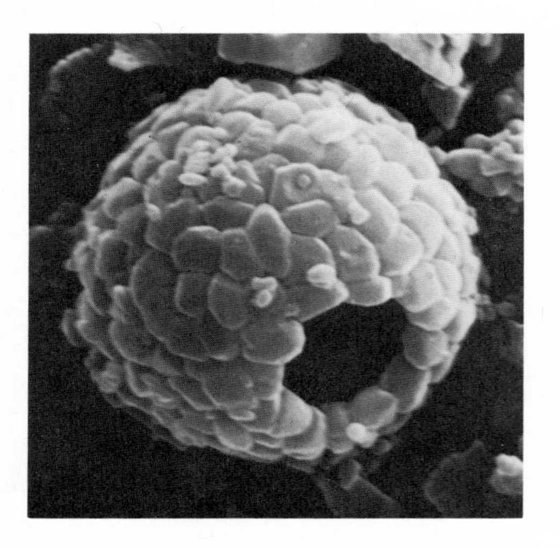

Fig. 12. *Thoracosphaera* sp. exhibits calcite overgrowths with dimpled surfaces (DSDP Sample 8-71-34-4, 48 cm), 5,000X.

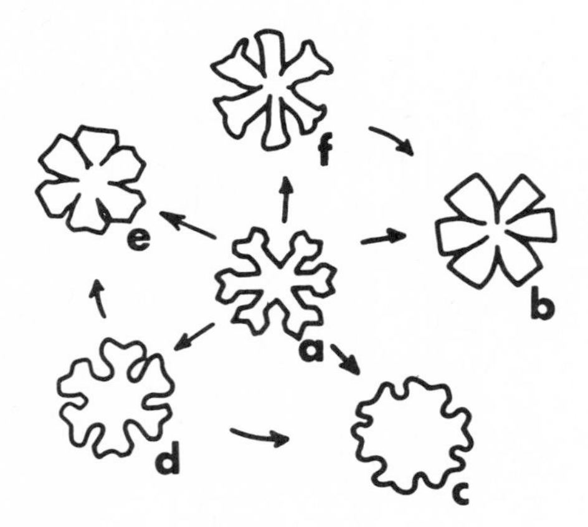

Fig. 13. Diagram showing modification of discoaster outlines as a result of secondary calcite overgrowths. Figure suggests that a number of forms formerly regarded as descrete biological species were derived diagenetically from generalized forms similar to the one shown in the center. Binonial names originally assigned to these forms are: (a) *Discoaster deflandrei* Bramlette and Riedel, (b) *Discoaster wooringi* Bramlette and Riedel, (c) *Discoaster lidzi* Hay, (d) *Discoaster nephados* Hay, (e) *Discoaster trinidadensis* Hay, (f) *Discoaster dialatus* Hay.

Fig. 14. Lower Cretaceous (Aptian) chalk composed largely of wedge-shaped particles (arrows) characteristic of relatively metastable braarudosphaerid/micrantolith type coccoliths. The chalk apparently was lithified by cementation early during diagenesis, thus allowing delicate forms such as the spoked specimens of *Corollithion* to be preserved within the interstities. Those fragile forms would probably otherwise have been destroyed by compaction (DSDP Sample 330-3-2, 115 cm); 3,500X.

discoasters, alterations may artificially produce a high morphologic diversity that does not represent a high biological diversity as has been generally assumed.

In dealing with the taxonomic dilemma created by diagenetic alterations of coccolith and foraminiferal assemblages, paleontologists have provided a wealth of observations which explain some of the detailed processes involved in chalk formation. Other valuable information has been provided by sedimentologists who have looked at broader aspects of the problem as well. *Schlanger and Douglas* (1974) found that a number of indurated chalk layers intercalated with post-Eocene carbonate oozes on plateaus and rises of the Equatorial Pacific are sufficiently lithified to form acoustical reflectors which can be traced continuously on profiles over hundreds of miles and correlated in sequences identifiable at drill sites located as much as 3800 kilometers apart. In the indurated units, not only were coccolith assemblages greatly altered as illustrated in this paper, but foraminiferal assemblages were even more radically affected. The planktonic foraminifera were highly fragmented and greatly reduced in numbers in relation to the more dissolution resistant benthonic foraminifera, an indication that planktonic foraminifera served as donors of a significant proportion of the calcite which formed the secondary overgrowths and cements on the calcareous nannofossils.

In seeking to explain the presence of indurated layers within otherwise unconsolidated ooze sequences, *Schalanger and Douglas* (1974) formulated the concept of "diagenetic potential". In their words (p. 133):

> Diagenetic potential may ... be defined as the length of the diagenetic pathway left for the original dispersed foraminiferal-nannoplankton assemblage to traverse before it reaches the very low free-energy level of a crystalline mosaic.
>
> The concept of the diagenetic potential can be used to explain deviations from a strict depth-of-burial/lithification dependence demanded by a simple gravitational compaction model of diagenesis. The diagenetic potential remaining in a sediment after it passes through the critical boundary between Realms IV and V (Table 2) will determine how far cementation will proceed per unit time. Thus, if layers of very different diagenetic potential are buried sequentially, the amount of cementation in a layer per unit time will be proportional to the diagenetic potential of that layer so that chalks can form above oozes and limestones above chalks in the sedimentary column.
>
> The diagenetic potential (DP) can be expressed as follows:
>
> DP=f (water depth, sed. rate, Temp (Surface), Productivity (Surface), foraminifer + coccolith: discoaster ratio; Size $_{(max)}$: Size $_{(min)}$ ratio, predation rate....)

Schlanger and Douglas' discussion of the various factors in the above equation are quite complete and need not be reviewed here. In simplified terms, the equation suggests that lithologic units of high diagenetic potential would have a higher percentage of well preserved metastable components than those of low diagenetic potential. In addition to the factors stipulated in the above formula, seemingly minor factors such as species composition within a major group (such as within the foraminifera or calcareous nannofossils) could be critically important in determining diagenetic potential and subsequent lithification history. An excellent example of this would be the Oligocene *Braarudosphaera* chalk of the South Atlantic. The parent ooze of that chalk was composed nearly exclusively of an exceptionally metastable coccolith species (*Braarudosphaera rosa*) which gave the ooze initially a high diagenetic potential and accelerated its subsequent cementation. Indeed, the ooze contained such a high percentage of metastable components that lithification appears to have occurred not long after deposition while the unit was still sufficiently close to the sediment water interface to allow isotopic equilibrium with bottom waters to be achieved (*Wise and Hsü*, 1971; *Lloyd and Hsü*, 1972; *Wise and Kelts*, 1972).

One factor which probably contributed toward the high instability of the *Braarudosphaera* ooze may be the wide disparity in particle size which results from the breakdown of coccoliths of *Braarudosphaera*. These coccoliths are finely laminated. With disaggregation they are dispersed into flat wedge-shaped segments of variable thickness, some only a few hundredths or tenths of a micron thick (see *Wise and Kelts*, 1972). The thinner segments have considerably higher surface area/volume ratios than the thicker segments. Thus there would be a wide range of stabilities within the sediment even though the ooze is composed predominantly of skeletal elements from one coccolith taxon. The thinner segments would be expected to supply calcite through dissolution for precipitation on the thicker, more stable segments.

It is interesting to note that the Oligocene *Braarudosphaera* chalk is not unique. A predominantly braarudosphaerid-micrantolith ooze was cored recently in Lower Cretaceous strata of the South Atlantic (DSDP Core 330-3). The ooze contained chalk laminae which were well cemented apparently during early diagenesis (Figure 14). It is quite similar in texture to its Oligocene counterpart (*Wise and Wind*, 1977). Another chalk in which a predominance of a relatively unstable (i.e., non-resistant to dissolution) species may have contributed to its lithification was cored at DSDP Site 329 where a chalk unit in the Paleocene-Eocene (Cores 32, 33) is dominated by the holococcolith *Zygrabdotus bijugatus*, all specimens of which exhibit exceptionally heavy secondary over-growths (*Wise and Wind*, 1977).

Another condition which apparently enhances diagenetic potential is the production during deposition of large amounts of fine skeletal particles which by their relatively high surface area/volume ratios are unstable. Sediment consisting of a large percentage of fine particles is produced in zones of high planktonic productivity, high lysocline and low calcite compensation depths such as in the eastern equatorial Pacific. Most of the illustrations in this paper are of samples from this region. Here high productivity in the photic zone contributes sufficient skeletal carbonate to the abyssal floor to depress the CCD to great depths while at the same time combustion of the abundant organic matter contributed to the water column produces enough CO_2 to raise the lysocline to abnormally high levels (*Berger*, 1971). The zone of rapid dissolution, therefore, is considerably thicker here than elsewhere, and calcareous fossils are subject to severe dissolution effects. A high percentage of fine skeletal rubble is generated in this environment particularly during times of exceptionally high carbonate productivity, as existed in the central Pacific during the Oligocene-early Miocene. Minor variations in productivity in such regions, however, could affect significantly the amount of fine skeletal rubble generated, and therefore the amount of variation in the diagenetic potential of successive layers within the sediment.

OTHER ACCELERATING AND DECELERATING FACTORS

In its simplest format, the concept of diagenetic potential assumes "that the amount of cementation in a layer per unit time will be proportional to the diagenetic potential of that layer" (*Schlanger and Douglas*, op. cit. p. 133). Therefore, rate of cementation is greater in lithologic units of high diagenetic potential than in those of low potential. All other factors being equal, units deposited with a larger amount of less stable elements (i.e., more unstable taxa, greater percentage of fine particles, etc.) will undergo lithification at a faster rate than those deposited with an initial mix of more stable components. From one locality to another or within a given sedimentary column, however, all other factors are not equal. Certain factors not related to diagenetic potential can serve to accelerate or decelerate the rate of conversion of an unconsolidated ooze layer to chalk. Important factors to consider here are the overall host rock lithology and carbonate content. Paleontologists have long noted that preservation of coccolith assemblages can vary significantly with sample lithology. Coccoliths in samples having higher amounts of clay, zeolite, volcanic ash or siliceous microfossils are generally far better preserved than those in pure carbonate oozes (*Bukry et al.*, 1971). Where coccoliths and siliceous organisms are mixed, the nannofossils may be slightly etched but show no overgrowths (example, Miocene of DSDP Site 329; see *Wise and Wind*, 1977, pl. 2), an indication that the presence of silica in solution may well inhibit the dissolution-reprecipitation reaction in carbonates. Significant reductions in secondary overgrowths are also noted where volcanic ash layers occur within otherwise pure carbonate oozes (*Bukry et al.*, 1971). Overgrowths are similarly reduced where clay contents are higher, apparently due to the insulating effects that clay particles have on individual coccoliths; this inhibits the diffusion necessary for the dissolution-diffusion-reprecipitation reactions to occur.[2] The above factors decelerate reactions necessary for cementation, and prolong the time necessary for the diagenetic potential of a sediment to be realized.

Other factors seem to accelerate the reactions. Depositional hiatuses expose surface sediment to prolonged burrowing activity and dissolution by bottom waters. Although the complete process is not totally understood, such hiatuses seem to promote early induration of the carbonate ooze. The end result may be a "hardground" (for examples and discussion, see *Fischer et al.*, 1967; *Milliman*, 1974; *Milliman*, 1977). Typical hardgrounds are present at and near the top of Maestrichtian chalk sections in many parts of the world. Sharp reductions in sedimentation rate and depositional hiatuses, therefore, may greatly accelerate the lithification of chalk units.

TIME STRATIGRAPHIC CONTROL OVER INDURATION

Schlanger and Douglas (1974) suggest that the close parallelism of the indurated reflectors widespread over much of the equatorial Pacific must result from a time-stratigraphic control of induration. This suggests that the reflectors provide a record of time-stratigraphic events which operated over a wide area. The exact cause for the existence of all reflectors observed, however, are certainly not all the same. Indeed, *Schlanger and Douglas* (1974) point to a variety of different events which correlate closely in time with the reflectors they discuss. Two of their reflectors ("c" and "d") correspond to temperature minima, major fluctuations in the calcite compensation depth, and periods of atoll emergence in the Pacific, all of which may correlate with glacio-eustatic regression. Other kinds of events which correspond to the major reflectors they observed are tectonic in nature, such as changes in the Pacific plate direction and velocity.

The Oligocene *Braarudosphaera* chalk of the South Atlantic also forms a reflector traceable over a wide regional extent (DSDP shipboard scientist labeled it the "Maxwell marker"). *Wise and Hsü* (1971) and *Wise and Kelts* (1972) suggested that unusual paleo-oceanographic conditions of major regional significance must have been responsible for the deposition and lithification of that unit. Among the hypotheses which have been offered is a suggestion by *Bukry* (1974) that unusually high rainfall in the South Atlantic region at various times during the Oligocene may have lowered surface water salinities sufficiently to favor the proliferation in the open ocean of *Braaruodsphaera*, a form tolerant of reduced salinities.

---

[2]Conversely, the poorest preserved coccolith assemblage are those in nearly pure carbonate oozes where there are no barriers to the diffusion of dissolved carbonate.

ULTIMATE FATE

From the above, it appears that a wide variety of factors may affect the diagenetic potential of carbonate oozes, the rates at which they may become lithified, and thereby their subsequent diagenetic history. Overall, the dissolution-diffusion-reprecipitation process seems to be quite effective in the redistribution of carbonate within the sediment of the shallow-burial and deep-burial diagenetic realms. It is also quite effective in conserving carbonate within the sediment pile. This is because donor-acceptor relationships between components within the sediment column promote *in situ* precipitation of carbonate liberated by dissolution in excess of that needed to establish equilibrium with pore waters. Pore water expulsion is a one time process which carries an insignificant amount of carbonate out of the sediment pile compared to that which remains. Calculations of alkalinities and pore water concentrations in the top meter of sediment by *Morse* (1977) suggest that loss of dissolved carbonate by diffusion or by advection from within the sediment across the sediment water interface is probably minimal although this is still an open question.

Effective recycling of carbonate or CO_2 out of deep sea sediment and into the hydrosphere or atmosphere, therefore, is dependent on tectonic activity which would expose the sediment to atmospheric weathering. *Hay and Southam* (1975), however, note that tectonic processes are extremely inefficient at recycling oceanic carbonate to the terrestrial environment. Most such carbonate is subducted in oceanic trenches where, according to calculations by Hay and Southam, it is absorbed into the mantle. Therefore, once locked into carbonate sediments of deep oceans, CO_2 is unlikely to be returned to the hydrosphere or atmosphere since diagenetic and tectonic processes tend to conserve it in the oceanic environment until recycled into the mantle.

ACKNOWLEDGEMENTS

Work on this topic began at the Swiss Federal Institute of Technology in collaboration with Professor Kenneth J. Hsü during the tenure of an NSF Postdoctoral Fellowship, and preliminary results were presented in abstract at a Southeastern Meeting of the Geological Society of America (*Wise*, 1972). Scanning electron micrographs were made at the Department of Geology, Florida State University, and at the ETH Laboratory for Electron Microscopy, Honneggberg, through the kindness of Drs. Hans M. Bolli (host professor) and Dr. Hans-Ude Nissen (Laboratory Director). I thank Dr. John W. Morse (Univ. of Miami) for critical review of the manuscript and Dr. Seymour O. Schlanger (University of Leiden) for helpful discussion. Study supported by NSF grant DES74-14161.

REFERENCES

Adelseck, C. G., G. W. Geehan, and P. H. Roth. 1973. Experimental evidence for the selective dissolution and overgrowth of calcareous nannofossils during diagenesis. *Bull. Geol. Soc. America, 84.* 2755.

Adelseck, C. G. and P. H. Roth. 1975. Test morphology and crystal size: controlling factors in the selective dissolution of Coccolithophorida and Foraminifera species. *Geol. Soc. America, Abstracts with Programs, 7.* 969.

Bé, A. W. H., J. W. Morse, and S. M. Harrison. 1975. Progressive dissolution and ultrastructural breakdown in planktonic foraminifera. *Cushman Found. Form Res. Spec. Publ., 13.* 27.

Berger, W. H. 1971. Sedimentation of planktonic foraminifera. *Marine Geol., 11.* 325.

Berger, W. H. 1973. Deep-sea carbonates: evidence for a coccolith lysocline. *Deep-Sea Res., 20.* 917.

Berger, W. H. and P. H. Roth. 1975. Oceanic micropaleontology: progress and prospect. *Reviews of Geophysics and Space Physics, 13.* 561-585, 624-635.

Bukry, D. 1971. Cenozoic calcareous nannofossils from the Pacific Ocean. *San Diego Soc. Nat. History Trans., 16.* 303.

Bukry, D. 1973. Coccolith stratigraphy, eastern equatorial Pacific, Leg 16, Deep Sea Drilling Project. *Deep Sea Drilling Proj. Initial Repts, 16.* 653.

Bukry, D. 1974. Coccoliths as paleosalinity indicators -- evidence from the Black Sea. In: *The Black Sea, Its Geology, Chemistry and Biology,* E. T. Degens and D. A. Ross (Eds.), Am. Assoc. Pet. Geol., Mem 20. 353.

Bukry, D., R. G. Douglas, S. A. Kling and V. Krasheninnikov. 1971. Planktonic microfossil biostratigraphy of the northwestern Pacific Ocean. In: *Initial Reports of the Deep Sea Drilling Project, Volume VI,* A. G. Fisher et al., Washington (U. S. Government Printing Office). 1253.

Fisher, A. G., S. Honjo, and R. E. Garrison. 1967. Electron micrographs of limestones and their nannofossils, Princeton, New Jersey, Princeton Univ. Press. 141.

Hakasson, E., R. Bromley, and K. Perch-Nielsen. 1974. Maastrichtian chalk of north-west Europe -- a pelagic shelf sediment. In: *Pelagic Sediments on Land and Under the Sea,* K. J. Hsü and H. C. Jenkyns (Eds.), *Internat. Assoc., Sedimentologist Spec. Publ., 1.* 211.

Hay, W. W. and J. Southam, Jr. 1975. Calcareous plankton and loss of *CaO* from the continents. *Geol. Soc. Amer., Abstracts with Programs,* 7. 1105.

Honjo, S. 1975. Dissolution of suspended coccoliths in the deep-sea water column and sedimentation of coccolith ooze. *Cushman Found., Foram. Res. Spec. Publ. 13.* 114-128.

Honjo, S. 1976. Coccoliths: production, transportation and sedimentation. *Marine Micropaleontology, 1(1).*

Lloyd, R. M. and K. J. Hsü. 1972. Stable-isotope investigations of sediments from the DSDP III Cruise to South Atlantic. *Sedimentology, 19.* 45.

Mapstone, N. B. 1975. Diagenetic history of a North Sea Chalk. *Sedimentology, 22.* 601.

Matter, A. and K. Perch-Nielsen. 1975. Fossil preservation, geochemistry, and diagenesis of pelagic carbonates from Shatsky Rise, Northwest Pacific. *Init. Repts. Deep Sea Drilling Project, 32.* 891.

McIntyre, A. and A. W. H. Bé. 1967. Modern Coccolithophoridae of the Atlantic Ocean -- I, placoliths and cyrtoliths. *Deep-Sea Res., 14.* 561.

McIntyre, A. and R. McIntyre. 1971. Coccolith concentrations and differential solution in oceanic sediments. In: *The Micropaleontology of Oceans*, F. M. Funnell and W. R. Riedel (Eds.). 253.

Milliman, J. D. 1974. Recent sedimentary carbonates. I. Marine carbonates. Springer Verlag, New York. 375.

Milliman, J. D. 1977. Characteristics and genesis of shallow-water and deep-sea limestones. In: *The Fate of Fossil Fuel CO_2*, N. R. Andersen and A. Malahoff (Eds.), Plenum, N. Y., (this volume).

Morse, J. W., F. H. Wind, and S. W. Wise. 1976. Selective dissolution sequences for calcareous nannofossils in deep sea carbonate ooze: emperical observations and experimental results. *Amer. Assoc. Petrol. Geol., Ann. Meet. AAPG-SEPM Ann. Meet. Abs., 3.*

Morse, J. W. 1977. The carbonate chemistry of North Atlantic Ocean deep-sea sediment pore water. In: *The Fate of Fossil Fuel CO_2*, N. R. Andersen and A. Malahoff (Eds.), Plenum, N.Y., (this volume).

Neugebauer, J. 1974. Some aspects of cementation in chalk. In: *Pelagic Sediments on Land and Under the Sea*, K. J. Hsü and H. C. Jenkyns (Eds.), *Internat. Assoc. Sedimentologists Spec. Publ., 1.* 149.

Roth, P. H. 1973. Calcareous nannofossils--Leg 17, Deep Sea Drilling Project. *Init. Repts. Deep Sea Drilling Project, 17.* 695.

Roth, P. H., and W. H. Berger. 1975. Distribution and dissolution of coccoliths in the South and Central Pacific. *Cushman Foundation Foram, Res., Special Publ., 13.* 87.

Roth, P. H., M. M. Mullin, and W. H. Berger. 1975. Coccolith sedimentation by fecal pellets: laboratory experiments and field observations. *Geol. Soc. Am., Bull., 86.* 1074.

Roth, P. H. and H. Thierstein. 1972. Calcareous nannoplankton: Leg 14 of the Deep Sea Drilling Project. *Init. Repts., Deep Sea Drilling Proj., 14.* 421-485.

Roth, P. H., S. W. Wise, and H. Thierstein. 1975. Early chalk diagenesis and lithification: sedimentological applications of paleontological approaches. *IXme Congres International de Sedimentologie Nice, Theme 7.* 187-192.

Schlanger, S. O. and R. G. Douglas. 1974. The pelagic-ooze-chalk-limestone transition and its implication for marine stratigraphy. In: *Pelagic Sediments on Land and Under the Sea,* K. J. Hsü and H. Jenkyns (Eds.), *Internat. Assoc. Sedimentologists, 1.* 117-148.

Schneidermann, N. 1973. Deposition of coccoliths in the compensation zone of the Atlantic Ocean. In: *Proc. Sympos. Calcareous Nannofossils, Gulf Coast Sect., Soc. Econ. Paleontologists Mineralogist,* L. A. Smith and J. Hardenbol (Eds.), Houston, Texas. 140-151.

Smayda, T. J. 1971. Normal and accelerated sinking of phytoplankton in the sea. *Mar. Geol., 11 (2).* 105.

Thierstein, H. 1974. Calcareous nannoplankton -- Leg 26, Deep Sea Drilling Project. *Deep Sea Drilling Proj., Init. Repts., 26.* 619.

Wise, S. W. 1972. Calcite overgrowths on calcareous nannofossils -- a taxonomic irritant and a key to the formation of chalk. *Geol. Soc. Amer. Prog. Abst., 4.* 115.

Wise, S. W. 1973. Calcareous nannofossils from cores recovered by Leg 18, Deep Sea Drilling Project: biostratigraphy and observations of diagenesis. *Deep Sea Drilling Proj. Init. Repts., 18.* 567.

Wise, S. W. and K. J. Hsü. 1971. Genesis and lithification of a deep sea chalk. *Eclog. Geol. Helvetiae, 64 (2).* 273.

Wise, S. W. and K. R. Kelts. 1972. Inferred diagenetic history of a weekly silicified deep sea chalk. *Gulf Coast Assoc. Geol. Soc. Trans. 22.* 177.

Wise, S. W. and F. H. Wind. 197 . Mesozoic and Cenozoic calcareous nannofossils recovered by DSDP Leg 36 to the Falkland Plateau, Atlantic Sector of the Southern Ocean. In: *Initial Reports of the Deep Sea Drilling Project, Volume 36,* P. F. Barker, I. W. D. Dalziel et al., Washington (United States Government Printing Office). In press.

Wolf, M. J. 1968. Lithification of a carbonate mud; Senonian chalk in Northern Ireland. *Sed. Geol., 2.* 263.

LIST OF CONTRIBUTORS AND PARTICIPANTS

NEIL R. ANDERSEN, Marine Chemistry, National Science Foundation, Washington, D. C. 20550

ROBERT BACASTOW, Scripps Institution of Oceanography, University of California, San Diego, La Jolla, California 92093

ROGER G. BATES, Department of Chemistry, University of Florida, Gainesville, Florida 32611

SAM BEN-YAAKOV, Institute of Electronics, Ben Gurion University of the Negev, P. O. Box 2053, Beer Sheva, 84 120, Israel

WOLFGANG H. BERGER, Scripps Institution of Oceanography, University of California, San Diego, La Jolla, California 92093

ROBERT A. BERNER, Department of Biology and Geophysics, Yale University, P. O. Box 2161, Yale Station, New Haven, Conn. 06520

PETER BETZER, Department of Marine Science, University of South Florida, St. Petersburg, Florida 33701

S. B. BETZER, University of Miami, Miami, Florida 33125

BERT BOLIN, Department of Meteorology, Arrhenius Laboratory, University of Stockholm, FACK, S-104-05, Stockholm, Sweden

EDWARD BOYLE, Grant Institute of Geology, University of Edinburgh, Edinburgh, Scotland

WALLACE S. BROECKER, Lamont-Doherty Geological Observatory, Columbia University, Palisades, New York 10964

R. W. BURLING, Institute of Oceanography, University of British Columbia, V6T 1W5, Canada

K. L. CARDER, University of South Florida, St. Petersburg, Florida 33701

KEITH CHAVE, University of Hawaii, Hawaii Institute of Geophysics, Honolulu, Hawaii 96822

C. H. CULBERSON, Department of Chemistry, University of Florida, Gainesville, Florida 32611

J. PAUL DAUPHIN, Graduate School of Oceanography, University of Rhode Island, Kingston, Rhode Island 02881

P. L. DONAGHAY, School of Oceanography, Oregon State University, Corvallis, Oregon 97331

JOHN M. EDMOND, Massachusetts Institute of Technology, Cambridge, Massachusetts 02129

D. W. EGGIMANN, Department of Marine Science, University of South Florida, St. Petersburg, Florida 33701

JORIS GIESKES, Scripps Institution of Oceanography, University of California, San Diego, La Jolla, California 92037

N. L. GUINASSO, JR., Department of Oceanography, Texas A & M University, College Station, Texas 77843

E. L. HAMILTON, Naval Undersea Center, San Diego, California 92132

K. HANSON, National Oceanic and Atmospheric Administration, 8060 13th Street, Silver Spring, Maryland 20910

WILLIAM HAY, Rosenstiel School of Marine and Atmospheric Research, University of Miami, Miami, Florida 33149

G. ROSS HEATH, Graduate School of Oceanography, University of Rhode Island, Kingston, Rhode Island 02881

SUSUMU HONJO, Woods Hole Oceanographic Institution, Woods Hole, Massachusetts 02543

CHARLES D. HOLLISTER, Woods Hole Oceanographic Institution, Woods Hole, Massachusetts 02543

THOMAS C. JOHNSON, Department of Geology and Geophysics, 108 Pillsbury Hall, University of Minnesota, Minneapolis, Minnesota 55455

CHARLES D. KEELING, Scripps Institution of Oceanography, University of California, San Diego, La Jolla, California 92037

D. R. KESTER, University of Rhode Island, Kingston, Rhode Island 02881

G. KIPPHUT, Lamont-Doherty Geological Observatory, Columbia University, Palisades, New York 10964

PETER KROOPNICK, Department of Oceanography, Hawaii Institute of Geophysics, University of Hawaii, Honolulu, Hawaii 96822

EDGAR R. LEMON, 1021 Bradfield Hall, Cornell University, Ithaca, New York 14853

L. MACHTA, Air Resources Laboratory, National Oceanic and Atmospheric Administration, 8060 - 13th Street, Silver Spring, Maryland 20910

ALEXANDER MALAHOFF, National Ocean Survey, National Oceanic and Atmospheric Administration, Rockville, Maryland 20852

STANLEY MARGOLIS, Department of Oceanography, University of Hawaii, Honolulu, Hawaii 96822

JOHN D. MILLIMAN, Woods Hole Oceanographic Institution, Woods Hole, Massachusetts 02543

THEODORE C. MOORE, Graduate School of Oceanography, University of Rhode Island, Kingston, Rhode Island 02881

JOHN W. MORSE, Rosenstiel School of Marine and Atmospheric Science, University of Miami, 4600 Rickenbacker Causeway, Miami, Florida 33149

JENS MÜLLER, Lehrstuhl für Geologie, Technische Universität, D-8 München 2, West Germany

ALLAN Z. PAUL, Lamont-Doherty Geological Observatory, Columbia University, Palisades, New York 10964

T. H. PENG, Lamont-Doherty Geological Observatory, Columbia University, Palisades, New York 10964

N. G. PISIAS, Graduate School of Oceanography, University of Rhode Island, Kingston, Rhode Island 02881

RICARDO M. PYTKOWICZ, School of Oceanography, Oregon State University, Corvallis, Oregon 97331

D. REINHARD-DERIE, Universite Libre de Bruxelles, Institut de Chimie Industrielle, Service Environment, Avenue F.-D. Roosevelt 50, 1050 Bruxelles, Belgium

RICHARD REZAK, Department of Oceanography, Texas A & M University, College Station, Texas 77843

D. K. RICHARDSON, Lamont-Doherty Geological Observatory, Columbia University, Palisades, New York 10964

RALPH ROTTY, Institute for Energy Analysis, P. O. Box 117, Oak Ridge, Tennessee 37830

DAVID R. SCHINK, Department of Oceanography, Texas A & M University, College Station, Texas 77843

N. J. SHACKLETON, University of Cambridge, 5, Salisbury Villas, Station Road, Cambridge CB1 2JF England

LAWRENCE F. SMALL, School of Oceanography, Oregon State University, Corvallis, Oregon 97331

JOHN R. SOUTHAM, Rosenstiel School of Marine and Atmospheric Research, University of Miami, Miami, Florida 33149

ERIC SUNDQUIST, Geology Department, Harvard University, Cambridge, Massachusetts 02138

TARO TAKAHASHI, Lamont-Doherty Geological Observatory, Columbia University, Palisades, New York 10964

JORN THIEDE, Universitetet i Oslo, Post Boks 1047, Blindern, Oslo 3, Norway

MARY-FRANCES THOMPSON, American Institute of Biological Sciences, 1401 Wilson Boulevard, Arlington, Virginia 22209

T. H. VAN ANDEL, Stanford University, Stanford, California 94305

GEORGE WEATHERLY, Department of Oceanography, Florida State University, Tallahassee, Florida 32306

RAY WEISS, Scripps Institution of Oceanography, University of California, San Diego, La Jolla, California 92037

GERALD WINTERER, Scripps Institution of Oceanography, University of California, San Diego, La Jolla, California 92037

SHERWOOD W. WISE, JR., Department of Geology, The Florida State University, Tallahassee, Florida 32306

ROLAND WOLLAST, Universite Libre de Bruxelles, Institut de Chimie Industrielle, Service Envrionment, Avenue F.-D. Roosevelt 50, 1050 Bruxelles, Belgium

C. S. WONG, Institute of Ocean Sciences, Victoria, British Columbia, Canada

INDEX